HYPERBOLIC MANIFOLDS

An Introduction in 2 and 3 Dimensions

Over the past three decades there has been a total revolution in the classic branch of mathematics called 3-dimensional topology, namely the discovery that most solid 3-dimensional shapes are hyperbolic 3-manifolds. This book introduces and explains hyperbolic geometry and hyperbolic 3- and 2-dimensional manifolds in the first two chapters, and then goes on to develop the subject. The author discusses the profound discoveries of the astonishing features of these 3-manifolds, helping the reader to understand them without going into long, detailed formal proofs. The book is heavily illustrated with pictures, mostly in color, that help explain the manifold properties described in the text. Each chapter ends with a set of Exercises and Explorations that both challenge the reader to prove assertions made in the text, and suggest further topics to explore that bring additional insight. There is an extensive index and bibliography.

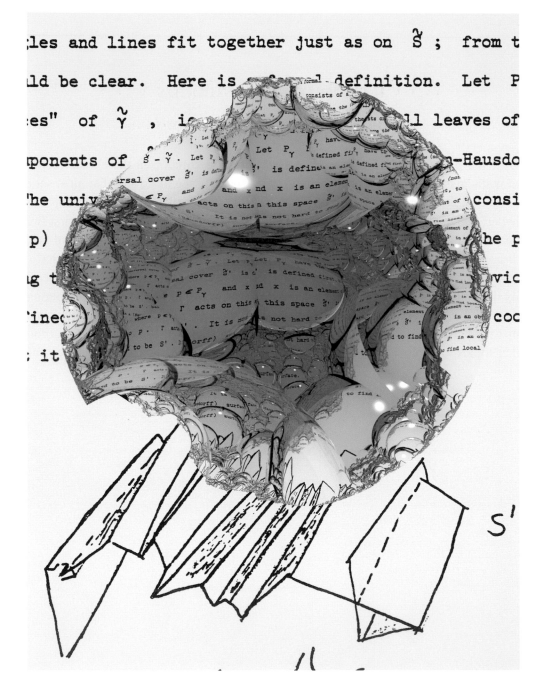

[Thurston's Jewel (JB)(DD)] **Thurston's Jewel**: Illustrated is the convex hull of the limit set of a kleinian group G associated with a hyperbolic manifold $\mathcal{M}(G)$ with a single, incompressible boundary component. The translucent convex hull is pictured lying over p. 8.43 of Thurston [1979a] where the theory behind the construction of such convex hulls was first formulated. This particular hyperbolic manifold represents a critical point of Thurston's "skinning map", as described in Gaster [2012].

This image was created by Jeffrey Brock and David Dumas; details about its creation can be found at http://dumas.io/convex.

The image of p. 8.43 was used with permission of Julian, Nathaniel, Dylan Thurston.

HYPERBOLIC MANIFOLDS

An Introduction in 2 and 3 Dimensions

ALBERT MARDEN

University of Minnesota

CAMBRIDGE
UNIVERSITY PRESS

CAMBRIDGE
UNIVERSITY PRESS

University Printing House, Cambridge CB2 8BS, United Kingdom

Cambridge University Press is part of the University of Cambridge.

It furthers the University's mission by disseminating knowledge in the pursuit of education, learning and research at the highest international levels of excellence.

www.cambridge.org
Information on this title: www.cambridge.org/9781107116740

First published 2007

Printed in the United Kingdom by Bell and Bain Ltd.

A catalogue record for this publication is available from the British Library

Library of Congress Cataloguing in Publication data
Marden, Albert.
[Outer circles]
Hyperbolic manifolds : an introduction in 2 and 3 dimensions / Albert Marden, University of Minnesota. – Second edition.
pages cm
Second edition of Outer circles, which has changed title to: Hyperbolic manifolds.
Includes bibliographical references and index.
ISBN 978-1-107-11674-0
1. Three-manifolds (Topology)   2. Hyperbolic spaces.   3. Complex manifolds.
4. Geometry, Hyperbolic.   I. Title.
QA613.2.M37 2016
515'.946–dc23
2015014690

ISBN 978-1-107-11674-0 Hardback

Cambridge University Press has no responsibility for the persistence or accuracy of URLs for external or third-party internet websites referred to in this publication, and does not guarantee that any content on such websites is, or will remain, accurate or appropriate.

To Dorothy

צו מיין פרוי דבורה, די אמת׳דיקע אשת-חיל.

and

In memory of William P. "Bill" Thurston
"God gave him the open book."—Jürgen Moser

Contents

List of Illustrations   *page* xi
Preface   xiii

1   **Hyperbolic space and its isometries**   1
    1.1   Möbius transformations   1
    1.2   Hyperbolic geometry   6
        1.2.1   The hyperbolic plane   8
        1.2.2   Hyperbolic space   8
    1.3   The circle or sphere at infinity   12
    1.4   Gaussian curvature   16
    1.5   Further properties of Möbius transformations   19
        1.5.1   Commutativity   19
        1.5.2   Isometric circles and planes 20
        1.5.3   Trace identities   23
    1.6   Exercises and explorations   24

2   **Discrete groups**   53
    2.1   Convergence of Möbius transformations   53
        2.1.1   Some group terminology   55
    2.2   Discreteness   55
    2.3   Elementary discrete groups   59
    2.4   Kleinian groups   62
        2.4.1   The limit set $\Lambda(G)$ 62
        2.4.2   The ordinary (regular, discontinuity) set $\Omega(G)$ 64
    2.5   Quotient manifolds and orbifolds   66
        2.5.1   Covering surfaces and manifolds   67
        2.5.2   Orbifolds   70
        2.5.3   The conformal boundary   73
    2.6   Two fundamental algebraic theorems   73
    2.7   Introduction to Riemann surfaces and their uniformization   75
    2.8   Fuchsian and Schottky groups   80
        2.8.1   Handlebodies   82
    2.9   Riemannian metrics and quasiconformal mappings   84

2.10 Teichmüller spaces of Riemann surfaces 87
    2.10.1 Teichmüller mappings 89
2.11 The mapping class group $\mathcal{MCG}(R)$ 91
    2.11.1 Dehn twists 91
    2.11.2 The action of $\mathcal{MCG}(R)$ on R and $\mathfrak{Teich}(R)$ 91
    2.11.3 The complex structure of $\mathfrak{Teich}(R)$ 94
2.12 Exercises and explorations 94
    2.12.1 Summary of group properties 114

3 Properties of hyperbolic manifolds **122**
3.1 The Ahlfors Finiteness Theorem 122
3.2 Tubes and horoballs 123
3.3 Universal properties in hyperbolic 3-manifolds and orbifolds 126
3.4 The thick/thin decomposition of a manifold 134
3.5 Fundamental polyhedra 135
    3.5.1 The Ford fundamental region and polyhedron 138
    3.5.2 Poincaré's Theorem 142
    3.5.3 The Cayley graph dual to tessellation 143
3.6 Geometric finiteness 144
    3.6.1 Finite volume 149
3.7 Three-manifold surgery 149
    3.7.1 Compressible and incompressible boundary 151
    3.7.2 Extensions $\partial \mathcal{M} \to \mathcal{M}$ 153
3.8 Quasifuchsian groups 155
    3.8.1 Simultaneous uniformization 157
3.9 Geodesic and measured geodesic laminations 158
    3.9.1 Geodesic laminations 158
    3.9.2 Measured geodesic laminations 161
    3.9.3 Geometric intersection numbers 162
    3.9.4 Length of measured laminations 164
3.10 The convex hull of the limit set 167
    3.10.1 The bending measure 169
    3.10.2 Pleated surfaces 170
3.11 The convex core 175
    3.11.1 Length estimates for the convex core boundary 177
    3.11.2 Bending measures on convex core boundary 178
3.12 The compact and relative compact core 180
3.13 Rigidity of hyperbolic 3-manifolds 182
3.14 Exercises and explorations 187

4 Algebraic and geometric convergence **219**
4.1 Algebraic convergence 219
4.2 Geometric convergence 225
4.3 Polyhedral convergence 226

4.4	The geometric limit	229
4.5	Sequences of limit sets and regions of discontinuity	232
	4.5.1 Hausdorff and Carathéodory convergence	232
	4.5.2 Convergence of groups and regular sets	234
4.6	New parabolics	237
4.7	Acylindrical manifolds	239
4.8	Dehn filling and Dehn surgery	241
4.9	The prototypical example	242
4.10	Manifolds of finite volume	245
	4.10.1 The Dehn Surgery Theorem	246
	4.10.2 Sequences of volumes	249
	4.10.3 Well ordering of volumes	250
	4.10.4 Minimum volumes	251
4.11	Exercises and explorations	253
5	**Deformation spaces and the ends of manifolds**	**276**
5.1	The representation variety	276
	5.1.1 The discreteness locus	279
	5.1.2 The quasiconformal deformation space $\mathfrak{T}(G)$	280
5.2	Homotopy equivalence	281
	5.2.1 Components of the discreteness locus	284
5.3	The quasiconformal deformation space boundary	287
	5.3.1 Bumping and self-bumping	289
5.4	The three conjectures for geometrically infinite manifolds	289
5.5	Ends of hyperbolic manifolds	290
5.6	Tame manifolds	291
5.7	The Ending Lamination Theorem	296
5.8	The Double Limit Theorem	303
5.9	The Density Theorem	304
5.10	Bers slices	305
5.11	The quasifuchsian space boundary	309
	5.11.1 The Bers (analytic) boundary	310
	5.11.2 The Thurston (geometric) boundary	314
5.12	Examples of geometric limits at the Bers boundary	317
5.13	Classification of the geometric limits	323
5.14	Cannon-Thurston mappings	326
	5.14.1 The Cannon-Thurston Theorem	326
	5.14.2 Cannon-Thurston mappings and local connectivity	329
5.15	Exercises and explorations	332
6	**Hyperbolization**	**371**
6.1	Hyperbolic manifolds that fiber over a circle	371
	6.1.1 Automorphisms of surfaces	371
	6.1.2 Pseudo-Anosov mappings	372

	6.1.3	The space of hyperbolic metrics	374
	6.1.4	Fibering	374
6.2		Hyperbolic gluing boundary components	375
	6.2.1	Skinning a bordered manifold	375
	6.2.2	Totally geodesic boundary	376
	6.2.3	Gluing boundary components	378
	6.2.4	The Bounded Image Theorem	381
6.3		Hyperbolization of 3-manifolds	382
	6.3.1	Review of definitions in 3-manifold topology	382
	6.3.2	Hyperbolization	385
6.4		The three big conjectures, now theorems, for closed manifolds	386
	6.4.1	Surface subgroups of $\pi_1(\mathcal{M}(G)) = G$	387
	6.4.2	Remarks on the proof of VHT and VFT: Cubulation	390
	6.4.3	Prior computational evidence	393
6.5		Geometrization	394
6.6		Hyperbolic knots and links	396
	6.6.1	Knot complements	396
	6.6.2	Link complements	397
6.7		Computation of hyperbolic manifolds	399
6.8		The Orbifold Theorem	401
6.9		Exercises and explorations	403
7		**Line geometry**	**425**
7.1		Half-rotations	425
7.2		The Lie product	426
7.3		Square roots	429
7.4		Complex distance	431
7.5		Complex distance and line geometry	432
7.6		Exercises and explorations	433
8		**Right hexagons and hyperbolic trigonometry**	**444**
8.1		Generic right hexagons	444
8.2		The sine and cosine laws	446
8.3		Degenerate right hexagons	448
8.4		Formulas for triangles, quadrilaterals, and pentagons	450
8.5		Exercises and explorations	453
Bibliography			472
Index			495

List of Illustrations

Credits: CG, Charlie Gunn; CM, Curt McMullen; CS, Caroline Series; DD, David Dumas; DW, Dave Wright; HP, Howard Penner; JB, Jeff Brock; JP, John Parker; JR, Jonathan Rogness; KS, Ken Stephenson; RB, Robert Brooks; SL, Silvio Levy; StL, Stuart Levy; YM, Yair Minsky.

1.1	Stereographic projection (DW)	2
1.2	Invariant spiral of a loxodromic (DW)	4
1.3	Disk and upper half-plane models of $\mathbb{H}^2$ (SL)	7
1.4	Ball and upper half-space model of $\mathbb{H}^3$ (SL)	9
1.5	Tubes around geodesics (SL)	14
1.6	Universal thinness of triangles (SL)	15
1.7	Outer Circles (DW)	21
1.8	Isometric circles (SL)	22
1.9	Ideal tetrahedron (SL)	34
1.10	Minkowski 3-space (JR)	39
2.1	Rolling up an octagon (SL)	79
2.2	Schottky group's generators and quotient (SL)	81
2.3	A Schottky group (DW)	83
2.4	Modular group (DW)	97
2.5	Thrice-punctured sphere group with Farey indexing (DW)	97
2.6	The (2,3,7) group (DW)	99
3.1	Seifert–Weber dodecahedral space (SL)	139
3.2	Solid pairing tube for a rank one cusp (JR)	145
3.3	Sharing a puncture (SL,JR)	146
3.4	Solid cusp torus for a rank two cusp (SL)	146
3.5	Cutting a solid torus along a compressing disk (SL)	151
3.6	A geodesic lamination (DW)	159
3.7	Contrasting torus laminations (DD)	160
3.8	Dome over a component of a quasifuchsian orthogonal set (YM, DW)	171
3.9	Relative compact core (JR)	181
3.10	The figure-8 knot (SL)	191
4.1	A cyclic loxodromic group near its geometric limit (DW)	242
4.2	A tripod (SL)	258
4.3	Computation of the modulus of a marked quadrilateral (RB, KS)	264
4.4	Circle packing an owl (KS)	267
4.5	Pairing punctures with tubes (SL)	273

xi

4.6	Earle–Marden coordinates for 4-punctured spheres and 2-punctured tori (DW)	274
4.7	Earle–Marden coordinates for genus-2 surfaces (DW)	275
5.1	A maximal Schottky boundary cusp (DW)	288
5.2	Simultaneous uniformization (SL)	306
5.3	The limit set of a once-punctured torus quasifuchsian group (DW)	307
5.4	Opening the cusp of Figure 5.3 (DW)	307
5.5	A Bers slice (DD)	309
5.6	Limit set of a two-generator group with elliptic commutator (DW)	311
5.7	A algebraic limit on the deformation space boundary of Figure 5.6 (DW)	312
5.8	A singly degenerate group (DW)	313
5.9	The Apollonian Gasket (DW)	317
5.10	An augmented Apollonian Gasket (DW)	318
5.11	Infinitely generated geometric limit(JR)	319
5.12	Iterating a Dehn twist on a Bers slice: algebraic and geometric limits (SL)	319
5.13	Limit set of an algebraic limit at a cusp (CM, JB)	320
5.14	Limit set of the geometric limit at the same cusp (CM, JB)	320
5.15	Iterating a partial pseudo-Anosov: the algebraic and geometric limits (SL)	323
5.16	Limit set of the algebraic limit of the iteration (CM, JB)	324
5.17	Limit set of the geometric limit of the iteration (CM, JB)	325
5.18	A Dehn twist in an annulus (SL)	339
5.19	The local structure of a train track (SL)	367
6.1	Sierpiński gasket limit set (DW), Helaman Ferguson's *Knotted Wye*, and Thurston's wormhole (SL)	379
6.2	Colored limit set of a 2-punctured torus quasifuchsian group (DW)	384
6.3	Cubulation of a dodecahedron (SL)	392
6.4	Borromean ring complement (CG,StL;The Geometry Center)	398
6.5	The Whitehead link and the Borromean rings (SL)	401
6.6	Extended Bers slice (DD)	414
7.1	Complex distance between lines (SL)	430
7.2	The limit set of a twice-punctured torus quasifuchsian group (DW)	440
8.1	The generic right hexagon (SL)	446
8.2	A right triangle or degenerate hexagon (HP, SL)	450
8.3	Planar pentagons with four right angles (HP, SL)	451
8.4	A quadrilateral with three right angles (HP, SL)	452
8.5	A generic triangle (HP, SL)	452
8.6	Planar quadrilaterals with two right angles (HP, SL)	454
8.7	Planar right hexagons (HP, SL)	455
8.8	A pair of pants with a seam (HP, SL)	456
8.9	Cylindrical coordinate approximation (SL)	458
8.10	A right hexagon expressing bending (JP, CS, SL)	467

Preface

To a topologist a teacup is the same as a bagel, but they are not the same to a geometer. By analogy, it is one thing to know the topology of a 3-manifold, another thing entirely to know its geometry—to find its shortest curves and their lengths, to make constructions with polyhedra, etc. In a word, we want to do geometry in the manifold just like we do geometry in euclidean space.

But do general 3-manifolds have "natural" metrics? For a start we might wonder when they carry one of the standards: the euclidean, spherical or hyperbolic metric. The latter is least known and not often taught; in the stream of mathematics it has always been something of an outlier. However it turns out that it is a big mistake to just ignore it! We now know that the interiors of "most" compact 3-manifolds carry a hyperbolic metric.

It is the purpose of this book to explain the geometry of hyperbolic manifolds. We will examine both the existence theory and the structure theory.

Why embark on such a study? Well after all, we do live in three dimensions; our brains are specifically wired to see well in space. It seems perfectly reasonable if not compelling to respond to the challenge of understanding the range of possibilities. For a while, it had even been considered that our own visual universe may be hyperbolic, although it is now believed that it is euclidean.

The twentieth-century history. Although Poincaré recognized in 1881 that Möbius transformations extend from the complex plane to upper half-space, the development of the theory of three-dimensional hyperbolic manifolds had to wait for progress in three-dimensional topology. It was as late as the mid-1950s that Papakyriakopoulos confirmed the validity of Dehn's Lemma and the Loop Theorem. Once that occurred, the wraps were off.

In the early 1960s, while 3-manifold topology was booming ahead, the theory of kleinian groups was abruptly awoken from its long somnolence by a brilliant discovery of Lars Ahlfors. Kleinian groups are the discrete isometry groups of hyperbolic 3-space. Working (as always) in the context of complex analysis, Ahlfors discovered their finiteness property. This was followed by Mostow's contrasting discovery that closed hyperbolic manifolds of dimension $n \geq 3$ are uniquely determined up to isometry by their isomorphism class. This too came as a bombshell as it is false for $n = 2$. Then came Bers' study of quasifuchsian groups and his and Maskit's fundamental discoveries of "degenerate groups" as limits of them. Along a different line, Jørgensen developed the methods for dealing with sequences of kleinian groups, recognizing the existence of two distinct kinds of convergence which he called "algebraic" and "geometric". He also discovered a key class of examples, namely hyperbolic 3-manifolds that fiber over the circle.

It wasn't until the late 1960s that 3-manifold topology was sufficiently understood, most directly by Waldhausen's work, and the fateful marriage of 3-manifold topology to the complex analysis of the group action on $\mathbb{S}^2$ occurred. The first application was to the classification and analysis of geometrically finite groups and their quotient manifolds.

During the 1960s and 1970s, Riley discovered a slew of faithful representations of knot and link groups in PSL$(2, \mathbb{C})$. Although these were seen as curiosities at the time, his examples pressed further the question of just what class of 3-manifolds did the hyperbolic manifolds represent? Maskit had proposed using his combination theorems to construct all hyperbolic manifolds from elementary ones. Yet Peter Scott pointed out that the combinations that were then feasible would construct only a limited class of 3-manifolds.

So by the mid-1970s there was a nice theory, part complex analysis, part three-dimensional geometry and topology, part algebra. No-one had the slightest idea as to what the scope of the theory really was. Did kleinian groups represent a large class of manifolds, or only a small sporadic class?

The stage (but not the players) was ready for the dramatic entrance in the mid-1970s of Thurston. He arrived with a proof that the interior of "most" compact 3-manifolds has a hyperbolic structure. He brought with him an amazingly original, exotic, and very powerful set of topological/geometrical tools for exploring hyperbolic manifolds. The subject of two- and three-dimensional topology and geometry was never to be the same again.

This book. Having witnessed at first hand the transition from a special topic in complex analysis to a subject of broad significance and application in mathematics, it seemed appropriate to write a book to record and explain the transformation. My idea was to try to make the subject accessible to beginning graduate students with minimal specific prerequisites. Yet I wanted to leave students with more than a routine compendium of elementary facts. Rather I thought students should see the big picture, as if climbing a watchtower to overlook the forest. Each student should end his or her studies having a personal response to the timeless question: What is this good for?

With such thoughts in mind, I have tried to give a solid introduction and at the same time to provide a broad overview of the subject as it is today. In fact today, the subject has reached a certain maturity. The characterization of those compact manifolds whose interiors carry a hyperbolic structure is complete, the final step being provided by Perelman's confirmation of the Geometrization Conjecture. Attention turned to the analysis of the internal structure of hyperbolic manifolds assuming only a finitely generated fundamental group. The three big conjectures left over from the 1960s and 1970s have been solved: Tameness, Density, and classification of the ends (ideal boundary components). And recently, entirely new and quite surprising structures to be described below have been discovered. If one is willing to climb the watchtower, the view is quite remarkable.

It is a challenge to carry out the plan as outlined. The foundation of the subject rests on elements of three-dimensional topology, hyperbolic geometry, and modern complex analysis. None of these are regularly covered in courses at most places.

I have attempted to meet the challenge as follows. The presentation of the basic facts is fairly rigorous. These are included in the first four chapters, plus the optional Chapters 7 and 8. These chapters include crash courses in three-manifold topology, covering surfaces and manifolds, quasiconformal mappings, and Riemann surface theory. With the basic information under our belts, Chapters 5 and 6 (as well as parts of Chapters 3 and 4) are expository, without most proofs. The reader will find there both the Hyperbolization Theorem and the newly discovered structural properties of general hyperbolic manifolds.

At the end of each chapter is a long section titled "Exercises and Explorations". Some of these are genuine exercises and/or important additional information directly related to the material in the chapter. Others dig away a bit at the proofs of some of the theorems by introducing new tools they have required. Still others are included to point out various paths one can follow into the deeper forest and beauty spots one can find there. Thus there are not only capsule introductions to big fields like geometric group theory, but presentations of other more circumscribed topics that I (at least) find fascinating and relevant.

The nineteenth-century history. For the full story consult Jeremy Gray's new book, particularly Gray [2013, Ch. 3], for a comprehensive treatment of this period.

The history of non-Euclidean geometry in the early nineteenth century is fascinating because of a host of conflicted issues concerning axiom systems in geometry, and the nature of physical space [Gray 1986; 2002].

Jeremy Gray [2002] writes:

Few topics are as elusive in the history of mathematics as Gauss's claim to be a, or even the, discoverer of Non-Euclidean geometry. Answers to this conundrum often depend on unspoken, even shifting, ideas about what it could mean to make such a discovery.... [A]mbiguities in the theory of Fourier series can be productive in a way that a flawed presentation of a new geometry cannot be, because there is no instinctive set of judgments either way in the first case, but all manner of training, education, philosophy and belief stacked against the novelties in the second case.

Gray goes on to quote from Gauss's 1824 writings:

...the assumption that the angle sum is less than 180° leads to a geometry quite different from Euclid's, logically coherent, and one that I am entirely satisfied with. It depends on a constant, which is not given a priori. The larger the constant, the closer the geometry to Euclid's....

The theorems are paradoxical but not self-contradictory or illogical. . . . All my efforts to find a contradiction have failed, the only thing that our understanding finds contradictory is that, if the geometry were to be true, there would be an absolute (if unknown to us) measure of length a priori. . . . As a joke I've even wished Euclidean geometry was not true, for then we would have an absolute measure of length a priori.

From his detailed study of the history, Gray's conclusion expressed in his Zurich lecture is that the birth of non-Euclidean geometry should be attributed to the independently written foundational papers of Lobachevsky in 1829 and Bólyai in 1832. As expressed in Milnor [1994, p. 246], those two were the first "with the courage to publish" accounts of the new theory. Still,

[f]or the first forty years or so of its history, the field of non-Euclidean geometry existed in a kind of limbo, divorced from the rest of mathematics, and without any firm foundation.

This state of affairs changed upon Beltrami's introduction in 1868 of the methods of differential geometry, working with constant curvature surfaces in general. He gave the first global description of what we now call hyperbolic space. See Gray [1986, p. 351], Milnor [1994, p. 246], Stillwell [1996, pp. 7–62].

It was Poincaré who brought two-dimensional hyperbolic geometry into the form we study today. He showed how it was relevant to topology, differential equations, and number theory. Again I quote Gray, in his translation of Poincaré's work of 1880 [Gray 1986, p. 268–9].

There is a direct connection between the preceding considerations and the non-Euclidean geometry of Lobachevskii. What indeed is a geometry? It is the study of a *group of operations* formed by the displacements one can apply to a figure without deforming it. In Euclidean geometry this group reduces to *rotations* and *translations*. In the pseudo-geometry of Lobachevskii it is more complicated. . . [Poincaré's emphasis].

As already mentioned, the first appearance of what we now call Poincaré's conformal model of non-Euclidean space was in his seminal 1881 paper on kleinian groups. He showed that the action of Möbius transformations in the plane had a natural extension to a conformal action in the upper half-space model.

Actually the names "fuchsian" and "kleinian" for the isometry groups of two- and three-dimensional space were attached by Poincaré. However Poincaré's choice more reflects his generosity of spirit toward Fuchs and Klein than the mathematical reality. Klein himself objected to the name "fuchsian". His objection in turn prompted Poincaré to introduce the name "kleinian" for the discontinuous groups that do not preserve a circle. The more apt name would perhaps have been "Poincaré groups" to cover both cases.

Recent history. A reader turning to this book may benefit by first consulting some of the fine elementary texts now available before diving directly into the theory of hyperbolic 3-manifolds. I will mention in particular: Jim Anderson's text [Anderson 2005], Frances Bonahon's text [Bonahon 2009], Jeff Weeks' text [Weeks 2002], as well as the classic book of W. Thurston [Thurston 1997]. Ours remains a tough subject to enter because of the range of knowledge involved—geometry, topology, analysis, algebra, number theory—-still, those who take the plunge find satisfaction in the subject's richness.

Within the past three years, our subject has gone well beyond the major accomplishments of the Thurston era, which include proofs of Tameness, Density, and the Ending Lamination, and, most importantly, Perleman's proof of the full Geometrization Conjecture. In March, 2012 was the dramatic announcement by Ian Agol of the solution of the Virtual Haken and Virtual Fibering Conjectures for hyperbolic manifolds. As the principal architects of the proof in drawing together new elements of hyperbolic geometry, cubical complexes, and geometric group theory, Ian Agol and Dani Wise shared the 2013 AMS Veblen prize.

Their proof required the Surface Subgroup Conjecture which had been confirmed to great acclaim by Kahn and Markovic a few years earlier. In addition, by the same method Kahn and Markovic confirmed another major longstanding question: the Ehrenpries Conjecture about Riemann surfaces. Jeremy Kahn and Vlad Markovic were awarded the 2012 Clay Prize for their accomplishments. Further remarkable consequences are duly reported in Chapter 6.

Mahan Mj resolved a different longstanding problem. He proved that connected limit sets of kleinian groups are locally connected. He proved this as a consequence of his existence proof of general "Cannon-Thurston maps". This work is discussed in Chapter 5.

The resolution of another important issue has been announced by Ken'ichi Ohtsuka and Teruhiko Soma with a paper in arXiv. They determined all possible geometric limits at quasifuchsian space boundaries. It too is discussed in Chapter 5.

Although the bibliography is extensive, it is hardly inclusive of all the papers that have contributed to the subject. Every signature accomplishment has been built on prior work of many others. I have tended to include references only to papers that are the most comprehensive and those which put down the last word. Upon referring to the referenced papers, one gets a better idea of the extent of the prior contributions culminating in a final result.

Acknowledgments. It is a great pleasure to thank the many people who have helped bring the book and its predecessor to fruition.

First I want to acknowledge the essential contributions of my friend and colleague Troels Jørgensen. Over more than 25 years we walked in the forest together discussing and admiring the landscape our studies revealed. In particular we discussed the "universal properties" of Chapter 3 for years, until it was too late to publish them. Chapters 7 and 8 are based on his private lectures.

David Wright kindly computed a number of limit sets of kleinian groups, some never before seen, others adapted from pictures created for *Indra's Pearls* [Mumford et al. 2002]. The extent of his contribution is evident from the list of figures. His pictures can be downloaded from www.okstate.edu/~wrightd/Marden together with computational details. In addition, David Dumas was willing to share his visualization of a Bers slice amidst the surrounding archipelago of discreteness components. Jeff Brock contributed his pictures of algebraic and geometric limits that originally appeared in Brock [2001b]; these too can be seen on www.math.brown.edu/~brock. The presence of the many artfully crafted pictures is a tangible expression of the mathematical beauty of the subject.

Jeff Brock and David Dumas created the stunning frontispiece "Thurston's Jewel"; its intricate rendering required 1425 CPU-hours. My colleague Jon Rogness created two clarifying diagrams for the text. The full list of figures and their creators is on p. xi.

The list of those providing timely advice on mathematical issues includes Scot Adams, Ian Agol, Shinpei Baba, Ara Basmajian, Brian Bowditch, Jeff Brock, Dick Canary, Daryl Cooper, David Dumas, Cliff Earle, David Epstein, Ren Guo, Misha Kapovich, Sadayoshi Kojima, Sara Maloni, Vlad Markovic, Howie Masur, Yair Minsky, Mahan Mj, Ken'ici Ohshika, Peter Scott, and Saul Schleimer.

I could not have completed the book in the present form without the expert guidance and participation of Silvio Levy. Not only did he help with the mathematics, he formatted much of the LaTeX, and led me out of numerous LaTeX jams.

I want to acknowledge the institutional support from the Forschungsinstitut für Mathematik at ETH in Zurich, the Maths Research Center, University of Warwick, and not least, from my own department, the School of Mathematics of the University of Minnesota. In my semester course Math 8380, I was able to present a solid introduction and overview of the subject based on the main points in the first six chapters.

My editor at Cambridge University Press is David Tranah. One could not have asked for a more supportive person, a wise advisor in all the publication issues. I am grateful to Caroline Series for introducing him.

So here we are today, nearly 130 years after Poincaré and approaching 200 after the initial ferment of ideas of Gauss. We are witnessing a full flowering of the vision and struggle for understanding of the nineteenth-century masters. Still, the final word remains an elusive goal.

Albert Marden
am@umn.edu
Minneapolis, Minnesota
July, 2015

1

Hyperbolic space and its isometries

In this chapter we gather together basic information about the geometry of two- and three-dimensional hyperbolic spaces and their isometries. This will set the stage for our study of quotient manifolds and orbifolds which begins in the next chapter.

1.1 Möbius transformations

A *Möbius transformation* in the unit sphere $\mathbb{S}^n$ of dimension n is, by definition, the result of a composition of reflections in $(n-1)$-dimensional spheres in $\mathbb{S}^n$. It will be orientation-preserving if it is the composition of an even number of reflections. A defining property is that Möbius transformations send $(n-1)$-dimensional spheres onto $(n-1)$-dimensional spheres. Automatically, a symmetric pair of points (with respect to reflection) about one sphere gets sent to a symmetric pair about the other.

*From now on, the unqualified term **Möbius transformation** will be reserved for those that preserve orientation.* The orientation-reversing kind will be called anti-Möbius transformations. For a discussion of the latter, see Exercise (1-31) at the end of the chapter.

The study of hyperbolic 3-manifolds is intimately connected with the study of Möbius transformations on the two-dimensional sphere $\mathbb{S}^2$. Via stereographic projection (Figure 1.1), $\mathbb{S}^2$ is homeomorphic to the extended plane $\mathbb{C} \cup \infty$, and we will freely use this fact to change points of view between the extended plane and the 2-sphere. Under stereographic projection, the collection of circles and straight lines in $\mathbb{C}$ corresponds to the collection of circles on $\mathbb{S}^2$; a straight line in $\mathbb{C}$ corresponds to a circle on $\mathbb{S}^2$ through the north pole. With this correspondence in mind, we can refer to the collection of circles and lines in $\mathbb{C}$ simply as "circles". Moreover stereographic projection is a conformal map, that is, it preserves angles between intersecting arcs—in particular, angles of intersection between circles.

Möbius transformations in two dimensions are *fractional linear transformations* of the extended plane. That is, a Möbius transformation acting on $\mathbb{C} \cup \infty$ has the form

$$z \mapsto A(z) = \frac{az+b}{cz+d}, \quad \text{with } a, b, c, d \in \mathbb{C} \text{ such that } ad - bc \neq 0. \tag{1.1}$$

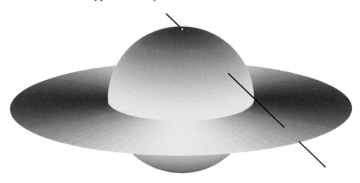

Fig. 1.1. Stereographic projection.

(When $ad - bc = 0$ the expression on the right is a constant, so the map is not a Möbius transformation.) As we will see shortly, a map of this form can indeed be expressed as the composition of an even number of reflections in circles (in fact, two or four circles: see Exercise (1-7)). The symmetry properties of such maps are established in Exercise (1-2).

Möbius transformations are conformal maps. In fact, the only conformal homeomorphisms of $\mathbb{C} \cup \infty$ are Möbius transformations.

We will generally assume that the representation in (1.1) is *normalized*, meaning that $ad - bc = 1$. Then we can identify the group of Möbius transformations with the quotient $\mathrm{PSL}(2, \mathbb{C}) := \mathrm{SL}(2, \mathbb{C})/ \pm I$, where $\mathrm{SL}(2, \mathbb{C})$ is the group of 2×2 matrices of determinant one and I is the identity matrix:

$$A(z) = \frac{az + b}{cz + d} \longleftrightarrow \pm \begin{pmatrix} a & b \\ c & d \end{pmatrix}, \qquad A^{-1}(z) \longleftrightarrow \pm \begin{pmatrix} d & -b \\ -c & a \end{pmatrix}.$$

The $\pm$ ambiguity cannot be avoided. We will not keep inserting it, unless it plays an essential role. In any case the value of changing from transformations to matrices lies mainly in the algebra of composition. If A, B are Möbius transformations, the Möbius transformation resulting from the application of A followed by B is written BA; the corresponding matrix is just the usual product BA of the component matrices, in the order written. The $\pm$ ambiguity follows along. We will hop from one to the other, the representation as a transformation to the representation as a matrix, depending on which best suits the situation, without changing the labeling.

Two Möbius transformations A, B are *conjugate* if there is a Möbius transformation U such that $B = UAU^{-1}$. Conjugate transformations have the same geometry: U effects transfer of the geometry of A to that of B.

The expression $ABA^{-1}B^{-1}$ is called the *commutator* of A and B and written as $[A, B]$. Two elements commute if and only if their commutator is the identity.[1]

The *trace* of a Möbius transformation A is, by definition, the trace of the normalized matrix of A:

---

[1] The alternative conventions $[A, B] = B^{-1}A^{-1}BA$ or $A^{-1}B^{-1}AB$ are preferred by some authors; they do the same job, but the formulas come out differently.

$$\tau_A = \operatorname{tr} A = \pm(a + d).$$

It is invariant under conjugation. The $\pm$ ambiguity can be avoided either by using τ_A^2 or by specifying $0 \le \arg \tau_A < \pi$.

By solving the equation $A(z) = z$, we find that a Möbius transformation $\ne$ id has one or two fixed points in $\mathbb{S}^2$, namely $(a - d \pm \sqrt{\tau_A{}^2 - 4})/2c$, when $c \ne 0$, or otherwise the points ∞ and $b/(d - a) = ab/(1 - a^2)$. Here $A = \begin{pmatrix} a & b \\ c & d \end{pmatrix}$, $ad - bc = 1$. Only the identity can have three fixed points.

Given three distinct points $(p_2, p_3, p_4) \in \mathbb{S}^2$, there exists a unique Möbius transformation T sending p_2 to 0, p_3 to 1, p_4 to ∞. It is given by

$$T(z) = \frac{(z - p_2)(p_3 - p_4)}{(z - p_4)(p_3 - p_2)} = (z, p_2, p_3, p_4),$$

when none of the points p_i is ∞. By taking the limit as some $p_i \to \infty$, we obtain the correct expression for $p_i = \infty$. The expression (z, p_2, p_3, p_4) is called the *cross ratio* of the four points.[2]

- Given distinct triples (p_2, p_3, p_4), (p_2', p_3', p_4') there is a unique Möbius transformation taking one triple onto the other.
- Cross ratios are invariant under Möbius transformations:

$$(Az, Ap_2, Ap_3, Ap_4) = (z, p_2, p_3, p_4) \quad \text{for any } A.$$

To establish the second item, note that both Möbius transformations $S(z) = (z, Ap_2, Ap_3, Ap_4)$ and $T \circ A^{-1}(z)$ take (Ap_2, Ap_3, Ap_4) to $(0, 1, \infty)$ so they are identical. Replace z by $A(z)$.

Apart from the identity, Möbius transformations fall into one of three types:

A is *parabolic* if the following equivalent properties hold.

- A is conjugate to $z \mapsto z + 1$.
- A has exactly one fixed point in $\mathbb{S}^2$.
- $\tau_A = \pm 2$ and $A \ne$ id.

A is *elliptic* if the following equivalent properties hold.

- A is conjugate to $z \mapsto e^{2i\theta} z$, with $2\theta \not\equiv 2\pi$.
- $\tau_A \in (-2, +2)$.
- A has exactly two fixed points, and the derivative of A has absolute value 1 at each of them.

A is *loxodromic* if the following equivalent properties hold.

- A is conjugate to $z \mapsto \lambda^2 z$, with $|\lambda| > 1$.
- $\tau_A \in \mathbb{C} \setminus [-2, +2]$.
- A has two fixed points; at one $|A'| < 1$ (attracting), at the other $|A'| > 1$ (repelling).

---

[2] In the induced direction on $\mathbb{R}$, UHP lies to the left; also $(z, 0, 1, \infty) = z$. A common alternate definition results in $(z, 1, 0, \infty) = z$.

We will use the term *standard forms* for the conjugates just listed. The geometry of a general normalized Möbius transformation A is most easily read off from the conjugate standard form. Note that the elliptic $z \mapsto 1/z$ is conjugate to $z \mapsto -z$.

A loxodromic Möbius transformation A has a collection of *loxodromic curves* or *invariant spirals* in $\mathbb{S}^2$. (In navigation, a *loxodromic curve* or *rhumb line* is a path of constant bearing: it makes equal oblique angles with all meridians, and so coils around the poles without ever reaching them.) For the standard form $z \mapsto \lambda^2 z$, one such spiral is given by

$$z(t) = \lambda^{2t}, \quad -\infty < t < \infty.$$

If σ denotes the segment $0 \leq t < 1$ of the spiral, the various images $\{A^n(\sigma)\}$ cover the spiral without overlap. See Figure 1.2.

For additional structure in special cases see Wright [2006].

The term *hyperbolic transformation* has historically been used to designate a loxodromic transformation whose trace is real. Such a transformation is conjugate to $z \mapsto \lambda^2 z$ with $\lambda > 1$. Nowadays the term "hyperbolic" is also used for a loxodromic element acting in hyperbolic 3-space.

The classification is proved by first conjugating A so that one fixed point lies at ∞ and the other, if there is one, at 0. The further conjugation $z \mapsto 1/z$ that interchanges 0 and ∞ may be needed to put the attracting fixed point at ∞.

If $p \in \mathbb{C}$ is a fixed point of $A \neq \mathrm{id}$, p is attracting if and only if $|A'(p)| < 1$ and repelling if and only if $|A'(p)| > 1$. The transformation A is parabolic if and only if $A'(p) = 1$; A is elliptic if and only if $|A'(p)| = 1$ but $A'(p) \neq 1$.

Upon referring to the normalized matrix $A = \left(\begin{smallmatrix} a & b \\ c & d \end{smallmatrix} \right)$, we find that the *eigenvalues* are $\lambda, \lambda^{-1} = \frac{1}{2}(\mathrm{tr}\, A \pm \sqrt{\mathrm{tr}^2 A - 4})$. The corresponding eigenvectors $\left(\begin{smallmatrix} \alpha \\ \beta \end{smallmatrix} \right)$ satisfy

$$\frac{\alpha}{\beta} = \frac{\lambda - d}{c} = p \quad \text{and} \quad \frac{\alpha}{\beta} = \frac{\lambda^{-1} - d}{c} = q,$$

where p, q are the fixed points. Like the trace, the eigenvalues are invariant under conjugation. The eigenvalues of an elliptic transformation have the form $e^{\pm i\theta}$ and the

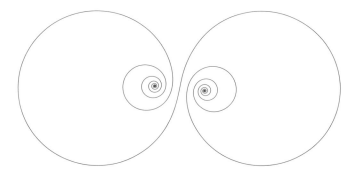

Fig. 1.2. Invariant spiral of a loxodromic with trace $\lambda + \lambda^{-1} = 1.976 + 0.005i$.

trace is $2\cos\theta$. A loxodromic transformation has eigenvalues $\lambda^{\pm 1}$ and trace $\lambda + \lambda^{-1}$. We can choose λ so that $|\lambda| > 1$, that is, so that λ is the expanding eigenvalue.

The expanding eigenvalue of a loxodromic element A can be expressed as a cross ratio by the formula

$$\lambda^2 = (z, p_+, A(z), p_-),$$

where p_+, p_- are the attracting and repelling fixed points. (It is enough to confirm this when $p_+ = \infty$ and $p_- = 0$.)

We can write $A = \begin{pmatrix} a & b \\ c & d \end{pmatrix}$ as

$$Az = \frac{1}{-c^2(z + d/c)} + \frac{a}{c} \quad \text{if } c \neq 0, \qquad Az = \frac{a}{d}\left(z + \frac{b}{a}\right) \quad \text{if } c = 0. \qquad (1.2)$$

This expresses A in terms of simple building blocks: maps in standard form, plus the map $z \mapsto 1/z$. Each of these has the property of preserving (generalized) circles. Therefore any Möbius transformation preserves circles, as mentioned earlier. Likewise each building block is easily seen to be a composition of two reflections, so a Möbius transformation is the composition of an even number of reflections.

Three distinct points p_2, p_3, p_4 uniquely determine a circle C, with an orientation determined by their order. When C is a proper circle or any closed curve, the *positive orientation* is determined by a sequence of points p_2, p_3, p_4 in counterclockwise direction, that is when the interior of the curve lies to the left. Let q_2, q_3, q_4 be another set of distinct points, and C' the circle through them. The Möbius transformation T that sends $p_i \to q_i$ automatically sends C onto C'. If both are proper circles, T sends the interior of C to the interior of C' if and only if the triples give both circles the same, positive or negative, orientations.

The transformation $T : z \to w$ can be expressed in terms of cross ratios as

$$(w, q_2, q_3, q_4) = (z, p_2, p_3, p_4).$$

But if we focus simply on sending C to C', and a designated side of C to a designated side of C', it is easier to find T by cross ratio using the symmetry property: A Möbius transformation sends points symmetric with respect to reflection in one circle, to a pair of points symmetric in its image (Exercise (1-2)). For a proper circle, the most conspicuous symmetric points are its center and ∞. If the circle is $\{z : |z - z_0| < r\}$, the points $|a| < 1$ and $|a^*| > 1$ are symmetric if and only if $a^* - z_0 = r^2/\overline{(a - z_0)}$.

A cross ratio (p, p_2, p_3, p_4) is real if and only if the four points lie on a circle in $\mathbb{S}^2$. The cross ratio is positive if and only if (p, p_3, p_4) gives the circle the same orientation as (p_2, p_3, p_4).

We are now ready to show that Möbius transformations in $\mathbb{C} \cup \infty$ can be extended to Möbius transformations acting in upper half-space (UHS) $\{\vec{x} = (z, t) : z \in \mathbb{C}, t > 0\}$ (and, correspondingly, lower half-space (LHS)). The simplest way to see this is by applying the following observation. Each Möbius transformation is the composition of an even number of reflections in circles or lines in $\mathbb{C}$. A reflection in a circle extends naturally to the reflection in the upper hemisphere bounded by that circle. Likewise the reflection in a straight line extends to the reflection in the vertical half-plane bounded by that line. (The same argument shows that Möbius transformations on $\mathbb{S}^n = \mathbb{R}^n \cup \{\infty\}$ extend to upper half $(n+1)$-space.)

A Möbius transformation acting on $\mathbb{C} \cup \infty$ sends a given circle to another circle or line. Its extension to upper half-space will therefore map the hemisphere bounded by the circle to the hemisphere or half-plane bounded by the image of the circle. We conclude that the extension to upper half-space maps the totality of hemispheres and vertical half-planes onto itself.

If two hemispheres intersect, or a hemisphere and a vertical half-plane intersect, the intersection is a semicircle which is orthogonal to $\mathbb{C}$. If two vertical half-planes intersect, they intersect in a vertical half-line orthogonal to $\mathbb{C}$. The extension of a Möbius transformation thus maps the totality of half-lines and semicircles orthogonal to $\mathbb{C}$ onto itself. The dihedral angles between intersecting hemispheres is the same as the angle of intersection between their bounding circles in $\mathbb{C}$.

It is useful to explicitly work out the formula for extension to upper half-space $\{\vec{x} = (z, t) : z \in \mathbb{C}, \, t > 0\}$. We first extend the building blocks. First,

$$z \mapsto az \qquad \text{becomes} \quad (z, t) \mapsto (az, |a|\, t);$$
$$z \mapsto z + b \quad \text{becomes} \quad (z, t) \mapsto (z + b, t).$$

The inversion $z \mapsto z^{-1}$ is most easily dealt with as the composition of two anti-Möbius transformations: $z \mapsto \bar{z}$ (reflection in a line) and $z \mapsto z/|z|^2 = \bar{z}^{-1}$ (reflection in the unit circle). Extending to reflections in a vertical plane and the unit hemisphere, we get respectively $(z, t) \mapsto (\bar{z}, t)$ and

$$\vec{x} \mapsto \frac{\vec{x}}{|\vec{x}|^2} \quad \text{or} \quad (z, t) \mapsto \left(\frac{z}{|z|^2 + t^2}, \frac{t}{|z|^2 + t^2} \right).$$

Therefore,

$$z \mapsto \frac{1}{z} \quad \text{becomes} \quad (z, t) \mapsto \left(\frac{\bar{z}}{|z|^2 + t^2}, \frac{t}{|z|^2 + t^2} \right).$$

Composing the building blocks we find that the extension of $\left(\begin{smallmatrix} a & b \\ c & d \end{smallmatrix} \right)$ is

$$(z, t) \mapsto \left(-\frac{\overline{z + d/c}}{c^2 \left(|z + d/c|^2 + t^2 \right)} + \frac{a}{c}, \, \frac{t}{|c|^2 \left(|z + d/c|^2 + t^2 \right)} \right) \qquad \text{when } c \neq 0.$$

$$(z, t) \mapsto \left(\frac{a}{d}(z + b/a), \, \left| \frac{a}{d} \right| t \right) \qquad \text{when } c = 0.$$

A Möbius transformation acting in UHS acts symmetrically in LHS under the correspondence $(z, t) \leftrightarrow (z, -t)$.

1.2 Hyperbolic geometry

In the euclidean plane, there is exactly one line through a given point and not meeting a given line disjoint from the point; this is the famous fifth postulate of Euclid. It gradually became clear in the nineteenth century that one can have a self-consistent and interesting geometry where this postulate is not valid—where "parallel" lines are not unique and indeed exist in uncountable abundance. This became known as *hyperbolic geometry*. Though the name was bestowed in connection with conics and projective geometry [Klein 1871, p. 72], it is a doubly felicitous choice, because the Greeks had named the hyperbola after the word for excess (compare "hyperbole", from the same Greek word). Hyperbolic geometric certainly has an excess of lines—and of "room"—compared to euclidean geometry!

Here are some of the salient features that distinguish hyperbolic geometry from the familiar euclidean and spherical geometry.

(i) The angle sum Σ of a hyperbolic triangle $\triangle$ satisfies $0 < \Sigma < \pi$; in fact, Σ equals $\pi - \text{area } \triangle$. The limiting case $\Sigma = 0$ is achieved by *ideal triangles* whose vertices are "at infinity": we will have more to say about such *ideal vertices* soon (p. 14). At the other extreme, the case $\Sigma = \pi$ is the limiting case of hyperbolic triangles of very small area. Indeed, on the infinitesimal scale, hyperbolic geometry is euclidean.

(ii) There are no similarities in hyperbolic space—one cannot scale a figure up or down without changing its angles and shape. It follows, for instance, that all hyperbolic triangles with the same angles are isometric (hyperbolic triangles are "rigid"), and also that the choice of a unit of length is not arbitrary, as in euclidean space; one can privilege a unit having some special property, say the side length of an equilateral triangle whose vertex angles are $\pi/4$.

(iii) For any $0 \leq \theta < \pi/(n-2)$ there is a regular n-sided hyperbolic polygon with vertex angles θ. More generally, a necessary and sufficient condition for the existence of an n-sided convex polygon with vertex angles θ_i (with $0 \leq \theta_i < \pi$) in clockwise order is that $\sum \theta_i < (n-2)\pi$. The polygon is uniquely determined up to isometry and its area is $(n-2)\pi - \sum \theta_i$.

We now discuss the most commonly used *conformal models* of the hyperbolic plane and of hyperbolic space with their metrics.

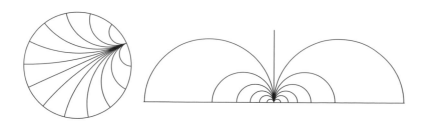

Fig. 1.3. Disk and upper half-plane models of $\mathbb{H}^2$ showing the same geodesics.

1.2.1 The hyperbolic plane

The *upper half-plane model* is $\{z \in \mathbb{C} : \operatorname{Im} z > 0\}$ with the metric

$$ds = \frac{|dz|}{\operatorname{Im} z}.$$

Here $\operatorname{Im} z$ is the notation for the imaginary part. The *unit disk model* is $\{z \in \mathbb{C} : |z| < 1\}$ with the metric

$$ds = \frac{2\,|dz|}{1 - |z|^2}.$$

The two models are equivalent under any Möbius transformation that maps the upper half-plane onto the unit disk. We will denote either one of these models by $\mathbb{H}^2$, the notation for the *hyperbolic plane*. These models have the following properties.

(i) The metrics are *infinitesimally euclidean*; at each point they equal a rescaled euclidean metric. Thus the angle between two curves in the disk or upper half-plane is the same whether measured in the hyperbolic or the euclidean geometry; as a result these models are often called *conformal*. (For other models see Exercise (1-25) and following.)

(ii) $\mathbb{H}^2$ is *complete* in its metric. Every arc tending to the "boundary" has infinite length.

(iii) The hyperbolic metric is invariant under any Möbius transformation that maps the upper half-space or disk model onto itself. These transformations comprise the full group of orientation-preserving isometries.

(iv) The hyperbolic lines (geodesics) in the upper half-plane model are semicircles orthogonal to $\mathbb{R}$ and vertical half lines. In the disk model they are diameters and circular arcs orthogonal to $\{|z| = 1\}$.

(v) Two hyperbolic triangles with the same vertex angles are isometric.

Proof of Property (v). Applying an isometry, we may assume that the base of the two triangles Δ, Δ' lie on the same geodesic ℓ, and that the left side of the two triangles also lie on the same geodesic ℓ^*. Denote the right edges by δ, δ'. If $\delta = \delta'$ then $\Delta = \Delta'$.

Now two distinct geodesics, which intersect a third geodesic at the same angle cannot themselves intersect. For then the angle sum would exceed π.

In view of this fact, assume $\delta \cap \ell$ separates $\delta' \cap \ell$ from the vertex $\ell^* \cap \ell$. δ' cannot intersect δ. Therefore δ' cannot intersect ℓ^*, in contradiction to the assumption that Δ' is a triangle. $\qquad\qquad\square$

1.2.2 Hyperbolic space

The *upper half-space model* is $\{(z, t) : z \in \mathbb{C},\ t > 0\}$ with the metric

$$ds = \frac{|d\vec{x}|}{t}, \qquad |d\vec{x}|^2 = |dz|^2 + dt^2.$$

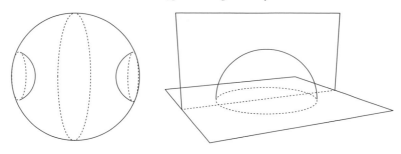

Fig. 1.4. Ball and upper half-space model of $\mathbb{H}^3$ showing geodesic planes.

The *ball model* is $\{\vec{x} \in \mathbb{R}^3 : |\vec{x}| < 1\}$ with the metric

$$ds = \frac{2\,|d\vec{x}|}{1 - |\vec{x}|^2}.$$

The two models are equivalent by a Möbius transformation that maps one to the other. Stereographic projection extends to such a Möbius transformation (Exercise (1-11)). We will refer to either of these models with its metric as *hyperbolic space* and denote it by $\mathbb{H}^3$.

Here is a further list of properties:

(i) The hyperbolic metric is infinitesimally euclidean and correctly represents the angles in $\mathbb{H}^3$.

(ii) $\mathbb{H}^3$ is complete in its metric: an arc tending to the "boundary" has infinite length.

(iii) The hyperbolic planes in the upper half-space model are hemispheres orthogonal to $\mathbb{C}$ and vertical euclidean half-planes. The lines (geodesics) are semicircles orthogonal to $\mathbb{C}$ and vertical euclidean half-lines. In the ball model the hyperbolic planes are spherical caps orthogonal to the unit sphere, and equatorial planes. The lines are circular arcs orthogonal to the unit sphere, and euclidean diameters.

(iv) The hyperbolic metric is invariant under any Möbius transformation that maps the upper half-space or the ball model onto itself. These transformations form the full group of orientation-preserving isometries.

(v) The volume V of a hyperbolic ball and the surface area S of its bounding sphere grow exponentially with the hyperbolic radius ρ. The ratio of the surface area to the volume approaches 2 as $\rho \to \infty$.

(vi) Two convex hyperbolic polyhedra that are combinatorially the same with the same dihedral angles and valence 3 at all vertices are isometric [Rivin 1996; Bobenko and Springborn 2004].

(vii) **Andreev-Thurston Theorem** [Roeder et al. 2007] and [Bowers and Stephenson 1996]. Also see Exercise (4-17). Let $\mathbb{T}$ be a triangulation of $\mathbb{S}^2$ and $E = \{e_i\}$ the set of edges. Assume that $\mathbb{T}$ is not equivalent to the edge set of a tetrahedron. Assign an angle $\Phi : E \to [0, \frac{\pi}{2}]$ to each edge e_i. The assignments must satisfy two conditions:

(a) If the sequence of edges (e_1, e_2, e_3) form a closed loop on $\mathbb{S}^2$ satisfying $\sum_{i=1}^3 \Phi(e_i) \geq \pi$, then these edges bound a triangle in $\mathbb{T}$.

(b) If the sequence (e_1, e_2, e_3, e_4) form a closed loop on $\mathbb{S}^2$ satisfying $\sum_{i=1}^4 \Phi(e_i) = 2\pi$ then these edges form the boundary of two adjacent triangles.

Then there is a realization of $\mathbb{T}$, (1) as a geodesic triangulation (in the spherical metric) $\mathbb{T}_{geod} \subset \mathbb{S}^2$, unique up to Möbius equivalence, together with (2) a family $C = \{C_i\}$ of circles, centered at the vertices of $\mathbb{T}_{geod}$, with the property that the two circles in C centered at the endpoints of an edge e_i intersect at angle $\Phi(e_i)$.

(viii) Erect the hyperbolic planes $\{P_{C_i}\} \subset \mathbb{H}^3$ rising from the circles $\{C_i\}$. Assume, say in the ball model, that 0 is in the "exterior" $Ext(P_i)$ of each plane. Then $\cap_i Ext(P_i)$ is a hyperbolic polyhedron with dihedral angles $\{\Phi(e_i)\}$ coming from the intersecting circles centered at the endpoints of the edges $\{e_i\}$.

For analysis of properties of ideal polyhedrons (all vertices at "infinity" in $\mathbb{H}^3$) see Rivin [1996].

Restricting the hyperbolic metric to a hyperbolic plane in the model yields the two-dimensional hyperbolic metric on that plane. Particular cases are the vertical half-plane rising from $\mathbb{R}$ in the upper half-space model and the equatorial plane in the ball model.

Proof of Property (iii). For the proof that the Möbius transformations are orientation-preserving isometries of the models, see Exercise 1-7. Here we show that there are no other such isometries, concentrating on the hyperbolic plane.

Given three positive distances d_1, d_2, d_3 satisfying the triangle inequality, and a point z on an oriented line $\ell \in \mathbb{H}^2$, there is a triangle Δ with a vertex at z, a side of length d_1 lying on the positive side of ℓ, a side of length d_2 lying in the right side of ℓ and sharing the vertex z, and a third side of length d_3. There is one other triangle with sides of these lengths: the reflection of Δ in ℓ. One of the two is uniquely determined if an ordering of the vertices is given and required to give the positive orientation of the triangle they bound.

Given an orientation-preserving isometry T, the T-images of three points not on a line are not on a hyperbolic line either. There is a Möbius transformation A such that $A \circ T$ fixes the three points. It then pointwise fixes the sides of the triangle they determine, and then fixes the whole triangle, say Δ. That is, $T(z) = A^{-1}(z)$, for $z \in \Delta$. If Δ' is a triangle sharing an edge with Δ, there is Möbius transformation A_1 such that $T(z) = A_1^{-1}(z)$ on Δ'. Necessarily $A_1 = A$. Continuing on, building up the whole plane $\mathbb{H}^2$ by adding in succession adjacent triangles, we conclude that $T \equiv A$. $\quad\square$

Proof of Property (iv). Working in the upper half-space model, in view of (iii), we need only prove that the vertical axis ℓ, rising from $z = 0$ is itself a geodesic. Fix a pair of points $t_1, t_2 \in \ell$. Let $\vec{x} = \vec{x}(s)$ be a smooth parameterized arc α between $t_1 < t_2 \in \ell$. Differentiating with respect to s,

$$\frac{d\vec{x}}{ds} = \frac{\sqrt{x'^2 + y'^2 + t'^2}}{t} \geq \frac{|t'(s)|}{t(s)} \geq \frac{t'(s)}{t(s)}.$$

Upon integrating we find for the length, $L(\alpha) \geq \log(t_2/t_1)$. Equality holds if and only if α is the vertical segment $[t_1, t_2]$.

This argument also shows that *orthogonal projection* of two points $x_1, x_2 \in \mathbb{H}^3$ onto ℓ strictly reduces the distance between them, unless they both lie on ℓ.

We conclude that the shortest path between any two points is given by a Möbius image of a vertical line segment; the geodesic lines in $\mathbb{H}^3$, are the Möbius images of ℓ.

Likewise in the vertical half-plane resting on $\mathbb{R}$ is a hyperbolic plane, the geodesic through any two points of the plane also lies in the plane. Therefore the totality of images under Möbius transformations gives the totality of hyperbolic planes. Any three distinct points, not on a line, uniquely determine a hyperbolic plane through them. □

Euclidean circles in $\mathbb{H}^2$ and euclidean circles and spheres in $\mathbb{H}^3$ are also hyperbolic circles and spheres. This is seen by starting with circles and spheres in the disk and ball models which are centered at the origin. The image of a circle or sphere under any Möbius transformation is again a euclidean circle or sphere, if no point on it gets sent to ∞. Conversely any circle or sphere can be sent by a Möbius transformation to one centered at the origin.

However, the hyperbolic center is *not* the euclidean center, except for the circles and spheres with center at the origin in the disk and ball models (Exercise (1-4)).

In short, in the hyperbolic plane and space there are more geometric shapes than in euclidean space, they have a tendency toward rigidity, and there is a lot more space in which to build them—in the estimate of Dick Canary, a baseball game played in the hyperbolic plane would require more than 10^{100} ballplayers to provide the same level of outfield coverage as in euclidean space!

Most two-dimensional abstract surfaces and three-dimensional manifolds can be modeled using hyperbolic geometry, but not euclidean or spherical geometry. Hyperbolic space is a good place to embed exponentially growing graphs, like a graph representing interconnected web sites. In fact PARC has patented an algorithm for laying out such graphs in $\mathbb{H}^2$ Lamping et al. [1995]. A different, unpatented, algorithm for laying out graphs in $\mathbb{H}^3$ is presented in Munzner [1997]. The change of focus from one site to another is effected by a hyperbolic isometry.

By studying the ancient microwave background radiation that pervades the universe, astrophysicists hope to get clues about the topology and large-scale curvature of our cosmic home. An earlier proposal that we live in a hyperbolic universe appears to be incompatible with recent data from the Wilkinson Microwave Anisotropy Probe (WMAP), which found the total density (matter plus vacuum energy) to have essentially the value expected for flat space. To the extent that there may be deviation, it is toward a spherical universe (positive curvature); see the discussion in Weeks [2004].

If the universe is a closed manifold with positive curvature, it can have one of only a few topological types.[3]

Such were earlier speculations. The results from the latest and much more accurate European space probe called Planck appeared in 2013. According to Jeff Weeks (personal communication), Planck found to 95% accuracy that

$$0.9926 < \Omega_{total} < 1.0057.$$

Here Ω_{total} is an astronomical measure, the ratio of the actual mass energy density to that required for a flat universe. The data rules out a dodecahedral universe which would have required the value 1.02. As Jeff writes "... the plausible hypotheses are that either

- Space really is flat, or
- Space is curved, but on such a large scale that we'll never be able to detect the curvature based on our observations of what we see within our horizon sphere".

To establish that the universe is not simply connected, which might be the case if $\Omega_{total} < 1$, would be astounding, but this now seems beyond our ability.

1.3 The circle or sphere at infinity

From the point of view of the hyperbolic metric, the models have no boundary: the metric is complete, and hyperbolic straight lines extend forever. Equal hyperbolic distances are represented by increasingly smaller euclidean distances in the model as one approaches "∞".

However, it is useful to regard the edge of the model as a sort of "conformal boundary" in a way that will be explained shortly. This boundary is denoted by $\partial \mathbb{H}^2$ ($= \mathbb{S}^1$ or $\mathbb{R} \cup \{\infty\}$ for the hyperbolic plane) and by $\partial \mathbb{H}^3$ ($= \mathbb{S}^2$ or $\mathbb{C} \cup \{\infty\}$ for hyperbolic space). Another common designation is $\mathbb{S}_\infty$, for the *circle or sphere at infinity*. If we fix a point in $\mathbb{H}^3$, we can also identify $\partial \mathbb{H}^3$ with the *visual sphere* of rays emanating from this point.

In $\mathbb{H}^2$ or $\mathbb{H}^3$, each hyperbolic line determines two "endpoints" on the boundary. Conversely, two distinct boundary points uniquely determine a line. Distinct lines may share an endpoint—indeed, a way to define the sphere at infinity *intrinsically*, without reference to a model, is by taking all oriented geodesics (parametrized by arclength) and defining as equivalent any two that remain within a bounded distance of each other as $t \to \infty$; the set of equivalence classes is $\mathbb{S}_\infty$.

Two distinct hyperbolic lines intersect in at most one point. In $\mathbb{H}^2$, they intersect if and only if their endpoints alternate on $\partial \mathbb{H}^2$. Given a line ℓ and a point $z \notin \ell$ in

---

[3] For example, it might conceivably be Poincaré dodecahedral space, the famous first example found by Henri Poincaré of a closed manifold with zero homology which is not homeomorphic to $\mathbb{S}^3$. He had initially believed that such a manifold must be $\mathbb{S}^3$; the example led him to the Poincaré Conjecture. A good explanation of this space and of the classification of spherical three-manifolds can be found in Thurston [1997].

$\mathbb{H}^2$, there are infinitely many lines through z which do not meet ℓ—unlike the case of the euclidean plane! These are the "parallel lines" of the hyperbolic plane. Among all these parallel lines, there are two that share an endpoint with ℓ.

In $\mathbb{H}^3$ each hyperbolic plane P is bounded by a circle on $\partial\mathbb{H}^3$ (which may appear as a euclidean line on the boundary of the upper half-space model). Conversely each circle on $\mathbb{S}^2 = \partial\mathbb{H}^3$ determines one such plane.

The isometries of $\mathbb{H}^3$ extend to $\partial\mathbb{H}^3$ as conformal automorphisms, that is, as Möbius or anti-Möbius transformations (depending on whether the isometry preserves or reverses orientation).

As mentioned earlier, the set of geodesic rays from a given point $\vec{x} \in \mathbb{H}^3$ can be identified with $\partial\mathbb{H}^3$. Any hyperbolic plane not through $\vec{x}$ subtends a *solid angle* at $\vec{x}$ of $< 2\pi$. That is, on a tiny sphere of radius ϵ about $\vec{x}$, the intersection of the sphere with the rays from $\vec{x}$ to the plane fill out a surface area strictly less than $2\pi\epsilon^2$, less than half the area of the sphere. In contrast, in euclidean space any plane subtends exactly a solid angle 2π.

If we lived in hyperbolic space, what we would see as flat lines and planes would automatically be the hyperbolic geodesics, since light would travel along hyperbolic geodesics. If we stood on a plane P, we would see the "circle at infinity" that supports a plane P as the *horizon* of P.

In practice, we have to view hyperbolic space from the outside, from euclidean space using one of our models. We then see the euclidean lines and planes as flat while most of the hyperbolic ones look curved. Looking at the disk or ball model from the outside, we also see the entire circle or sphere at infinity.

An elliptic transformation T has an *axis of rotation* inside $\mathbb{H}^3$. It is the hyperbolic line connecting its fixed points on $\mathbb{S}^2$. The axis is pointwise fixed by T.

A loxodromic transformation T likewise has an *axis* in $\mathbb{H}^3$. It too is the hyperbolic line connecting the fixed points. T maps the line onto itself, moving each point toward the attracting fixed point. If T is in standard form, $\lambda^2 z$, with $|\lambda| > 1$, the axis is the vertical half-line $z = 0$ in upper half-space. The hyperbolic distance between any pair of points z, $T(z)$ on the axis is $d = 2\log|\lambda|$, or $2\cosh(d/2) = |\lambda| + |\lambda^{-1}| \geq |\tau_T|$.

Both elliptic and loxodromic transformations in $\mathbb{H}^3$ leave invariant not just the axis, but also each of a family of surfaces equidistant from the axis. These surfaces are particularly easy to visualize in upper half-space when the transformation is in standard form: the surface is a euclidean cone with vertex at $(z, t) = (0, 0)$ and a vertical axis (the half-line $z = 0$). When both endpoints of the axis line on the plane $t = 0$, the euclidean shape of the surfaces is a tube, tapering, like a banana, to a cone at each endpoint (Figure 1.5). Note that though we often describe features of hyperbolic space by talking about their euclidean shapes in the model—and this mixture is almost inevitable—you should strive to visualize each object both intrinsically (the tube has constant diameter) and in terms of the model (the tube looks like a cone or a crescent).

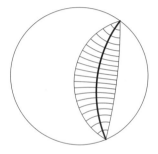
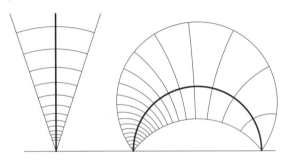

Fig. 1.5. Invariant tubes viewed in cross section, in the ball model and the upper half-space. The axis of the transformation is shown thicker in each case. The transversal lines represent discs orthogonal to the axis; all these disks (within the same tube) are congruent.

A parabolic transformation P has no axis, since there is only one fixed point. P does have invariant surfaces, each mapped to itself (in the spirit of the tubes of the previous paragraph); they *are* euclidean spheres, but *not* hyperbolic spheres! (See Exercise (1-33).) They are called *horospheres*. All horospheres of a parabolic transformation P are tangent to one another and to the sphere at $\mathbb{S}_\infty$ at the fixed point ζ of P. The region of $\mathbb{H}^3$ cut off by a horosphere is called a *horoball*. In the upper half-space model, there is an exceptional case, when the fixed point ζ is at infinity (say for $P(z) = z+1$): then the horospheres are euclidean planes $\{(z,t) : t = \text{constant}\}$, and the horoballs are the half-spaces above these planes. For parabolic transformations of $\mathbb{H}^2$ the corresponding objects are called *horocycles* and *horodisks*.

At a parabolic fixed point $\zeta \in \mathbb{S}^2$, there is a family of double horocircles and horodisks disks at ζ, namely one on each "side"; in the case that $\zeta = \infty$ and $P(z) = z + 1$, the family of double horodisks is $\{|\operatorname{Im} z| > h > 0$. The family consists of mutually tangent circles at ζ that bound disjoint open disks which are invariant under the parabolic transformation.

The prefix "horo" comes from the Greek word for "limit". Fix a point $O \in \mathbb{H}^3$. Take the *hyperbolic* sphere σ_x centered at $x \in \mathbb{H}^3$ and passing through O. As $x \to \zeta \in \partial\mathbb{H}^3$, the limit of σ_x is the horosphere at ζ passing through O.

We now examine the simplest of polygons, the triangles. As already mentioned the area of a triangle is equal to the "angle deficit" $\pi - \sum \theta_i$, where the θ_i are the vertex angles; see Exercise (1-6) for a proof. Thus the greatest area a triangle can have is π, which happens when all vertices have "angle zero"—this is a limiting case, when the vertices are no longer points in hyperbolic space but in the sphere at infinity. A point in the sphere at infinity is also called an *ideal point*, and so triangles whose vertices are at infinity are *ideal triangles*. Given two ideal triangles and a labeling of

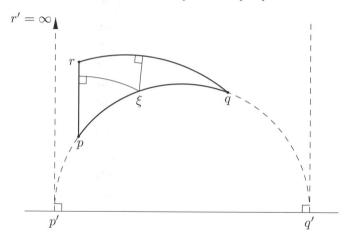

Fig. 1.6. Universal thinness of triangles.

the respective vertices in the positive direction, there is a unique isometry that takes one to the other, matching the designated labeling.

Theorem 1.3.1 (All triangles are thin). *Any point ξ on a side of a hyperbolic triangle Δ is within distance* $\log(1 + \sqrt{2}) = \operatorname{arcsinh} 1$ *from one of the two other sides. The distance attains its maximum only for an ideal triangle, with ξ of equal distance from the two other sides.*

Any point inside a hyperbolic triangle is within distance $\log(1 + \sqrt{2}) = \operatorname{arcsinh} 1$ *of one of the sides.*

Proof. We will work in the upper half-plane model. We may assume by changing the position of Δ in $\mathbb{H}^2$ by an orientation-preserving isometry that the side $[p, q]$ of $\Delta = (p, q, r)$ containing ξ lies on the unit semicircle centered at the origin, Δ lies above this semicircle, and the side $[p, r]$ lies on the vertical euclidean line through p. Here take p to be the left vertex on the semicircle. Assume at the start that none of the vertices is ideal. (See Figure 1.6.)

We start by showing that we may replace r by $r' = \infty$. As r goes up along the vertical line from p, the distance of ξ to $[p, r]$ increases or (once r is no longer the closest point to ξ) remains the same. The distance of ξ to the side $[q, r']$, too, either equals the distance $q\xi$ or the length of the perpendicular from ξ to $[q, r']$; therefore it exceeds the distance from ξ to $[q, r]$, since $[q, r]$ separates $[q, r']$ from ξ in the new triangle (p, q, r'). Thus the minimum distance of ξ to the sides increases when we move r to $r' = \infty$, so that Δ becomes a triangle with an ideal vertex.

Next consider what happens as q slides down the semicircle to its right endpoint $q' \in \mathbb{R}$. The distance of ξ to $[q, r' = \infty]$ strictly increases as q changes to q'; in the new triangle (p, q', ∞) the side $[q, \infty]$ separates $[q', \infty]$ from ξ. Similarly the distance of ξ to $[p, \infty]$ strictly increases as p slides down the semicircle to its left endpoint $p' \in \mathbb{R}$.

So now we have an ideal triangle with ξ on the side that is now the full semicircle. The minimal distance of ξ to the two vertical sides is greatest when ξ is the symmetric point $\xi = i$. Finally we have to compute the distance from ξ to one of the vertical sides. There is exactly one semicircle C' through $\xi = i$ with center at $z = p'$ which is orthogonal to the vertical line $[p', \infty]$. In polar coordinates at p', the orthogonal segment is the arc $0 \leq \theta \leq \pi/4$ of C', if θ is measured from the vertical. The length of this segment is

$$\int_{\pi/4}^{\pi/2} \frac{\rho \, d\theta}{\rho \sin \theta} = \log(\sqrt{2} + 1) = \operatorname{arcsinh} 1.$$

Here the radius ρ of C' doesn't enter—the map $z \mapsto kz, k > 0$ (a euclidean similarity) is a hyperbolic isometry.

Given a point z in a triangle Δ and a pair of sides, divide Δ by a geodesic arc between the designated sides and passing through z. Application of what we just proved shows that z is within distance $\operatorname{arcsinh} 1$ of the two designated sides. Then repeat the argument with an arc through z from the third side. $\qquad \square$

A related fact is described in Exercise (1-17).

1.4 Gaussian curvature

The hyperbolic plane is a simply connected surface with a complete riemannian metric of constant negative gaussian curvature. It is usually taken (by multiplying the metric by the appropriate constant) to be -1, as we have done in Section 1.2. The purpose of this section is to explain the meaning of the expression "gaussian curvature -1".

Using the disk model of $\mathbb{H}^2$ and polar coordinates (r, θ) based at the origin, we begin with the following computations: the hyperbolic radius ρ of the circle of euclidean radius $R < 1$ centered at the origin, its hyperbolic area A, and its hyperbolic circumference C. Our results are as follows:

$$\rho = \int_0^R \frac{2 \, dr}{1 - r^2} = \log \frac{1 + R}{1 - R}, \quad R = \tanh \frac{\rho}{2} = \frac{\cosh \rho - 1}{\sinh \rho},$$

$$A(\rho) = \int_{\theta=0}^{2\pi} \int_{r=0}^{R} \frac{4r \, dr \, d\theta}{(1 - r^2)^2} = \frac{4\pi R^2}{1 - R^2} = 2\pi (\cosh \rho - 1),$$

$$C(\rho) = \int_0^{2\pi} \frac{2R \, d\theta}{1 - R^2} = \frac{4\pi R}{1 - R^2} = 2\pi \sinh \rho.$$

$$C(\rho)^2 = A(\rho)^2 + 4\pi A(\rho).$$

Thus both the area and the circumference grow exponentially with the hyperbolic radius; more than 63% of the surface of any hyperbolic disk is within 1 unit of the boundary; the ratio $A(\rho)/C(\rho)$ equals $R < 1$ and in particular approaches 1 as $\rho \to \infty$.

For the record, the analogous formulas for the volume $V(\rho)$ of a hyperbolic ball of hyperbolic radius ρ and its surface area $S(\rho)$ are:

$$V(\rho) = 8 \int_0^{2\pi} d\theta \int_0^{\pi} \sin\phi \, d\phi \int_0^R \frac{r^2 \, dr}{(1-r^2)^3} = \pi(\sinh 2\rho - 2\rho),$$

$$S(\rho) = 4 \int_0^{2\pi} \int_0^{\pi} \frac{R^2 \sin\phi \, d\phi \, d\theta}{(1-R^2)^2} = 4\pi \sinh^2 \rho.$$

Thus $2V(\rho) < S(\rho)$ and $\lim_{\rho\to\infty} 2V(\rho)/S(\rho) = 1$. The value of the ratio is 1 only in the limiting case of a horoball.

Now consider a smooth riemannian surface, a point z on the surface, and, for $\rho > 0$ variable, the disk of radius ρ around that point (in the given metric, say hyperbolic). The gaussian curvature K_0 at z can be characterized by the following properties involving the limiting behavior of the area $A(\rho)$ and circumference $C(\rho)$ of such disks, compared with their euclidean counterparts [Struik 1950, §4.3]:

$$K_0 = -3\frac{d^2}{d\rho^2}\left(\frac{C(\rho)}{2\pi\rho}\right)_{\rho=0} = \frac{3}{\pi}\lim_{\rho\to0}\frac{2\pi\rho - C(\rho)}{\rho^3},$$

$$K_0 = -6\frac{d^2}{d\rho^2}\left(\frac{A(\rho)}{\pi\rho^2}\right)_{\rho=0} = \frac{12}{\pi}\lim_{\rho\to0}\frac{\pi\rho^2 - A(\rho)}{\rho^4}.$$

In particular negative curvature is characterized by the property that $C(\rho) > 2\pi\rho$ for all small values of R. Or by the property that $A(\rho) > \pi\rho^2$. This is confirmed for $\mathbb{H}^2$ from the formulas for area and circumference above. Contrast this with the corresponding properties of euclidean space.[4]

Here is a construction of a surface with discrete negative curvature: Now equilateral (euclidean) triangles with unit side lengths tessellate the euclidean plane when six are arranged about each vertex. Instead form a polyhedral surface in $\mathbb{R}^3$ by placing seven triangles about each vertex. In a polyhedral surface, each vertex v has a discrete curvature defined by $2\pi - \sum\theta_i$, where the θ_i is the sum of the vertex angles of the triangles sharing the vertex v_i. In our case the curvature at each vertex is $-\pi/3$. The "circle" of radius $R = 1$ about a vertex has circumference 7, larger than the euclidean circumference of 6 when there are six triangles about each vertex.

Kale is a lettuce-like vegetable with hyperbolic-like leaves.

The expression $u|dz|$, $u(z) > 0$, is a conformal metric on the plane, more generally on a Riemann surface. Its gaussian curvature expressed terms of the laplacian is

$$K_u = -\frac{\Delta \log u}{u^2} = -\frac{4}{u^2}\frac{\partial^2 \log u}{\partial z \, \partial \bar{z}}. \tag{1.3}$$

Ahlfors had great success early in his career 1973 with applications involving singular conformal metrics having this form.

---

[4] While $\mathbb{C}$ has zero gaussian curvature in the euclidean metric, it is known that a complete, noncompact riemannian surface with positive gaussian curvature is also conformally equivalent to $\mathbb{C}$.

If instead we want a model of $\mathbb{H}^2$ with gaussian curvature $-c < 0$, replace u in Equation (1.3) by $\frac{u}{\sqrt{c}}$. For notational simplicity, one often sees the disk model metric chosen to have curvature -4, namely

$$ds = \frac{|dz|}{1 - |z|^2}.$$

Gauss originally defined the curvature as follows. Suppose $S \subset \mathbb{R}^3$ is an embedded surface and $p \in S$. Draw a simple closed curve $c \subset S$ enclosing a region $D \subset S$ containing p. Interpret each exterior unit normal vector $\vec{N}$ (determined by the right-hand rule) at a point of D as a vector from $(0, 0, 0)$ to the unit 2-sphere $\mathbb{S}^2$. As $\vec{N}$ ranges over all possibilities, a certain region $\Omega \subset \mathbb{S}^2$ is filled out. Gauss defined the *total curvature* of D to have absolute value $A(\Omega)$, the area of Ω. The sign is determined as follows. As $\vec{N}$ runs over c in the positive direction (D to its left), use $+$ if the corresponding $\vec{N}$ runs over $\partial\Omega$ also in the positive direction (Ω to its left); otherwise use $-$. Thus the total curvature of a region in a plane is zero, while the total curvature of a hemisphere is $2\pi \sin(\pi/2)$. Gauss defined the *curvature* of S at the point p as

$$\lim_{D \searrow \{p\}} \frac{\pm A(\Omega)}{A(D)}.$$

Gaussian curvature is an intrinsic property of a surface—although the definition just given is for surfaces embedded in $\mathbb{R}^3$, Gauss's famous *Theorema Egregium* is that isometric surfaces have the same gaussian curvature at corresponding points. So we can define the curvature for a metric defined on an abstract surface.

Hilbert proved that there is no C^2 surface in $\mathbb{R}^3$ whose metric induced from $\mathbb{R}^3$ is a *complete* metric of constant negative gaussian curvature. There do exist smooth surfaces embedded in $\mathbb{R}^3$ with constant negative curvature, but they cannot be extended to a complete surface; see Thurston [1997, §2,1] for example.

The most famous example is the *pseudosphere*, a surface of revolution about the x-axis in $\mathbb{R}^3$ described by the parametric equations

$$(u - \tanh u, \ \mathrm{sech}\, u \cos v, \ \mathrm{sech}\, u \sin v) \quad \text{for } u \geq 0, \ v \in \mathbb{R}.$$

If we take the subset $\{z : \mathrm{Im}\, z \geq 1\}$ of $\mathbb{H}^2$ and quotient it by the cyclic group of hyperbolic isometries generated by $z \to z + 2\pi$, we get a riemannian surface isometric to the pseudosphere (with its inherited metric from $\mathbb{R}^3$); see Coxeter [1961, p. 378]. Conformally, this is a once-punctured disk.

The *Gauss–Bonnet formula* for a *simply connected* surface element S of constant gaussian curvature K, bounded by the union of n smooth arcs meeting with interior angles θ_i at the vertices, is

$$\iint_S K \, dS + \int_{\partial S} \kappa_g \, ds = 2\pi - \sum_{i=1}^{n} (\pi - \theta_i) = \pi(2 - n) + \sum_{i=1}^{n} \theta_i, \qquad (1.4)$$

where κ_g is the geodesic curvature of the arcs [Struik 1950, §4.8]. On a geodesic arc, the geodesic curvature κ_g vanishes; thus, for example, if $S \subset \mathbb{H}^2$ is a hyperbolic triangle Δ, the formula becomes

$$-\text{area } \Delta = -\pi + \theta_1 + \theta_2 + \theta_3.$$

In particular the area of a hyperbolic triangle cannot exceed π. It is π only in the limiting case of an ideal triangle.

The Gauss–Bonnet formula, and indeed the area formula for triangles directly, can be verified by using Green's formula (see also Exercise (1-6)). By breaking more general surfaces into simply connected regions one can apply the formula further. See Exercise (3-1) for the details.

See Exercise (1-33) for computations of the hyperbolic curvature of horocycles, equidistant arcs to geodesics, and circles.

For hyperbolic 3-space $\mathbb{H}^3$ (or n-space more generally), the normalized metric is characterized by having *sectional curvature* -1: all two-dimensional planes through a given point have gaussian curvature -1 in the metric induced from that of $\mathbb{H}^3$.

1.5 Further properties of Möbius transformations

The following facts are part of the repository of basic knowledge of hyperbolic geometry.

1.5.1 Commutativity

Lemma 1.5.1. *Let A, B be Möbius transformations $\neq$ id.*

(i) *A and B share a fixed point if and only if*

$$\text{tr}(ABA^{-1}B^{-1}) = +2.$$

(ii) *Assume that A and B do not share a fixed point. Then $ABA^{-1}B^{-1}$ is parabolic if and only if*

$$\text{tr}(ABA^{-1}B^{-1}) = -2.$$

Proof. The second statement follows directly from the characterization on p. 3. The proof of the first is not hard if one of the transformations is taken to be in standard form. □

Lemma 1.5.2. *Let A and B be Möbius transformations distinct from* id. *The following statements are equivalent:*

(i) *A and B commute.*

(ii) *A and B share the same axis.*

(iii) *A and B have the same set of fixed points, whether one or two, OR*

(iv) *A and B have order two and each interchanges the fixed points of the other, that is, their axes intersect orthogonally in $\mathbb{H}^3$.*

Again, this can be checked by assuming A to be in standard form, and observing that the general elliptic transformation of order two exchanging 0 with ∞ is $z \mapsto a^2/z$.

Lemma 1.5.3. *Let $k \geq 1$ and let A and B be Möbius transformations with $B^k \neq$ id. Then $A^k = B^k$ if and only if either $A = B$, or $A = EB$ for an elliptic E whose order divides k and whose set of fixed points is the same as that of B.*

Proof. Sufficiency is obvious, since E and B commute by Lemma 1.5.2.

To prove necessity, we can assume that k is minimal with the property that $A^k = B^k$. Note that taking powers preserves both type and fixed points, except that an elliptic can become the identity. Thus, if A is parabolic, so is A^k, and hence so is B; but the parabolic elements fixing a given point of S_∞ form a torsion-free abelian group (see again Lemma 1.5.2), so $A = B$ in this case. If instead A fixes two points, the same argument (since $A^k \neq$ id) again shows that A and B commute; hence $(AB^{-1})^k =$ id, so $E = AB^{-1}$ is elliptic and shares the fixed points of A, B. $\qquad\square$

1.5.2 Isometric circles and planes

Consider a Möbius transformation A on $\mathbb{C} \cup \infty$. If A does not fix ∞, it has the form

$$A(z) = \frac{az + b}{cz + d}, \quad ad - bc = 1, \ c \neq 0.$$

One may ask, at what points does A preserve the length of (euclidean) tangent vectors? Since $A'(z) = 1/(cz + d)^2$, the set of such points, denoted by

$$\mathcal{I}(A) = \{z \in \mathbb{C} : |A'(z)| = 1\} = \{z \in \mathbb{C} : |cz + d| = 1\},$$

is a circle. It is called the *isometric circle* of A. Its center and radius satisfy

$$\text{center } \mathcal{I}(A) = -\frac{d}{c} = A^{-1}(\infty), \quad \text{radius } \mathcal{I}(A) = \frac{1}{|c|}.$$

Because A maps circles to circles, the restriction of A to $\mathcal{I}(A)$ is a euclidean isometry onto the circle $\mathcal{I}(A^{-1})$ of the same radius. Also, $|A'(z)| > 1$ for z in the interior of $\mathcal{I}(A)$, and $|A'(z)| < 1$ for z in the exterior.

Now consider the same transformation A, regarded as an isometry of $\mathbb{H}^3$ in the upper half-space model (recall that $\mathbb{C} \cup \infty = \partial\mathbb{H}^3$). The *isometric plane* of A is likewise defined as the set of points where the jacobian preserves length:

$$\{\vec{x} : |A'(\vec{x})| = 1\}. \tag{1.5}$$

Explicitly, it is the hemisphere

$$\{\vec{x} = (z, t), \ t > 0 : |cz + d|^2 + |c|^2 t^2 = 1\}.$$

rising from the isometric circle $\mathcal{I}(A) \subset \partial\mathbb{H}^3$. When the context is clear, we will use the notation $\mathcal{I}(A)$ interchangeably for the isometric circle and the isometric plane.

There is also an isometric circle and plane for the ball model, with the same defining Equation (1.5) as for the upper half-space model. However, the isometric circle

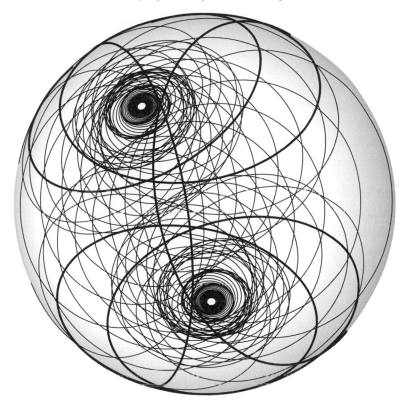

Fig. 1.7. Outer Circles: Isometric circles on $\mathbb{S}^2$ of numerous elements of a cyclic group generated by a loxodromic with approximate trace $1.92 + 0.03i$.

in one model does not usually map to the isometric circle in the other under a hyperbolic isometry conjugating the two models; in fact, even within the same model, a conjugating isometry U need not map $\mathfrak{I}(A)$ to $\mathfrak{I}(UAU^{-1})$, as is the case with the axis and fixed points. Although not intrinsic, the notion of isometric circles and planes is nonetheless useful because of its metric properties. It was introduced by L. R. Ford.

Thus looking from the outside, $\partial\mathbb{H}^3$ is full of circles corresponding to the elements of discrete groups of isometries, the *Outer Circles* of the former book title, as seen in Figure 1.7. The action of isometries on geodesic planes in $\mathbb{H}^3$ is paired with the corresponding action on the outer circles.

Here is another description of the isometric planes (see Exercises (1-10, 3-4)).

Lemma 1.5.4.

(i) *In the ball model, the isometric plane for A is the perpendicular bisector of the line segment $[0, A^{-1}(0)]$, where 0 denotes the origin of the ball.*

(ii) *In the upper half-space model, if ∞ is not a fixed point of A, the isometric plane results from the following construction. There is exactly one horosphere $\mathcal{H}$ at ∞ such that the horosphere $A^{-1}\mathcal{H}$ at $A^{-1}(\infty)$ is tangent to $\mathcal{H}$. The line ℓ between ∞ and $A^{-1}(\infty)$ goes through the point of tangency and is orthogonal to the two horospheres. The isometric plane is the unique plane through the point of tangency and orthogonal to ℓ.*

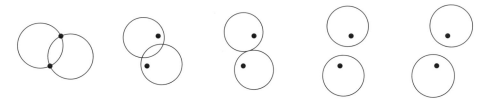

Fig. 1.8. Isometric circles for various transformations A that have the same fixed points. From left to right, the elliptic case ($\tau_A \in (-2, 2)$), the loxodromic case with $\tau_A \notin \mathbb{R}$ and respectively $|\tau_A| < 2$, $|\tau_A| = 2$ and $|\tau_A| > 2$, and finally the loxodromic case with τ_A real.

We summarize here the properties of isometric planes and circles in the upper half-space model. Refer to Figure 1.8 for examples.

Proposition 1.5.5. *Let A be a Möbius transformation of the upper half-space model, not fixing ∞. Let $\mathcal{B}(A)$ be the closed disk bounded by the isometric circle $\mathfrak{I}(A)$ in $\mathbb{C}$, and let $\mathcal{E}(A)$ be the closure of its exterior (including ∞).*

(1) *A sends $\mathfrak{I}(A)$ to $\mathfrak{I}(A^{-1})$, $\mathcal{B}(A)$ to $\mathcal{E}(A^{-1})$, and $\mathcal{E}(A)$ to $\mathcal{B}(A^{-1})$. If $\vec{x} = (z, t)$ lies on the isometric plane $\mathfrak{I}(A)$, then $A(\vec{x}) = (A(z), t)$ lies on $\mathfrak{I}(A^{-1})$.*
(2) *$\mathfrak{I}(A) = \mathfrak{I}(A^{-1})$ if and only if $\tau_A = 0$.*
(3) *The intersection of circles $\mathfrak{I}(A) \cap \mathfrak{I}(A^{-1})$ consists of two points if and only if $0 < |\tau_A| < 2$. If A is elliptic, these intersection points are the fixed points and the corresponding isometric planes intersect in the axis of rotation.*
(4) *$\mathfrak{I}(A)$ and $\mathfrak{I}(A^{-1})$ are tangent if and only if A is parabolic, in which case the tangency point is the fixed point.*
(5) *$\mathfrak{I}(A)$ and $\mathfrak{I}(A^{-1})$ are disjoint if and only if $|\tau_A| > 2$.*
(6) *If A is loxodromic, $\mathcal{B}(A) \cap \mathcal{E}(A^{-1})$ contains the repelling fixed point of A, and $\mathcal{B}(A^{-1}) \cap \mathcal{E}(A)$ its attracting fixed point.*
(7) *If U fixes ∞, then $\mathfrak{I}(UAU^{-1}) = U(\mathfrak{I}(A))$. If U is a euclidean translation, $\mathfrak{I}(UA) = \mathfrak{I}(A)$ and $\mathfrak{I}(AU^{-1}) = U\mathfrak{I}(A)$.*
(8) *If A preserves a circle in $\mathbb{S}^2$, $\mathfrak{I}(A)$ is orthogonal to that circle.*

Proof. (1) The jacobian determinants or derivatives are related:

$$\left(BA^{-1}\right)'(Az) = B'(z)/A'(z).$$

(2) A normalized matrix $A = \pm A^{-1}$ if and only if either $A = \pm \mathrm{id}$ or $\tau_A = 0$. The isometric circle $|-cz + a| = 1$ of A^{-1} is identical to that of A if and only $a + d = 0$.

(3)–(5) The distance between the centers of the isometric circles is

$$\left| \frac{a}{c} + \frac{d}{c} \right| = \frac{|a + d|}{|c|}.$$

Since the radius of the circles is $1/|c|$ they intersect whenever the distance between centers is less than $2/|c|$ and are tangent when there is equality.

Now the distance between the centers is exactly $2/|c|$ when A is parabolic, less than $2/|c|$ when A is elliptic, and can have any positive value when A is loxodromic. Only when the loxodromic satisfies $|\tau_A| > 2$ does the distance between centers exceed $2/|c|$ so that the circles are disjoint.

(6) The derivative $|A'|$ is greater than 1 at the repelling fixed point, and less than 1 at the attracting one, when these points are finite. (In contrast, at a finite elliptic or parabolic fixed point ζ, $|A'(\zeta)| = 1$.)

(7) This is a direct computation, or an application of the chain rule.

(8) If A preserves $\mathbb{R} \cup \infty$, its normalized form has real or purely imaginary entries, the latter case if A interchanges the upper and lower half-planes. Therefore the center of the isometric circle is real, so $\mathfrak{I}(A)$ is orthogonal to $\mathbb{R}$. If A maps the unit disk onto itself, it has the following form (Exercise (1-2)):

$$A = e^{i\theta} \frac{z - a}{1 - \bar{a}z}.$$

From this we compute that $\mathfrak{I}(A)$ has center $1/\bar{a}$ and squared radius $(1-|a|^2)/|a|^2$. This implies that $\mathfrak{I}(A)$ is orthogonal to the unit circle. If A interchanges the two sides of the unit circle, it can be expressed by replacing z by $1/z$ in the formula and proceeding in the same way. The general transformation A is conjugate to one we have considered via a transformation that fixes ∞.

$\square$

1.5.3 Trace identities

Here we present the common trace identities that help form the bridge between the algebra of matrices and hyperbolic geometry. See also Exercise (1-20), Exercise (1-39), Exercise (2-13).

Lemma 1.5.6. *Let X and Y be 2×2 complex matrices of determinant one.*

(i) $\operatorname{tr}(XY^{-1}) = \operatorname{tr}(X)\operatorname{tr}(Y) - \operatorname{tr}(XY)$.

(ii) $\operatorname{tr}(XYX^{-1}Y^{-1}) + 2 = \operatorname{tr}^2(X) + \operatorname{tr}^2(Y) + \operatorname{tr}^2(XY) - \operatorname{tr}(X)\operatorname{tr}(Y)\operatorname{tr}(XY)$.

(iii) $\operatorname{tr}(XYX^{-1}Y^{-1}) - 2 = \big(\operatorname{tr}(X) - \operatorname{tr}(Y)\big)^2 - \big(\operatorname{tr}(XY) - 2\big)\big(\operatorname{tr}(XY^{-1}) - 2\big)$.

(iv) *If* $\operatorname{tr}^2(X) \neq 4$ *then*

$$\frac{\operatorname{tr}(X^m Y X^{-m} Y^{-1}) - 2}{\operatorname{tr}^2(X^m) - 4} = \frac{\operatorname{tr}(XYX^{-1}Y^{-1}) - 2}{\operatorname{tr}^2(X) - 4}.$$

(v) *If* $[X, Y] = XYX^{-1}Y^{-1}$ *is parabolic and* X, Y *do not share a fixed point, so that* $\operatorname{tr}(XYX^{-1}Y^{-1}) = -2$, *then*

$$\operatorname{tr}(X)\operatorname{tr}(Y)\operatorname{tr}(XY) = \operatorname{tr}^2(X) + \operatorname{tr}^2(Y) + \operatorname{tr}^2(XY),$$
$$\operatorname{tr}(XY)\operatorname{tr}(XY^{-1}) = \operatorname{tr}^2(X) + \operatorname{tr}^2(Y).$$

Conversely, either of these two identities implies $\operatorname{tr}[X, Y] = -2$.

Remark 1.5.7. The first equation in (v) is called the *Markov identity*. Markov proved that for the equation $xyz = x^2 + y^2 + z^2$, the only integer solutions (called Markov triples) are provided by the traces of group elements $X, Y, Z = XY$ in the modular group (Exercise (2-10)), with tr$[X, Y] = -2$. If (u, v, w) is a Markov triple, so are $(u, v, uv - w)$, $(u, uw - v, w)$, $(vw - x, v, w)$. A famous unsolved problem in number theory is Markov's Conjecture that if (x, y, z), (x', y', z') are Markov triples, with $x \leq y \leq z$ and $x' \leq y' \leq z'$, then $x = x'$ and $y = y'$. See Bowditch [1998] and Goldman [2003] for more detail.

Proof. To verify (i), (ii) and (iii), apply a conjugacy to convert Y to standard form:

$$Y = \begin{pmatrix} \lambda & 0 \\ 0 & \lambda^{-1} \end{pmatrix} \text{ or } \begin{pmatrix} 1 & \lambda \\ 0 & 1 \end{pmatrix}, \qquad X = \begin{pmatrix} a & b \\ c & d \end{pmatrix}, \ ad - bc = 1.$$

The identities are now easily verified. In particular we find that for Y normalized we have, depending on whether Y is nonparabolic and parabolic,

$$\text{tr}(XYX^{-1}Y^{-1}) = 2 - bc(\lambda - \lambda^{-1})^2 \text{ or } 2 + c^2\lambda^2.$$

Thus the commutator cannot have trace $+2$ unless $b = 0$, $c = 0$, or $X = \pm I$. All three possibilities are excluded by the hypotheses.

The Markov identity in (v) can be regarded as a quadratic equation for $w = \text{tr}(XY)$ in terms of the coefficients tr(X) and tr(Y). The two solutions are $w = \text{tr}(XY)$ and $w = \text{tr}(XY^{-1})$. This is the reason for the second identity in (v).

If the transformations corresponding to X and Y are loxodromic and preserve the upper half-plane, and their matrix representations are chosen so that the traces are positive, then tr(XY) will automatically be positive as well.

For item (iv), put X instead of Y in standard form. It is enough to verify the formula for $m = 2$; the general case follows by induction. The ratio has the constant value $-bc$ if $Y = \begin{pmatrix} a & b \\ c & d \end{pmatrix}$. $\qquad \square$

1.6 Exercises and explorations

1-1. For a Möbius transformation T of $\mathbb{C} \cup \infty$ with normalized matrix $\begin{pmatrix} a & b \\ c & d \end{pmatrix}$, prove:

(i) $|T(z) - T(w)| = |z - w| \sqrt{|T'(z)|} \sqrt{|T'(w)|}$.

(ii) T preserves the upper half-plane if and only all of a, b, c, d are real. (Thus the group of orientation-preserving isometries of the hyperbolic plane is PSL$(2, \mathbb{R})$.) If such a T is loxodromic or parabolic, its fixed points lie in $\mathbb{R}$. If T is elliptic, its fixed points are symmetric about $\mathbb{R}$ under reflection, but do not lie in $\mathbb{R}$. Moreover, Im $T(z) = (\text{Im } z) \cdot |T'(z)|$.

(iii) T preserves the right half-plane if and only if $T = \begin{pmatrix} a' & ib' \\ -ic' & d' \end{pmatrix}$ with $a', b', c', d' \in \mathbb{R}$ and $a'd' - b'c' = 1$.

(iv) Find conditions on a, b, c, d for T to preserve the unit disk. (*Hint:* Conjugate by a Möbius transformation taking $-1, 1, \infty$ to $-1, 1, i$.) Prove an alternative characterization: T preserves the unit disk if and only if it can be written as

$$T(z) = e^{i\theta} \frac{z - z_0}{1 - \bar{z}_0 z}, \quad |z_0| < 1.$$

Moreover, such a T satisfies $|T'(z)|(1 - |z|^2) = (1 - |T(z)|^2)$.

1-2. Two points are said to be symmetric in a circle or straight line if reflection in the circle or line carries one point to the other. Thus z and $\bar{z}$ are symmetric in $\mathbb{R}$, while z and $1/\bar{z}$ are symmetric in the unit circle centered at the origin. Verify that the formula for symmetric points ζ, ζ^* with respect to the circle $\{|z - a| = R\}$ is

$$\zeta^* - a = \frac{R^2}{\bar{\zeta} - \bar{a}}.$$

This map extends to a reflection about the corresponding hyperbolic plane in $\mathbb{H}^3$.

(i) Prove that a Möbius transformation maps points symmetric in a circle/line to points symmetric in the image circle/line. Hence the extension to $\mathbb{H}^3$ preserves symmetry in planes. (See Ahlfors [1978].)

(ii) If C_1, C_2 are disjoint circles and Δ is the region they bound on $\mathbb{S}^2$, show how to find a Möbius transformation that sends Δ to an annulus centered at $z = 0$ with C_2 sent to the outer circle of radius 1.

(iii) Suppose $D_2 \subset D_1$ are disks centered at $z = 0$ of radii $r_2 < r_1$ respectively. Let T be any Möbius transformation such that $T^{-1}(\infty)$ is not in the closure of D_1. Denote the radii of $T(D_2) \subset T(D_1)$ by r_2', r_1'. Show that

$$\frac{r_1'}{r_2'} \geq \frac{r_1}{r_2}.$$

When is there equality?

1-3. *Liouville measure.* Suppose ℓ is a hyperbolic line in the disk model, with endpoints a, b. Suppose p, q are two points on ℓ so labeled that p separates a and q. Show that the hyperbolic distance $d(p, q)$ between p and q is given in terms of the cross ratio by

$$d(p, q) = \log(a, q, b, p) = \log \frac{(a - q)(b - p)}{(a - p)(b - q)}.$$

Suppose ℓ_1, ℓ_2 are hyperbolic lines which intersect in $\mathbb{H}^2$ with angle θ. Suppose their endpoints a_i, b_i, $i = 1, 2$, are arranged clockwise $[a_1, a_2, b_1, b_2]$ around $\partial \mathbb{H}^2$. Prove that

$$(a_1, b_1, a_2, b_2) = \cos^2 \frac{\theta}{2} = \frac{\cos \theta + 1}{2}.$$

Denote the space of *unoriented* geodesics in $\mathbb{H}^2$ by $G(\mathbb{H}^2)$. In terms of endpoints on $\mathbb{S}^1 = \partial \mathbb{H}^2$,

$$G(\mathbb{H}^2) = \left(\mathbb{S}^1 \times \mathbb{S}^1 \setminus \{\text{diagonal}\} \right) / \mathbb{Z}_2,$$

which is topologically a Möbius band (Exercise (4-15)).

The *Liouville measure L* is a measure on the space $G(\mathbb{H}^2)$ with the property that for any isometry T acting on disjoint intervals $I_1 = [e^{i\alpha_1}, e^{i\alpha_2}]$, $I_2 = [e^{i\beta_1}, e^{i\beta_2}]$ on $\mathbb{S}^1 \equiv \partial \mathbb{H}^2$.

$$L(T(I_1) \times T(I_2)) = L(I_1 \times I_2).$$

The infinitesimal form of the measure is defined as

$$\frac{d\alpha \, d\beta}{|e^{i\alpha} - e^{i\beta}|^2} = -\frac{d\alpha \, d\beta}{4\sin^2\left(\frac{\alpha-\beta}{2}\right)}, \quad (e^{i\alpha}, e^{i\beta}) \in (\mathbb{S}^1 \times \mathbb{S}^1 \setminus \{\text{diagonal}\})/\mathbb{Z}_2.$$

The measure of two disjoint arcs $(e^{i\alpha_1}, e^{i\beta_1})$, and $(e^{i\alpha_2}, e^{i\beta_2})$, similarly directed, is then

$$\int_{\beta_1}^{\beta_2} \int_{\alpha_1}^{\alpha_2} \frac{d\alpha \, d\beta}{4\sin^2 \frac{\alpha-\beta}{2}} = \left| \log \left| \frac{(e^{i\alpha_1} - e^{i\beta_1})(e^{i\alpha_2} - e^{i\beta_2})}{(e^{i\alpha_1} - e^{i\beta_2})(e^{i\alpha_2} - e^{i\beta_1})} \right| \right|.$$

For details see Bonahon [1988].

1-4. *Euclidean and hyperbolic circle centers.* Suppose ℓ is the vertical half-line rising from the origin in the upper half-space or upper half-plane model. Given $d > 0$, show that the locus of the points of hyperbolic distance d from ℓ consists of the cone of angle 2ϕ; its edges are the two euclidean lines of angle ϕ from ℓ, where $\sec \phi = \cosh d$.

The corresponding neighborhood about a geodesic, which is a semicircle, looks like a banana (Figure 1.5).

Next construct a sphere with euclidean center on ℓ which is tangent to the cone of distance d. Find its hyperbolic center which by symmetry also lies on ℓ. *Hint:* Construct the hyperbolic line segment between two opposite points of tangency of the sphere with the cone. It is orthogonal to both the sphere and ℓ. Show that it is a hyperbolic diameter and its intersection $(c, 0)$ with ℓ is the hyperbolic center.

Denote by $(0, a)$, $(0, b)$ the north and south pole of the sphere with hyperbolic center $(0, c)$. Show that $c^2 = ab$.

Show that in the disk model $\mathbb{D}$, the hyperbolic center of a circle coincides with the euclidean center if and only they both are euclidean circles centered at $(0, 0) \in \mathbb{D}$. In the upper half-plane model the centers never coincide. The corresponding statements hold in three dimensions.

1-5. Let ℓ be a line in the upper half-space model, which we assume without loss of generality to have endpoints $\pm 1/\beta \in \mathbb{R}$ on the boundary plane $\mathbb{C}$. Choose two other points on ℓ. Their projections to $\mathbb{C}$ lie on the line segment joining the endpoints and so are the form $\lambda/\beta, \mu/\beta$, with $-1 < \lambda < \mu < 1$. Let their heights above $\mathbb{C}$ be s and t, respectively. Verify the cross ratio identity:

$$\left(-\frac{1}{\beta}, \frac{\lambda}{\beta} + is, \frac{1}{\beta}, \frac{\mu}{\beta} + it \right) = \sqrt{\frac{(1+\lambda)(1-\mu)}{(1-\lambda)(1+\mu)}} < 1. \tag{1.6}$$

Specialize to the case that ℓ is the axis of a loxodromic T with $T(\lambda/\beta + is) = \mu/\beta + it$.

1-6. *Areas and side lengths of triangles.* Prove directly (without Gauss–Bonnet) that the angle sum Σ of a hyperbolic triangle Δ satisfies $\Sigma = \pi -$ Area Δ. *Hint:* First prove this for a triangle in the upper half-plane model with a vertex at ∞ and the other two above the points $0, 1 \in \mathbb{R}$. Then show that the area of a general triangle is the difference of the areas of two such ideal triangles. To find the area of the ideal triangle, you can use Green's formula from advanced calculus, plus the fact that the hyperbolic length of the horizontal segment $\{y = t, \; 0 \le x \le 1\}$ goes to 0 as $t \to \infty$.

Go on to prove, as in Epstein and Marden [1987, A.6.1,2], that the area A of a hyperbolic triangle Λ with a side of finite hyperbolic length s is strictly less than s. (*Hint:* The area increases when the other two sides of Δ have infinite length. Then show, still in the upper half-plane model, that given $a > 0$, the hyperbolic area of the rectangular strip $\{z : 0 < \operatorname{Re} z < s, a < \operatorname{Im} y\}$ is s/a. Use this to show that

$$\left(\frac{dA}{ds}\right)_{s=0} = \sin\theta < 1,$$

where θ is the angle between the short side and one of the vertical sides.)

Deduce that the area of a hyperbolic polygon P with one ideal vertex is less than s, where s is the sum of the lengths of the finite sides. Moreover, the sum of the exterior angles of P is less than $s + 2\pi$.

1-7. *A Möbius transformation is the composition of two or four reflections* Let $T = \left(\begin{smallmatrix} a & b \\ c & d \end{smallmatrix}\right)$ be a Möbius transformation of $\mathbb{C} \cup \infty$ such that $T^2(\infty) \ne \infty$ (so that the isometric circles $\mathfrak{I}(T)$ and $\mathfrak{I}(T^{-1})$ are distinct). Show that T is the composition of reflection in $\mathfrak{I}(T)$, followed by reflection in the perpendicular bisector of the line joining the center of $\mathfrak{I}(T)$ to the center of $\mathfrak{I}(T^{-1})$, followed by a rotation about the center of $\mathfrak{I}(T^{-1})$ of angle ϕ, where $e^{i\phi} = (\overline{a+d})/(a+d)$. What about the case that $\mathfrak{I}(T) = \mathfrak{I}(T^{-1})$?

If the trace of T is real, the rotation step is not needed.

In this we see that every Möbius transformation is the composition of two or four reflections in circles on $\mathbb{S}^2$. (A rotation is the composition of two reflections.) If $\operatorname{tr}(T)$ is real, only two reflections are needed.

1-8. (i) Prove that a Möbius transformation that has a real trace leaves invariant some circle in $\partial \mathbb{H}^3$ (which can be taken as $\mathbb{R} \cup \{\infty\}$).

   (ii) If, in addition, the transformation is loxodromic, it maps every hyperbolic plane that contains its axis onto itself.

  (iii) [Van Vleck 1919] Prove that the composition $F = E_2 \circ E_1$ of two elliptics has real trace if and only if there is a circle σ in $\mathbb{S}^2$ containing the fixed points of E_1 and E_2. If the fixed points of E_2 separate the fixed points of E_1 on σ then F is elliptic. If the fixed points do not so separate, then $E_1 E_2 E_1^{-1} E_2^{-1}$ is loxodromic with real trace.

1-9. *Möbius transformaions in ≥ 3 dimensions.* Prove the formula for the extension to upper half-space:

$$\left|A\vec{x}_1 - A\vec{x}_2\right| = \frac{1}{|c|^2} \frac{\left|\vec{x}_1 - \vec{x}_2\right|}{\left|\vec{x}_1 - A^{-1}(\infty)\right|\left|\vec{x}_2 - A^{-1}(\infty)\right|}, \qquad c \neq 0, \qquad (1.7)$$

and a corresponding formula for $c = 0$. Deduce that the hyperbolic metric is invariant under Möbius transformations and that the designation "isometric hemisphere" is justified as $|A'(\vec{x})| = 1$, where here $|A'|$ denotes the jacobian determinant.

The extension to upper half-space—in fact to all $\mathbb{R}^3 \cup \infty$—is conformal. Conversely if $F : D \subset \mathbb{R}^n \to \mathbb{R}^n$ is a conformal (angle-preserving) mapping, then F is the restriction to D of a Möbius transformation, *provided $n \geq 3$*. This striking result is called *Liouville's Theorem*. Liouville proved it under the assumption that the third partial derivatives of F are continuous; it is now known to be true under much weaker hypotheses on F; see Vuorinen [1988].

1-10. *Symmetry in isometric circles and planes.* Suppose S preserves the upper half-plane UHP, $S(\infty) \neq \infty$, while T preserves the unit disk $\mathbb{D}$. Prove that the isometric circle $\mathcal{I}(S)$ is characterized by the property that $\mathcal{I}(S) \cap$ UHP is the (hyperbolic) perpendicular bisector of $[i, S^{-1}(i)]$, that is, i and $S^{-1}(i)$ are symmetric about $\mathcal{I}(S)$. Correspondingly prove that $\mathcal{I}(T) \cap \mathbb{D}$ is the perpendicular bisector of $[0, T^{-1}(0)]$, that is, 0 and $T^{-1}(0)$ are symmetric about $\mathcal{I}(T)$.

Deduce that if $A(\mathbb{D}) =$UHP, then A maps $\mathcal{I}(T)$ to $\mathcal{I}(S)$.

Show that the corresponding facts are true for the isometric planes of transformations that preserve the upper half-space and ball models of $\mathbb{H}^3$.

Returning to the upper half-space assertion of Lemma 1.5.4, suppose $p \in \partial\mathbb{H}^3$ is not a fixed point of A. Given $\vec{x} \in \mathbb{H}^3$, let $e(\vec{x}, A)$ denote the plane which is the perpendicular bisector of the line segment $[\vec{x}, A^{-1}(\vec{x})]$. Then p lies on the circle bounding $e(\vec{x}, A)$ if and only if $\vec{x}$ lies on the plane $e(p, A)$, which has the expression (with $A = \left(\begin{smallmatrix} a & b \\ c & d \end{smallmatrix}\right)$)

$$\left|z - \left(p + \frac{(p^2c + pd - pa - b)(\overline{a - pc})}{|a - pc|^2 - 1}\right)\right|^2 + t^2 = \frac{|p^2c + pd - pa - b|^2}{(|a - pc|^2 - 1)^2},$$

if $|a - pc| \neq 1$. If $p = \infty$, then $e(p, A)$ reduces to the isometric plane for A. Also the plane $e(\vec{x}, A)$ converges to $e(p, A)$ as $\vec{x} \to p$.

Choose A so that $A^{-1}(\infty) = 0$. Then when $\vec{x} = (0, t)$, the vertical coordinate of $A(\vec{x})$ is $1/(t|c|^2)$. This takes the value $1/|c|$ when $t = 1/|c|$. Thus the horosphere σ at $z = 0$ of euclidean diameter $1/|c|$ is tangent to the horizontal plane P of height $t = 1/|c|$. Moreover A maps $(0, 1/|c|) \in \mathbb{H}^3$ to $(a/c, 1/|c|)$. Therefore A maps σ onto P. The hemisphere centered at $z = 0 \in \mathbb{C}$ of radius $1/|c|$ is the isometric plane.

1-11. *Stereographic projection.* Confirm that stereographic projection from the north pole of the unit sphere $\mathbb{S}^2$ to the complex plane $\mathbb{C}$ containing the equator of $\mathbb{S}^2$ is given by the following formulas: If $(x_1, x_2, x_3) \in \mathbb{S}^2$ and $z = x + iy$ is the corresponding point in $\mathbb{C}$ then

$$x = \frac{x_1}{1 - x_3}, \quad y = \frac{x_2}{1 - x_3}, \quad |z|^2 = \frac{1 + x_3}{1 - x_3}.$$

This can be extended to map the interior $\mathbb{B}$ of $\mathbb{S}^2$ to upper half-space as follows. Reflection in the unit circle $z \mapsto 1/\bar{z}$ extends to the reflection in $\mathbb{H}^3$ given by $\vec{x} \mapsto \vec{x}/|\vec{x}|^2$. Take however the reflection in the sphere S_N of radius $\sqrt{2}$ about the north pole of $\mathbb{S}^2$:

$$I_1(\vec{x}) = 2\frac{\vec{x} - \vec{k}}{|\vec{x} - \vec{k}|^2} + \vec{k}, \quad \vec{k} = (0, 0, 1).$$

S_N intersects $\mathbb{S}^2$ in its equator and I_1 pointwise fixes that. Also fixed are the vertical planes through the origin.

Because I_1 sends $(0, 0, 1)$ to ∞, $(1, 0, 0)$ to $(1, 0, 0)$, $(0, 1, 0)$ to $(0, 1, 0)$, and $(0, 0, -1)$ to $(0, 0, 0)$, we see that the image of $\partial \mathbb{B}$ is the plane $\{x_3 = 0\}$ and the image of $\mathbb{B}$ is lower half-space.

Follow I_1 by reflection in the horizontal plane $\{x_3 = 0\}$:

$$I_2 : (x_1, x_2, x_3) \mapsto (x_1, x_2, -x_3).$$

The required extension of stereographic projection is $I = I_2 \circ I_1$. The collection of euclidean half-planes and hemispheres in upper half-space corresponds to the collection of spherical caps in $\mathbb{B}$ orthogonal to $\partial \mathbb{B}$. We know this once we know that stereographic projection maps the collection of circles/lines in $\mathbb{C}$ to circles on $\mathbb{S}^2$. From this we can also deduce that I preserves the dihedral angles between intersecting hyperbolic planes.

The group of isometries of $\mathbb{B}$ is then the conjugate of the group of the upper half-space by the Möbius transformation I. The formulas are best found by the method given in the next exercise.

1-12. *Formulas for the ball model.* We will follow the elegant treatment presented by Ahlfors [1981]. The notation $x^* = \vec{x}/|x|^2$ for reflection of $\vec{x}$ in the unit sphere will be useful. More generally, given $\vec{a} \in \mathbb{B}^3$, the sphere with center a^* orthogonal to $\partial \mathbb{B}^3$ has radius $(|a^*|^2 - 1)^{1/2}$. The formula for reflection $x \in \mathbb{B}^3 \mapsto y$ is

$$y = a^* + (|a^*|^2 - 1)(x - a^*)^*.$$

The group of Möbius transformations preserving $\mathbb{B}^3$ is generated by an even number of such reflections. A Möbius transformation that sends $\mathbb{B}^3$ onto itself and a given point $a \in \mathbb{B}^3$ to the origin 0 is

$$T_a x = -a + (1 - |a|^2)(x^* - a)^*.$$

The general transformation that sends $a \to 0$ is the composition of T_a followed by a euclidean rotation about 0. The jacobian determinant $|T_a'(x)|$ of T_a represents the local stretch of T_a, the same in all directions. It remains unchanged if T_a is followed by a rotation about 0, so it represents the jacobian determinant of any element that sends $a \to 0$ and preserves $\mathbb{B}^3$.

The jacobian satisfies

$$|T_a'(x)| = \frac{(1 - |a|^2)}{|x|^2 |x^* - a|^2} = \frac{1 - |a|^2}{\left|x - a|x|^2\right|^2}.$$

In addition confirm the formulas

$$|T_a x| = |T_a x - T_a a| = |x - a| \sqrt{T_a'(x)} \sqrt{T_a'(a)},$$

and

$$|T_a x| = \frac{|x - a|}{|a|\,|x - a^*|},$$

so that

$$1 - |T_a x|^2 = \frac{(1 - |x|^2)(1 - |a|^2)}{|a|^2 |x - a^*|^2}.$$

Conclude that

$$\frac{|T_a'(x)|}{1 - |T_a x|^2} = \frac{1}{1 - |x|^2}.$$

In other words, the hyperbolic metric is invariant under any transformation that sends some point $a \to 0$; therefore is invariant under all Möbius transformations preserving the ball.

In fact, for any Möbius transformation T in 3-space, whether or not it preserves a ball or half-space,

$$|T\vec{x} - T\vec{y}| = |T'(\vec{x})|^{1/2} |T'(\vec{y})|^{1/2} |\vec{x} - \vec{y}|.$$

This follows from the fact that any nontrivial Möbius transformation T is the composition of similarity mappings $\vec{x} \mapsto m\vec{x} + \vec{b}$ and the reflection $\vec{x} \mapsto \vec{x}/|x|^2$. The jacobian matrix for the similarity is simply mI. For the reflection it is $(I - 2Q(\vec{x}))/|x|^2$, where the matrix $Q(\vec{x}) = (1/|x|^2)\,(x_i x_j)$ satisfies $Q^2 = Q$ and $(I - 2Q)^2 = I$. In other words, at each point $x \in \mathbb{B}^3$, the jacobian $T'(\vec{x})$ is a scalar multiple $|T'(x)|$ of an orthogonal matrix.

1-13. Show the existence of regular n-sided hyperbolic polygons as follows. In the disk model of $\mathbb{H}^2$ start at the origin with a tiny regular n-sided euclidean polygon. Radially expand the polygon insuring by rotational symmetry that all sides remain equal in length. Show that the vertex angle decreases monotonically from the euclidean $(n - 2)\pi/n$ to zero when the vertices are on the unit circle. Use the same argument for regular polyhedra.

1-14. If g is a Möbius transformation acting in $\mathbb{B}^3$ with center 0 (really $(0, 0, 0)$) and $d(\cdot, \cdot)$ denotes hyperbolic distance, show that

$$e^{-d(0, g(0))} = \frac{1 - |g(0)|}{1 + |g(0)|}.$$

Also show that for any two points x, y in $\mathbb{B}^3$,

$$\frac{1 - |g(x)|}{1 - |g(y)|} \leq 2e^{d(x,y)}.$$

Let G be a countable group of Möbius transformations. Show that for $\alpha > 0$,

$$\sum_{g \in G} e^{-\alpha d(0,g(0))} < \infty \quad \text{if and only if} \quad \sum_{g \in G} e^{\alpha d(x,g(x))} < \infty,$$

and correspondingly

$$\sum_{g \in G} (1 - |g(0)|)^{\alpha} < \infty \quad \text{if and only if} \quad \sum_{g \in G} (1 - |g(x)|)^{\alpha} < \infty.$$

Referring back to Exercise (1-12) or Nicholls [1989, Ch. 1] show that

$$\sum_{g \in G} e^{-\alpha d(0,g(0))} < \infty \quad \text{if and only if} \quad \sum_{g \in G} |g'(0)|^{\alpha} < \infty.$$

Confirm the analogous formulas for groups acting instead in the unit disk.

1-15. *Iwasawa decomposition.* Prove that $\mathrm{SL}(2, \mathbb{R}) = KAN$, where

$$K = \begin{pmatrix} \cos\theta & -\sin\theta \\ \sin\theta & \cos\theta \end{pmatrix}; \quad A = \begin{pmatrix} r & 0 \\ 0 & 1/r \end{pmatrix}, \quad r > 0; \quad N = \begin{pmatrix} 1 & x \\ 0 & 1 \end{pmatrix}.$$

K consists of orthogonal matrices, A of diagonal matrices with positive entries, and N of unipotent matrices, upper triangular matrices with 1s on the diagonal. Every matrix $g \in \mathrm{SL}(2, \mathbb{R})$ can be written as a product $\{k \times a \times n\}$.

Conclude that $\mathrm{SL}(2, \mathbb{R})$ is homotopy equivalent to a circle, and its fundamental group is $\mathbb{Z}$. $\mathrm{SL}(2, \mathbb{R})$ has infinitely many coverings each determined by $n\mathbb{Z}$, for $n \geq 1$; its universal cover is $\mathbb{R}$.

Let G_n be the degree $n \geq 1$ covering group of $\mathrm{PSL}(2, \mathbb{R})$. Bill Goldman [1988] found an exact formula for the number of connected components in the representation space $\mathrm{Hom}(\pi_1(S), G_n)$ obtained by group homomorphisms $p_1(S) \to G_n$ where S is a closed surface of genus $g > 1$. In particular for the degree two cover $\mathrm{SL}(2, \mathbb{R})$, there are $2^{2g+1} + 2g - 2$ components.

In contrast, the representation space for $G = \mathrm{PSL}(2, \mathbb{C})$ has two connected components and $G = \mathrm{SL}(2, \mathbb{C})$ has only one, no matter what the value of $g \geq 2$.

1-16. Given four distinct points $z_1, z_2, w_1, w_2 \in \mathbb{C} \cup \infty$, show that there exists a Möbius transformation A such that $A(z_1) = -1$, $A(z_2) = 1$, $A(w_1) = -u$, and $A(w_2) = u$ for some $u \in \mathbb{C}$. Clearly $(z_1, z_2, w_1, w_2) = (-1, 1, -u, u)$. A is uniquely determined if it is required that $|u| \geq 1$. *Hint:* Take $z_1 = -1$, $z_2 = 1$, $w_1 = i$, $w_2 = \zeta$. Find an equation for the coefficients of A. For there to be a nonzero solution, the determinant of the coefficients must vanish.

Consider the hyperbolic line ℓ with endpoints z_1, z_2 and the line m with endpoints w_1, w_2. Show that there is a uniquely determined common perpendicular to ℓ and m. Show that the hyperbolic distance between the lines is $\log |u|$.

1-17. *The horizon in* $\mathbb{H}^2$. Prove that there is a unique largest disk in an ideal triangle and that its hyperbolic radius is $\frac{1}{2}\log 3$. Deduce that any hyperbolic disk in $\mathbb{H}^2$ that meets three mutually disjoint open half-planes must have radius exceeding $\frac{1}{2}\log 3$. (*Hint:* Put $z = 0$ and the vertices of the ideal triangle at equally spaced points on the circle.)

The *horizon* of a line ℓ subtended at say $z = 0$ is the angle subtended by the rays from 0 to the endpoints of ℓ. In euclidean space, this angle is π for any ℓ not through 0. In hyperbolic space the angle is always $< \pi$, but approaches π as ℓ approaches 0.

In particular a line ℓ in $\mathbb{H}^2$ at hyperbolic distance $\frac{1}{2}\log 3$ from a point $z \in \mathbb{H}^2$ covers exactly one third of the horizon. The union of three hyperbolic lines, each at distance at least $\frac{1}{2}\log 3$ from z and at least one at distance strictly greater than that, do not separate z from $\partial\mathbb{H}^2$.

Verify that in $\mathbb{H}^3$, a hyperbolic ball B that meets three mutually disjoint open hyperbolic half-spaces must likewise have radius exceeding $\frac{1}{2}\log 3$.

(*Hint:* In the ball model assume that the origin is the center of B. The half-spaces determine three mutually disjoint disks $D_i \subset \partial\mathbb{H}^3$. Denote their spherical radii by r_i. A great circle has length 2π so $\sum 2r_i \leq 2\pi$. For at least one index, $2r_j \leq 2\pi/3$. The distance to the origin of the plane rising from ∂D_j is therefore at least $\frac{1}{2}\log 3$.)

1-18. Given a point $z \in \mathbb{H}^2$ and a geodesic γ not through z show that

$$\sinh d(z, \gamma) = \cot(\theta_z/2),$$

where $d(z, \gamma)$ is the distance from z to γ and $\theta_z < \pi$ is the visual angle γ subtends at z.

Hint: In the disk model take $z = 0$. Then γ is an arc of a circle of euclidean circle of radius r, say. Set $a = d(0, \gamma)$. Show that $\sinh a = 1/r$ and $\sinh d(0, a) = 2a/(1-a^2)$. Use the right triangle with euclidean sides of length $1, r, d+r$.

1-19. In the upper half-space model let γ be the geodesic with endpoints $\pm 1 \in \mathbb{C}$ so that it lies in the vertical plane rising from $\mathbb{R}$. Show that in the hyperbolic arc length parameter s with basepoint at i,

$$\gamma(s) = \frac{1 + ie^{-s}}{1 - ie^{-s}} = \tanh s + i\,\text{sech}\,s.$$

1-20. *Trace calculations for cyclic groups.* We are here reporting on work of Jørgensen [Jørgensen 1973]. Let $\tau = \text{tr}\,A$ be normalized so that $0 \leq \arg\tau < \pi$. Prove (by putting A in standard form and using induction) that there is a sequence $\{\beta_n\}, -\infty < n < \infty$, such that $\beta_0 = 0, \beta_1 = 1$,

$$A^n = -\beta_{n-1}I + \beta_n A, \quad \text{and} \quad \beta_{n+1} = -\beta_{n-1} + \tau\,\beta_n.$$

Show that $\beta_{-n} = -\beta_n$. Furthermore, $\beta_n = 0$ for some n if and only if $A^n(z) = \text{id}$.
    Set $\tau_n = \text{tr}\,A^n$, so that $\tau_0 = 2$ and $\tau_1 = \tau$. Then $\tau_{-n} = \tau_n$. Prove that

(i) $\tau_n = -\beta_{n-1} + \beta_{n+1}$;
(ii) if $\tau = \lambda + \lambda^{-1}$, then $\tau_n = \lambda^n + \lambda^{-n}$ and $\beta_n = (\lambda^n - \lambda^{-n})/(\lambda - \lambda^{-1})$;

(iii) $\tau_m \tau_n = \tau_{m+n} + \tau_{m-n}$, $\beta_m \tau_n = \beta_{m+n} + \beta_{m-n}$, and $\beta_m \beta_n = (\tau_{m+n} - \tau_{m-n})/(\tau^2 - 4)$.

Show also that $\lim_{|n| \to \infty} \beta_n = \lim_{|n| \to \infty} \tau_n = \infty$, if $|\tau| > 2$.

Finally show that β_n is a polynomial of degree $|n| - 1$ in τ, and τ_n is of degree $|n|$ in τ. Furthermore

$$\frac{d}{d\tau} \tau_n = n\beta_n, \qquad \frac{d}{d\tau} \beta_n = \frac{n\tau_n - \tau\beta_n}{\tau^2 - 4}.$$

The isometric circles $\mathfrak{I}(A^{\pm 1})$ are symmetric about the midpoint of line segment joining their centers. Replace A by a conjugate so that the midpoint $A(\infty) + A^{-1}(\infty)$ becomes $z = 0$. Show that A then has the form

$$A = \begin{pmatrix} \frac{1}{2}\tau & \frac{1}{4}(\tau^2 - 4) \\ 1 & \frac{1}{2}\tau \end{pmatrix}. \tag{1.8}$$

Jørgensen used this form to study the behavior of the cyclic group $\langle A \rangle$ as a function of its trace.

Show that

$$A^n = \begin{pmatrix} \frac{1}{2}\tau_n & \frac{1}{4}(\tau_n^2 - 4)\beta_n^{-1} \\ \beta_n & \frac{1}{2}\tau_n \end{pmatrix}. \tag{1.9}$$

Also $A^k(-z) = -A^{-k}(z)$, for $-\infty < k < \infty$.

1-21. *Tangent bundle of a closed surface.* Consider with Tukia 1985c the 3-manifold

$$K = \{(x_1, x_2, x_3) : x_i \in \mathbb{R} \text{ are distinct and induce the positive orientation.}\}$$

Define a map $\rho : K \to \text{UHP}$ as follows. Let ℓ be the geodesic in UHP between x_1 and x_2. Let x_3^* be the foot of the perpendicular from x_3 to ℓ. Set $\rho(x_1, x_2, x_3) = x^*$. Prove:

(i) For $z \in \text{UHP}$ the set $\rho^{-1}(z)$ is homeomorphic to the circle $\mathbb{S}^1$, and hence that K is homeomorphic to $\text{UHP} \times \mathbb{S}^1$.

(ii) If A is loxodromic and preserves UHP with axis $\ell \in \text{UHP}$, the set $\mathcal{S}(A) := \rho^{-1}(\ell)$ is homeomorphic to $\mathbb{R} \times \mathbb{S}^1$.

(iii) If B is another Möbius transformation preserving UHP, then $B(\mathcal{S}(A)) = \mathcal{S}(BAB^{-1})$.

(iv) $\mathcal{S}(A)$ and $B(\mathcal{S}(A))$ are either disjoint, identical, or have intersection $\rho^{-1}(z)$ for some $z \in \text{UHP}$.

Suppose $R = \text{UHP}/G$ is a closed hyperbolic surface. Show that there is a natural discrete action of G on K. Show that K/G is homeomorphic to the unit tangent bundle $T(R)$ of R.

Next, show that any orientation-preserving homeomorphism (automorphism) $\alpha : R \to R$ induces an automorphism $\hat{\alpha} : T(R) \to T(R)$ of the 3-manifold $T(R) = K/G$. Moreover, homotopic automorphisms α, α_1 of R correspond to homotopic

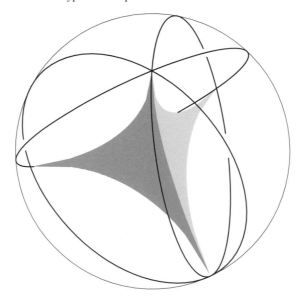

Fig. 1.9. An ideal tetrahedron in the ball model. The dihedral angles are the same as the angles of intersection of the circles on $\partial\mathbb{H}^3$ determined by the faces.

automorphisms $\hat{\alpha}, \hat{\alpha}_1$ of $T(R)$. This result is attributed to Cheeger and Gromov— see Casson and Bleiler [1988, pp. 54–55] for details. (*Hint*: Set $x = \rho(x_1, x_2, x_3)$ and $p(x) = (x, \sigma_x)$ where $\sigma_x \subset \ell$ is the oriented segment of length two, centered at x.)

1-22. *Ideal tetrahedra.* On $\partial\mathbb{H}^3$, choose any four distinct points z_1, z_2, z_3, z_4. Then draw the six hyperbolic lines obtained by connecting pairs of points. Each triple of points lies on the edge of a uniquely determined hyperbolic plane. The four hyperbolic planes so obtained pairwise intersect in the six lines. The common exterior of these four planes is a four-sided solid called an ideal tetrahedron. It is uniquely determined up to isometry by its four "ideal" vertices z_1, z_2, z_3, z_4.

Now using the upper half-space model, send any one of the vertices to ∞. The three faces meeting at ∞ now become vertical planes. The cross section obtained by intersecting with any sufficiently high horizontal plane $\{t = N\}$ is a euclidean triangle. The three angles α, β, γ of that triangle are exactly the dihedral angles formed by the intersection of the corresponding two planes. And of course $\alpha + \beta + \gamma = \pi$.

Label the three other dihedral angles so that δ is opposite β, ϵ is opposite γ, and ρ is opposite α. Any of the four ideal vertices can be sent to ∞. As a consequence the six dihedral angles satisfy four equations. From this, deduce that $\alpha = \rho$, $\beta = \gamma$, $\gamma = \epsilon$. That is, the dihedral angles at opposite edges are the same. In addition, the sum of all the dihedral angles is 2π.

Show that the ideal tetrahedron is uniquely determined by the three angles α, β, γ at the vertex ∞ up to similarity ($z \mapsto az + b$).

Taking thus one of its vertices at ∞, denote the other three ideal vertices by t, u, v, all of which lie in $\mathbb{C}$. These are the vertices of the ideal triangle forming the base of the tetrahedron. Orthogonal projection to $\mathbb{C}$ takes this to a euclidean triangle with vertices

t, u, v. Assume the labeling is chosen so that t, u, v in order give the clockwise orientation. Define

$$z_1 = (t, u, v, \infty) = z,$$
$$z_2 = (u, v, t\infty) = 1/(1 - z),$$
$$z_3 = (v, t, u, \infty) = (z - 1)/z.$$

Each z_i determines the other two. Consequently,

$$\text{(i)} \quad z_1 z_2 z_3 = -1, \quad \text{hence} \quad \sum \arg(z_i) = \pi, \quad \text{and} \tag{1.10}$$

$$\text{(ii)} \quad z_1 z_3 - z_1 + 1 = 0.$$

Assign the numbers z_1, z_2, z_3 to the vertical edges of the tetrahedron through v, t, u, that is, the three edges at ∞. Now apply a Möbius transformation to the tetrahedron sending a different ideal vertex to ∞ and correspondingly obtain three numbers, using the same clockwise ordering. Show that the same three numbers appear as before and an edge which runs between the original vertex and the new one placed at ∞ is assigned the same number.

Conclude that ideal tetrahedra are uniquely determined up to isometry by three complex numbers z_1, z_2, z_3 that satisfy Equations (1.10): Starting at any vertex the three edges there are labeled in clockwise order z_1, z_2, z_3. Then the three opposite edges are given the same labeling.

Show that an ideal tetrahedron is also determined up to isometry by the three dihedral angles (which sum to π) along the three edges ending at an ideal vertex. Conversely, given three positive angles which sum to π, there is an ideal tetrahedron with these as dihedral angles at an ideal vertex. The dihedral angles at opposite edges of an ideal tetrahedron are the same.

It may happen that the four ideal vertices lie on a circle in $\mathbb{S}^2$. In this case the "ideal tetrahedron" is degenerate: it lies in a plane. There are three patterns (up to reordering) in which a proper ideal tetrahedron may degenerate: The possible orders of the ideal vertices on the circle are (1234), (1342), (1423).

1-23. *Volume of tetrahedra.* Show that the volume of tetrahedra, like the area of triangles, is uniformly bounded above.

An exact formula for the volume of an ideal tetrahedron is derived in Milnor [1994, §3] and Ratcliffe [1994, §10.4]. The basic function involved is what Milnor calls the *Lobachevsky function*,

$$Л(\theta) = -\int_0^\theta \log |2 \sin u|\, du = \theta\left(1 - \log 2\theta + \sum_1^\infty \frac{2^{2n} B_n}{2n(2n+1)!} \theta^{2n}\right), \tag{1.11}$$

where B_n denotes the n-th Bernoulli number. The series, which is obtained by twice integrating $d^2 Л(\theta)/d\theta^2 = -\cot\theta$, converges for $|\theta| < \pi$ although $Л(\theta)$ itself is periodic with period π. For computations, one generally works with the infinite series.

The volume of the ideal tetrahedron with dihedral angles $\alpha, \alpha, \beta, \beta, \gamma, \gamma$ (the opposite dihedral angles of an ideal tetrahedron are equal) is

$$\textrm{Л}(\alpha) + \textrm{Л}(\beta) + \textrm{Л}(\gamma). \tag{1.12}$$

One can also compute the volumes of the regular hyperbolic polyhedra.

Of all hyperbolic tetrahedra, ideal or not, there is a one with the largest volume, which is uniquely determined up to Möbius equivalence [Milnor 1994, p. 200]. It is the ideal tetrahedron whose vertices are the vertices of a regular euclidean tetrahedron inscribed in $\mathbb{S}^2$. All its dihedral angles are $\pi/3$ and its group of orientation-preserving hyperbolic symmetries is the group of rotations preserving the euclidean tetrahedron. Its volume is $1.0149\ldots$. (The area of the ideal triangle is π.)

There is a classical variational formula useful in studying deformations of hyperbolic polyhedra. It is called the *Schläfli formula* [Milnor 1994, p. 281]:

$$dV(P) = -\frac{1}{2} \sum_e L(e)\, d\theta_e.$$

Here P is a hyperbolic polyhedron of volume $V(P)$, the sum is over all edges e of P; $L(e)$ is the length of the edge e and θ_e is the interior dihedral angle along e.

1-24. *Thinness of tetrahedra.* Recall from Theorem 1.3.1 that if p is a point on a side of a triangle, there is a point on at least one of the other two sides which is of distance at most $\log(1 + \sqrt{2}) = \operatorname{arcsinh} 1$ away.

Show that something similar holds for hyperbolic tetrahedra in $\mathbb{H}^3$: there exists $C > 0$ such that if $\vec{x}$ is a point on an edge of the tetrahedron, then the minimum distance from $\vec{x}$ to the union of the other edges does not exceed C. Can you find the optimal C? An analogous property is that there is a constant $C > 0$ such that if $\vec{x}$ is any point in a tetrahedron, the minimum distance of $\vec{x}$ to an edge does not exceed C. See Exercise (1-17).

1-25. *Hyperboloid and projective (Klein) models.* We have worked with the most common conformal models: The "(unit) disk model" or "Poincaré disk" and the upper half-plane (UHP) model. We will build two more models of the hyperbolic plane, closely related to one another, and explain their relationship to the conformal models introduced in this chapter. For another approach, see Cannon et al. [1997, Section 7].

We start with the plane $\mathbb{C} = \mathbb{R}^2$, containing the unit circle S^1 and the real line $\mathbb{R}$, which we complete to $\mathbb{R} \cup \infty$ (the boundary of the upper half-plane model). Let U be the unique Möbius transformation of $\mathbb{C} \cup \infty$ fixing -1 and 1 and taking ∞ to i; you found its expression, $U(z) = (z - i)/(1 - iz)$, in Exercise (1-1(iv)). When restricted to the real line, U can be thought of as stereographic projection from the point i, that is, it maps $x \in \mathbb{R} \cup \infty$ to

$$U(x) = (u, v) = \left(\frac{2x}{x^2 + 1}, \frac{x^2 - 1}{x^2 + 1} \right) \in \mathbb{R}^2 = \mathbb{C};$$

the inverse stereographic projection from the circle to the line (compare Exercise (1-11)) is $(u, v) \mapsto u/(1 - v)$.

U conjugates the upper half-plane model and the disk model; in particular, the orientation-preserving isometries of the disk model can be thought of as elements of $U \, \mathrm{PSL}(2, \mathbb{R}) \, U^{-1} \subset \mathrm{PSL}(2, \mathbb{C})$—since we know from Exercise (1-1) that the orientation-preserving isometries of the upper half-plane model are the elements of $\mathrm{PSL}(2, \mathbb{R})$.

The setup is completed by considering $\mathbb{E}^{1,2}$, which is $\mathbb{R}^3$ with the inner product

$$\langle \vec{x}, \vec{y} \rangle = x_1 y_1 + x_2 y_2 - x_3 y_3. \tag{1.13}$$

(This is studied in detail in Exercises (1-26) and (1-27).) The set of vectors in $\mathbb{E}^{1,2}$ having length 0—that is, satisfying $\langle \vec{x}, \vec{x} \rangle = 0$—is the *light cone*. The light cone corresponds to the unit circle in $\mathbb{R}^2$ via the usual projectivization map

$$u = \frac{x_1}{x_3}, \quad v = \frac{x_2}{x_3}.$$

The name "light cone" comes from relativity. Vectors of "imaginary length" ($\langle \vec{x}, \vec{x} \rangle < 0$) are called *timelike*, those lying on the light cone are *lightlike*, and those of positive length are *spacelike*.

Now take a Möbius transformation A preserving $\mathbb{R} \cup \infty$ and having normalized matrix $\left(\begin{smallmatrix} a & b \\ c & d \end{smallmatrix} \right)$; recall that a, b, c, d are real. We associate to A the linear map $\widehat{A}$ of $\mathbb{E}^{1,2}$ given by the matrix

$$\frac{1}{2} \begin{pmatrix} 2(ad + bc) & 2(ac - bd) & 2(ac + bd) \\ 2(ab - cd) & a^2 - b^2 - c^2 + d^2 & a^2 + b^2 - c^2 - d^2 \\ 2(ab + cd) & a^2 - b^2 + c^2 - d^2 & a^2 + b^2 + c^2 + d^2 \end{pmatrix}.$$

Check that:

(i) $\widehat{A}$ preserves the inner product of $\mathbb{E}^{1,2}$, and thus leaves the light cone invariant. (*Hint:* Taking the inner product (1.13) of the columns of the matrix of $\widehat{A}$ yields zero if the columns are distinct and ± 1 if the columns are the same—the minus sign appearing for the third column only. Thus the matrix is "orthonormal" for the given inner product.)

(ii) The map from the light cone onto itself defined by $\widehat{A}$ induces on S^1 a map that coincides with $U A U^{-1}$.

(iii) $\widehat{A}$ also induces a map on the unit disk bounded by S^1 (since the set of timelike vectors in $\mathbb{E}^{1,2}$ is also preserved by $\widehat{A}$ and corresponds to the unit disk). This map on the disk takes straight line segments to straight line segments.

The *hyperboloid model* of the hyperbolic plane is the upper sheet of the hyperboloid $x_1^2 + x_2^2 - x_3^2 = -1$, $x_3 > 0$, with the metric induced from the ambient space $\mathbb{E}^{1,2}$.

The *projective model* or *Klein model* of the hyperbolic plane is the unit disk in $\mathbb{R}^2$, with the metric transported from the hyperboloid model by the central projection map $(x_1, x_2, x_3) \mapsto (u, v) = (x_1/x_3, x_2/x_3)$. We now justify these metrics and study the basic properties of these models.

(iv) We first define standard maps from the upper half-plane (UHP) to the hyperboloid and projective disk. Given a point z in the upper half-plane, take any nontrivial $A \in \mathrm{PSL}(2, \mathbb{R})$ that fixes z; show that the corresponding linear map $\widehat{A}$ has a timelike one-dimensional eigenspace in $\mathbb{E}^{1,2}$ that does not depend on the choice of A. We take the intersection of this eigenspace with the hyperboloid as the image of z in the hyperboloid model; likewise we take the projection of this eigenspace onto the projective disk as the image of z in the projective model.

(v) Show that this standard map from UHP to the hyperboloid is an isometry between the hyperbolic metric on UHP and the metric induced on the hyperboloid from the ambient space $\mathbb{E}^{1,2}$. Thus the hyperboloid really is a model of the hyperbolic plane. So is, trivially, the projective disk (since we defined the metric by pullback from the hyperboloid).

(vi) Show that hyperbolic lines are straight line segments in the projective model. What are they in the hyperboloid model? What are the horocycles in the projective and hyperbolic models?

(vii) The orientation-preserving isometries of the hyperboloid and projective models are induced by the linear maps $\widehat{A}$, as A ranges over $\mathrm{PSL}(2, \mathbb{R})$. Work out the special cases for A in standard form: $x \mapsto x + 1$, $x \mapsto \lambda^2 x$, and $x \mapsto (x \cos \varphi + \sin \varphi)/(-x \sin \varphi + \cos \varphi)$. For each case show that there is a one-dimensional fixed eigenspace and determine where it is. Identify what corresponds to the axes and fixed points of loxodromic transformations, and to horocycles for parabolics.

(viii) The map U and the map constructed in (iv) take UHS to the disk model and the projective model, respectively. Composing one with the inverse of the other we get a map that fixes the boundary—see (ii) above. Show that its action of this map on the interior of the disk—that is, the map that conjugates the disk and projective models—is radial and corresponds to stereographic projection onto a hemisphere, followed by orthogonal projection back to the unit circle.

(xi) Consider a hyperbolic polygon in the disk model of $\mathbb{H}^2$ and then, as Poincaré before you, take its counterpart in the projective model. Show that the property that a vertex angle be $< \pi$ is preserved, although the angle itself is not. Recover Poincaré's proof that all of the interior vertex angles of a hyperbolic polygon are $< \pi$ if and only if the polygon is hyperbolically convex.

1-26. *Minkowski 3-space.* [Cannon et al. 1997, p. 66] The space $\mathbb{E}^{1,2}$ of the previous exercise is called *Minkowski space*. Consider the situation in one dimension lower, in $\mathbb{R}^2 = \mathbb{E}^{1,1}$. Here the inner product is

$$\langle x, y \rangle = x_1 y_1 - x_2 y_2, \quad x = (x_1, x_2), y = (y_1, y_2).$$

Suppose $\vec{p}(t) = (x(t), y(t))$ describes the motion of a car on the upper sheet of the hyperbola $x^2 - y^2 = -1$. Differentiating, we find that $xx' - yy' = 0$, so that the vectors $\vec{p} = (x, y)$ and $\vec{p}' = (x', y')$ are orthogonal: $\langle \vec{p}(t), \vec{p}'(t) \rangle = 0$. Consequently there is a scalar function $k(t)$ such that $\vec{p}' = k(t)(y(t), x(t))$. If $\vec{p}'$ moves at constant unit speed so $|\vec{p}'| = 1 = |k|\sqrt{y^2 - x^2}$, and $k = \pm 1$. We will take $k = 1$.

Therefore we also have at hand the coupled pair of differential equations $x'(t) = y(t)$, $y'(t) = x(t)$. These can be solved by infinite series; in fact for suitable initial values, $y = \cosh t$, $x = \sinh t$. The parameter t can be identified with hyperbolic arc length on the hyperbola $x^2 - y^2 = -1$ since $\vec{p}(t) = (x, y)$ travels along at unit speed.

Figure 1.10 is a visual interpretation of $\mathbb{E}^{1,2}$ and how it corresponds to the disk model $\mathbb{D}$ of $\mathbb{H}^2$:

Let $\vec{x}$ be the vector

$$\vec{x} = (x_1, x_2, x_3), \qquad \langle \vec{x}, \vec{x} \rangle = x_1^2 + x_2^2 - x_3^2.$$

Set $u = \dfrac{x_1}{x_3}$ and $v = \dfrac{x_2}{x_3}$. First comes the hyperboloid of two sheets in $\mathbb{R}^3$,

$$\mathcal{H}_\pm = \{x_1^2 + x_2^2 = x_3^2 - 1\} \equiv \{u^2 + v^2 = 1 - \frac{1}{x_3^2}\}.$$

Each of the upper and lower sheets $\mathcal{H}_+$ and $\mathcal{H}_-$ corresponds to the unit disk $\mathbb{D}$. The hyperboloid is asymptotic to the *light cone*,

$$\mathcal{L}^+ = \{x_1^2 + x_2^2 = x_3^2\} \equiv \{u^2 + v^2 = 1\},$$

each half of which corresponds to $\partial\mathbb{D}$. As the vertical coordinate x_3 sweeps through $-\infty < x_3 < +\infty$ the circles sweep out both napes of the cone. Finally there is the one-sheeted hyperboloid in $\mathbb{R}^3$,

$$\mathcal{H} = \{x_1^2 + x_2^2 = 1 + x_3^2\} = \{u^2 + v^2 = 1 + \frac{1}{x_3^2}\},$$

which is also asymptotic to the light cone. The values $\pm x_3 \neq 0$ correspond to two circles of radius > 1; $x_3 = 0$, corresponds the unit circle. The hyperboloid $\mathcal{H}$ is

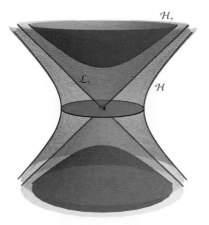

Fig. 1.10. Minkowski 3-space: The upper half $\mathcal{L}_+$ of the light cone corresponding to $\partial\mathbb{D}$, the upper sheet $\mathcal{H}_+$ of the hyperboloid corresponding to $\mathbb{D}$, and the enveloping hyperboloid $\mathcal{H}$ each half corresponding to the exterior of $\mathbb{D}$.

asymptotic to the full light cone from the outside. Both its upper and lower half corresponds to the exterior of $\mathbb{D}$ as $0 \leq x_3 < +\infty$.

1-27. *Minkowski space, dimension-n.* We now consider Minkowski space in arbitrary dimension. The inner product in $\mathbb{E}^{1,n}$ is

$$\langle \vec{x}, \vec{y} \rangle = \left(\sum_1^n x_i y_i \right) - x_{n+1} y_{n+1}, \quad \| \vec{x} \|^2 = \langle \vec{x}, \vec{x} \rangle,$$

where $\vec{x} = (x_1, \ldots, x_{n+1})$. Two nonzero vectors are called orthogonal if $\langle \vec{x}, \vec{y} \rangle = 0$.

A vector $\vec{x}$ is called *timelike, lightlike, or spacelike* according to whether $\| \vec{x} \|^2$ is negative, zero, or positive.

The collection of lightlike vectors forms the *light cone* $\{\vec{x} : \| \vec{x} \| = 0\}$. The upper sheet of the light cone is denoted by L^+; it is asymptotic to the upper sheet of the hyperboloid

$$\mathcal{H}^n = \{\vec{x} \in \mathbb{R}^{n+1} : |\vec{x}|^2 = -1, \ x_{n+1} > 0\}.$$

A ray from the origin in L^+ corresponds to a point on $\partial \mathcal{H}^n = \mathbb{S}^{n-1}$.

An n-dimensional plane P with a *spacelike* normal vector $\vec{v}$, namely $P = \{\vec{x} \in \mathbb{E}^{1,n} : \langle \vec{x} - \vec{\zeta}, \vec{v} \rangle = 0\}$ for some $\vec{\zeta} \in P$, intersects $\mathcal{H}^n$ in a $(n-1)$-dimensional hyperbolic subspace.

A plane P with a *timelike* normal vector $\vec{v} \neq 0$ intersects $\mathcal{H}^n$ in a (possibly degenerate) $(n-1)$-sphere.

For the borderline case, a plane P with a *lightlike* normal vector $\vec{v} \neq 0 \in L^+$ intersects $\mathcal{H}^n$ in an $(n-1)$-dimensional horosphere. There is a unique lightlike vector $\vec{v}$ such that $P = \{\vec{x} \in \mathbb{E}^{1,n} : \langle \vec{x}, \vec{v} \rangle = -1\}$. The ray of L^+ from the origin through $\vec{v}$ corresponds to the point at ∞ of the horosphere. The corresponding horoball is

$$\{\vec{x} \in \mathbb{H}^n : 0 \geq \langle \vec{x}, \vec{v} \rangle \geq -1\}.$$

As **v** increases along its ray, the horoball contracts to its point at ∞. Because of the peculiarities of the metric, the normal vector is simultaneously orthogonal to and parallel to the plane P.

Now set $n = 3$. $\mathbb{E}^{1,3}$ is the space/time of relativity theory. Parabolic, elliptic, and loxodromic Möbius transformations correspond to linear transformations:

$$\begin{pmatrix} 1 & 0 & 1 & -1 \\ 0 & 1 & 0 & +0 \\ 1 & 0 & \frac{3}{2} & -\frac{1}{2} \\ 1 & 0 & \frac{1}{2} & +\frac{1}{2} \end{pmatrix}, \begin{pmatrix} \cos\theta & -\sin\theta & 0 & 0 \\ \sin\theta & +\cos\theta & 0 & 0 \\ 0 & +0 & 1 & 0 \\ 0 & +0 & 0 & 1 \end{pmatrix}, \begin{pmatrix} \cos\theta & -\sin\theta & 0 & 0 \\ \sin\theta & +\cos\theta & 0 & 0 \\ 0 & +0 & \cosh\lambda & \sinh\lambda \\ 0 & +0 & \sinh\lambda & \cosh\lambda \end{pmatrix}.$$

A parabolic transformation has only one eigenvalue, which is 1, and preserves a single ray in L^+. A loxodromic has two eigenvalues not on the unit circle, which are $\cosh\lambda \pm \sinh\lambda$ (its other two are $e^{\pm i\theta}$), and preserves two rays in L^+. All the eigenvalues of an elliptic lie on the unit circle; the eigenvalue 1 is repeated twice, the other two are $e^{\pm i\theta}$.

Back to $\mathbb{E}^{1,2}$ *and* $\mathbb{E}^{1,3}$. Show that if $x \mapsto Ax$ is parabolic with fixed point $u \in \mathbb{R} \cup \infty$, the corresponding linear transformation fixes every point on the corresponding ray of the light cone L^+ in $\mathbb{E}^{1,2}$:

$$x_1 = \frac{2u}{u^2 + 1}x_3, \quad x_2 = \frac{u^2 - 1}{u^2 + 1}x_3.$$

If instead $x \mapsto Ax$ is loxodromic with fixed points p, q, show that the corresponding linear transformation fixes each point of the ray from $(0, 0, 0) \in L^+$ orthogonal to the plane spanned by the two rays in the light cone determined by the fixed points. There is expansion by the factor ρ^2 along the ray for the attracting fixed point, and contraction by ρ^{-2} along the ray for the repelling one. In fact, it is best to work out first the geometry for $\mathbb{E}^{1,2}$. For more detail see Hodgson and Weeks [1994].

Explain why elliptic, parabolic, and hyperbolic transformations of $\mathbb{H}^2$ are associated with ellipses, parabolas, and hyperbolas, respectively. (Each has an invariant plane that cuts the light cone in the respective conics.)

Here is an alternate way to view $\mathcal{H}^2$. Associate column vectors in $\mathbb{E}^{1,2}$ with real symmetric matrices as follows:

$$\begin{pmatrix} x_1 \\ x_2 \\ x_3 \end{pmatrix} \longleftrightarrow \begin{pmatrix} x_3 - x_2 & x_1 \\ x_1 & x_3 + x_2 \end{pmatrix}.$$

Again let A be a normalized 2×2 matrix with real coefficients. Consider the action

$$\begin{pmatrix} x_3 - x_2 & x_1 \\ x_1 & x_3 + x_2 \end{pmatrix} \mapsto A \begin{pmatrix} x_3 - x_2 & x_1 \\ x_1 & x_3 + x_2 \end{pmatrix} A^t,$$

which preserves the determinant $-(x_1^2 + x_2^2 - x_3^2)$. Relate this action to the action of $\widehat{A}$ in Exercise (1-25).

Passing on to $\mathbb{E}^{1,3}$, associate vectors with hermitian matrices as follows:

$$\begin{pmatrix} x_1 \\ x_2 \\ x_3 \\ x_4 \end{pmatrix} \longleftrightarrow \begin{pmatrix} x_4 - x_1 & x_2 + ix_3 \\ x_2 - ix_3 & x_4 + x_1 \end{pmatrix}.$$

Let A be a normalized 2×2 matrix now with complex coefficients. It acts on $\mathbb{E}^{1,3}$ by

$$\begin{pmatrix} x_4 - x_1 & x_2 + ix_3 \\ x_2 - ix_3 & x_4 + x_1 \end{pmatrix} \mapsto A \begin{pmatrix} x_4 - x_1 & x_2 + ix_3 \\ x_2 - ix_3 & x_4 + x_1 \end{pmatrix} \bar{A}^t,$$

which leaves the determinant $-(x_1^2 + x_2^2 + x_3^2 - x_4^2)$ invariant. Show that the action preserves $\mathcal{H}^3$ (the upper sheet of the hyperboloid). Show that this is the action brought over from $\mathbb{C} \cup \infty$ to $\mathbb{E}^{1,3}$ by stereographic projection and homogeneous coordinates.

See Weeks [1993] or Greenberg [1962] for more details.

1-28. *The quaternion description.* Upper half-space can be neatly described by the division ring of quaternions. Quaternions can be identified with the group of matrices

$$Q = \left\{ \begin{pmatrix} u & v \\ -\bar{v} & \bar{u} \end{pmatrix} : u, v \in \mathbb{C} \right\}$$

as follows. Set

$$\mathbf{1} = \begin{pmatrix} 1 & 0 \\ 0 & 1 \end{pmatrix}, \quad \boldsymbol{i} = \begin{pmatrix} i & 0 \\ 0 & -i \end{pmatrix}, \quad \boldsymbol{j} = \begin{pmatrix} 0 & 1 \\ -1 & 0 \end{pmatrix}, \quad \boldsymbol{k} = \begin{pmatrix} 0 & i \\ i & 0 \end{pmatrix}.$$

Then $\boldsymbol{ij} = \boldsymbol{k}$, $\boldsymbol{jk} = \boldsymbol{i}$, $\boldsymbol{ki} = \boldsymbol{j}$. Also $\boldsymbol{i}^2 = \boldsymbol{j}^2 = \boldsymbol{k}^2 = -\mathbf{1}$. Writing $u = u_1 + iu_2$, $v = v_1 + iv_2$, set

$$\begin{pmatrix} u & v \\ -\bar{v} & \bar{u} \end{pmatrix} = u_1 + u_2\boldsymbol{i} + v_1\boldsymbol{j} + v_2\boldsymbol{k} = u + v\boldsymbol{j}.$$

For $z = u + v\boldsymbol{j} \in Q$, define $\bar{z} = \bar{u} - v\boldsymbol{j}$, and $|z|$ by $|z|^2 = |u|^2 + |v|^2$. Note that for $c \in \mathbb{C}$, $c\boldsymbol{j} = \boldsymbol{j}\bar{c}$.

Points in $\mathbb{R}^3$ can now be denoted as the special quaternion $\mathbf{z} = z + t\boldsymbol{j}$, $z \in \mathbb{C}$, $t \in \mathbb{R}$.

Given $\begin{pmatrix} a & b \\ c & d \end{pmatrix}$, with $ad - bc = 1$ and $a, b, c, d \in \mathbb{C}$, show that the action of the corresponding Möbius transformation in upper half-space is described by

$$A(\mathbf{z}) = (a\mathbf{z} + b)(c\mathbf{z} + d)^{-1} = (\mathbf{z}c + d)^{-1}(\mathbf{z}a + b),$$

where the quaternion $A(\mathbf{z})$ is of the same special type as $\mathbf{z}$.

For more information about the use of quaternions see [Fenchel 1989; Ahlfors 1981; Beardon 1983].

1-29. Let $T : \mathbb{E}^{1,n} \to \mathbb{E}^{1,n}$ be a (not necessarily orientation-preserving) linear transformation that (i) maps the (upper sheet) hyperboloid $\mathcal{H}^n$ onto itself and (ii) preserves the inner product:

$$\langle T\mathbf{x}, T\mathbf{y} \rangle = \langle \mathbf{x}, \mathbf{y} \rangle.$$

Prove that a necessary and sufficient condition for T satisfy conditions (i) and (ii) is that

$$M^t J M = J \quad \text{and} \quad M_{n+1,n+1} > 0,$$

where M is the matrix corresponding to T.

Then prove that T acts isometrically on $\mathcal{H}^n$, and that the only isometries of $\mathcal{H}^n$ are of this type. See Cannon et al. [1997] if you need help.

1-30. *Relation of hyperbolic and euclidean metrics.* Suppose $f : \mathbb{D} \to \Omega$ is a conformal mapping from the unit disk. The *hyperbolic* or *Poincaré* metric in Ω is defined to be

$$\rho(w)\,|dw| = \frac{2|f'(z)||dz|}{1 - |f(z)|^2}.$$

The Riemann map f is uniquely determined only up to postcomposition by Möbius transformations; such compositions do not change the metric in Ω. Prove using the Koebe $\frac{1}{4}$-Theorem from conformal mapping theory (see Pommerenke [1992, 4.6(6)]) that

$$\frac{1}{2d(w, \partial\Omega)} \le \rho(w) \le \frac{2}{d(w, \partial\Omega)},$$

where $d(w, \partial\Omega)$ is the shortest euclidean distance from w to the boundary of Ω. Equality on the left holds if and only if the complement of Ω is a ray to ∞ from w. Equality on the right holds if and only if Ω is a round disk centered at w.

Suppose instead that $\Omega \subset \mathbb{C}$ is a multiply, possibly infinitely, connected region whose complement contains at least two points (plus ∞). That is, assume that Ω carries a hyperbolic metric $\rho(w)\,|dw|$ that arises from projection from the unit disk, its universal cover (see Chapter 2). According to Beardon and Pommerenke [1978] and Pommerenke [1984] there exists $C = C(\Omega) > 0$ such that

$$\frac{C}{d(w, \partial\Omega)} \le \rho(w) \le \frac{2}{d(w, \partial\Omega)}, \tag{1.14}$$

if and only if $X = \partial\Omega$ has the property called *uniformly perfect*.

The notion of uniformly perfect is really directed to multiply connected regions as when Ω is simply connected $\ne \mathbb{C}$ we already know that Equation (1.14) holds with $C = 1/2$. Yet it simplifies terminology to simply declare that closed, connected sets $X \in \mathbb{S}^2$ with more than two points are automatically uniformly perfect (their complementary components are simply connected).

So consider closed sets $X \subset \mathbb{S}^2$ that are *not* connected. Equivalently, $X \subset \mathbb{S}^2$ is *uniformly perfect* [Beardon and Pommerenke 1978; Pommerenke 1984] if and only if there exists a constant $M < \infty$ such that any annular region $A \subset \mathbb{S}^2 \setminus X$ that separates the components of X has modulus $\mathrm{mod}(A) < M$. Here $\mathrm{mod}(A) = (\log r)/2\pi$ where $f : A \to \{1 < |z| < r\}$ is a conformal map. The uniformly perfect condition relates to a requirement that X be "uniformly thick" at each of its points, independent of scaling by Möbius transformations. For example if X contains an isolated point, there would be a separating annular region of arbitrarily large modulus.

If X is uniformly perfect, the boundary of any multiply connected complementary component is uniformly perfect as well.

A third definition, equivalent to the others, is that X is uniformly perfect if and only if for every $a, b \in X$ and $w \notin X$ there exists $c \in X$ such that for some constant M the cross ratio satisfies

$$\frac{1}{M} \le |(a, b, c, w)| \le M.$$

Most importantly for us, the limit set $\Lambda(G)$ of any finitely generated (nonelementary) kleinian group G is uniformly perfect [Pommerenke 1984]! Therefore each component Ω of the complement $\Omega(G)$ satisfies Equation (1.14) for some $C = C(\Omega)$. On the other hand, there exist infinitely generated Schottky groups whose limit sets are not uniformly perfect. See Chapter 2 for the basic properties of kleinian groups.

1-31. *Anti-Möbius transformations.* An anti-Möbius, that is orientation-reversing, transformation A can be expressed as $A = B \circ J$, where J is complex conjugation and B is orientation preserving. Show that in $\mathbb{S}^2$, A either pointwise fixes a circle, or it has zero, one or two fixed points. Examples:

$$z \mapsto \frac{1}{\bar{z}}, \quad z \mapsto -\frac{1}{\bar{z}} + 1, \quad z \mapsto \bar{z} + 1, \quad z \mapsto 2\bar{z}.$$

The extension of A to open upper half-space pointwise fixes a plane, a line, or a point.

A finer classification [Fenchel 1989, pp. 48–53] has the possibilities grouped in three conjugacy (by an orientation-preserving transformation) classes. The first two are called *involutive* since the elements have order two. The third conjugacy class consists of elements T with the property that T^2 is loxodromic or parabolic. These arise as described below.

We will have to use Lemma 7.3.1 in that a loxodromic or parabolic transformation has a square root which is a Möbius transformation of the same type. From Lemma 7.1.2 we take the result that any Möbius transformation that interchanges two distinct points in $\mathbb{S}^2$ is elliptic of order two. Fenchel's classification is as follows:

Reflection in a plane: Reflection J in a plane in $\mathbb{H}^3$.

Reflection in a point $c \in \mathbb{H}^3$: Suppose $c = (0, 0, t)$ in the upper half-space model of $\mathbb{H}^3$. Let ℓ denote the vertical axis and P be a plane through c and orthogonal to ℓ. The point-reflection in c has the form $J \circ E$ where E is the elliptic of order two $E : (x, y, t) \mapsto (-x, -y, t)$ with rotation axis ℓ and J is reflection in P.

Noninvolutive anti-Möbius transformations: Let T^2 be loxodromic or parabolic and pick a square root T. Choose any $x \in \mathbb{S}^2$ distinct from its fixed points and set $y = T^{-1}(x)$, $z = T(x)$ so that the three points x, y, z are distinct. Let $\tau \subset \mathbb{H}^3$ be the line with endpoints y, z and let P be the plane orthogonal to τ whose boundary passes through x. Let J denote reflection in P. Then $T \circ J = E$ interchanges x, z so that it is elliptic of two with rotation axis $\ell \subset P$. Thus $T = E \circ J$.

Show that an elliptic of order two is itself the composition of reflections in two orthogonal planes intersecting in its rotation axis.

Given two distinct circles $C_1, C_2 \in \mathbb{S}^2$, show that there is a circle C^* such that reflection in C^* interchanges C_1 and C_2.

1-32. *Hilbert metric.* Let Ω be a bounded, euclidean convex domain in $\mathbb{R}^n$. Then every euclidean straight line that contains a point of Ω intersects its boundary $\partial\Omega$ in exactly two points. Given two points $x, y \in \Omega$, denote by $x', y' \in \partial\Omega$ the two points of intersection of the line L through x, y with $\partial\Omega$, so labeled that x separates x' from y along L. Consider the expression

$$d(x, y) = \log \frac{|x - y'||y - x'|}{|x - x'||y - y'|}.$$

Show that Ω is a complete metric space with metric $d(\cdot, \cdot)$. The geodesics in this space are the euclidean line segments. If $\Omega = \mathbb{D}^n$, $d(\cdot, \cdot)$ is the hyperbolic metric.

In the projective model of $\mathbb{D}^n$ (Exercise (1-25)), $d(\cdot, \cdot)$ is the hyperbolic distance; projective transformations preserve cross ratios. If Ω is not convex, the Hilbert metric is not a metric.

An affine map of R^n sends the metric to the Hilbert metric of the image domain. It turns out that $d(\cdot, \cdot)$ is a riemannian metric on Ω if and only if Ω is an ellipsoid and d is the hyperbolic metric on Ω, here considering Ω as the Klein model of $\mathbb{H}^n$.

Show that if $\partial\Omega$ contains a straight line segment, then two rays from any point $O \in \Omega$ to different points on the line are of uniformly bounded distance apart. For a survey see Guo [≥ 2015].

1-33. *Hyperbolic curvature of arcs.* Suppose $z = z(t) = x(t) + iy(t)$ is a parametrized arc in $\mathbb{H}^2$, $-m < t < m$, with continuous second derivative. In euclidean geometry, the parameter t is an *arc length* parameter if and only if $|z'(t)| = 1$. In hyperbolic geometry, in the upper half-plane model, t is arc length parameter if and only if

$$\frac{|z'(t)|}{y(t)} = 1, \qquad -m < t < m.$$

In euclidean geometry with arc length parameter s the curvature of $z = z(s)$ is defined as

$$\kappa_e(s) = \lim_{\Delta s \to 0} \frac{|\Delta\varphi|}{|\Delta s|}.$$

Here $\Delta\varphi = \varphi(s + \Delta s) - \varphi(s)$, where the slope of the tangent line at the point $z(s)$ is $\tan\varphi(s)$ and correspondingly $\tan\varphi(s + \Delta s)$ at $z(s + \Delta s)$.

In hyperbolic geometry, the curvature is defined by exactly same formula but the meaning of the terms is hyperbolic: s is the hyperbolic arc length parameter and $\Delta\varphi$ is the angle between the *hyperbolic tangent lines* at $z(s)$ and $z(s + \Delta s)$.

Another characterization valid in both the euclidean and hyperbolic situations is this: About the point $s = 0$ take the local coordinate system determined by the tangent vector $\vec{\alpha}$ to the curve at $s = 0$ and the normal $\vec{\beta}$ to the curve at that point. In this new coordinate system the curve $\zeta = \zeta(s)$ has the expansion

$$\zeta(s) = s\vec{\alpha} + \kappa(0)\frac{s^2}{2}\vec{\beta} + O(s^3).$$

Using Exercise (1-19) confirm that the curvature of a geodesic is zero.

Curvature of a horocycle. In the upper half-plane model, consider the horocycle $z(t) = ai + t$, $a > 0$ (which is already in euclidean arc length parameter). In hyperbolic arc length parameter, the equation is

$$z(s) = ai + as.$$

Along the horocycle, $a\Delta s = \Delta x$. So it suffices to compute the limit of $\dfrac{a\Delta\varphi(s)}{\Delta x}$. Actually it suffices to take the case $a = 1$.

The hyperbolic tangent lines to $\{z : \operatorname{Im} z = 1\}$ are the semicircles with center on $\mathbb{R}$ and unit radius. Because the horocycle is invariant under the continuous group of translations, it suffices to make the computation at say $s = 0$. To do this take the unit semicircle centered at $z = 0$ and the semicircle centered at $z = \Delta x$. Find their point of intersection, and then find the angle between them at this point (choose the angle so that it would be zero if the two tangents coincided); this is our $\Delta\varphi$. Taking the limit as $\Delta x \to 0$ we find (in sharp contrast to the euclidean case) that

$$\text{curvature of a horocycle} = 1.$$

Curvature of an equidistant arc. In the upper half-plane model measure distances from the vertical half line. Consider the line $y = cx$, $c > 0$, making angle θ with the vertical. In its hyperbolic arc length parameter,

$$z(s) = e^{s\cos\theta} e^{i(\pi/2-\theta)},$$

where $c = \cot\theta$. The line is invariant under the continuous group $z \to kz$, $k > 0$, so it suffices to make the computation at (m, mc) where $m = \sin\theta$. Find the angle of intersection $\Delta\varphi$ between the hyperbolic tangent lines at (m, cm) and $e^{s\cos\theta}(m, cm)$. Show that at the point $s = 0$,

$$\lim_{\Delta s \to 0} \left| \frac{\Delta\varphi}{\Delta s} \right| = \frac{\cos\theta}{c}.$$

Conclude that

$$\textbf{curvature of a line of distance } d \textbf{ from a geodesic} = \tanh d = \sin\theta. \qquad (1.15)$$

For example, the $45°$ line has curvature $1/\sqrt{2}$.

Curvature of a circle. In the unit disk model, consider a circle of euclidean radius R about the origin. In hyperbolic arc length coordinates starting from $(R, 0)$ its equation is

$$z(s) = re^{is(1-R^2)/(2R)}.$$

Take the geodesics tangent to the circle at $z = R$ and at $z = Re^{i\theta}$. Find their point of intersection (x_0, y_0) within the unit disk and then their angle $\varphi = \varphi(\theta)$ of intersection. After a long calculation, for example by calculating $\lim_{\theta\to 0}\varphi/\theta$, one finds

$$\left.\frac{d\varphi}{d\theta}\right|_{\theta=0} = \frac{1+R^2}{1-R^2}.$$

Because the circle is invariant under the continuous group of rotations, it suffices to make this calculation at a single point.

Note that

$$\frac{d\varphi}{ds} = \frac{1-R^2}{2R}\frac{d\varphi}{d\theta}.$$

The hyperbolic radius ρ satisfies $e^\rho = (1+R)/(1-R)$ so that $\coth\rho = (1+R^2)/2R$. We end up with the following formula (compare Section 1.4):

$$\textbf{curvature of a circle of hyperbolic radius } \rho = \coth\rho. \qquad (1.16)$$

In both 1.15 and 1.16, the curvature approaches 1 as $R \to 1$ or $\rho \to \infty$. Why?

Summary

$$\text{constant curvature} \begin{cases} < 1 & \Longleftrightarrow \quad \text{curves of finite distance from a geodesic,} \\ = 1 & \Longleftrightarrow \quad \text{horocycles,} \\ > 1 & \Longleftrightarrow \quad \text{circles.} \end{cases}$$

In $\mathbb{H}^3$ consider a surface of distance d from a hyperbolic plane. The nearest point map that projects the surface to the plane scales hyperbolic distances by a factor $1/\cosh d$. Conclude from this that:

gaussian curvature of surface of distance d from a plane $= -\operatorname{sech}^2 d$. (1.17)

Also deduce:

gaussian curvature of surface of distance d from a line $= 0$. (1.18)

1-34. *Conjugation by involution: Wada's Lemma* 2003. Suppose A, B are two Möbius transformations which do not share a fixed point. If A and B are conjugate, show that there is a Möbius transformation Q, with $Q^2 = \text{id}$, such that $A = QBQ$. That is, show that A and B are conjugate by an involution.

Hint: For the parabolic case take $A = \left(\begin{smallmatrix} 1 & 1 \\ 0 & 1 \end{smallmatrix}\right)$. If $B = XAX^{-1}$, set $Y = XT$ where $T = \left(\begin{smallmatrix} 1 & t \\ 0 & 1 \end{smallmatrix}\right)$. Note that $YAY^{-1} = B$.

1-35. *Parametrization of two generator groups with parabolic commutator* [Jørgensen 2003]. Assume that the matrices X, Y satisfy the relation $[X, Y^{-1}] = \left(\begin{smallmatrix} -1 & -2 \\ 0 & -1 \end{smallmatrix}\right)$. Introduce the notation

$$a = \operatorname{tr} X, \quad b = \operatorname{tr} Y, \quad c = \operatorname{tr} XY^{-1} = \operatorname{tr} Y^{-1}X, \quad c' = \operatorname{tr} XY.$$

Applying Lemma 1.5.6, these quantities are connected by the relations

$$abc = a^2 + b^2 + c^2, \quad cc' = a^2 + b^2.$$

These relations are symmetric with respect to c, c', that is, $abc' = a^2 + b^2 + c'^2$. If one of the traces a, b, c is zero, that element represents an elliptic of order two. If two of the traces are zero, say $a = b = 0$, then $c = 0$. If none of the traces are zero,

$$\frac{a}{bc} + \frac{b}{ac} + \frac{c}{ab} = 1 = \frac{a}{bc'} + \frac{b}{ac'} + \frac{c'}{ab}. \quad (1.19)$$

Complex numbers u, v, w that lie on the hyperplane

$$\mathfrak{P} = \{(u, v, w) \in \mathbb{C}^3 : u + v + w = 1\}$$

are called *complex probabilities*. From our perspective, the *singular subset* $\Sigma_0 \in \mathfrak{P}$ is the union of the three coordinate lines:

$$\Sigma_0 = \{(u, v, w), u + v + w = 1 : u = 0 \text{ or } v = 0 \text{ or } w = 0\}.$$

Note that $\mathfrak{P} \setminus \Sigma_0$ is connected.

We are now ready to parametrize two generator groups with parabolic commutator by complex probabilities. Confirm the following facts.

Suppose u, v, $w \in \mathbb{C}$ are nonvanishing numbers such that $u + v + w = 1$. Set

$$d = \frac{1}{uvw}, \quad a = \sqrt{ud}, \quad b = \sqrt{vd},$$

where the arguments are chosen so that $-\pi < \arg d$, $\arg a$, $\arg b \leq \pi$. Choose $\arg c$ so that

$$c = \sqrt{wd}$$

satisfies

$$abc = \sqrt{ud}\sqrt{vd}\sqrt{wd} = d.$$

Set

$$X = \begin{pmatrix} a - b/c & a/c^2 \\ a & b/c \end{pmatrix}, \quad Y^{-1} = \begin{pmatrix} b - a/c & -b/c^2 \\ -b & a/c \end{pmatrix}. \tag{1.20}$$

Then

$$XY^{-1}X^{-1}Y = \begin{pmatrix} -1 & -2 \\ 0 & -1 \end{pmatrix}, \quad XY^{-1} = \begin{pmatrix} c & -1/c \\ c & 0 \end{pmatrix}, \quad Y^{-1}X = \begin{pmatrix} c & 1/c \\ -c & 0 \end{pmatrix}$$

and $\operatorname{tr} X = a$, $\operatorname{tr} Y = b$, and $\operatorname{tr} XY^{-1} = c$ with $a, b, c \neq 0$. Conversely, if X and Y satisfy $[X, Y^{-1}] = \begin{pmatrix} -1 & -2 \\ 0 & -1 \end{pmatrix}$, and the numbers $a = \operatorname{tr} X$, $b = \operatorname{tr} Y$ and $c = \operatorname{tr} XY^{-1}$ are nonzero, and $XY^{-1}(0) = \infty$, then X and Y^{-1} have the matrix representations (1.20). In summary,

$$\operatorname{tr} X = a, \quad \operatorname{tr} Y = b, \quad \operatorname{tr} XY^{-1} = c,$$

$$u = \frac{a}{bc}, \quad v = \frac{b}{ac}, \quad w = \frac{c}{ab},$$

$$a^2 = \frac{1}{vw}, \quad b^2 = \frac{1}{uw}, \quad c^2 = \frac{1}{uv}, \quad abc = \frac{1}{uvw},$$

$$a^2 + b^2 + c^2 = abc, \quad u + v + w = 1.$$

And in the opposite direction:

Lemma 1.6.1. *Set $K = \begin{pmatrix} -1 & -2 \\ 0 & -1 \end{pmatrix}$. Suppose the matrices X, Y satisfy*

$$XY^{-1}X^{-1}Y = K,$$

and the transformation corresponding to X does not fix ∞. Then X and Y^{-1} have the form

$$X = \begin{pmatrix} * & * \\ \tau_X & * \end{pmatrix}, \quad Y^{-1} = \begin{pmatrix} * & * \\ -\tau_Y & * \end{pmatrix}.$$

Proof. As usual, all matrices have determinant one. Set

$$X = \begin{pmatrix} a & b \\ c & d \end{pmatrix}, \quad Y^{-1} = \begin{pmatrix} \alpha & \beta \\ \gamma & \delta \end{pmatrix}.$$

Replace X and Y^{-1} by the conjugates WXW^{-1} and $WY^{-1}W^{-1}$, where $W = \begin{pmatrix} 1 & d/c \\ 0 & 1 \end{pmatrix}$, so that $WXW^{-1}(0) = \infty$; conjugation by the translation W leaves K unchanged and it also leaves the entries c and γ in the matrix for X and Y^{-1} unchanged. The identity $XY^{-1} = KY^{-1}X$ gives the four equations

$$c\beta = -b\gamma, \quad c\alpha = -a\gamma - c\delta, \quad a\alpha + b\gamma = -a\alpha - c\beta - 2a\gamma - 2c\delta,$$
$$a\beta + b\delta = -b\alpha - 2b\gamma.$$

Substituting the first and second into the third we wind up with either $c = a = \tau_X$ or $\alpha = 0$. In the former case, $-\gamma = \alpha + \delta = \tau_Y$ and $\beta = -\tau_Y/\tau_X^2$, since $bc = -1$. With the normalization $X(0) = \infty$, leaving aside the formulas for α, δ, we have

$$X = \begin{pmatrix} \tau_X & -1/\tau_X \\ \tau_X & 0 \end{pmatrix} \quad Y^{-1} = \begin{pmatrix} \alpha & -\tau_Y/\tau_X^2 \\ -\tau_Y & \delta \end{pmatrix}. \tag{1.21}$$

Suppose instead $\alpha = 0$. Using the fact $bc = -1$ and $\beta\gamma = -1$, we find from the first equation that $c^2 = -\gamma^2$, or $\gamma = \pm ci$. The second equation becomes $a\gamma + c\delta = 0$, or $\delta = \mp ai$; in other words $\tau_Y = \mp\tau_X i$. The fourth equation becomes $a\beta + b\delta = -2b\gamma$ and upon rewriting in terms of c and a yields $c = a = \tau_X$. Putting it all together we actually have a special case of (1.21),

$$X = \begin{pmatrix} \tau_X & -1/\tau_X \\ \tau_X & 0 \end{pmatrix}, \quad Y^{-1} = \begin{pmatrix} 0 & 1/\tau_Y \\ -\tau_Y & \tau_Y \end{pmatrix}.$$

In this case $Y^{-1}X$ and XY^{-1} represent the order two elliptics $z \mapsto -z$ and $z \mapsto -z + 2$. □

The bottom line is that given the traces of generators X, Y, plus the trace of XY^{-1}, plus the condition that the trace of the commutator is -2, the group is uniquely determined up to conjugation. Moreover, by the trace identities, the trace of any element of $\langle X, Y \rangle$ is a polynomial in the initial traces. The systematic method for doing this uses the modular diagram and the Farey sequence; it is spelled out in Mumford et al. [2002].

1-36. Suppose the Möbius transformations A, B are such that $[A, B]$ is parabolic. Assume that $A, B, C = BA$ all have real traces. Show that for some Möbius T, the entries of the normalized matrices for TAT^{-1}, TBT^{-1} are real. Consequently the group $T\langle A, B \rangle T^{-1}$ preserves the upper half-plane. What if $[A, B]$ is instead elliptic? Show by modifying Jørgensen's method of complex probabilities that the same conclusion holds.

1-37. *Two-jets of locally injective analytic functions.* Suppose $f(z)$ is a locally injective analytic function (that is, $f'(z) \neq 0$) in a neighborhood of $z_0 \in \mathbb{C}$. Show that there is a Möbius transformation $M(f; z_0)$ uniquely determined by the three properties:

$$M(f; z_0)(z_0) = f(z_0), \quad M(f; z_0)'(z_0) = f'(z_0), \quad M(f; z_0)''(z_0) = f''(z_0).$$

The value of the first two derivatives of f at z_0 is called the *two-jet* of f at z_0. If A is any Möbius transformation, $M(Af; z_0) = AM(f; z_0)$.

Thurston 1986d showed that for any $v \neq 0 \in \mathbb{C}$,

$$vS_f(z_0) = v\left(\left(\frac{f''}{f}\right)' - \frac{1}{2}\left(\frac{f''}{f'}\right)^2\right)_{z_0} = \frac{\partial^2}{\partial z^2} M(f; z)^{-1} M(f; z + tv)\Big|_{t=0, z=z_0}.$$

Here $S_f(z_0)$ is called the *schwarzian derivative* (Exercise (6-9)).

Suppose now that $f : \text{UHP} \to \mathbb{C}$ is a conformal map. There is an extension F of f to $\mathbb{H}^3$, taken as UHS, determined as follows.

Denote by $P \subset \mathbb{H}^3$ the vertical half-plane rising from $\mathbb{R}$. Identify P with UHP by orthogonal projection. Given $x \in \mathbb{H}^3$, let ℓ_x be the geodesic through x that is orthogonal to P. Denote by $r(x)$ its intersection with $P \equiv \text{UHP}$.

Set

$$F(x) = M(f; r(x)).$$

Show that F is continuous. It is also equivariant if f is so: Suppose G is a fuchsian group acting on UHP, and on $\mathbb{H}^3$, and f satisfies for all $\gamma \in G$, $f \circ \gamma = \phi(\gamma) \circ f$, where ϕ is a homomorphism of G to another group, not necessarily discrete. Then

$$M(f; r(\gamma x)) = \phi(\gamma)M(f; r(x))\gamma^{-1},$$

and consequently $F \circ \gamma = \phi(\gamma) \circ F$. Although F is not necessarily a local homeomorphism, there exists $d > 0$ such that F is a local homeomorphism outside of a distance-d neighborhood of P. See Bromberg [2000] for details.

1-38. *The hyperbolic Gauss map.* Suppose $S \subset \mathbb{H}^3$ is a smoothly immersed, *oriented* surface in the ball model. Given $\zeta \in S$ let n^ζ denote the geodesic ray normal to S at ζ and denote its endpoint on $\mathbb{S}^2$ by $n(\zeta)$. Epstein 1986 defined the hyperbolic Gauss map by $G(\zeta) = n^+(\zeta)$, for $\zeta \in S$ There is a uniquely determined horosphere $\sigma(\zeta)$ based at $n(\zeta)$ that is tangent to S at ζ.

If we reverse the orientation of S we get another Gauss map $n^-(\zeta)$ that sends ζ to the "other" side of $\mathbb{S}^2$. When S is smoothly embedded and its principal curvatures satisfy $k_1, k_2 < 1$, the maps n^+, n^- are diffeomorphisms to disjoint open sets in $\mathbb{S}^2$. The composition $n^- \circ (n^+)^{-1}$ is a kind of "reflection" in $\partial S \subset \mathbb{S}^2$. The situation is studied in detail in Epstein [1986].

1-39. *Fricke's Lemma* (see Magnus [1980] for the history and a proof). Let A_1, A_2, A_3 be Möbius transformations. With Magnus introduce the notation

$$x_\nu = \text{tr}\, A_\nu, \quad y_{\nu\mu} = \text{tr}\, A_\nu A_\mu, \quad z_{\nu\mu\sigma} = \text{tr}\, A_\nu A_\mu A_\sigma,$$
$$P = x_1 y_{23} + x_2 y_{13} + x_3 y_{12} - x_1 x_2 x_3,$$
$$Q = x_1^2 + x_2^2 + x_3^2 + y_{12}^2 + y_{13}^2 + y_{23}^2 + y_{12}y_{13}y_{23} - x_1 x_2 y_{12} - x_1 x_3 y_{13} - x_2 x_3 y_{23} - 4.$$

Prove Fricke's Lemma, namely the formula

$$P = z_{123} + z_{132}, \quad Q = z_{123}z_{132}.$$

In other terms, z_{123} and z_{132} are the roots of the equation $z^2 - Pz + Q = 0$.

1-40. *Finer properties of isometric circles.* The following properties have proved very useful in Jørgensen's hands in analyzing one- and two-generator groups.

Lemma 1.6.2. *Let A and B be Möbius transformations on S^2 and let $\mathfrak{I}$, $\mathcal{B}$ be as in Section 1.5.2.*

(i) $\mathcal{B}(B)$ *covers a set σ on the isometric plane or circle $\mathfrak{I}(A)$ if and only if $\mathcal{B}(BA^{-1})$ covers $A(\sigma)$ on the isometric plane or circle $\mathfrak{I}(A^{-1})$.*

(ii) *If the circle $\mathfrak{I}(B)$ is internally tangent to $\mathfrak{I}(A)$ at the point x, then $\mathfrak{I}(BA^{-1})$ is externally tangent to $\mathfrak{I}(A^{-1})$ at $A(x)$.*

(iii) $\mathfrak{I}(A), \mathfrak{I}(AB), \mathfrak{I}(B)$ *have a common point x if and only if $\mathfrak{I}(A), \mathfrak{I}(AB^{-1})$, and $\mathfrak{I}(B^{-1})$ have a common point $B(x)$.*

(iv) *Suppose $\mathfrak{I}(B_1), \ldots, \mathfrak{I}(B_n)$, $n \geq 3$, go through a point x, and that $\bigcup_1^n \mathcal{B}(B_i)$ covers a neighborhood of x. Then for each k, the circle $\mathfrak{I}(B_k^{-1})$ and every circle $\mathfrak{I}(B_i B_k^{-1})$, for $i \neq k$, pass through $B_k(x)$, and the union of their interiors covers a neighborhood of $B_k(x)$.*

(v) *The sum of the excesses determined by the three pairs of isometric circles (A, B), (A^{-1}, BA^{-1}) and (B^{-1}, AB^{-1}) is $12\pi - 2(\lambda_1 + \lambda_2 + \lambda_3)$, where $\lambda_1, \lambda_2, \lambda_3$ are the exterior angles of intersection of each of the three pairs of circles.*

If two circles bound overlapping open disks, the overlap is bounded by an arc of each one. The *excess* of the pair is defined as the sum of the (euclidean) central angles subtended at the center of the two circles by the *complements* of the arcs. If the circles do not intersect at all, the excess is defined as 4π.

Proof. Properties (1)–(4) follow from the chain rule,

$$(BA^{-1})'(Az) = \frac{B'(z)}{A'(z)}.$$

For $z \in \sigma$ we have $|A'(z)| = 1$ and $|B'(z)| > 1$, so $|(BA^{-1})'(Az)| > 1$. Conversely, if $|(BA^{-1})'(Az)| > 1$ and $|A'(z)| = 1$, then $A(z) \in \mathfrak{I}(a)$ and necessarily $|B'(z)| > 1$. Assertions (3) and (4) are consequences and (2) is a limiting case.

For (5), the key step here is to remember a theorem from high school geometry. Consider a closed arc σ on a circle. Draw the rays from the center to its endpoints. Let 2θ denote the angle they subtend. Choose a point ζ on the circle, but not on σ and draw the lines from ζ to the endpoints of σ. Then the lines from ζ subtend the angle θ with σ. In the limiting case that one of the endpoints of σ approaches ζ, the angle θ approaches the angle between the remaining line from ζ and the tangent to the circle at ζ.

Next, suppose $\mathcal{B}(A) \cap \mathcal{B}(B) \neq \emptyset$. Let σ_A denote the arc $\mathfrak{I}(A) \cap \mathcal{B}(B)$ and σ_B the arc $\mathfrak{I}(B) \cap \mathcal{B}(A)$. Let θ_1 denote the angle subtended at the center $c(A)$ by the *complementary arc* $\mathfrak{I}(A) \setminus \sigma_A$ and θ_2 subtended at $c(B)$ by $\mathfrak{I}(B) \setminus \sigma_B$. The excess at this intersection is, by definition, $\theta_1 + \theta_2$.

Draw the straight line l through the two points $\mathfrak{I}(A) \cap \mathfrak{I}(B)$. For ease of reference assume l is a vertical line. Choose one of the points of intersection ζ and draw there the tangent lines to the two circles. Let λ_3 denote the *exterior* angle of intersection of the two circles, that is, the angle between the two tangents that lies exterior to both circles. We claim that the angle between l and the tangent line to $\mathfrak{I}(A)$ is $\theta_1/2$ and correspondingly that between l and $\mathfrak{I}(B)$ is $\theta_2/2$. This is a consequence of the limiting case of the high school theorem presented above.

Summing the angles at ζ shows that $\frac{1}{2}\theta_1 + \frac{1}{2}\theta_2 + \lambda_3 = 2\pi$.

Now pass on to the next pair of intersecting circles, $\mathfrak{I}(A^{-1})$ and $\mathfrak{I}(BA^{-1})$. Let $\sigma_{A^{-1}}$ and $\sigma_{BA^{-1}}$ denote the arcs $\mathfrak{I}(A^{-1}) \cap \mathcal{B}(BA^{-1})$ and $\mathfrak{I}(BA^{-1}) \cap \mathcal{B}(A^{-1})$. Recall that $\mathfrak{I}(A)$ has the same radius as $\mathfrak{I}(A^{-1})$ and that A is a euclidean isometry from the former to the latter. This means that the angle subtended at $c(A^{-1})$ by the complementary arc $\mathfrak{I}(A^{-1}) \setminus \sigma_{A^{-1}}$ is again θ_1. Let θ_3 denote the angle subtended at $c(BA^{-1})$ by the complementary arc $\mathfrak{I}(BA^{-1}) \setminus \sigma_{BA^{-1}}$. Denote the exterior angle of intersection of the two circles by λ_2. Again we find that $\frac{1}{2}\theta_1 + \frac{1}{2}\theta_3 + \lambda_2 = 2\pi$.

Once more, carry out this construction for the intersecting circles $\mathfrak{I}(B^{-1})$ and $\mathfrak{I}(AB^{-1})$. The angle subtended at $c(B^{-1})$ is now θ_2 while the angle at $c(AB^{-1})$ is θ_3. If λ_1 denotes the exterior angle of intersection of the circles, we find as before that $\frac{1}{2}\theta_2 + \frac{1}{2}\theta_3 + \lambda_1 = 2\pi$.

Putting the three calculations together we conclude that the *total excess* is

$$2(\theta_1 + \theta_2 + \theta_3) = 12\pi - 2(\lambda_1 + \lambda_2 + \lambda_3).$$

In the limiting case that the circles do not cross each other, the total excess is $3 \times 4\pi = 12\pi$. $\qquad\qquad\square$

2

Discrete groups

This chapter introduces the related notions of discreteness and discontinuity, limit set and ordinary set. We establish the connection between discrete groups of Möbius transformations and hyperbolic manifolds and orbifolds. Some classical special cases of discrete groups are presented: elementary groups (which we classify), fuchsian and Schottky groups. The chapter includes crash courses on covering surfaces, Riemann surfaces, and quasiconformal mappings. The first two of these topics help us understand the boundaries of the 3-manifolds, while the latter shows us how to make controlled deformations of them. Along the way we introduce the Uniformization Theorem, Teichmüller spaces, and mapping class groups.

2.1 Convergence of Möbius transformations

Proposition 2.1.1 (General convergence theorem). *Suppose* $\{T_n\}$ *is an infinite sequence of distinct Möbius transformations such that the corresponding fixed point(s)* p_n, q_n *converge to* $p, q \in \mathbb{S}^2$; *here either* $p_n = q_n$, *or* T_n *is elliptic, or* p_n *is the repelling and* q_n *the attracting fixed point of* T_n. *There is a subsequence* $\{T_k\}$ *with one of the following properties.*

(i) *There exists a Möbius transformation* T *such that* $\lim T_k(z) = T(z)$ *uniformly on* $\mathbb{H}^3 \cup \mathbb{S}^2$ *(considered with the euclidean metric), or equivalently,* $T_k \to T$ *for suitable choices of the associated matrices.*

(ii) $\lim T_k(z) = q$ *for all* $z \neq p$, *uniformly on compact subsets of* $\mathbb{H}^3 \cup (\mathbb{S}^2 \setminus \{p\})$. *Also* $\lim T_k^{-1}(z) = p$ *for all* $z \neq q$, *uniformly on compact subsets of* $\mathbb{H}^3 \cup (\mathbb{S}^2 \setminus \{q\})$. *Possibly* $p = q$.

Examples: $\{z + n\}, \quad \{k^n z\}, \quad \{e^{i/n} z\}, \quad \{a^2 z + n(1 - a^2)\}$.

Before proving the lemma, we state as a corollary a stronger form of Montel's theorem on "normal families" (the original requires three omitted values).

Corollary 2.1.2. *Suppose* $\{T_n\}$ *is an infinite sequence of distinct Möbius transformations and* $U \subset \mathbb{S}^2$ *is a connected open set. Suppose there are two distinct points*

ζ_1, ζ_2 in $\mathbb{S}^2$ such that $T_n(U)$ avoids ζ_1 and ζ_2, for all n. Then there is an infinite sub-sequence $\{T_m\}$ which converges on U, uniformly on compact subsets, to a Möbius transformation or to a constant.

Proof of Proposition 2.1.1. Assume that $\{T_n\}$ is a sequence whose fixed points converge as described in Lemma 2.1.1, and assume it has no subsequence which converges to a Möbius transformation.

Case 1: $p \neq q$. Choose $\zeta \in \mathbb{C}$ distinct from p, q, p_n, q_n for all n. Set $R_n(z) = (z, p_n, \zeta, q_n)$ so that $\lim R_n(z) = R(z) = (z, p, \zeta, q)$, uniformly on $\mathbb{S}^2$.

The transformation $S_n(z) = R_n T_n R_n^{-1}(z)$ fixes $0, \infty$ and has the same convergence properties as $\{T_n\}$. We have for large indices $S_n(z) = a_n z$ with $|a_n| \geq 1$. If $|a_n|$ is bounded for infinitely many indices then a subsequence converges to a Möbius transformation. Otherwise there exists a subsequence $\{S_m\}$ for which $\lim a_m = \infty$. In this case, $\{S_m\}$ converges uniformly to ∞ outside any given neighborhood of $z = 0$.

Case 2: $p = q$. Choose $\zeta_1, \zeta_2 \neq q_n, q$ and $\zeta_1 \neq \zeta_2$. Set $R_n(z) = (z, \zeta_1, \zeta_2, q_n)$. Again $\lim R_n(z) = R(z) = (z, \zeta_1, \zeta_2, q)$. Set $S_n(z) = R_n T_n R_n^{-1}(z)$. This fixes ∞ and has the same convergence properties as $\{T_n\}$. So $S_n(z) = a_n z + b_n$; the other fixed point of S_n is $-b_n/(a_n - 1)$. If for a subsequence $\lim b_m = b \neq \infty$, then $\lim a_m = 1$. In this case $\lim S_m(z) = z + b$. If instead $\lim b_m = \infty$, rewrite S_m as

$$S_m(z) = b_m \left(\frac{(a_m - 1)z}{b_m} + 1 \right) + z.$$

Since $\lim(a_m - 1)/b_m = 0$, we have $\lim S_m(z) = \infty$ for all z. As for the inverse,

$$S_m^{-1}(z) = \frac{b_m}{a_m} \left(\frac{z}{b_m} - 1 \right).$$

Because

$$\lim \frac{a_m - 1}{b_m} = \lim \left(\frac{a_m}{b_m} - \frac{1}{b_m} \right) = 0$$

and $\lim b_m = \infty$, we find $\lim a_m/b_m = 0$. Therefore $\lim S_m^{-1}(z) = \infty$ as well, for all $z \in \mathbb{C}$. $\qquad\square$

Proof of Corollary 2.1.2. Let $\{T_m\}$ be a convergent sequence as in Proposition 2.1.1. Suppose the limit is not a Möbius transformation. Then $\lim T_m(z) = q$ uniformly outside any given neighborhood of p.

Case 1: $p \neq q$. We may assume that $p = 0$, $q = \infty$. If $0 \notin U$, Corollary 2.1.2 is true. If $0 \in U$ then $p_m \in U$ for all large indices. We show that then, for all large indices, the equation $T_m(z) = 0$ has a solution in U. For choose a disk $D \subset U$ centered at 0. Given a smaller concentric disk D_0, the fixed points $\{p_m\}$ are contained in D_0 for all large indices, while the $\{q_m\}$ are in the exterior of D. The image disk $T_m(D)$ contains p_m but not q_m. And $\lim_{m \to \infty} T_m(\partial D) = \infty$. Therefore $T_m(D)$ covers 0, and in fact any given point $\zeta \in \mathbb{C}$, for all large indices.

So if $p \in U$, then $p \neq \zeta_1, \zeta_2$. Moreover if say $\zeta_1 \neq \infty$, $T_m(D)$ would cover ζ_1 for all large indices, again a further contradiction to our assumption that $D \subset U$. In any case, $p \notin U$. There is a subsequence converging uniformly on compact subsets of U to q.

Case 2: p=q. We may assume that $p = q = \infty$. If $\infty \notin U$, then Corollary 2.1.2 holds. Suppose instead that $\infty \in U$. Fix a disk $D \subset U$ centered at ∞. Because $\lim_{m\to\infty} T_m^{-1}(\zeta_1) = \infty$, ultimately $T_m^{-1}(\zeta_1) \in D$. That is for large indices, $\zeta_1 \subset T_m(D) \subset T_m(U)$, in contradiction to the hypothesis. $\square$

The example of the powers of a loxodromic transformation acting on the complement U of its attracting fixed point shows that the hypotheses of Corollary 2.1.2 are best possible.

We also include in this section the following elementary fact.

Lemma 2.1.3. *If g is loxodromic and h exchanges the fixed points of g, then $h^2 = \mathrm{id}$ ($\mathrm{tr}(h) = 0$).*

Proof. In any case h^2 fixes the fixed points of g. But h has its own fixed point or points which h^2 fixes as well. $\square$

2.1.1 Some group terminology

Two groups G, H of Möbius transformations are said to be *conjugate* if there is a Möbius transformation T such that $G = THT^{-1}$; in other words G is the group consisting of the elements ThT^{-1}, for $h \in H$. As we did with a single Möbius transformation in Chapter 1, we will often find it convenient to "normalize" a group of transformations, replacing it by the representative of its conjugacy class for which we stipulate some propitious property.

If A, B are Möbius transformations, $\langle A_1, A_2, \ldots, A_n \rangle$ denotes the group generated by $A_1, A_2, \ldots, A_n$. In particular $\langle A \rangle$ denotes the cyclic group generated by A.

A group is *torsion-free* if no element apart from the identity has finite order.

If a group G acts on a set X, the *stabilizer* of a subset $\Sigma \subset X$ under G is the set

$$\mathrm{Stab}(\Sigma) = \mathrm{Stab}_G(\Sigma) = \{g \in G : g(\Sigma) = \Sigma\}.$$

The case that interests us is where G is a group of Möbius transformations and Σ is a subset of $\mathbb{S}^2$.

2.2 Discreteness

In this section we begin our study of groups of Möbius transformations. A group G of Möbius transformation is *discrete* if there is no infinite sequence of distinct elements in the group that converges to the identity. Using Proposition 2.1.1 we see that each of the following conditions is equivalent to discreteness.

(i) No infinite sequence of distinct elements of G converges to a Möbius transformation.

(ii) G acts *properly discontinuously* in $\mathbb{H}^3$: Given any closed ball $B \subset \mathbb{H}^3$, the set $\{g \in G : g(B) \cap B \neq \varnothing\}$ is finite.

(iii) G has no limit points in $\mathbb{H}^3$: Given $\vec{x} \in \mathbb{H}^3$, there is no point $\vec{y} \in \mathbb{H}^3$ with an infinite sequence of distinct elements $\{g_n\}$ in G such that $\lim g_n(\vec{y}) = \vec{x}$.

Proper discontinuity implies that discrete groups have at most a countable number of elements. To see this, exhaust $\mathbb{H}^3$ by a countably many closed balls $V_1 \subset V_2 \subset \cdots$ centered at some point $\mathcal{O} \in \mathbb{H}^3$. For each i, enumerate the at most finitely many elements $g \in G$ for which $g(O) \in V_i$.

A group G is called *elementary* if and only if it preserves one point or a pair of points on $\mathbb{S}^2$, or a point in $\mathbb{H}^3$. An equivalent definition is that a group is elementary if and only if any two elements of infinite order have a common fixed point; see Exercise (2-1).

It is difficult to determine whether the group generated by a given set of elements is discrete. An algorithm for deciding discreteness of two-generator groups in PSL(2, $\mathbb{R}$) is presented in Gilman [1995; 1997]. The best general result is the following necessary condition.

Jørgensen's Inequality [Jørgensen 1974b]. *If $G = \langle A, B \rangle$ is discrete then*

$$\left| \mathrm{tr}^2(A) - 4 \right| + \left| \mathrm{tr}(ABA^{-1}B^{-1}) - 2 \right| \geq 1, \tag{2.1}$$

except in the following three cases, which are elementary groups:

(i) *G cyclic or a finite abelian extension of a cyclic group and $|\mathrm{tr}^2(A) - 4| < 1$.*

(ii) *A is loxodromic or elliptic with $|\mathrm{tr}^2(A) - 4| < \frac{1}{2}$ while B interchanges the fixed points of A.*

(iii) *A is parabolic while B is parabolic or elliptic of order 2, 3, 4 or 6 and fixes the fixed point of A.*

Note that the left side of (2.1) depends continuously on the Möbius entries. For cases that equality occurs see Exercise (2-13).

The inequality is often applied to show the impossibility of a situation that $\langle A_n, B_n \rangle$ remains nonelementary while both A_n, B_n converge with $\lim A_n = \mathrm{id}$.

Jørgensen went on to draw the following conclusion [1977b].

Corollary 2.2.1. *A nonelementary group G is discrete if and only if every two-generator subgroup is discrete.*

If G preserves a disk in $\mathbb{S}^2$, then G is discrete if and only if every one-generator subgroup is discrete, that is, if and only if there are no elliptic transformations of infinite order.

In contrast, if a *nonelementary group H* is not discrete, Leon Greenberg 1962 proved that its closure in PSL(2, $\mathbb{C}$), that is the set of all Möbius transformations which are limits of elements of H, is either the full group PSL(2, $\mathbb{C}$) or it is the group of all Möbius transformations which preserve some round disk in $\mathbb{S}^2$, possibly augmented by the inversion in the bounding circle.

Proof of Corollary 2.2.1. We will see later (Theorem 4.1.5) that if all elements in G are elliptic, then G is elementary with a common fixed point in $\mathbb{H}^3$.

Assume G is not elementary. Then there is a loxodromic or parabolic element B_1 in G and an element C which neither shares a fixed point with B_1 nor interchanges the fixed points p, q of B_1, if B_1 is loxodromic. Then the fixed point(s) $C(p), C(q)$ of the parabolic or loxodromic $B_2 = CB_1C^{-1}$ are distinct from p, q. For all n we have $B_1^n(C(p)) \neq C(p)$ and $B_1^n(C(q)) \neq C(q)$, yet these points are as close as we please to one of p, q. Therefore for some n the fixed points of $B_3 = B_1^n B_2 B_1^{-n}$ are distinct from those of both B_1, B_2. That is, the three loxodromic or parabolic transformations B_1, B_2, B_3 have mutually distinct fixed points.

We claim that B_1, B_2, B_3 can be replaced if necessary by three other transformations with distinct fixed points, so that all of them are loxodromic. To confirm the claim, it is enough to show that B_1 and B_2 can be chosen not to be parabolic. This is a consequence of the following argument.

If X, Y are parabolic without a common fixed point we may assume $X = \left(\begin{smallmatrix} 1 & 1 \\ 0 & 1 \end{smallmatrix}\right)$ and $Y = \left(\begin{smallmatrix} a & b \\ c & d \end{smallmatrix}\right)$ in normalized matrices. Then $\operatorname{tr}(X^n Y) = a + d + nc$, so that $Z = X^n Y$ will be loxodromic for all large n and will not share a fixed point with Y. Consequently YZY^{-1} is loxodromic as well and does not share a fixed point with Z.

Now assume that G is not discrete. We will show the existence of a two-generator subgroup that is not discrete. This will show that if every two-generator subgroup of a group is discrete, the group itself must be discrete.

So assume that there is an infinite sequence $\{A_n\}$ of distinct elements of G with $\lim A_n = \operatorname{id}$. For n sufficiently large,

$$\left| \operatorname{tr}^2(A_n) - 4 \right| + \left| \operatorname{tr}(A_n B_i A_n^{-1} B_i^{-1}) - 2 \right| < 1,$$

for $i = 1, 2, 3$. For each n, at least one element of B_1, B_2, B_3 does not share a fixed point with A_n. Passing to a subsequence if necessary we may assume say B_1 does not share a fixed point with any A_n. The group $\langle A_n, B_1 \rangle$ is not elementary provided that A_n does not exchange the fixed points of B_1. Since $\lim A_n = \operatorname{id}$, the order of A_n, if finite must increase to ∞. For large enough n, A_n, B_1 do not satisfy Jørgensen's inequality; therefore $\langle A_n, B_1 \rangle$ cannot be discrete.

Now assume that G is a nonelementary group preserving the unit disk $\mathbb{D}$. Suppose G is not discrete. We claim that G then contains an elliptic element of infinite order.

By the previous result we may assume that G is a two-generator group. By Selberg's lemma below (p. 73), there is a finitely generated subgroup G_0 of finite index without elliptic transformations of finite order. This too is nonelementary and nondiscrete. Now we call on a theorem of C. L. Siegel repeated in Lehner [1964, III.3J] and that we will ask the reader to prove in Exercise (2-3), that establishes that G_0 must contain elliptic elements of arbitrarily high order. In fact then, G_0 must contain elliptic elements of infinite order. $\square$

Proof of Jørgensen's inequality. We will follow the original proof. Assume the inequality fails to hold, so that for $A, B \neq \operatorname{id}$ generating a discrete group,

$$\mu = \left|\text{tr}^2(A) - 4\right| + \left|\text{tr}(ABA^{-1}B^{-1}) - 2\right| < 1.$$

We study the sequence obtained by setting $T_0 = B$ and define inductively

$$T_n = T_{n-1}AT_{n-1}^{-1} = \begin{pmatrix} a_n & b_n \\ c_n & d_n \end{pmatrix}, \quad a_n d_n - b_n c_n = 1.$$

Case 1: $A = \begin{pmatrix} 1 & 1 \\ 0 & 1 \end{pmatrix}$ *is parabolic.* Write

$$B = \begin{pmatrix} a_0 & b_0 \\ c_0 & d_0 \end{pmatrix},$$

where $a_0 d_0 - b_0 c_0 = 1$. We may assume B does not fix ∞—otherwise $\langle A, B \rangle$ would be elementary and to be discrete, B would have to be parabolic or be elliptic of order 2, 3, 4 or 6 by Lemma 2.3.1(iii). Therefore $c_0 \neq 0$. Furthermore,

$$ABA^{-1}B^{-1} = \begin{pmatrix} 1 + a_0 c_0 + c_0{}^2 & 1 - a_0 c_0 - a_0{}^2 \\ c_0{}^2 & 1 - a_0 c_0 \end{pmatrix}, \quad \text{tr}(ABA^{-1}B^{-1}) - 2 = c_0^2.$$

Therefore $\mu = |c_0^2| < 1$. We find for the sequence of conjugates $\{T_n\}$ that

$$\begin{pmatrix} a_n & b_n \\ c_n & d_n \end{pmatrix} = \begin{pmatrix} 1 - a_{n-1}c_{n-1} & a_{n-1}{}^2 \\ -c_{n-1}{}^2 & 1 + a_{n-1}c_{n-1} \end{pmatrix}.$$

From this we deduce that

$$c_n = -c_0{}^{2^n}, \quad |c_n| = \mu^{2^{n-1}} < 1,$$

$$|a_{n+1} - 1| \leq |c_n|(n + |a_0|), \quad |d_{n+1} - 1| \leq |c_n|(n + |a_0|), \quad |b_{n+1} - 1| = |a_n{}^2 - 1|.$$

So $\lim c_n = 0$, $\lim a_n = \lim d_n = \lim b_n = 1$, and hence $\lim T_n = A$. Since $0 < |c_0| < 1$ we see from $c_n = -c_0{}^{2^n}$ that the elements of the sequence $\{c_n\}$, hence of the sequence of transformations $\{T_n\}$, are distinct. Therefore $\langle A, B \rangle$ is not discrete, a contradiction.

Case 2: $A = \begin{pmatrix} \rho & 0 \\ 0 & 1/\rho \end{pmatrix}$, *with* $|\rho| \geq 1$. Take B as before. We find that

$$\text{tr}(ABA^{-1}B^{-1}) - 2 = -b_0 c_0 (\rho - \rho^{-1})^2, \quad (\rho - \rho^{-1})^2 = \text{tr}^2(A) - 4.$$

Note in particular that $|\rho - \rho^{-1}|^2 \leq \mu < 1$.

Now

$$T_{n+1} = \begin{pmatrix} a_n d_n \rho - b_n c_n \rho^{-1} & a_n b_n (\rho^{-1} - \rho) \\ c_n d_n (\rho - \rho^{-1}) & a_n d_n \rho^{-1} - b_n c_n \rho \end{pmatrix}.$$

Consequently,

$$b_{n+1}c_{n+1} = -a_n b_n c_n d_n (\rho - \rho^{-1})^2 = -b_n c_n (1 + b_n c_n)(\rho - \rho^{-1})^2.$$

Inserting the formula for $\text{tr}(ABA^{-1}B^{-1})$,

$$|b_1 c_1| = |b_0 c_0| \, |(1 + b_0 c_0)(\rho - \rho^{-1})^2|$$
$$= |b_0 c_0| \, |\text{tr}^2(A) - 4 - \text{tr}(ABA^{-1}B^{-1}) + 2| \leq |b_0 c_0| \mu \leq |b_0 c_0|,$$

since we are assuming $\mu < 1$. Using induction starting with the case $n = 1$ we just investigated, we find that $|b_n c_n| \le |b_0 c_0| \mu^n$. Moreover the analysis shows that sequence $\{|b_n c_n|\}$ strictly decreases to 0, unless it equals zero after some point.

Note that $b_{n+1}/b_n = a_n(\rho^{-1} - \rho)$ and $c_{n+1}/c_n = d_n(\rho - \rho^{-1})$. Consequently if $b_{n+1} = c_{n+1} = 0$ while $b_n \ne 0$ and $c_n \ne 0$, necessarily $a_n = d_n = 0$ at $\text{tr}(T_n) = 0$.

Case 2a: $b_n c_n \ne 0$, for all n. Since the sequence $\{b_n c_n\}$ is strictly decreasing, the elements $\{T_n\}$ are distinct. Because $\lim a_n d_n = 1$, from the formula for T_{n+1} we see that $\lim a_n = \rho$ and $\lim d_n = \rho^{-1}$. Again using the formula for T_{n+1} we find that $\lim(b_{n+1}/b_n) = \rho(\rho^{-1} - \rho)$ and $\lim c_{n+1}/c_n = \rho^{-1}(\rho - \rho^{-1})$.

If A is elliptic, that is if $|\rho| = 1$, the ratios $|b_{n+1}|/|b_n|$ and $|c_{n+1}|/|c_n|$ are approximately $|\rho - \rho^{-1}| < 1$ and therefore $\lim b_n = \lim c_n = 0$. Consequently $\lim T_n = A$, contradicting discreteness.

Consider more generally the transformations

$$S_n = A^{-n} T_{2n} A^n = \begin{pmatrix} a_{2n} & \rho^{-2n} b_{2n} \\ \rho^{2n} c_{2n} & d_{2n} \end{pmatrix}.$$

Again from the formula for T_{n+1}, the ratios $|b_{2n}|/|b_{2n-2}|$ and $|c_{2n}|/|c_{2n-2}|$ are approximately $|\rho - \rho^{-1}|^2 < 1$. Therefore $\lim S_n = A$, again a contradiction to discreteness.

Case 2b: $b_n c_n = 0$, $n \ge N$. For $n \ge N$, A and T_n share a fixed point.

If A is elliptic, its order exceeds 6. This is because $\mu < 1$ implies $\sin \theta < \frac{1}{2}$ since $\text{tr}^2(A) - 4 = 4 \sin^2 \theta$ and A has the form $z \mapsto e^{2i\theta} z$. If A and T_n share exactly one fixed point, then by Lemma 2.3.1 G is not discrete, a contradiction.

If A is loxodromic and shares exactly one fixed point with T_n then $\langle A, T_n \rangle$ cannot be discrete.

Therefore A and T_n, $n \ge N$, have the same pair of fixed points $0, \infty$.

If $N = 0$, $G = \langle A, B \rangle$ is a discrete elementary group.

If $N = 1$, then $a_0 = d_0 = 0$ and G is elementary.

Suppose $N \ge 2$ so that T_{N-1} is conjugate to A. Then $\text{tr}(T_{N-1}) = \text{tr}(A) = 0$. But then $\mu \ge 4$, contrary to our assumption.

Further analysis yields the itemization of elementary groups for Jørgensen's inequality. □

In particular the group $\langle \left(\begin{smallmatrix} 1 & 1 \\ 0 & 1 \end{smallmatrix}\right), \left(\begin{smallmatrix} a & b \\ c & d \end{smallmatrix}\right) \rangle$, where $ad - bc = 1$, is not discrete when $0 < |c| < 1$.

2.3 Elementary discrete groups

A loxodromic or elliptic element g in a discrete group G is called *primitive* if g is a generator of the cyclic subgroup consisting of all loxodromic or elliptic elements in G having the same fixed points (and axis) as g.

The purpose of this section is to present the classical classification [Ford 1929] of the elementary discrete groups, these being the discrete groups with one or two fixed points on $\mathbb{S}^2$, or one in $\mathbb{H}^3$. Our analysis will show:

A discrete group is elementary if and only if it is either finite, abelian, or it contains an abelian subgroup of finite index (virtually abelian).

Finite groups

If G is a finite group, it consists only of elliptic transformations and there is a common fixed point in $\mathbb{H}^3$ (Corollary 4.1.8). If the group is not cyclic there is exactly one fixed point. The common fixed point may be taken as the origin of the ball model so that G becomes a group of rotations of $\mathbb{S}^2$. Thus G is the group of orientation-preserving symmetries of a regular figure inscribed in $\mathbb{S}^2$: one of the platonic solids, or else an equatorial regular polygon. More specifically, we have the following cases: the *tetrahedral group* of order 12, which preserves (collectively) the set of vertices of a tetrahedron; the *octahedral group* of order 24, which preserves the vertices of an octahedron, or those of its dual cube; the *icosahedral group* of order 60, which preserves the vertices of an icosahedron or dodecahedron; and the *dihedral group* of order $2n$, for $n \geq 2$, which preserves the dihedron, the degenerate "solid" consisting of two coincident faces in the shape of an n-sided regular polygon inscribed in the equator. The dihedral group contains, in addition to rotations by $2\pi i/n$ about the center, rotations of order two about any diameter from a vertex or the midpoint of a side. Each of these groups is generated by two elements. See Exercise (2-29) for more detail.

For any finite G, the sphere $\mathbb{S}^2$ is a branched covering of $\mathbb{S}^2/G \cong \mathbb{S}^2$ with branching orders $r_i \geq 2$. If G is cyclic, there are two branch points, the fixed points of G. If G is not cyclic, the branching is over three points having the following orders (r_1, r_2, r_3): $(2, 3, 3)$ for the tetrahedral group, $(2, 3, 4)$ for the octahedral group, $(2, 3, 5)$ for the icosahedral group, and $(2, 2, n)$ for the dihedral group. See Exercises (3-1, 2-29).

More generally, we will show in Theorem 4.1.5 that if H is an arbitrary group consisting entirely of elliptic transformations, then H is conjugate to a group of rotations of $\mathbb{S}^2$.

Infinite groups

An elementary discrete group G that is not finite has one of two properties: (1) G fixes a single point ζ on $\mathbb{S}^2$; or (2) G fixes a pair of points on $\mathbb{S}^2$.

(1) *One fixed point.* Here G contains only parabolic transformations and elliptic transformations all sharing the same fixed point, say ∞. The parabolic subgroup G_0 is either cyclic and conjugate to $\langle z \mapsto z + 1 \rangle$, or it is a free abelian group of rank two and conjugate to $\langle z \mapsto z + 1, \ z \mapsto z + \tau \rangle$, for some $\tau \in \mathbb{C}$ with $\operatorname{Im} \tau > 0$. See Exercise (2-4).

In the cyclic case G itself can be the finite extension of G_0 by an elliptic of order two.

In the rank two case, G can be a finite extension of G_0 by elliptics fixing ∞, of order not exceeding six by Lemma 2.3.1(iii). The possibilities are $(2, 2, 2, 2)$ and $(3, 3, 3)$, $(2, 3, 6)$, $(2, 4, 4)$, meaning these are the orders of primitive elliptic elements, non-conjugate under G_0, which generate the four possible extensions. Each of the triples is possible only for special choices of τ in G_0. For details see Ford [1929] and Exercise (2-30).

(2) *Two fixed points.* Here G is a finite extension of a cyclic loxodromic group with axis ℓ. It can be extended by an elliptic of finite order with rotation axis ℓ and extended once again by an elliptic of order two which exchanges the endpoints. All these groups preserve ℓ.

Lemma 2.3.1. *Let G be an infinite group of Möbius transformations.*

 (i) *If G is discrete, G is elementary if and only if it is a finite extension of an abelian group.*
 (ii) *If $g \in G$ is loxodromic and $h \in G$ has exactly one fixed point in common with g then G is not discrete.*
 (iii) *If $g_1 \in G$ is elliptic of order exceeding six and $g_2 \in G$ has exactly one fixed point in common with g_1, then G is not discrete.*
 (iv) *If $g \neq$ id is an element of a nonelementary discrete or nondiscrete group G, there is a loxodromic element in G without a common fixed point with g.*
 (v) *A nonelementary discrete or nondiscrete group contains two loxodromic elements with no fixed points in common, and hence it contains infinitely many loxodromic elements with mutually distinct fixed points.*

Proof. Item (i) follows from the discussion above.

For (ii), we may assume that $g = \begin{pmatrix} \rho & 0 \\ 0 & 1/\rho \end{pmatrix}$ and $h = \begin{pmatrix} a & b \\ 0 & 1/a \end{pmatrix}$. We find that

$$g^n h g^{-n} h^{-1} = \begin{pmatrix} 1 & -ab(1 - \rho^{2n}) \\ 0 & 1 \end{pmatrix}.$$

If $|\rho| < 1$ let $n \to +\infty$. If $|\rho| > 1$, let $n \to -\infty$. In either case, G cannot be discrete.

To prove (iii) suppose g_1 and g_2 have the common fixed point ∞. According to Lemma 1.5.2, their commutator $g_1 g_2 g_1^{-1} g_2^{-1}$ is parabolic, also with fixed point ∞. If $G = \langle g_1, g_2 \rangle$ is to be discrete, then the subgroup G_∞ of parabolic transformations fixing ∞ has a generator K whose period ω satisfies $|\omega| \leq |\omega'|$ in comparison to the periods ω' of other elements of G_∞ (see Ahlfors [1978]). Write $g_1(z) = az+b$, $|a| = 1$ and $K(z) = z + \omega$. Then $g_1 K g_1^{-1}(z) = z + a\omega$. In particular $|\omega| \leq |a\omega - \omega|$ or $1 \leq |a - 1|$.

Now $a = e^{i\theta}$, where $\theta = 2\pi k/m$ for some relatively prime $m, k \in \mathbb{Z}$, since if G is to be discrete the elliptic elements have finite order. We may choose g_1 so that $\theta = 2\pi/m > 0$ and then $|a - 1| = 2\sin(\pi/m)$. If $|a - 1|$ is to be ≥ 1, then we must have $m \leq 6$, where $m = 6$ gives equality.

The proof of (iv) involves three cases. We will show later in Theorem 4.1.5 that a nonelementary group, discrete or not, contains nonelliptic elements.

Case 1. $g = \left(\begin{smallmatrix} 1 & 1 \\ 0 & 1 \end{smallmatrix}\right)$ is parabolic. There exists $h \in G$ without a common fixed point with $g\colon h = \left(\begin{smallmatrix} a & b \\ c & d \end{smallmatrix}\right)$ with $c \neq 0$. We find that $\mathrm{tr}(g^n h) = (a + d) + nc$. Thus for all large $|n|$, $g^n h$ is loxodromic and does not share a fixed point with g.

Case 2. $g = \left(\begin{smallmatrix} \rho & 0 \\ 0 & \rho^{-1} \end{smallmatrix}\right)$ is loxodromic, $|\rho| > 1$. We have to show there is an element $h \in G$ which does not share one of the fixed points p, q of g. Not all elements of G can fix say p, but perhaps there is one h_p which fixes only p and another h_q that fixes only q. But then $h = h_q h_p$ fixes neither. In addition h does not exchange p, q.

Since $\mathrm{tr}(g^n h) = a\rho^n + d\rho^{-n}$, $g^n h$ is loxodromic for most n, and does not share a fixed point with g.

Case 3. $g = \left(\begin{smallmatrix} \rho & 0 \\ 0 & \rho^{-1} \end{smallmatrix}\right)$, $|\rho| = 1$, is elliptic with fixed points $p = 0, q = \infty$. If there is a loxodromic $h \in G$ which does not share a fixed point with g we are done. If there is a parabolic $h \in G$ which does not share a fixed point, then $h^n g$ is loxodromic for all large $|n|$. Moreover it does not share a fixed point with g. Finally if $h \in G$ shares exactly one fixed point with g then either h is parabolic or $ghg^{-1}h^{-1}$ is parabolic. So assume h_p is parabolic and fixes p while h_q is parabolic and fixes q. Then for all large $|n|$, $h_p^n h_q$ is loxodromic and fixes neither.

Item (v) follows from (iv). For given $g \neq \mathrm{id}$ let $h \in G$ be loxodromic without a common fixed point with g. Then the fixed points of the loxodromic element ghg^{-1} are $g(p), g(q)$ where p, q are the fixed points of h. Unless g is elliptic of order two and exchanges p, q, the fixed points of ghg^{-1} will be distinct. If g exchanges the fixed points of h the subgroup $\langle g, h \rangle$ is elementary. Yet there is some element $g_1 \in G$ which does not fix or exchange the fixed points of h. Now we can use $g_1 h g_1^{-1}$. Once we have two, we can keep conjugating so as to get infinitely many. $\qquad \square$

2.4 Kleinian groups

Discrete groups of Möbius transformations are called *kleinian groups*. To avoid special cases, unless stated otherwise we will take the term to mean *nonelementary* groups; this is the usual practice.

Perhaps the smallest noncyclic kleinian group is the *Klein 4-group*, $\mathbb{Z}^2 \times \mathbb{Z}^2$. It is the direct product of two elements of order two, with presentation $\langle a, b : a^2 = b^2 = (ab)^2 = 1 \rangle$. In nature, it is found as the group of automorphisms of the 1-skeleton of the unit cube in 3D.

A kleinian group that preserves the interior (hence also the exterior) of a round disk on $\mathbb{S}^2$ is called a *fuchsian group*. Typically, a fuchsian group is taken to act on the unit disk $\mathbb{D} = \{z \in \mathbb{C} : |z| < 1\}$ or on the upper half-plane $\mathrm{UHP} = \{z \in \mathbb{C} : \mathrm{Im}\, z > 0\}$.

We know that a group is discrete if and only if it is properly discontinuous on $\mathbb{H}^3$. Therefore we focus our attention on $\mathbb{S}^2$ and make the following definition.

2.4.1 The limit set $\Lambda(G)$

A point $\zeta \in \mathbb{S}^2$ is a *limit point* of the discrete group G if there exists $\xi \in \mathbb{S}^2$ such that $\lim T_n(\xi) = \zeta$, for an infinite sequence of distinct elements $\{T_n\} \in G$. The set

$$\Lambda(G) = \{\zeta \in \mathbb{S}^2 : \zeta \text{ is a limit point}\}$$

is called the *limit set* of G. It contains all loxodromic and parabolic fixed points. It is automatically invariant under G. If $\Lambda(G)$ contains no, one, or two points, G is an elementary group.

Lemma 2.4.1 (Properties of the limit set). *Suppose G is nonelementary, so $\Lambda(G)$ contains at least three points.*

(i) *The G-orbit of any $\zeta \in \Lambda(G)$ is dense in $\Lambda(G)$.*
(ii) *$\Lambda(G)$ is the closure of the set of loxodromic fixed points, and if there are parabolics, it is the closure of the set of parabolic fixed points as well.*
(iii) *$\Lambda(G)$ is a closed set.*
(iv) *The G-orbit of any point $x \in \mathbb{H}^3 \cup \mathbb{S}^2$ accumulates onto $\Lambda(G)$.*
(v) *If $D_1, D_2 \in \mathbb{S}^2$ are two open disks with disjoint closures, each of which meets $\Lambda(G)$, there exists a loxodromic element in G with a fixed point in D_1 and in D_2.*
(vi) *$\Lambda(G)$ is a perfect set (it has no isolated points).*
(vii) *Either $\Lambda(G) = \mathbb{S}^2$ or its interior is empty.*
(viii) *If G_0 has finite index in G, or if G_0 is a normal subgroup of G, then $\Lambda(G_0) = \Lambda(G)$.*

Proof. Given $\zeta \in \Lambda(G)$ let $\zeta^* \neq \zeta$ be another limit point. If ζ^* is the fixed point of a loxodromic or parabolic A, then $\lim A^n(\zeta) = \zeta^*$, as $n \to +\infty$, or to $-\infty$, and (i) is established.

Otherwise, for some $x \in \mathbb{S}^2$, and a sequence $\{A_n\} \subset G$, $\lim_{n \to \infty} A_n = \zeta^*$. We can assume all elements are either loxodromic or are parabolic, and with convergent fixed points $(p_n, q_n) \to (p, q)$.

In the loxodromic case with limit attracting point $q = \zeta^*$ so that $\lim A^n(\zeta) = \zeta^*$, unless $\zeta = p$, the repelling point. But then $\lim A_n(\zeta_1) = \zeta^*$, for a limit point $\zeta_1 \neq \zeta, \zeta^*$. In the parabolic case, $\lim A^n(\zeta) = \zeta^*$, unless $\zeta = p$, the limiting fixed point. Otherwise, $\lim A^n(\zeta_1) = \zeta^*$.

Thus every $\zeta^* \in \Lambda(G)$ is the limit of other limit points.

The argument also proves (ii), (iii), and (vi).

For (iv), if $x \in \mathbb{S}^2$, and $T \in G$ is loxodromic with attracting fixed point q and repelling p, then $\lim T^n(x) = q$, if $x \neq p$. There are no limit points in $\mathbb{H}^3$, but for any $x \in \mathbb{H}^3$, $\lim T^n(x) = q \in \mathbb{S}^2$, for example.

If $\Lambda(G)$ is not all of $\mathbb{S}^2$ there is an open set U in its complement. Every loxodromic fixed point is a limit point of the G-orbit of U, and then so is every point of $\Lambda(G)$. Therefore $\Lambda(G)$ can have no interior, for points there could not also be limit points. This establishes (vii).

To prove (v) (after Beardon [1983, Theorem 5.3.8]), choose disjoint closed disks, $\overline{D_1}, \overline{D_2} \subset \mathbb{S}^2$ such that $D_i \cap \Lambda(G) = \varnothing$, $i = 1, 2$. Suppose there exists loxodromic A_i, $i = 1, 2$, with attracting fixed point $q_i \in D_i$ but its repelling p_i *not* in D_j, $i = 1, 2$. If there is not such A_i or A_j we are done.

Choose a loxodromic $h \in G$, attractive fixed point q, repelling p, both being different than the fixed points of A_1, A_2. For all large m, $A^m(p), A^m(q) \in D_1$. Fix m.

Choose $\overline{D'_1} \subset D_1$ with $A_1^m(p) \in D'_1$, and $\overline{D'_2} \subset D_1$. Choose them so $\overline{D'_1} \cap \overline{D'_2} = \varnothing$. Then choose n so large that (a) $A_2^n(D'_2) \subset D_2$, while (b) $A_2^{-n}(\overline{D'_1}) \cap D_1 = \varnothing$.

Set $B = A_1^m h A_1^{-m}$. Then pick $r > 0$ so large that (c) $B^r(\overline{D'_2}) \subset D'_2 \subset D_1$, and (d) $B^{-r}(\mathbb{S}^2 \setminus D_1) \subset D'_1 \subset D_1$.

Finally, set $T = A_2^n B^r$. Applying (c) and (a) we find

$$T(\overline{D_2}) = A_2^n B^r(\overline{D_2}) \subset A_2(D'_2) \subset D_2$$

and, using (b) and (d),

$$T^{-1}(D'_1) = B^{-r} A_2^{-n}(\overline{D'_1}) \subset B^{-r}(\mathbb{S}^2 \setminus D_1) \subset D'_1 \subset D_1.$$

We conclude that T is loxodromic, with attracting fixed point in D_2, while its repelling is in D_1.

It remains to prove item (viii). Suppose G_0 has finite index in G. For any loxodromic $A \in G$, for some k, $A^k \in G_0$. Therefore $\Lambda(G_0) = \Lambda(G)$. If instead G_0 is a normal subgroup, $g G_0 g^{-1} = G_0$ for any $g \in G$. If p is a fixed point of $A \in G$, then $g(p)$ is a fixed point of $g A g^{-1} \in G_0$. Now $\Lambda(G_0)$ is a closed set and $\{g(p) : g \in G\}$ is dense in $\Lambda(G)$. Therefore $\Lambda(G_0) = \Lambda(G)$. $\qquad\square$

Each component of $\Lambda(G)$ which is not a circle or a point is a fractal set; see Exercises (2-15), (3-20).

2.4.2 The ordinary (regular, discontinuity) set $\Omega(G)$

The complementary open set,

$$\Omega(G) = \mathbb{S}^2 \setminus \Lambda(G),$$

is called the *ordinary set*. Alternate names in use include *regular set*, or *set of discontinuity*. Like $\Lambda(G)$, $\Omega(G)$ is preserved by G. It is the largest open subset of $\mathbb{S}^2$ on which G acts properly discontinuously.

Lemma 2.4.2 (Properties of the ordinary set). *Assume that G is finitely generated and not elementary, and that $\Omega(G) \neq \varnothing$.*

(i) *$\Omega(G)$ has one, two, or infinitely many components.*

(ii) *Each component of $\Omega(G)$ is either simply or infinitely connected.*

(iii) *If each of two components Ω_1, Ω_2 of $\Omega(G)$ is preserved by G, then each one is simply connected and $\Omega(G) = \Omega_1 \cup \Omega_2$.*

(iv) *If one component Ω of $\Omega(G)$ is preserved by G, all the others are simply connected.*

Proof. To prove (ii), assume a component Ω is finitely but not simply connected. At this point we have to anticipate the Ahlfors Finiteness Theorem (p. 122) to assert that Ω is preserved by an element $g \in G$ of infinite order (such a g may not exist if G

is not finitely generated). Choose a simple loop $\sigma \subset \Omega$ that separates the boundary components. The simple loops $\{g^k(\sigma)\} \subset \Omega$ converge to the fixed points or point of g. But each simple loop $g^k(\sigma)$ separates boundary components of Ω. Hence the fixed points are limits of infinitely many boundary components of Ω, a contradiction.

To prove (i), suppose there are a finite number of components $\Omega_1, \ldots, \Omega_m$ of $\Omega(G)$; we may assume that $\infty \in \Omega_m$. There is a subgroup G_0 of finite index and with the same limit set that preserves each of them.

Choose a loxodromic transformation $g \in G_0$. Since g in particular preserves Ω_1 and Ω_2, we can find simple arcs $\sigma_i \in \Omega_i$, $i = 1, 2$, such that $\sigma_i^* = \bigcup_{k=-\infty}^{\infty} g^k(\sigma_i)$ forms a simple arc in Ω_i between the two fixed points of g. This is most easily done by using the quotient surface $\Omega_i/\mathrm{Stab}(\Omega_i)$. Then $\sigma^* = \sigma_1^* \cup \sigma_2^* \cup \{p, q\}$ forms a simple closed curve meeting $\Omega(G_0)$ only in Ω_1 and Ω_2. Consider the two components of $\mathbb{S}^2 \setminus \sigma^*$. One of them, say U, contains Ω_m and ∞. The other, U', contains points of $\Lambda(G_0)$, for otherwise σ_1^* and σ_2^* could be connected by an arc that does not meet $\Lambda(G_0)$. Therefore we can find a loxodromic element $h \in G_0$ with attracting fixed point in U'. Connect ∞ to $h(\infty)$ by an arc $\tau \subset \Omega_m$ and set $\tau^* = \bigcup_{k=0}^{+\infty} h^k(\tau)$. Now τ^* is an arc in Ω_m connecting $\infty \in U$ to the attracting fixed point of h in U', so τ^* must cross σ^*, giving a contradiction.

Item (iii) also depends on Ahlfors' theorem. Using that the simplest proof involves three-dimensional topology. We will present it in Section 3.8.

Item (iv) is a consequence of the fact that $\Lambda(G) = \partial\Omega$. The analysis in terms of three-dimensional topology is suggested in Exercise (3-11). $\qquad\square$

It is relevant to refer again to L. Greenberg's theorem 1962, which has the following consequence. Suppose $\Omega \neq \mathbb{S}^2$ is a connected open set which is not a round disk. Then the group of all Möbius transformations which map Ω onto itself is either discrete or elementary, as in the case of a horizontal strip. Usually it will consist only of the identity.

The term *function group* is usually reserved for a group G with the property that $\Omega(G)$ has an infinitely connected component Ω that is invariant under G. The term arises from the fact that functions invariant under G can be constructed on Ω. The finitely generated function groups can be completely classified [Maskit 1988] or by topology; see Marden [1977] and Exercise (3-11).

When G is not finitely generated, if two of the components of $\Omega(G)$ are invariant under G then as before they are both simply connected. Yet there may also be other components; each of these is also simply connected, but its stabilizer consists only of the identity (see Accola [1966] or apply three-dimensional topology as in Section 3.8). An example of Accola shows that indeed there can be infinitely many other components, which he called "atoms". (The situation is reminiscent of the classical construction in point set topology known as the *Lakes of Wada*: a family of three—or any number up to countably infinite—simply connected open sets on $\mathbb{S}^2$ each of which has the same boundary, namely the complement in $\mathbb{S}^2$ of the union of the open sets.

See Hocking and Young [1961, pp. 143–145].) However, when G is finitely generated, atoms cannot occur, as we will see from the Ahlfors Finiteness Theorem.

Here is an answer to the question of the $\pm$ ambiguity as we go from a group of Möbius transformations to a set of associated matrices in $SL(2, \mathbb{C})$. In short, for discrete groups, the signs can be chosen unambiguously, except if there are elements of order two.

Theorem 2.4.3 [Culler 1986]. *A discrete group G can be lifted to an isomorphic group of matrices in $SL(2, \mathbb{C})$ if and only if G has no elements of order two.*

This result is the best that can be hoped for, since the matrices corresponding to Möbius transformations of order two have order four—for example the normalized matrix that corresponds to $z \mapsto 1/z$ has order four. On the other hand, one can ask, with John Fay, whether any group can be lifted to an unnormalized matrix group in $GL(2, \mathbb{C})$. (For example, the unnormalized matrix $\left(\begin{smallmatrix} 0 & 1 \\ 1 & 0 \end{smallmatrix}\right)$ does have order two.) The answer is not known, to my knowledge.

2.5 Quotient manifolds and orbifolds

A kleinian group G is usually best studied in conjunction with its quotient spaces:

$$\mathcal{M}(G) = \mathbb{H}^3 \cup \Omega(G)/G, \quad \partial \mathcal{M}(G) = \Omega(G)/G,$$

namely, the set of equivalence classes

$$\{\{x\} : x \in \mathbb{H}^3 \cup \Omega \text{ with } x \equiv x_1 \text{ if and only } x_1 = g(x), g \in G\}.$$

The projection $x \to \{x\}$ is denoted by π.

We will often switch between thinking of a situation in $\mathbb{H}^3 \cup \Omega(G)$ and thinking of it in the quotient $\pi : \mathbb{H}^3 \cup \Omega \to \mathcal{M}$.

If G is torsion-free (no elliptics), then $\mathcal{M}(G)$ is an oriented[1] manifold with boundary $\partial \mathcal{M}(G)$, which may be empty. The projection π is a local homeomorphism $\mathbb{H}^3 \to \mathbb{H}^3/G$ and $\Omega \to \partial \mathcal{M}(G)$, because of proper discontinuity of the group action. The interior $\mathcal{M}(G)^{int} = \mathbb{H}^3/G$ has a complete hyperbolic structure arising from the projection of the hyperbolic metric in $\mathbb{H}^3$. Its fundamental group $\pi_1(\mathcal{M}(G))$ is isomorphic to G.

Fix $x, y \neq x \in \mathcal{M}(G)$, and fix x^* over x in $\mathbb{H}^3$. There are infinitely many lifted rays from x^* to the set of lifts $\{\pi^{-1}(y)\}$ of y. We would actually see this picture by standing at x in $\mathcal{M}(G)$. There, we would see our friend y infinitely often, from different angles; she would appear dimmer and dimmer as the rays become longer. This phenomenon is strikingly visualized in the video *Not Knot* [Gunn and Maxwell 1991].

The name "hyperbolic manifold" is reserved for those $\mathcal{M}(G)$ arising from groups G without elliptics.

---

[1] For the record we point out that if G were a group with orientation-reversing elements, the subgroup of orientation-preserving elements would form a normal subgroup of index two. The corresponding nonorientable quotient manifold would have a two-sheeted cover which is orientable and an orientation-reversing isometry which interchanges the sheets.

On the other hand if G contains elliptics, $\mathcal{M}(G)$ is called an *orbifold*. The additional structure of orbifolds will be described in Section 2.5.2 below.

2.5.1 Covering surfaces and manifolds

We will briefly review salient aspects of the theory of coverings of oriented surfaces and 3-manifolds thereby giving more insight to the nature of quotient spaces as well. Our applications will be to Riemann surfaces (Section 2.6) and hyperbolic manifolds, so our discussion will be carried out with these cases in mind.

We will start by focusing on surfaces.[2]

Associated with a surface S and a given basepoint $O \in S$ is the *fundamental group* $\pi_1(S; O)$ of homotopy classes of closed paths from O. If $O_1 \neq O$ is a different basepoint, $\pi_1(S; O)$ is isomorphic to $\pi_1(S; O_1)$ via any arc α from O to O_1 sending each $\gamma \subset \pi_1(S; O)$ to $\alpha\gamma\alpha^{-1} \subset \pi_1(S; O_1)$.

Choose a subgroup H of $F = \pi_1(S; O)$. For example, H may be the cyclic group generated by a single loop, or it may be the identity. A more interesting example is the commutator subgroup, which is the subgroup generated by commutators of pairs of elements of G (the subgroup is also called the homology group since it corresponds to the elements in $\pi_1(S; O)$ which are homologous to zero).

Corresponding to H is the regular[3] covering surface S_H constructed as follows. Consider equivalence classes of pairs $\{(z; \alpha_z)\}$, where $z \in S$ and α_z is a path from O to z. The equivalence relation is that two paths α_1, α_2 from O to z are equivalent if the homotopy class of $\alpha_2^{-1}\alpha_1$ is in H. That is, $(z, \alpha_z) \equiv (z, \alpha_z h)$ if $h \in H$. The surface S_H is the set of equivalence classes $\{(z, \alpha_z)\}$ with the topology determined from S as follows: A neighborhood N^* of (z, α_z) consists of the pairs $\{(w, \sigma_w \alpha_z)\}$, where w lies in a neighborhood N of z and σ_w is a path in N from z to w.

The map $\pi : (z, \alpha_z) \in S_H \mapsto z \in S$, called the projection, is a local homeomorphism of S_H onto S. The points in $\{\pi^{-1}(z)\}$ are said to *lie over* $z \in S$. If H has finite index[4] n in $G = \pi_1(S; O)$, S_H is n-sheeted over S—there are exactly n distinct points of S_H lying over each point of S.

The point $O^* \in S_H$ determined by the class (O, γ), $\gamma \sim 1$, can be taken as basepoint of S_H; it (and many others) lies over O. *The fundamental group $\pi_1(S_H; O^*)$ is isomorphic to H*. If $H \neq \mathrm{id}$ is cyclic, so is the fundamental group of S_H; in this case S_H is homeomorphic to an annulus. If $H = \mathrm{id}$, then S_H is simply connected and is called the *universal covering surface* of S, as it covers all other covering surfaces of S. If $H = G$ then $S_H = S$.

Suppose

$$f : (z, \alpha_z) \to (f(z), f(\alpha_z)), \quad f(O) = O',$$

---

[2] Formally, a surface is a connected two-dimensional manifold, that is, a Hausdorff space with an open covering of sets homeomorphic to open sets in $\mathbb{C}$.

[3] A regular covering S^* is one with the property that if α is a closed arc in S from $x \in S$ to some x_1, and $x^* \in S^*$ lies over x then α can be lifted to an arc $\alpha^* \in S^*$ from x^* to a point x_1^* over x_1.

[4] H has index n in G if there are n distinct cosets $\{Hg_k\}$, $g_k \in G$, such that $G = \bigcup_k Hg_k$. In this case $\bigcap_{g \in G} gHg^{-1} = \bigcap_k g_k Hg_k^{-1}$ is a normal subgroup of finite index in G.

is an automorphism $S \to S$. Then f lifts to S_H if and only if for an arc τ from O to O', $\tau f(\gamma)\tau^{-1} \in H$ whenever the closed curve γ at O lies in H.

A *cover transformation* (also called a *deck transformation*) is a fixed-point-free, orientation-preserving homeomorphism h^* of S_H onto itself with the property that $\pi(h^*(z^*)) = \pi(z^*)$; that is for each point $z \in S$, h^* interchanges the points lying over z. The group of cover transformations of S^* over S is isomorphic to the quotient group $N(H)/H$. Here $N(H) = \{g \in G : gHg^{-1} = H\}$ is called the *normalizer* of H in G. An element $g \neq \mathrm{id} \in N(H)$ induces the cover transformation $(z, \alpha_z) \mapsto (z, \alpha_z g)$.

If $N(H) = G$ then H is a called a *normal subgroup* of G and S_H is called a *normal covering*. In this case the group of cover transformations is isomorphic to G/H: Given any two points O_1^*, O_2^* over O, in a normal covering S_H there is a cover transformation taking $O_1^* \to O_2^*$. In particular, when $H = \{\mathrm{id}\}$, the group of cover transformations of the universal cover is isomorphic to the fundamental group G. Another normal covering is generated by the commutator subgroup. In this case the group of cover transformations is isomorphic to G/H and to the first homology group of S, this is a free abelian group of rank $2g$ if S is a closed surface.

In general, however, there may or may not be cover transformations; for example if H is cyclic and S is a closed surface of genus exceeding one, then $N(H) = H$ and there are none, yet S_H is infinite-sheeted over S.

In contrast, we will construct the universal cover of $G = \mathrm{PSL}(2, \mathbb{R})$.

To do this, following [Labourie 2013, p. 97], start by noting that the group $G = \mathrm{PSL}(2, \mathbb{R})$ can be continuously shrunk onto the subgroup $\mathrm{SO}(2, \mathbb{R})$ of rotations (a deformation retraction), as in the disk model $\mathbb{H}^2$ can be shrunk to a circle about the origin. The quotient $\mathrm{PSL}(2, \mathbb{R})/\mathrm{SO}(2, \mathbb{R})$, is homeomorphic to $\mathbb{H}^2$. So

$$\pi_1(\mathrm{PSL}(2, \mathbb{R})) = \pi_1(\mathrm{SO}(2, \mathbb{R})) = \mathbb{Z}.$$

Now the universal cover of a circle C is $\mathbb{R}$ with projection $\pi : \mathbb{R} \to C$. Denote by $\widetilde{G}$ the subgroup of diffeomorphisms of $\mathbb{R}$ that is the lift of the G-action on $\mathbb{R}$:

$$\widetilde{G} := \{f \in \mathrm{Diffeo}(\mathbb{R}) : \exists g \in \mathrm{PSL}(2, \mathbb{R}) \text{ such that } \pi \circ f = g \circ \pi.$$

This gives a group homomorphism $\widetilde{\pi} : \widetilde{G} \mapsto G$.

The preimage of $\mathrm{SO}(2, \mathbb{R})$ in $\widetilde{G}$ is the subgroup of translations; $\widetilde{G}$ retracts onto this subgroup. Consequently $\widetilde{G}$ is connected and simply connected. It is the universal cover of $\mathrm{PSL}(2, \mathbb{R})$ with cover projection $\widetilde{\pi}$.

We will have need of extending our definition to *branched covers* S^* of S. The difference here is that a *discrete* set of points $\{\zeta_i\} \subset S^*$ is distinguished. We then have regular coverings of the punctured surfaces $S^* \setminus \{\zeta_i\} \to S \setminus \{\pi(\zeta_i)\}$ on which π is a local homeomorphism. But in a small neighborhood Δ of each ζ_i, $\pi : \Delta \to S$ is not a homeomorphism, rather it can be taken as the map $z \mapsto z^r$, where ζ_i corresponds to 0: In Δ the projection is r-to-1. The geometry can be viewed either as (i) $\Delta \setminus \zeta_i$ is an r-sheeted cover of $\pi(\Delta) \setminus \pi(\zeta_i)$, (ii) a sector of angle $2\pi/r$ of Δ is mapped into a cone of vertex angle $2\pi/r$. The point $\zeta_i \in S^*$ is referred to as a *branch point* and its projection $\pi(\zeta_i)$ is the *branch value* or *cone point* (sometimes it is called a branch point as well). The integer $r = r(\zeta_i) \geq 1$ is the order of ramification ($r = 1$ stands for a regular point). Paths in S lift to S^* provided they avoid the cone points.

That the branched cover has N sheets implies that if $\{x_i^*\} \subset S^*$ are the distinct points lying over $x \in S$, then

$$\sum r(x_i^*) = N. \tag{2.2}$$

In particular, if $N = 2$, there is at most one branch point over x; if there is one, its order is two.

More details are provided in Exercise (3-1).

The *Euler characteristic* of an oriented, compact, triangulated surface of genus $g \geq 0$ and $b \geq 0$ punctures or boundary components is

$$\chi(S) = V - E + T = 2 - 2g - b, \tag{2.3}$$

where V is the number of vertices, E the number of edges and T the number of triangles.

The precise relationship between the topologies of a surface and its covering is governed by the Riemann–Hurwitz formula (R–H). Suppose S^* is an N-sheeted cover of the compact surface S, where S has genus $g \geq 0$ and $b \geq 0$ boundary components. Triangulate S so that all the branch values are vertices, and assume there are no branch values on the boundary components. Over each branch value, there may be a number of branch points. Lifting the triangles to S^*, we can compute $\chi(S^*)$ in terms of $\chi(S)$. The result is the *Riemann–Hurwitz formula*,

$$\chi(S^*) = N\chi(S) - \sum_{x \in S} \sum_{x_i^* \, \text{over} \, x} (r(x_i^*) - 1),$$

where the sum is over the branch points in S^* lying over each branch value $x \in S$. In a more useful form,

$$2g^* + b^* - 2 = N(2g + b - 2) + \sum_{x \in S} \sum_{x_i^* \in S^*} (r(x_i^*) - 1). \tag{2.4}$$

This should be coupled with Equation (2.2).

If $S = \mathbb{S}^2$, the (finite) coverings are closed surfaces[5] satisfying $2g^* + 2(N - 1) = \sum (r(x_i^*) - 1)$; a closed surface of any genus can be so constructed. For a torus, $g = 1, b = 0$, the corresponding formula is $g^* - 1 = \frac{1}{2} \sum (r(x_i^*) - 1)$. If each $r_i = 2$, the formula is $g^* - 1 + N = n/2$, where there are n branch points. From the formula we learn, for example, that there is no covering with $g^* = 1$ and six branch points of order 2. But a torus is covered by a genus two surface.

A close study of orientable and nonorientable branched coverings of closed surfaces is made in Edmonds et al. [1984]. In terms of Euler characteristics, there is a degree N branched cover S^* of the orientable S if and only if

$$\chi(S^*) \geq N\chi(S) \quad \text{and} \quad \chi(S^*) - N\chi(S) \text{ is } even.$$

---

[5] A closed surface or manifold is one which is compact, without boundary.

When $S = \mathbb{S}^2$, R–H alone is not always sufficient. But if the total branching is even, $\neq 4$, and satisfies $\sum \sum (r(x_i{}^*) - 1) \geq 3(N - 1)$, R–H is sufficient for the existence of a cover.

We now turn to the case of hyperbolic 3-manifolds $\mathcal{M}(G)$. A covering manifold (unbranched) is a regular covering of $\mathcal{M}(G)$. The group of cover transformations is isomorphic to $N(H)/H$ where H is a subgroup of $G \equiv \pi_1(\mathcal{M}(G))$. The cover transformations are fixed-point-free, orientation-preserving isometries.

On the other hand, the group of orientation-preserving isometries $\mathcal{M}(G) \to \mathcal{M}(G)$ is isomorphic to $N(G)/G$. Here $N(G)$ is the normalizer of G in the full group of all Möbius transformations. Now elements of $N(G)$ can be elliptic. However very often $\mathcal{M}(G)$ has no orientation-preserving "symmetries" at all, that is, $N(G) = G$. For finer details of coverings see Exercise (3-14).

2.5.2 *Orbifolds*

When G contains elliptic transformations, $\mathcal{M}(G)$ is called a hyperbolic three-dimensional *orbifold*.[6] Still, it is useful to define "orbifold" more abstractly, by specifying how such a structure is constructed in an underlying manifold M. We will only discuss the case of three-dimensional orbifolds.

The definition is enlarged from the example that the unit disk $\mathbb{D}$ becomes an orbifold $\mathcal{O}$ with *singular set* $z = 0$ by means of introducing, say, the map $\phi : \mathbb{D} \to \mathbb{D}/\langle z \mapsto e^{2\pi i/8} z \rangle$ which can be viewed as a map to a cone of cone angle $\pi/4$.

For the record we will give a formal definition of an orbifold below; much more extensive discussion of orbifolds is found in Thurston [1979a, §13.2], Boileau et al. [2003], Cooper et al. [2000], and Kapovich [2001, Chapter 6].

Definition [Boileau et al. 2003]. A 3-orbifold $\mathcal{O}$ is a metrizable topological space M, called the underlying space, with a collection of charts $\{U_i, \widetilde{U}_i, \phi_i, \Gamma_i\}$. Here (i) $\{U_i\}$ is an open cover of $\mathcal{O}$; (ii) $\widetilde{U}_i$ is open in $\mathbb{R}^3$; (iii) Γ_i is a finite group of automorphisms of $\mathbb{R}^3$ acting on $\widetilde{U}_i$; (iv) $\phi_i : \widetilde{U}_i \to U_i$ is a continuous map that factors through a homeomorphism between $\widetilde{U}_i / \Gamma_i$ and U_i.

Finally the charts must be compatible with each other: For every $x \in \widetilde{U}_i$ and $y \in \widetilde{U}_j$ for which $\phi_i(x) = \phi_j(y)$, there is a diffeomorphism ψ from a neighborhood V of x to a neighborhood W of y satisfying $\phi_j(\psi(z)) = \phi_i(z)$ for all $z \in V$.

If $\Gamma_i = \text{id}$ for all indices, $\mathcal{O}$ is a manifold. The obifold $\mathcal{O}$ is *orientable* if the maps $\{\phi_i\}$ as well as the elements of $\{\Gamma_i\}$ are orientation preserving.

However, in this book, we are more interested in starting with a hyperbolic orbifold, not the underlying space. When G has elliptic elements $\mathcal{M}(G)$ is called an *orbifold*. The hyperbolic structure of $\mathcal{M}(G)$, which remains oriented, has mild singularities (see Exercise (2-2)) along the projection of the rotation axes of the elliptic elements. The projection of these axes is the *singular set* or *branch locus* of the orbifold.

---

[6] The term, meaning "many folds", was established by vote of the students in Thurston's famous 1976-77 Princeton class.

The projection of an elliptic axis ℓ is often called a *cone axis* as it is reminiscent of the paper-and-scissors construction of a cone by wrapping up a wedge of angle $< 2\pi$; correspondingly the points on the cone axis are called *cone points*. Locally the projection has the form $(z, t) \mapsto (z^r, t)$, where t is a coordinate along the rotation axis and z is a complex coordinate in a plane orthogonal to the axis. The *cone angle* $2\pi/r$ assigned to $\pi(\ell)$ is the angle of rotation of a primitive element—a generator of the cyclic subgroup that has rotation axis ℓ. The Isolation of Cone Axes property in Theorem 3.3.4 gives additional information about the separation of cone axes in $\mathcal{M}(G)$.

A rotation axis γ^* may also be the axis of a loxodromic in G. In this case ℓ will project to a simple loop γ in the singular set of the quotient. There may also be other elliptic axes that intersect γ^*. If so, the common point of intersection is stabilized by a finite elliptic group.

To better understand the structure of the singular set, we will start with the case of a finite group G.

Lemma 2.5.1. *For a finite group H associated with a regular solid, $\mathcal{M}(H)$ is topologically a closed ball, and there exists a point $O \in \mathbb{H}^3/H$ from which exactly three cone axes emanate. Each cone axis from O ends at one of three branch points on $\partial\mathcal{M}(H) = \mathbb{S}^2/H$. The angles at O between the axes are uniquely determined by H.*

Conversely given three distinct points on $\mathbb{S}^2$ there are conjugates of H whose cone axes end at those points in any prescribed order.

In a general an elliptic rotation axis either ends at a point of $\Omega(G)$, or at a parabolic fixed point $\zeta \in \partial\mathbb{H}^3$. In a discrete group, the subgroup of parabolics that fix ζ is either cyclic, or it is free abelian of rank two, as we found when we examined the elementary groups. Correspondingly, we will refer to ζ as a rank one or rank two parabolic fixed point. The conjugacy classes of parabolic fixed points give rise to certain structures in $\mathcal{M}(G)$ called cusps, to be described in detail in Section 3.2. Here it suffices to say that a geodesic ray in $\mathcal{M}(G)$ ends at a rank one or rank two cusp if and only if every lift to $\mathbb{H}^3$ ends at a rank one or rank two parabolic fixed point.

Here is a description of the singular set as a graph in the quotient orbifold:

Proposition 2.5.2. *In any kleinian $\mathcal{M}(G)$, the singular locus is a graph, and a component can be compact in $Int(\mathcal{M}(G))$ or not. Each edge has an order $r \geq 2$. Emanating from each interior vertex are three edges of orders $(2, 3, 3)$, $(2, 3, 4)$, $(2, 3, 5)$, or $(2, 2, n)$, for $n \geq 2$.*

If an edge does not end at a vertex, it may end at a point on $\partial\mathcal{M}(G)$, or at a rank one or rank two cusp. If it ends at a rank one cusp, it must have order two. If it ends at a rank two cusp it also has order two unless two or three additional edges end there as well, in which case they have orders $(3, 3, 3)$, $(2, 3, 6)$, $(2, 4, 4)$, or $(2, 2, 2, 2)$.

Note that for the various cases, quotient of a small horoball or euclidean ball about the common fixed point results in a euclidean or spherical orbifold (compare with Section 3.2 and Exercise (3-1)).

For an example, consider three mutually orthogonal lines ℓ_1, ℓ_2, ℓ_3 intersecting at a point, for example the three coordinate axes at the origin in the ball model. Consider the group G generated by 180° rotations about each line.

Next take a point $x \neq 0$ on say ℓ_1, and take an orthogonal system ℓ_1, ℓ_2', ℓ_3' through x. Let G' be the group generated by the 180° rotations about the three lines ℓ_1, ℓ_2', ℓ_3'. Consider the group $H_0 = \langle G, G' \rangle$. Now choose the loxodromic T with axis ℓ_1 such that distance $[x, T(0)] = [0, x]$. Set $H = \langle H_0, T \rangle$. The branch lines form a trivalent graph in $\mathbb{H}^3$ with two interior vertices. A fundamental set with respect to the action of H consists of the half-lines $\ell_2^+, \ell_3^+, \ell_2'^+, \ell_3'^+$ and, in addition the segment of ℓ_1 from 0 to x. Down in $\mathbb{H}^3/H$, the projection $\pi(\ell_1)$ is a line segment from $\pi(0)$ to $\pi(x)$, and back, a degenerate simple loop. From each of $\pi(0)$ and $\pi(x)$, there are two rays ending at points on the boundary.

The boundary of $\mathbb{H}^3/\langle T \rangle$ is a torus and on it are eight distinguished points, the endpoints of the projection of $\ell_2, \ell_3, \ell_2', \ell_3'$. There is an automorphism of order two of the torus, that has no fixed points, that takes four of these points to the other four. Also acting on the torus is a group of order four generated by two automorphisms of order two, each with four of the distinguished points as fixed points. The quotient of the torus with respect to this group of order four is the sphere, and the torus is a four-sheeted cover, branched over four points on the sphere. The boundary of the orbifold $\mathbb{H}^3/H$ is the sphere: The four branch values are the endpoints of the four singular loci $\pi(\ell_2), \pi(\ell_3), \pi(\ell_2'), \pi(\ell_3')$.

Proof of Proposition 2.5.2. The quotient of $\mathbb{S}^2$ under the groups of the regular solids is again $\mathbb{S}^2$ with exactly three branch values. These have orders $(2, 3, 3)$ for the symmetries of a regular tetrahedron, $(2, 3, 4)$ for the symmetries of a cube or octahedron, $(2, 3, 5)$ for the symmetries of a icosahedron or dodecahedron, and $(2, 2, n)$ for a dihedral group. All of the groups are *triangle groups*, which have three generators each of finite order (Exercise (2-5)). These statements follow from the formula of Exercise (3-1) with details given in Exercise (2-29).

As in the example above, a given rotation axis ℓ may be intersected by other rotation axes at succession of distinct points. Each intersection point is the fixed point of one of the standard finite groups.

The second statement also follows from our itemization of elementary groups. Applying Equation (2.4), we see that the torus is a two-sheeted cover of $\mathbb{S}^2$ in the case $(2, 2, 2, 2)$, a three-sheeted cover in the case $(3, 3, 3)$, or a four-sheeted cover in the cases $(2, 4, 4)$ and $(2, 3, 6)$. □

For kleinian groups with elliptics, $\mathbb{H}^3$ is a simply connected branched cover of the orbifold $\mathbb{H}^3/G$ and $\Omega(G)$ may or may not be branched over $\partial\mathcal{M}(G)$. Actually 3-orbifolds are manifolds too, but new local coordinates need to be introduced in neighborhoods of the singular edges and vertices that map them to euclidean balls.

We will reserve the term orbifold for the cases that a singular set—cone axes—exists. Some authors use it to include both manifolds and orbifolds. We have not considered the case of nonorientable orbifolds. Such an orbifold would result, for example, from a reflection in a plane in $\mathbb{H}^3$. For further details about euclidean orbifolds see Exercise (2-30).

In the case above an underlying manifold is the manifold which results by "ignoring" the orbifold structure, for example by removing a tubular neighborhood of the singular set and replacing the components by solid tubes $\mathbb{D} \times (0, 1)$ as required.

It is easier if we start, for example, with $\mathbb{S}^3$ and a knot or link $K \subset \mathbb{S}^3$. Assign positive integers to each component of K. Then we can ask, is there a hyperbolic orbifold, or a manifold in some other geometry, whose *underlying space* is $\mathbb{S}^3$, that models the data? For an answer see the discussion of the Orbifold Theorem, Section 6.8.

An orbifold $\mathcal{O}$ can itself have an orbifold cover $\widetilde{\mathcal{O}}$. There is then a projection between underlying manifolds $p : \widetilde{M} \to M$ with the following property. Every $x \in M$ has a neighborhood $D_0 = D/\Gamma$, $D \subset \mathbb{R}^3$, such that each component $\widetilde{V}$ of $p^{-1}(D_0)$ is isomorphic to D/Γ_0 with $\Gamma_0 \subset \Gamma$ (respecting the projection p). Thus $\mathbb{D}/\langle z \mapsto e^{\pi i/2} z \rangle$ covers $\mathbb{D}/\langle z \mapsto e^{\pi i/4} z \rangle$.

The orbifold cover is a manifold if all the groups Γ_0 are the identity.

Finally, if M has a boundary, we can put mirrors on the boundary. Reflection in the mirror is an orientation-reversing homeomorphism of M to itself; we have introduced an orientation-reversing orbifold structure with underlying space M. Yet if we lived in M what we would see is the manifold resulting from reflecting M in its boundary.

2.5.3 *The conformal boundary*

The "boundary" $\partial \mathcal{M}(G)$ is infinitely far away from any interior point in the hyperbolic metric on $\mathcal{M}(G)^{\text{int}}$, yet it is intimately related to the interior structure. The isometries and the geodesics extend to it. The infinitesimal three-dimensional coordinate frame at each point in $\Omega(G)$, with one direction the interior normal to $\mathbb{S}^2$, projects to the corresponding frame in $\partial \mathcal{M}$, also with one direction the interior normal to $\partial \mathcal{M}$. The boundary $\partial \mathcal{M}(G)$ has a conformal structure induced from $\Omega(G) \subset \mathbb{S}^2$: it is a union of Riemann surfaces (see next section). For this reason it is often called the *conformal boundary* of $\mathcal{M}(G)$.

If G is not elementary, no component of $\partial \mathcal{M}$ can be a sphere or a torus. Tori (without cone points) are excluded by the Uniformization Theorem (see next section): tori can arise only if a component Ω of $\Omega(G)$ over the torus is Möbius equivalent to $\mathbb{C}$, if Ω is simply connected, or Möbius equivalent to $\mathbb{C} \setminus \{0\}$, if Ω is not.

2.6 Two fundamental algebraic theorems

Using the following purely algebraic fact, the quotient orbifolds can often be analyzed by analyzing manifolds. For every orbifold obtained from a finitely generated group has a finite-sheeted cover which is a manifold:

Selberg's Lemma [1960]. *A finitely generated group of matrices in* SL(2, $\mathbb{C}$) *has a finitely generated normal subgroup of finite index which contains no element $\neq$ id of finite order.*

For a proof see Matsuzaki and Taniguchi [1998] or Ratcliffe [1994].

To obtain the corresponding result for a finitely generated kleinian group, choose a set of N generators, and then pass to the matrix group generated by the $2N$ pairs of matrices $\{\pm A_i\}$.

There is a useful inequality relating the rank (minimum number of generators) of a kleinian group G to the rank of an index $N < \infty$ subgroup H of G [Lyndon and Schupp 1977, p.164]:

$$r(H) - 1 \leq N(r(G) - 1) \tag{2.5}$$

See Exercise (5-23) for a proof.

Group presentation. Let G be a group generated by elements $g_1, g_2, \ldots$; we write $G = \langle g_1, g_2, \ldots \rangle$. A *word* in the chosen generators is a finite sequence (of length ≥ 0) whose letters are of the form g_i or g_i^{-1}; any such word gives rise, by composition, to an element of G. A word (of length > 0) giving rise to the identity of G is called a *relator* in G; it is called a *trivial relator* if it collapses to the identity by successive pairwise cancellations of the form $g_i g_i^{-1}$ or $g_i^{-1} g_i$.

Suppose $R_1, R_2, \ldots$ are relators in G. A word W is *derivable* from the relators $\{R_i\}$ if repeated application of the following operations changes W to the empty word in finitely many steps: Insertion or deletion of one of the relators $R_1, R_1^{-1}, \ldots$, or of one of the trivial relators, between any two consecutive letters of W, or before or after the word W. If every relator is so derivable from the relators on the list $R_1, R_2, \ldots$ (plus the empty word), we say that the generators $g_1, g_2, \ldots$ and the relators $R_1, R_2, \ldots$ constitute a *presentation* of G, and we write

$$G = \langle g_1, g_2, \cdots \mid R_1, R_2, \ldots \rangle.$$

A *free group* is one that has a presentation $\langle g_1, g_2, \cdots \mid \rangle$; that is, if (for appropriately chosen generators) there are no nontrivial relators. A group is *finitely presented* if it has a presentation where both the generators g_i and the relators R_i are finite in number.

Suppose G is generated by $g_1, \ldots, g_N$ and $F_N = \langle f_1, \ldots, f_N \mid \rangle$ is a free group in the same number of generators. The map $\phi : f_i \to g_i$ extends to a homomorphism $\phi : F_N \to G$. The elements in the kernel of ϕ are exactly the relators of G; any generating set for this kernel is a set of relators for a presentation of G.

The fundamental group of a closed surface of genus g has a presentation

$$\langle a_1, b_1, a_2, b_2, \ldots, a_g, b_g \mid \textstyle\prod_{i=1}^{g}[a_i, b_i] \rangle,$$

where the generators come from appropriately chosen loops, as in Figure 2.1, p. 79.

For example, a closed surface of genus g has the single relation $\prod_{i=1}^{g}[a_i, b_i] = 1$. Here $[a, b]$ denotes the commutator $aba^{-1}b^{-1}$. On the other hand, the fundamental group of a closed surface with punctures is a free group.

It is a basic property of hyperbolic 3-manifolds (and 3-manifolds more generally) that finitely generated fundamental groups are automatically finitely presented. The proof, due to Scott and Shalen, is a formal consequence of the existence of a compact core in the quotient manifold (see Section 3.9).

For orbifolds, Selberg's lemma can be applied.

Theorem 2.6.1 (Scott, Shalen). *Finitely generated kleinian groups are finitely presented.*

The finite presentation property is automatically true for compact manifolds, as we will see in Section 3.5.

This is in sharp contrast to the case of 4-manifolds, where any countable group, finitely presented or not, can be a fundamental group. There even exist finitely presented groups which have finitely generated subgroups which are not finitely presented [Scott 1973b]. In fact, in [B. H. Bowditch and G. Mess 1994], a torsion-free, finitely generated group of isometries of four-dimensional hyperbolic space is constructed which is not finitely presented and has no parabolics. (Thus the quotient hyperbolic manifold cannot be geometrically finite.)

2.7 Introduction to Riemann surfaces and their uniformization

A Riemann surface is a one-dimensional complex analytic manifold: It is defined by coordinate coverings $\{U_\alpha, \phi_\alpha\}$ (of a connected Hausdorff space), where $\phi_\alpha : U_\alpha \to \mathbb{C}$ is such that the transition mappings

$$\phi_\beta \phi_\alpha^{-1} : \phi_\alpha(U_\alpha \cap U_\beta) \to \phi_\beta(U_\alpha \cap U_\beta$$

associated with overlapping coordinate neighborhoods are analytic homeomorphisms (conformal mappings).[7] These structures will be explored in Exercise (6-9). If the transition mappings are affine mappings, i.e., of the form $Az + B$, then the manifold is called a *complex affine manifold*. Riemann surfaces are orientable and have countable bases.

The collection of local homeomorphisms into $\mathbb{C}$, together with their conformal transition mappings provide angles on R and allow us to declare:

A Riemann surface is an oriented two-dimensional surface with a "rule" for measuring angles.

A homeomorphism $f : R \to S$ between Riemann surfaces is a *conformal mapping* if $\phi_\beta f \phi_\alpha^{-1}$ is conformal where defined; the two surfaces are then *conformally equivalent*. Of course, angles are preserved. Usually one does not distinguish between conformally equivalent surfaces.

The most familiar cases are regions $\Omega \subset \mathbb{C}$ where the euclidean angles are taken. Make a new rule for measuring angles by defining the angle between two rays at $z \in \Omega$ to be the angle resulting after applying the affine transformation $T : (x, y) \mapsto (x', y')$ with $x' = x$, $y' = 2y$. This determines a new Riemann surface structure on the same underlying point set. With its new structure, Ω is conformally equivalent to $T(\Omega)$, the latter with the natural structure from $\mathbb{C}$.

Another common situation is a smoothly embedded surface in $\mathbb{R}^3$ with the "rule" that is induced by the ambient euclidean metric. However to find the local

---

[7] If the transition mappings are instead required to be the restriction of Möbius transformations, the additional structure is called a *complex projective structure*.

homeomorphisms into $\mathbb{C}$ with conformal transition functions, it is necessary to solve the *Beltrami equation* to be discussed in Section 2.9. A polyhedral surface is made into a Riemann surface with the rule given by the euclidean metric in the polygons, except the neighborhood around each vertex must be flattened out so the angles add to 2π (the vertices can be viewed as cone points with the cone angle being the sum of the vertex angles of the triangles sharing the vertex). Flattening is done by $w = z^\alpha$ for some $\alpha > 0$,

In this connection, mention of the following theorem [Rüedy 1971] is irresistible: Any abstract Riemann surface can be *conformally* embedded as a smooth (C^∞) surface, or even a polyhedral surface, in $\mathbb{R}^3$. Here the angles on the embedded surface are to be measured by restricting the ambient euclidean metric. If the surface is not compact, the ends of the embedded conformal equivalent go off to ∞. A conformal embedding can be found in an arbitrarily small neighborhood of a smoothly embedded model surface by deforming it in the normal direction.

In the same vein, the famous paper Nash [1956, Theorems, 2,3] contains the proof that, in particular, (i) a closed hyperbolic surface can be smoothly and *isometrically* embedded in $\mathbb{R}^{17}$, while a noncompact hyperbolic surface can be smoothly and *isometrically* embedded in $\mathbb{R}^{51}$. In each case, the embedding can be effected in an arbitrarily small open set!

However, for most applications one works with Riemann surfaces that are not naturally embedded in any ambient space. Such an example is given below in terms of algebraic curves.

A Riemann surface may be of any genus $g \geq 0$ (number of "handles"), and with any number of "ends" (or "ideal boundary components"), countable or uncountable—like the Riemann surface which is the complement of the Cantor set. A *puncture* is an isolated ideal boundary component which has a neighborhood conformally equivalent to the once-punctured unit disk. To put it another way, a *puncture* is obtained by removing a point from a Riemann surface. One can also speak of Riemann surfaces with borders—like the closed unit disk—but we will not be using them here.

A regular (unbranched) covering surface $\widetilde{R}$ of a Riemann surface R is also a Riemann surface. By means of the locally injective projection map $\pi : \widetilde{R} \to R$ the local complex structure can just be lifted. Cover transformations automatically become conformal automorphisms of $\widetilde{R}$.

A branched cover $\widetilde{R}$ of R is also a Riemann surface. It has a discrete set of special points called branch points. If $\xi^* \in \widetilde{R}$ is a branch point of order $r \geq 2$ and $\pi(\xi^*) = \xi$ is its projection in R, then given a small V neighborhood of ξ there is a neighborhood U of ξ^* such that $\pi(U) = V$ and each point $\neq \xi$ of V is covered exactly r times in U. If the branch values are removed from R and preimages from $\widetilde{R}$, one is left with a regular covering, that can be described by a subgroup of the fundamental group of the base surface.

Conversely, if G is a discontinuous group acting on R then R/G is also a Riemann surface with R a possibly branched cover, depending on whether G has fixed points in R. A typical example is $\mathbb{H}^2/G$.

Good references are Ahlfors and Sario [1960], Donaldson [2011], Farkas and Kra [1991] especially for closed surfaces, and Springer [1957] for an elementary introduction. For a survey of recent work on Riemann surfaces with singular conformal metrics see Bonk [2002].

What is uniformization?

An abstract Riemann surface is described only in terms of local coordinates. Wouldn't it be nice if there were a global coordinate system $w = \phi(t)$, in terms of a complex parameter t, that served *uniformly* at all points? By way of analogy, the unit circle $\{w : |w| = 1\}$ is uniformized by the real line via the projection map $w = e^{it}, -\infty < t < \infty$.

For example, suppose P is an irreducible polynomial of two complex variables. Then $R = \{(x, y) \in \mathbb{C}^2 : P(x, y) = 0\}$ is a Riemann surface. To suggest why, suppose m is the degree of P in y. For most $x \in \mathbb{S}^2$ there will be m distinct values $y_k(x)$ that satisfy $P(x, y) = 0$; the m points $(x, y_k(x)) \in R$ lie over x. A small neighborhood N about such an x determines m disjoint neighborhoods $N_k \subset R$ and the map $x : (x, y) \mapsto x$ is a homeomorphism of each back down to N. The complex structure can be extended over the other points as well. As a result, R is a closed Riemann surface. Conversely, it is a famous classical theorem that every closed Riemann surface is associated with such an algebraic curve.

A noteworthy class of examples are the *Fermat curves* $x^n + y^n = 1, n \geq 2$, which represent closed Riemann surfaces of genus $\frac{1}{2}(n - 1)(n - 2)$. The world now knows that when $n \geq 3$, there are no solution pairs of nonzero rational numbers. For algebraic curves more generally, Mordell's Conjecture is known to be true too: For curves $P(x, y) = 0$ of genus at least two, there are at most a finite number of solution pairs (x, y) where both x and y are rational numbers.

In short, for closed Riemann surfaces in particular, it would be nice if we could find a single complex parameter t such that $x = x(t), y = y(t)$ for all points $(x, y) \in R$, that is, for all solution pairs of an associated $P = 0$.

Uniformization Theorem A. *A simply connected Riemann surface can be conformally mapped onto exactly one of: the Riemann sphere $\mathbb{S}^2$, the complex plane $\mathbb{C}$, the unit disk $\mathbb{D}$.*

Ahlfors [1973, p. 136] wrote about the Uniformization Theorem, "This is perhaps the single most important theorem in the whole theory of analytic functions of one variable". A proof can be found in that same reference, or better in Hubbard [2006, Theorem 1.8.8]. For simply connected regions properly embedded in $\mathbb{C}$ it reduces to the Riemann Mapping Theorem. The famous application is to the universal covering surface $\widetilde{R}$ of a Riemann surface R, which as defined is not embedded anywhere.

Uniformization Theorem B. *The universal cover $\widetilde{R}$ is both determined by R and imposes its geometry on R as follows:*

(i) $\widetilde{R} = \mathbb{S}^2$ *if and only if R is itself conformally* $\mathbb{S}^2$; *spherical geometry,*

(ii) $\widetilde{R} = \mathbb{C}$ *if and only if R is conformally equivalent to* $\mathbb{C}$, *to* $\mathbb{C} \setminus \{0\}$, *or to a torus; euclidean geometry,*

(iii) $\widetilde{R} = \mathbb{D}$ *in all other cases; hyperbolic geometry.*

The group Γ of cover (or deck) transformations is isomorphic to the fundamental group $\pi_1(R)$. Cover transformations are conformal automorphisms of the universal covering, that is, Möbius transformations when one of the standard models are used. A cover transformation cannot have a fixed point in the cover, hence cannot be elliptic. Furthermore, Γ is *properly discontinuous* in $\widetilde{R}$: $N \cap \gamma(N) = \varnothing$ for any $\gamma \in \Gamma$ distinct from id, and any small neighborhood N of any point in $\widetilde{R}$. So the cover transformations form a discrete group.

Parabolic cover transformations are associated with *punctures* on R: recall that a puncture is an isolated "ideal boundary component" with the property that it has a "neighborhood" in R conformally equivalent to the once-punctured unit disk. The lift of a small loop surrounding a puncture determines a parabolic transformation, and conversely, every parabolic transformation is associated with a puncture in this manner. If R has no punctures then Γ contains only loxodromic transformations (plus the identity).

Once we know that the abstract $\widetilde{R}$ is conformally equivalent to, say, the concrete $\mathbb{D}$, we can replace it by (identify it with) $\mathbb{D}$. Likewise, R is conformally equivalent to—and we can replace it by—the quotient surface $\mathbb{D}/\Gamma$ with Γ the group of cover transformations. The complex structure descends automatically from $\mathbb{D}$ to $\mathbb{D}/\Gamma$. The coordinate coverings are just $\{U_\alpha, z\}$, where the $U_\alpha \subset \mathbb{D}$ are small enough to project injectively, via the identity map, into R. In fact the group Γ is uniquely determined up to conjugation: if $\mathbb{D}/\Gamma$ is conformally equivalent to $\mathbb{D}/\Gamma_1$, then $\Gamma_1 = T\Gamma T^{-1}$ for some $\mathbb{D}$-preserving Möbius transformation T.

The third case of Theorem B is operative in particular whenever R is a closed Riemann surface of genus exceeding one. Such Riemann surfaces are nonelliptic algebraic curves, yet uniformization is not an algebraic process. One of the mysteries concerns the precise relation between an explicit polynomial that generates the surface and the uniformizing function.

The group of conformal automorphisms, if $\neq$ id, of a Riemann surface is discrete if and only if its fundamental group is nonabelian. For the record, the group of conformal automorphisms of a closed surface of genus $g \geq 2$ has at most $84(g-1)$ elements; see Exercise (3-1). The lowest genus for which this number can be attained is $g = 3$ and the surface that attains it is the *Klein surface*. In $\mathbb{D}$ it can be represented by fitting together 24 isometric regular hyperbolic heptagons with interior angles $2\pi/3$ (and area $\pi/3$); see Figure 2.6, on p. 99. After making the appropriate pairwise identification of the consequent free edges, the configuration "rolls up" to form a closed surface composed of 24 regular heptagons arranged in triples around 56 vertices. A model was sculpted by Helaman Ferguson and is on the terrace at the Mathematical Sciences Research Institute in Berkeley. The Klein surface as an algebraic curve is $x^3 y + y^3 + x = 0$.

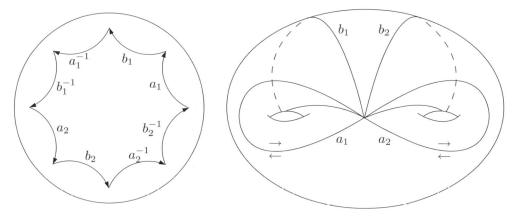

Fig. 2.1. Rolling up a regular octagon: The four transformations mapping edges $a_i \to a_i^{-1}$, $b_i \to b_i^{-1}$ and sending the octagon into its exterior generate the fuchsian covering group.

Even though it does not include the cases of simply and doubly connected plane regions, often one proclaims:

Hyperbolization Theorem for Riemann Surfaces. *Every Riemann surface with a nonabelian fundamental group carries a hyperbolic metric compatible with its complex structure.*

The Uniformization/Hyperbolization Theorem expresses the fact that when the universal cover $\tilde{R}$ can be taken as $\mathbb{D}$, we can bring to bear the dual role of $\mathbb{D} = \mathbb{H}^2$ having both a complex structure and a hyperbolic structure. The group Γ of cover transformations consists of conformal automorphisms of $\mathbb{D}$; therefore the complex structure of $\mathbb{D}$ induces a complex structure on $R = \mathbb{D}/\Gamma$. The group Γ is also a group of isometries of $\mathbb{H}^2$. Therefore the hyperbolic structure on $\mathbb{D} = \mathbb{H}^2$ induces a hyperbolic structure on $R = \mathbb{H}^2/\Gamma$. That is, if z denotes the coordinate in $\mathbb{D}$ and $w = \pi(z)$ the corresponding coordinate in R, define the hyperbolic metric on R by the equation $\lambda(w)\,|dw| = \rho(z)\,|dz|$, where $\rho\,|dz|$ is the hyperbolic metric in $\mathbb{D}$. Usually it is not possible to compute $\lambda(w) = \rho(z)/|\pi'(z)|$ with $z = \pi^{-1}(w)$ explicitly. Notable exceptions are the once-punctured disk $\{0 < |w| < 1\}$ and annulus $\{1 < |w| < R\}$—see Exercise (2-2).

The surface R has finite hyperbolic area if and only if it is a closed Riemann surface of genus $g \geq 0$ with $n \geq 0$ points removed (punctures) satisfying $2g + n \geq 3$. Its area is $2\pi(2g + n - 2)$. Examples include the n-punctured spheres when $n \geq 3$.

On the one hand we can study analytic and meromorphic functions on R, in terms of its complex structure. On the other hand we can do geometry on R, talking about geodesics, triangles, etc. It is often easier to study the analysis and geometry in the universal cover, taking account of the covering group Γ. As explained in Section 3.5, there is a concrete model of R within $\mathbb{D}$ as a convex hyperbolic polygon called a Dirichlet region. Its sides are organized in pairs; when the polygon is "rolled up" by identifying the side pairs by Γ, a surface results and it is conformally equivalent to R.

The Uniformization Theorem can be divided into a topological part and an analytic part. The topological part says in particular that every orientable surface S with nonabelian fundamental group is homeomorphic to $\mathbb{H}^2/G$ for some fuchsian group G. Hence S carries a hyperbolic metric. This can be proven directly by modeling each surface type by a fuchsian group.

The analytic part says that S can be made into a Riemann surface with its conformal structure compatible with its hyperbolic metric.

It is the topological part that has an analogue for 3-manifolds, as is realized in the Hyperbolization Theorem, p. 385. This too is proved by finding geometric models.

Branched covering surfaces. There is a generalization of uniformization theory to Riemann surfaces with a discrete set of points designated as *cone points*. Assign to each of these points a rational cone angle of the form $2\pi/r$, where r is a positive integer ≥ 2. The choice $r = \infty$ means that the point should become a puncture. One requires now a *branched simply connected cover* with the following property. If ξ is a cone point with cone angle $\frac{2\pi}{r}$ then at each point ξ^* over ξ, the stabilizer of ξ^* in the cover group G is generated by an elliptic transformation of order r. A branched, simply connected covering corresponding to the assigned data exists as $\mathbb{S}^2$, $\mathbb{C}$, or most commonly $\mathbb{H}^2$ according to the possibilities described in Exercise (3-1).

We emphasize there are two aspects to the consideration of branch points. Consider the cyclic group $H = \langle z \mapsto e^{2\pi i/6} z \rangle$. A fundamental region for H in $\mathbb{D}$ is the sector $\{z : 0 \leq \arg z < 2\pi/6\}$. There are two ways to consider the quotient $R = \mathbb{D}/\langle z \mapsto e^{2\pi i/6} \rangle$. To a complex analyst, $\mathbb{D}/H$ is made into a Riemann surface by defining the complex structure in $\mathbb{D}/H$ in terms of the map $w = z^6 : \mathbb{D}/H \to \mathbb{D}$. Then $R = \mathbb{D}$ and $\mathbb{D}$ is a branched covering of itself with projection map $w = z^6$.

On the other hand, a geometer sees $\mathbb{D}/H$ as the cone with cone angle $2\pi/6$ obtained when the fundamental sector $\{z \in \mathbb{D} : 0 \leq \arg z \leq 2\pi/6\}$ is rolled up to identify the edges. To make the cone into a Riemann surface at the cone point, it must be flattened out there. This is what is done by interpreting the map $z \mapsto z^6$ as a homeomorphism of the sector of central angle $2\pi/6$ with the edge identifications onto the full disk $\mathbb{D}$. The point $\{z = 0\}$ is called a *cone point*. The situation is analogous to that encountered by three-dimensional orbifolds. An orbifold is actually a manifold, but that involves "flattening out" the cone points.

2.8 Fuchsian and Schottky groups

Fuchsian groups

Suppose G is a nonelementary, discrete group preserving the upper UHP and lower half-plane LHP. Each element $A \in G$ is symmetric in $\mathbb{R}$; it satisfies $\overline{A(z)} = A(\bar{z})$. Each elliptic transformation has one fixed point in UHP and the other at the symmetric point in LHP. The limit set is contained in $\mathbb{R}$. Classically, G is said to be of the *first kind* if $\Lambda(G) = \mathbb{R}$, otherwise it is said to be of the *second kind*.

Suppose G is of the first kind. Then $\Omega(G) = \text{UHP} \cup \text{LHP}$ and if G is finitely generated, $R_{\text{top}} = \text{UHP}/G$ is a closed surface with at most a finite number of punctures and a finite number of branch values (cone points); see Marden [1967]; Casson and Bleiler [1988]. The quotients of the upper and lower half-plane are symmetric surfaces under reflection $z \mapsto \bar{z}$. The 3-manifold $\mathcal{M}(G) = \mathbb{H}^3 \cup (\text{UHP} \cup \text{LHP})/G$ is homeomorphic

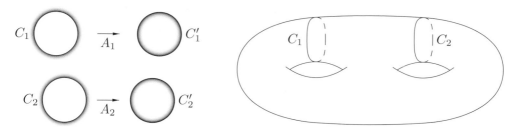

Fig. 2.2. A Schottky group's generators (left) and the group's quotient (right).

to $R_{\text{top}} \times [0, \pi]$; that is $\mathcal{M}(G)$ is an "I-bundle" with top surface R_{top}. This can be seen explicitly as follows. Let H_θ be the euclidean half-plane bordering $\mathbb{R}$, inclined at angle $0 < \theta < \pi$ to $\mathbb{C}$. G maps each half-plane H_θ onto itself. Their quotients H_θ/G are the cross sections in the I-bundle. In addition the orientation-reversing involution $z \mapsto \bar{z}$ extends to all $\mathbb{H}^3$ and projects to an orientation-reversing involution of $\mathcal{M}(G)$, interchanging its top and bottom boundary components, pointwise fixing the middle surface $R_{\pi/2}$.

Suppose instead G is of the second kind and nonelementary. Its ordinary set $\Omega(G)$ contains the countable number of open intervals $\Lambda_+ = (\mathbb{R} \cup \{\infty\}) \setminus \Lambda(G)$ and is connected and infinitely connected. The quotient $R_{\text{top}} = \text{UHP} \cup \Lambda_+/G$, if G is finitely generated, is a *compact bordered* Riemann surface containing at most a finite number of punctures and branch values. Its boundary consists of the finite number of simple closed curves Λ_+/G which are pointwise fixed by the involution of $\Omega(G)$, $z \mapsto \bar{z}$. Equally we have an involution of the surface $\Omega(G)/G$, which is called the *double* of R_{top}. The 3-manifold $\mathcal{M}(G)$ has a connected boundary and the product structure $R_{\text{top}} \times [0, 1]$, where the "top" and "bottom" pieces are joined across ∂R_{top}.

Schottky groups

This is the simplest class of function groups. Take $g \geq 1$ pairs of mutually disjoint circles in $\mathbb{C}$, $\{C_1, C'_1, \ldots, C_g, C'_g\}$, with mutually disjoint interiors. For each index, choose any Möbius transformation A_i that maps C_i to its partner C'_i and sends the interior of C_i to the exterior of its partner. The group generated by $\{A_i\}$ is called a *Schottky group of genus* g. It is the archetypical free group on g generators. The G-orbit of the circles nest down on the limit set $\Lambda(G)$ which is totally disconnected (every component of $\Lambda(G)$ is a point) as shown in Figure 2.3 (p. 83). If G is not cyclic (which is an exceptionally simple special case), the limit set is a perfect set. In Mandelbrot's terminology it is "fractal dust", since it is known to have a positive Hausdorff dimension (Exercise (3-20)). The ordinary set $\Omega(G)$ is connected and infinitely connected. The quotient surface $R = \Omega(G)/G$ is a closed surface of genus g. From the point of view of R, $\Omega(G)$ is a *planar covering surface*. There is a wonderful, thorough discussion of the two-generator case in Mumford et al. [2002].

The Schottky construction works equally well if we replace the circle pairs by pairs of Jordan curves that are known to be associated with Möbius transformations sending the interior of one to the exterior of its partner. To reflect this distinction the Schottky groups generated by circles are known to aficionados as *classical Schottky groups*, whereas groups with the less restrictive requirement are known merely as Schottky groups. The more general situations arise naturally in planar uniformizations of surfaces. It is known [Marden 1974c] that not every Schottky group in the general sense can be generated by circles, no matter how the generators are chosen—for examples see Gilman and Waterman [2006]. In any case Schottky groups form that class of kleinian groups for which $\mathcal{M}(G)$ is a handlebody. The handlebodies of genus g obtained from classical groups are characterized by the property of containing g mutual disjoint hyperbolic planes that are bounded by simple loops which are not retractable to points in the boundary.

Conversely, a discrete, finitely generated, purely loxodromic group, $\Omega(G) \neq \varnothing$, that is a free group is automatically a Schottky group [Maskit 1988, X.H.6]. A special case is a finitely generated fuchsian group of the second kind, without elliptics or parabolics. In fact this is a classical Schottky group; it is an illuminating exercise to verify this fact directly. See Exercise (2-22).

Any finitely generated subgroup of a classical Schottky group is also a classical Schottky group [Anderson $\geq$ 2015].

In the opposite direction, we have:

Maskit Planarity Theorem [Maskit 1988]. *Suppose R is a closed Riemann surface with at most a finite number of punctures and the covering surface $\widehat{R}$ determined by a normal subgroup N of $\pi_1(R; O)$ is planar. Then there is a finite set of mutually disjoint simple loops $\{\alpha_i\}$ in R and a corresponding set of integers $\{r_i \geq 1\}$ with the following property: N is the smallest normal subgroup[8] of $\pi_i(R; O)$ determined by $\{\alpha_i{}^{r_i}\}$, or equivalently, $\widehat{R}$ is the highest normal covering surface of R with the property that all lifts of the curves $\{\alpha_i{}^{r_i}\}$ are simple loops.*

A *planar* Riemann surface $\widehat{R}$ is one which is conformally equivalent to a region in $\mathbb{C}$. Because the covering $\widehat{R} \to R$ corresponds to a normal subgroup $N \subset \pi_1(R)$, the cover transformations are conformal automorphisms of $\widehat{R} \subset \mathbb{C}$. If the result of cutting R along the simple loops $\{\alpha_i\}$ is itself a planar surface, then the cover transformations of $\widehat{R}$ are known to consist of the restrictions of Möbius transformations—see Ahlfors and Sario [1960, IV.4B, IV.19F]. Otherwise, as shown in Maskit [1968], there is a conformal map of $\widehat{R}$ onto another representation $\widehat{R}'$ of the covering for which the cover transformations become restrictions of Möbius transformations (Exercise (2-17)).

2.8.1 Handlebodies

For a Schottky group G, the quotient manifold $\mathcal{M}(G)$ is called a *handlebody of genus g*. The common exterior of the circles serves as a fundamental region for $\Omega(G)$.

---

[8] That is, N is generated by $\{\gamma \alpha_i{}^{r_i} \gamma^{-1}\}$ for all $\gamma \in \pi_1(R; O)$ with each α_i joined to O by an auxiliary path.

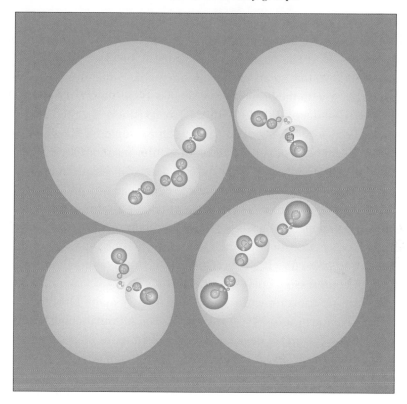

Fig. 2.3. A two-generator Schottky group showing the orbit of the Schottky circles nesting to the limit set.

The common exterior of the hyperbolic planes rising from the circles serves as a fundamental region in $\mathbb{H}^3$. For $g = 1$ the handlebody is called a *solid torus*—a bagel! More generally $\mathcal{M}(G)$ is homeomorphic to the result of g bagels being stuck together.

A handlebody M of genus $g \geq 1$ is characterized by the following property. Its boundary ∂M is a closed surface of genus g. There exist g mutually disjoint simple curves on ∂M called *compressing curves*, no two of which are freely homotopic nor any one of which homotopic to a point, which bound mutually disjoint *compressing disks* within M. Furthermore, when M is cut along these disks what results is connected and homeomorphic to a ball. In our Schottky construction these disks can be taken to be the planes rising from the circles.

If the handlebody M is embedded in $\mathbb{R}^3$, its exterior in $\mathbb{S}^3$ is either a handlebody, with its own, distinct, collection of compressing curves on ∂M, or it is knotted.

There is a generalization of handlebody called a *compression body*. Suppose M is a compact 3-manifold such that one boundary component S, of genus $g \geq 2$, has the following property: There are g mutually disjoint compressing disks, bounded by compressing curves in S, such that when M is cut along these disks, S becomes a 2-sphere. If $\partial M \setminus S = \varnothing$, then M is a handlebody. Otherwise M is called a *compression body*. All the other boundary components are incompressible. In kleinian terms, M arises from a function group. There is more discussion in Exercise (3-11).

Suppose X_1 and X_2 are two handlebodies of the same genus and $h : \partial X_1 \to \partial X_2$ is an orientation-reversing homeomorphism. Glue X_1 to X_2 by identifying each point $x \in \partial X_1$ to $h(x) \in \partial X_2$. The result is a closed orientable 3-manifold M. Conversely it has long been known that in *every* closed, orientable 3-manifold M^3, one can find embedded surfaces S with the property that $M^3 \setminus S$ is the union of two handlebodies; see Hempel [1976] or Jaco [1980]. Such a decomposition is called *Heegaard split-ting*. In the 3-sphere $\mathbb{S}^3$, an embedded torus is an example. A Heegaard splitting of a closed manifold can be obtained by taking a tubular neighborhood about the union of 1-simplices of a triangulation. There are two sets of simple loops S_1, S_2 on S such that each loop in S_1 bounds a disk in one component of $M^3 \setminus S$ and each loop in S_2 bounds a disk in the other. There is an orientation-reversing homeomorphism between the two components that interchanges the two sets. The seeming simplicity of the splitting is very deceptive; all efforts to decipher the topology of the resulting manifold enough to prove the Poincaré Conjecture from the homeomorphism h and its interplay with compressing curves have failed.

It is known that, up to isotopy, there are only finitely many Heegaad splittings of each genus in a given manifold $\mathcal{M}(G)$ [Li 2007]. The minimal genus of all Heegaard surfaces is called the *Heegaard genus* of $\mathcal{M}(G)$. Even more, there is an algorithm to construct a finite list of all possible Heegaard splittings of $\mathcal{M}(G)$, up to isotopy [Li 2011]. Recently Tao Li [Li $\geq$ 2007] proved that, in a closed hyperbolic manifold, the Heegaard genus can be strictly larger, in fact arbitrarily larger, than the rank of its fundamental group[9]. It is easily seen that the rank cannot exceed the Heegaard genus. This in answer to a longstanding question of Waldhausen who asked if there is always equality between the rank and genus. If this were the case when the rank is zero, the Poincaré Conjecture would follow.

It is a conjecture of Agol that every two-generator, closed kleinian manifold is homeomorphic to the result of so gluing two genus-2 handlebodies.

If instead $\mathcal{M}(G)$ has a boundary, a Heegaard splitting is a oriented surface S that divides $\mathcal{M}(G)$ into two compression bodies C_1, C_2 bounded by S, $\mathcal{M}(G) = C_1 \cup_h C_2$. The two sides of S are glued together by an orientation-reversing homeomorphism h.

2.9 Riemannian metrics and quasiconformal mappings

A riemannian surface S carries (we will assume) a smooth, nonsingular riemannian metric which is associated with a form $ds^2 = E\,dx^2 + 2F\,dx\,dy + G\,dy^2$. This is positive definite and invariant under change of local parameters (x, y). It determines angles between intersecting arcs at each point. But to make S into a Riemann surface we have find possibly new local parameters $\{w\}$ and maps $\{f\}$ into $\mathbb{C}$ with the property that, on overlapping neighborhoods, $f_1 \circ f^{-1}$ is conformal.

The fundamental form can be written in complex form in terms of $z = x + iy$, $\bar{z} = x - iy$ as

---

[9] The rank is the minimal number of generators.

$$ds^2 = \lambda(z)|\,dz + \mu(z)\,d\bar{z}|^2,$$

with $\lambda(z) > 0$ and $0 \leq |\mu(z)| \leq k < 1$. In the special case that $\mu = 0$, $ds^2 = |dw|^2 = \lambda(z)|dz|^2$ is a conformal metric: a tiny circle $|z| = \varepsilon$ becomes a tiny circle $|w| = \lambda(0)\varepsilon$ and the angle measure is the same as in the z-coordinate.

Given the riemannian metric, we introduce new local coordinates so that the new metric becomes conformal and the surface becomes a Riemann surface. This is a classical procedure called introducing *isothermal coordinates*.

This is done in each coordinate patch by solving the *Beltrami equation*

$$\frac{\partial f}{\partial \bar{z}} = \mu(z) \frac{\partial f}{\partial z}. \tag{2.6}$$

If $\mu = 0$, (2.6) reduces to the Cauchy–Riemann equations for analyticity of f. A solution f of Equation (2.6) satisfies the infinitesimal equation

$$|df| = |f_z|\,|dz + \mu(z)\,d\bar{z}| = |f_z||dw|.$$

The mapping $w = f(z, \bar{z})$, sending a neighborhood of S into $\mathbb{C}$, sends angles in the w-coordinate to the same angles in $\mathbb{C}$.

Isothermal coordinates have the property that if f is isothermal and if h is a conformal mapping of the range of f, then $h \circ f$ also satisfies (2.6) so is isothermal. This implies that if f, f_1 are each isothermal on overlapping neighborhoods $f_1 \circ f^{-1}$ is conformal where defined.

The solutions $\{w = f(z, \bar{z})\}$ determine a complex structure on the underlying pointset S.

For example, consider the metric $ds^2 = |dz + k d\bar{z}|^2$ in $\mathbb{C}$. The map $f(z, \bar{z}) = z + k\bar{z}$, $z \in \mathbb{C}$, $0 < k < 1$, solves the Beltrami equation with $\mu = k$. This is an orientation-preserving (since $k < 1$), nonsingular (since $k \neq 1$), homeomorphism sending circles about $z = 0$ to ellipses with major and minor axes in the ratio $K = (1+k)/(1-k) \geq 1$. Introduce a new angle measure at $z = 0$ by defining the angle between two rays to be the angle between the image of the rays. This is the angle measure determined by the new metric ds^2. In this case the new Riemann surface is also conformally $\mathbb{C}$ but with respect to the new w-coordinates, it has the conformal metric $ds^2 = |dw|^2$.

In the general theory, $\mu(z)$ needs only to be measurable on its domain, say $\Omega \subset \mathbb{C}$ with essential supremum $\|\mu\|_\infty = k < 1$. The Beltrami equation has a solution f which is a *K-quasiconformal mapping*[10]. Very near a point, say $z = 0$, where f is differentiable (which it is almost everywhere), f is approximately affine: $z \mapsto az + b\bar{z}$. A solution f is uniquely determined up to postcomposition with conformal mappings of its range. The number $K = (1 + k)/(1 - k)$, where $k = \|\mu\|_\infty$, is called the *maximal dilatation* of f, and $\mu = f_{\bar{z}}/f_z$ is called its *complex dilatation*. The maximal dilatation measures the maximal distortion of the mapping in the sense that

---

[10] f is an orientation-preserving homeomorphism with locally integrable distributional derivatives f_z, $f_{\bar{z}}$.

infinitesimal circles are sent to infinitesimal ellipses with ratio of major to minor axis uniformly bounded by K; $K = 1$ if and only if f is conformal[11].

The inverse of a K-quasiconformal mapping is also a K-quasiconformal mapping.[12]

It is often better to define the complex dilatation μ on all $\mathbb{C}$—which can be regarded as $\mathbb{S}^2$ as the values of μ at isolated points do not matter. For example, set $\mu = 0$ on the complement of Ω. When μ is defined on $\mathbb{S}^2$, except perhaps for a set of zero spherical area, and satisfies $\|\mu\|_\infty < 1$ then there is a unique solution of the Beltrami equation up to postcomposition by Möbius transformations. It is a homeomorphism $\mathbb{S}^2 \to \mathbb{S}^2$. The easiest way of normalizing the solution is to require that it fix three prescribed points.

Now we will indicate how to use the Beltrami equation to deform kleinian groups. Suppose $\Omega \subset \mathbb{C}$ is preserved by a kleinian group G. We want to consider quasi-conformal mappings f of Ω that when projected, map the quotient surface Ω/G onto itself or onto another Riemann surface. For this to happen, μ must be a *Beltrami differential* with respect to G, that is, μ must imply, for any $g \in G$, that both f and $f \circ g$ satisfy the same Beltrami equation. The condition that this be the case is

$$\mu(g(z))\frac{\overline{g'(z)}}{g'(z)} = \mu(z) \quad \text{for all } g \in G \text{ and (almost) all } z \in \Omega. \tag{2.7}$$

Often we will extend μ to $\mathbb{C}$ by setting it equal to zero in the complement of Ω, so μ will automatically become a Beltrami differential for G in $\mathbb{S}^2$. Any solution f will a quasiconformal map when restricted to Ω and a conformal mapping when restricted to the complement of $\overline{\Omega}$.

Since it is now known that $\Lambda(G)$ has zero area, it suffices to require μ to be a Beltrami differential on $\Omega(G) = \mathbb{S}^2 \setminus \Lambda(G)$.[13] If μ satisfies (2.7) on $\mathbb{C}$, then both F and $F \circ g$ are solutions of (2.6) for any $g \in G$.

Now the solution of a Beltrami equations is determined only up to postcomposition with at conformal map. To choose which one, start by normalizing the solution F by requiring it to fix, say, $(0, 1, \infty)$. Then, given $g \in G$, choose the unique Möbius transformation g^* so that for the three points, $F \circ g = g^* \circ F$. Then the map defined as $\varphi(g) = g^*$ becomes an isomorphism $\varphi : G \to H = \{g^* : g \in G\}$, $F \circ g = \varphi(g) \circ F$ for all $g \in G$. The group $H = FGF^{-1}$ acting on $F(\Omega)$ is called a *quasiconformal deformation* of G. It is a *trivial deformation*, really not a deformation at all, if φ is a conjugation: for some Möbius U, $\varphi(g) = UgU^{-1}$ for all $g \in G$.

Suppose G is a fuchsian group acting in the upper half-plane UHP and μ is a Beltrami differential for G. There is a way of arranging things so that the quasiconformal

---

[11] The equivalent geometric definition that generalizes to arbitrary metric spaces is that a homeomorphism f of Ω is quasiconformal if there is some constant $H < \infty$ such that for every $z \in \Omega$,

$$\limsup_{r \to 0} \frac{\sup_{|w-z|=r} |f(w) - f(z)|}{\inf_{|w-z|=r} |f(w) - f(z)|} \leq H.$$

[12] An L-bilipschitz map is L^2-quasiconformal, but a quasiconformal map is in general not bilipschitz.

[13] Application of Sullivan's Theorem (page 183) also implies that we do not have to worry about $\Lambda(G)$.

deformation of G is fuchsian as well. This is done by extending μ by symmetry to the lower half-plane LHP: $\mu(z) = \overline{\mu(\overline{z})}$. It will remain a Beltrami differential for G. Normalize F so as to fix, for example, $(0, 1, \infty)$. Then F maps each of UHP and LHP onto itself. The quasiconformal deformation $H = FGF^{-1}$ is fuchsian.

Another example is a Schottky group G. Take a Beltrami differential μ in its ordinary set $\Omega(G)$. We don't need to bother with the limit set because it has zero area. A normalized solution f of the Beltrami equation will induce an isomorphism φ onto another Schottky group H. Even if G is a classical Schottky group it is unlikely that the F-images of the Schottky circles are round circles. But the pairing geometry of these F-images will remain the same.

Return to the case that Ω is simply connected. At the quotient level: F induces a quasiconformal mapping $F_* : R = \Omega/G \to S = F(\Omega)/H$. In each homotopy class $[F_*]$ of a quasiconformal map between the two surfaces, there will be uncountably many quasiconformal mappings. One of them may even be conformal. If so, the deformation, or deformation class $[F_*]$, is said to be *trivial*—there has been no real deformation at all. One of Teichmüller's basic contributions is a characterization of trivial classes.

There is a longstanding, compelling classical conjecture in complex analysis called the *Ehrenpreis Conjecture*: Let R_1, R_2 be any two closed Riemann surfaces of genus ≥ 2, not necessarily homeomorphic. Given $\varepsilon > 0$, there exist finite-sheeted, unbranched covers R_1^* of R_1 and R_2^* of R^* which are homeomorphic and are close to each other in the sense that there is a (smooth) quasiconformal map $f : R_1^* \to R_2^*$ with maximal dilatation $K < 1 + \varepsilon$. The conjecture was finally confirmed in deep work jointly by Jeremy Kahn and Vlad Markovic [Kahn and Markovic 2014] (see p. 387).

2.10 Teichmüller spaces of Riemann surfaces

We will use the notation $\mathfrak{Teich}(R)$ to designate the Teichmüller space of the Riemann surface R.

We will give three definitions. Each is commonly used, depending on the context. The space for the 3-punctured sphere is a single point, but in all other cases the space has positive dimension.

Suppose $R = \mathbb{H}^2/G$ is a closed Riemann surface of genus $g \geq 0$ with $n \geq 0$ punctures such that $3g + n - 3 \geq 1$.

(1) *The mapping definition.* The *Teichmüller space* $\mathfrak{Teich}(R)$ is the quotient space of pairs

$$\mathfrak{Teich}(R) = \{(S, f) \mid f : R \to S \text{ is quasiconformal}\}/\equiv,$$

with the equivalence

$$(S, f) \equiv (S', f') \text{ if and only if } f' \circ f^{-1} \text{ is homotopic to a conformal map.}$$

If R is a closed surface, we can use the term "orientation-preserving[14] homeomorphism" rather than "quasiconformal". The latter is needed only to insure that the punctures are not opened up to holes. We emphasize that $\mathfrak{Teich}(R)$ is the space of "marked" Riemann surfaces: each equivalence class is associated with a particular homotopy class of maps $R \to S$, or given basepoints O, $f(O)$, an isomorphism $\pi_1(R; O) \to \pi_1(S; f(O))$. (R, id) can be taken as basepoint of $\mathfrak{Teich}(R)$. In particular there are infinitely many distinct points $\{(R, h_i)\}$ where $\{h_i\}$ is set of mutually non homotopic automorphisms of R onto itself.

The advantage of this definition is its appeal to our geometric intuition.

(2) *The metric definition.* $\mathfrak{Teich}(S)$ is the space of isotopy classes of hyperbolic metrics (curvature -1) on a fixed C^∞-surface S. That is,

$$\mathfrak{Teich}(S) = \{\text{Hyperbolic metrics on } (S)\}/\mathrm{Diff}_0(S)$$

Here two hyperbolic metrics $\rho(z)|dz|$, $\rho_1(w)|dw|$ on S are identified if $\rho(z)|dz|$ is the result of pulling back the metric $\rho_1(w)|dw|$ under an element $w = f(z) \in \mathrm{Diff}_0$, namely a diffeomorphism f isotopic to the identity (see Exercise (3-25)). Via the uniformization theorem, choosing distinct hyperbolic metrics on S is equivalent to changing the complex structures—the "rules" for measuring angles—on S. Instead of changing surfaces, change the metrics on a fixed surface.

(3) *The fuchsian definition.* $\mathfrak{Teich}(G)$ is the deformation space of fuchsian groups G, $R = \mathbb{H}^2/G$:

$$\mathfrak{Teich}(G) = \{\theta \mid \theta : G \to G', \text{ type-preserving isomorphism to fuchsian } G'\}/\equiv .$$

Type preserving means that parabolics correspond to parabolics (and any elliptics to elliptics of the same order). Here θ corresponds to a homotopy class of quasiconformal maps $\mathbb{H}^2/G \to \mathbb{H}^2/G'$. Isomorphisms θ, θ' represent the same point (are equivalent) if and only if they are conjugate: $\theta(g) = U \circ \theta'(g) \circ U^{-1}$ for some Möbius U and all $g \in G$.

The fuchsian definition gives $\mathfrak{Teich}(G)$ a real analytic structure as it involves matrices with real entries. Suppose R is a closed surface of genus g and $n \geq 0$ punctures. Each loxodromic corresponds to a real matrix normalized having a unit determinant, thus the matrix depends on 3 parameters. Each parabolic has trace ± 2 so the corresponding matrix depends on 2 real parameters. Counting up the parameters gives $3(2g) + 2(n)$. But then

$$[a_1, b_1] \cdots [a_g, b_g]c_1 \cdots c_n$$

when properly oriented surrounds a cell. The matrix representing this again has 3-real parameters. The final equation is another matrix which itself can be normalized. We conclude that the *fuchsian deformation space* has **real** dimension

$$6g + 2n - 6.$$

---

[14] All maps we consider will be orientation-preserving unless stated otherwise.

It can be parameterized explicitly by any suitable choice of *Fenchel-Nielson coordinates*.

It is irresistible to go a step further, deforming into PSL$(2, \mathbb{C})$ instead of just PSL$(2, \mathbb{R})$. This is eminently possible and it leads us into the space of normalized complex matrices. If R has genus g and n punctures this space of matrices has *complex*-dimension $6g + 2n - 6$, by the same computation as above. The reason for having exactly double the dimension is that the space we have now constructed is *not* $\mathfrak{Teich}(R)$ but

$$\mathfrak{Teich}(R) \times \mathfrak{Teich}(\overline{R}),$$

taking account of both, say, UHP/G and LHP/G which can be deformed independently of each other.

2.10.1 Teichmüller mappings

Teichmüller's famous theorem tells us that in the homotopy class of a quasiconformal map $g : S \to S_1$ there is a unique extremal mapping, called a *Teichmüller mapping*, that minimizes the maximal dilatation K among all quasiconformal mappings in the homotopy class. The *Teichmüller distance* between two points (S, f), $(S_1, f_1) \in \mathfrak{Teich}(R)$ is defined as $\log K$, which for the space of tori, coincides with the hyperbolic metric $d(\tau_1, \tau_2)$. (Instead, the Teichmüller metric is often taken as $\frac{1}{2} \log K$.) Here $K = (1 + k)/(1 - k)$ is the minimal maximal dilatation of all quasiconformal mappings in the homotopy class $[f_1 \circ f^{-1} : S \to S_1]$: There is exactly one such mapping H whose maximal dilatation achieves the value K. Likewise $h = H^{-1} : S \to R$ is the extremal mapping in the homotopy class of its inverse, and also has the maximal dilatation K.

A *quadratic differential* on R is an invariant form (that is, it does not depend on the choice of local coordinates) $\phi(z)dz^2$. Up in the universal cover with covering group G the invariance is expressed as

$$\phi(Az)A'(z)^2 = \phi(z), \quad A \in G.$$

A *Beltrami differential* μ on R is also a form independent of the choice of local coordinates. Up in the universal cover $\mathbb{H}^2$ with covering group G the invariance is expressed as

$$\mu(Az)\frac{\overline{A'(z)}}{A'(z)} = \mu(z).$$

On R, in a neighborhood with overlapping local coordinates z_i, z_j and corresponding μ_i, μ_j, the invariance is expressed as

$$\mu_i \frac{\overline{dz_i}}{dz_i} = \mu_j \frac{\overline{dz_j}}{dz_j},$$

or more simply one just states that $\mu \frac{\overline{dz}}{dz}$ is an invariant Beltrami form.

Some integral expressions are

$$\| \phi \| := \int \int_{\mathbb{H}^2} |\phi| dx dy, \text{ and } \int \int_{\mathbb{H}^2} |\phi\mu| := \int \int_{\mathbb{H}^2} |\phi dz^2 \mu \frac{\overline{dz}}{dz}| = \int \int_{\mathbb{H}^2} |\phi\mu||dz|^2.$$

The theory [Strebel 1984] shows that each such extremal pair H, h corresponds to Beltrami equations of a special form: There are uniquely determined (up to positive constant multiples) holomorphic quadratic differentials $\phi\, dz^2$ on R and $\psi\, dz^2$ on S. The Beltrami equations for the corresponding maps are:

$$H_{\bar{z}} = k \frac{\overline{\varphi}}{|\varphi|} H_z, \quad h_{\bar{w}} = k \frac{\overline{\psi}}{|\psi|} h_w, \quad 0 \le k = \frac{K-1}{K+1} < 1.$$

We can also use ϕ and ψ to introduce local coordinates in R and S, away from the zeros of the differentials, which correspond under H and h. This is by means of the local maps $z = \int \sqrt{\varphi}, w = \int \sqrt{\psi}$. Using them, the horizontal and vertical lines in $\mathbb{C}$ are locally pulled back to R and S. Away from the zeros, in the new coordinates, the extremal mappings $H : R \to S, h : S \to R$, become affine maps. When normalized to have unit jacobian $|f_z|^2 - |f_{\bar{z}}|^2$,

$$H : z \mapsto w = \frac{z + k\bar{z}}{\sqrt{1 - k^2}}, \quad h : w \mapsto z = \frac{w - k\bar{w}}{\sqrt{1 - k^2}}.$$

The local pull-back of the horizontal and vertical lines then have the form

$$\text{Re}(z) \mapsto \sqrt{K}\, \text{Re}(z), \quad \text{Im}(z) \mapsto \frac{1}{\sqrt{K}} \text{Im}(z), \quad K \ge 1.$$

Distance between vertical lines is stretched while distance between horizontals is compressed. This is in agreement with the torus case, where the extremal mapping are affine mappings.

With the Teichmüller metric, $\mathfrak{Teich}(R)$ becomes a metric space, homeomorphic to $\mathbb{R}^{6g+2n-6}$. Each point lies on a uniquely determined geodesic ray from a given basepoint. The ray has the parametric form $F_{\bar{z}} = t(\overline{\phi}/|\phi|)F_z, 0 \le t < 1$. The map F is the solution to a Beltrami equation of Teichmüller type. See also Exercise (5-33).

Possible extensions of quasiconformal maps beyond their original domain is an important issue in the theory. A basic result is the following, which has particular importance to the study of the fuchsian Teichmüller spaces:

Lemma 2.10.1 ([Ahlfors 1966], Ch. 1V). *A K-quasiconformal map $f : \mathbb{H}^2 \to \mathbb{H}^2$ has a continuous extension, also denoted by f, to a "quasisymmetric" mapping $f : \partial\mathbb{H}^2 \to \partial\mathbb{H}^2$.*

That the extension f to $\mathbb{R}$ is *quasisymmetric* means that there is a constant $M = M(K) < \infty$, such that

$$\frac{1}{M} \le \frac{f(x+t) - f(x)}{f(x) - f(x-t)} \le M. \tag{2.8}$$

Such maps are also called *1-quasiconformal*. Conversely, every homeomorphism f of R satisfying Equation (2.8) for some M extends to a quasiconformal mapping of UHP.

For introductions to the theory of quasiconformal mappings and Teichmüller spaces see Ahlfors [1966], Hubbard [2006]; Lehto [1987], Imayoshi and Taniguchi [1992] and Fletcher and Markovic [2007].

2.11 The mapping class group $\mathcal{MCG}(R)$

2.11.1 Dehn twists

The simplest class of surface automorphisms are the Dehn twists. In an annulus $A = \{z : 1 < |z| < R\}$ they are defined as follows (see Exercise (5-10)): A (positive) Dehn twist $\tau : A \to A$ is a map obtained by holding the inner contour fixed while rotating the outer one by 2π in the positive direction. Equivalently, by holding the outer fixed and rotating the inner in the negative direction. In polar coordinates a formula for it is

$$T : (r, \theta) \mapsto (r, \theta + 2\pi \frac{r-1}{R-1}), \quad 1 \le r \le R.$$

If A is embedded in a larger surface R the twist is extended by setting it equal the identity in $R \setminus A$.

The Dehn twist about a simple loop γ, which we may assume is a simple closed geodesic, is the homotopy class determined by the following mapping. Choose a collar neighborhood C about γ and a conformal map f of it to the annulus: $C \to A$. Pull T back to C, $T_\gamma := f^{-1} \circ T \circ f : C \to C$. Extend T_γ by setting it equal to the identity on $R \setminus C$. We speak of "the" Dehn twist about γ to be any map in the homotopy class of T_γ just constructed, and/or any choice of simple curve in the free homotopy class of γ.

When applied to a geodesic δ crossing γ, $T_\gamma(\delta)$ is not freely homotopic to δ.

2.11.2 The action of $\mathcal{MCG}(R)$ on R and $\mathfrak{Teich}(R)$

The mapping class $\mathcal{MCG}(R)$ of the surface R is the group of *isotopy classes* of automorphisms that extend to any punctures (or quasiconformal maps) of R onto itself; a chosen element α of $\mathcal{MCG}(R)$ must be regarded as a representative of its isotopy[15] class $[\alpha]$. Yet in the context of homeomorphisms of surfaces, the terms "isotopy" and "homotopy" can be used interchangeably (see footnote below). For a torus T^2, each isotopy class contains a unique element of SL(2, $\mathbb{Z}$).

Note that an element of the mapping class group of R determines an automorphism of $\pi_1(R)$ only up to conjugacy, because the base point is typically not preserved.

---

[15] While two homeomorphisms f, f_1 are homotopic if there is a continuous family $\{g_t\}$, $0 \le t \le 1$, of *continuous* mappings with $g_0 = f$, $g_1 = f_1$, to be isotopic, the maps $\{g_t\}$ must be *homeomorphisms*. However in Epstein [1966] it is shown that for surfaces, these notions are equivalent: homotopic homeomorphisms are in fact isotopic.

Each α determines the automorphism $\alpha : (S, f) \mapsto (S, f \circ \alpha^{-1})$ of Teichmüller space $\mathfrak{Teich}(R)$. The totality of isotopy classes $\{[\alpha]\}$ of orientation-preserving automorphisms of R form the *mapping class group* (or *Teichmüller modular group*) $\mathcal{MCG}(R)$.[16]

The action of $\mathcal{MCG}(R)$ on a surface of type $(g, n) = (1, 1), (1, 2), (2, 0)$ is special in that the kernel of the action—the elements that fix the free homotopy class of every nontrivial simple closed curve—is the group of order 2 generated by the hyperbolic involution. One can think of a rod stuck through a roasting marshmallow so that a 180° rotation sends the surface to itself. The other exception is $(g, n) = (0, 4)$ when the kernel is the *klein 4-group* $\mathbb{Z}_2 \times \mathbb{Z}_2$ generated by hyperbolic involutions as well. When $(g, n) = (0, 3)$, $\mathfrak{Teich}(R)$ is a point. In all remaining cases (when $2g + n - 3 \geq 1$), the kernel is just the identity. See Farb and Margalit [2012].

A point (S, f) is fixed by $[\alpha] \in \mathcal{MCG}(R)$ if and only if $f\alpha f^{-1} : S \to S$ is homotopic to a conformal map, that is if S has a conformal symmetry in the homotopy class of $f\alpha f^{-1}$. If α has a fixed point it has finite order—it can be thought of as an "elliptic" element. It is a famous theorem of Kerckhoff 1983, resolving a longstanding conjecture called the Nielsen Realization Problem: Corresponding to every *finite* subgroup $F \subset \mathcal{MCG}(R)$, there exists a point $(S, f) \in \mathfrak{Teich}(R)$ which is fixed by F. There, F acts as a finite group of conformal automorphisms of S.

Besides the elements of finite order, $\mathcal{MCG}$ has elements analogous to the parabolics (Dehn twists) and loxodromics (pseudo-Anosov transformations) of kleinian groups. We will return to these matters in Exercise (5-12).

In the four cases mentioned above, $\mathcal{MCG}(R)$ is exceptional: When R is the sphere with ≤ 3 punctures, $\mathcal{MCG}(R)$ is finite and there are no pseudo-Anosovs). When R is a 4-punctured sphere or a torus with ≤ 1-puncture, $\mathcal{MCG}(R)$ is commensurate with the modular group $\mathrm{SL}(2, \mathbb{Z})$. Typically, to avoid special situations, these exceptional cases are automatically excluded.

It has been an old question of whether, not just for finite subgroups, but $\mathcal{MCG}(R)$ itself can be realized by a group of homeomorphisms. There would then be no need to bother with homotopy classes. While the answer is affirmative for tori, it is negative for closed surfaces of genus $g \geq 2$; [Markovic 2007] (for the cases $g \geq 5$), [Le Calvez 2012] (for the remaining cases). However the question is still unresolved for subgroups of finite index in the mapping class group.

For closed surfaces of genus ≥ 3 the mapping class group is generated by six elements, each of order two [Brendle and Farb 2004]. It is also generated by two elements: a Dehn twist Section 2.11.1 and a torsion element, or, two torsion elements [Korkmaz 2005].

The mapping class group is also known to be finitely presented.

---

[16] As defined here, $\mathcal{MCG}(R)$ consists of orientation-preserving homeomorphisms; it has index two in the *extended mapping class group* $\mathcal{MCG}^{\pm}(R)$ consisting in addition of orientation-reversing mappings. For a torus the homotopy classes $\mathcal{MCG}(T^2)$ are isomorphic to $\mathrm{SL}(2, \mathbb{Z})$ while $\mathcal{MCG}^{\pm}(T^2)$ is isomorphic to $\mathrm{GL}(2, \mathbb{Z})$.

The mapping class group acts discontinuously on $\mathfrak{Teich}(R)$. Namely given $\chi \in \mathcal{MCG}(R)$,

$$\chi : (S, f) \in \mathfrak{Teich}(R) \mapsto (S, f \circ \chi).$$

In other terms, χ sends the marking of S to the same Riemann surface but marked differently by the χ-image. The quotient orbifold **Mod**(R) is called *moduli space*. We will give two formal definitions:

(1) Mod$(R) = $ **Mod**$_{g,n}(R) := \mathfrak{Teich}(R)/\mathcal{MCG}(R).$

Namely two points of Teichmüller space are identified if they differ by the action of an element of $\mathcal{MCG}$.

Parallel to the metric definition of Teichmüller space, an equivalent definition is:

(2) Mod$(R) = $ **Mod**$_{g,n}(R) := \{$hyperbolic metrics on $S\}/\text{Diff}^+(R).$

Namely two hyperbolic metrics on R determine the same point of Mod(R) if one is the pullback of the other under a diffeomorphism $f : S \to S$ that extends to any punctures.

In other words, **Mod**(R) consists of Riemann surfaces without designated markings: Two surfaces are identified if they are conformally (thus isometrically) equivalent.

$\mathcal{MCG}(R)$ acts on **Mod**(R). A surface S with conformal symmetries is fixed by the corresponding elements of $\mathcal{MCG}$. These fixed points form subvarieties of codimension ≥ 1 of Mod(R), each fixed by certain elements of $\mathcal{MCG}(R)$. The subvarieties are the singular sets making moduli space an orbifold.[17]

On a surface with punctures, there are no finite order elements that pointwise fix the punctures [Farb and Margalit 2012, Cor. 7.3]. Often in the literature, one finds that the mapping class group is defined so as to have the property that the set of punctures is fixed under the group. With such a definition, when there are punctures, the corresponding moduli space is then a manifold. This version of the mapping class group has finite index in our $\mathcal{MCG}(R)$ and so is a finite-sheeted, manifold cover.

It is an interesting fact that in parallel to Selberg's Lemma for Möbius transformations, there is a torsion-free, normal subgroup $\mathcal{MCG}_0(R) \subset \mathcal{MCG}_{g,0}(R)$ of finite index, see Ivanov [1992].

An excellent introduction to mapping class groups is Farb and Margalit [2012].

It is a celebrated theorem of Royden [1971] that $\mathcal{MCG}(R)$ constitutes the full group of *orientation-preserving* isometries of $\mathfrak{Teich}(R)$, in the Teichmüller metric, provided R is not one of the exceptional surfaces listed in the footnote.

The extended modular group $\mathcal{MCG}^{\pm}(R)$, which also contains the orientation-reversing elements, contains $\mathcal{MCG}(R)$ as an index-2 subgroup. The corresponding quotient

---

[17] The mapping class group does not act faithfully if the kernel of the action consists of more than the identity. The exceptional cases $\{(g, b)\}$ are: $(2, 0)$, $(1, 1)$, $(0, 4)$, $(0, 3)$. The first three have an involution of order two fixing each point of the space.

$$\mathfrak{Teich}(R)/\mathcal{MCG}^{\pm}(R)$$

can be interpreted as the space of all isotopy classes of hyperbolic metrics on R, not just the orientation-preserving ones. The moduli space **Mod**(R) is a 2-sheeted orbifold cover [Farb and Margalit 2012, p. 344].

2.11.3 The complex structure of $\mathfrak{Teich}(R)$

$\mathfrak{Teich}(R)$ is a complex analytic manifold of dimension $(3g + b - 3)$, where g is the genus of R and $b \geq 0$ is the number of punctures. The complex structure is not at all apparent from the fuchsian deformations.[18] Compared to the real analytic structure, it took a long time to discover the complex structure; the first complete proof was by Ahlfors [1960]. Later Bers greatly simplified the proof, making use of the quasiconformal deformations into PSL(2, $\mathbb{C}$); from this point of view, the complex structure arises naturally.

With respect to the analytic structure, the mapping class group $\mathcal{MCG}(R)$ consitities the full group[19] of biholomorphic automorphisms of $\mathfrak{Teich}(R)$ [Earle and Kra 1974]. Thus **Mod**(R) $= \mathfrak{Teich}(R)/\mathcal{MCG}(R)$) is an analytic orbifold.

Corresponding to the normal, torsion-free, finite index subgroup $\mathcal{MCG}_0(R)$ of $\mathcal{MCG}(R)$ is the finite-sheeted covering $\mathfrak{Teich}(R)/\mathcal{MCG}_0(R)$ of **Mod**(R). This is an analytic manifold, in fact an open subset of a (compact) algebraic variety.

Global coordinate systems for $\mathfrak{Teich}(R)$ exist: A much used set of real analytic coordinates, called "Fenchel Nielsen" coordinates consists of (i) the lengths of $3g + b - 3$ simple, mutually disjoint geodesics, comprising a pants decomposition, and (ii) $3g + n - 3$ associated real numbers called "twist parameters" measured by transverse arcs about the geodesics, see Hubbard [2006, §7.8]. Direct complex analytic coordinates are described in Earle and Marden [$\geq$ 2015]. These project to give complex coordinates on the orbifold **Mod**(R).

We will present a "concrete" realization of $\mathfrak{Teich}(R)$ in Section 5.10 in which its complex structure is more apparent. Unlike $\mathfrak{Teich}(R)$, the moduli space **Mod**(R) is a Zariski open subset of a projective algebraic variety. There is an algebraic compactification, called the Deligne-Mumford compactification, which is a projective variety. For an equivalent natural analytic compactification, see Earle and Marden [2012] and further references there.

2.12 Exercises and explorations

2-1. *Elementary and reducible groups.*
(i) Prove that a (not necessarily discrete) group G is elementary if and only if any two elements of infinite order have a common fixed point.

Hint: Clearly the definition given in Section 2.2 implies this one. If all elements are elliptic, we must appeal to the fact, to be proven in Theorem 4.1.5, that either the

---

[18] Fuchsian representations of $\pi_1(R)$ depend on $6g + 2b - 6$ real numbers, including one relation costing 3 real numbers and normalization costing another 3.

[19] There are four exceptional cases of low genera where the kernel is nonempty, see Section 2.11.

group is cyclic, or it has a common fixed point in $\mathbb{H}^3$. All the parabolic elements of G must share the same fixed point. If A and B share the fixed point ζ and at least one of them is not parabolic, then their commutator $[A, B] = ABA^{-1}B^{-1}$ is either parabolic or the identity; it is the identity if and only if A and B have the same set of fixed points. Thus for $B \in G$ with two fixed points, there cannot exist two other transformations, A, C such that A, B share one fixed point of B and C, B share the other. In particular all parabolic and loxodromic elements of G must have a common fixed point ζ. Then any elliptic element must fix ζ as well, unless all the loxodromic elements have the same pair of fixed points and the elliptic element interchanges the two fixed points.

Conclude as in Lemma 2.3.1(v) that every nonelementary group contains two loxodromic elements without a common fixed point.

(ii) A Möbius group H is called *reducible* if there is a fixed point common to all elements of H. A reducible group is, in particular, elementary.

Suppose that H is not reducible. Show that there exist two elements without a common fixed point. As a consequence show that a nonabelian group is reducible if and only if the trace of every commutator is $+2$ (Lemma 1.5.1).

Hint: Assume that H is not reducible. Choose an element $h \neq$ id. If h is parabolic, there is an element which does not fix the fixed point of h. Instead suppose h is loxodromic or elliptic with fixed points $\zeta_1, \zeta_2 \in \mathbb{S}^2$. If there is an element with distinct fixed points we are finished. Otherwise there is an element h_1 which fixes ζ_1 but not ζ_2, and h_2 which fixes ζ_2 but not ζ_1. If h_1 and h_2 have distinct fixed points we are done. Otherwise h_1 and h_2 have a common fixed point $\zeta_3 \neq \zeta_1, \zeta_2$. But $h_2 \circ h_1$ fixes neither ζ_1 nor ζ_2.

2-2. Show that $w = e^{2\pi i z}$ is a conformal mapping of the quotient space

$$\mathbb{H}^2/\langle z \mapsto z + 1 \rangle$$

onto the punctured disk $0 < |w| < 1$. Find a corresponding mapping from

$$\mathbb{H}^2/\langle z \mapsto kz \rangle,$$

where $k > 1$, to some annulus $1 < |w| < R$. Then show that the hyperbolic metrics $\lambda(w)|dw|$ in the punctured disk and annulus are given by

$$\lambda(w) = \frac{1}{|w| \log \dfrac{1}{|w|}}, \qquad \lambda(w) = \left(\frac{\pi}{\log R}\right) \frac{1}{|w| \sin \dfrac{\pi \log |w|}{\log R}}.$$

In the annulus, the geodesic is the circle $\{|w| = \sqrt{R}\}$. It is fixed by the involution switching the boundary components. Its hyperbolic length is $\frac{2\pi^2}{\log R}$.

In contrast, verify the following formula for the mildly singular metric that results from pulling the hyperbolic metric down to $\mathbb{H}^2/\langle E \rangle$, where E is elliptic of order n:

$$\lambda(w) = \frac{2}{n|w|^{(n-1)/n}(1 - |w|^n/2)}.$$

Hint: Use the disk model and the map $w = z^n$.

The hyperbolic metric $\rho|dz|$ for the 3-punctured sphere was discovered by Agard [1968]; see also Gardiner and Lakic [2001], McMullen [2014, Thm4.13]. When the punctures are at $(0, 1, \infty)$ it is,

$$\frac{1}{\rho(\zeta)} = \frac{|\zeta(\zeta - 1)|}{2\pi} \int \int_{\mathbb{C}} \frac{dx\, dy}{|z(z - 1)(z - \zeta)|}.$$

2-3. Prove the result of C. L. Siegel repeated in Lehner [1964, Theorem III.J] that a nonelementary group that preserves the upper half-plane and which is not discrete contains an elliptic element of arbitrarily high order.

Hint: Suppose $A = \begin{pmatrix} \lambda & 0 \\ 0 & 1/\lambda \end{pmatrix}$ is an element of G and there is a sequence $\{B_n\}$ with $\lim B_n = \mathrm{id}$. Compute the trace of the commutators $C_n = AB_nA^{-1}B_n^{-1}$ and $D_n = AC_nA^{-1}C_n^{-1}$. Writing the normalized matrix $B_n = \begin{pmatrix} a_n & b_n \\ c_n & d_n \end{pmatrix}$, show first that $\lim b_nc_n = 0$ so that $\lim a_nd_n = 1$ and $a_nd_n > 0$ for large indices. Conclude that for infinitely many indices either $\mathrm{tr}^2(C_n) < 4$, so that C_n is elliptic, or $\mathrm{tr}^2(D_n) < 4$.

2-4. Suppose $\{T_n\}$ is a sequence of loxodromic or elliptic transformations such that $\{(\mathrm{tr}\, T_n)^2\}$ has limit 4. Show that there is a subsequence if conjugates $\{U_kT_kU_k^{-1}\}$ such that $\lim U_kT_kU_k^{-1}$ is a parabolic transformation. One example is the sequence $\{z \mapsto e^{2\pi i/n}z\}$. (For an application, see the video *Not Knot* [Gunn and Maxwell 1991].)

2-5. *The modular group.* For this exercise it may be helpful to refer, for example, to Ahlfors [1978]. The group M of normalized matrices with integer entries is called the *modular group* and is often denoted by $\mathrm{Mod} = \mathrm{SL}(2, \mathbb{Z})$.[20] This is an object of fundamental importance in number theory, in particular in the proof of Fermat's Last Theorem by Wiles. It is also involved in the theory of quadratic forms: If an integer N can be represented as $N = ax^2 + 2bxy + cy^2$, where a, b, c are given integers and x, y are integer variables, then replacing x, y by $mx + ny, px + qy$, where $\begin{pmatrix} m & n \\ p & q \end{pmatrix} \in \mathrm{Mod}$, gives a new representation of N.

The action of Mod on the upper half-plane UHP is given by the corresponding Möbius transformations $\mathrm{PSL}(2, \mathbb{Z})$.

Show that in its action on UHP, Mod is generated by $z \mapsto z + 1$ and $z \mapsto -1/z$. It can also be expressed as the free product $\mathbb{Z}_2 * \mathbb{Z}_3$, that is, it is generated by elliptics of order two and three with parabolic commutator. Confirm that the following is a fundamental polygon F for the action of Mod in the upper half-plane: $\{z : -\frac{1}{2} < \mathrm{Re}\, z \le \frac{1}{2}, |z| > 1\}$ with the boundary segment $\{|z| = 1, 0 \le \mathrm{Re}\, z \le \frac{1}{2}\}$.

Denote by M_2 the subgroup of the modular group M, called the *level 2 congruence subgroup* of M, consisting of normalized matrices which satisfy

$$\begin{pmatrix} a & b \\ c & d \end{pmatrix} \equiv \pm \begin{pmatrix} 1 & 0 \\ 0 & 1 \end{pmatrix} \mod 2.$$

---

[20] Often the modular group Mod is simply taken as $\mathrm{PSL}(2, \mathbb{Z})$.

Fig. 2.4. Tessellations of the upper half-plane and disk by the orbits of the standard fundamental polygon under the modular group.

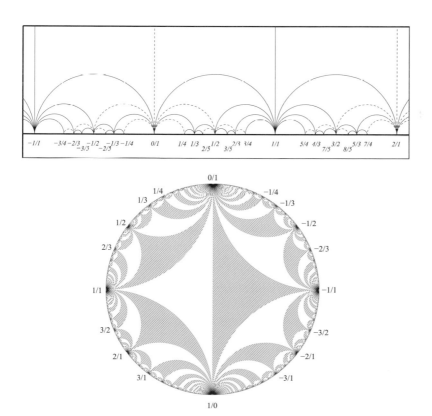

Fig. 2.5. Tessellation of the upper half-plane and disk by the orbit of an ideal quadrilateral, the union of two adjacent ideal triangles, under the 3-punctured sphere group. The ideal vertices are labeled by the Farey sequence (Exercise (2-10)), which are their coordinates on $\mathbb{R}$. If the endpoints of the outer edge of an ideal triangle are $p/q < r/s$, then $ps - qr = \pm 1$ and the coordinate of the third vertex is $(p + r)/(q + s)$.

Show that M_2 is generated by $z \mapsto z + 2$ and $z \mapsto z/(2z + 1)$. Furthermore M_2 has index 6 in M; show this by showing that the ideal quadrilateral

$$F_2 = \left\{ z : -1 < \operatorname{Re} z \le 1, \ |z + \tfrac{1}{2}| > \tfrac{1}{2}, \ |z - \tfrac{1}{2}| \ge \tfrac{1}{2} \right\}$$

is a fundamental polygon for M_2 and that it contains 6 copies of F. The quotient surface $\mathbb{H}^2/M_2$ is conformally equivalent to the triply punctured sphere.

Conversely suppose all three transformations $A, B, C = AB$ are parabolic with distinct fixed points. Prove that $\langle A, B \rangle$ is a discrete group preserving some round disk in $\mathbb{S}^2$. In fact, it is conjugate to M_2. In short, there is only one triply punctured sphere, up to Möbius equivalence.

More generally, a fuchsian group Γ is called a (hyperbolic) *triangle group* of *signature* (p, q, r), $2 \le p, q, r \le \infty$, if it is generated by elements A, B such that $A, B, C = BA$ are elliptic of orders p, q, r—or parabolic if the corresponding order is infinite. Such a fuchsian group exists if and only if $\frac{1}{p} + \frac{1}{q} + \frac{1}{r} < 1$ (use trace identities or refer to Exercise (3-1)). The triangle group Γ arises from a hyperbolic triangle with vertex angles $(\pi/p, \pi/q, \pi/r)$ by first taking the group $\langle a, b, c \rangle$ generated by the reflections in the sides and then passing to the index two, orientation-preserving, subgroup generated by $A = ab$, $B = bc$, $C = ca$. Its presentation is $\Gamma = \langle A, B, C : A^p = B^q = C^r = ABC = 1 \rangle$.

A triangle group is unique up to conjugacy, in view of that fact that a Möbius transformation that sends three distinct points in $\mathbb{S}^2$ another three is uniquely determined.

The modular group has signature $(2, 3, \infty)$ and M_2 has signature (∞, ∞, ∞). Up to Möbius equivalence, there is only one group for each admissible signature.

While we are dealing with the fundamental polygon for M_2 we will take the opportunity of pointing out the following phenomenon. If we move the fundamental polygon to the unit disk $\mathbb{D}$, it is bounded by a chain of four circular arcs orthogonal to $\partial \mathbb{D}$ and mutually tangent at their points of intersection. The group $\mathcal{M}_2$ is generated by pairing successive arcs, sending the exterior of one to the interior of its partner. The four points of tangency correspond to the three punctures on the quotient 3-punctured sphere. But we can equally pair the opposite arcs instead of the adjacent ones. Show that this results in a quotient which is a once-punctured torus!

In particular we have shown that the same fundamental polygon can serve for two entirely different groups.

The nonprojectivized group $\mathrm{SL}(2, \mathbb{Z})$ itself is generated by $A = \begin{pmatrix} 0 & -1 \\ 1 & 0 \end{pmatrix}$ and $B = \begin{pmatrix} 0 & -1 \\ 1 & 1 \end{pmatrix}$, and also by $\begin{pmatrix} 1 & 1 \\ 0 & 1 \end{pmatrix}$ and $V = \begin{pmatrix} 1 & 0 \\ 1 & 1 \end{pmatrix}$. It has the presentation $\langle A, B \mid A^2 = B^3, A^4 = \mathrm{id} \rangle$. See Magnus [1974, p. 108].

2-6. *A crash course on tori.* Take $\omega_1, \omega_2 \in \mathbb{C}$ with $\operatorname{Im}(\omega_2/\omega_1) > 0$. Consider the rank two parabolic group

$$G = \langle z \mapsto z + \omega_1, z \mapsto z + \omega_2 \rangle,$$

associated with the lattice in $\mathbb{C}$ of the points $\{m\omega_1 + n\omega_2\}$, $m, n \in \mathbb{Z}$. The parallelogram with vertices $(0, \omega_1, \omega_2, \omega_1 + \omega_2)$ is a fundamental parallelogram: its

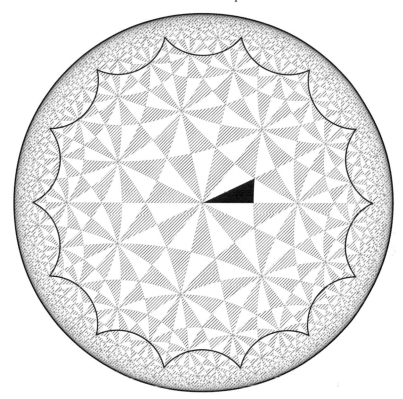

Fig. 2.6. Tessellation by the (2,3,7) group. Two adjacent triangles form a fundamental polygon. A single triangle is a fundamental polygon for the reflection group. A fundamental polygon for the subgroup associated with the Klein surface (p. 78) is also indicated.

G-orbit covers $\mathbb{C}$ without overlap. The quotient $T = \mathbb{C}/G$ is a torus. The euclidean metric in $\mathbb{C}$ projects to T and the sides $[0, \omega_1]$, $[0, \omega_2]$ project to a pair of simple loops which cross each other only at the projection of 0.

Set $\omega_2' = a\omega_2 + b\omega_1$, $\omega_1' = c\omega_2 + d\omega_1$, where a, b, c, d are integers satisfying $ad - bc = 1$ (so that ω_1, ω_2 likewise can be expressed as an integral combination of ω_1', ω_2'). Thus $\{\omega_1', \omega_2'\}$ is a new basis of the lattice. Every change of basis arises in this manner.

The only conformal mappings of $\mathbb{C}$ are the affine mappings $z \mapsto az + b$. An affine mapping takes one lattice (ω_1, ω_2) to another; it projects to a conformal mapping of one quotient torus to the other, sending one pair of loops to the other. Since we do not want to distinguish two lattices so related, we can normalize by focusing instead on ratios $\tau = \omega_2/\omega_1$, with Im $\tau > 0$, and lattices $\{m + n\tau\}$.

In this convention $\tau' = \omega_2'/\omega_1'$ determines the same lattice if and only if there exists a normalized Möbius transformation with integer entries a, b, c, d such that

$$\tau' = A(\tau) = \frac{a\tau + b}{c\tau + d}, \quad \tau = \frac{\omega_2}{\omega_1}.$$

Such a transformation A is called a *modular transformation*. It sends each torus to a conformally equivalent one: Instead of the parallelogram $(1, \tau)$ there is a new fundamental parallelogram $P' = (1, \tau')$ for the same lattice. The group of modular transformations is of course the modular group of Exercise (2-5).

We can subject τ to the following additional normalization: $\mathrm{Im}\, \tau > 0$, $|\tau| \geq 1$, $-\frac{1}{2} < \mathrm{Re}\, \tau \leq \frac{1}{2}$ but $\mathrm{Re}\, \tau > 0$ if $|\tau| = 1$ [Ahlfors 1978, §6.2.3]. This object so determined serves as a fundamental polygon P for the modular group: its orbit covers $\mathbb{C}$ without overlap. Each torus has a unique conformal representative $\tau \in P$. (Exercise (2-5)).

The space $\mathfrak{Teich}$ of all tori can thus be taken to be the upper half-plane $\{\tau : \mathrm{Im}\, \tau > 0\}$. Two points τ, τ' represent conformally equivalent tori if and only if they differ by a modular transformation. The modular transformations are of course isometries of $\mathfrak{Teich}$. The distance between two points $\tau, \tau' \in \mathfrak{Teich}$ is taken as the hyperbolic distance in UHP between the two points, $d(\tau, \tau')$.

A fundamental parallelogram P "rolls up" to give the quotient torus T. The specific choice of fundamental parallelogram gives a marking of the torus, in that the two pairs of edges project to an specific ordered pair of simple loops on the quotient torus. (The two loops are geodesics in the euclidean metric of the torus and cross each other exactly once.) Different choices for the fundamental parallelogram correspond to different choices for this pair of simple loops and different choices in the orbit of the initial τ under the modular group.

The change of marking arising from changing τ to τ' is induced by a conformal automorphism of the underlying torus if and only if $\tau' = \tau$ is a fixed point of $A \in \mathrm{PSL}(2, \mathbb{Z})$, which is then necessarily elliptic. This can happen only for special lattices—special values of τ, namely for $\tau = i$ (square), $\tau = e^{2\pi i/3}$, or $\tau = e^{2\pi i}/6$. The modular transformation that fixes the point $\tau = i$ is $\tau' = -1/\tau$.

There is a continuous group that maps every torus unto itself, without fixed points: there is a unique element that maps a given point to any other. This group is the projection to the torus of the group $\{z \mapsto z + c : c \in \mathbb{C}\}$ of translations of $\mathbb{C}$. By fixing say $z = 0$ as a lattice point, we prevent this group from acting on the quotients.

In addition the map $z \mapsto -z$ projects to every torus T. It becomes a conformal automorphism of order two with exactly four fixed points. The quotient $T/\langle z \mapsto -z\rangle$ is conformally equivalent to $\mathbb{C} \cup \infty$. and T is branched of order two over four distinct points. Apart from the group of translations above, this is the only affine map that induces a conformal automorphism of all tori. Since it has order two, it is called the *hyperelliptic involution*. All closed surfaces of genus 2 also have such an conformal involution which bynecessity has six fixed points (see Exercise (2-14)), but relatively few closed Riemann surfaces of each $g > 2$ support such an automorphism.

The euclidean line segment from 0 to $p + q\tau$, with (p, q) relatively prime integers, projects to a simple loop, and conversely every simple loop on T from the projection of 0 is determined in such a fashion. In other words, there is a one-to-one correspondence between rational numbers q/p and unoriented simple loops from a given point $0 \in T$. The fraction $0/1$ corresponds to α and $1/0$ corresponds to β.

Actually it is quite artificial to choose the base point 0. Given a slope p/q with p, q relatively prime, consider the family of all parallel euclidean lines with this slope. The projection of parallel, mutually disjoint simple loops (geodesics in the euclidean metric) that fills up T. Conversely, each simple loop determines such a family. In short, given any torus T and a basis α, β the set of free homotopy or homology classes of simple closed curves on T is in one-to-one correspondence with the rational numbers.

Let T be the square torus $\tau = i$ and $\alpha \in T$ be the simple geodesic loop coming from a line with slope p/q. If (p, q) are relatively prime, which of course we will always assume, there are relatively prime positive integers (r, s) for which $ps - qr = \pm 1$. Choose $\beta \in T$ to be the simple loop coming from a line with slope r/s. Show that the number of times that α crosses β on T is exactly

$$i(\alpha, \beta) = \left| \det\left(\begin{smallmatrix} p & r \\ q & s \end{smallmatrix}\right) \right|.$$

Here $i(\alpha, \beta)$ is called the *geometric intersection number* of the loops α, β. It is the least number of intersections that any pair of curves α' in the free homotopy class of α and β' in the free homotopy class of β can have. If $h : T \to T$ is a homeomorphism, $i(\alpha, \beta) = i(h(\alpha), h(\beta))$.

Let p, q be relatively prime integers, $0 < p < q$ and consider the line $L : y = (p/q)x$ in the (x, y)-plane. It projects to a simple loop α on the quotient torus T. On T, there is a shortest distance d from one side of α to the other side. Remembering that there is a pair of integers (p', q') with $pp' - qq' = \pm 1$, show that

$$d = \frac{1}{\sqrt{p^2 + q^2}}.$$

Hint: Minimize the distance of the lattice point $m + ni$ to the line L. Show that the line $L' : y = (p/q)x + 1/2q$ is as close as possible to L while projecting to the simple geodesic on T parallel to α and halfway between its two sides.

Suppose $\tau \to p/q \in \mathbb{R}$ along a vertical segment ℓ ending at p/q. It passes through a finite number of elements of the orbit of the fundamental polygon F. The torus becomes pinched along its (p, q) curve ending up as a twice-punctured sphere. Analyze instead the case that $\tau \to p/q$ along a horocycle.

What happens if τ approaches an irrational number ζ along a vertical segment ℓ? In this interesting case τ runs through a sequence of polygons $\{A_j(P)\}$ with $A_j \in \text{Mod}$ and $\lim A_j = \zeta$. If we fix a point $\tau \in P$, the sequence of tori, are all conformally all the same. Yet in the limit they collapse to the lines of slope ζ, which is the "ending lamination" for the sequence. We will formally study this idea in a more general context in Chapter 5.

Again work with the square torus T. Any (orientation-preserving) automorphism $\neq$ id of the torus T is homotopic to the projection of an affine map f described in terms of column vectors by

$$\begin{pmatrix} x \\ y \end{pmatrix} \mapsto \begin{pmatrix} a & b \\ c & d \end{pmatrix} \begin{pmatrix} x \\ y \end{pmatrix},$$

where a, b, c, d are integers satisfying $ad - bc = 1$. Let λ, λ^{-1} denote the eigenvalues of the coefficient matrix (which therefore is conjugate to $\begin{pmatrix} \lambda & 0 \\ 0 & \lambda^{-1} \end{pmatrix}$). If $\lambda > 1$ show that $\lim_{n \to \infty} i(f^n(\alpha), \beta)) = \infty$. Such a map on a torus, which is area preserving, is called an *Anosov mapping*. It preserves two lines through $z = 0$, ℓ_a and ℓ_r, both at angles which are irrational multiples of π. It stretches one by a factor of λ and compresses the other by a factor λ^{-1}. The projection of each line to T is a geodesic of infinite length, never intersecting itself and dense on T. In contrast, if $|\lambda| = 1, \lambda \neq \pm 1$, then f has finite order ($f^n = $ id for some n). If $\lambda = \pm 1$, f satisfies $f(\alpha) = \alpha$ for some simple geodesic loop α on T.

An affine map $A(z) = \alpha z + \beta$ maps the lattice $\Lambda = (1, \tau)$ onto itself if and only if $\alpha \in \Lambda$ and $\alpha \tau \in \Lambda$. Of course this is satisfied if $\alpha \in \mathbb{Z}$. For particular values of τ it can be that $\alpha \notin \mathbb{Z}$. For these cases show that τ and likewise α satisfy quadratic equations with integer coefficients.

The jacobian of $A(z)$ is $|\alpha|^2$. That means the fundamental parallelogram $P = (1, \tau)$ is sent to a parallelogram of $|\alpha|^2$-times the area of P; the image covers the torus T $|\alpha|^2$-times, which is necessarily an integer. Alternately P can be subdivided into $|\alpha|^2$ subparallelograms so that the image of each covers the torus once. The induced (analytic) mapping A_* of the torus onto itself has degree $|\alpha|^2$.

How many fixed points does A_* have? How many distinct solutions mod(Λ) does $(\alpha - 1)z = 0$ have? Well $|\alpha - 1|^2$ is the jacobian of the map $z \mapsto (\alpha - 1)z$ so it induces a map of the torus that covers itself $|\alpha - 1|^2$-times, and by necessity this too is an integer. So A_* has exactly $|\alpha - 1|^2$-fixed points. Likewise the n-th iterate $A^n(z) = \alpha^n z + \beta_n$ has $|\alpha^n - 1|^2$ fixed points. These become dense in T as $n \to \infty$, if $|\alpha| \neq 1$, that is, if A_* is a $n \geq 2$ to 1 analytic mapping of T onto itself.

For example, if $\alpha = i$ then we must have $\tau = i$. The map A_* has degree one and has two fixed points, namely the points on T corresponding to $z = 0$ and $z = (1 + i)/2$.

2-7. *Two tori.* Set $A(z) = z + 1$. Suppose $B = (az + b)/(cz + d)$, given in normalized form, has trace -2, but does not commute with A. If $[A, B]$ is also parabolic, find c and the possible fixed points of B.

Suppose A, S are commuting parabolics so that $\langle A, S \rangle$ is a torus group. Show that there exists a parabolic T which commutes with B such that the two tori $\langle A, S \rangle$, $\langle B, T \rangle$ are equivalent.

2-8. *Discrete extension of parabolic groups.* Suppose G is a noncyclic discrete group all of whose elements fix ∞. Suppose the subgroup G_0 of parabolic transformations is a rank two parabolic group. Then G is the extension of G_0 by at least one of the following: an element of order two (possible in all cases), an element of order four (possible only if a fundamental parallelogram P for G_0 is a square), or an element of order three or six (P is a rhombus with a $\pi/3$ vertex angle).

2-9. *The complex conjugate of a group.* Suppose G is discrete. Take all the matrices corresponding to G and replace them by their complex conjugates. Show that the resulting group G' is also discrete and $\Omega(G') = J(\Omega(G))$, $\Lambda(G') = J\Lambda(G)$, where $J(z) = \bar{z}$. In fact $G' = JGJ$.

If two elements A, B in a discrete group without elliptics satisfy $A^p = BA^q B^{-1}$, $p, q \neq 0$, then B preserves the fixed point set of A. Furthermore $p = q$ and either A, B lie in a one- or two-generator parabolic subgroup or both are powers of a loxodromic element C.

2-10. *Punctured tori and the Farey sequence.* A once-punctured torus has a hyperbolic metric but a torus has only a euclidean metric. Yet topologically and analytically there is a close relationship as a once-punctured torus corresponds to a choice of basepoint on the torus.[21]

A group G representing a once-punctured torus in $\mathbb{H}^2$ is given by two loxodromic generators X, Y without common fixed point and with parabolic commutator $K = XYX^{-1}Y^{-1}$ that we may assume fixes ∞. Prove that XY cannot fix ∞. *Hint:* If XY fixes ∞ show that $YX = X^{-1}(XY)X$ also fixes ∞ and must be parabolic. Then show that X and Y must fix ∞.

A once-punctured torus is called *square* if it has two simple geodesics α, β which have the same length and cross each other orthogonally. Show that these curves are the *systoles* for the surface—the geodesics that have the minimum length among all geodesics on the surface. Terik Aougab showed that this statement is false if the interesection is not orthogonal.

Automatically $\alpha\beta\alpha^{-1}\beta^{-1}$ is freely homotopic to a simple loop that is retractable to the puncture.

Show that the matrices

$$\Lambda = \begin{pmatrix} -1 + \sqrt{2} & 0 \\ 0 & 1 + \sqrt{2} \end{pmatrix}, \quad B = \begin{pmatrix} \sqrt{2} & 1 + \sqrt{2} \\ -1 + \sqrt{2} & \sqrt{2} \end{pmatrix}$$

determine a square torus. Find a fundamental polygon in UHP.

Find the matrix generators for the once-punctured torus that corresponds to a regular euclidean hexagon with puncture at the center. What are its symmetries?

The *Farey sequence* $\mathcal{F}$ is very useful in studying once-punctured tori. It is based on the modular diagram. The Farey sequence is the orbit of the boundary ∂F_2 of the fundamental polygon for M_2 presented in Exercise (2-5). What is interesting is its labeling.

Note that the orbit of the ideal vertex $\{\infty\}$ under Mod is the set of all rational numbers $\mathbb{Q}$ (plus ∞). Prove:

(i) The rational numbers m/n and p/q are ideal vertices of the tile $g(F_2)$, $g \in M_2$ if and only if $mq - np = \pm 1$.

---

[21] The period ratio τ determining a torus, and parameters determining the fuchsian uniformization of the punctured torus are related by a differential equation involving a mysterious "accessary parameter" [Keen et al. 1979].

(ii) The ideal vertex x/y of a tile $g(F_2)$ separates the other two ideal vertices $m/n <$
p/q of $g(F_2)$ if and only if

$$\frac{x}{y} = \frac{m+p}{n+q}.$$

Each irrational number ζ is the limit of a nested sequence of geodesics in the orbit of
∂F_2. For example the sequence

$$\ldots, [p/q, m/n], [p/q, (p+m)/(q+n)], [(2p+m)/(2q+n)], \ldots$$

The sequence can be described in terms of a left-right pattern in the edges of the tessellation $\{M_2(F_2)\}$. There is a wonderful description of the Farey sequence in Mumford
et al. [2002].

What we want to point out here is that the geodesic with endpoints m/n, p/q represents a pair of simple closed geodesics α, β on the punctured torus with the property
that α crosses β exactly once. Here α has slope m/n and β has slope p/q—see
Exercise (2-6).

2-11. *Volume and area computations.* G is a nonelementary kleinian group.

(i) Consider first the case G contains the subgroup $P_2 = \langle z \mapsto z+1, z \mapsto z+\tau \rangle$,
where $\mathrm{Im}\,\tau > 0$. The horoball $\mathcal{H} = \{(z, t) \in \mathbb{H}^3 : t > a > 0\}$ is invariant
under P_2. Assume that its projection $C(\tau)$ to $\mathcal{M}(G)$ is embedded; this is a solid
cusp torus. Using the hyperbolic volume form $dV = dx\,dy\,dt/t^3$ show that for
$\theta = \arg \tau$,

$$\mathrm{Vol}(C(\tau)) = \frac{|\tau| \sin \theta}{2a^2}, \quad \mathrm{Area}(\partial C(\tau)) = \frac{|\tau| \sin \theta}{a^2}, \quad \frac{\mathrm{Vol}(C(\tau))}{\mathrm{Area}(\partial C(\tau))} = \frac{1}{2}.$$

For the universal horoball, $a = 1$. By Exercise (2-5) a fundamental parallelogram
for P_2 can be chosen so that $|\tau| \geq 1$ and $\frac{\pi}{3} \leq \theta \leq \frac{2\pi}{3}$. Thus the volume of the
solid cusp torus is not less than $\frac{\sqrt{3}}{4}$.

(ii) Next assume that the equidistant tube $T(r)$ of hyperbolic radius r about a closed
geodesic of length L is embedded in $\mathcal{M}(G)$. Show by working in the upper half-
space model that

$$\mathrm{Vol}\,T(r) = \pi L \sinh^2 r, \quad \mathrm{Area}\,\partial T(r) = 2\pi L \sinh r \cosh r,$$

$$\frac{\mathrm{Vol}\,T(r)}{\mathrm{Area}\,\partial T(r)} = \tfrac{1}{2} \tanh r \nearrow \tfrac{1}{2} \quad \text{as } r \nearrow \infty,$$

$$\mathrm{Vol}\,T(r) = \tfrac{1}{2} \mathrm{Area}\,\partial T(r) + \tfrac{1}{2} \pi L (e^{-2r} - 1).$$

(iii) In the case that $\mathcal{M}(G)$ has finite volume, borrowing terminology from Section 3.4, deduce that

$$\mathrm{Vol}(\mathcal{M}(G)^{\mathrm{thick}}) \leq \mathrm{Vol}(\mathcal{M}(G)) \leq \mathrm{Vol}(\mathcal{M}(G)^{\mathrm{thick}}) + \tfrac{1}{2}\mathrm{Area}(\partial \mathcal{M}(G)^{\mathrm{thick}}). \quad (2.9)$$

In fact $\mathcal{M}(G)^{\mathrm{thick}}$ is a compact submanifold whose complement consists of a
finite number of cusp tori and tubes about short geodesics.

(iv) Suppose $\mathcal{M}(G)$ is a closed manifold (compact, without boundary). Show that there is a shortest closed geodesic γ. If γ has length L, show that it has an embedded tubular neighborhood of radius $L/4$. *Hint:* Expand the tubular neighborhood until at radius r it first touches itself at a point p. The two orthogonals of length r from p to γ, with a segment of γ of length $\leq L/2$, form a closed loop. Its length must be $\geq L$.

2-12. A *gaussian integer* is a number of the form $p + iq$, where p, q are integers. The group Γ of matrices in $\mathrm{SL}(2, \mathbb{C})$ whose entries are gaussian integers is called the *Picard group* $\mathrm{SL}(2, \mathbb{Z}(i)$.[22] Show that it is a discrete group. In its action on $\mathbb{C} \cup \{\infty\}$, it is generated by the four parabolic transformations,

$$S(z) = z + 1, \quad T(z) = \frac{z}{-z+1}, \quad U(z) = z + i, \quad V(z) = \frac{z}{iz+1}.$$

However $T = ASA$, and $V = AUA$, where $A(z) = -1/z$. Furthermore, $A = TST$ and $B = UAU^{-1}AUA$, where $B(z) = -z$. In fact, the Picard group is generated by three elements $\langle S, U, A \rangle$. For details, including an explicit fundamental polyhedron see Wielenberg [1978], Maskit [1988, p. 82].

The former reference discusses a variety of subgroups of finite index without elliptic elements of Γ. These interesting subgroups give rise to quotient spaces which are homeomorphic to a variety of knot and link complements, including the Borromean rings.

2-13. *Equality in Jørgensen's inequality.* There are continuous families of geometrically finite groups (Section 3.6) as well as uncountably many nonconjugate, nonelementary, geometrically infinite two-generator discrete groups that give equality in Jørgensen's inequality [Jørgensen et al. 1992]. In these cases the generator A must be elliptic or parabolic [Jørgensen 1976]. The examples are typically extensions of the modular or other triangle groups. The Picard group is one such extreme group.

However in the class of fuchsian groups, only the triangle groups G (Exercise (2-5)) with signature $(2, 3, q)$ with $7 \leq q \leq \infty$ give equality [Jørgensen and Kiikka 1975]. Confirm that this is the case for the first few examples.

Prove that if $\langle A, B \rangle$ gives equality, then $\langle A, B_1 = BAB^{-1} \rangle$ is also a nonelementary, discrete group which gives equality in Jørgensen's inequality. *Hint:* You will need the identity,

$$\mathrm{tr}(AB_1A^{-1}B_1^{-1}) = [\mathrm{tr}(ABA^{-1}B^{-1}) - 2][\mathrm{tr}(ABA^{-1}B^{-1}) - \mathrm{tr}^2(A) + 2],$$

and its consequence,

$$\left| \mathrm{tr}(AB_1A^{-1}B_1^{-1}) - 2 \right| \leq \left| \mathrm{tr}(ABA^{-1}B^{-1}) - 2 \right|. \tag{2.10}$$

See also Lemma 1.5.6.

Using the property proved in the preceding paragraph, prove that if $\langle A, B \rangle$ gives equality in Equation (2.10), then A is either elliptic of order at least 7, or is parabolic.

---

[22] Often the Picard group is simply taken as $\mathrm{PSL}(2, \mathbb{Z})$.

Prove that if two Möbius transformations A, B with equal traces generate a nonelementary discrete group, then [Jørgensen 1981]

$$\left|\text{tr}(ABA^{-1}B^{-1}) - 2\right| > \tfrac{1}{8}.$$

2-14. *Genus-two surfaces.* In the disk model find a regular hyperbolic octagon with vertex angles $\pi/4$. *Hint:* Start with a tiny octagon centered at $z = 0$. It is nearly a regular euclidean octagon with vertex angles $3\pi/4$. Now increase the distance of the vertices from the origin; the vertex angles strictly decrease to zero, as they become ideal vertices.

Next, in the positive direction label sides of the octagon $\mathcal{P}$ as $a_1, b_1, a_1^{-1}, b_1^{-1}$, $a_2, b_2, a_2^{-1}, b_2^{-1}$, as in Figure 2.1 on p. 79. Find an isometry A_i that maps a_i onto a_i^{-1} but sends the positive direction along a_i to the negative direction along a_i^{-1}, that is $A_i(\mathcal{P})$ is adjacent to $\mathcal{P}$ along the exterior side of a_i^{-1}. Similarly find B_i. By starting with a vertex p of side a_1, show that the commutator product (starting from the right) satisfies $B_2^{-1}A_2^{-1}B_2A_2B_1^{-1}A_1^{-1}B_1A_1 = \text{id}$. This is called a *vertex relation*; it says that in the orbit of $\mathcal{P}$, successive images of $\mathcal{P}$ are arranged in the indicated cyclic order about a vertex. Show that the orbit of $\mathcal{P}$ under the group $\langle A_1, A_2, B_1, B_2 \rangle$ covers $\mathbb{H}^2$ without overlap. Show that the quotient surface R is a genus 2 surface, the eight vertices project to a single point O, each side projects to a simple loop from O, the loops are mutually disjoint except at O, and they bound a simply connected region Δ on R. The vertex relation is a consequence of the fact that if you make a complete circuit of $\partial \Delta$, the resulting loop is contractible to a point.

Does the rotation by $\pi/4$ of the octagon induce a conformal mapping of R onto itself? How about rotation by π? How about reflection about the geodesic between two opposite vertices. What fixed points on R do the induced mappings have?

Now label the edges in the sequence $a_1, b_1, a_2, b_2, a_1^{-1}, b_1^{-1}, a_2^{-1}, b_2^{-1}$ and repeat the process, pairing the opposite sides of $\mathcal{P}$ as before. Find the vertex relation. The quotient gives another surface of genus 2 but for which the simple loops are arranged in a different pattern. Consider the rotation of $\mathcal{P}$ by π. Confirm that on the quotient, it maps each simple loop to its inverse, and has exactly six fixed points. This involution is called the *hyperelliptic involution J*; every closed surface of genus two has one. The quotient $R/\langle J \rangle$ is a sphere with six branch values of order two, that is, R is a two-sheeted cover of $\mathbb{S}^2$, branched over six points.

Conversely, given six distinct points in $\mathbb{S}^2$, the two-sheeted cover branched over the six points is a closed surface of genus two. The covering surface is determined by a normal subgroup of index two in the fundamental group of the 6-punctured sphere; can you find the subgroup?

Every closed genus-2 surface has a hyperelliptic involution. *Hint*: Using fuchsian groups, express (following Jørgensen) the basic relation

$$ABA^{-1}B^{-1}CDC^{-1}D^{-1} = 1$$

as $ABA^{-1}B^{-1} = DCD^{-1}C^{-1}$. Set $U = A^{-1}B^{-1}C$ and $V = DC^{-1}BA$. Conclude that $ABUV = VUBA$. Think then of a double bagel with A, B simple loops around the holes and U, V through the holes. Skewer the double bagel, puncturing it at 6 points, and then rotate by 180°. The involution is given by $A \to A^{-1}$, $B \to B^{-1}$, $C \to C^{-1}$, $D \to D^{-1}$. See also the example of Section 7.2.

2-15. *No tangents at loxodromic fixed points.* Show that a limit set $\Lambda(G)$ cannot have a tangent line at a fixed point of a loxodromic $g \in G$ with $\mathrm{tr}(G) \notin \mathbb{R}$ unless $\Lambda(G)$ is Möbius equivalent to a circle. Therefore there are no smooth limit sets except circles and euclidean lines.

Outline of proof. [Lehto 1987, Lemma 4.2]. Assume $\Lambda = \Lambda(G)$ is not Möbius equivalent to a circle. Suppose to the contrary that $\mathbb{R}$ is the tangent line to Λ at $z = 0$ and there is a loxodromic $g \in G$ of the form $z \mapsto ke^{i\varphi}z$, $0 < k < 1$, $0 \le \varphi < 2\pi$.

If $\varphi = 0$, find $z \in \Lambda$, $\mathrm{Im}\, z \ne 0$. Then for all n, $\arg g^n(z) = \arg z \ne 0, \pi$. There is a contradiction as $n \to \infty$. If instead $\varphi = \pi$, upon working with g^2 we likewise get a contradiction.

More generally set $\phi = \min\{\varphi, |\pi - \varphi|, 2\pi - \varphi\}$ so that $0 < \phi \le \pi/2$. Construct the symmetric wedges of angle ϕ centered along $\mathbb{R}$: $V = \{re^{i\theta} : \theta \in (-\phi/2, \phi/2)\}$, $V' = \{re^{i\theta} : \theta \in (\pi - \phi/2, \pi + \phi/2)\}$. Thus if z lies in $V \cup V'$, then $g(z) \notin V \cup V'$. Choose a sufficiently small disk D about 0 so that $\Lambda \cap D \subset (V \cup V') \cap D$. This is possible since $\mathbb{R}$ is a tangent line. Then choose $z \ne 0$ in $\Lambda \cap D$ sufficiently small so that $g(z) \in D$. But then $g(z) \in \Lambda \cap D$ yet $g(z) \notin V \cup V'$, a contradiction.

2-16. Prove a Möbius transformation of the form $A = \begin{pmatrix} a & b \\ b & d \end{pmatrix}$, $ad - b^2 = 1$, satisfies $JAJ = A^{-1}$, where $J(z) = -z$. Conclude that if a discrete group G is generated by elements of this form, then $\Lambda(G)$ is invariant under J.

2-17. *Modeling conformal groups by Möbius groups.* Suppose G is a group of Möbius transformations preserving $\Omega \subset \mathbb{C}$ and $\phi : G \to H$ is an isomorphism onto a group of conformal automorphisms H that map another region Ω' onto itself. Suppose $\Phi : \Omega \to \Omega'$ is a quasiconformal mapping such that $\Phi g \Phi^{-1} = \phi(g)$ for all $g \in G$. Prove that there is a conformal mapping $F : \Omega' \to \Omega^*$ such that FhF^{-1} is a Möbius transformation for all $h \in H$.

Once F is constructed, it induces an isomorphism from the conformal group H to a Möbius group H^*; the action of H on Ω' is conformally equivalent the action of H^* on Ω^* [Maskit 1968].

Hint: Confirm that the Beltrami differential $\mu = (\partial\Phi/\partial\bar{z}) \big/ (\partial\Phi/\partial z)$ satisfies

$$\mu(gz)\frac{\overline{g'(z)}}{g'(z)} = \mu(z) \text{for all } g \in G.$$

Extend μ to $\mathbb{S}^2$ by setting it equal to zero in the complement of Ω. Then solve the corresponding Beltrami equation on $\mathbb{S}^2$; the solution $\Psi : \Omega \to \Psi(\Omega) = \Omega^*$ is uniquely determined if we require it to fix three prescribed points. Show that $\psi : g \mapsto \Psi g \Psi^{-1}$ is an isomorphism of G to a group of Möbius transformations H^* preserving Ω^*. Show that $F = \Psi \circ \Phi^{-1} : \Omega' \to \Omega^*$ is a conformal mapping inducing the isomorphism $\psi \circ \phi^{-1} : H \to H^*$.

2-18. *Geometric group theory.* An abstract finitely generated group G can be investigated *geometrically* by analyzing its action on its *Cayley graph* $\mathcal{G}$. Good introductions to the theory are Bowditch [2006]; Ohshika [2002]. The theory has become fundamentaly important in the study of hyperbolic 3-manifolds.

The graph is constructed as follows. Select for G a generating set $C = \{g_1, g_2, \ldots, g_r\}$. We will assume that if $g_k \in C$ then also $g_k^{-1} \in C$; the identity is not put in C. The *vertices* of the graph $\mathcal{G}$ are the distinct elements of G. The (oriented) *edges* $\{e\}$ eminating from the vertex $g \in G$ terminate at the vertices $g \cdot c$, $c \in C$; write $e = (g, c; g \cdot c)$. The edge $e^{-1} = (g, c^{-1}; g \cdot c^{-1})$ goes in the opposite direction to e.

There is a special case when the element $c \in C$ has order two: then the edge e can be regarded as unoriented.

If O is the vertex that corresponds to id $\in G$, then any word in the designated generators is uniquely represented by a path of oriented edges starting from O. The word is the identity if and only if the corresponding path is a closed loop.

For example the word $g_1 g_2 g_1^{-1}$ reading from the left corresponds to the path composed of the successive edges from O:

$$e_1 = (\text{id}, g_1; g_1), \quad e_2 = (g_1, g_2; g_1 g_2), \quad e_3 = (g_1 g_2, g_1^{-1}; g_1 g_2 g_1^{-1}).$$

Two graphs are called isomorphic if there is a one-one mapping of the vertices and edges of one onto the vertices and edges of the other which preserves orientations.

To embed $\mathcal{G}$ in a particular space, for example in $\mathbb{R}^2$, $\mathbb{H}^2$ or $\mathbb{H}^3$, we have to represent its vertices by distinct points and its edges by smooth arcs or geodesic arcs which are mutually disjoint except for common endpoints.

The Cayley graph $\mathcal{G}$ is connected (why?). If there are no (nontrivial) closed loops the graph is called a *tree*. A closed loop is trivial if it is a succession of edges followed by the succession of the edges with the opposite orientations.

Show that for a free group G on one generator with generating set $\{g, g^{-1}\}$, $\mathcal{G}$ is isomorphic to the graph on $\mathbb{R}$ whose vertices are integers and whose edges are directed segments between them. Show that graph of the free group on two generators with a generating set C of four elements is a tree and draw an embedding in $\mathbb{R}^2$. Use the model of a n-generator Schottky group to embed the Cayley graph of an n-generator free group in $\mathbb{C}$.

A Cayley graph can be made into a metric space by mapping each edge with distinct endpoints onto the unit interval thereby assigning it unit length and proportionally giving smaller lengths to each segment of the edge. If both endpoints of the edge are the same, map it onto the circle of unit length. There is at least one geodesic between any two vertices; its length is the number of edges in a shortest chain in the graph that connects them. This metric determines a topology on the graph.

The Cayley graph for one choice of group generators of G is quasiisometric (see Exercise (3-19)) to the graph based on a different set of generators. It is also quasiisometric to the graph of any finite index subgroup.

The *ends* of the graph are defined as follows. For any compact subset K, count the number of unbounded components of $\mathcal{G} \setminus K$. The number of ends of $\mathcal{G}$ is defined to be the supremum of the number of such components over all K. Show that the number of ends does not depend on the generating set. It is known that a Cayley graph has either zero, one, two, or an infinity of ends.

The Cayley graph allows "visualization" of the group G, generalizing the following classical representations: To construct the Cayley graph of a fuchsian group or kleinian group we can use the tiling by a fundamental polygon or polyhedron centered at a point O, which is not an elliptic fixed point (Section 3.5). Use the generating set determined by the face pairing transformations. Show that the Cayley graph is represented by drawing geodesic segments from O the successive points in the orbit of O under words in the face pairing transformations. The graph then appears as "dual" to the tiling by the orbit of the fundamental region and combinatorially reflects that tiling—see Section 3.4. As the graph gets closer to the boundary, more and more it looks like hyperbolic space itself, especially if you are very farsighted and cannot see the edges clearly. Analogy with this concrete situation often inspires the intuition for finding "geometry" in abstract Cayley graphs and looking for a "sphere at infinity". The Cayley graph has turned out to be a powerful tool to study particular classes of abstract groups.

The group G acts on its Cayley graph. Each $f \in G$ sends a vertex v to fv and an edge $e = (g, c; g \cdot c)$ to $fe = (fg, c; fg \cdot c)$. The group action is an isometry in the path metric. In the classical cases at least, we can find a connected finite subgraph that serves as a fundamental set for the action.

Even the family of finitely presented groups is too general to deal with; for example, the question of deciding whether a given element of the group is the identity is known to be undecidable; such groups do not seem amenable to a geometric approach. It was Gromov's work, followed by Thurston's, that brought modern combinatorial group theory back to its historic, geometrical roots. In his famous 1987 paper, Gromov presented a condition on the groups that would make possible an effective geometric theory. His definition models a certain property of isometry groups of hyperbolic space.

An abstract, infinite, finitely generated group $\mathcal{G}$ is said to be *Gromov hyperbolic, δ-hyperbolic, word-hyperbolic, negatively curved,* or simply *hyperbolic* if it has the following property: There exists a constant $\delta > 0$ such that for *any* geodesic triangle[23] Δ in the Cayley graph $\mathcal{G}$, a point on one side of Δ lies within distance δ of the union of the other two sides, however long the sides are. This property is called *δ-thinness* or the *Rips thin triangle property*. We know it holds for hyperbolic triangles by Theorem 1.3.1. The thinness property is independent of the chosen generating set of G. (What does "zero thinness" mean?) The condition of thinness is a global condition that suggests that the Cayley graph in the large "looks" like hyperbolic space. The theory was first outlined in Gromov [1987]. It now occupies a large place in combinatorial

---

[23] A geodesic triangle consists of three vertices and three geodesic segments connecting each vertex pair.

group theory (see expositions Cannon et al. [1997], Cannon [2002], which have many explicit examples, and also Cannon [1991] or Ohshika [2002]; Bowditch [2006]). The study of hyperbolic groups involves a kind of discrete hyperbolic geometry in the large.

A word-hyperbolic group has a "space at infinity", a "boundary" $\partial\mathcal{G}$, which for kleinian groups is $\mathbb{S}^2$. The boundary serves a fundamental role in analyzing the group. The group $\mathcal{G}$ acts on $\partial\mathcal{G}$. In studying this action, it is assumed that the action is *effective*: that if $g \neq$ id then the action of g on $\partial\mathcal{G}$ is also $\neq$ id. Within the theory the most pressing question is how to determine whether a hyperbolic group is isomorphic to a kleinian group, more specifically,

The Cannon Conjecture: *Assume that $\mathcal{G}$ is a Gromov hyperbolic group that acts effectively and is orientation preserving on its boundary $\partial\mathcal{G}$. Suppose $\partial\mathcal{G}$ is homeomorphic to $\mathbb{S}^2$. Then, if there is torsion in $\mathcal{G}$, it is isomorphic to a closed kleinian orbifold. Otherwise, $\mathcal{G}$ is isomorphic to a kleinian group corresponding to a closed manifold.*

It is known [Markovic 2013] that the positive solution of the Cannon Conjecture would give a purely algebraic/topological proof of the Hyperbolization Theorem for closed manifolds (6.3.2)(2). This paper also gives a criterion for its solution akin to the following property of a kleinian group representing a closed manifold: Any two distinct points on $\mathbb{S}^2$ are separated by the limit set of a quasifuchsian subgroup.

An equivalent formulation (by Sullivan) is that if the Gromov boundary of a hyperbolic group is homeomorphic to $\mathbb{S}^2$, then the boundary with its visual metric is quasiisometric to $\mathbb{S}^2$.

Hyperbolic groups include finite groups, finitely generated closed hyperbolic surface groups and fuchsian triangle groups, finitely generated free groups, and more generally, fundamental groups of compact n-dimensional riemannian manifolds of negative sectional curvature. See also Exercise (5-20).

The (Gromov) boundary $\partial\mathcal{G}$ of the Cayley graph of a hyperbolic group $\mathcal{G}$ is defined as follows. Given a vertex O, consider the set of geodesic rays from O parameterized by distance $t \in [0, \infty)$. Two rays $\alpha(t)$, $\beta(t)$ from O represent the same point on $\partial\mathcal{G}$ if the supremum $\sup\{d(\alpha(t), \beta(t))\} < \infty$. The two rays are said to be "close" if $\alpha(t)$ is close to $\beta(t)$ for t in a long interval $[0, T]$. The boundary is independent of the choice of basepoint O. It turns out that the boundary is finite dimensional, compact, and metrizable.

The Cannon Conjecture requires that $\partial\mathcal{G}$ be a topological sphere. The Conjecture is known to be true if and only if the Hausdorff dimension of the boundary is two (not greater than two). Current work [Cannon 2011] reduces the requirement of Hausdorff dimension (see Exercise (3-20)) to arbitrarily close to two.

According to Gromov, randomly chosen groups are hyperbolic. Measure the complexity of a group given by n generators and a finite number of relations in the generators $\{r_k\}$ by $N = n + \sum_k \text{length}(r_k)$. The relations are chosen as random words in the generators. Let A_N denote the number of groups with complexity $\leq N$ and let H_N be the number of *hyperbolic* groups with complexity $\leq N$. It is a theorem [Ol'shanskiĭ 1992] that $\lim_{N \to \infty} H_N / A_N = 1$!

Yet not all groups are hyperbolic, for example, two-generator abelian groups. For these, the Cayley graph is a square lattice in $\mathbb{C}$. In turn, hyperbolic groups are a special class of *automatic groups* as in finite state automata. This is the class of groups that can be effectively analyzed by computer, see Epstein et al. [1992]; Ohshika [2002]. The theory originated with a paper of Cannon as distilled by Thurston, and was extensively developed in The Geometry Center (1988–1994). Automatic groups are finitely presented, and have a solvable word problem. An extension to the theory admits the fundamental groups of finite volume hyperbolic manifolds which are not closed.

A finitely presented group is hyperbolic if and only if its Cayley graph satisfies a *linear isoparametric inequality*, while if it is automatic it satisfies a *quadratic isoparametric inequality*.

Actually the notion of "hyperbolic" is not restricted to graphs. Any metric space with the property that there is a geodesic between any two points can be considered from the point of view of hyperbolicity—see Ohshika [2002]; Bowditch [2006].

Hyperbolic groups have a certain "negative curvature" while abelian groups have more of a zero curvature (think tori!). Interesting groups may have a negative-like structure yet may also include some special abelian subgroups. An example is the mapping class group. Another example is Teichmüller space where the flatness is associated with boundary cusps. Such groups are not Gromov hyperbolic. To remedy this situation in many cases, Farb [1998] introduced the concept of *relatively hyperbolic groups*.

Here is his definition in the simplest situation. Suppose G is a finitely generated group and $\mathcal{G}$ is its Cayley graph. Let $H \subset G$ be a finitely generated (for example a rank two) subgroup. Form a new graph $\widehat{\mathcal{G}}$ as follows. For each $g \in G$ identify all the vertices of $\mathcal{G}$ that correspond to elements lying in the left coset gH. For each coset gH, one new vertex is then added to $\mathcal{G}$ together with a connection of length 1/2 to each element of gH. (If H is a rank two parabolic with fixed point $\zeta \in \partial \mathbb{H}^3$, this process is akin to grouping together as one point, the G-orbit of ζ.) The process is called *electrification*. The corresponding graph $\widehat{\mathcal{G}}$ is said to be *relatively hyperbolic with respect to* H if $\widehat{\mathcal{G}}$ is Gromov hyperbolic. If so, then from properties of $\widehat{\mathcal{G}}$, information about $\mathcal{G}$ can be deduced.

One can extend this definition to a finite number of finitely generated subgroups H_i.

With this new definition, the mapping class group and Teichmüller space itself become relatively hyperbolic [Masur and Minsky 1999]. Also finite area fuchsian groups and hyperbolic knots are relative hyperbolic groups. One might think there is an analog to the Cannon Conjecture that covers this case as well.

An interesting reference that applies these ideas to random walks on the mapping class group is Maher [2011].

The following three exercises draw attention to deeper properties that abstract groups, especially kleinian groups and mapping class groups, may or may not possess. Such properties can be useful in constructing covering manifolds. See Section 6.4.

2-19. *Residual finiteness.* Suppose G is a finitely generated group. The group G is called *residually finite* if it satisfies any of the following equivalent conditions [Farb and Margalit 2012]:

  (i) In intersection of all finite index subgroups and finite index normal subgroups $\{H\}$ of G satisfies $\bigcap_{H \subset G} H = 1$.
 (ii) Corresponding to each $g \neq \mathrm{id} \in G$, there exists a finite index normal subgroup $H_g \subset G$ with $g \notin H_g$; the homomorphism $\phi : G \to G/H$ satisfies $\phi(g) \neq 1$.
(iii) The *profinite* completion

$$\widehat{G} = \lim_{\leftarrow} G/H, \quad H \in \Sigma,$$

    exists and G injects into $\widehat{G}$ as a dense subgroup; see Exercise (2-20).

Many finitely generated groups are known to be residually finite. Known examples include finitely generated matrix groups, in particular fuchsian and kleinian groups, and also mapping class groups $\mathcal{MCG}(R)$ of a compact surface, see Magnus [1969], Farb and Margalit [2012, p. 179]. Subgroups of residually finite groups are residually finite Hempel [1976].

Even more strongly, suppose S is a surface of possibly infinite topological type and $G \subset \pi_1(S)$ is a finitely generated proper subgroup. Choose $g \in \pi_1(S) \setminus G$. There exists a finite-sheeted covering surface S^* of S such that $G \subset \pi_1(S^*)$ with injective inclusion $G \hookrightarrow \pi_1(S^*)$, but that $g \notin \pi_1(S^*)$ [Scott 1978].

For 3-manifolds, Hempel [1987] proved that the fundamental group of a (compact) Haken manifold is residually finite. Because finitely generated matrix groups are residually finite so is the fundamental group of every geometric 3-manifold proclaimed in the Geometrization Conjecture / Theorem Section 6.4 [Thurston 1982, Theorem 3.3].

There is a hierarchy of terms associated with **separability**:

  (i) A subgroup $H \subset G$ is **separable** if $H = \cap H_i$ where $\{H_i\}$ are finite index subgroups of G.
 (ii) If $\{\mathrm{id}\}$ itself is separable then G is called **residually finite**.
(iii) If every finitely generated subgroup is separable then G is "subgroup separable" or **LERF**.

2-20. *Inverse limits and profinite completion of a group.* Given a finitely generated group G, consider the set $\Sigma = \{H\}$ of all finitely generated normal subgroups of finite index in G. The set Σ is partially ordered by inclusion.

To understand item (iii) in Exercise (2-19) we have to know (1) what the inverse limit is, and (2) what the profinite completion of G is.

(1) *Inverse limits of finite quotients of G.* This involves the totality of normal subgroups $\{H\}$ of finite index in G, and their finite quotient groups $\{F = G/H\}$. If $H = \text{id}$, then $F = G$, and if $H = G$, then $F = \text{id}$. These examples are the extremes in a partial ordering by inclusions: if $H_1 \supseteq H_2$, then $F_1 \subseteq F_2$. There is an associated homomorphism $h_{1,2} : F_1 \to F_2$ by inclusion of elements.

Let I denote the appropriate index set so that the set of all quotient groups can be written $\{F_i : i \in I\}$. We write $i < j$ if $F_i \subseteq F_j$ ($G/H_i \subseteq G/H_j$, $H_i \supseteq H_j$) and $h_{i,j} : F_i \to F_j$ is the associated homomorphism.

The projections

$$\pi_i : G \to G/H_i = F_i, \quad g \in G \mapsto \pi_i(g) \in G/H_i.$$

are homomorphisms. Moreover, if $i < j$, $\pi_j(g) = h_{i,j} \circ \pi_i(g)$.

The homomorphisms $\{h_{i,j}\}$ have the properties

(i) $h_{i,i} = \text{id}$ on F_i,
(ii) $h_{i,k} = h_{i,j} \circ h_{j,k}, \quad i \leq j \leq k$.

(2) The *profinite completion* $\widehat{G}$ is the *inverse limit* of the finite subgroups $\{F_i\}$. It is a subgroup of the product $P(F) = \prod_{i \in I} F_i$ as follows.

The components of a "vector" $\vec{\mathbf{a}} = (\ldots, a_i, \ldots) \in P(F)$ are denoted by $\{a_i : a_i \in F_i\}$.

The *inverse limit* is defined in terms of the product $P(F)$:

$$\widehat{G} := \varprojlim_{i \in I} F_i = \{\vec{\mathbf{a}} \in \prod_i F_i : a_i = h_{i,j}(a_j), \forall j, i \leq j \in I\}.$$

The homomorphism $h^* : G \to \widehat{G}$ is defined by

$$g \in G \mapsto (\ldots, \pi_i(g), \ldots) \subset \widehat{G}.$$

The homomorphism h^* is injective, each $g \in G$ is sent to a distinct point of $\widehat{G}$: $h^*(g_1) \neq h^*(g_2)$ when $g_1 \neq g_2$. For G is residually finite so that $\bigcap_j F_j = \text{id}$.

The *profinite completion* $\widehat{G}$ is the closure in $P(F)$ of the image of G. Inside $\widehat{G}$, G is a totally disconnected, dense subset. In some sense, we are approximating G by finite groups; G is the "ideal boundary" of the graph of partial orderings of finite quotients.

2-21. *Groups which are LERF.* The acronym LERF is short for "locally extended residually finite". Compared to the notion of residually finite in Exercise (2-19), LERF is more difficult for a group to satisfy. A finitely generated group G is LERF if given *any* finitely generated subgroup H and an element $g \notin H$, there is a normal subgroup H_g of finite index with the following property: Under the homomorphism $\phi : G \to G/H_g$, $\phi(g) \notin \phi(H)$. In other words, finitely generated subgroups H can be "residually separated" from G.

LERF groups G are automatically residually finite. Using the criterion given above, given $g \neq \text{id} \in G$ there is a finitely generated normal subgroup H_g such that $\phi : G \to G/H_g$ satisfies $\phi(g) \notin \phi(H)$, that is, $g \notin H_g$. For otherwise, $\phi(g) \in \phi(H_g) = \text{id}$.

The following groups are known to be LERF:

 (i) Free groups (M. Hall Jr.),
 (ii) Closed surface groups (Peter Scott),
(iii) Finitely generated kleinian groups ([Agol 2012, Cor. 9.4]).
(iv) The group of automorphisms of a handlebody boundary that extend to the handlebody [Leininger and McReynolds 2007], also see Exercise (5-25).

The proof of item (iii) uses in addition to the Virtual Fibering Theorem, p. 387 two facts:

 (i) Any finitely generated kleinian group is isomorphic to a geometrically finite kleinian group by the Tameness Theorem 5.6.6, p. 293.
(ii) Any geometrically finite group is arbitrarily close to an isomorphic subgroup of a finite volume group, Theorem 4.11.8, p. 266.

On the other hand, the mapping class group of a closed surface $g \geq 1$ is not LERF [Leininger and McReynolds 2007].

2.12.1 *Summary of group properties*

Let G be a finitely generated group. It may or may not have one of the following properties:

Separable. A subgroup $H < G$ is separable if it can be expressed as an intersection of finite index subgroups $\{H_i\}$ of G: $H = \bigcap H_i$.

Residually finite. Given any $g \in G$, $g \neq 1$, there exists a finite index normal subgroup $H_g \subset G$ with $g \notin H_g$. Equivalently, $\{id\}$ itself is separable.

LERF. (Also called "subgroup separable".) Given *any* finitely generated subgroup H of G and any $g \notin H$ there is a normal subgroup of finite index $H_g \subset G$, such that $g \notin H_g$. Equivalently, *every* finitely generated subgroup of G is separable.

2-22. *More on Schottky groups.* Prove that a freely generated, purely loxodromic fuchsian group G (which is necessarily a group of the second kind) acting in the upper and lower half-planes, normalized so that ∞ is a limit point, is a classical Schottky group, where the Schottky circles are orthogonal to $\mathbb{R}$. *Hint:* Start with the simplest case that UHP$/G$ is a torus with one boundary component so that G has two-generators. There are four mutually exterior circles orthogonal to $\mathbb{R}$. The opposite, not adjacent, circles are paired. (If instead the adjacent circles are paired the quotient is a 2-holed disk.) When does the converse hold?

What happens when the 4-Schottky circles form a chain of mutually tangent circles with respect to $\mathbb{S}^2$? (Answer: When the pairing is again opposite there results a once-punctured torus and the group becomes fuchsian of the first kind; the fractal dust of a Schottky groups congeals to $\mathbb{R}$.)

Really, in talking about Schottky groups, to a large degree it makes little matter, in describing the construction, if there are tangencies of circles, so long as they are arranged in pairs such that the pairing elements are loxodromic or parabolic sending the exterior of one circle onto the interior of its partner. When a point of tangency is fixed by a parabolic it becomes a puncture, however if the point is not so fixed it does not necessarily become a puncture—see Gilman and Waterman [2006]. In particular, any finitely generated fuchsian group such that the quotient is a finitely punctured (≥ 1) closed surface is such a limiting case of a circle-Schottky group.

Of course the conditions can be weakened further so that instead of Schottky circles there are Jordan curves. Such will be the case we take general quasiconformal deformations of these fuchsian groups. There are explicit examples given in Mumford et al. [2002].

Using Ahlfors' Finiteness Theorem (Section 3.1) and Maskit's Planarity Theorem, prove that any finitely generated, free, purely loxodromic kleinian group G with $\Omega(G) \neq \varnothing$ is a Schottky group [Maskit 1967]. A much shorter proof makes use of the convex core to be introduced in Section 3.10: A Schottky group G is characterized by the fact that the convex core of $\mathcal{M}(G)$ is a handlebody (in the case G is fuchsian, we have to take an ε-neighborhood of the convex core). This is the case for a geometrically finite group that is free and purely loxodromic.

Bringing in the notions of ends and tameness from Section 5.3 for $\mathcal{M}(G)$, together with the Covering Theorem (5.6.2), we can state the following which is particularly interesting when $\Omega(G) = \varnothing$ [Canary 1996, Corollary D]:

Theorem 2.12.1. *Assume G is a finitely generated kleinian group such that $\mathbb{H}^3/G$ has infinite volume. Suppose $H \subset G$ is a finitely generated subgroup of infinite index which is purely loxodromic and free. Then H is a Schottky group.*

Given a set of Schottky circles, consider the group generated by reflections in them. How is this group related to the Schottky group?

Given a set of Schottky circles (mutually disjoint) let U denote their common exterior in $\mathbb{S}^2$. Show that any conformal automorphism of U is the restriction of a Möbius transformation. (*Hint:* Consider the group of reflections in the circles and correspondingly reflect each automorphism g to get a conformal map on the complement of the limit set Λ; the map also extends to be an automorphism of Λ. You will need to use the fact that this set has area zero so that the functions are analytic on this set as well.) Go on to prove that this group of automorphisms is finite—if there are more than two circles. (*Hint:* Erect the hyperbolic plane in $\mathbb{H}^3$ on each circle and consider the set of hyperbolic distances between every two of them.) What is the effect of the group of automorphism on the quotient surface $\Omega(G)/G$?

2-23. If g, h generate a discrete group without elliptics, prove that at least one of the four transformations is loxodromic: g, h, gh, gh^{-1}. See the proof of Theorem 4.1.1 for related results.

2-24. *Rational billiards.* Here is a Riemann surface construction that has been used extensively to study the dynamics of "rational" billiards on a euclidean polygon $P \subset \mathbb{C}$, that is not necessarily convex or even simply connected.

Corresponding to each side e_i of P, place a parallel line e_i', through the origin. Denote the reflection in e_i' by σ_i. By definition, a *rational billiard table* is one with the property that the group Γ generated by the reflections $\{\sigma_i\}$ is finite. If ∂P is connected, this condition is satisfied if and only if each interior vertex angle is a rational fraction of 2π. If ∂P is not connected, this requirement is only a necessary condition for a rational table.

Under the assumption that P is a rational table, here is how to glue copies of P together to get a closed Riemann surface.

Let N be the number of distinct elements $\{\gamma_k\}$ of Γ, including the identity. Take N copies of P each with the labeled edges; denote the copies by $\{P_{\gamma_k}\}$, $1 \le k \le N$.

Suppose $\gamma_j = \gamma_i \sigma_m$ for some index m. Then identify the edge e_m of P_{γ_j} with the edge e_m of the reflection $\sigma_m(P_{\gamma_i})$: attach the reflected polygon $\sigma_m(P_{\gamma_i})$ to the polygon P_{γ_j} along the common edge e_m.

Show that with this rule for attachment, an abstract polygon S can be built up from the N tiles. The end result will have no free edges. The vertex angles will be integer multiples of 2π.

Another description of gluing is as follows. Consider the normal subgroup Γ_0 of even index $2M$ in Γ consisting of even numbers of reflections in the lines e_i'. Interpreted as the product of reflections in the edges of P, Γ_0 consists orientation-preserving euclidean motions $z \mapsto e^{i\varphi}z + c$. The cosets of $\Gamma_0 \subset \Gamma$ are $\{\Gamma_0\sigma_j\}$, $0 \le j \le m$. Now take the polygon P, and the reflected copies of P, $\sigma_1(P), \sigma_2(P), \ldots$, and glue the edges together using the elements of Γ_0.

The complex structure on S is given by the euclidean coordinates on the polygons, but the vertices must be flattened out by use of $z^{\frac{1}{p}}$ at a vertex with angle sum $2\pi p$.

When P is a rectangle, the group has order 4 and the Riemann surface is a torus; when P is an equilateral triangle, the group has order 10 and the Riemann surface is also a torus. For the theory, see Masur and Tabachnikov [2002]. The point is that on the surface, a ball starting at a point of P, instead of bouncing off the edges of P runs in a straight line on S, except a billiard path that hits a vertex must end since there is no unique continuation.

2-25. Starting with the finite group of a euclidean polyhedron, can you adjoin other such finite polyhedral groups to obtain a nonelementary kleinian group with singular set forming a specified trivalent graph with the properties specified by Proposition 2.5.2?

2-26. *Homology and simple loops.* Confirm the following folk theorem: Suppose S is a closed, oriented surface of genus $g \ge 1$. Fix a "canonical homology basis" A_i, B_i, $1 \le i \le g$. This means $\{A_i, B_i\}$ are simple loops generating the first homology, and A_i crosses B_i once but is disjoint from A_j, B_j for $j \ne i$; each pair (A_i, B_i) corresponds to a "handle". An element γ of the first integral homology group can be written, $\gamma \sim \sum(a_i A_i + b_i B_i)$, where each a_i, b_i is an integer.

Prove that the homology class of γ contains a simple closed curve if and only if the greatest common denominator of $\{a_1 \ldots a_g, b_1 \ldots b_g\}$ is one. For a proof see Schafer [1976].

2-27. *Belyĭ functions on Riemann surfaces.* A Belyĭ function on the closed Riemann surface R is a meromorphic (rational) function $f : R \rightarrow \mathbb{S}^2$ such that each of its critical values is at one of the points $0, 1, \infty$. Here a critical point is a point x where the derivative vanishes, $f'(x) = 0$; the corresponding critical value is $f(x)$. Not every closed surface supports such a function; not every Riemann surface can be realized as the branched cover of $\mathbb{S}^2$ with all branch values in $\{0, 1, \infty\}$ (for the topological possibilities see p. 67).

Each of the following conditions is necessary and sufficient for R to support a Belyĭ function:

(i) There exists a finite set of points $\{x_i\} \subset R$ such that the Riemann surface $R' = R \setminus \{x_i\}$ is uniformized by a finite index subgroup Γ of the modular group Mod $=$ PSL$(2, \mathbb{Z})$: $R' = \mathbb{H}^2 / \Gamma$.
(ii) R' carries a horodisk packing such that the complement is a union of triangular regions.
(iii) **Belyĭ's Theorem.** R is the Riemann surface determined by an irreducible polynomial equation $P(x, y) = 0$ whose coefficients are algebraic numbers.

A horodisk on R' is the projection of a horodisk at a parabolic fixed point of Γ.

For example, the Riemann surface given by $x^m + y^n = 1$ has the property that the projection $f : (x, y) \mapsto x$ has critical values in $\{1, \infty\}$, and so is a Belyĭ function.

The Riemann surfaces carrying Belyĭ functions are dense in all Riemann surfaces. Markovic asks: In the moduli space of a closed Riemann surface R, could it be that, corresponding to any two Belyĭ surfaces, there a Riemann surface which is an unbranched cover of each?

We will only prove item (i) which has the simplest proof: Let $\pi : \mathbb{H}^2 \rightarrow \mathbb{H}^2/M_2$ be the projection to the thrice-punctured sphere $S_3 = \mathbb{S}^2 \setminus \{0, 1, \infty\}$ as in Exercise (2-5). If a Belyĭ function f exists on R then $R \setminus f^{-1}\{0, 1, \infty\}$ is a covering surface of S_3 and therefore corresponds to a finite index subgroup of Mod.

Conversely suppose $R = \mathbb{H}^2 / \Gamma$. Then R is a covering surface of $S = \mathbb{H}^2/\text{Mod}$, which is the sphere punctured at ∞ with two branch values. Let $f : R \rightarrow S$ be the projection. Let R' denote the result of removing from R the inverse images of branch values on S. Let S' denote the sphere punctured at the two branch values and at ∞. Then $f : R' \rightarrow S'$ is an unbranched covering. The points on R that we removed are the critical points of f.

For thorough studies of this subject and its relation to oriented trivalent graphs and Grothendieck's "dessins d'enfants", see the beautiful papers Jones and Singerman [1978; 1996], and also Brooks [1999].

See also Exercise (8-12).

2-28. *Fuchsian subgroups of finite index.* Suppose G is a fuchsian group representing a closed surface $R = \mathbb{H}^2/G$. Show that there are subgroups of finite index k (not necessarily normal subgroups) for any $k \geq 2$. In other words, show that there are k-sheeted, unbranched covering surfaces of R. If H has index k in G, the subgroup $H^* \supset H$ generated by $\{ghg^{-1} : g \in G, \ h \in H\}$ is a normal subgroup of G, of index at most k.

The topological possibilities are described by the Riemann–Hurwitz relation

$$g^* - 1 = k(g - 1),$$

where g^* is the genus of the k-sheeted cover. (*Hint:* If you can topologically find finite-sheeted cover S' of R, you can lift the hyperbolic metric from R to S' to get a conformal cover of R. To find a topological cover, cut R along a simple geodesic, and join two copies of the cut surface by cross identifying along the cuts.) In fact there are only a finite number of index-n subgroups (why?).

Let G_n^* be the *intersection* of all subgroups $\{G_k\}$ of G of index $k \leq n$. Show that G_n^* also has finite index; $\mathbb{H}^2/G_n^*$ is at most a finite-sheeted cover of R and of any other k-sheeted cover of R, where $k \leq n$. (*Hint:* The intersection $G_k \cap G_k'$ has index at most n^2.) Show that $G_{n+1} \subset G_n$.

Define a metric $\rho(\cdot, \cdot)$ on G as follows. Given two elements $A, B \in G$ set

$$\rho(A, B) = \min \left\{ \frac{1}{n} : AB^{-1} \text{ lies in a subgroup of index } n \right\}.$$

Thus $\rho(A, B) \leq 1$ and $\rho(A_n, \mathrm{id}) \to 0$ if and only if $A_n \in G_n$ with $n \to \infty$. The completion of G with respect to the metric ρ (called the *profinite completion*) is a compact topological group $\widehat{G}$ homeomorphic to a Cantor set. One then works with the space $\mathbb{D} \times \widehat{G}$. The action of G on this space is $T(z, t) = (Tz, tT^{-1})$, where $T \in G$, $t \in \widehat{G}$. For an exposition and further development of this subject, which leads to an infinite-dimensional Teichmüller-like space called the *universal hyperbolic solenoid*, see Markovic and Sarić [2006].

2-29. *The groups of regular polyhedra: spherical orbifolds.* Here we will follow the treatment of Ford [1929]. Let $\mathcal{P}$ be a regular euclidean polyhedron inscribed in the unit sphere $\mathbb{S}^2$. Denote the number of its faces, edges and vertices by F, E, V respectively. By Euler's formula, $F - E + V = 2$. Let v denote the number of faces at each vertex. When ∂P is projected on $\mathbb{S}^2$ there results a tessellation of $\mathbb{S}^2$ by F regular spherical polygons of vertex angles $2\pi/v$. Let μ denote the number of edges bounding each face.

We are interested in the group G of symmetries of $\mathcal{P}$. This is a group of rotations of $\mathbb{S}^2$. There are $2E$ of them: for given an edge $[a_0, b_0]$ and another $[a, b]$, there is a symmetry that sends $[a_0, b_0]$ to $[a, b]$ and to $[b, a]$ in either order.

Here is how to construct a fundamental domain for the action of G on $\mathbb{S}^2$. It will be a spherical triangle (with one exceptional case to be included below).

Choose an edge, to be called the outer edge, and an adjacent face. Join the ends of the edge to the midpoint of the face by two lines we will call inner edges. We have then a euclidean triangle with central angle $2\pi/\mu$.

Project the triangle to $\mathbb{S}^2$, for example by stereographic projection from the plane. We have an spherical triangle σ of central angle $2\pi/\mu$ and angle π/ν at the other two vertices.

Let S be the elliptic of order μ that fixes the inner vertex of σ. Locate the midpoint of the outer edge and let T be the elliptic of order two that fixes it. The axes of S and T pass through the center of the ball; S and T rotate $\mathcal{P}$ onto itself.

Prove that $G = \langle S, T \rangle$ with $(T \circ S)^\nu = \mathrm{id}$, and that σ is a fundamental region for its action on $\mathbb{S}^2$.

All the groups in this class are two-generator groups. The rays from the origin of the ball to the fixed point of S, T, TS on the boundary of σ are pointwise fixed by these three elliptics, and their G-orbit gives the complete set of rotation axes for G.

The possibilities are listed in the following table from Ford [1929]:

	F	V	E	μ	ν	Order(G)
Tetrahedron	4	4	6	3	3	12
Cube	6	8	12	4	3	24
Octahedron	8	6	12	3	4	24
Dodecahedron	12	20	30	5	3	60
Icosahedron	20	12	30	3	5	60
Dihedron	2	n	n	n	2	$2n$

The dihedron is special in that it has zero volume and two faces which are regular $n \geq 2$-sided polygons inscribed in the equatorial plane.

2-30. *Euclidean orbifolds.* Show that any rank two parabolic group G_0 is a subgroup of a group generated by four elliptics of order two (whose fundamental domain is half a fundamental parallelogram of G_0). This is the $(2, 2, 2, 2)$-group.

Show that the rank two group of the square torus can be generated by two elliptics of order four and one of order two (its fundamental domain is $1/4$ of the fundamental square of the rank two parabolic subgroup). This is the $(2, 4, 4)$-group.

Consider the rank two group G_0 whose fundamental parallelogram P is spanned by the vectors 1 and $e^{\pi i/3}$. Show that G_0 is a subgroup of (i) the group generated by two elliptics of order three, fixed points at the two centers of the equilateral triangles T_1, T_2 formed by the diagonal $[1, e^{\pi i/3}]$, and (ii) the group generated by an elliptic of order two with fixed point the midpoint of $[1, e^{\pi i/3}]$, and an elliptic of order three with fixed point the center of T_1. These are the $(3, 3, 3)$-group and $(2, 3, 6)$-group. A fundamental domain of (i) is the equal-sided 60° parallelogram with vertices 1, $e^{\pi i/3}$ and the centers of T_1, T_2, and of (ii) is the 30° isosceles triangle formed by the diagonal and the center of T_1.

2-31. *Variation of length under a Dehn twist.* This is to display a wonderful formula discovered by Scott Wolpert.

Suppose on a Riemann surface $R = \mathbb{H}^2/\Gamma$, α is an oriented simple geodesic and β is a geodesic that crosses α at least once.

Theorem 2.12.2 ([Wolpert 1981]). *Denote the length of β by $\ell(\beta)$. Apply to β an infinitesimal Dehn twist τ_t about α. Then*

$$\frac{d\,\tau_t(\ell(\beta))}{d\,t}\Big|_{t=0} = \sum_{p\in\alpha\cap\beta} \cos\theta_p.$$

Here θ_p is the angle at $p \in \alpha \cap \beta$ between the tangent of α and the arc of β lying on the right side of α.

What happens if we replace α by β and twist about β? Orient β so the angle between β and the right-hand segment of α at p is now $(\pi - \theta_p)$ at each intersection. Then

$$\frac{d\,\tau_t(\ell(\beta))}{d\,t}\Big|_{t=0} = -\sum_{p\in\alpha\cap\beta} \cos\theta_p,$$

displaying the interesting antisymmetry between effect of a Dehn twist on each geodesic.

Wolpert's proof proceeds by analyzing the following quasiconformal deformation that fixes α, and its Beltrami differential. We may assume that a lift of α is the vertical imaginary axis in the universal cover UHP. There, set $\theta = \arg z$ and choose a smooth function $\phi(\theta) \geq 0$ with compact support in $(0, \pi)$, and $\int_0^\pi \phi\,d\theta = \frac{1}{2}$. The function

$$w := z e^{2t\Phi(\theta)}, \quad \Phi(\theta) := \int_0^\theta \phi\,d\theta,$$

is a quasiconformal map of UHP onto itself, maps the lift of α onto itself, and is the identity when $t = 0$. Although it does not project to R it determines a twist about α in R.

For a generalization of the formula, see Series [2001].

2-32. *Images of horizontal lines under a Möbius transformation.* This exercise repeats an observation of Clifford Earle. Consider a normalized transformation $w = T(z) = (az+b)/(cz+d)$ with $c \neq 0$. The collection of all images under T of horizontal lines can be described as follows.

Parameterize the family of all horizontal lines by $\rho \neq 0 \in \mathbb{R}$:

$$L_\rho := \{z \in \mathbb{C} : \operatorname{Im}(z + \frac{d}{c}) = \rho\} \cup \{\infty\}.$$

The images under T form the family of circles

$$C_\rho = \left\{ w \in \mathbb{C} \ : \ \left| w - \left(\frac{a}{c} + \frac{i}{2c^2\rho} \right) \right| = \frac{1}{2|c|^2|\rho|} \right\}.$$

These circles are tangent to each other at the point $w = \frac{a}{c} = T(\infty)$. If the group $\langle T(z), z \mapsto z+1 \rangle$ is discrete, the radius r of C_ρ satisfies $r \le \frac{1}{2|\rho|}$ (see Theorem 3.3.4).

3

Properties of hyperbolic manifolds

In this chapter we gather together basic properties of hyperbolic 3-manifolds. We start with a characterization of their (conformal) boundaries. Then the universality of key elements of their internal geometry and their thick/thin decomposition are described. After that, we study the global structure as revealed by their fundamental polyhedra, and by their convex and compact cores. We introduce the class of manifolds which are essentially compact (geometrically finite); this class is at the core of our studies.

As we progress through Chapter 3, we introduce the class of quasifuchsian groups. Returning to geometry, we present crash courses in 3-manifold surgery and the theory of geodesic laminations on surfaces. The chapter ends with a description of the rigidity of hyperbolic 3-manifolds of finite volume.

3.1 The Ahlfors Finiteness Theorem

The beginning of the modern theory of hyperbolic manifolds can be pinpointed at the appearance in 1964 of a transformational result:

Ahlfors Finiteness Theorem ([Ahlfors 1964]). *If G is a finitely generated kleinian group, $\partial \mathcal{M}(G) = \Omega(G)/G$ is the union of a finite number of surfaces. Each of them is a closed surface with at most a finite number of punctures and elliptic cone points.*

Punctures[1] arise from rank one parabolic fixed points, and cone (branch) points from elliptic fixed points. I have taken the liberty of including cone points in the statement although the proof of this was established later as described in Exercise (3-15).

The hyperbolic area formula (Exercise (3-1)) implies that a Riemann surface R of genus $g \geq 0$ with $m \geq 0$ cone points of (finite) orders $r_1, \ldots, r_m \geq 2$ and $n \geq 0$ punctures appears as a boundary component of $\partial \mathcal{M}(G)$, for finitely generated, nonelementary G, only if

$$2g + n - 2 + \sum_i \left(1 - \frac{1}{r_i}\right) > 0.$$

---

[1] Ahlfors had initially forgotten to include a proof that there are also at most a finite number of triply punctured spheres. This omission was quickly rectified by Bers and by Greenberg.

In fact the inequality is a necessary and sufficient condition for R to be represented as $R = \mathbb{H}^2/H$ for some fuchsian group H.

Good estimates can be found for the genus and number of punctures of the boundary of kleinian manifolds $\mathcal{M}(G)$ in terms of the number N of generators, and the number of rank one and rank two parabolic conjugacy classes. In particular $\sum g_i \leq N$ where g_i is the genus of the i-th component of $\partial\mathcal{M}(G)$. If G is purely loxodromic, then $\partial\mathcal{M}(G)$ has at most $N/2$ components. This calculation is made using the homology considerations of Remark 3.7.2.

The deepest part of Ahlfors' theorem is the assertion that the ideal boundary of a component of $\partial\mathcal{M}(G)$ consists only of punctures—that, in particular, there are no simply connected components. The proof is based on the following idea. The group G, being finitely generated, depends on the finite number of complex parameters in its generating matrices. On the other hand, if a boundary surface R were not of the "finite analytic type" indicated above, then that surface would have an infinite-dimensional space of distinct deformations. This results in a contradiction. Recent proofs (see Kapovich [2001, 4.9] or Marden [2006], for example) are much easier than Ahlfors' original; in particular, the finiteness assertions on total genus, number of punctures and cone points can be deduced by topological methods. The Ahlfors theorem also follows from the solution of the tameness conjecture (Section 5.4); see Exercise (5-31).

Conjugacy classes of parabolic and elliptic subgroups are not necessarily represented by punctures and cone points in $\partial\mathcal{M}(G)$. Yet these classes are finite too (Exercise (3-15)), rounding out Ahlfors' theorem.

3.2 Tubes and horoballs

Consider the axis γ^* of a primitive loxodromic element $g \in G$ (g is a generator of the cyclic loxodromic group fixing γ^*). Suppose first that γ^* is not also the rotation axis of an elliptic element, and that G contains no elliptic element (of order two) that interchanges its endpoints. Then γ^* projects to a closed geodesic γ in $\mathcal{M}^{\text{int}}$; the full collection of lifts of γ is the orbit $\{G(\gamma^*)\}$. The length of γ is the length of any segment $[x, gx]$ of γ^*. The loop γ is a simple loop if and only if the orbit of γ^* consists of mutually disjoint geodesics. Conversely, every closed geodesic in $\mathcal{M}(G)$ is the projection of a loxodromic axis γ^*.

Suppose γ^* is taken as the vertical axis from $0 \in \mathbb{C}$ in the upper half-space model. Given r consider the *tubular neighborhood* of radius r about γ^*,

$$N_r(\gamma^*) = \{\vec{x} \in \mathbb{H}^3 : d(\vec{x}, \gamma^*) < r\}.$$

This appears as a euclidean cone with central angle 2θ given by the equation $r = \log(\sec\theta + \tan\theta)$. Alternate expressions of the equation are

$$\tanh r = \sin\theta, \quad \cosh r = \sec\theta, \quad \sinh r = \tan\theta. \tag{3.1}$$

The image of N_r under a Möbius transformation A such that $A(0)$, $A(\infty) \neq \infty$ looks like a banana.

If it is embedded, the projection $N_r(\gamma) = \pi(N_r(\gamma^*)) \subset \mathcal{M}$ is called the *tubular neighborhood of radius* r about γ. The volume and surface area of tubular neighborhoods in $\mathcal{M}(G)$ are presented in Exercise (2-11).

If γ^* is also the axis of rotation of an elliptic element, the projection γ is a closed curve and a cone axis. If there is an elliptic element that interchanges the endpoints of γ^*, then γ^* projects to a finite geodesic segment of length $[x, g(x)]/2$ with endpoints on cone axes, the degenerate case of a closed curve as we go forth and return along the segment.

We now turn to the structure at a parabolic fixed point $\zeta \in \mathbb{S}^2$ of G. Let σ denote any horosphere at ζ. The horosphere has an intrinsic euclidean metric $d_\sigma(\cdot, \cdot)$. There is a parabolic pair $T^{\pm 1} \in G$ for which

$$d_\sigma(x, T^{\pm 1}(x)) \leq d_\sigma(x, T_1(x))$$

for all parabolic $T_1 \in \mathrm{Stab}_\zeta$ and all $x \in \sigma$. The same inequality is true for $T^{\pm 1}$ on any horosphere at ζ. We will refer to either $T^{\pm 1}$ as a *least (translation) length parabolic* in Stab_ζ. For the parabolic $x \mapsto x + 1$, $d_\sigma(x, x + 1) = 1/h$ on the horosphere $\sigma = \{(z, t) : t = h\}$.

We can replace G by a conjugate so that $\zeta = \infty \in \mathbb{S}^2$ and that $z \mapsto z + 1$ is a least length parabolic. Suppose for simplicity, ζ is not also fixed by an elliptic element. Then Stab_ζ is either a cyclic parabolic group or a free abelian parabolic group of rank two; we may assume that Stab_ζ is either $\langle z \mapsto z + 1 \rangle$ or $\langle z \mapsto z + 1, z \mapsto z + \tau \rangle$, where $\mathrm{Im}\, \tau > 0$ and $|\tau| \geq 1$.

In the former case, the doubly infinite strip $\{z : 0 < \mathrm{Re}\, z \leq 1\}$ forms a fundamental domain for its action in $\mathbb{C}$. In the upper half-space model, the slab rising vertically from the strip is a fundamental region for its action in $\mathbb{H}^3$. We see that $\mathbb{C}/\mathrm{Stab}_\zeta$ can be viewed as a doubly infinite cylinder; $w = e^{2\pi i z}$ conformally maps it onto $\mathbb{C} \setminus \{0\}$. The quotient $\mathbb{H}^3/\mathrm{Stab}_\zeta$ is homeomorphic to $\{z : 0 < |z| < 1\} \times (-\infty, \infty)$, since the quotient of each vertical slice of the slab is conformally equivalent to the punctured disk.

In the latter case, the parallelogram with vertices $\{0, 1, \tau, \tau + 1\}$ with two adjacent sides included is a fundamental parallelogram for the action of Stab_ζ on $\mathbb{C}$. The quotient is a torus. The vertical chimney rising from the parallelogram is a fundamental region for the action in the upper half-space model of $\mathbb{H}^3$. The quotient is homeomorphic to $\{z : 0 < |z| < 1\} \times \{z : |z| = 1\}$, where the first factor comes as before from the quotient of vertical slices.

The projection of the horoball $\mathcal{H}_s = \{(z, t) \in \mathbb{H}^3 : t \geq s\}$ may or may not be embedded in $\mathcal{M}(G)$. Once it is embedded for some $t = s$ it will be embedded for all larger values of t. If Stab_ζ is cyclic and $\pi(\mathcal{H}_s)$ is embedded, it is homeomorphic to $\{0 < |z| \leq 1\} \times \mathbb{R}$. We refer to this as a *solid cusp cylinder*, a curtain rod with its axis removed. It has infinite volume and surface area. Its boundary, $\pi(\partial\mathcal{H}_s)$ is called a *cusp cylinder*.

If Stab_ζ has rank two and $\pi(\mathcal{H}_s)$ is embedded, it is homeomorphic to the product $\{0 < |z| \leq 1\} \times \mathbb{S}^1$ and is called a *solid cusp torus*, homeomorphic to the inside of a bagel with its core curve removed. Its boundary is called a *cusp torus*.

Yet it is worthwhile to consider the difference between a bagel embedded in $\mathbb{R}^3$ and a solid cusp torus. The former has a pair of simple loops on its boundary, neither null-homotopic on the surface, such that one bounds a disk on the inside of the bagel, the other bounds a disk on the outside. Neither such curve exists on the boundary of a solid cusp torus; for this reason the boundary is said to be "incompressible".

A solid cusp cylinder has finite volume and surface area by Exercise (2-11). We have defined the solid objects to be closed sets, but we will not always be fastidious in distinguishing one from its interior.

Here is a summary of terms used for cusp tubes and cylinders:

- cusp torus, (for rank two cusp; $\{(|z| = 1) \times \mathbb{S}^1\}$)
- solid cusp torus, (for filled in cusp torus; $\{(0 < |z| \leq 1) \times \mathbb{S}^1\}$)
- cusp cylinder, (for rank one cusp; $\{(|z| = 1) \times \mathbb{R}\}$)
- solid cusp cylinder, (for filled in rank-1 cusp; $\{(0 < |z| \leq 1) \times \mathbb{R}\}$
- solid pairing tube, (for rank one cusp: small disks around two punctures are paired; $\{(0 < |z| \leq 1) \times [0, 1]\}$

Remark 3.2.1. It is important to understand that a cusp cylinder, solid cusp cylinder, cusp torus and solid cusp torus all arise from the action on a small horoball of a rank one or rank two parabolic. Consequently in each case, the object can be retracted back to the associated parabolic fixed point within the manifold. As we go forward in understanding 3-manifold topology, we must remember that none of these objects can be intertwined with the internal topology of the manifold: No *geodesic* in the manifold can interfere with its retraction to the parabolic fixed point, nor can an object obtained from a horoball at a parabolic fixed point in a different conjugacy class interfere.

Least length parabolics. As pointed out earlier, each parabolic subgroup of a group G has a "least length" element as measured by the intrinsic metric of any of its horospheres. We can also measure the length by conjugating the subgroup so that its common fixed point is at ∞. The least length is determined by a generator with the least translation length in the subgroup.[2]

By Feighn and Mess [1991], a finitely generated group G has at most a finite number of conjugacy classes of parabolic subgroups. Given a manifold G, there is a minimum translation length among each of its finite number of conjugacy classes of parabolic subgroups: among translations $\{z \to z + u\}$ there is a minimum value for $u \neq 0$.

---

[2] The intrinsic distance and the hyperbolic distance d between $(-1/2, a)$ and $(1/2, a)$ on the horosphere $\{(z, t) : t = a > 0\}$ are $1/a$ and $d = 2\log((1 + \sqrt{4a^2 + 1})/(2a))$. To have $d = 2\varepsilon$, say, we must have $a = 1/(2\sinh\varepsilon)$.

Moreover, among all the parabolic subgroups there is a subgroup with the least such translation length.

3.3 Universal properties in hyperbolic 3-manifolds and orbifolds

We will record some important internal properties of the quotient. While the properties are stated for 3-manifolds and orbifolds, they have analogues for hyperbolic surfaces as well. During the proof we will often rely on the characterization of limits of nonelementary groups to be presented in Theorem 4.1.1. For our use here we will draw from it the following special case.

Lemma 3.3.1. *Suppose* $\{\langle A_n, B_n \rangle\}$ *is a sequence of nonelementary, discrete groups such that* $\lim A_n = A$, $\lim B_n = B$. *Then* $\langle A, B \rangle$ *is also a nonelementary, discrete group. The corresponding conclusion holds as well for a sequence of three generator nonelementary, discrete groups.*

The injectivity radius at a point. Given a discrete group G and $x \in \mathbb{H}^3$, for $r > 0$ set

$$\delta_x(r) = \{A \neq \mathrm{id} \in G : d(x, Ax) \leq 2r\}.$$

Define the *injectivity radius* at x as

$$r_x = \mathrm{Inj}(x) = \mathrm{Inj}(G; x) = \inf\{r : \delta_x(r) \neq \varnothing\}.$$

Thus $d(x, Ax) \geq 2r_x$, for any $A \neq \mathrm{id} \in G$; that is, the G-orbit of the ball $\{y : d(x, y) < r_x\}$ has no overlaps. Interpreted at the projection $\pi(x) \in \mathcal{M}(G)$, $\mathrm{Inj}(\pi(x))$ is the radius of the largest embedded open ball centered at $\pi(x)$. On the other hand there exist elements $A \in \delta_x(r_x)$ such that the points $A^{\pm 1}(x)$ lie on the boundary of the ball of radius $2r_x$ about x.

The injectivity radius is infinite at a point $x \in \mathbb{H}^3$ only when $G = \{\mathrm{id}\}$. As long as x is not on a rotation axis, the radius is positive by the discreteness of G. As $\pi(x)$ approaches a cusp, $\mathrm{Inj}(\pi(x)) \to 0$.

Lemma 3.3.2. *Given* $\delta > 0$ *there exists* $N = N(\delta)$ *such that for any* $x \in \mathbb{H}^3$ *and for any kleinian group* G *with* $r_x = \mathrm{Inj}(G; x) < \delta$, *the set* $\delta_x(r_x)$ *has at most* N *elements.*

Proof. If $A, B \in \delta_x(r_x)$, $A \neq B$, then $d(Ax, Bx) = d(x, A^{-1}Bx) \geq 2r_x$. Therefore the points Ax, Bx on the sphere of radius $2r_x$ about x are of distance $\geq 2r_x$ apart. For fixed $r = r_x$, only finitely many such points are possible. As $r \to 0$, it is approximately the same number as in the euclidean case and this number is uniformly bounded, independent of $x \in \mathbb{H}^3$. $\qquad\square$

In the same vein:

Lemma 3.3.3. *Suppose* M *is a hyperbolic surface or manifold with the property that there is a constant* $\delta > 0$ *such that all geodesics have length at least* 2δ *(and there are no punctures or cusps). Then* M *can be covered by embedded* δ-*balls.*

Given $V > 0$, there exists $N(V)$ such that all M with area or volume not exceeding V can be covered by $N(V)$ embedded δ-balls.

The universal horoball/horosphere. Corresponding to every parabolic subgroup $H \subset G$ there exists a uniquely determined horosphere at its fixed point ζ with the following property. Let $T \in H$ denote an element of least translation length. Find the horosphere σ, $d_\sigma(x, Tx) = 1$. Here d_σ is the intrinsic euclidean metric on σ.

If ∞ is the fixed point of H, then d_σ is just euclidean length in the horizontal plane $\sigma = \{(z, t) : t = 1\}$. Instead, if $T(z) = z + u$ is a least length element, then the universal horoball is $\{(z, t) : t > |u|\}$ as the flat metric is $\frac{|d\vec{x}|}{|u|}$.

Theorem 3.3.4 (Universal properties of kleinian groups). *There exist universal constants in terms of which the manifold* $\mathcal{M}(G)$, *corresponding to any nonelementary, finitely generated, kleinian group* G, *has the following properties.*

Universal ball. There exists $\delta > 0$ *such that* $\mathcal{M}(G) \setminus \{cone\ axes\}$ *contains an embedded hyperbolic ball of radius* δ.

Universal horoball. Suppose $\mathcal{H}_p$ *is the universal horoball for a parabolic subgroup* $H_p \subset G$ *with fixed point* p. *Then for* **any** $A \in G$ *with* $A(p) \neq p$, $A(\mathcal{H}_p) \cap \mathcal{H}_p = \varnothing$.

- *The projection to* $\mathcal{M}(G)$ *of the universal horoballs of* G *comprise the horoballs in* $\mathcal{M}(G)$, *which are mutually disjoint and finite in number.*
- *If* $p = \infty$ *and* $T(z) = z + 1 \in H_p$, *then any* $A = \left(\begin{smallmatrix} * & * \\ c & * \end{smallmatrix} \right) \in G$ *with* $c \neq 0$, *satisfies* $|c| \geq 1$.

Tubular neighborhoods about short geodesics. There exist $r > 0$ *and* $L_0 > 0$ *such that in any* $\mathcal{M}(G)$:

(i) *The radius* r *tubular neighborhood about any closed geodesic of length* $\leq L_0$ *is embedded; any geodesic of length* $< L_0$ *is simple.*
(ii) *The* r-*tubular neighborhoods about different geodesics of length* $< L_0$ *are mutually disjoint.*
(iii) *The* r-*tubular neighborhoods about geodesics of length* $< L_0$ *do not intersect the universal horoballs.*

Universal elementary neighborhood. There exists $\varepsilon > 0$ *such that for any* $x \in \mathbb{H}^3$, *the subgroup generated by* $\{A \in G : d(x, Ax) < \varepsilon\}$ *is elementary. When* G *is torsion free:*

- *if a generator loxodromic, it is a tubular neighborhood about a simple geodesic,*
- *otherwise it is a solid cusp cylinder or a solid cusp torus.*

Isolated cone (rotation) axes. There exists $\delta > 0$ *such that the distance between any two nonintersecting rotation axes in* $\mathcal{M}(G)$ *is at least* δ, *except if the axes have a common endpoint at a rank two cusp, or perhaps if the axes are both rotation axes of order two.*

Proof: Universal horoballs. The existence of the universal constant is an immediate consequence of Jørgensen's inequality: If $A = \left(\begin{smallmatrix} 1 & 1 \\ 0 & 1 \end{smallmatrix} \right)$ and $Y = \left(\begin{smallmatrix} a & b \\ c & d \end{smallmatrix} \right)$, $ad - bc = 1$, then we find that $\text{tr}(AYA^{-1}Y^{-1}) = 2+c^2$. Jørgensen's inequality with $A = X$ implies that

$$|c| \geq 1 \tag{3.2}$$

since $\text{tr}^2 X = 4$, if $c \neq 0$.

The formula for extension of Y to upper half-space $(z, t) \mapsto (z', t')$ then yields

$$t' = \frac{t}{|cz + d|^2 + |c|^2 t^2} \leq \frac{t}{|c|^2 t^2} \leq \frac{1}{t}. \tag{3.3}$$

For the horoball $\mathcal{H} = \{t > 1\}$, necessarily $t' < 1$. That is, $Y(\mathcal{H}) \cap \mathcal{H} = \varnothing$.

If there is another parabolic subgroup with fixed point say $\zeta \neq \infty$ we may conjugate G by a translation so that the fixed point is moved to $\zeta = 0$. A least length parabolic then has the form $Y' : z \mapsto z/(uz + 1)$. Applying Equation (3.2) we find that $|u| \geq 1$.

Applying in turn Equation (3.3), we find

$$t' \leq \frac{1}{|u|^2 t} \leq \frac{1}{t}; \quad t' < 1 \text{ when } t > 1.$$

So $Y'(\mathcal{H}) \cap \mathcal{H} = \varnothing$.

If we conjugate Y' by $z \to 1/z$ it becomes $Y^*(z) = z + u$. Since $|u| \geq 1$, its universal horoball $(z, t) : t \geq |u| \geq 1$. Map the horoball for $Y^*(z)$ back to that for Y' by $\begin{pmatrix} 0 & -1 \\ 1 & 0 \end{pmatrix}$. Equation (3.3) yields

$$t' = \frac{t}{|z^2| + t^2} < \frac{1}{t} \leq \frac{1}{|u|} \leq 1.$$

The universal horoball for Y' is disjoint from that of Y.

Now a universal horoball an invariant of a parabolic subgroup, not depending on the location of its fixed point. We conclude that the G-orbits of the conjugacy classes of universal horoballs in $\mathbb{H}^3$ are mutually disjoint. In the projection to $\mathcal{M}(G)$, the finite collection of universal horoballs are mutually disjoint.

It is possible that a parabolic fixed point, say $\zeta = \infty$, is also fixed by elliptics $E \in G$. Such elliptics have the form $E(z) = e^{2i\theta} z + a$ and clearly preserve the horoballs at ∞ as well.

A similar argument applies to horodisks in $\mathbb{H}^2$. See also Exercise (3-3) on p. 190. See Exercise (3-46) and Proposition 3.5.3 for much larger universal horodisks when there are no elliptics. □

Proposition 3.5.3 also presents a version of the horodisk theorem that applies in simply connected regions $\Omega \subset \mathbb{C}$ more general than just $\mathbb{H}^2$.

Tubular neighborhoods about short geodesics. Suppose property (i) does not hold. We may assume there are sequences $r_n \to 0$, $L_n \to 0$ and a corresponding sequence of groups G_n and geodesics γ_n such that the radius r_n-tube about γ_n of length $\leq L_n$ is not embedded.

Let ℓ denote the vertical half-line rising from $z = 0$ in the upper half-space model. We may replace each G_n by a conjugate so that γ_n is the projection of ℓ and the corresponding primitive transformation is $A_n : z \mapsto a_n z$, $|a_n| > 1$ where $\log |a_n| \to 0$ is the length of γ_n.

Our hypothesis insures that there is no elliptic of order two in G_n that interchanges the fixed points of A_n. However our proof will still work if we allow elliptics with the same axis as A_n, although the tubular neighborhood will then have a singular axis.

Let C_n denote the euclidean cone about ℓ, which is the radius-r_n tubular neighborhood of ℓ. Let $F_n = \{ \vec{x} \in C_n : 1 \leq |\vec{x}| \leq |a_n| \}$ be a fundamental chunk of C_n. There is an element $B_n^* \in G_n$, which does not preserve ℓ, with $B_n^*(C_n) \cap C_n \neq \emptyset$. For some k, m, $B_n = A_n^k B_n^* A_n^m \subset G_n$ has the property that $B_n(F_n) \cap F_n \neq \emptyset$. Therefore for some $x_n \in F_n$, $B_n(x_n) \in F_n$. Furthermore $\langle A_n, B_n \rangle$ is not elementary.

After passing to a subsequence if necessary, $\lim A_n = A$ and $\lim B_n = B$ exist. But A, B fix the point $p \in \ell$ with $|p| = 1$ so that $\langle A, B \rangle$ is elementary, a contradiction to Lemma 3.3.1.

Exactly the same proof shows that there cannot be a sequence of groups $\{G_n\}$ in which there are two loxodromics A_n, B_n with translation lengths satisfying $L_n \to 0$, such that the closure of their r_n-tubes intersects while $r_n \to 0$. (This is possible for two elliptic axes of order two.)

To prove (iii), suppose that $A : z \mapsto z + 1$ is an element of G and $T \in G$ is loxodromic with translation length L, which is the length of the corresponding geodesic in $\mathcal{M}(G)$. Conjugate the group by a translation so that the fixed points of T are symmetric about $z = 0$. Then a normalized matrix for T has the form $T = \begin{pmatrix} a & bd \\ d/b & a \end{pmatrix}$, $a^2 - d^2 = 1$, $\mathrm{tr}(T) = 2a$, and its fixed points are $+b$. The r-tube about the axis of T will not intersect the horoball $\mathcal{H} = \{(z, t) : t > 1\}$ provided that $|b| \leq e^{-r}$.

We claim that for all sufficiently small $L > 0$, it will be true that $|b| \leq e^{-r}$. For otherwise, there is a sequence of groups G_n containing A and a loxodromic T_n with fixed points $\pm b_n$ symmetric to $z = 0$ such that $\lim L_n = 0$, while $\lim b_n = b^* \geq e^{-r}$. If $b^* = \infty$, then a long segment of the axis of T_n penetrates $\mathcal{H}$ in which case $T_n(\mathcal{H}) \cap \mathcal{H} \neq \emptyset$ in contradiction to the universal horoball property. On the other hand, since $b^* \neq 0$, $\lim T_n = T$ exists with T either elliptic or the identity. This violates Lemma 3.3.1. (Alternatively, such a sequence of r-tubes converges to a horoball.) We conclude that there exists $L' \leq L$ for which the r-tube about any geodesic in any $\mathcal{M}(G)$ of length $\leq L'$ does not intersect any universal solid cusp torus or cusp cylinder.

With the proper interpretation, the case that the loxodromic axis is also the rotation axis of an elliptic in G, or is preserved by an element of order two, is included in our analysis. $\qquad\square$

For the record we also point out the following interesting inequality [Meyerhoff 1987]. By the universal horoball property $|b| \leq |d|$ so that $4|b|^2 \leq 4|d|^2 = |\mathrm{tr}^2(T) - 4|$. Now T is conjugate to $\begin{pmatrix} k & 0 \\ 0 & k^{-1} \end{pmatrix}$, where $k = e^{\frac{L}{2}} e^{i\varphi}$, $0 \leq \varphi < \pi$, and $\mathrm{tr}^2(T) - 4 = (k - k^{-1})^2$. Consequently

$$|b|^2 \leq \sinh^2 \frac{L}{2} + \sin^2 \varphi. \tag{3.4}$$

With the help of a sophisticated computer search, D. Gabai, R. Meyerhoff, and N. Thurston [Gabai et al. 2003] proved that with a few exceptions, if G has no parabolic or elliptic transformations, there is a geodesic in $\mathcal{M}(G)$ with an embedded tubular neighborhood of radius $r = (\log 3)/2$.

Isolated rotation axes. Confirmation of this property runs along the same lines. In a sequence of groups $\{G_n\}$, suppose that $\{d_n, d_n > 0\}$ is a sequence with $\lim d_n = 0$, that $E_n \in G_n$ is an elliptic with rotation axis ℓ, and that $F_n \in G_n$ is an elliptic whose axis ℓ_n does not intersect ℓ but comes within distance d_n of ℓ. We may replace G_n by a conjugate so that for some $p_n \in \ell_n$, $\lim p_n = p \in \ell$.

For all large n we may assume that either $\langle E_n, F_n \rangle$ is nonelementary, it is an infinite dihedral group, that is, each of E_n, F_n is of order two and $E_n F_n$ is loxodromic with axis orthogonal to the axes of E_n and F_n, or ℓ, ℓ_n have a common endpoint at a rank two cusp at ∞ with $\langle E_n, F_n \rangle$ a subgroup of $\text{Stab}_n(\infty)$. In the former case for a subsequence, both $\lim E_n = E$, $\lim F_n = F$ are Möbius transformations fixing p. But then $\langle E, F \rangle$ is elementary, again a violation of Lemma 3.3.1. $\qquad \square$

The universal elementary neighborhood. Denote by $G_x(r)$ the subgroup of G generated by the set $\delta_x(r)$, that is by the elements A for which $d(x, Ax) \leq 2r$. We claim that there exists $r > 0$ such that, for any $x \in \mathbb{H}^3$, and any kleinian group G, the subgroup $G_x(r)$ is elementary. In other words, for the ball $B_x(r)$, the subgroup generated by the elements $\{g\}$ for which $g(\overline{B_x(r)}) \cap \overline{B_x(r)} \neq \varnothing$ is elementary.

For a fixed G and $x \in \mathbb{H}^3$, there must be some $r > 0$ for which $G_x(r)$ is elementary. For as $r = r_n \to 0$, any infinite sequence of distinct elements $A_n \in \delta_x(r_n)$ converges either to an elliptic transformation fixing x or to the identity. No such sequence can exist! So for all sufficiently small r, the set of elements $\delta_x(r)$ is independent of r and either contains no elements or consists of elliptic transformations fixing x.

Now assume that for some $x \in \mathbb{H}^3$, there is no universal elementary neighborhood. Then there is a sequence of kleinian groups $\{G_n\}$ and a sequence $r_n \to 0$ such that $G_{n,x}(r_n)$ is not elementary. On the other hand, for fixed n, $G_{n,x}(\rho)$ is elementary for some ρ with $0 < \rho < r_n$. As ρ increases to r_n, the elementary groups $G_{n,x}(\rho)$ are nested. There is a first number $\tau_n < r_n$ for which $G_{n,x}(\rho) = G_{n,x}(\tau_n)$ for $\tau_n \leq \rho < r'_n \leq r_n$, and is elementary but $G_{n,x}(r'_n)$ not. We may take $r_n = r'_n$.

If $G_{n,x}(\tau_n)$ is finite but not cyclic, there are elements $A_n, B_n \in \delta_x(\tau_n)$ with distinct, yet intersecting, axes of rotation. The set $\delta_x(r_n)$ must contain an element X_n which does not fix the common fixed point of A_n, B_n. Hence $\langle A_n, B_n, X_n \rangle$ is not elementary.

If $G_{n,x}(\tau_n)$ is finite cyclic, let $A_n \in \delta_x(\tau_n)$ be a generator. We can find an $X_n \in \delta_x(r_n)$ that does not fix the axis of A_n. If X_n is elliptic with axis intersecting that of A_n, there must be another element $B_n \in \delta_x(r_n)$ that does not fix this common point. Thus one of $\langle A_n, X_n \rangle$ and $\langle A_n, B_n, X_n \rangle$ is not elementary.

Next suppose $G_{n,x}(\tau_n)$ is an infinite group that keeps invariant a line $\ell \subset \mathbb{H}^3$. Either $\delta_x(\tau_n)$ contains a loxodromic A_n, or it contains two elliptics A_n, B_n of order two that interchange the endpoints of ℓ. There must be an element $X_n \in \delta_x(r_n)$ which does not leave ℓ invariant. Again, $\langle A_n, X_n \rangle$ or $\langle A_n, B_n, X_n \rangle$ is not elementary.

Finally, suppose $G_{n,x}(\tau_n)$ fixes a point $\zeta \in \mathbb{S}^2$, but does not fall into one of the previous cases. Then $\delta_x(\tau_n)$ contains a parabolic transformation A_n or two elliptics A_n, B_n such that $B_n A_n$ is parabolic (e.g., $z \mapsto -z$, $z \mapsto -z + 1$). The set $\delta_x(r_n)$ contains an element X_n that does not fix ζ. Hence $\langle A_n, X_n \rangle$ or $\langle A_n, B_n, X_n \rangle$ is not elementary.

In all cases we have found a nonelementary two- or three-generator subgroup generated by elements of $\delta_x(r_n)$. As $n \to \infty$, convergent subsequences converge to elements which fix x and are therefore elliptic or the identity. Once again we draw on Lemma 3.3.1 to reach the contradiction.

To complete the argument we claim that, if $A \in G_x(r)$ is loxodromic, it represents a simple geodesic in the quotient. Otherwise the projection of $[x, Ax]$ into $\mathcal{M}(G)$ would contain two simple subloops of shorter length. (The projection $\pi([x, Ax])$ is a closed loop which is a geodesic except for a likely corner at $\pi(x)$.) We could find two other loxodromics A_1, A_2 with different axes, and satisfying $d(x, A_i x) < 2r$. Both elements would be in $G_x(r)$ which then could not be elementary.

Meyerhoff proved [1987] that $2\delta > 0.104$. $\qquad\qquad\qquad\qquad\square$

The universal ball. Given an $x \in \mathbb{H}^3$ that is not an elliptic fixed point of G, denote by G_x the subgroup of G generated by the set $\delta_x(r_x)$, where $r_x = \mathrm{Inj}(G; x)$; in other terms, $\delta_x(r_x) = \{A \in G : d(x, Ax) = 2r_x\}$.

Given any nonelementary, discrete group G we will find somewhere in $\mathbb{H}^3$ a ball of radius δ, where δ is the universal elementary neighborhood constant, with the property that there is no overlapping in its G-orbit. To do this we will find a point $x \in \mathbb{H}^3$ for which G_x is not elementary. The universal elementary neighborhood property then assures us that $r_x \geq \delta$, and as a consequence $g(B_x(\delta)) \cap B_x(\delta) = \varnothing$ for all $g \neq \mathrm{id} \in G$, for the δ-ball about x. In searching for such a point, we may restrict our attention to groups whose injectivity radii are uniformly bounded above (no elementary discrete groups have this property).

Start with a point $x \in \mathbb{H}^3$ which is not a fixed point. Suppose G_x is elementary. We will find a polygonal line along which the injectivity radius strictly increases until the terminal point y where G_y is nonelementary. To find this line we have to examine various classes of elementary groups separately.

Case 1: The elements of G_x have a single common fixed point $\xi \in \partial \mathbb{H}^3$. We may take $\xi = \infty$ in the upper half-space model. Then G_x is a finite extension of a rank one or rank two parabolic group. For each $A \in \mathrm{Stab}_\infty(G)$, $A \neq \mathrm{id}$, the perpendicular bisectors of the segments $[x, A^{\pm 1} x]$ are vertical half-planes. If $A \in \delta_x(r_x)$, they are tangent to the ball D_x of radius r_x about x. For $A \notin \delta_x(r_x)$, the perpendicular bisectors are uniformly bounded away from D_x.

Let the point y move down the vertical line ℓ through x. For a certain open interval near x, $\mathrm{Inj}(y) = r_y$ is determined by the same vertical planes that determine $\mathrm{Inj}(x)$. However r_y is strictly increasing since y is moving closer to $\partial \mathbb{H}^3$, away from ∞.

Since we are assuming r_y is uniformly bounded there must be a first point w with the following property. For some $B \in \delta_w(r_w)$, $B \notin \delta_x(r_x)$, the perpendicular bisector of $[w, Bw]$ is tangent to the ball D_w of radius r_w about w. This cannot be a vertical plane since $B \notin \mathrm{Stab}_\infty$. Because $\langle G_x, B \rangle \subset G_w$, the group G_w is not elementary.

Case 2: G_x is a finite group but not a cyclic group nor a $\mathbb{Z}_2$ extension of a cyclic group. We now use the ball model and take the common fixed point of G_x to be the origin. Then G_x is a subgroup of the finite group $\mathrm{Stab}_0 \subset G$ of euclidean rotations. The ball D_x of radius r_x centered at x is inscribed in a convex cone with flat faces and vertex at the origin; its faces are contained in the perpendicular bisecting planes of $[x, A^{\pm 1}x]$, for $A \in \delta_x(r_x)$. These are equatorial planes of the ball model $\mathbb{H}^3$.

Let now the point y move away from x along the ray from the origin through x. There is a first point w for which the ball D_y of radius r_y hits a new plane, the perpendicular bisector of some $[w, Bw]$, $B \notin \delta_x(r_x)$. This new plane does not pass through the origin so that $B \notin \mathrm{Stab}_0$. Therefore $\langle G_x, B \rangle \subset G_w$ is not elementary.

Case 3: G_x is cyclic loxodromic or a finite extension of a cyclic loxodromic group. In preparation for the analysis of this case, we take note of the following situation. Suppose in the ball model $\mathbb{B}$, we have a closed ball B with center on the positive radius of $\mathbb{B}$ that does not contain 0. Let P denote the equatorial plane through 0 and orthogonal to the vertical diameter of $\mathbb{H}^3$; it contains the center of B. Let ρ denote the diameter of B that lies in P and is orthogonal to the positive radius. Consider two planes β_1, β_2 tangent to B at the ends of ρ. Compare these two planes with two planes X_1, X_2 containing the vertical diameter of $\mathbb{B}$ through 0 that are tangent to B, as the pages of a book with spine the vertical diameter. Necessarily the planes X_1, X_2 intersect the planes β_1, β_2 in two lines. Consequently we see that the pages of the book grow apart more quickly than do β_1, β_2, as we head toward $\partial \mathbb{H}^3$ along the positive radius.

Now we can deal with Case 3. Let ℓ denote the axis of a primitive loxodromic element $T \in \delta_x(r_x)$; we may assume that it is the vertical diameter. The ball $B_x(r_x)$ of radius r_x about x is tangent to two planes $\beta_\pm$ orthogonal to ℓ, namely, the perpendicular bisectors of the segments $[x, T^{\pm 1}(x)]$.

If $\delta_x(r_x)$ contains a rotation $z \mapsto e^{2i\theta}z$ with axis ℓ then $B_x(r_x)$ is also supported by two planes passing through ℓ and opened at angle 2θ.

If $\delta_x(r_x)$ has an element of order two that interchanges the endpoints of ℓ, then $T = E_2 E_1$, where each E_i is such a half-rotation. There are two other planes $\beta'_\pm$ orthogonal to ℓ and tangent to $B_x(R_x)$. One contains the axis of E_1 and the other the axis of its conjugate $E_2 E_1 E_2$.

Construct the ray ρ orthogonal to ℓ and passing through x. Follow a point $y \in \rho$ as y moves from x towards $\partial \mathbb{B}$. The thrust of our initial observation is that $B_y(r_y)$ will eventually no longer be supported by the pages of the open book whose spine is ℓ but it might still be by the planes $\beta_\pm$. If G_y is still elementary, it will continue to be cyclic or an infinite dihedral group. As y continues to move along ρ towards $\partial \mathbb{B}$, there will be a first point w such that $\delta_w(r_w)$ contains an element $X \notin G_x$. The perpendicular bisecting plane of $[w, X(w)]$ cannot then be orthogonal to ℓ. Consequently $\langle G_w, X \rangle$ is not elementary.

Case 4: G_x is a finite cyclic group or the extension of one by an elliptic of order two that exchanges the fixed points. Once again in the ball model, we can assume the vertical diameter ℓ is the axis of rotation. If G_x is cyclic, the ball $B_x(r_x)$ of radius r_x about x is supported by two vertical planes containing ℓ, as between the pages of an open book. If in addition there is an elliptic of order two $E \in G_x$ exchanging the north and south poles, then $E \in \delta_x(r_x)$ and we can assume in addition that $B_x(r_x)$ is also tangent to the horizontal equatorial plane.

In the latter case, let the point y, as before, move toward $\partial \mathbb{H}^3$ from x along the ray orthogonal to ℓ passing through x. There is a first point w at which the ball $B_w(r_w)$ hits the perpendicular bisecting plane P of $[w, Xw]$ for $X \notin G_x$. The new plane P does not contain ℓ. If $\langle G_w, X \rangle$ is still elementary, we must return to Cases 1–3.

If instead G_x is cyclic, follow the same procedure. There is a first point w at which the ball $B_w(r_w)$ hits the perpendicular bisecting plane of $[w, Xw]$ for $X \notin G_x$. The group $\langle G_w, X \rangle$ is not cyclic but may fall into any of the cases 1–4.

For numerical information about the size of the universal ball in hyperbolic manifolds, see Shalen [2011]. □

There has been much recent work studying tubular neighborhoods, especially as a way of better understanding the volume of manifolds [Gabai et al. 2001; Meyerhoff 1987; Przeworski 2003]. The radius ρ of the tube about a closed geodesic γ can be chosen as a function of the length so that as the length of γ shrinks to 0 (in a sequence of groups), and a primitive loxodromic generator converges to a parabolic transformation, $N(\gamma^*)$ converges to the corresponding universal horoball. Explicit estimates are given in Meyerhoff [1987].

For torsion-free Haken manifolds (see Section 6.3), it is shown in Culler and Shalen [2010] that the universal elementary constant 2δ can be taken as 0.286. In Meyerhoff [1987] it is shown that the lower bound for the universal elementary constant in any finite volume manifold can be taken as ~ 0.104.

In the case of fuchsian groups, the tubular neighborhood property is called the *collar lemma*. The first paper on it was by Linda Keen. The sharp statement is this:

On any Riemann surface, the length L of any nonsimple closed geodesic α satisfies

$$L > 4 \sinh 1.$$

If instead α is a simple closed geodesic then it has a collar neighborhood of width

$$2 \operatorname{arcsinh}((\sinh L/2)^{-1}).$$

See Exercise (8-8)); for a complete discussion see Buser [1992, Chapter 4].

Historical remarks

The universal horoball property seems to have been discovered by Fatou 1930, p. 159 though, as pointed out by Alan Beardon, his proof was incomplete. Apparently the first complete proof in the literature is in Shimizu [1963] and in some papers the property is referred to as "Shimizu's lemma".

The universal elementary constant is today usually called the *Margulis constant*. Existence for the case without elliptics, appears in Kazhdan and Margulis [1968]. Existence for the general case appears in Wang [1969]. These early results were proved in the context of general Lie groups. Following an entirely different track, in the context of hyperbolic geometry in $\mathbb{H}^3$, the property was independently discovered in 1973 in discussion with Jørgensen. It was one of a number of universal properties that followed from Jørgensen's inequality. This discovery was motivated by the fuchsian analogue in Marden [1974d], and independently [Sturm and Shinnar 1974]. Jørgensen's lemma brings the analysis closer to the actual phenomena allowing, in principle, estimates for the optimal value.

In this book we have chosen to call the universal constants by descriptive names.

3.4 The thick/thin decomposition of a manifold

Assume that G has no elliptics; the only elementary subgroups of G are then rank one and two parabolic cusp groups, and cyclic loxodromic groups.

The *ε-thin part* $\mathcal{M}^{\text{thin}}(G)$ of $\mathcal{M}(G)$ is defined in terms of the universal elementary constant ε

$$\{x \in \text{Int}(\mathcal{M}(G)) : \text{Inj}(x) < \varepsilon\},$$

where $\text{Int}(\mathcal{M}(G))$ denotes the interior.

For each component C of $\mathcal{M}^{\text{thin}}(G)$, given $x \in C$, the subgroup generated by $\{g \in G : d(x, g(x)) < \varepsilon\}$ is either cyclic loxodromic, or it is either a cyclic rank one cusp subgroup or a rank two cusp subgroup. The subgroups generated do not depend on the choice of $x \in C$.

The components of $\mathcal{M}^{\text{thin}}(G)$ are of three types.

- A tubular neighborhood of a short geodesic.
- A solid cusp cylinder associated with a rank one cusp.
- A solid cusp torus associated with a rank two cusp.

These neighborhoods are mutually disjoint, and disjoint from the universal horoballs. For if a point $x \in \mathbb{H}^3$ is common to two of them, there in each component C_1, C_2 there is an element g_1, g_2 which together do not generate an elementary group. This is a contradiction to our choice of ε.

We can choose $\varepsilon_0 \leq \varepsilon$ if needed so that the lengths of the geodesics, and the translation length of generators of the cusp groups are all $< \varepsilon_0$.

The complement of $\mathcal{M}(G)^{\text{thin}}$ in the interior of $\mathcal{M}(G)$ is denoted $\mathcal{M}(G)^{\text{thick}}$. This is an open, connected submanifold. The reason behind this decomposition of the interior of $\mathcal{M}(G)$ is that the thin parts are more troublesome in the analysis of the manifold.

It is shown in Meyerhoff [1987] that one can choose $\varepsilon = 0.104$.

3.5 Fundamental polyhedra

Fundamental polyhedra provide "concrete" models of the manifolds $\mathcal{M}$. Suppose we are standing at an interior point $\pi(\mathcal{O}) \in \mathcal{M}(G)$ and blow up a balloon. If it keeps growing without ever touching itself, we must be living in $\mathbb{H}^3$ itself. Otherwise at some point the balloon will meet itself. We blow some more, and keep blowing until the balloon fills the whole manifold (ignoring the fact that this may require an infinite volume of air). The balloon will then be the projection of the Dirichlet region centered at $\mathcal{O}$; the faces comprise the balloon surface and form a *spine* for the manifold.

The Dirichlet regions, or Poincaré fundamental polyhedra (Poincaré first used them to study kleinian groups), are constructed as follows. Given a kleinian group G, choose a base point $\mathcal{O} \in \mathbb{H}^3$ which is not a fixed point of G. For each element $g \in G$, $g \neq \text{id}$, construct the hyperbolic plane which is the perpendicular bisector P_g of the geodesic segment $[\mathcal{O}, g^{-1}(\mathcal{O})]$. Denote by H_g the relatively closed half-space which is bounded by P_g and contains $\mathcal{O}$. The labeling is such that $g(P_g) = P_{g^{-1}}$, which is the boundary of $g(H_g)$.

The *Dirichlet region* or *Dirichlet fundamental polyhedron* $\mathcal{P}_{\mathcal{O}}$ with center $\mathcal{O}$ is defined as the closed, convex hyperbolic polyhedron

$$\mathcal{P}_{\mathcal{O}} = \mathcal{P}_{\mathcal{O}}(G) = \bigcap_g H_g \subset \mathbb{H}^3.$$

(The use of the word "region", and also "domain", is traditional in this context although the set in question is not open.)

If $h \in G$, $\mathcal{P}_{h(\mathcal{O})} = h(\mathcal{P}_{\mathcal{O}})$.

The relative boundary of $\mathcal{P}_{\mathcal{O}}$ in $\mathbb{H}^3$ is the union of possibly an infinite number of faces $\{f\}$ (a face is a polygonal region in some P_g), edges $\{e\}$ (an edge is a geodesic segment that lies in the boundary of two adjacent faces), and vertices $\{v\}$. At most a finite number of faces, edges and vertices meet any given compact subset of $\mathbb{H}^3$. Moreover, since $\mathcal{P}_{\mathcal{O}}$ is convex, its intersection with any hyperbolic plane is connected.

Proposition 3.5.1. *$\mathcal{P}_{\mathcal{O}}$ has the following properties*:

(i) *The faces are arranged in pairs (σ, σ'). To each pair corresponds an element $g \in G$, called a face pairing transformation, such that*

$$g(\sigma) = \sigma' \quad \text{and} \quad g(\mathcal{P}_{\mathcal{O}}) \cap \mathcal{P}_{\mathcal{O}} = \sigma'.$$

(ii) *If a face pairing transformation is elliptic, there is an edge contained in its rotation axis.*

(iii) *To each edge e corresponds an edge relation: $g_1 g_2 \ldots g_n = g_e$ where either g_e is elliptic with rotation axis containing e and $g_e^m = \text{id}$ for some $m > 1$, or $g_e = \text{id}$. Each g_i is a face pairing transformation. The polyhedra*

$$\mathcal{P}_{\mathcal{O}}, \ g_1(\mathcal{P}_{\mathcal{O}}), \ g_1 g_2(\mathcal{P}_{\mathcal{O}}), \ldots, \ g_1 g_2 \cdots g_e(\mathcal{P}_{\mathcal{O}})$$

are arranged cyclically about e, each sharing a face with the previous and the succeeding. If $g_e = \text{id}$ then $g_1 g_2 \cdots g_e(\mathcal{P}_{\mathcal{O}}) = \mathcal{P}_{\mathcal{O}}$. Otherwise the full cycle is *completed by applying in succession* $g_e, g_e^2, \ldots, g_e^m = $ id *to the union of the listed polyhedra.*

(iv) *The orbit of* $\mathcal{P}_\mathcal{O}$ *under G fills* $\mathbb{H}^3$ *without overlap on interiors.*

(v) *The face pairing transformations generate G; the edge relations generate the relations in G.*

(vi) $\overline{\mathcal{P}}_\mathcal{O} \cap \Omega(G)$ *is a fundamental region for the action of G on* $\Omega(G)$. *Here* $\overline{\mathcal{P}}_\mathcal{O}$ *denotes the closure of* $\mathcal{P}_\mathcal{O}$ *in* $\Omega(G) \cup \mathbb{H}^3$.

(vii) *Let* $B_R(\mathcal{O})$ *be the closed ball of radius R centered at* $\mathcal{O}$. *Then the intersection* $\mathcal{P}_\mathcal{O} \cap B_R(\mathcal{O})$ *projects to a compact submanifold of* $\mathcal{M}(G)$ *bounded by the projection of* $\mathcal{P}_\mathcal{O} \cap \partial B_R(\mathcal{O})$.

Proof. (a) The polyhedron $\mathcal{P}_\mathcal{O}$ is characterized by the property that a point $y \in \mathbb{H}^3$ lies in its interior if and only if $d(\mathcal{O}, y) < d(y, h^{-1}(\mathcal{O})) = d(\mathcal{O}, h(y))$ for all $h \neq$ id $\in G$. Thus Int($\mathcal{P}_\mathcal{O}$)) $\cap h(\mathcal{P}_O) = \varnothing$ since y is closer to $\mathcal{O}$ than any $h(\mathcal{O})$.

In particular, g maps P_g to $P_{g^{-1}}$ and H_g into the closure of $\mathbb{H}^3 \setminus H_{g^{-1}}$. For $x \in P_g$, $d(\mathcal{O}, x) = d(x, g^{-1}(\mathcal{O})) = d(g(x), \mathcal{O})$.

The argument shows that the interior of $\mathcal{P}_\mathcal{O}$ cannot contain points of a rotation axis of G; also that there cannot be any overlap in the interiors in the G-orbit of $\mathcal{P}_\mathcal{O}$.

(b) If x is an interior point of a face $f' \subset P_g$, we have $d(g(x), \mathcal{O}) = d(x, \mathcal{O}) < d(x, g^{-1}h(\mathcal{O}))$, so long as $h \neq g$, id. Thus $g(x) \in P_{g^{-1}}$ also lies in a face.

On the other hand, no conjugate hgh^{-1} can also be a face pairing transformation. Instead, hgh^{-1} is a face pairing transformation of $\mathcal{P}_1 = h(\mathcal{P}_\mathcal{O})$.

(c) There cannot be different faces f_1, f_2 with the property that $g_1(f_1) = g_2(f_2) = f$. For $g_1(\mathcal{P}_\mathcal{O})$ is exterior to $\mathcal{P}_\mathcal{O}$ but adjacent to f. The transformation g_2^{-1} maps f to f_2 and necessarily sends $g_1(\mathcal{P}_\mathcal{O})$ back to $\mathcal{P}_\mathcal{O}$. Thus $h = g_2^{-1}g_1$ maps $\mathcal{P}_O$ onto itself so $g_1 = g_2$ and hence $f_1 = f_2$.

We conclude that the faces of $\mathcal{P}_\mathcal{O}$ are arranged in mutually disjoint (except for perhaps a common edge) isometric pairs.

(d) *The edge relations.* Choose an edge e_1 and then one of the two faces sharing e_1, say f_1. A face pairing transformation g_1 sends the partner face f_1' to $f_1 = g_1(f_1')$ and $g_1(\mathcal{P}_\mathcal{O})$ is adjacent to $\mathcal{P}_\mathcal{O}$ along f_1. An edge e_2 of f_1' is sent by g_1 to e_1. A special case is when g_1 is elliptic and e_1 is contained in its axis of rotation. Then the partner face f_1' and f_1 both share the edge e_1. If g_1 has order m, the m polyhedra $\mathcal{P}_\mathcal{O}, g_1(\mathcal{P}_\mathcal{O}), \cdots, g_1^{m-1}(\mathcal{P}_\mathcal{O})$ form a complete cycle of polyhedra, sharing the edge e_1, each sharing a face with the adjacent polyhedra. In this case the *edge relation* determined by e_1 is simply $g_1^m = $ id.

Otherwise there is a face $f_2 \neq f_1$ that shares with f_1' the edge e_2. Its partner face f_2' is sent by some g_2' to $f_2 = g_2(f_2')$. There is an edge e_3 of f_2' that g_2 sends to e_2. Note that the three polyhedra $\mathcal{P}_\mathcal{O}, g_1(\mathcal{P}_\mathcal{O}), g_1g_2(\mathcal{P}_\mathcal{O})$ are arranged in cyclic order about the edge e_1. Successive polyhedra share a face.

Next take the face $f_3 \neq f_2'$ that also shares e_3 and find its partner and the face pairing map $g_3(f_3') = f_3$. To our cyclic arrangement about e_1 we can add one more,

$g_1 g_2 g_3(\mathcal{P}_\mathcal{O})$. Keep going. The process will necessarily end after a finite number of steps. We will arrive at f_k with the property that f_k' shares e_1 with f_1. At this point the polyhedra $\mathcal{P}_\mathcal{O}, g_1(\mathcal{P}_\mathcal{O}), \ldots, g_1 g_2 \ldots, g_k(\mathcal{P}_\mathcal{O})$ are arranged in cyclic order about e_1. Furthermore the transformation $h = g_1 g_2 \ldots, g_k$ fixes the edge e_1. There are two possibilities.

The first is that $g_1 \cdots g_k = \mathrm{id}$, that is, the final polyhedron in the cycle, namely $g_1 g_2 \ldots, g_k(\mathcal{P}_\mathcal{O})$, coincides with $\mathcal{P}_\mathcal{O}$. The *edge relation* determined by e_1 is $h = \mathrm{id}$. The sequence of edges $e_1, \ldots, e_k = e_1$, is called an *edge cycle*. Had we started instead with a different edge e_j in the cycle, its edge relation is conjugate to that for e_1. The dihedral angles corresponding to the edges in the cycle sum to 2π.

The second possibility is that $h = g_1 \cdots g_k$ is an elliptic transformation fixing the edge e_1, and $k \geq 1$ is the smallest number with this property. If h has order m then for $\mathcal{P}^* = \mathcal{P}_\mathcal{O} \cup g_1(\mathcal{P}_\mathcal{O}) \cup \cdots \cup g_1 g_2 \cdots g_k(\mathcal{P}_\mathcal{O})$, the collection $\mathcal{P}^*, h(\mathcal{P}^*), \ldots, h^{m-1}(\mathcal{P}^*)$ is a nonoverlapping cyclic ordering of km polyhedra about e_1. The edge relation associated with e_1 is $h^m = \mathrm{id}$. The sequence of edges $e_1, e_2, \ldots, e_k = e_1$ forms an *elliptic edge cycle*. Each edge in the cycle is contained in the rotation axis of an elliptic element conjugate to h. The sum of the dihedral angles about e_1 of the polyhedra $\mathcal{P}_\mathcal{O}, g_1(\mathcal{P}_\mathcal{O}), \ldots, g_k(\mathcal{P}_\mathcal{O})$ must be $2\pi/m$.

By adjoining to $\mathcal{P}_\mathcal{O}$ the polyhedra which share a face with $\mathcal{P}_O$, and then those that share just an edge, we can completely surround $\mathcal{P}_\mathcal{O}$ by other polyhedra of its orbit. A vertex v of $\mathcal{P}_\mathcal{O}$ will be shared exactly by those polyhedra that are part of the edge cycles about the edges of $\mathcal{P}_\mathcal{O}$ that end at v.

(e) *The G-orbit of $\mathcal{P}_\mathcal{O}$ covers* $\mathbb{H}^3$. For suppose to the contrary that the orbit does not cover $y \in \mathbb{H}^3$. Consider the geodesic segment $[\mathcal{O}, y]$. At most a finite number of elements in the orbit can intersect this segment. There is a point $w \in [\mathcal{O}, y]$ such that w lies on the boundary of some element of the orbit, but no point closer to y does. But we know we can completely surround any element $h(\mathcal{P}_\mathcal{O})$ of the orbit by other neighbors sharing a face or edge. Therefore $w = y$ and y is covered, after all.

(f) As a consequence we can assert that the rotation axis of each elliptic in G contains a segment which is conjugate to an edge of $\mathcal{P}_\mathcal{O}$. For if not, the rotation axis of some conjugate g would meet the interior of $\mathcal{P}_\mathcal{O}$. But then g could not send $\mathcal{P}_\mathcal{O}$ into its exterior, a contradiction.

The rotation axis of an elliptic g is the line of intersection of the two planes $P_{g^{\pm 1}}$. If the rotation axis of a primitive elliptic contains an edge e of $\mathcal{P}_\mathcal{O}$, the two faces sharing e must necessarily be contained in the planes $P_{g^{\pm 1}}$. Therefore g is a face pairing transformation, and no conjugate can also be face pairing.

It is time to bring up a special case: Suppose f' is contained in $P_{g^{-1}}$ for g elliptic of order two. Then $g(P_{g^{-1}}) = P_g = P_{g^{-1}}$, and $g(f') = f'$. The face f' is divided in two parts by the rotation axis of g and application of g interchanges the two parts. To incorporate this special case into our general theory, we must allow any segment of the rotation axis that meets $P_\mathcal{O}$ to be counted as an edge of $\mathcal{P}_\mathcal{O}$, and regard f' itself as the union of two adjacent faces.

(g) *The presentation of G*. In the G-orbit of $\mathcal{P}_\mathcal{O}$, the first generation of polyhedra consists of those that share an edge with $\mathcal{P}_\mathcal{O}$. The second generation consists of those which share an edge with a member of the first generation. The n-th generation consists of polyhedra that share an edge with the $(n-1)$ generation but not with a member of an earlier generation. It is clear that any given compact subset of $\mathbb{H}^3$ is covered by the polyhedra in a sufficiently high generation. This shows that the face pairing transformations of $\mathcal{P}_\mathcal{O}$ generate G: any $g \in G$ can be written as a composition of face pairing transformations by following a connected union of polyhedra in the orbit, beginning with $\mathcal{P}_\mathcal{O}$ and ending with $g(\mathcal{P}_\mathcal{O})$.

A small sphere about a vertex v is subdivided into circular polygons by its intersection with the polyhedra sharing v.

Consider the graph Γ_e formed by the union of the edges of the polyhedra in the orbit of $\mathcal{P}_\mathcal{O}$. This may or may not be connected. But any simple loop in $\mathbb{H}^3 \setminus \Gamma_e$ is homotopic to a finite product of tiny circles about edges, connected by an arc to the base point of the fundamental group. Furthermore each edge is conjugate to an edge of $\mathcal{P}_\mathcal{O}$. This translates into the statement that all relations in G are generated by the edge relations of $\mathcal{P}_\mathcal{O}$. For any relation in the generators $g_1 g_2 \cdots g_k = \text{id}$ corresponds to a loop in the complement of Γ_e.

(h) If $\Omega(G) \neq \varnothing$, set $\mathcal{P}_* = \overline{\mathcal{P}} \cap \Omega(G)$. We claim that the G-orbit of $\mathcal{P}_*$ covers $\Omega(G)$ without overlap on the interiors. But this is clear from the fact the orbit of $\mathcal{P}_\mathcal{O}$ covers $\mathbb{H}^3$ without overlap. In general $\mathcal{P}_*$ is not connected. The sides of $\mathcal{P}_*$ are outer edges of faces of $\mathcal{P}_\mathcal{O}$, and the vertices of $\mathcal{P}_*$ are endpoints of edges. Thus the sides of $\mathcal{P}_*$ are arranged in pairs where the side pairing transformations also generate G.

(i) Finally if $\mathcal{P}_\mathcal{O}$ is truncated by intersection with $B_R(\mathcal{O})$, the ball of radius R about $\mathcal{O}$, the truncated faces of $\mathcal{P}_\mathcal{O}$ are still arranged in pairs, with the same pairing transformations as before. This is because if a point x in a face σ is distance R from $\mathcal{O}$ and $g : \sigma \to \sigma'$ is the face pairing transformation, then $g(x) \in \sigma'$, being equidistant from $\mathcal{O}$ and $g(\mathcal{O})$, is also distance R from $\mathcal{O}$.   □

3.5.1 The Ford fundamental region and polyhedron

In this section we will work with the upper half-space model. For the basic facts about isometric circles and planes we refer back to Section 1.5. They are defined for all elements $\neq$ id in a group provided ∞ is not a fixed point. So long as $\Omega(G) \neq \varnothing$, we can replace G by a conjugate if necessary so that $\infty \in \Omega(G)$. Then every element has a well-defined isometric circle and isometric plane which is the hemisphere that rises from the isometric circle.

For $g \in G$, let $\mathcal{E}(g)$ and $\mathcal{E}^*(g)$ denote the closure of the exterior of the isometric circle for g and the isometric hemisphere rising from that circle, respectively. In line with our penchant to define "fundamental regions" as relatively closed sets, we define the *Ford region* or *isometric fundamental region F* and the *Ford polyhedron* or *isometric fundamental polyhedron $\mathcal{F}$* as the following relatively closed sets:

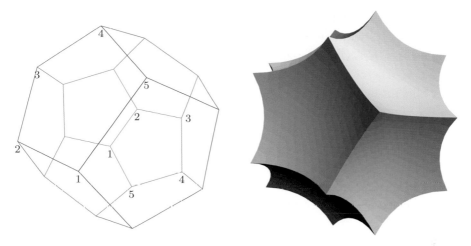

Fig. 3.1. A regular hyperbolic dodecahedron with 72° dihedral angles (right). There is a Möbius transformation that maps each face to the opposite face with a $\frac{3}{10}$ clockwise twist. These generate a kleinian group. The quotient manifold is called the Seifert–Weber dodecahedral space. Its first homology group vanishes. The combinatorial pattern of the identifications is shown on the left.

$$F = \left(\textstyle\bigcap_{g \in G} \mathcal{E}(g)\right) \cap \Omega(G), \quad \mathcal{F} = \textstyle\bigcap_{g \in G} \mathcal{E}^*(g).$$

The isometric polyhedron is a limiting case of Dirichlet polyhedra. For if g does not fix ∞, as $\mathcal{O} \to \infty \in \partial \mathbb{H}^3$, H_g converges to the complement of the isometric hemisphere for g (see Exercise (3-4)). From this we see that $\mathcal{P}_\mathcal{O}$ converges to $\mathcal{F}$ uniformly on compact subsets of $\mathbb{H}^3$.

The polyhedron $\mathcal{F}$ (as well as F) has all the properties listed in Proposition 3.5.1:

Lemma 3.5.2. *If $\infty \in \Omega(G)$, the isometric fundamental polyhedron $\mathcal{P}_\infty(G) = \mathcal{F}$ in the upper half-space model is well defined and is the limit of the Dirichlet polyhedra $\mathcal{P}_\mathcal{O}$ as $\mathcal{O} \to \infty$.*

If instead $\mathcal{O} = \infty$ is a parabolic fixed point of G, the convex polyhedron $\widetilde{\mathcal{P}}_\infty$ exterior to the isometric planes of all $g \in G$, $g(\infty) \neq \infty$, is periodic with respect to the stabilizer $\mathrm{Stab}(\infty)$, while the elements of the G-orbit of $\widetilde{\mathcal{P}}_\infty$ correspond to the G-cosets of Stab_∞.

The second statement follows from Proposition 1.5.5(7). It is sometimes a very useful object to consider, in spite of the periodicity. This is especially true when there is only one cusp.

The intersection with $\Omega(G)$ of the euclidean closure $\overline{\mathcal{P}}_\infty$ is the isometric region $F = \overline{\mathcal{P}}_\infty \cap \Omega(G)$. It may have isolated points, as a church steeple rising toward $\Omega(G)$ from $\mathcal{P}_\infty$. This subtlety is of concern only if one desires precise information about F itself because a neighborhood of an isolated point is covered by a finite number of elements in the orbit of F. The Ford region itself is not necessarily connected and its intersection with a component of $\Omega(G)$ may not be connected. Certainly it is not connected if $\Omega(G)$ is not connected. In any case the orbit of F tiles the region of discontinuity $\Omega(G)$ without interior overlap.

The interior of F is characterized by the property that for any $g \neq \mathrm{id} \in G$, $|g'(z)| < 1$ for $z \in F$. Therefore among all tiles in the orbit of F, it is F itself that is largest, in view of the formula

$$\iint_{g(F)} du \, dv = \iint_F |g'(z)|^2 dx \, dy < \iint_F dx \, dy = \infty.$$

Since $|g'(z)|^2 = O(|z|^{-2})$ as $|z| \to \infty$, the intermediate integral is automatically finite. The inequality becomes more meaningful if $\Omega(G)$ has a bounded component Ω and we replace F by $F \cap \Omega$ and G by $\mathrm{Stab}(\Omega)$.

Fuchsian groups. When G is a fuchsian group acting in UHP, the Ford region is easier to see. If $i\infty$ is not a fixed point then the Ford region F is well defined. It will have a finite or infinite number of sides according to whether G is finitely generated or not. If instead $i\infty$ is fixed by a parabolic element, say by $T(z) = z + 1$, which does not have an isometric circle, F is invariant under T, all the isometric circles will lie in the strip $\{z : \mathrm{Im}\, z \leq 1\}$; F is not a fundamental region. Instead, it is the intersection with any vertical strip such as $F \cap \{0 \leq \mathrm{Re}(z) \leq 1\}$ that serves as a fundamental region.

The following extended form of the universal horodisk theorem is due to Naomi Halpern, and augmented in Earle and Marden [$\geq$ 2015]. It can be quite useful.

Suppose $\Omega \subset \mathbb{C}$ is invariant under a nonelementary group G without torsion and containing the translation $T(z) = z + 1$. Assume that T is determined by a puncture $\zeta \in S = \Omega(G)/G$ corresponding to $i\infty$ in $\partial\Omega(G)$.

Proposition 3.5.3 ([Halpern 1981]). *In the hyperbolic distance $d_\Omega(\cdot, \cdot)$, the region*

$$H_\zeta = \{z \in \Omega : d_\Omega(z, T(z)) < \log\left(\frac{\sqrt{2} + 1}{\sqrt{2} - 1}\right)\}$$

is the maximal horodisk at $i\infty$. Its projection to S is the maximal embedded horodisk at ζ.

Proof. First suppose Ω is the upper half-plane so that G is fuchsian. Following N. Halpern, let $A = \frac{az+b}{cz+d}$, $ad - bc \neq 0$, be any element of G with $c \neq 0$. We find

$$\mathrm{tr}(TATA^{-1}) = 2 - c^2.$$

Since there are no elliptics, either $-2 \geq 2 - c^2$ or $+2 \leq 2 - c^2$. It is the first inequality that must hold. Therefore $|c| \geq 2$.

The isometric circle for A, namely $|cz + d| = 1$, has radius $1/|c| \leq \frac{1}{2}$. Therefore the half-plane

$$H_\Omega = \{z \in \Omega = \mathrm{UHP} : \mathrm{Im}(z) > \tfrac{1}{2}\}$$

is the maximal horodisk at ∞. For the line $\{z \in \mathbb{C} : \mathrm{Im}(z) = \frac{1}{2}\}$ is characterized by

$$\partial H_\Omega = \{z \in \Omega : d_\Omega(z - \tfrac{1}{2}, z + \tfrac{1}{2}) = \log\left(\frac{\sqrt{2}+1}{\sqrt{2}-1}\right)\}. \tag{3.5}$$

The common exterior of the isometric circles, modulo action by $\langle T \rangle$, forms a fundamental domain for G.

Invariance of the hyperbolic metric under Riemann mappings carries this result over to any Ω as described above. $\qquad\square$

The maximal horodisk has the disadvantage that it is usually not a round disk in $\mathbb{S}^2$. Instead, a good substitution in Ω which is round and is associated with the puncture ζ is found as follows:

The Jordan arc ∂H_Ω is periodic with period one. Therefore there is a uniquely determined maximum

$$m = m(\zeta) = \max\{\mathrm{Im}(z); z \in \partial H_\Omega\}.$$

The round disk $D_{max} = \{z \in \Omega : \mathrm{Im}(z) > m\}$ is certainly invariant under $z \mapsto z + 1$. But we will choose the slightly smaller disk,

$$D_\zeta := \{z \in \Omega : \mathrm{Im}(z) > m + \tfrac{1}{2}\}.$$

In terms of the hyperbolic metric d' on the half-plane H_Ω,

$$d_\Omega(m i - \tfrac{1}{2}, \ m i + \tfrac{1}{2}) < d'(m i - \tfrac{1}{2}, \ m i + \tfrac{1}{2}) = \log\left(\frac{\sqrt{2}+1}{\sqrt{2}-1}\right).$$

Proposition (3.5.3) now tells us that D_ζ lies inside the maximal horodisk in Ω associated with ζ.

Therefore D_ζ, while invariant under T, also has the crucial property $A(D_\zeta) \cap D_\zeta = \varnothing$ for all $A \in G$, $A(\infty) \neq \infty$. Furthermore, the projection of D_ζ/G to S is embedded as a neighborhood of the puncture ζ.

Such a round disk $D_\zeta \subset \Omega$ can be called a *G-horodisk*.

Each puncture on S can be dealt with in the same way. Only the number $m(\zeta) + \frac{1}{2}$ changes. The number

$$h = h(S) := \max_{\zeta \in S}\{m(\zeta) + \tfrac{1}{2}\}.$$

is an invariant of S. The corresponding disks in S have mutually disjoint closures.

If we do not assume the kleinian group G acting on Ω is without elliptics then Equation (3.5) charactizing the horodisk at $+i\infty$ in Ω =UHP is as follows. The relevant line is now $\{z : \mathrm{Im}(z) = +1\}$,

$$\partial H_\Omega = \{z \in \Omega : d_\Omega(i - \tfrac{1}{2}, \ i + \tfrac{1}{2}) = \log\left(\frac{\sqrt{5}+1}{\sqrt{5}-1}\right)\}.$$

The presence of elliptics that also share ∞ as a fixed point, would complicate things a bit. But without the assumption that $\infty \subset \partial\Omega$ corresponds to a puncture the result would be false.

3.5.2 Poincaré's Theorem

A particular consequence of Proposition 3.5.1 is the *local finiteness* of the G-orbits of $\mathcal{P}_\mathcal{O}$ in $\mathbb{H}^3$ and $\overline{\mathcal{P}}_\mathcal{O} \cap \Omega(G)$ in $\Omega(G)$: Any neighborhood of a point intersects only a finite number of elements of the orbit.

It is possible to have a polygon or polyhedron that seems to have the properties of a fundamental region, yet it does not have the local finiteness property. A nice example is presented in Mumford et al. [2002, Project 7.1] (another example is Beardon [1983, 9.2.5]): Consider the group generated by the two parabolics $A(z) = z + 3$ and $B(z) = 2z/(3z + 2)$, which acts in the upper and lower half-plane. The element A maps the circle $C_1 = \{|z + 1/2| = 1/2\}$ onto $C_2 = \{|z - 1| = 1\}$, sending the inside of C_1 onto the outside of C_2. The element B maps the line $C_3 = \{\operatorname{Re} z = -1\}$ onto $C_4 = \{\operatorname{Re} z = 2\}$ sending the right side of C_3 onto the right side of C_4. The group $G = \langle A, B \rangle$ is discrete and preserves the upper and lower half-planes. In fact G is a variation on the modular group M_2 of Exercise (2-5). The element $A^{-1}B$ is loxodromic with fixed points $-2, -1$. Therefore $\lim_{n \to +\infty}(A^{-1}B)^n(C_3)$ is the circle $\{|z + 3/2| = 1/2\}$. The quotient $\mathbb{H}^2/G$ is conformally equivalent to a twice-punctured disk. The region exterior to C_1, C_2 and between C_3 and C_4 has the properties of a fundamental region, except it is not locally finite. It has an edge which ends at a fixed point of a loxodromic element but which is not itself preserved by that element; the projection of the edge to the quotient spirals into the corresponding geodesic without meeting it.

It is also possible to have a polyhedron that seems to be a fundamental polyhedron but the face pairing transformations do not generate a discrete group. Take a convex euclidean quadrilateral Q with no two sides parallel. Find the two affine mappings $A_i(z) = a_i z + b_i$ that map one side to its opposite side and send Q to a polygon $A_i(Q)$ that does not overlap Q except along a side. The two elements generate a nondiscrete group in $\mathbb{C}$. In the upper half-space model, above Q rises a chimney Q^*. The transformations A_i act in $\mathbb{H}^3$ and are hyperbolic isometries pairing opposite faces of Q^*, as required of face pairing transformations. Yet the group they generate is not discrete. What went wrong? This example is from Epstein and Petronio [1994].

Still, if we start with a convex polyhedron Q^* with the properties (i), (iii) of Proposition 3.5.1, the face pairing transformations will in general generate a discrete group for which Q^* is a (locally finite) fundamental region. This is called *Poincaré's Theorem*. One must be particularly careful in understanding the orbit of the ends of the polyhedron on $\partial \mathbb{H}^3$. While this is often self-evident if there are a finite number of faces, in the presence of infinitely many faces special care must be taken. For the definitive analysis, valid in all dimensions, see Epstein and Petronio [1994].

Note that there are perfectly good fundamental regions that are neither Dirichlet nor isometric fundamental regions. Simple examples are most fundamental parallelograms for discrete, rank two groups of translations. Another example is the modular group M_2 of Exercise (2-5); there one pair of circles are isometric circles, but the other pair are not. However one might think of the fundamental region as a truncated Ford polygon because ∞ is a parabolic fixed point.

3.5.3 The Cayley graph dual to tessellation

The dual graph Λ associated with $\mathcal{P}_{\mathcal{O}}$ is constructed as follows. Draw a geodesic from $\mathcal{O}$ to each point $g(\mathcal{O})$ where $g(\mathcal{P}_{\mathcal{O}})$ shares a face with $\mathcal{P}_{\mathcal{O}}$. Then draw geodesics to the centers of the polyhedra of the G-orbit that share faces with the first generation, and so on. We get an infinite connected graph Λ, embedded in $\mathbb{H}^3$. If each edge cycle has length 3, which is true in the generic case, then there is a geodesic triangle transverse to each edge. These geodesic triangles are the 2-simplices of the graph.

The graph Λ is equivariant under G. Its projection is therefore a graph $\Lambda_* \subset \mathcal{M}(G)$. The edges of the graph project to simple loops from $\pi(\mathcal{O})$. These loops generate the fundamental group $\pi_1(\mathcal{M}(G); \mathcal{O})$. The 2-cells generate the relations in $\pi_1(\mathcal{M}(G); \mathcal{O})$.

Let $\mathcal{S}$ denote the generator set coming from the face pairing transformations of $P_{\mathcal{O}}$. The graph Λ can then be described as follows. The vertices of Λ are the elements of G. Two elements g, h form an edge if $g^{-1}h \in \mathcal{S}$. That is, Λ is a *Cayley graph* for the group G.

For the general definition of Cayley graphs see Exercise (2-18). If $\mathcal{P}_{\mathcal{O}}$ has an infinite number of faces, Λ has not been found to be useful since there is a finite set of generators whereas there is an infinite number of face pairing transformations of $\mathcal{P}_{\mathcal{O}}$. The abstract Cayley graph for G does not suffer under the same handicap.

Additional remarks

Wielenberg [1981] has given examples showing that a polyhedron may be the fundamental polyhedron for more than one group; different pairings of faces give rise to different groups. An example of this phenomenon for fuchsian groups is in Exercise (2-14).

R. Riley over many years developed a computer program to test whether a group given by generating matrices is discrete [Riley 1983]. In effect, it tests for discreteness using Jørgensen's inequality and the universal horoball property, and then it tries to construct an isometric fundamental polyhedron. If successful, the program can read off the presentation of the group.

Jørgensen [1973] has completely analyzed the isometric fundamental polyhedron for cyclic loxodromic groups $\langle T \rangle$ in terms of the trace parameter, using the normalization of Exercise (1-35). The polyhedron can have an arbitrarily large number of faces; large numbers of faces arise when the trace with $|\mathrm{tr}(T)| < 2$ tangentially approaches 2. (When $|\mathrm{tr}(T)| > 2$, the isometric circles of $T^{\pm 1}$ are disjoint.) The combinatorial arrangement of faces is completely described in terms of $\mathrm{tr}(T)$. Moreover, either F

is the region bounded by the isometric circles of $T^{\pm 1}$ (when $|\mathrm{tr}(T)| \geq 2$) or it is the closure of a simply connected domain with either four or six sides.

Jørgensen also analyzed the Ford fundamental polyhedra of once-punctured torus groups G in terms of the combinatorics of the faces. He shows how, starting with the side pairing transformations of the Ford regions (typically bounded by six circular arcs) on the two components of $\Omega(G)$, the sequence of face pairing transformations of the Ford polyhedron can be read off. This study has been important to this day because this class is the simplest nontrivial class of groups, depending on only two complex parameters. Besides important applications in its own right, especially to two-bridge knots [Akiyoshi et al. 1999], it serves as a test bed for more general situations. For details of Jørgensen's analysis see [Jørgensen 2003; Akiyoshi et al. 2003], Akiyoshi et al. [2005]. However, except for the work of Jørgensen on once-punctured tori, no-one has been able to decipher patterns that may be intrinsic in an infinite number of face pairing transformations.

3.6 Geometric finiteness

Geometrically finite groups is the central class of kleinian groups. They correspond to the manifolds $\mathcal{M}(G)$ which are "essentially" compact [Marden 1974a]. They are dense in the class of all finitely generated groups (This will be discussed in some detail in Sections 5.4–5.6.)

The precise definition and characterization is as follows. The terms "pairs of punctures" and "solid pairing tubes" will be explained below:

Definition. A kleinian group G and corresponding manifold or orbifold $\mathcal{M}(G)$ are said to be *geometrically finite* if G has a Dirichlet region with a finite number of faces.

Assume now that G is nonelementary and is without elliptics.

Theorem 3.6.1 [Marden 1974a; 1977]. *Given a base point* $\mathcal{O} \in \mathbb{H}^3$, $\mathcal{P}_\mathcal{O}(G)$ *has a finite number of faces if and only if the quotient manifold* $\mathcal{M}(G)$ *is compact except perhaps for a finite number of rank one and rank two cusps, and the rank one cusps correspond to pairs of punctures on* $\partial \mathcal{M}(G)$.

If the condition holds for one base point $\mathcal{O}$, *it holds for any choice of base point.*

Corollary 3.6.2. *A manifold* $\mathcal{M}(G)$ *is geometrically finite if and only if (i) the punctures on* $\partial \mathcal{M}(G)$ *are arranged in pairs such that each pair determines a solid pairing tube, and (ii) the result of removing the interiors of all solid pairing tubes and solid cusp tori is a compact manifold* $\mathcal{M}_0(G)$.

Schottky groups and finitely generated fuchsian groups (Section 2.7) are examples of geometrically finite groups with $\Omega(G) \neq \varnothing$.

The term was coined by Leon Greenberg. After Ahlfors' announcement of his finiteness theorem, the next thought was that a Dirichlet region in $\mathbb{H}^3$ for a finitely generated kleinian group had to have a finite number of faces, as is the analogous case in $\mathbb{H}^2$ for fuchsian groups. This hope was decisively dashed when Greenberg pointed out that this is not the case for the "degenerate" groups (Chapter 5) discovered by Bers on the boundary of Teichmüller space [Greenberg 1966; Marden 1974a]. This was the first indication that $\mathbb{H}^3$ really matters.

In contrast, consider the following interesting fact, a consequence of the Ahlfors Finiteness Theorem. For proofs see Beardon and Jørgensen [1975], Greenberg [1977], and Exercise (3-34).

When G is finitely generated, the boundary $\overline{\mathcal{P}}_O \cap \Omega(G)$ "at ∞" of a Dirichlet polyhedron $\mathcal{P}_O(G)$, or the Ford fundamental region $F(G) \subset \mathbb{S}^2$ has a finite number of sides.

There are three other definitions that will be discussed later. Because these are all used in particular situations we will list them here:

Equivalent definitions of geometric finiteness of $\mathcal{M}(G)$

- As above, $\mathcal{M}(G)$ is "essentially" compact.
- There is a neighborhood of each end (see Section 5.5) of $\text{Int}(\mathcal{M}(G))$ which intersects no closed geodesics.
- There is a neighborhood of each end of $\text{Int}(\mathcal{M}(G))$ which does not intersect the convex core of $M(G)$.
- The convex core of $\mathcal{M}(G)$ has finite volume except, perhaps, when G is fuchsian (see Section 3.11).
- All points of $\Lambda(G)$, other than parabolic fixed points, are conical limit points (see Exercise (3-18)).
- The Hausdorff dimension $\dim(\Lambda(G))$ satisfies $0 < \dim(\Lambda(G)) < 2$ (see Exercise (3-20)).

Fig. 3.2. Solid pairing tube for a rank one cusp.

The special role of parabolics. Let $\zeta \in \partial\mathbb{H}^3$ be a parabolic fixed point of G and Stab_ζ the parabolic subgroup fixing ζ. We have called ζ a *rank one* or *rank two cusp* if Stab_ζ has one or two generators respectively. Associated with ζ is its universal horoball $\mathcal{H}$, whose "size" depends only on a least length generator (Section 3.2). For $T \notin \text{Stab}_\zeta$, $T(\mathcal{H}) \cap \mathcal{H} = \varnothing$ while $T(\mathcal{H}) = \mathcal{H}$ for $T \in \text{Stab}_\zeta$.

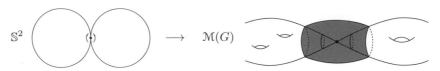

Fig. 3.3. Fragments of one or two surfaces on $\partial\mathcal{M}(G)$ sharing a puncture. Punctures on geometrically finite manifolds come in pairs: A pair really consists of two sides of a single puncture corresponding to a parabolic fixed point: more precisely a pair of small disjoint neighborhoods as indicated here and separated in Figure 3.6. An associated solid pairing tube is also indicated here. Within $\mathcal{M}(G)$, the configuration can be retracted back to the puncture without intersecting any closed geodesic or other pairing tube.

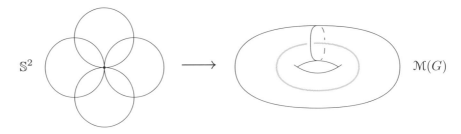

Fig. 3.4. A solid cusp torus. Inside $\mathcal{M}(G)$, it can be retracted to the associated rank two cusp without intersecting any closed geodesic, cusp cylinder or a different cusp torus.

The geometric structure associated with a rank two cusp is the same for all hyperbolic manifolds, even those with nonfinitely generated fundamental groups. Embedded in $\mathcal{M}$ is a one-parameter family of solid cusp tori (Section 3.2) for every conjugacy class of rank two parabolic subgroups. The universal horoball property assures us that if we choose the solid cusp tori to come from horoballs properly contained in the universal horoballs, those corresponding to different conjugacy classes have mutually disjoint closures.

If the interiors of the solid cusp tori interiors are removed from $\mathcal{M}(G)$, there results a manifold with the same fundamental group but with a number of torus boundary components. These are in addition to the components of $\partial\mathcal{M}(G)$, none of which can be tori. Every noncyclic abelian subgroup of $\pi_1(\mathcal{M}) \cong G$ arises by an injection into $\pi_1(\mathcal{M})$ of the fundamental group of a cusp torus. A particular horoball associated with a rank two cusp can be chosen to be of maximal size in that its boundary torus is just tangent to itself; this is not necessarily true of the universal horoball. The set of volumes of these maximal solid cusp tori is an invariant of the particular hyperbolic structure.

If there are elliptics sharing the fixed point ζ then instead of the solid cusp torus there will be an object homeomorphic to $S' \times [0, \infty)$ where S' is a sphere with three or four cone points.

Rank one cusps in a geometrically finite group are associated with a very particular geometric structure that may not appear in a general group. The geometric structure is much stronger than the mere existence of a horosphere and solid pairing tube (Section 3.2); the solid pairing tube must be directly related to two punctures on the boundary of the manifold. In a geometrically finite group, there corresponds a *pair of punctures* p_1, p_2 on $\partial \mathcal{M}(G)$, uniquely associated with the conjugacy class of the cusp: If c_1, c_2 are small circles in $\partial \mathcal{M}(G)$ retractable to p_1, p_2, there is a *pairing cylinder C* in $\mathcal{M}(G)$, which is a cylinder, closed in $\mathcal{M}(G)$, and bounded by c_1 and c_2. It bounds a subregion of $\mathcal{M}(G)$, called a *solid pairing tube*, which is homeomorphic to $C \times (0, 1]$ (and retractable to a cusp). The solid pairing tubes corresponding to the different conjugacy classes of rank one cusps can be chosen to be mutually disjoint in the geometrically finite manifold $\mathcal{M}(G)$.

Let T be a parabolic generator of an element of the conjugacy class that represents the cusp. The circles c_1, c_2 can be chosen so that the pair lifts to round circles in $\Omega(G)$ mutually tangent at the fixed point ζ of T; see Corollary 3.5.3. Such a pair of circles is called a *double horocycle* at ζ, even though this is an abuse of terminology if the components of $\Omega(G)$ containing them are not round disks on $\mathbb{S}^2$.

Suppose the fixed point ζ is shared by an order two elliptic. then instead of the solid pairing cylinder there will be an object of the form $D^* \times [0, \infty)$. The subset $D^* \times \{0\}$ of $\mathrm{Int}(\mathcal{M}(G))$ is a disk with one puncture or cone point.

Consider a fuchsian manifold $\mathcal{M}(G)$ with G acting on the upper and lower half-planes; every puncture on one component of $\partial \mathcal{M}$ is paired with a puncture on the other. Suppose $T : z \mapsto z + 1$ is a least length generator of a rank one parabolic subgroup. For $b > 1$, $\{z \in \mathbb{C} : \mathrm{Im}\, z = \pm b\}$ is a pair of horocycles at the fixed point ∞. These project to "circles" about a pair of punctures. Let $P_\pm \subset \mathbb{H}^3$ denote the vertical planes rising from them and consider the vertical slab $Q = \{(z, t) \in \mathbb{H}^3 : -b \leq \mathrm{Im}\, z \leq b, \ t > 0\}$ they bound. Truncate Q by the half-space $K = \{(z, t) : t \geq a > 1\}$. The relative boundary in $\mathbb{H}^3$ of the resulting tunnel $Q \setminus K$ projects to a pairing tube. This explicit construction suggests how solid pairing tubes can be created in general—there does not seem to be a canonical construction.

Proof of Theorem 3.6.1, Corollary 3.6.2. We continue to assume that G has no elliptics—elliptics will be dealt with at the end of the proof. Assume that $\mathcal{P}_\ominus$ has a finite number of faces. Then every edge is surrounded by—lies in the interior of the closure of—a finite number of polygons in the orbit $G(\mathcal{P}_O)$.

The set $F = \overline{\mathcal{P}_\ominus} \cap \mathbb{S}^2$ consists of a finite number of finite-sided circular polygons and perhaps a finite number of isolated points. The points of F we need to consider are on the boundary of a face of $\mathcal{P}_\ominus$; vertices are such points.

A point $x \in F$ contained in the closure of only finitely many elements $\mathcal{P}_1 = \mathcal{P}_\ominus, \mathcal{P}_2, \ldots, \mathcal{P}_k$ of the orbit $G(\mathcal{P}_\ominus)$, must lie in $\Omega(G)$. To verify this statement (as in Greenberg [1966]), construct a horosphere $\mathcal{H}_x$ at x so small in spherical diameter that it intersects only the polyhedra $\{\mathcal{P}_j\}$. This means that $\mathcal{H}_x$ is partitioned into sectors, each of which lies in some $\mathcal{P}_j$. A neighborhood of x on $\mathbb{S}^2$ is likewise partitioned. Therefore $x \notin \Lambda(G)$.

If interior points of F are adjacent to x, then x cannot be a limit point.

A point of $x \in F$ cannot be the attractive fixed point of a loxodromic $A \in G$. For then the plane containing a face $f \subset \mathcal{P}_0$ has x on its boundary. If P does not also have the other limit point of A on its boundary, the A-orbit of P accumulates to the full axis of A which is impossible. If P contains the axis of A, both the A-axis and the face f meet $\mathbb{S}^2$ orthogonally at x. The A-orbit of f accumulates to x within f which is also impossible.

Consider more generally a limit point $x \in F \cap \Lambda$. Because there are only a finite number of faces, x lies in the boundary of infinitely many elements $\{\mathcal{P}_j\}$ of the G-orbit $G(\mathcal{P}_0)$. Of the infinitely many faces of the $\{\mathcal{P}_j\}$ that contain x on their boundary, infinitely many are images of the same face of $\mathcal{P}_0$ by elements of $\text{Stab}_x \subset G$. All these transformations fix x so are necessarily parabolic. So x is the common fixed point of a rank one or two parabolic subgroup Stab_x. Its Dirichlet region $\mathcal{P}_x = \mathcal{P}_0(\text{Stab}_x)$ contains $\mathcal{P}_0$. If Stab_x has rank one, $\mathcal{P}_x$ is the region bounded by two hyperbolic planes which are tangent at x. If Stab_x has rank two, $\mathcal{P}_x$ is a chimney of four or six faces rising to x.

We have to consider in more detail the case where Stab_x is rank one. In $\mathbb{S}^2$, choose two circles tangent at x that bound a strip S_x whose Stab_x-orbit is all $\mathbb{C}$. For example, if $x = \infty$ and Stab_x is generated by $z \mapsto z + 1$, we can choose the strip $S_x = \{0 \leq \text{Re } z \leq 1\}$. The intersection with S_x of a small neighborhood of x must lie in F since boundaries of faces of $\mathcal{P}_0$ cannot accumulate to x within S_x. This shows that there is a double horocycle at x with respect to Stab_x. In other words with respect to $\partial \mathcal{M}(G)$, x supports a pair of punctures.

We conclude that $\mathcal{M}(G)$ is compact except for a finite number of solid cusp tori and solid cusp tubes with respect to pairs of punctures.

Conversely, if $\mathcal{M}(G)$ has the "essential compactness" just described, we claim that $\mathcal{P}_0$ has a finite number of faces. Otherwise, where in $\mathcal{M}(G)$ would the projection of an infinite number of faces $\{\pi(f_j)\}$ accumulate? We know there can be no accumulation point within $\mathcal{M}(G)$.

Suppose infinitely many $\pi(f_j)$ were in the interior C of a solid cusp torus. A face $\pi(f_i) \subset C$ does not separate C. Therefore there is a simple loop in C, not retractable to a point, joining one side to the other. This loop determines an element of the fundamental group $\pi_1(C)$, which is a rank two abelian group. Because not more than one pair of faces can be paired by elements of a cyclic subgroup, the projection of at most two faces can lie inside C, a contradiction. The same argument applies to the interior of a solid pairing tube. We conclude that $\mathcal{P}_O$ has a finite number of faces.

Corollary 3.6.2 follows from our argument.

We will indicate how the corresponding theorem for orbifolds can be derived from the theorem for manifolds. By Selberg's Lemma (p. 73), there is a torsion-free normal subgroup H of finite index. Let $G = \bigcup_{i=1}^N g_i H = \bigcup_{i=1}^N H g_i$ be a decomposition by distinct cosets. Then $\mathcal{P}^* = \bigcup_{i=1}^N g_i(\mathcal{P}_0(G))$ serves as a fundamental domain for H. Although it may not be connected, it has the properties of $\mathcal{P}_0$, in particular the faces are arranged in pairs with respect to H. For example, if $(f, g(f))$ is a pair of faces of $\mathcal{P}_0(G)$ then the $2N$ faces $\{g_i(f), g_i g(f)\}$ are arranged in N pairs under H. Now

$g_i g = h g_j$ for some j and $h \in H$—because $G = Gg = \bigcup g_i H g = \bigcup g_i g H = \bigcup H g_i$. Therefore the faces $g_i(f)$ and $g_j(f)$ are paired by $h \in H$. Also we know that $h_1 g_j \neq h_2 g_k$ for $k \neq j$, $h_1, h_2 \in H$. In effect, $\mathcal{P}_0(H)$ is made up of N copies of $\mathcal{P}_0(G)$. We conclude that G is geometrically finite if and only if H is as well. (The picture at orbifold cusps is more complicated if there are elliptics that share the parabolic fixed points.) $\qquad \square$

Lemma 3.6.3 [Thurston 1986b]. *If G is geometrically finite and $\Omega(G)$ is nonempty, every finitely generated subgroup is also geometrically finite.*

A proof is indicated in Exercise (3-7). Without the assumption that $\Omega(G) \neq \varnothing$, the statement would be false in general, as we will later see in Section 6.1.

3.6.1 Finite volume

Lemma 3.6.4 ([Wielenberg 1977]). *If $\mathrm{Vol}(\mathcal{M}(G)) < \infty$, then G is geometrically finite, $\Lambda(G) = \mathbb{S}^2$, $\partial \mathcal{M}(G) = \varnothing$, there are no rank one cusps, and at most a finite number of rank two cusps.*

Proof. Again we may assume that G has no elliptics. Consider the ε-thick part $\mathcal{M}(G)^{\mathrm{thick}}$ (with ε chosen as in Section 3.4). The surface area of a cusp cylinder coming from a rank one cusp is infinite. Therefore a small neighborhood in the thick part would have infinite volume. So G cannot have any rank one cusps. On the other hand the volume of each ε-solid cusp torus is not less than $2\varepsilon^2 |\tau| \sin \theta \geq \sqrt{3}\varepsilon^2$ by Exercise (2-11), so there are at most a finite number of them. If the thick part were not compact there would be an infinite sequence $x_n \in \mathcal{M}(G)^{\mathrm{thick}}$ which are centers of mutually disjoint ε balls. Therefore the volume of $\mathcal{M}(G)$ would have to be infinite, which is not the case. $\qquad \square$

3.7 Three-manifold surgery

In this section we will present what is needed from 3-manifold topology for direct application to hyperbolic manifolds. For a rigorous treatment of the aspects of topology that we are using, we refer to Hempel [1976] or Jaco [1980].

Dehn's Lemma and the Loop Theorem. *Let S be a boundary component of an orientable 3-manifold M^3. Suppose $\gamma \subset S$ is a simple loop homotopic to a point within M^3 but not within S. Then γ is the boundary of an essential disk in M^3.*

Suppose a nonsimple loop $\gamma \subset S$ is homotopic to a point in M^3 but not in S. Given a neighborhood $N_\gamma \subset S$ of γ, there a simple loop $\gamma_0 \subset N_\gamma$ that bounds an essential disk $D \subset \mathcal{M}(G)$.

An *essential disk* is an embedded closed disk $D \subset M^3$ such that $D \cap \partial M^3 = \partial D$, where ∂D is not homotopic to a point in ∂M^3. We call a loop $\gamma \subset S$ *nontrivial* if it is not homotopic to a point within S. When obtaining a disk from application of Dehn's Lemma and the Loop Theorem, we will automatically choose one that is essential. A boundary component that supports an essential disk is called *compressible*.

The equivariant version, formulated here for kleinian manifolds $\mathcal{M}(G)$, is also useful:

Equivariant Dehn's Lemma and the Loop Theorem. ([Meeks and Yau 1981]). *Suppose* $\Gamma = \{\gamma_1, \ldots, \gamma_n\}$ *are mutually disjoint and pairwise nonisotopic smooth simple loops on* $\partial\mathcal{M}(G)$, *such that each* γ_i *is contractible to a point in* $\mathrm{Int}(\mathcal{M}(G))$ *but not on* $\partial\mathcal{M}(G)$. *Assume* X *is a finite group of orientation-preserving diffeomorphisms acting freely on the set* Γ. *Then there exists a corresponding* X-*invariant set of mutually disjoint disks* $\Delta = \{D_1, \ldots, D_n\}$ *in* $\mathcal{M}(G)$ *with boundary curves* Γ.

The Loop version is as follows.

Start instead with a finite group X *of orientation-preserving diffeomorphisms of* $\mathcal{M}(G)$. *Denote the kernel of the inclusion map* $\pi_1(\partial\mathcal{M}(G)) \hookrightarrow \pi_1(\mathcal{M}(G))$ *by* K. *Then there exists an* X-*invariant set* $\Delta = \{D_1, \ldots, D_n\}$ *of essentially embedded, mutually disjoint disks in* $\mathcal{M}(G)$ *with the following property:* K *is the kernel of the normal subgroup of* $\pi_1(\partial\mathcal{M}(G))$ *generated by the set* $\Gamma = \{\partial D_1, \ldots, \partial D_n\}$.

Here is an important extension of Dehn's lemma:

Cylinder Theorem. *Suppose* $\gamma_1, \gamma_2 \subset \partial\mathcal{M}(G)$ *are disjoint nontrivial simple loops that are freely homotopic in* $\mathcal{M}(G)$ *but not within* $\partial\mathcal{M}(G)$. *There is an essential cylinder embedded in* $\mathcal{M}(G)$ *bounded by* γ_1 *and* γ_2.

Suppose instead that the freely homotopic loops are not simple but $\gamma_i \subset N_i \subset \partial\mathcal{M}(G)$, *where the neighborhoods* N_1 *and* N_2 *are disjoint. There are simple loops* $\gamma_i' \subset N_i$ *that bound an essential cylinder in* $\partial\mathcal{M}(G)$.

That two loops are *freely homotopic* means that there is a continuous mapping of an annulus A into M^3 sending the boundary components of A to the two loops. Another way of describing free homotopy is as follows: γ_1 is freely homotopic to γ_2 if and only if there is an arc α from any given point $O_1 \in \gamma_1$ to any given point $O_2 \in \gamma_2$ such that γ_1 is homotopic to $\alpha^{-1}\gamma_2\alpha$ (here we are composing curves from right to left).

Two disjoint simple loops that are freely homotopic in ∂M^3, but neither is homotopic to a point in ∂M^3, bound a (topological) annulus in ∂M^3.

An *essential cylinder* is a closed cylinder $C \subset M^3$ such that $C \cap \partial M^3 = \partial C$, the boundary components of C are not homotopic to points in M^3, and C cannot be homotoped (relative to $\partial\mathcal{M}(G)$—it is allowed to slide along $\partial\mathcal{M}^3$) to an annulus in ∂M^3. When obtaining a cylinder from application of the Cylinder Theorem, we will automatically choose one that is essential.

The words "cylinder" and "annulus" can be used interchangeably; one is more likely to think of an annulus as lying in a surface and a cylinder as lying in a manifold.

In the case of a kleinian manifold $\mathcal{M}(G)$ we will add the following requirement to the definition: For C to be called an essential cylinder, *it cannot bound a solid pairing tube*—or even a solid cusp cylinder. Here we are regarding these cusp tubes as homotopic into the boundary.

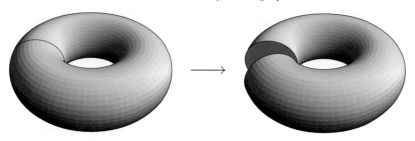

Fig. 3.5. Cutting a solid torus along a compressing disk results in a topological ball.

It is possible that a simple loop $\gamma \in \partial \mathcal{M}(G)$ may be a boundary component of two or more homotopically distinct essential cylinders which are disjoint, except for sharing the common boundary γ. On the other hand,

A simple nontrivial loop on a cusp cylinder or cusp torus cannot be freely homotopic to a loop either on a cusp cylinder or cusp torus corresponding to a different cusp.

3.7.1 Compressible and incompressible boundary

If a component Ω of $\Omega(G)$ is not simply connected, there is a simple loop $\gamma^* \in \Omega$ which separates its boundary components. Of course γ^* is homotopic to a point if we move it into $\mathbb{H}^3$. Its projection $\gamma \subset R = \Omega / \operatorname{Stab}(\Omega)$ is a closed loop, perhaps not a simple loop, which is not homotopic to a point in R, but is homotopic to a point in $\mathcal{M}(G)$. Dehn's Lemma and the Loop Theorem say that there is a simple loop $\gamma' \in R$ which bounds an essential disk in $\mathcal{M}(G)$.

A component R of $\partial \mathcal{M}(G)$ is *incompressible* if the inclusion $\pi_1(R) \hookrightarrow \pi_1(\mathcal{M}(G))$ is injective. Our argument shows that R is incompressible if and only if all the components of $\Omega(G)$ which lie over R are simply connected. Otherwise R is called *compressible*. If all the boundary components are incompressible, the manifold $\mathcal{M}(G)$ is called *boundary incompressible*.

More generally, an orientable surface S embedded in $\mathcal{M}(G)$ is called *incompressible*, if it is not a topological disk and if the inclusion $\pi_1(S) \hookrightarrow \pi_1(\mathcal{M}(G))$ is injective. This means that every loop in S which is homotopic to a point in $\mathcal{M}(G)$ is already homotopic to a point in S. Otherwise there is a simple loop in S bounding an essential disk whose interior lies in $\mathcal{M}(G) \backslash S$ [Jaco 1980, III.8]. The surface S is incompressible if and only if each lift over it in $\mathbb{H}^3$ is simply connected.

An essential disk D in $\mathcal{M}(G)$, $\partial D \subset \partial \mathcal{M}(G)$, is called a *compressing disk*. It either divides $\mathcal{M}(G)$ into pieces M_1, M_2 or $M_1 = \mathcal{M}(G) \setminus D$ is connected. In the first case the fundamental group of $\mathcal{M}(G)$ splits into a free product: $\pi_1(\mathcal{M}) = \pi_1(M_1) * \pi_1(M_2)$ and correspondingly G splits: $G = G_1 * G_2$ (van Kampen's Theorem). In the second case let γ be a simple loop from an origin $O \in M_1$ that crosses D once. Then $\pi_1(\mathcal{M}(G)) = \langle \pi_1(M_1), \gamma \rangle$ or $G = \langle G_1, T \rangle$ where $T G_1 T^{-1} = G_1$, $T \notin G_1$ (G is an HNN-extension of G_1).

The subgroup $\pi_1(M_i)$ corresponds to a conjugacy class of subgroups of G—take a lift M_i^* of M_i to $\mathbb{H}^3$ and let G_i denote its stabilizer. There are one or more copies of the compressing disk D in the relative boundary of M_i^* in $\mathbb{H}^3$. These lifted disks bound a topological half-space of $\mathbb{H}^3$ not containing M_i^*. Adding these half-spaces to M_i^* gives back all $\mathbb{H}^3$. Moreover the half-spaces project injectively into $\mathcal{M}(G_i)$.

Now starting with some manifold $\mathcal{M}(G)$, the process of repeated insertion of compressing disks, which we can take to be mutually disjoint, terminates after a finite number of steps. We end up with a union of manifolds that are either balls or are boundary incompressible (see Hempel [1976] or Jaco [1980]). For example, if we start with a handlebody of genus g coming from a Schottky group, after cutting it along g mutually disjoint disks none of which divide the handlebody we will end up with a topological ball. See Exercise (3-11).

Here is a way of reversing the process of cutting $\mathcal{M}(G)$ by an essential disk: Choose disjoint, closed, round disks D_1, D_2 in $\Omega(G)$. Choose them small enough that each projects injectively into $\partial\mathcal{M}(G)$ and they remain disjoint there. Let σ_1, σ_2 denote the hyperbolic planes rising from the circles $\partial D_1, \partial D_2$. Let σ_i^- denote the half-space adjacent to D_i and σ_i^+ the other half-space. Choose any Möbius transformation T which has the property that $T(\sigma_1) = \sigma_2$ and $T(\sigma_1^+) = \sigma_2^-$. We see that $G^* = \langle G, T \rangle$ is a discrete group: T conjugates all the action of G in σ_1^+ to the action of TGT^{-1} in σ_2^-. Of course the operation is duplicated over the full orbit $G(D_i)$. The associated manifold $\mathcal{M}(G^*)$ is obtained from $\mathcal{M}(G)$ as follows. Down in $\mathcal{M}(G)$ we have the disks $D_i \subset \partial\mathcal{M}(G)$, and the planes σ_i which lie in the interior of $\mathcal{M}(G)$ except for their boundaries and bound balls (here we are using the same notation for the projections). Let M denote the result of removing from $\mathcal{M}(G)$ the two half-spaces σ_i^-. The action by T forms a new hyperbolic manifold $\mathcal{M}(G^*)$ from M by gluing σ_1 to σ_2. In $\mathcal{M}(G^*)$, $\sigma_1 \equiv \sigma_2$ is an essential disk which does not separate.

The procedure works equally well if we have two manifolds $\mathcal{M}(G_i)$ and take a disk in each boundary. In this case the new essential disk will divide the manifold. This process we have described is an example of *Klein–Maskit combination theory*, developed by Klein and refined and extended by Maskit 1988; see Marden [1974a] for the manifold interpretation. See also Exercise (3-8).

For the following result, see for example Waldhausen [1968].

Proposition 3.7.1. *Suppose M^3 is a compact, orientable and irreducible 3-manifold. If $\pi_1(M^3) = A * B$, $A, B \neq$ id, is a free product of subgroups, there exists a compressing disk bounded by a simple loop in ∂M^3.*

Remark 3.7.2. In calculating the genus of the boundary of a 3-manifold in terms of its fundamental group the following simple fact is very useful. Suppose $\alpha, \beta \in \partial M^3$ are two 1-cycles with nonzero intersection number. Then at most one of them can be homologous to zero, or, in particular, homotopic to a point in M^3. Thus if the fundamental group has N generators so that its first homology group has at most N generators, the total genus of the boundary is at most N. In particular, if G has no parabolics, $\partial\mathcal{M}(G)$ has at most $N/2$ components. Similar arguments give useful estimates for the topology of the boundary [Marden 1971; 1974a]. For example, if G is a g-generator free group ($g \geq 2$) and $\mathcal{M}(G)$ is compact with $\partial\mathcal{M}(G)$ a closed surface of genus g, then $\mathcal{M}(G)$ is a handlebody.

3.7.2 Extensions $\partial \mathcal{M} \to \mathcal{M}$

Often we will be in the position of having a group G and a quasiconformal deformation $F : \Omega(G) \to \Omega(H)$ that induces an isomorphism $\varphi : G \to H$. Such a map is called equivariant; it is the lift of a quasiconformal map $f : \partial \mathcal{M}(G) \to \partial \mathcal{M}(H)$ which (i) sends puncture pairs to puncture pairs, and (ii) sends compression loops to compression loops.

We will spell out in terms of given basepoints how the boundary map f respects the isomorphism $\varphi : \pi_1(\mathcal{M}(G); O) \to \pi_1(\mathcal{M}(H); O')$. On each boundary component R of $\partial \mathcal{M}(G)$, choose a basepoint p, and then choose the basepoint $f(p) \in f(R)$. To each loop $\alpha \subset R$ with basepoint p corresponds a loop $f(\alpha) \subset f(R)$ with basepoint $f(p)$. Upon joining the loops to the basepoints O, O' by auxiliary arcs, we get inclusion homomorphisms $\pi_1(R; p) \hookrightarrow \pi_1(\mathcal{M}(G); O)$ and $\pi_1(f(R); f(p)) \hookrightarrow \pi_1(\mathcal{M}(H); O')$ with kernels $K = \pi_c(R)$, $K' = \pi_c(f(R))$. There are a finite number of mutually disjoint simple compression loops on R such that the kernel K is the least normal subgroup $\pi_c(R) \subset \pi_1(R)$ generated by these (see the Maskit Planarity Theorem, p. 82). In turn the map f induces an isomorphism between the images of the inclusions.

We want to find a quasiconformal extension to $f : \mathcal{M}(G) \to \mathcal{M}(H)$. Although no "canonical" method seems available, the extension can be done by topological means (extension is not always possible in the geometrically infinite case).

Suppose first that $\mathcal{M}(G)$ is compact. According to Hempel [1976, Theorem 13.9 and Corollary 13.7], f is homotopic on $\partial \mathcal{M}(G)$ to a homeomorphism f_1 which has an extension to a homeomorphism between the manifolds $f_1 : \mathcal{M}(G) \to \mathcal{M}(H)$. In turn f_1 is homotopic to a diffeomorphism $f_2 : \mathcal{M}(G) \to \mathcal{M}(H)$, [Munkres 1960]; f_2 is automatically quasiconformal. We can choose a lift F_2 of the new f_2 to $\mathbb{H}^3 \cup \Omega(G)$ so that its restriction to $\Omega(G)$ induces φ and is homotopic to F. But now, applying Gehring [1962], F_2 has an φ-equivariant quasiconformal extension to all of $\mathbb{S}^2$.

If there are parabolics we have to replace the manifolds by the compact manifolds resulting from the removal of the solid pairing tubes and the solid cusp tori and extend the extension back to the original manifolds.

For applications it suffices to replace (F, f) by (F_2, f_2). However it is nicer to apply the stronger result Theorem 3.7.3 below.

It is shown in Exercise (3-37) that the original F itself has a homeomorphic extension to $\mathbb{S}^2$ satisfying $F(\zeta) = F_2(\zeta)$ for all $\zeta \in \Lambda(G)$. In fact, the extension of F is quasiconformal on all $\mathbb{S}^2$ by Theorem 3.14.7. This puts us in a position to apply Theorem 3.7.4(iii) below. We end up with a most satisfying result as follows:

Theorem 3.7.3. *Assume that G is geometrically finite and F is a quasiconformal mapping $\Omega(G)$ onto $\Omega(H)$ that induces an isomorphism $\varphi : G \to H$. Then F is the restriction of an equivariant quasiconformal map of $\mathbb{S}^2$ which extends to an equivariant quasiconformal mapping $F : \mathbb{H}^3 \cup \mathbb{S}^2 \to \mathbb{H}^3 \cup \mathbb{S}^2$. The mapping F then projects to a quasiconformal mapping $f : \mathcal{M}(G) \to \mathcal{M}(H)$.*

Now suppose $\mathcal{M}(G)$ is not necessarily geometrically finite. We start afresh with a quasiconformal mapping $F : \mathbb{S}^2 \to \mathbb{S}^2$ that induces an isomorphism $\varphi : G \to H$

satisfying $F(g(z)) = \varphi(g)F(z)$ for all $g \in G$, $z \in \mathbb{S}^2$. If the restriction of F is conformal $\Omega(G) \to \Omega(H)$, or if $\Omega(G) = \varnothing$, then F is Möbius and the two groups are conjugate. Here we are applying Theorem 5.6.8 or, if $\Omega(G) = \varnothing$, Corollary 3.13.6.

One general approach is the following. It is based on a canonical method of Douady and Earle 1986, see also Abikoff et al. [2004]. They discovered that a quasiconformal automorphism F of the $(n-1)$-sphere (when $n = 2$ such a map is called *quasisymmetric*) to can be extended to a surjective mapping of n-ball for $n \geq 2$. The extension is equivariant under a group G if F is so. On the other hand the extension is guaranteed to be a homeomorphism only when $n = 2$, or when the complex dilatation of the boundary mapping is sufficiently small (see McMullen [1996, p. 231]). Fortunately in the case $n = 3$ a modification suggested by Pekka Tukia (personal communication) allows one to get a homeomorphism of the ball without any restrictions. This modification is based on the fact [Ahlfors 1966, p. 100] that in dimension 2, given $\varepsilon > 0$, a quasiconformal mapping can be factored into the composition $F = F_n \circ F_{n-1} \circ \cdots \circ F_1$ of a finite number of equivariant quasiconformal mappings each of whose complex dilatations satisfies $\|\mu_k\|_\infty < \varepsilon$. This is done by taking $\mu_k = (k/n)\mu$ for sufficiently large n, where μ is the complex dilatation of F. In consistent normalizations, denote the solution of the corresponding Beltrami equation by g_k. Then set $F_k = g_k \circ g_{k-1}^{-1}$. The Douady-Earle extension is then applied to each factor F_k resulting in an extension to a equivariant homeomorphism of $\mathbb{H}^3$. The final result is stated below as item (ii).

The weakness of this approach is that the extension is not known to be quasiconformal or even bilipschitz. There is an alternate approach by integrating an extension of a vector field on $\mathbb{S}^n$. This method is suggested in Thurston [1979a, Chapter 11]; Reimann [1985] and carried out in McMullen [1996, Corollary B.23]. Martin Reimann's paper drew from this work the strong statement (i) listed below.

Theorem 3.7.4 (Basic Extension Theorems). *Suppose G, H are arbitrary kleinian groups and $F : \mathbb{S}^2 \to \mathbb{S}^2$ is a K-quasiconformal mapping that induces an isomorphism $\varphi : G \to H$. Then:*

(i) [Reimann 1985, §8] *F extends to a K^3-quasiconformal, φ-equivariant map of $\mathbb{H}^3$.*

(ii) [Abikoff et al. 2004; Douady and Earle 1986; Tukia 2005] *The map F has a φ-equivariant extension to $\mathbb{H}^3$ that is a homeomorphism.*

(iii) [McMullen 1996, Corollary B.23] *The map F has a φ-equivariant $K^{3/2}$-quasiisometric extension to $\mathbb{H}^3$.*

(iv) [Tukia 1985c] *The map F has a φ-equivariant (L, a)-quasiisometric extension for some $L = L(K)$, $a = a(K)$.*

(v) [Markovic 2014] *The map F has a φ-equivariant, quasiisometric, harmonic extension to $\mathbb{H}^3$.*

The extensions obtained in (iii), (iv), (v) are not necessarily homeomorphisms of $\mathbb{H}^3$. Nevertheless, they all project to mappings $\mathcal{M}(G) \to \mathcal{M}(H)$.

A mapping f of $\mathbb{H}^3$ is (L, a)-quasiisometric if there exist finite constants $1 \le L$ and $a \ge 0$ such that in the hyperbolic metric

$$\frac{1}{L}d(x, y) - a \le d(f(x), f(y)) \le Ld(x, y) + a.$$

Thus a quasiisometric map need not be continuous, but at long range it is essentially bilipschitz. Like quasiconformal maps of $\mathbb{H}^3$ [Gehring 1962], quasiisometric maps can be extended to $\partial \mathbb{H}^3 \equiv \mathbb{S}^2$ and the extension is a quasiconformal map of $\mathbb{S}^2$. If the quasiisometric map is a homeomorphism it will automatically be quasiconformal (but quasiconformal maps are not automatically bilipschitz). See Exercise (3-19).

For another version of extension, see Theorem 4.7.1.

The case of harmonic extensions of quasisymmetric maps $\partial \mathbb{H}^2 \to \partial \mathbb{H}^2$ was recently solved by Vlad Markovic (2015), finally resolving a long standing question.

Theorem 3.7.5 (The Schoen Conjecture). *Every quasisymmetric homeomorphism* $f : \partial \mathbb{H}^2 \to \partial \mathbb{H}^2$ *has a uniquely determined harmonic quasiconformal extension* $F : \mathbb{H}^2 \to \mathbb{H}^2$. *It is automatically equivariant under the action of a fuchsian group.*

This result complements the Douady-Earle extension recorded above. Uniqueness was proved earlier.

3.8 Quasifuchsian groups

A *quasifuchsian group* G is the quasiconformal deformation (p. 86) of a fuchsian group Γ. The purpose of this section is to characterize this class of groups by the topology of $\mathcal{M}(G)$.

Assume that G is a finitely generated kleinian group such that $\Omega((G)$ has two G-invariant components Ω_1, Ω_2. We will first show that $\Omega(G)$ has only the two components Ω_1, Ω_2 as claimed in Lemma 2.4.2(iii). And further, that there is an orientation-reversing homeomorphism between the Riemann surfaces $R_1 = \Omega_1/G$ and $R_2 = \Omega_2/G$.

Each component must be simply connected. Otherwise there would a simple loop α in Ω_1, say, that separates its boundary. This would force Ω_2 to make a choice of which component of $\mathbb{S}^2 \setminus \alpha$ to lie in. Whichever it chose, its boundary could not be the full limit set, a contradiction. In addition the Ahlfors Finiteness Theorem tells us each surface R_1, R_2 is a closed surface with at most a finite number of punctures and cone points.

From the perspective of $\partial \mathcal{M}(G) = R_1 \cup R_2$ there is an "identity" isomorphism $j : \pi_1(R_1) \to \pi_1(R_2)$ that comes from identification of the action of $g \in G$ on Ω_1 with its action on Ω_2. Loxodromics, parabolics and elliptics in Ω_1 correspond to transformations of the same type in Ω_2, as is seen by their fixed points on the boundaries.

Fix basepoints $O_i^* \in \Omega_i$ and a geodesic τ^* in $\mathbb{H}^3$ that connects them. Given $g \in G$, a simple arc $\gamma_i^* \subset \Omega_i$ from O_i^* to $g(O_i^*)$ projects to a loop $\gamma_i \in R_i$ from the point $O_i = \pi(O_i^*)$, $i = 1, 2$. Then $\gamma_1^* \sim g(\tau^*)\gamma_2^*\tau^{*-1}$ and the projections to $\mathcal{M}(G)$ satisfy $\gamma_1 \sim \tau\gamma_2\tau^{-1}$; γ_1 is freely homotopic to γ_2 in $\mathcal{M}(G)$. This gives the isomorphism j.

A change of basepoints will give the same isomorphism j if the connecting arc τ^* is correspondingly adjusted.

Applying the Cylinder Theorem, given a simple loop $\gamma_1 \subset R_1$ there is a simple loop $\gamma_2 \subset R_2$ that bounds with γ_1 an essential cylinder within $\mathcal{M}(G)$.

A very similar situation arises for punctures and cone points. If $g \in G$ is parabolic, since each Ω_i is simply connected, its fixed point corresponds to a puncture in each of R_i. According to Corollary 3.5.3, its fixed point supports a horocycle in both Ω_1 and Ω_2. From this, we can construct a solid pairing tube in $\mathcal{M}(G)$ pairing the two punctures. Also each elliptic transformation has one fixed point in each component and its axis of rotation extends from one to the other, analogous to the situation for a parabolic.

The argument proceeds as follows. Suppose first that there are no elliptics or parabolics. Consider a simple closed geodesic c in the hyperbolic metric on R_1. There is a corresponding geodesic c' in R_2 such that c, c' are the boundary components of a cylinder in $\mathcal{M}$. If d is a simple geodesic in R_1 crossing c exactly once and $d' \subset R_2$ corresponds to d, the two cylinders can be adjusted so that they are transverse to each other within $\mathcal{M}(G)$—they intersect in a single arc.

Now take a chain of $2g$ simple geodesics $\{c_i\}$ in R_1, where g is the genus, such that c_i crosses c_{i-1} and c_{i+1} while c_{2g} crosses c_{2g-1} and c_1, but otherwise the geodesics are mutually disjoint. The complement of their union in R_1 is simply connected. Insert cylinders so that within $\mathcal{M}(G)$ each is transverse to its neighbors but disjoint from the others. Let M denote the complement in $\mathcal{M}(G)$ of the union of the cylinders. The interior of M can only be a ball because it is bounded by a topological 2-sphere. This establishes the product structure for this case.

In the general case, choose mutually disjoint solid pairing tubes for the pairs of punctures and solid tubes about the rotation axes. Connect the union of the geodesics $\{c_i\}$ with the circles about the punctures and branch points in R_1 so the result bounds a simply connected region. Then extend this to connect within $\mathcal{M}$ the union of cylinders about punctures and branch points to once again get a complementary region M bounded by a topological 2-sphere. The argument is completed as before.

We have shown that $\Omega(G) = \Omega_1 \cup \Omega_2$. Moreover, Ω_1 and Ω_2 share the same boundary $\Lambda(G)$ and the same group G.

Select a fuchsian group Γ, acting say in UHP and LHP, for which there is a type-preserving isomorphism $\varphi : \Gamma \to G$. Now there is a reflection in $\mathbb{R}$, $J^* :$ UHP $\leftrightarrow$ LHP. It projects to an involution

$$J : X_1 = \text{UHP}/\Gamma \longleftrightarrow X_2 = \text{LHP}/\Gamma.$$

Corresponding to J there is a "reflection" $\chi : R_1 \leftrightarrow R_2$ that commutes with $\varphi : \varphi \circ J = \chi \circ \varphi$.

Choose quasiconformal maps $F_i : X_i \to R_i$, $i = 1, 2$ so that $F_2 \circ J = \chi \circ F_1$ and $F_1 \circ J = \chi \circ F_2$ thereby becoming fiber preserving. Lift to a map $\widehat{F} : \Omega(\Gamma) \to \Omega(G)$. The complex dilatation $\mu(z)$ of $\widehat{F}$ on $\Omega(X)$ can be taken so that $\| \mu \| \leq k < 1$,

$i = 1, 2$. Solving the Betrami equation for μ gives us a quasiconformal map $\widehat{F}'$: $\mathbb{S}^2 \to \mathbb{S}^2$. The solution may give rise to a new group G' but isomorphic to G. The restriction to $\Omega(\Gamma)$ projects to a map $F : \partial \mathcal{M}((\Gamma) \to \partial \mathcal{M}(G')$. If we appeal in addition to Theorem 3.7.3, we can take $F : \mathcal{M}(\Gamma) \to \mathcal{M}(G')$.

The bottom line is that G' (and G as well) is a quasiconformal deformation of Γ. For another proof see Maskit [1970].

3.8.1 Simultaneous uniformization

Suppose Γ is a fuchsian group acting again in the upper and lower half-planes and then in upper half-space. The orientation-reversing involution $J_0 : (z, t) \in \mathbb{H}^3 \mapsto (z, t)$ interchanges the upper and lower half-plancs, UHP and LHP, and pointwise fixes the vertical plane P rising from $\mathbb{R}$. It satisfies $J_0 \circ \gamma = \gamma \circ J_0$ for all $\gamma \in \Gamma$. The projection J_{0*} is an anticonformal, fiber-preserving mapping of $\mathcal{M}(\Gamma)$ that exchanges the two boundary components, and pointwise fixes the midplane P/Γ. It exchanges the outer normal at a point on the bottom surface and the outer normal at the corresponding point on top.

A quasifuchsian deformation G of Γ is induced by a quasiconformal map (see Section 3.6.3) $f : \mathbb{H}^3 \cup \partial \mathbb{H}^3 \to \mathbb{H}^3 \cup \partial \mathbb{H}^3$ that satisfies $f \circ \gamma(x) = \theta(\gamma) \circ f(x)$ for all $x \in \mathbb{H}^3 \cup \partial \mathbb{H}^3$, for an isomorphism $\theta : \Gamma \to G$. The map f projects to a homeomorphism $f_* : \mathcal{M}(\Gamma) \to \mathcal{M}(G)$. Hence $J' = f_* \circ J \circ f_*^{-1}$ is an orientation-reversing involution of $\mathcal{M}(G)$, exchanging its two boundary components and fiber preserving.

$$
\begin{array}{ccc}
R^{\text{top}} & \xrightarrow{\text{Conf}} & \partial^{\text{top}} \mathcal{M}(G) \\
{\scriptstyle J}\downarrow & & {\scriptstyle J'}\downarrow \\
R_{\text{bot}} & \xrightarrow{\text{Conf}} & \partial_{\text{bot}} \mathcal{M}(G)
\end{array}
$$

It is customary to refer to the boundary component UHP/Γ as the *top* boundary component of $\mathcal{M}(\Gamma)$ and LHP/Γ as the *bottom* and correspondingly for any quasifuchsian deformation of Γ. The following generative result is due to Bers.

Simultaneous uniformization ([Bers 1960]). *Suppose R_{bot}, R^{top} are two Riemann surfaces of finite hyperbolic area and $J : R_{\text{bot}} \leftrightarrow R^{\text{top}}$ is an orientation-reversing involution. There exists a quasifuchsian group G, uniquely determined up to Möbius equivalence, such that the top boundary component of $\mathcal{M}(G)$ is conformally equivalent to R^{top}, the bottom conformally equivalent to R_{bot}, and such that the fiber-preserving "reflection" J' of $\mathcal{M}(G)$ is homotopic on $\partial \mathcal{M}(G)$ to the restriction of J.*

In other words, the two components of $\Omega(G)$ serve as the universal covering surfaces of R_{bot} and R^{top} respectively.

3.9 Geodesic and measured geodesic laminations

In this section we will introduce the notions of geodesic and measured geodesic laminations in $\mathbb{H}^2$ which are needed to understand not only the full structure of simple geodesics on a surface, but also the internal structure of hyperbolic manifolds. General references are [Fathi et al. 1979], [Canary et al. 1987], [Bonahon 2001].

3.9.1 Geodesic laminations

It is helpful to think in terms of the disk model of $\mathbb{H}^2$. Let G be a fuchsian group such that $\mathbb{H}^2/G = R$ is a surface of finite hyperbolic area (no elliptics). We will assume that R is not the triply punctured sphere so that we are not distracted by this simple case.

Draw any simple closed curve γ from a basepoint $O \in R$ which is not retractible to a point or a puncture. Choose $O^* \in \mathbb{H}^2$ over O. A lift γ_0^* of γ beginning at O^* terminates at $g(O^*)$ for some $g \neq \mathrm{id} \in G$. The orbit γ^* of γ_0^* under the cyclic group $\langle g \rangle$ is a simple arc in $\mathbb{H}^2$ with endpoints at the fixed points of g. Since γ^* projects to a simple loop, it will have the property that its orbit under the full group G consists of mutually disjoint arcs.

Now consider the axis α^* of g, namely the hyperbolic line between the endpoints of γ^*; α^* projects to a closed loop α on R which is necessarily a simple loop and a geodesic. Furthermore, γ is freely homotopic to α. In Exercise (3-3) we will find that α^* does not penetrate the universal horoballs at parabolic fixed points.

Fix a fundamental polygon P for G (for examples, see Figure 3.6, and Exercise (2-14)). Consider a sequence of simple closed geodesics $\{\alpha_n\}$ that are getting longer and longer, say in terms of a fixed set of generators for $\pi_1(R; O)$. Choose a point $p_n \in \alpha_n$ and a lift so that $p_n^* \in \alpha_n^*$ lies in P; the corresponding axes α_n^* all intersect P. What happens as $n \to \infty$? Since all the axes intersect P (but do not enter the universal horoballs at the cusps of P) we can find a subsequence that converges to a geodesic σ^* in $\mathbb{H}^2$. Necessarily neither endpoint of σ^* is a parabolic fixed point. What about the projection σ to R?

First of all σ, can have no self-intersections, since the orbit of σ^* under G consists of mutually disjoint geodesics. Second, it cannot be a closed geodesic, for $\lim g_n$ cannot exist as a proper Möbius transformation. Therefore σ is a simple geodesic of infinite length on R. As such it has limit points in R—that is, there are sequences of points $\{p_n^*\} \in \sigma^*$ which converge to an endpoint so that the projections $\{p_n = \pi(p_n^*)\} \in R$ converge to a point $p \in R$. Such a geodesic σ is called *recurrent*; it keeps returning to a compact set in R.

However we cannot say that in R, $\{\alpha_n\}$ "converges" to σ. For if a different point $p_n' \in \alpha_n$, increasingly far away along α_n from p_n, was lifted to P, the sequence of lifts will not necessarily converge to σ^*, or even to a leaf in the G-orbit of σ^*.

A *geodesic lamination* $\Lambda^* \subset \mathbb{H}^2$ is a *closed* set of mutually disjoint geodesics. Each component of Λ^* is called a *leaf*. Two leaves are allowed to have a common endpoint on $\partial \mathbb{H}^2$. The components of $\mathbb{H}^2 \setminus \Lambda^*$ are called *gaps*. The gaps are ideal

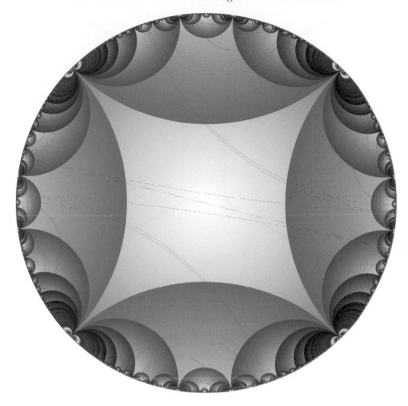

Fig. 3.6. A discrete geodesic lamination consisting of the lifts of a long simple geodesic on a once-punctured torus. A large fundamental polygon is in the foreground, and its iterates are shown to cover $\mathbb{H}^2$.

polygons, possibly infinite-sided, possibly bounded by arcs of $\partial\mathbb{H}^2$. If there are no punctures, the minimal gaps are idea triangles.

The space $\mathcal{GL}(\mathbb{H}^2)$ of all geodesic laminations on $\mathbb{H}^2$ is given the topology of Hausdorff convergence (see p. 232): A sequence converges $\Lambda_n^* \to \Lambda^*$ if and only if on every compact subset $K \subset \mathbb{H}^2$, $\{\Lambda_n^* \cap K\}$ converges to $\Lambda^* \cap K$ in the Hausdorff topology. That is, a neighborhood of $\Lambda^* \subset \mathbb{H}^2$ contains all but a finite number of $\{\Lambda_n^*\}$, and if $U \subset \mathbb{H}^2$ is an open set containing all but a finite number of $\{\Lambda_n^*\}$ then $\Lambda^* \subset U$. With this topology, $\mathcal{GL}(\mathbb{H}^2)$ becomes a compact Hausdorff space.

In fact, there is a natural topology on the space of geodesics in $\mathbb{H}^2$ so that it becomes a Möbius band (Exercises (1-3, 4-15)). An individual geodesic becomes a point in the Möbius bandwhile a geodesic lamination becomes a closed pointset.

Assume now that Λ^* is G-invariant. The projection Λ to $R = \mathbb{H}^2/G$ is a closed set of mutually disjoint simple (but not in general closed) geodesics in R which cover a set of zero area. The same Hausdorff topology applies to sequences. On the finite surface R, $\mathcal{GL}(R)$ is a compact Hausdorff space.

Here we will typically consider not all geodesic laminations on R, but, when R has punctures, only compactly supported ones—the closed subset $\mathcal{GL}_0(R) \subset \mathcal{GL}(R)$ consisting of those whose leaves do not end at a puncture (a simple closed geodesic cannot penetrate a universal horodisk—Exercise (3-3)).

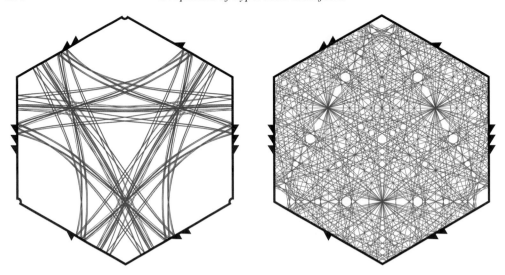

Fig. 3.7. At left, a picture of the union of all simple closed hyperbolic geodesics for the hexagonal once-punctured torus, conformally mapped to a euclidean hexagon with side identifications. At right, a smaller collection of simple closed geodesics for the euclidean structure of this torus with a distinguised point. Note the dramatic difference in their density in the two metrics.

A leaf $\ell \subset \Lambda$ is *isolated* if every point $z \in \ell$ has a neighborhood whose intersection with Λ consists of a segment of ℓ through z. For example, suppose α, β are disjoint simple closed geodesics on R. There is an isolated geodesic ℓ one end of which spirals infinitely often around one side of α and the other end infinitely often around one side of β. A lift of γ in $\mathbb{H}^2$ will connect one fixed point of a loxodromic over α to a fixed point of a loxodromic over β. Yet ℓ is not a geodesic lamination since it is not closed in the space of geodesics. The lamination is $\gamma \cup \alpha \cup \beta$.

In R, simple closed geodesics are dense in the subspace $\mathcal{GL}_0(R)$. See for example Canary et al. [1987] for more discussion.

A well-known result of Birman and Series [1985] is that the set of all simple (but not necessarily closed) geodesics on a finite surface R form a set of Hausdorff dimension one—see Exercise (3-20). An interesting consequence is that almost every geodesic arc $[a, b] \subset R$ is *generic* with respect to simple geodesics in the sense that it is transverse to *every* simple closed geodesic on R [Bonahon 2001, p. 19].

Another consequence is that if Λ has no isolated leaves, then Λ has uncountably many leaves: For any transverse segment τ, $\tau \cap \Lambda$ is totally disconnected; $\tau \cap \Lambda$ is a Cantor set [Bonahon 2001, Prop. 7].

Mirzakhani [2008] found the precise growth of the number $S_X(L)$ of simple *closed* geodesics of length $\leq L$ on a hyperbolic surface X of genus g and n punctures:

$$S_X(L) \sim nL^{6g+2n-6}, \quad \text{as } L \to \infty,$$

where $n = n(X)$ is a constant depending on X.

Any lamination $\Lambda^* \subset \mathbb{H}^2$, invariant under G, can be augmented by additional leaves if necessary so that the closures of the gaps $\mathbb{H}^2 \setminus \Lambda^*$ are ideal triangles. The projection to R of a leaf of Λ^* is either a closed curve, ends at a puncture, or spirals around R without intersecting itself or other gap projections. Since each ideal triangle has area π, there are exactly $2(2g + n - 2)$ different gaps $R \setminus \Lambda$ (Exercise (3-1)).

A lamination $\Lambda \subset R$ is called *minimal* if it has no closed sublaminations. Each geodesic lamination on R can be decomposed into (i) the union of finitely many isolated leaves whose ends spiral to a minimal sublamination or end at a puncture, and (ii) the union of finitely many minimal sublaminations with the property that every half-leaf is dense in the sublamination. A closed geodesic is in the second category, and a geodesic whose ends are at punctures is in the first.

Finally, for later application, we note the following. A given Λ can be covered with open sets $\{U_i\}$ with continuous maps $\phi_i : U_i \cap \Lambda \to X_i \times (0, 1) \subset \mathbb{R}^2$, taking leaves to vertical line segments indexed by $X_i \subset \mathbb{R}$, so that $\phi_j \circ \phi_i^{-1}(x, y) = (f(x), g(x, y))$ preserves verticality for overlapping neighborhoods.

3.9.2 Measured geodesic laminations

A geodesic lamination Λ on a finite area hyperbolic surface R (and hence its lift $\Lambda^* \subset \mathbb{H}^2$) is called a *measured lamination* and written $(\Lambda; \mu)$ if there is a Borel measure μ with support contained in (usually it will be equal to) Λ. More precisely, each transverse segment τ, with endpoints in gaps, has finite, positive (occasionally zero) measure $\mu(\tau)$ where the measure depends only on the equivalence class of τ, that is, $\tau_1 \equiv \tau$ if and only if the endpoints of τ_1 are in the same gaps as the endpoints of τ.

We will be working with the class $\mathcal{ML}(R)$ of measured laminations $\{(\Lambda; \mu)\}$ with measures $\{\mu\}$ that are *uniformly bounded* in the sense that for each μ, there is a constant $C < \infty$ such that $\mu(\tau) < C$ for all transversals τ of unit length.

These are compactly supported measures on R: No leaf λ can end at a puncture. For the closer to the puncture a transversal of unit length gets, the higher its intersection number with λ. Likewise no leaf λ can spiral into a closed geodesic. This would occur if up in $\mathbb{H}^2$, a leaf of Λ^* shares an endpoint with the axis of a loxodromic representing a simple closed geodesic. For this reason, the only isolated leaves of a measured lamination $\Lambda \in \mathcal{ML}(R)$ are simple closed geodesics.

The topology imposed on $\mathcal{ML}$ is that $(\Lambda_n; \mu_n) \to (\Lambda; \mu)$ if and only if (i) $\Lambda_n \to \Lambda$ in the Hausdorff topology, and (ii) for any transverse segment τ with endpoints in the gaps of Λ, we have $\lim \mu_n(\tau) = \mu(\tau)$. We also allow the zero-lamination with no leaves and zero measure. Of course the measures are assumed to be uniformly bounded.

$\mathcal{ML}(R)$ is a Hausdorff space but it is not compact (just multiply the measure by positive constants). If $(\Lambda; \mu) \in \mathcal{ML}(R)$, the support $|\Lambda|$ of μ can be referred to as the "unmeasured lamination". However the topology imposed on $|\Lambda|$ alone is less restrictive. Unmeasured laminations do not form a Hausdorff space.

The Hausdorff topology induced on the unmeasured laminations $\{|\Lambda|\}$ is less restrictive than just the Hausdorff topology since $\mathcal{ML}$ topology requires the existence of uniformly bounded transverse measures. In fact with the induced topology, unmeasured laminations do not form a Hausdorff space.

The minimal gaps determined by a measured lamination in R are ideal triangles, ideal bigons containing one puncture, and ideal monogons containing one puncture.[3] An *ideal bigon* is the union of two ideal triangles and an *ideal monogon* one, after it is slit from the puncture to the ideal point. Each ideal triangle has area π, each ideal bigon has area 2π and each monogon has area π. An n-sided ideal polygon made up of $n-2$ ideal triangles, or n triangles if it contains one puncture, can also appear as a gap.

Finally we introduce the quotient space of *projective measured laminations*

$$\mathcal{PML}(R) = (\mathcal{ML}(R) \setminus 0)/\text{multiplication by positive scalars.}$$

In this space two measured laminations with identical support are equivalent, $\mu_1 \equiv \mu_2$, if and only if $\mu_2 = c\mu_1$ for some constant $c > 0$. Unlike the space $\mathcal{ML}(R)$, $\mathcal{PML}(R)$ is compact. In fact $\mathcal{PML}(R)$ is homeomorphic to the sphere $\mathbb{S}^{6g-7}$, or $\mathbb{S}^{6g+2n-7}$ if there are n-punctures.

3.9.3 Geometric intersection numbers

Given simple loops α, β, their *geometric intersection number* $\iota(\alpha, \beta) = \iota(\beta, \alpha)$ is defined to be the minimum number of crossings of simple loops in their respective free homotopy classes. This minimum is achieved by the geodesics in the free homotopy classes. The intersection number $\iota(\alpha, \beta) = 0$, if and only if, in their free homotopy classes, there are representatives which are disjoint, or if $\beta = \alpha$.

The geometric intersection number should not be confused with the algebraic intersection number from combinatorial topology in which the crossings of a pair of oriented curves or chains are counted according to the direction of the crossings.

Equally we can define the geometric intersection number of two multicurves (a *multicurve* is a collection of mutually disjoint, nonparallel simple closed curves). If we take all the elements of a multicurve $\{\alpha_i\}$ to be geodesics, the intersection number is the total number of crossings $\cup_{i,j} \iota(\alpha_i, \alpha_j)$, remembering to identify $\iota(\alpha_i, \alpha_j) \equiv \iota(\alpha_j, \alpha_i)$.

The geometric intersection number of a geodesic arc τ with endpoints in gaps and a multicurve is defined likewise (see Exercise (2-6) for the torus case).

The most general transverse measure on a finite system Λ of mutually disjoint closed geodesics is obtained by assigning an atomic measure $\alpha(\ell)$ to each leaf $\ell \in \Lambda$. Then we define a measure μ on the lamination Λ by linearity

---

[3] A bigon B (monogon M) on S is a annular region about a puncture whose boundary in S consists of two (one) infinite geodesics whose ends are asymptotic to each other "at ∞". Each lift of B or M to $\mathbb{H}^2$ is an infinite-sided ideal polygon invariant under a cyclic parabolic subgroup.

$$\mu(\tau) = \int_\tau d\mu = \iota(\tau; \mu) = \sum_{\ell \in \Lambda} \alpha(\ell)\iota(\tau; \ell). \tag{3.6}$$

The measured lamination is the pair $(\Lambda; \mu)$.

Thurston (see Bonahon [2001]) proved that the measured laminations with support on a simple closed geodesic are dense in $\mathcal{ML}(R)$ [Thurston 1988; Fathi et al. 1979]. Bonahon describes this process as being akin in 2-dimensions to the passage from the lattice $\mathbb{Z}^2$ to $\mathbb{R}^2$.

Specifically, given $(\Lambda; \mu) \in \mathcal{ML}(R)$ there exists (i) a sequence of simple closed geodesics $\{\alpha_n\}$ which converge to Λ in the Hausdorff topology, and (ii) a corresponding sequence $\{a_n\}$ of strictly positive numbers such that for any simple loop or arc τ transverse to Λ,

$$\mu(\tau) = \lim_{n \to \infty} \frac{\iota(\tau, \alpha_n)}{a_n} = \int_\tau d\mu. \tag{3.7}$$

Conversely, given a sequence $\{\alpha_n\}$ of simple closed geodesics on a closed surface (or contained in a compact part of a punctured surface), there exists a subsequence $\{\alpha_m\}$ which, with associated positive numbers $\{a_m\}$, there is convergence to a measure μ on the Hausdorff limit $\Lambda = \lim \alpha_m$:

$$\lim_{m \to \infty} \frac{\iota(\tau, \alpha_m)}{a_m} = \mu(\tau) = \int_\tau d\mu = \iota(\tau, \mu) \neq \infty,$$

for all simple loops or transverse segments τ and $\mu(\tau) \neq 0$ for at least one τ. Typically, good choices are $a_m = \text{Len}(\alpha_m)$ (Exercise (3-38)), or $a_m = \iota(\tau_0, \alpha_m)$ if τ_0 is a fixed geodesic transverse to all α_m. Here $1/a_m$ is the atomic measure assigned to α_m.

The sequence $\{a_m\}$ and therefore μ is uniquely determined asymptotically, up to a positive multiplicative constant. Indeed, if $\{a_m'\}$ is another sequence that determines the measure μ' on Λ, there exists a constant $C \neq 0, \infty$ such that $\lim a_m/a_m' = C$. To see why, suppose for transversals τ, τ', $C(\tau) = \lim \iota(\tau, \alpha_m)/a_m \neq 0$ and $C'(\tau') = \lim \iota(\tau', \alpha_m)/a_m' \neq 0$. Then

$$\lim a_m/a_m' = C'(\tau')/C(\tau') = C'(\tau)/C(\tau) \neq 0, \infty.$$

The support of a measure μ may not be the whole Hausdorff limit Λ. To illustrate what can happen, take the lamination consisting of two disjoint simple geodesics α_1, α_2 with assigned integer multiplicities m_1, m_2. We can construct a sequence of simple geodesics $\{\alpha_n\}$ that go nm_1 times around α_1, and nm_2 times around α_2. The Hausdorff limit $\Lambda = \lim \alpha_n$ is a union $\alpha_1 \cup \alpha_2 \cup \ell_1 \cup \ell_2$ where ℓ_1, ℓ_2 are infinite length geodesics each spiraling around one side each of α_1 and of α_2. Up in $\mathbb{H}^2$, each endpoint of a lift of ℓ_i is a fixed point of a transformation determined by one of the closed leaves. If τ is a simple loop transverse to α_1 but not α_2, $\iota(\tau, \alpha_n)/n \to m_1$ and similarly the limit is m_2 if it is transverse to α_2 but not α_1. If τ cuts $\ell_1 \cup \ell_2$ but not $\alpha_1 \cup \alpha_2$, the limit is zero so $\ell_1 \cup \ell_2$ is not in the support of the measure we defined.

The bottom line is that every geodesic lamination Λ has a transverse measure whose support consists of all the minimal sublaminations of Λ [Bonahon 2001;

Otal 1996]. Two minimal laminations with the same (nonzero) transverse measure are identical. In the above example the minimal laminations are α_1 and α_2. The spiraling geodesic is not in the support of μ.

3.9.4 Length of measured laminations

The sequence of geodesic lengths $\{\text{Len}(\alpha_n)\}$ also has a limit if it is scaled by the same $\{a_n\}$ that determines μ, namely

$$\mu(\tau) = \lim_{n \to \infty} \frac{\iota(\tau, \alpha_n)}{a_n}, \quad \text{Len}_\mu(\Lambda) = \lim_{n \to \infty} \frac{\text{Len}(\alpha_n)}{a_n}. \tag{3.8}$$

$L_\mu(\Lambda)$ is called the *length* of the measured lamination $(\Lambda; \mu)$. Unlike intersection numbers, the value of L_μ depends on the particular hyperbolic surface R where the measurement is made; it is known to change continuously as the underlying surface is deformed [Kerckhoff 1985, Lem. 2.4].

Another way of looking at the length is to cover the support Λ of μ by a finite number of "rectangles" such that the "horizontal" sides are disjoint from Λ and the "vertical sides" are transverse arcs $\{\tau_i\}$ that cut Λ so that each leaf that crosses one vertical side also crosses the opposite. Then in each rectangle integrate with respect to $d\ell$, the hyperbolic length is followed by integration with respect to $d\mu$ along a vertical side. From this point of view, the continuity of the length function on $\mathcal{ML}(R)$ in terms of change of hyperbolic metric on R follows ([Kerckhoff 1985, p. 26], [Bonahon 2001, p. 21]). This reasoning leads to the intrinsic expression for length.

$$\text{Len}_\mu(\Lambda) := \iint_R d\ell \times d\mu. \tag{3.9}$$

Thus if Λ is a single simple closed geodesic, and μ is the atomic measure of unit weight, $\text{Len}_\mu(\Lambda)$ is just the geodesic length on R. The integral is obtained from the local product structure of the lamination determined by an open cover, using a partition of unity.

Thurston showed that a finite family of "generic" arcs $\{\tau_i\}_{i=1}^n$ can be chosen so that, corresponding to $(\Lambda; \mu)$, is the map

$$\mu \in \mathcal{ML}(R) \mapsto (\mu(\tau_1), \dots, \mu(\tau_n)) \in \mathbb{R}^n.$$

This induces a homeomorphism to a piecewise linear (PL) submanifold[4] of $\mathbb{R}^n$ of real dimension $(6g+2b-6)$. An arc τ is "generic" if it is transverse to every simple closed geodesic in R. For each τ there exists a PL function which expresses $\mu(\tau)$ in terms of $\{\mu(\tau_i)\}$ for every $\mu \in \mathcal{ML}(R)$.

Using a (quasiconformal) homeomorphism f from one finite area surface R to another S, the space $\mathcal{ML}(R)$ can be transferred to $\mathcal{ML}(S)$. Namely let f^* be a lift of f to $\mathbb{H}^2 \equiv \mathbb{D}$. The homeomorphism extends to a homeomorphism also denoted f^* on

---

[4] A PL submanifold M is a space with a family of charts which locally model M by open subsets of $\mathbb{R}^n$ so that transition maps are restrictions of PL maps $\mathbb{R}^n \to \mathbb{R}^n$.

$\partial \mathbb{D}$. By taking images of endpoints, f^* induces a map of the (lifted) measured lamination over one surface to the other. Then project back to R and S. For this reason, if you have seen $\mathcal{ML}$ on one surface, you have seen it on all surfaces in the deformation space. However metric properties will differ, one surface to the other.

Start with the intersection number $\iota(\sigma, \mu)$ defined for an atomic measure μ on a simple closed geodesic σ. The intersection number ι can be extended by continuity to any measured lamination $(\Lambda; \mu) \in \mathcal{ML}$. It can be extended again by continuity to $\iota(\mu, \nu)$ for any pair of measured laminations [Rees 1981]. Specifically, if we write $(\Lambda, \mu), (\Lambda', \nu)$ as limits of simple closed geodesics $\mu = \lim \alpha_n/a_n, \nu = \lim \beta_n/b_n$, then

$$\iota(\mu, \nu) = \lim_{n \to \infty} \frac{\iota(\alpha_n, \beta_n)}{a_n b_n};$$

see Bonahon [1986]. If $\iota(\mu, \nu) = 0$, any component of the support of μ is either identical to a component of the support of ν or disjoint from all of its components.

This generalization of the geometric intersection number remains a topological entity, independent of any particular complex structure the underlying surface R may have.

There is a theory of measured foliations $\mathcal{MF}(R)$ on a surface R which is the topological version of measured laminations. Roughly, a measured foliation is a (necessarily singular) foliation with a measure of distances between leaves. Measured foliations are modeled by quadratic differentials; see Exercise (5-33). Every measured foliation comes from a measured lamination by lifting to $\mathbb{H}^2$, first showing (noncritical) leaves have endpoints on $\partial \mathbb{H}^2$, then replacing the leaves by geodesics with the same endpoints. The converse is also true. In fact there is a homeomorphism between $\mathcal{MF}(R)$ and $\mathcal{ML}(R)$. The map is the identity on weighted simple closed curves, and it preserves intersection numbers [Kerckhoff 1992].

The quantitative results depend on taking a pants decomposition of R and classifying the intersection of a measured foliation with each pants (for a taste, see Exercise (3-38)). One way of getting lots of nontrivial examples of measured foliations is by means of interval exchange maps, see Exercise (3-40).

As with measured laminations, a measured foliation is uniquely determined up to a positive multiplicative constant. For this reason, one often deals with the quotient space $\mathcal{PMF}(R)$ of projective measured foliations.

Formal introduction to this beautiful and essential subject can be found, for example, in Thurston [1988]; Fathi et al. [1979]; Canary et al. [1987]; Casson and Bleiler [1988]; Bonahon [2001]; Otal [1996, Appendix]; Marden and Strebel [1984]; Matsuzaki and Taniguchi [1998].

Summary

(i) Each infinite length leaf $\ell \subset R$ of a lamination with compact support in R is *recurrent*: there is a sequence of points $\{\zeta_n\} \subset \ell$ such that along ℓ, $\zeta_n \to \infty$ yet there exists $\zeta \in \ell$ such that in a neighborhood of ζ in R, $\lim \zeta_n = \zeta$. Up in $\mathbb{H}^2$, this says that given a lift ℓ^*, there is a sequence of (mutually disjoint) lifts ℓ_n^* which converge to ℓ^* as euclidean circular arcs.

(ii) Two measured laminations in $\mathbb{H}^2$ whose set of leaves have the same combinatorics and the same transverse measures are usually not Möbius equivalent. For example, they may be lifts of finite laminations on two different surfaces where the distances between leaves differ (compare with Theorem 3.11.6).

(iii) We have seen how a lamination consisting of two or more mutually disjoint simple geodesics has many projectively inequivalent transverse measures. Yet there are geodesic laminations which support *only one* projective class of measures [Masur 1982]; such measured laminations are called *uniquely ergodic*. Uniquely ergodic laminations Λ have the minimality property that Λ is not the union of proper sublaminations. Uniquely ergodic laminations are dense in all measured laminations. Yet it is a subtle business to determine if a particular lamination is uniquely ergodic. The pair of laminations fixed by a pseudo-Anosov automorphism of a surface (see Exercise (5-12)) does have this property [Thurston 1988]. The analogous result on a square torus is a famous theorem of Hopf, which says that the projection to the quotient torus of a line of irrational slope in the square lattice in $\mathbb{C}$ is equally distributed on the torus.

(iv) A measured lamination $(\Lambda; \mu)$ on a finite area surface S is called *arational* or *filling* if each complementary component of Λ is an ideal polygon, possibly containing a single puncture. Consequently there are at most $4g + 2n - 4$ gaps for an arational lamination, where $n \geq 0$ is the number of punctures. An arational lamination is cut by every simple closed geodesic; more generally, if ν is any measured lamination on R with support different than Λ, then the geometric intersection number satisfies $\iota(\mu, \nu) \neq 0$. Arational laminations Λ also have the property that every half-leaf is dense; in particular Λ is minimal. Uniquely ergodic laminations are arational (see Otal [1996] for details).

(v) The discussion works as well on compact surfaces with boundary. However usually the simple geodesics one works with are not allowed to be parallel to boundary components, as in the case of cusps.

We close this section by listing adjectives attached to geodesics Λ or measured geodesic laminations $(\Lambda; \mu)$ on a finite area hyperbolic surface R:

maximal lamination Λ is not a proper subset of another lamination; each component of $R \setminus \Lambda$ is an ideal triangle.

minimal or connected lamination The support Λ has no sublaminations; either Λ consists of a single closed geodesic, or every leaf λ has infinite length and each half-leaf is dense in Λ. Every lamination is the union of finitely many minimal sublaminations and, if Λ is not measurable, possibly finitely many isolated[5] leaves whose ends spiral in to the minimal laminations or end at a cusp.

---

[5] A leaf λ is isolated if every point $p \in \lambda$ has a neighborhood U with $U \cap \lambda$ an arc through p.

filling or arational lamination[6] Each component of $R \setminus \Lambda$ is an ideal polygon possibly containing one puncture of R. Λ has positive intersection number with every closed geodesic. Every half-leaf is dense so Λ is also minimal.

filling or binding pair Two laminations Λ_1 and Λ_2 form a *filling pair*, and Λ_1 and Λ_2 *fill up* R if every nontrivial, nonperipheral closed curve on R cuts at least one of Λ_1, Λ_2. A filling pair satisfies $\iota(\gamma, \Lambda_1) + \iota(\gamma, \Lambda_2) > 0$ for every simple closed geodesic γ. Each component of $R \setminus (\Lambda_1 \cup \Lambda_2)$ is a (simply connected) polygon possibly containing a puncture of R.

uniquely ergodic lamination There is one and only one measure μ with support Λ, up to positive multiples. The support of a uniquely ergodic measured lamination is minimal, but not necessarily maximal.

3.10 The convex hull of the limit set

Fenchel had long advocated using the convex hull construction in $\mathbb{H}^3$ to study kleinian groups, since in his work with Nielsen he had found the corresponding construction in $\mathbb{H}^2$ for fuchsian groups very useful. However the difficulty was not in the construction, but in the analysis of the convex hull boundary. It was Thurston who taught us how to use the convex hull as an effective tool. The application required prior development of the theory of measured laminations.

In describing the theory, we will stick with the upper half-space model. We start with a closed set $\Lambda \subset \mathbb{C} \cup \infty \equiv \mathbb{S}^2$, with a nonempty complement $\Omega = \mathbb{S}^2 \setminus \Lambda$. The hyperbolic convex hull of Λ is defined as follows.

Let $C \subset \overline{\Omega}$ be a round circle in $\mathbb{S}^2$ that bounds an open disk $\Delta \subset \Omega$. If Λ is connected so that each component of Ω is simply connected, any circle in $\overline{\Omega}$ will determine such a disk. The circle C in turn determines a hyperbolic plane $C^* \in \mathbb{H}^3$. Denote by $H(C)$ the relatively closed half-space bounded by C^* that abuts the *exterior* of Δ. The (hyperbolic) *convex hull* of Λ is the relatively closed set

$$\mathcal{CH}(\Lambda) = \bigcap_{C \subset \overline{\Omega}} H(C). \tag{3.10}$$

In constructing $\mathcal{CH}(\Lambda)$ it suffices to restrict attention to maximal disks Δ—those that are not proper subsets of larger disks in Ω. The circle bounding a maximal disk meets $\partial \Omega$ in at least two points.

Since $\mathcal{CH}(\Lambda)$ is convex, the (hyperbolic) line segment joining any two of its points lies in the set. In fact any geodesic with endpoints in $\partial \Omega = \Lambda$ is contained in $\mathcal{CH}(\Lambda)$. With Peter Storm one can define $\mathcal{CH}(\Lambda)$ as the package obtained by "shrink wrapping" the set of all geodesics with endpoints in Λ.

The relative boundary $\partial \mathcal{CH}\Lambda) \subset \mathbb{H}^3$ is the union of *flat pieces* and *bending lines*.

---

[6] The two terms mean the same thing and are both widely used.

A flat piece, if not a whole plane, is a noncompact hyperbolic polygon contained in one of the hyperbolic planes C^* used to form the convex hull. It lies in the plane determined by a maximal disk that is bounded by a circle that meets $\partial\Omega$ in at least three points.

The complement in $\partial\mathcal{CH}(\Lambda)$ of the union of open flat pieces is the closed set of bending lines. A bending line ℓ is a geodesic whose endpoints lie in $\partial\Omega$. Distinct bending lines are disjoint but they possibly have a common endpoint. There are in general an uncountable number of them. An infinite sequence of bending lines has a subsequence which either converges to a bending line or to a point in the common boundary $\partial\mathcal{CH} = \partial\Omega$.

An isolated bending line ℓ is the common boundary of adjacent flat pieces. The *bending angle* at ℓ is taken to be the *exterior bending angle* (that is the angle taken with respect to the complement of Ω) α so that $\alpha = 0$ corresponds to no bending at all and $\alpha = \pi$ corresponds to one flat piece folded over the other.

Each component S of the relative boundary $\partial\mathcal{CH}(\Lambda) \cap \mathbb{H}^3$ faces a component Ω_S of Ω. It helps to keep in mind the picture of a domed stadium, such as one used to find in Minneapolis. The *floor* of the stadium is Ω_S and the *dome* is S.

In particular, if all components of Ω represent closed surfaces, then all sufficiently small horoballs at rank two parabolics, if any, lie in $\mathcal{CH}(\Lambda)$.

There is a continuous map $r : \Omega_S \to S$ called the *nearest point retraction*. This is defined as follows: Given $z \in \Omega_S$ examine the family of horospheres tangent to $\partial\mathbb{H}^3$ at z. This family depends on a parameter, for example the euclidean diameter. Exactly one of these spheres just touches S, necessarily at a single point, without crossing S. This point of first touching is called the *nearest point* and is denoted by $r(z)$. If $r(z)$ is in a flat piece, then there is a geodesic ray from $r(z)$, where it is orthogonal to $\partial\mathcal{CH}(\Lambda)$, ending at z. An isolated bending line $\ell \in S$ with bending angle α will be the image under r of a crescent $C_\ell \in \Omega_S$ with vertices in $\partial\Omega_S$.

The crescent C_ℓ is constructed as follows. There are two planes C_1^*, C_2^* rising from maximal circles C_1, C_2 and intersecting with exterior angle α. The angle interior to $\mathcal{CH}(\Lambda)$ is $\pi - \alpha$. The sides of C_ℓ are orthogonal to C_1, C_2. Therefore the interior vertex angles of C_ℓ are

$$\alpha = 2\pi - \left(\frac{\pi}{2} + (\pi - \alpha) + \frac{\pi}{2}\right).$$

In particular, C_ℓ is not the crescent formed by $C_1 \cap C_2$ unless C_1 and C_2 are orthogonal.

Distinct isolated bending lines correspond to nonoverlapping crescents in Ω_S. If there are no isolated bending lines, r is a homeomorphism.

The nearest point retraction r fixes the points on the common boundary $\partial\Omega_S = \partial S$. Convex hulls are studied in detail in Epstein et al. [2004]; see also Bridgeman [1998], Bridgeman [2003], Bridgeman and Canary [2010].

Examples

In the degenerate case that Λ has exactly two points, the convex hull is simply the geodesic between the two points.

If Λ is the half infinite line $[0, +\infty]$ the convex hull is the vertical wall arising from the line. Like the above example, this is a degenerate case: the interior of the convex hull is empty. However in this case, its boundary is regarded as the union of the two sides of the wall, with exterior bending angle π.

If Λ is the exterior of the unit circle, its convex hull is the half-plane underneath the hyperbolic plane rising from the circle; the dome is the plane. How about two round disks with angle of intersection α measured exterior to one disk and interior to the other? The dome over the union consists of two flat pieces meeting with exterior bending angle α. There are two flat pieces and one bending line.

The dome over the region bounded by an ellipse is a half-ellipsoid. There is a continuous family of bending lines which sweep out the dome which is a smooth surface. There are no flat pieces and the dome is a smooth, ruled surface.

Next consider a wedge $W = \{z \in \mathbb{C}, \; 0 \le \arg z < \alpha \le \pi\}$. If $\Lambda = W$, the convex hull boundary consists of the two flat pieces rising from the edges of W and one bending line. The exterior bending angle is $\pi - \alpha$. If instead Λ is the closure of the complement of W, then the dome over W is a half cone. Again it is swept out by the bending lines; there are no flat pieces.

The dome over a convex euclidean triangle contains one flat piece which is contained in the plane rising from the maximal inscribed circle, and parts of three cones near the vertices. The dome is a smooth C^1-surface. In fact the dome over any euclidean convex region is a smooth surface Epstein et al. 2006; 2004].

3.10.1 The bending measure

Each component $S = \mathrm{Dome}(\Omega_S)$ which is not a whole plane carries a nonzero *bending measure*. At an isolated bending line, it is just the atomic measure with support on the line given by the exterior bending angle. In general, the bending measure is constructed by a process akin to Riemann integration, that is, by approximating the dome by a sequence of finitely bent surfaces. The basic result is the following theorem of Thurston; the detailed proof appears in Epstein and Marden [1987].

Theorem 3.10.1. *Suppose Ω is a simply connected region whose complement Λ in $\mathbb{S}^2$ has at least three points.*

(i) *The hyperbolic metric in $\mathbb{H}^3$ restricts to give a path metric on $\mathrm{Dome}(\Omega)$ referred to as its hyperbolic metric.*

(ii) *There is an isometry in the respective hyperbolic metrics $\Upsilon : \mathrm{Dome}(\Omega) \to \mathbb{H}^2$.*

(iii) *Under Υ, the set of bending lines is carried to a geodesic lamination Λ in $\mathbb{H}^2$ and the bending measure on $\mathrm{Dome}(\Omega)$ is carried to a (bounded) transverse measure on Λ.*

(iv) *If Ω is invariant under a kleinian group G, then $\mathrm{Dome}(\Omega)$ and the set of bending lines are also G-invariant. The corresponding measured lamination in $\mathbb{H}^2$ is invariant under the fuchsian group $\Upsilon G \Upsilon^{-1}$.*

The hyperbolic metric on the simply connected region Ω is carried over from $\mathbb{H}^2$ by a Riemann mapping. In terms of the hyperbolic metrics on Ω and its dome, the nearest point retraction $r : \Omega \to \text{Dome}(\Omega)$ satisfies $d(r(z_1), r(z_2)) \leq 2d(z_1, z_2)$; that is, r is 2-Lipschitz [Epstein et al. 2004]. If $\text{Dome}(\Omega)$ is instead infinitely connected, one can pass to its universal cover and map that and its measured bending lamination to $\mathbb{H}^2$.

Now suppose $\Lambda(G)$ is the limit set of a kleinian group G. Its convex hull $\mathcal{CH}(\Lambda)$ is G-invariant. Each relative boundary component S of $\mathcal{CH}(\Lambda)$ is the dome over a component Ω_S of $\Omega(G)$ and is invariant under $\text{Stab}(\Omega_S)$.

The convex hull $\mathcal{CH}(\Lambda)$ necessarily contains the axes of all loxodromics of G since these have endpoints in the limit set. Can the axis of a $g \in G$ be a bending line? Only if the trace of g is real with $|\text{tr } g| > 2$. Otherwise the angular part of the trace would force a rotation about the axis, and therefore could not preserve the convex hull.

The section cannot be closed without mentioning the following remarkable fact described by Dennis Sullivan. For a full discussion and proof see Epstein and Marden [1987] or Epstein et al. [2004].

Theorem 3.10.2. (Sullivan Convex Hull Theorem). *There exists a universal constant* $2 < K < 14$ *with the following property. Given any simply connected region* $\Omega \subset \mathbb{C}$, $\Omega \neq \mathbb{C}$, *there exists a K-quasiconformal mapping* $F : \Omega \to \text{Dome}(\Omega)$ *which extends continuously and pointwise fixes every point on the common boundary* $\partial\Omega$.

If Ω is invariant under a group Γ of Möbius transformations, F can be chosen to satisfy additionally $F \circ \gamma = \gamma \circ F$ *for all* $\gamma \in \Gamma$.

The precise constant K is not known. The corresponding result is false in general for multiply connected regions. It only holds for the multiply connected hyperbolic regions Ω whose ideal boundary components are bunched sufficiently closely together—the region is uniformly perfect (Exercise (1-30), [Marden and Markovic 2008]). In fact the nearest point retraction from Ω to its dome is Lipschitz if and only if Ω is uniformly perfect [Bridgeman and Canary 2010].

3.10.2 Pleated surfaces

We have spoken of the structure of a convex hull boundary component, especially the dome over a simply connected region. Now consider the reverse process. That is, given a measured lamination (Λ, μ) in $\mathbb{H}^2$, can we construct a surface in $\mathbb{H}^3$ whose bending measure is μ?

Let's start with the simplest cases. Take the equatorial plane $\mathbb{H}^2$ (the unit disk) in the ball model and fix a diameter ℓ. Bend $\mathbb{H}^2$ along ℓ with exterior bending angle $0 < \theta < \pi$. Here $\theta = 0$ corresponds to no bending at all. The other extreme $\theta = \pi$ corresponds to two situations: (i) folding $\mathbb{H}^2$ in half along ℓ, or (as commonly used) (ii) pushing ℓ out to ∞ to become a single point ξ. This forces $\mathbb{H}^2$ in the limit to become two hyperbolic planes whose boundaries are tangent at ξ so that one plane is the image of the other under a designated parabolic with fixed point ξ.

To normalize the direction of bending, bend so that the result lies in the upper half of the ball. The resulting "pleated surface" S bounds on one side a convex region

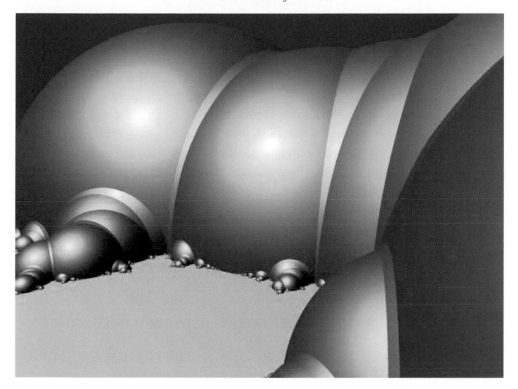

Fig. 3.8. A section of the dome over a component of a quasifuchsian ordinary set.

whose floor is bounded by two circular arcs with interior bending angle $\pi - \theta$. The dome has only one bending line.

The construction is easily generalized to a finite system of ordered, mutually disjoint hyperbolic lines, possibly with common endpoints, $\ell_1, \ldots, \ell_k \subset \mathbb{H}^2$. Assign an exterior bending angle $0 < \theta_i < \pi$ to each line. Then systematically bend the plane $\mathbb{H}^2$. For example we may assume that first bend along ℓ_1 results in $P_1 = P$ constructed above. Then in P_1 locate the copy of ℓ_2, say it lies to the right of ℓ_1. Then bend the half-plane in P_1 lying to the right of ℓ_1 along ℓ_2 with exterior angle θ_2. And so on for all the lines. We end up with what is called a *pleated surface* P_k. It is locally convex but is not necessarily embedded in $\mathbb{H}^3$—it may well have self-intersections. It has k bending lines, the images of the $\{\ell_i\}$. In any case there is a hyperbolic isometry $\Upsilon : \mathbb{H}^2 \to P_k$—such that Υ^{-1} is just unbending. The finite measured lamination is carried to the bending lines and bending measure on P_k.

The same construction can be carried out given a general lamination Λ in $\mathbb{H}^2$ and a positive transverse Borel measure by using finite approximations. In fact it equally works for a real valued transverse Borel measure. In the general case the pleated surface has both positive and negative bending. It may not be locally embedded and may even be dense in all $\mathbb{H}^3$. The construction is such that if $(\Lambda; \mu)$ is invariant under a fuchsian group G, a deformation of G, corresponding to the bended image is a homomorphism to a Möbius group H. H acts in $\mathbb{H}^3$ by mapping the image pleated surface onto itself reflecting the action of G in $\mathbb{H}^2$; H is unlikely to be discrete. The details are carried out in Epstein and Marden [1987].

Another way of constructing a pleated surface from a geodesic lamination $\Lambda \subset \mathbb{H}^2$ is as follows. Suppose Λ is such that all gaps are ideal triangles; this is the generic case. If there is an injection $f : \partial\mathbb{H}^2 \to \partial\mathbb{H}^3$ (for example, the restriction of a quasiconformal deformation of a fuchsian group) then each leaf $\ell \subset \Lambda$ can be mapped to the line determined by the f-images of the endpoints of ℓ and the ideal triangles can then be filled in. If in addition Λ is invariant under a fuchsian group G and f conjugates G to a quasifuchsian group H then the lamination Λ/G gives rise to a pleated surface in $\mathbb{H}^3/H$. For an example, see Exercise (6-13).

Since the convex hull contains all geodesics, a flat piece of a pleated surface that is bounded by two or more geodesics lies in the convex hull: Pleated surfaces lie in the convex hull of the limit set.

Yet another possibility is to construct a pleated surface by earthquaking along a geodesic lamination followed by bending (Exercise (3-35)). In this case the bending determines a pleated surface on the new hyperbolic structure resulting from the earthquake.

Formally, a *pleated surface* is determined by a *pleating map* $f : S \to \mathcal{M}$ of a hyperbolic surface S into a hyperbolic 3-manifold $\mathcal{M}$ with these properties:

  (i) f takes any rectifiable path in S to a path in $\mathcal{M}$ of the same length.
 (ii) Every point $z \in S$ lies in an open geodesic arc which f maps to a geodesic arc in $\mathcal{M}$.
(iii) f sends cusps to cusps: it sends a small neighborhood of a cusp of S into a small neighborhood of a cusp of $\mathcal{M}$; the homomorphism $f_* : \pi_1(S) \to \pi_1(\mathcal{M})$ sends parabolics to parabolics.

Assumption (i) can replaced by (i'): geodesic paths in S are sent to rectifiable paths of the same length in $\mathcal{M}$. The apparently stronger definition is equivalent [Canary et al. 1987, II.5.2.6 or Canary et al. 2006, I.5.2.6]. We may equally work with a lift of f to the universal covers.

The pleated surface is called *incompressible* if $f_* : \pi_1(S) \to \pi_1(\mathcal{M})$ is injective.

The *pleating locus* is the set $\Lambda \subset S$ consisting of those points $z \in S$ with the following property. There is one and only one geodesic in S through z which f maps onto a geodesic in the 3-manifold $\mathcal{M}$. These are the bending lines described earlier. The pleating locus Λ is a closed subset of S and is in fact a geodesic lamination. Its image $f(\Lambda)$ is often referred to as the *pleating locus* as well, or as the *bending lines*. The map f is an isometry of the complementary gaps onto polygons in $\mathcal{M}$ that in general are infinitely sided.

Given such a general pleated surface, there is likely to be a great deal of positive and negative bending. Yet by associating a transverse segment τ to the set of positive endpoints on $\partial\mathbb{H}^3$ of the oriented leaves through τ and then a continuum in $\partial\mathbb{H}^3$, it is possible to construct a kind of bending measure which however is only finitely additive. This measure and the pleating locus characterize the pleated surface. For the details see Bonahon [1996; 2001].

Given a lamination $\Lambda \subset S$ and a hyperbolic manifold $\mathcal{M}$, the lamination Λ is said to be *realizable* in $\mathcal{M}$ if there is a pleating map $f : S \to \mathcal{M}$ whose pleating locus contains Λ.

Given a closed topological surface S with at most a finite number of punctures, a homotopy class $[f]$ of maps to some geometrically finite $\mathcal{M}(G)$ and a maximal lamination $\Lambda \subset S$, is there a hyperbolic metric on S such that some $h \in [f]$ is a pleating map $h : S \to \mathcal{M}(G)$ whose pleating locus is contained in Λ? The answer if affirmative if $\Omega(G) \neq \varnothing$, the image S' of S is incompressible, and under f, cusps correspond to cusps [Canary et al. 1987, II.5.3.11 or Canary et al. 2006, I.5.3.11].

Uniform injectivity of pleated surfaces

Proposition 3.10.3 [Thurston 1986b, Prop. 5.3]. *Suppose $\varepsilon > 0$ determines a thick/thin decomposition of $\mathcal{M}$ and $A > 0$ is a given constant. Any incompressible pleated surface $f : S \to \mathcal{M}$ from a hyperbolic surface satisfying* $\mathrm{Area}(S) \leq A$ *has the following property for some constant $C = C(A) > 0$:*

$$\mathrm{Inj}_{\mathcal{M}}(f(x)) \leq \mathrm{Inj}_S(x) \leq C\,\mathrm{Inj}_{\mathcal{M}}(f(x)),$$

provided the distance of $f(x)$ from any closed geodesic in $\mathcal{M}$ of length not exceeding ε is at least 1.

That the injectivity radius is $r = \mathrm{Inj}_{\mathcal{M}}(f(x))$ means there is a hyperbolic ball in $\mathcal{M}$ of radius r, centered at $f(x)$, whose interior is embedded in $\mathcal{M}$, and no larger ball has this property. Proposition 3.10.3 guarantees that the injectivity radius in $\mathcal{M}$ at $f(x)$ is not substantially different from the injectivity radius on S at x, provided that $f(x)$ is not too close to a short geodesic in $\mathcal{M}$. The proof uses the fact that there is an upper bound for the injectivity radii on S in terms of A.

Consider now for simplicity the case of a closed hyperbolic surface S. The pleated surface $f : S \to f(S) \subset \mathcal{M}$ is called *doubly incompressible* if, in addition to being incompressible, (i) two loops on $f(S)$ which are freely homotopic in $\mathcal{M}$ come from loops which are already freely homotopic in S, and (ii) under f_*, maximal cyclic subgroups of $\pi_1(S)$ are sent to maximal subgroups of $\pi_1(\mathcal{M})$ (primitive elements are preserved).

There is an important injectivity property for such pleated surfaces as follows (see Minsky [2000] for the statement when there are parabolics and the application to the proof of the Ending Lamination Conjecture).

Theorem 3.10.4 (Uniform injectivity of pleated surfaces ([Thurston 1986b, Theorem 5.2])). *Let S be a closed hyperbolic surface with its ε thick/thin parts. Given $\delta > 0$ there exists $\delta^* > 0$ such that the following property holds for any doubly incompressible pleated surface $f : S \to \mathcal{M}$: Denote by $\Lambda \subset S$ the lamination that corresponds to the pleating locus $f(\Lambda)$. Then if $x, y \in \Lambda$ are in the ϵ-thick part of S,*

$$d_{T(\mathcal{M})}(v_x, v_y) \leq \delta \quad \text{implies} \quad d(x, y) \leq \delta^*.$$

Here v_x and v_y denote unit tangent vectors to the leaves of $f(\Lambda)$ at $f(x)$ and $f(y)$, respectively. Their distance apart is measured in the projectivized tangent bundle $T(\mathcal{M})$. The theorem says that when the tangent vectors are not too far from being parallel, the initial points x, y (and the leaves of Λ containing them) are δ-close in S. In particular the bit of $f(S)$ bounded by the lines containing $f(x)$, $f(y)$ is not wildly oscillating.

Compactness of pleated surfaces

There is another useful fact about pleated surfaces. In a mapping $f : \mathbb{H}^2 \to \mathbb{H}^3$, we are free to fix base points $x_0 \in \mathbb{H}^2$ and $y_0 \in \mathbb{H}^3$ and require that $f(x_0) = y_0$. We may even require that a vector direction at x_0 be sent to a given vector direction at y_0. From Canary et al. [1987, II.5.2.18] or Canary et al. [2006, I.5.2.18]] we find,

Theorem 3.10.5 (Compactness of pleated surfaces). *Suppose we are given $A < \infty$, $\varepsilon > 0$ and points $x_0 \in \mathbb{H}^2$, $y_0 \in \mathbb{H}^3$ such that:*

(i) *$S = \mathbb{H}^2/G$ has area $< A$ and $\Lambda \subset \mathbb{H}^2$ is a G-invariant geodesic lamination,*
(ii) *The map f induces an isomorphism $\phi : G \to \Gamma$ to a kleinian group acting on $\mathbb{H}^3$, parabolics corresponding to parabolics,*
(iii) *$P = f(\mathbb{H}^2)$ is a Γ-invariant pleated surface with pleating locus $f(\Lambda)$ satisfying $f(x_0) = y_0 \in P$,*
(iv) *For $x_0 \in S$, $f(x_0) = y_0 \in \mathbb{H}^3$, the injectivity radius $\mathrm{Inj}_{y_0}(\mathbb{H}^3/\Gamma) > \varepsilon$.*

Then for given A, ε, x_0, y_0, the set of all triples $\{(G, \Gamma; f)\}$ is compact.

Sketch of proof. We know the set of hyperbolic 3-manifolds with given basepoint y_0 and injectivity radius at y_0 bounded above zero is compact in the topology of geometric convergence. In addition the set of surfaces with uniformly bounded area is also compact. So we can pick simultaneously geometrically convergent subsequences of $\{G_n\}, \{\Gamma_n\}$, and $\{f_n\}$, the latter by Ascoli's theorem. And such that the corresponding sequence of laminations $\{\Lambda_n\} \subset \mathbb{H}^2$ converges to a lamination as well. □

It is tough to give conditions on a measured lamination $(\Lambda; \mu)$ with μ a positive measure, so that the corresponding pleated surface is the dome over a simply connected region. The best result is in terms of the norm $\|\mu\| = \sup_\sigma \mu(\sigma)$, where σ ranges over all transverse segments of unit length. In Epstein et al. [2004] it is shown that there exists a constant $0 < c \leq 2 \arcsin \tanh(\frac{1}{2}) \cong 0.96$, with the following property. If $\|\mu\| < c$ then $(\Lambda; \mu)$ is the bending measure of $\mathrm{Dome}(\Omega)$ for some simply connected region Ω. It is conjectured that the upper bound given is best possible; in any case it is known that it cannot be larger. On the other hand it is known that if the pleated surface is a dome, then $\|\mu\| < 4.88$ [Bridgeman 2003].

3.11 The convex core

The quotient of the convex hull

$$\mathcal{CH}(\Lambda)/G = \mathcal{CC}(G) \subset \mathcal{M}(G)^{\text{int}}$$

is hyperbolically convex and is called the *convex core of* $\mathcal{M}(G)$. Every closed geodesic in $\mathcal{M}(G)$ lies in $\mathcal{CC}(G)$. Indeed the convex core can be defined to be the smallest convex set with this property. At the level of fundamental groups, the inclusion $\pi_1(\mathcal{C}) \hookrightarrow \pi_1(\mathcal{M}(G))$ is an isomorphism. Thus the convex core is representative of the full manifold. At one extreme, for fuchsian groups the convex core is flat without interior. At the other extreme, if $\Omega(G)$ is empty then the convex core is the full manifold $\mathcal{M}(G)$.

The boundary components of the convex core are incompressible if and only if the limit set is connected.

Here are three additional facts about convex cores.

(1) *The nearest point retraction projects to the quotient and is a continuous map from each component of $\partial\mathcal{M}(G)$ to the component of $\partial\mathcal{CC}(G)$ that it faces.*
(2) *G is geometrically finite if and only if $\mathcal{CC}(G)$ has finite volume.*
(3) *If G is geometrically finite without rank one cusps then associated with any rank two cusps are solid cusp tori contained in $\mathcal{CC}(G)$.*

The fuchsian group case is trivial for its three-dimensional convex core has zero volume, even if the group is not finitely generated. The infinite generated case would be excluded by instead requiring that an ε-neighborhood of the core be of finite volume.

For fuchsian groups, Fenchel and Nielsen made good use of the fact that the group is finitely generated if and only if its convex core with respect to $\mathbb{H}^2$ has finite area.

Proof of item (2). We begin with a lemma.

Lemma 3.11.1. *Suppose G is nonelementary, has no elliptics and $\zeta \in \partial\mathbb{H}^3$ is a parabolic fixed point.*

(i) *If* $\operatorname{Stab}_\zeta$ *is a rank two parabolic group, then $\mathcal{CH}(\Lambda)$ contains a horoball $\mathcal{H}_\zeta$ at ζ.*
(ii) *Suppose* $\operatorname{Stab}_\zeta$ *is rank one and supports a double horocycle σ_1, σ_2 at ζ. For some horoball $\mathcal{H}_\zeta$, $\mathcal{CH}(\Lambda)$ contains $\mathcal{H}_\zeta \cap (H_1 \cap H_2)$. Here H_i is the half-space, bounded by the plane rising from the horocycle $\sigma_i = \partial\Delta_i$, which abuts the exterior of its horodisk Δ_i.*

Proof. We may assume that in the upper half-space model $\zeta = \infty$ and $T_1(z) = z+1$ is a generator of $\operatorname{Stab}_\zeta$.

In the rank two case, there is another generator $T_2(z) = z + a$ and we may assume that $|a| \geq 1$. We claim there is a maximal diameter d for circles $C \subset \overline{\Omega(G)}$ that bound disks in $\Omega(G)$. Suppose otherwise. When C has a sufficiently large diameter the disk Δ that it bounds will have the property that $\Delta^* = \Delta \cup T_1(\Delta) \cup T_2(\Delta) \cup T_1T_2(\Delta)$ is simply connected. But then the orbit of Δ^* under $\operatorname{Stab}_\zeta$ covers $\mathbb{C}$. This means that

$\Omega(G) = \mathbb{C}$ which is impossible if G is not Stab_ζ itself. Consequently the horoball $\{(z,t) : t > d/2\}$ is contained in $\widehat{C}(\Lambda)$.

In the rank one case, because there is a double horodisk at ∞, $\Lambda(G)$ is contained inside a minimal width strip $S = \{z : b_1 \leq \text{Im}\, z \leq b_2\}$ where both horizontal lines $\text{Im}\, z = b_1, b_2$ contain limit points. In fact their intersection with the vertical strip $V = \{z : 0 \leq \text{Re}\, z \leq 1\}$ also *contains* limit points. We see that there is a maximal diameter $d < \infty$ for circles $C \subset \overline{\Omega(G)}$ centered in V that bound disks in $\Omega(G)$. Now the maximal horocycles σ_1, σ_2 are bounded by $\text{Im}\, z = b_1, b_2$. These two observations translate into the second statement. $\qquad\square$

We continue our proof of item (2). If $\mathcal{M}(G)$ is geometrically finite, parallel to each component of $\partial\mathcal{M}(G)$ is a component of $\partial\mathcal{CC}(G)$. Each cusp is taken care of by Lemma 3.11.1. The convex hull has finite volume.

Conversely assume the convex hull has finite volume. Ahlfors' Finiteness Theorem implies that $\mathcal{M}(G)$ has a finite number of boundary components and each is a closed surface with at most a finite number of punctures. Each component is parallel to a boundary component of the convex hull with the same property. Consequently $\mathcal{CC}(G)$ has a finite number of boundary components.

In view of the universal horoball theorem, in $\mathcal{M}(G)$ there are only a finite number of cusp tori/solid cusp tori corresponding to rank two cusps, and a finite number of cusp cylinders/solid cusp cylinders corresponding to rank one cusps: There are at most a finite number of cusps.

While a solid cusp torus has finite volume Lemma 3.11.1, unless a rank one cusp ζ supports a double horodisk, it contributes infinite volume to $\mathcal{CC}(G)$. For there is universal horoball at ζ in any case, but unless there is a double horodisk its local T quotient will have infinite volume in $\mathcal{M}(G)$, where T generates the rank one group.

To see this erect two planes P_1, P_2 which are tangent at ζ and $T(P_1) = P_2$. These bound two disks $D_1, D_2 \subset \mathbb{S}^2$. No matter how small we take these disks, at least one of them will contain limit points in its interior. Hence $\partial\mathcal{CC}(G)$ will contain limit points giving $\mathcal{CC}(G)$ infinite volume. $\qquad\square$

A recent theorem of Brian Bowdich shows that the "thickness" of the convex core is bounded (even if it is compressible or comprises the whole manifold). A prior unpublished result of R.A. Evans [Evans 2005] proved this for geometrically finite, boundary incompressible manifolds.

Theorem 3.11.2 [Bowditch 2013]. *Let G be a finitely generated, torsion-free kleinian group, and $\mathcal{CC}(G)$ its convex core. There exists $0 < \rho < \infty$ such that the radius of any embedded hyperbolic ball in $\mathcal{CC}(G)$ is $\leq \rho$.*

The same bound ρ holds for any kleinian group isomorphic to G.

A bound ρ is called a *Bowditch constant* for $\mathcal{CC}(G)$, or *the* Bowditch constant if it is minimized over all groups. We emphasize that it is measured with respect to the convex core, not the whole manifold—unless the two are the same. Compare with Exercise (4-20).

In contrast we note:

Proposition 3.11.3. *A sequence of manifolds* $\{\mathcal{M}(G_n)\}$ *has no upper bound on the injectivity radius if and only if* $\{G_n\}$ *can be replaced by a subsequence of conjugates* $H_m = A_m G A_m^{-1}$ *for which the geometric limit of* $\{M(H_m)\}$ *is* $\mathbb{H}^3$.

Corollary 3.11.4. *Suppose* $\{G_n\}$ *is a sequence of isomorphic, torsion-free, finitely generated kleinian groups which converge algebraically to G and geometrically to H. Then H also has a Bowditch constant.*

Proof. If H is finitely generated it has a Bowditch constant ρ_H and so the constant for the whole sequence can be taken as $\max(\rho_H, \rho)$. But even if H is infinitely generated it has a Bowditch constant. For if not, we can apply Proposition 3.11.3 to get a sequence of conjugates of $\{G_n\}$ that converge geometrically to $\mathbb{H}^3$ (cf. Exercise (4-6)). $\qquad\square$

It is the requirement that the injectivity radius be measured in the convex hull that protects a hyperbolic manifold from being so destroyed. For an important consequence of Theorem 3.11.2 see Theorem 4.5.6.

Convex cocompact groups. This term is frequently used in the literature to refer to a geometrically finite group G without parabolics. The term arises because such groups are characterized by the property that the intersection of the convex hull of the limit set with a fundamental polyhedron is compact. Another defining property [Sullivan 1985] is that the limit set $\Lambda(G)$ has an expanding property: For each $\zeta \in \Lambda$ there exists $\gamma = \gamma_\zeta \in G$ such that $|\gamma'(\zeta)| > 0$ (in the spherical metric).

3.11.1 Length estimates for the convex core boundary

Suppose $\widehat{S}$ is a simply connected boundary component of $\mathcal{CH}(\Lambda)$, and S is the corresponding boundary component of the convex core $\mathcal{CC}(G)$ of $\mathcal{M}(G)$. The surface S faces a component R of $\partial\mathcal{M}(G)$. If γ_S is a closed geodesic in S, there is a uniquely determined geodesic γ_R in the hyperbolic metric on R which is freely homotopic to γ_S.

Theorem 3.11.5. [Epstein et al. 2004]. *In the respective hyperbolic metrics,*

$$\frac{1}{14} < \frac{\ell(\gamma_S)}{\ell(\gamma_R)} \leq 2.$$

The same bounds hold for the lengths of corresponding measured laminations.

For the proof, the lower bound is obtained from the best current estimate for the equivariant "K" in Theorem 3.10.2 and the fact that the minimal Lipschitz constant in the same homotopy class does not exceed this "K". The upper bound follows directly from the fact that the Lipschitz constant of the nearest point retraction does not exceed 2.

The inequality shows that the hyperbolic geometry of the two surfaces is tightly bound together.

The length of the geodesic γ_M in the interior of $\mathcal{M}(G)$ freely homotopic to γ_R is likewise bounded by $2\ell(\gamma_R)$ as shown in Exercise (5-5)—γ_M will be identical to γ_S if γ_S is a bending line.

According to Bridgeman [1998], there exists a universal constant B with the following property. If S is a component of $\partial\mathcal{CC}(G)$ as above, then

$$\ell(\beta_S) \leq B\pi^2|\chi(S)|, \tag{3.11}$$

where $\ell(\beta_S)$ is the length on S of the bending lamination β_S and $\chi(S)$ is the Euler characteristic. In particular if β_S is supported on a single geodesic of length L_β with bending angle θ, $\ell(\beta_S) = L_\beta \cdot \theta$ is bounded; the longer L_β is, the smaller θ must be.

3.11.2 Bending measures on convex core boundary

There is a beautiful recent result of Bonahon and Otal [2004], completed by Lecuire [2003], characterizing geometrically finite groups by the bending laminations of their convex core boundaries. See [Lecuire 2004, Lecuire 2006, [Lecuire 2008]] for further applications of this subject.

Start with an orientable, compact manifold M^3 other than a solid torus T^2 or a thickened torus $T^2 \times [0, 1]$, and whose interior has a hyperbolic structure. Thus M^3 is a model for a geometrically finite manifold with solid cusp cylinders and cusp tori removed. We assume that ∂M^3 has some nontorus components, which may or may not be incompressible, and we may as well assume each has a hyperbolic structure. Let (Λ, μ) be a measured lamination on the nontorus components of ∂M^3. We allow that on some boundary components, (Λ, μ) may be the zero lamination (no leaves).

On a closed leaf γ of Λ, μ has atomic measure $\mu(\gamma) > 0$ which we will think of as a bending angle. Let D and C be an essential disk and cylinder in M^3. As we know, the *geometric intersection number* $\iota(\partial D, \Lambda)$ or $\iota(\partial C, \Lambda)$ is the generalization of the case that Λ consists of a finite number of closed leaves and μ is the unit atomic measure on each. In the finite case, $\iota(\partial D, \Lambda)$ or $\iota(\partial C, \Lambda)$ is the minimum number of times that simple loops freely homotopic to ∂D cross the leaves of Λ or the minimum number of times simple loops freely homotopic to the components of ∂C cross the leaves of Λ. We are assuming that Λ has at least one leaf, yet it is possible that one or more nontorus components of $\partial\mathcal{M}$ carry no leaves.

Theorem 3.11.6 Existence of bending measures [Bonahon and Otal 2004; Lecuire 2003]. *Given the measured lamination $(\Lambda; \mu)$ on ∂M^3, there exists a geometrically finite, nonfuchsian, $\mathcal{M}(G_\mu)$ whose convex core boundary has the bending lamination $(\Lambda; \mu)$ if and only if the following conditions are satisfied:*

(i) *On each closed leaf α, $\mu(\alpha)$ satisfies $0 < \mu(\alpha) \leq \pi$.*
(ii) *For each essential disk $D \subset M^3$, $\iota(\partial D, \Lambda) > 2\pi$.*
(iii) *There exists $\eta > 0$ such that $\iota(\partial C, \Lambda) \geq \eta$ for each essential cylinder $C \subset M^3$.*

If Λ consists of a finite number of closed leaves then the kleinian group G_μ is uniquely determined up to Möbius equivalence.

The closed leaves γ with $\mu(\gamma) = \pi$ will correspond to the rank one cusps of $\mathcal{M}(G_\mu)$. Of course the torus boundary components of ∂M^3 will correspond to the rank two cusps of $\mathcal{M}(G_\mu)$. The proof of uniqueness is outlined in Exercise (6-3). Uniqueness for all laminations is known for the once-punctured torus quasifuchsian case [Series 2006] and conjectured for the general case.

Suppose in addition that M^3 is compact, boundary incompressible, and has no essential cylinders (see Exercise (3-17)). Assume we are given a maximal finite lamination of $\sum(3g_i - 3)$ simple closed geodesics $\Lambda = \cup\beta_j$ on ∂M^3, g_i the genus of the i-th component of ∂M^3, and an atomic measure $0 < \mu(\beta_j) < \pi$ for each index. According to Theorem 3.11.6, there exists a uniquely determined (up to isometry) hyperbolic structure $\mathcal{M}(G_\mu)$ on M^3 whose convex core $\mathcal{CC}(G)$ has exactly the bending lamination $(\Lambda; \mu)$. In Choi and Series [2006] it is shown that the $\sum(3g_i - 3)$-complex lengths in $\mathcal{M}(G_\mu)$ (see Section 7.4) of the geodesics $\{\beta_j\}$ serve as local coordinates for the local deformations of $\mathcal{M}(G_\mu)$ in the representation variety $\mathfrak{R}(G_\mu)$ (see Section 5.1).

If the lamination is finite, condition (ii) on the geometric intersection number ι requires that the boundary of each essential disk has at least three essential crossings with Λ. Condition (iii) insures that if one boundary curve of C is a leaf of Λ then the other must be transverse to Λ.

If a nontorus component of ∂M^3 carries no leaves, Theorem 3.11.6 provides that the corresponding component R of $\partial\mathcal{M}(G_\mu)$ is *totally geodesic*. This means that every component Ω_R of $\Omega(G_\mu)$ lying over R is a round disk; the convex hull boundary component that faces R is a hyperbolic plane. A compressible boundary component cannot become totally geodesic; in line with this fact condition (ii) requires compressible components to contain leaves of Λ.

To understand why condition (ii) is necessary, suppose we have a compact convex core with bending lines as simple loops $\{\sigma_i\}$ with exterior bending angles $\{\beta_i\}$. Consider a compressing disk D with $\partial D \subset \partial\mathcal{CC}(G)$. We may assume ∂D is piecewise geodesic and D is piecewise flat. The Gauss–Bonnet formula (1.4) tells us that

$$\sum \beta_i \iota(\partial D, \sigma_i) = \text{Area}(D) + 2\pi > 2\pi.$$

For (iii), suppose C with $\partial C \subset \partial\mathcal{CC}(G)$ is an essential cylinder, also piecewise geodesic. We find that $\sum \beta_i \iota(\partial C, \sigma_i) = \text{Area}(C)$ so that we must have $\iota(C, \bigcup\sigma_i) > 0$.

The proof in the finite case starts by showing that there exists a geometrically finite $\mathcal{M}(G)$ homeomorphic to M^3 whose convex core is bent along $\sigma = \bigcup\sigma_i$. Then the manifold and bending angles are continuously deformed until they match the assigned angles. To establish existence the following argument is used. Remove half-tubular neighborhoods of the $\{\sigma_i\}$ and double the resulting manifold. This gives a compact manifold with tori boundary components. Make assumptions on $\{\sigma_i\}$ so the manifold is irreducible and atoroidal. As a consequence it has a complete hyperbolic structure of finite volume. One then uses the theory of cone manifolds (Exercises (4-7, 6-3)) to deform the rank two cusps to get a symmetric cone manifold with small cone angles. Undoubling results in the required convex hull.

A typical application is the following. Consider quasifuchsian groups representing a pair of surfaces of genus 2, say. For Λ, take a simple loop γ_{bot} on the "bottom" component and a finite number of mutually disjoint, nonparallel, simple loops $\{\beta_i\}$ on the top. To fulfill condition (iii) of Theorem 3.11.6 we must assume that *every* β_i *is freely homotopic to a loop on the bottom component which is transverse to* γ_{bot}. Assign positive atomic measures each less than π to all the simple loops. According to Theorem 3.11.6 there is a unique quasifuchsian group representing genus 2 surfaces whose convex hull boundary has the prescribed bending measure. By varying the measure on γ_{bot} while leaving the measures on $\{\beta_i\}$ fixed, we obtain a "slice" of the deformation space.

Parker and Series [1995] have an explicit construction for bending along one geodesic in the case of once-punctured torus quasifuchsian groups; see their bending formulas Equations (8.39), (8.41).

3.12 The compact and relative compact core

There is another important "core" in a hyperbolic manifold, and this one is always compact but is not hyperbolic. It was discovered by Peter Scott [1973a] and independently by Peter Shalen for the freely indecomposable case.

In the interior of any hyperbolic manifold $\mathcal{M}(G)$ *with* G *finitely generated there is a compact, connected, submanifold* $C = C(G)$ *such that* (i) *inclusion of the fundamental group* $\pi_1(C) \hookrightarrow \pi_1(\mathcal{M})$ *is an isomorphism, and* (ii) *each component of* ∂C *is the full boundary of a noncompact component of* $\text{Int}(\mathcal{M}(G)) \setminus C$.

Property (ii) follows from (i). For if a complementary component were bounded by two components S_1, S_2 of ∂C, there would exist a simple loop in $\mathcal{M}(G)$ that crossed each of S_1, S_2 exactly once. Such a loop cannot be homotopic to a curve within C.

If $\mathcal{M}(G)$ is geometrically finite without parabolics, each component of ∂C is parallel to a component of $\partial \mathcal{M}(G)$. In the general case of no parabolics, each complementary component E of the core C is a neighborhood of exactly one *end* (see Section 5.5) of $\mathcal{M}(G)$.

The submanifold C is called a *compact core* of $\mathcal{M}(G)$. A core is uniquely determined up to homeomorphism: Two cores C_1, C_2 of $\mathcal{M}(G)$ are homeomorphic [McCullough et al. 1985]. An immediate consequence of its existence is that $\pi_1(\mathcal{M})$ is finitely presented, as stated in Theorem 2.6.1. Cores are a fundamental structure in studying geometrically infinite manifolds. According to Bonahon [1986] (see also Exercise (3-11)), each core can be cut along incompressible surfaces to result in a finite union of compression bodies and submanifolds with incompressible boundaries.

When there are parabolics, there is a useful refinement of the compact core that takes account of the cusps. Namely, McCullough [1986] chooses a system of mutually

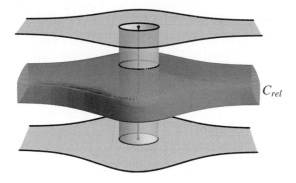

C_{rel}

Fig. 3.9. Two boundary components of a fragment of a relative compact core lying on a solid pairing tube.

disjoint horoballs in $\mathbb{H}^3$, associated with the parabolic fixed points, having the property that the union $\mathcal{H}$ is invariant under the action of G and the "parabolic locus" $\mathcal{P} = \mathcal{H}/G$ is embedded in $\mathcal{M}(G)$. The components of $\mathcal{P}$ are solid cusp cylinders and solid cusp tori. Then $\mathcal{M}_{nc} = \mathcal{M}(G) \setminus \mathcal{P}$ (the "noncuspidal part" of $\mathcal{M}(G)$)) has the property that each component of the relative boundary $\partial\mathcal{M}_{nc}$ in $\mathcal{M}(G)$ is either a component of $\partial\mathcal{M}(G)$, a doubly infinite cusp cylinder, or a cusp torus.

There exists a compact, connected, submanifold $C_{rel} \subset \text{Int}(\mathcal{M}(G))$ such that

  (i) *The inclusion $\pi_1(C_{rel}) \hookrightarrow \pi_1(\mathcal{M}(G))$ is an isomorphism,*
  (ii) *Each torus component of $\partial\mathcal{P}$ is a component of ∂C_{rel},*
 (iii) *Each cylinder component of $\partial\mathcal{P}$ intersects ∂C_{rel} in a closed annular region, and*
 (iv) *Each component of $\Omega(G)/G$ is parallel to a component of $\partial C_{rel} \setminus C_{rel} \cap \partial\mathcal{P}$.*

The submanifold C_{rel} is called a *relative compact core*.

When $\mathcal{M}(G)$ is geometrically finite, each doubly infinite solid cusp cylinder can be replaced by a solid pairing tube each of whose boundary components surround a puncture on $\partial\mathcal{M}(G)$. In this case we can choose $\mathcal{M}_{nc}$ to be bounded by closed surface components of $\partial\mathcal{M}(G)$, surface components with neighborhoods of punctures removed plus pairing cylinders connecting the boundaries of these punctured neighborhoods, and cusp tori. The submanifold $C_{rel} \subset \text{Int}(\mathcal{M}(G))$ is then bounded by closed surfaces and cusp tori.

If $\mathcal{M}(G)$ is quasifuchsian with k pairs of punctures, $\mathcal{M}_{nc}$ has two components each of which is a subsurface of $\partial\mathcal{M}(G)$ bounded by k simple loops. On the other hand C_{rel}, or the compactified $\mathcal{M}_{nc}$, has one boundary component which is a closed surface.

Note that in the presence of rank one parabolics, $\partial\mathcal{M}(G)$ might be incompressible while ∂C_{rel} is compressible; it could be a handlebody.

However if a boundary component S of $\partial\mathcal{M}(G)$ is compressible a boundary component of the compact core and of C_{rel} will be compressible as well.

For a full discussion of geometrically infinite manifolds see Section 5.5 and for relative cores, p. 293.

3.13 Rigidity of hyperbolic 3-manifolds

As mentioned in Chapter 1, hyperbolic polygons or convex polyhedra tend to be rigid—uniquely determined up to isometry by their angles. In dimensions larger than two, the same is true of finite volume hyperbolic manifolds.[7] Yet finite area surfaces are not rigid, except for the thrice-punctured sphere. For this reason it was a wake-up call for the then kleinian world when Mostow first came up with his rigidity theorem for closed, hyperbolic n-manifolds. Later, Mostow's theorem was extended to finite volume manifolds by Marden 1974a (for $n = 3$), and independently by Prasad [Prasad 1973] (for $n \geq 3$ dimensions). This was before the Thurston era, when we knew a lot less than we thought we did.

Theorem 3.13.1 (**Mostow Rigidity Theorem** [1973]). *Suppose we have a hyperbolic manifold or orbifold* $\mathcal{M}(G)$ *of finite volume, with dimension* $n \geq 3$, *and an isomorphism* $\phi : G \to H$ *onto another kleinian group* H. *Then* ϕ *is determined by an isometry* $\mathcal{M}(G) \to \mathcal{M}(H)$

In other words, $\mathcal{M}(G)$ is uniquely determined in the isomorphism class of G, up to orientation-preserving or -reversing Möbius equivalence. As already pointed out, rigidity does not hold in the hyperbolic plane: all surfaces with the same genus and the same number of punctures have the same area but are not usually isometric to each other. Surprisingly, the reason behind this state of affairs is that the dimension of the limit set of a finite area fuchsian group is one, while for $n \geq 3$ the limit set of a finite volume manifold has dimension ≥ 2.

There is a homotopy analogue of Mostow's theorem as follows. This is a deep result by Gabai, Meyerhoff and N. Thurston that required a sophisticated computer program to complete. Later we will restate and discuss this theorem from a different point of view (p. 283).

Theorem 3.13.2 ([Gabai 2001a; Gabai et al. 2003]). *Suppose* M^3 *is a closed, irreducible manifold (for this notion see Section 6.3) and* $\mathcal{M}(G)$ *a closed hyperbolic manifold. Assume there is an isomorphism[8]* $\phi : \pi_1(M^3) \to \pi_1(\mathcal{M}(G)) \cong G$. *Then*

(i) ϕ *is induced by a homeomorphism* $\Phi : M^3 \to \mathcal{M}(G)$,

(ii) *If* Φ_1, Φ_2 *are homotopic homeomorphisms, then* Φ_1 *is isotopic to* Φ_2,

(iii) *If* ρ_1, ρ_2 *are two hyperbolic metrics on* M^3, *there exists an isometry between* M^3 *with the metric* ρ_1 *and* M^3 *with* ρ_2 *which is homotopic to the identity on* M^3.

(iv) *The inclusion* $\mathrm{Isom}(\mathcal{M}(G)) \to \mathrm{Diff}(\mathcal{M}(G))$ *is a homotopy equivalence.*

Of course if M^3 is also hyperbolic, the first item is just Mostow's theorem with Φ an isometry. Even in this case the second item is new.

We recall that a homotopy is a continuous map $F : M^3 \times [0, 1] \to \mathcal{M}(G)$ such that $F(\,\cdot\,, 0) = \Phi_1$, $F(\,\cdot\,, 1) = \Phi_2$. For it to be an isotopy, each intermediate map $F(\,\cdot\,, t)$

---

[7] In fact $\mathbb{H}^3$ itself is rigid in the sense that there is no nonconstant harmonic map of $\mathbb{H}^3$ into any riemannian 3-manifold of nonpositive sectional curvature [Leung and Wan 2001].

[8] The precise assumption here is that there is a homotopy equivalence between M^3 and $\mathcal{M}(G)$; this is the same as an isomorphism if the higher homotopy groups of M^3 vanish—cf. Section 5.2.

must also be a homeomorphism. For example, a homotopy can send a geodesic α to a simple loop α' whose intersection with a tiny ball is knotted there. An isotopy cannot cause this effect. There is a subtle but important distinction between homotopy and isotopy. See Exercise (3-25). Isotopies are in particular homotopies, but in 3D having an isotopy between two mappings is a much stronger condition.

Theorem 3.13.2 implies that, for a closed manifold $\mathcal{M}(G)$, the mapping class group $\mathcal{MCG}(\mathcal{M}(G))$, defined as the group of isotopy classes, is a finite group. In contrast, Theorem 3.13.4 implies only that the group of *homotopy classes* of automorphisms of $\mathcal{M}(G)$ is finite.

For a further discussion of topological rigidity see Section 5.2.

Corollary 3.13.3. *If $\mathcal{M}(G)$ has finite volume, every orientation-preserving (and orientation-reversing) homeomorphism σ of $\mathcal{M}(G)$ onto itself is homotopic (isotopic if $\mathcal{M}(G)$ is closed) to an isometry.*

Unlike the case for finite area surfaces, the mapping class group of a finite volume hyperbolic 3-manifold is finite! We are implicitly assuming that a homotopy class contains at most one isometry, see Exercise (3-25).

In our study of quasifuchsian manifolds we have already made use of an analogue of Mostow rigidity for manifolds with boundary; in recent literature this is referred to as Marden's isomorphism (or rigidity) theorem [1974a]. See also Tukia [1985b, Theorems 4.2, 4.7].

Theorem 3.13.4 (Marden Isomorphism Theorem). *Suppose G is a geometrically finite group without elliptics. Assume either (i) $\Phi : \mathcal{M}(G) \to \mathcal{M}(H)$ is a homeomorphism, or (ii) $\pi_1(\mathcal{M}(G)) \to \pi_1(\mathcal{M}(H))$ is an isomorphism that restricts to an isomorphism $\phi_b : \pi_1(\partial\mathcal{M}(G)) \to \pi_1(\partial\mathcal{M}(H))$. Suppose in case (i) the restriction of Φ to $\partial\mathcal{M}(G) \to \partial\mathcal{M}(H)$ is homotopic to a conformal mapping, or in case (ii) ϕ_b is induced by a conformal mapping. Then in case (i) Φ and in case (ii) ϕ_b is induced by an isometry $\mathcal{M}(G) \to \mathcal{M}(H)$; the groups G, H are conjugate via a Möbius transformation.*

Thus $\mathcal{M}(G)$ is uniquely determined up to isometry by the isomorphism type of its fundamental group and the conformal structure of its boundary. In a sense, Mostow's theorem is a special case.

The boundary rigidity theorem first appears in Marden [1974a].

Sullivan [1981] established the most general form of rigidity that does not require geometric finiteness at all:

Theorem 3.13.5 (Sullivan Rigidity Theorem). *Suppose $\mu(z)$, $z \in \mathbb{S}^2$, is a Beltrami differential with respect to a finitely generated kleinian group G. Then $\mu(z) = 0$ for a.e. $z \in \Lambda(G)$.*

Corollary 3.13.6. *Geometrically infinite (as well as geometrically finite) manifolds $\mathcal{M}(G)$ with $\partial\mathcal{M}(G) = \varnothing$ are rigid under quasiconformal deformation.*

This result does not require any knowledge of the area of the limit set, nor does it give any information about its area. It says that the limit set can only support the zero

Beltrami differential so that from the point of view of quasiconformal deformations, its area has no consequence. In the case that $\Omega(G) \neq \varnothing$, we now know that $\Lambda(G)$ has zero area—Ahlfors' Conjecture is confirmed! (See Section 5.5.1.) So Sullivan's theorem for this case follows from the fact that a Beltrami differential needs only to be defined up to a set of zero measure for the Beltrami equation to have a solution (Section 2.8), uniquely determined up to postcomposition with a Möbius transformation. On the other hand when $\Lambda(G) = \mathbb{S}^2$, Sullivan's theorem still comes to the fore:

If $f : \mathbb{S}^2 \rightarrow \mathbb{S}^2$ is quasiconformal, induces an isomorphism $\phi : G \rightarrow H$, and, if $\Omega(G) \neq \varnothing$, restricts to a conformal map $\Omega(G) \rightarrow \Omega(H)$, then f is a Möbius transformation.

There is a remarkable analogue of Mostow's theorem for Teichmüler space itself due to [Bowditch and Epstein 1988, Thm. 1.7] and, independently, to Eskin et al. [2015] as follows. Suppose R is a Riemann surface with $(3g + n - 3) > 1$, and $\mathfrak{T}(R)$ its Teichmüller space.

Theorem $\mathfrak{T}(R)$ *is* **rigid** *in the following sense: If $\phi : \mathfrak{T}(R) \rightarrow \mathfrak{T}(R)$ is a quasi-isometry there exists an element $\sigma \in \mathcal{MCG}(R)$, that is, an isometry, such that the Teichmüller distance of $\sigma(x)$ to $\phi(x)$ is uniformly bounded for all $x \in \mathfrak{T}(R)$.*

Outline of the proof of the Mostow Rigidity Theorem. The theorem holds as well in n-dimensional hyperbolic space but here we will stick to three.

The orbifold case of the theorem can be reduced to the manifold case by Selberg's lemma. Namely, given G, there is a torsion-free, normal subgroup $G_0 \subset G$ of finite index; thus $\mathcal{M}(G_0)$ has finite volume, being finite-sheeted over $\mathcal{M}(G)$. The isomorphism ϕ restricts to the isomorphism $G_0 \rightarrow \phi(G_0) := H_0 \subset \phi(G) = H$. Assuming the manifold case, $\phi : G_0 \rightarrow H_0$ is a conjugation $G_0 \mapsto A G_0 A^{-1} = H_0$. We can assume A to be Möbius, rather than anti-Möbius, by replacing G if necessary with the group JGJ, where J is a reflection in some hyperbolic plane.

Now the cover transformations form a finite group C of isometries of $\mathcal{M}(G_0)$ and likewise the cover transformations of $\mathcal{M}(H_0)$ over $\mathcal{M}(H)$. We conclude that H itself is A-conjugate to G.

We will base our argument on the following theorem of Tukia:

Theorem 3.13.7 ([Tukia 1985a]). *Suppose G is any nonelementary kleinian group, $\zeta \in \Lambda(G)$ is a conical limit point, and $f : \mathbb{S}^2 \rightarrow \mathbb{S}^2$ is a homeomorphism which is differentiable at ζ with nonzero derivative. Assume $\phi : G \rightarrow H$ is a homomorphism to another kleinian group given by $f \circ g(z) = \phi(g) \circ f(z)$ for all $g \in G$, $z \in \mathbb{S}^2$. Then f is a Möbius transformation!*

Proof. We will give the proof reported in Kapovich 2001 [Theorem 8.34]. We will assume the homeomorphism is orientation preserving, although this is not necessary.

The limit point $\zeta \in \Lambda(G)$ is a *conical limit point*, also called *point of approximation*, if it has the following property.

Let $\gamma(t)$, $0 \leq t < \infty$, be a geodesic ray ending at ζ. Given a point $O \in \mathbb{H}^3$, there exists $r > 0$ such that there is an infinite subsequence of the orbit $G(O)$ that lies in the

r-tubular neighborhood about γ (and hence converges to ζ). In the quotient manifold, the condition means that the projection of the ray $\gamma(t)$ is *recurrent* in the sense that it meets a ball of radius r about the projection of O for a sequence $\{t_n\}$, $t_n \to \infty$. A loxodromic fixed point is always a conical limit point but a parabolic fixed point is not. Beardon and Maskit [1974] proved that a kleinian group is geometrically finite if and only if all limit points, except parabolic fixed points, are conical limit points; see Exercise (3-18).

We may assume that $\zeta = 0 = f(0)$ and that O lies on the vertical axis rising from $z = 0$ in the upper half-space model. Let γ be the vertical segment descending from $O \in \mathbb{H}^3$ to $z = 0$. There is an infinite sequence $g_n \in G$ such that for some $r > 0$ and each large index n, the (hyperbolic) distance $d(g_n(O), \gamma) < r$. Find the point $y_n \in \gamma$ that is closest to $g_n(O)$; it is within distance r. Then find $a_n > 0$ such that the transformation $A_n : \vec{x} \mapsto a_n\vec{x}$ takes O to y_n; $\lim a_n = 0$. Passing to a subsequence if necessary we may also assume that $\lim g_n^{-1} A_n = B$ exists as a Möbius transformation (because the distance of $g_n^{-1} A_n(O)$ to O is uniformly bounded by r). Set

$$f_n(z) \;=\; a_n^{-1} f(a_n z) \;=\; A_n^{-1} \circ f \circ A_n(z), \quad z \in \mathbb{C}.$$

That the complex valued function $f(z)$ is differentiable at $z = 0$ with nonzero derivative means that there is a linear transformation $L : \mathbb{R}^2 \to \mathbb{R}^2$ (i.e., a 2×2 real matrix operating on $z \in \mathbb{C}$ as a vector), with nonzero determinant, such that

$$f(\Delta z) \;=\; L(\Delta z) + \epsilon(\Delta z)\Delta z, \qquad \lim_{\Delta z \to 0} \epsilon(\Delta z) = 0.$$

(Alternatively $L(z, \bar{z}) = az + b\bar{z}$, for some $a, b \in C$, $|a|^2 - |b|^2 > 0$.) Treating A_n as a linear transformation on vectors $z \in \mathbb{R}^2$ and setting $\Delta z = A_n(z)$, we obtain for f_n that

$$f_n(z) \;=\; L(z) + \epsilon(A_n(z))z.$$

We have used that the real diagonal matrix A_n commutes with L. Consequently

$$\lim_{n \to \infty} f_n(z) = \lim A_n^{-1} f A_n(z) = L(z) \quad \text{uniformly on compact subsets of } \mathbb{C}.$$

In short, L is nothing but the "blow-up" of f at $z = 0$.

It now follows that

$$\lim A_n^{-1} G A_n \;=\; \lim A_n^{-1} g_n G g_n^{-1} A_n \;=\; B^{-1} G B.$$

This implies that the sequence of groups $\{A_n^{-1} G A_n\}$ *converges geometrically*. In the next chapter, we will study this notion in detail; it suffices to say here that every $B^{-1} g B$ is the limit of elements of the approximants $\{A_n^{-1} G A_n\}$, namely given $g \in G$, $B^{-1} g B = \lim(A_n^{-1} g_n) g(g_n^{-1} A_n)$. Conversely, the limit of any convergent sequence of elements of $\{A_n^{-1} G A_n\}$ lies in $B^{-1} G B$, namely,

$$h := \lim A_n^{-1} h_n A_n = \lim A_n^{-1} g_n (g_n^{-1} h_n g_n) g_n^{-1} A_n = B^{-1}(\lim g_n^{-1} h_n g_n) B.$$

Recall that for any $g \in G$, $f \circ g = \phi(g) \circ f$. Given $g \in G$,

$$L \circ B^{-1}gB \circ L^{-1} = \lim_{n \to \infty} A_n^{-1} f \circ g_n g g_n^{-1} \circ f^{-1} A_n$$
$$= \lim_{n \to \infty} A_n^{-1} \phi(g_n g g_n^{-1}) \circ f \circ f^{-1} A_n$$
$$= \lim_{n \to \infty} A_n^{-1} \phi(g_n g g_n^{-1}) A_n.$$

The element on the left is therefore a Möbius transformation. We have established that $L \circ h \circ L^{-1}$ is a Möbius transformation for any $h \in B^{-1}GB$.

Since not all elements of G fix $B(\infty)$, there exists $h \in B^{-1}GB$ with $h(\infty) \neq \infty$. We claim that LhL^{-1} being a Möbius transfomation this forces L itself to be a Möbius transformation, necessarily fixing 0 and ∞.

For let ℓ be a euclidean line in $\mathbb{C}$. Then $L^{-1}(\ell)$ is again a straight line. Choose ℓ such that $L^{-1}(\ell)$ does not go through $h^{-1}(\infty)$. Then $h \circ L^{-1}(\ell) = C$ is a proper circle. Therefore $LhL^{-1}(\ell) = L(C)$ is a circle as well, since on the one hand L maps bounded sets to bounded sets, and on the other, LhL^{-1} is Möbius. But an affine mapping L that by definition fixes 0 and ∞ cannot send a circle onto a circle unless it can be expressed as $z \mapsto az$ (or $z \mapsto a\bar{z}$, if we allowed f and hence L to be orientation reversing). So L is a Möbius transformation, as claimed, and it has the simple form $z \mapsto az$.

Pick three distinct points p_1, p_2, $p_3 \in \mathbb{S}^2$. For any homeomorphism $F : \mathbb{S}^2 \to \mathbb{S}^2$ set $N(F) = F^\sharp \circ F$ where the Möbius transformation $F^\sharp$ is uniquely chosen so that $N(F)$ fixes each p_i. Upon setting $u_n = g_n^{-1} A_n$ so that $\lim u_n = B$,

$$N(f_n) = N(A_n^{-1} f A_n) = N(f A_n) = N(f g_n u_n) = N(\phi(g_n) f u_n) = N(f u_n).$$

Going to the limit,

$$N(L) = \lim_{n \to \infty} N(f_n) = N(fB).$$

Since L and B are Möbius transformations, f must be one as well. $\square$

Mostow's Rigidity Theorem follows from this result. We have to construct, given the isomorphism $\phi : G \to H$, a quasiconformal automorphism of $\mathbb{S}^2$ that induces it. If $\mathcal{M}(G)$ has finite volume, results of Waldhausen 1968, Lemma 6.3, Theorem 6.1 applied and extended to the noncompact case in Marden [1974a], or of Tukia [1985b, Theorem 4.7] show that there is an orientation-preserving or -reversing quasiconformal mapping Φ_* between the manifolds, inducing ϕ on the fundamental group, so that $\mathcal{M}(H)$ has finite volume as well. By replacing H by JHJ if necessary we may assume it is an ordinary quasiconformal mapping. A lift Φ to $\mathbb{H}^3$ is a quasiconformal mapping, and quasiconformal mappings of, say, upper half-space extend to be quasiconformal on $\mathbb{C} \cup \infty$ [Gehring 1962]. Also $\Phi G \Phi^{-1} = H$. Quasiconformal maps of $\mathbb{S}^2$ are differentiable with nonzero derivative almost everywhere [Ahlfors 1966]. Since G is geometrically finite and $\partial \mathcal{M}(G) = \varnothing$, every point on $\mathbb{S}^2$ is a limit point, and all those except the countable number of parabolic fixed points are conical limit points. All that remains is to apply Tukia's result at one point of differentiability which is not a parabolic fixed point.

If $\mathcal{M}(G)$ is a closed manifold the following alternate argument can be used: It follows from topology that there is a *homotopy equivalence* (see the discussion in

Section 5.1) between the manifolds: continuous maps $f_1 : \mathcal{M}(G) \to \mathcal{M}(H)$ and $f_2 : \mathcal{M}(H) \to \mathcal{M}(G)$ such that $f_1 \circ f_2$ and $f_2 \circ f_1$ are homotopic to the identity. Working in terms of a piecewise linear structure (subdividing into hyperbolic tetrahedra) on the manifolds, the mappings can be taken to be Lipschitz. It turns out [Mostow 1973, Lemma 9.2; Thurston 1979a, p. 5.39] that their lifts F_1, F_2 to $\mathbb{H}^3$ are quasiisometries; see Exercise (3-19). Consequently each one extends to $\partial \mathbb{H}^3$ and is quasiconformal there. Using the cusp tori, this approach too extends to the finite volume case; see Prasad [1973]; Thurston [1979a], p. 5.39; Tukia [1985b, Lemma 3.4]. □

It is interesting to compare the situation we just considered to the case of a quasiconformal map $f : \mathbb{H}^2 \to \mathbb{H}^2$. Likewise f can be extended to $\mathbb{S}^1$ and is again a homeomorphism there, necessarily having a derivative almost everywhere. The extension to $\partial \mathbb{H}^2$ is either the restriction of a Möbius transformation, or its derivative is zero wherever it exists. Now one knows in advance that most fuchsian groups have nontrivial deformations. The corresponding homeomorphisms of the circle $\partial \mathbb{H}^2$ are therefore examples of totally singular functions: their derivatives are zero almost everywhere. This seems to be the simplest construction of singular functions.

It is perhaps surprising that the mapping class group $\mathcal{MCG}(S)$ of any finite area hyperbolic Riemann surface S is itself rigid [Hamenstädt 2007]:

Theorem 3.13.8 (Rigidity of mapping class group). *Suppose the finitely generated group Γ is quasiisometric to $\mathcal{MCG}(S)$, in the sense of the corresponding Cayley graphs. Then there exists a finite index subgroup $\Gamma' \subset \Gamma$ and a homomorphism $\varphi : \Gamma' \to \mathcal{MCG}(S)$ with finite kernel and finite index range.*

3.14 Exercises and explorations

3-1. (a) Prove the area formula for a surface S of constant gaussian curvature $K = 0, \pm 1$, the area being given in the euclidean, spherical, or hyperbolic metric. Here S is a closed surface of genus $g \geq 0$, with $n \geq 0$ punctures and $m \geq 0$ cone points of orders $\{2 \leq r_i < \infty\}$.

$$K \, \text{Area}(S) = 2\pi \left(2 - 2g - n - \sum_{i=1}^{m} \left(1 - \frac{1}{r_i} \right) \right). \tag{3.12}$$

Hint: The Euler characteristic formula for a triangulated closed surface is $\chi(S) = T - E + V = 2 - 2g$, where T is the number of triangles, E the number of edges, V the number of vertices and g the genus. If S is a closed surface of genus g with n punctures or boundary components, $\chi(S) = 2 - 2g - n$. In calculating $\chi(S)$, the punctures are not counted as vertices—they are not in the surface—and any boundary components are treated like punctures.

To compute the area, cut the surface into small geodesic triangles. Each puncture and cone point should be a vertex. Think of how the neighborhood of each arises by projection from the branched universal cover. Since each triangle has three edges each of which is shared by the adjacent triangle, $2E = 3T$. The area of each triangle satisfies $K \, \text{Area}(\Delta) = \theta_1 + \theta_2 + \theta_3 - \pi$. If there are no punctures or cone points, summing the triangles we find that $K\,\mathrm{Area}(S) = 2\pi V - \pi T = 2\pi\chi(S)$. If there are cusps (cone points of order ∞ on a negatively curved surface) the area is too great because the angle sum about a cusp is 0 instead of 2π, so $2\pi n$ must be subtracted. At a cone point the angle sum is instead $2\pi/r_i$ rather than 2π, so we must subtract the difference $2\pi(1 - 1/r_i)$.

Thus if S is a closed surface of genus g with n-punctures, and $K = -1$, no matter how many cone points there are, we have

$$\mathrm{Area}(S) \geq 2\pi|\chi(S)|. \qquad (3.13)$$

(b) Conversely, choose a discrete set of ≥ 3 points $\{x_i\}$ on a Riemann surface R, and a corresponding sequence of integers $\{r_i \geq 2\}$. Assume as per Equation (3.12) that R satisfies $2g - 2 + n + \sum(1 - 1/r_i) > 0$. Show that there exists a fuchsian group G such that $\mathbb{H}^2/G$ is a branched covering surface of R with branch points of order r_i exactly over the points x_i (for a proof, see Farkas and Kra [1991, Theorem IV.9.12]).

Suppose instead, we are given a finite group H. There exists a closed Riemann surface $R = H^2/G$, and a normal subgroup N of G such that G/N is isomorphic to H. In short, every finite group H is the group of automorphisms of some Riemann surface R, see Jones [1999].

In fact, Jones begins with an finitely generated fuchsian group G (typically with torsion) and counts the number of finite index normal subgroups N all of which have the property that G/N is isomorphic to the given finite group H. That is, he counts the number of equivalence classes of regular coverings of the orbifold $R = \mathbb{H}^2/G$ with covering group H.

(c) For a closed, oriented surface S of genus g with riemannian metric h and Gaussian curvature $K(h)$ the Gauss–Bonnet formula reads

$$2\pi\chi(S) = 2\pi(2 - 2g) = \iint_S K(h)\,dA_h,$$

where $\chi(S)$ is the Euler characteristic and dA_h is the element of surface area. In the hyperbolic case $K(h) = -1$ and the surface area is $4\pi(g - 1)$, $g \geq 2$. Hyperbolic metrics are best: If $K(h) \geq -1$ (resp. ≤ -1), then $\mathrm{Area}_h(S) \geq \mathrm{Area}_{\mathrm{hyp}}$ (resp. $\leq$), with equality only when h is the hyperbolic metric. For generalizations to 3-manifolds, see Besson et al. [1999]; Storm [2002; 2007].

Equation (3.12) holds for cone angles $2\pi/r_i \leq 2\pi$. But it is sometimes applied to cone manifolds with arbitrary cone angles. If in an n-punctured surface S of genus g the cone points become cone points with angles satisfying $2\pi/r_i \geq 2\pi$, then Equation (3.13) becomes instead

$$\mathrm{Area}(S) \leq 2\pi\,|\chi(S)|.$$

A common application of the area formula is to find the possibilities that a closed surface of genus g and n punctures with designated cone points carries the spherical metric $K = +1$, euclidean metric ($K = 0$), or hyperbolic ($K = -1$).

In other words, the surface is covered by $\mathbb{S}^2$, $\mathbb{C}$ or $\mathbb{H}^2$. This gives rise to three inequalities:

$$2g + n + \sum_{i=1}^{m}\left(1 - \frac{1}{r_i}\right) < 2 \qquad \text{spherical case,} \qquad (3.14)$$

$$2g + n + \sum_{i=1}^{m}\left(1 - \frac{1}{r_i}\right) = 2 \qquad \text{euclidean case,} \qquad (3.15)$$

$$2g + n + \sum_{i=1}^{m}\left(1 - \frac{1}{r_i}\right) > 2 \qquad \text{hyperbolic case.} \qquad (3.16)$$

Inequality (3.14) requires $g = 0$ and $n = 0, 1$. If $n = 1$ then $m = 1$ and $2 \leq r_1 < \infty$. For $n = 0$, if $m = 3$ the possibilities for the cone points are $(2, 3, 5)$, $(2, 3, 4)$, $(2, 3, 3)$, $(2, 2, n)$; if $m = 2$ then $2 \leq r_1, r_2 < \infty$; if $m = 1$, then $2 \leq r_1 < \infty$.

Equality (3.15) requires $g = 0, 1$. If $g = 1$ then $m, n = 0$. For $g = 0$, we can have $n = 2$ and $m = 0$; otherwise if $n = 1$, then $m = 2$ and $r_1 = r_2 = 2$; if $n = 0$, the cone points are given by $(2, 2, 2, 2)$, $(3, 3, 3)$, $(2, 3, 6)$, or $(2, 4, 4)$.

Inequality (3.16) is satisfied by all combinations except those listed already.

(d) Show that as $\Gamma \subset \mathrm{PSL}(2, \mathbb{R})$ ranges over all fuchsian groups (that may have elliptics and/or parabolics),

$$\mathrm{Area}(\mathbb{H}^2/\Gamma) \geq \frac{\pi}{21}.$$

The minimum value $\frac{\pi}{21}$ is uniquely attained for the $(2, 3, 7)$-triangle group (and its conjugates). Conclude that for a group R representing a closed surface R of genus $g \geq 2$, the order of the group $C(R)$ of conformal automorphisms of R satisfies

$$\mathrm{Order}(C(R)) \leq \frac{4\pi(g-1)}{\pi/21} = 84(g-1).$$

The group $C(R)$ is isomorphic to $N(G)/G$, where $N(G)$ is the normalizer of G, and the area of $\mathbb{H}^2/N(G)$ is not less than $\pi/21$.

(e) Show that if the fuchsian group G is of finite index n in the fuchsian group H, the area of $\mathbb{H}^2/H$ is n times that of $(\mathbb{H}^2/G)$. Because n cannot become too large, conclude that every fuchsian group of finite area is contained in a *maximal fuchsian group*, one that has finite area and is not a subgroup of any other fuchsian group [Greenberg 1974]

For example the triangle groups (genus zero, no punctures) are maximal except when the signatures have the form (a, a, b), $(2, a, 2a)$, $(3, a, 3a)$. In particular the triangle group with signature $(2, 7, 14)$ is not maximal.

Does the same argument work for finite volume kleinian groups (Section 4.11.1)?

(f) (Wiman's theorem, [Farb and Margalit 2012, Theorems 7.5, 7.14].) Show that on a closed surface of genus $g \geq 2$, the order of an automorphism cannot exceed

$(4g+2)$. Equality holds for the surface formed from a regular polygon of $(4g+2)$ sides with opposite sides identified. If there is an element of order $(4g + 2)$ on R, up to conjugacy, R has only one of them.

3-2. If A is loxodromic prove that in the hyperbolic metric, $\min_{\vec{x} \in \mathbb{H}^3} d(\vec{x}, A(\vec{x}))$ is achieved only when $\vec{x}$ lies on the axis of A.

Exercise (1-4) showed that the set $V = \{\vec{x} \in \mathbb{H}^3 : d(\vec{x}, A(\vec{x})) < \epsilon\}$ is a radius ϵ tube about the axis of A.

If A is parabolic, what is the set $\{\vec{x} \in \mathbb{H}^3 : d(\vec{x}, A(\vec{x})) < \epsilon\}$?

3-3. Prove that for a fuchsian group G, the universal horodisk at a parabolic fixed point is not penetrated by the axis of any loxodromic element that represents a simple geodesic on $\mathbb{H}^2/G$. Is the same statement true for the universal horoball in a kleinian group? (*Hint*: Apply $z \mapsto z + 1$ to the axis.)

Prove that for a kleinian group G the corresponding statement is as follows. Suppose $T(z) = z+1 \in G$ and $\mathcal{H} = (z, t) : t > 1$ is its universal horoball. No hyperbolic plane C with $\partial C \in \mathbb{C}$, and whose projection is embedded in $\mathcal{M}(G)$, can penetrate $\mathcal{H}$. (*Hint*: The euclidean height of C exceeds 1, the diameter of ∂C exceeds 2, resulting in $T(C) \cap C \neq \varnothing$.)

Prove that a discrete group with all real traces is conjugate to a fuchsian group.

3-4. Show that if $x \in \mathbb{H}^3$ approaches $\zeta \in \partial \mathbb{H}^3$, then the limit of the half-space H_g of Section 3.5 is the half-space determined as follows. There is a unique horosphere σ at ζ such that σ is tangent to the horosphere $g^{-1}\sigma$ at $g^{-1}(\zeta)$. Take the hyperbolic plane tangent to both horospheres at their point of tangency; it is orthogonal to the geodesic with endpoints ζ, $g^{-1}(\zeta)$. Choose the half-space determined by this plane that is adjacent to ζ. If $\zeta = \infty$, this half-space is the exterior of the isometric hemisphere for g. Also see Lemma 1.5.4.

3-5. *Figure-8 knot.* Find a Dirichlet region for the rank two parabolic group

$$G = \langle z \mapsto z + 1, \ z \mapsto z + \tau; \ \mathrm{Im}\,\tau > 0 \rangle.$$

Show that generically it has six edges, but in some situations it has only four. The square and the regular hexagon provide the associated torus with symmetries of order four and order six.

Compute the hyperbolic volume of the part of the polyhedron lying above a horosphere (a horizontal plane, in the present situation). Show that the quotient $\mathbb{H}^3 \cup \mathbb{C}/G$ is homeomorphic to $\{0 < |z| \leq 1, \ z \in \mathbb{C}\} \times \mathbb{S}^1$, that is, the complement of the core circle in the solid torus. This is the prototype of the local structure about a knot when the knot complement has a hyperbolic structure—as all of them do, except torus knots and satellite knots (Section 6.3). The parabolic fixed point is "stretched" into the knot.

For example, the figure-8 knot complement can be constructed as follows. In the upper half-space model, choose an ideal tetrahedron with one vertex at ∞, as in Exercise (1-22). Each of the four faces is an ideal triangle. Thinking of the ball model, the ideal vertices of each face lie on a circle in $\mathbb{S}^2$. The circles corresponding to adjacent faces intersect, and their angle of intersection is the dihedral angle between the faces. Arrange it so that the six dihedral angles are all 60° so as to become the regular ideal tetrahedron (compare (Exercise (1-23))). In fact the dihedral angles of any ideal tetrahedron add up to 360°.

Back in UHP, line up two such ideal tetrahedra T_1 and T_2, one next to the other so they share a face and the ideal vertex ∞. There are six free faces on the union of the two tetrahedra. The faces can be paired and the face identification via isometries precisely given so that the tetrahedral union is a fundamental polyhedron for the group G generated by the face pairing transformations. If down properly, $\mathbb{H}^3/G$ is homeomorphic to the complement of the figure-8 knot in $\mathbb{S}^3$. The five ideal vertices become parabolic fixed points which are in the single parabolic conjugacy class of G. For details see Thurston [1997, pp. 39–42], Ratcliffe [1994, 10.5], or Neumann [1999].

The figure-8 knot complement is $\mathbb{H}^3/G$, where G can be taken to be generated by

$$\begin{pmatrix} 1 & 0 \\ 1 & 1 \end{pmatrix}, \quad \begin{pmatrix} 1 & e^{\pi i/3} \\ 0 & 1 \end{pmatrix}.$$

For a discussion of hyperbolic knots see Section 6.3.

3-6. *Volume of maximal solid cusp tori* [Adams 1987]. We will use the fact that the densest circle packing in the plane with circles of the same radius is the hexagonal packing: each circle is surrounded by six others. This is applied as follows.

Suppose P is a parallelogram. Place a disk of radius r centered at each of the four vertices. Choose r maximally subject to the constraint that the interiors of the disks be mutually disjoint. If $|P|$ denotes the area of P, then $|P| \geq 2r^2\sqrt{3}$. Equality occurs if and only if all the sides of P have the length $2r$. If the sides of P have length ≥ 1, then $r \geq 1/2$. If one side has length one, $r = 1/2$.

Now consider a hyperbolic 3-manifold $\mathcal{M}(G)$ such that G has a rank two parabolic subgroup. We can conjugate so that $G_\infty = \langle z \mapsto z + 1, z \mapsto z + \tau \rangle$ with $\tau = u + iv$, $-\frac{1}{2} \leq u \leq \frac{1}{2}$, $y \geq \sqrt{3}/2$ (Exercise (2-5)). A *maximal horoball* at the fixed

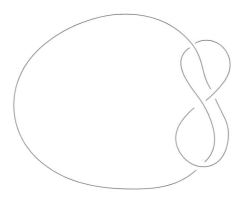

Fig. 3.10. The figure-8 knot.

point $z = \infty$ is the largest horoball $\mathcal{H}_\infty$ with the property that $g(\mathcal{H}_\infty) \cap \mathcal{H}_\infty = \varnothing$ for all $g \in G$, $g \notin G_\infty$. In our setup with ∞ the fixed point, this means that $\mathcal{H}_\infty = \{(z, t) \in \mathbb{H}^3 : t > s\}$ with the smallest possible $1 \geq s > 0$.

Let σ denote the horosphere $\{(z, t) \in \mathbb{H}^3 : t = s\}$. Because this bounds the maximal horoball at ∞ there will be an element $g \in G$, $g \notin G_\infty$ such that the euclidean sphere $g(\sigma)$ is tangent to σ. We may assume that $g(\sigma)$ is based at $z = 0 = g(\infty)$.

Prove that $g^{-1}(\infty)$ cannot lie in the orbit $G_\infty(0)$. *Hint*: Suppose otherwise so that for some $h \in G_\infty$, $h^{-1}g^{-1}(\infty) = 0$. Also $h^{-1}g^{-1}(0) = \infty$. Therefore $h^{-1}g^{-1}$ fixes a point on the vertical half line, a contradiction since G has no elliptic elements.

In $\mathbb{C}$, choose the fundamental parallelogram P for G_∞ to have vertices at 0, 1, τ, and $1 + \tau$. The horoball $g(\mathcal{H}_\infty)$ is tangent to $\mathbb{C}$ at $z = 0$ and its G_∞-orbit contains horoballs tangent to $\mathbb{C}$ at all the vertices of P. Its G_∞-orbit is also disjoint from the G_∞-orbit of the horoball $g^{-1}(\mathcal{H}_\infty)$. There will be at least one point $\zeta \in P$ which is a tangent point of the latter G_∞-orbit.

All these horoballs tangent to $\mathbb{C}$ have the same euclidean radius $r = 1/(2s)$. The vertices of P have distance at least $2r = 1$ apart, and also must be distance at least 1 from ζ. Place disks of radius $1/2$ centered at the vertices of P and at ζ. Consider their G_∞-orbit. Their interiors are mutually disjoint and P has to be covered by the equivalent of two disks. Deduce that $|P| \geq \sqrt{3}$.

The volume of $\mathcal{H}_\infty / G_\infty$ is $|P|/2s^2$.

Conclude that *the volume of the maximal solid cusp torus is at least $\sqrt{3}/2$ in any hyperbolic manifold $\mathcal{M}(G)$ with nonelementary G.*

Compare with Exercise (2-11).

Now show that the number of primitive lattice points of the orbit $G_\infty(0)$ whose distance from $z = 0$ is less than 2π is ≤ 48 [Bleiler and Hodgson 1996]; there are at most 24 simple closed geodesics in the quotient torus of length $< 2\pi$. Primitive means that the ray from 0 to the lattice point does not pass through any other lattice points.

3-7. *Subgroups of geometrically finite groups* [Thurston 1986b]. Suppose $\mathcal{M}(G)$ is a geometrically finite hyperbolic 3-manifold such that its convex core $\mathcal{CC}(G) \neq \mathcal{M}(G)$. Prove that for every finitely generated subgroup G_1 of G, $\mathcal{M}(G_1)$ is also geometrically finite.

Hint: Consider the case that G has no parabolics. Then $\mathcal{CC}(G)$ is compact. There exists d such that every point $x \in \mathcal{CC}(G)$ has distance at most d from $\partial\mathcal{CC}(G)$. Let $\widehat{\mathcal{CC}}(G)$ denote the lift to $\mathbb{H}^3$. Every point in the manifold $\widehat{\mathcal{CC}}(G)$ has distance at most d from $\partial\widehat{\mathcal{CC}}(G)$.

Now $\widehat{\mathcal{CC}}(G)/G_1$ is a convex submanifold of $\mathcal{M}(G_1)$, thus containing $\mathcal{CC}(G_1)$. Each point $x \in \mathcal{CC}(G_1)$ therefore has distance at most d from $\partial\widehat{\mathcal{CC}}(G)/G_1$ and then also from $\partial\mathcal{CC}(G_1)$.

The Ahlfors Finiteness Theorem implies that $\partial\mathcal{CC}(G_1)$ has a finite number of components and each component is a compact surface without boundary (there are no parabolics in G_1). Consequently a finite number of d-balls centered on $\partial\mathcal{CC}(G_1)$ cover it. A finite number of d-balls will then cover $\mathcal{CC}(G_1)$ as well. Thus G_1 is geometrically finite.

The proof in the general case also requires use of the thick/thin decomposition of $\mathcal{CC}(G)$.

3-8. *Klein–Maskit combination theory.* Here we will display only the classical situations. In Maskit [1988] the reader will find extensive generalizations.

(i) Suppose G is a kleinian group. Select two mutually disjoint closed disks D_1, D_2 in $\Omega(G)$ such $g(D_i) \cap D_j = \varnothing$ for $i, j = 1, 2$ and all $g \neq$ id $\in G$. Let T be any Möbius transformation that maps the exterior of D_1 onto the interior of D_2. Prove that $G^* = \langle G, T \rangle$ is also discrete, as claimed in Section 3.7.

Topologically show that what you have done is the following. The projection $\pi : D_i \rightarrow \Delta_i$ to the quotient $\mathcal{M}(G)$ is a homeomorphism. Remove Δ_1, Δ_2 from $\partial\mathcal{M}(G)$ and identify the resulting boundaries $\partial\Delta_1$ and $\partial\Delta_2$. If the two disks lie on the same component of $\partial\mathcal{M}(G)$ what you have done is create a new handle. If they lie in different boundary components, you have connected the two components. In either case, the simple loop $\partial\Delta_1 = \partial\Delta_2$ on $\partial\mathcal{M}(G^*)$ bounds a disk within $\mathcal{M}(G^*)$. In either case this disk does not divide $\mathcal{M}(G^*)$.

Exactly the same process can be used to connect two manifolds $\mathcal{M}(G_1)$ and $\mathcal{M}(G_2)$. In this case $G^* = G_1 * G_2$ is a free product since the new disk divides.

Show that you can adjoin a solid torus and/or a solid cusp torus to $\mathcal{M}(G)$.

What happens if you make the following alternative combination? Given a small closed disk $D \subset \Omega(G)$, let J denote reflection in the circle ∂D, and equally in the plane in $\mathbb{H}^3$ rising from ∂D. Form the new group $G^* = \langle G, JGJ \rangle$. This too will be discrete. Describe $\mathcal{M}(G^*)$.

And here is another possibility. Suppose two components D_1, D_2 of $\Omega(G)$ are round disks, where G is not quasifuchsian. Denote the stabilizing subgroups G_1, G_2. Suppose T is a Möbius transformation (i) sending the exterior of D_2 to the interior of D_1, and (ii) $TG_2T^{-1} = G_1$. Show that the augmented group $G^* = \langle G, T \rangle$ is discrete. How is $\mathcal{M}(G^*)$ related to $\mathcal{M}(G)$? A similar combination is used in Section 6.2.2.

(ii) Suppose ζ_1, ζ_2 are two parabolic fixed points of G. Suppose $\Delta_1 \in \Omega(G)$ is a relatively closed horodisk associated with ζ_1 and Δ_2 is a one associated with ζ_2 so that $g(\Delta_i) \cap \Delta_i = \varnothing, i = 1, 2$, unless $g \in \text{Stab}_{\zeta_i}$. Possibly $\zeta_1 = \zeta_2$ and then the disks are externally tangent at the fixed point. Let B be any Möbius transformation that maps the exterior of Δ_1 onto the interior of Δ_2 *and* conjugates $\text{Stab}(\zeta_1)$ to $\text{Stab}(\zeta_2)$. Prove that $G^* = \langle G, B \rangle$ is discrete. Topologically what has happened is this: We have chosen two circles c_1, c_2 about two distinct punctures on $\partial\mathcal{M}(G)$, the projections of the two horocycles. Remove the once-punctured disks bounded by these two circles from $\partial\mathcal{M}(G)$ and identify the two circles. If the two circles are on the same boundary component of $\mathcal{M}(G)$, that component loses two punctures but gains a handle. If they are on different components, the two components become connected and lose a puncture each. In G^* the two cyclic parabolic groups become conjugate; if $\zeta_1 = \zeta_2$ the cyclic parabolic group becomes a rank two parabolic group. See also Exercise (4-28).

Algebraically G^* is the free product with amalgamation of the two cyclic parabolic groups. Likewise the construction can be carried out to join two different manifolds.

It was originally hoped that with the two classical combination techniques, (i) and (ii), all kleinian groups could be constructed. Peter Scott, in the mid 1970s, showed (personal communication) that this cannot be the case. It is a key part of the Thurston theory, specifically the skinning lemma (Section 6.2), that allows general forms of combination to be effectively and generally applied—it shows that the group can be deformed so that there exist Möbius transformations that do the job required of hyperbolic gluing. Armed with the skinning lemma, most kleinian groups can be formed from simpler ones using hyperbolic gluing—the combination theorems.

3-9. *Extended quasifuchsian groups.* Let G be a fuchsian or quasifuchsian group with $\Omega(G) = \Omega_1 \cup \Omega_2$, the components of the regular set. Suppose there is a Möbius transformation T that maps Ω_1 onto Ω_2 and such that if $g \in G$ then also $TgT^{-1} \in G$. Show that the extended group $G^* = \langle G, T \rangle$ is discrete. Describe the topology of $\mathcal{M}(G^*)$ (it has only one boundary component).

The group G^* is called an *extended fuchsian* or *extended quasifuchsian* group. It has the same limit set as G and an index two subgroup which is fuchsian or quasifuchsian. Can you construct an extended fuchsian group by adjoining $z \mapsto -z$ or $z \mapsto 1/z$ to the modular group?

3-10. Suppose G is a finitely generated kleinian group and Ω is a simply connected component of $\Omega(G)$ with the following properties:

(i) Ω is invariant under G and is a proper subset of $\Omega(G)$.
(ii) Every simple loop in $S = \Omega/G$ that determines a parabolic element of G is retractable in S to a puncture.

Apply the Ahlfors Finiteness Theorem and the Cylinder Theorem to prove that G is a fuchsian or quasifuchsian group.

3-11. *Function and Schottky groups; compression bodies.* Assume that a component Ω of $\Omega(G)$ is invariant under G. Traditionally, complex analysts have called such a group a *function group* because by using Poincaré series, differentials and functions can be constructed on it. We will however reserve the name for the cases that Ω is not simply connected. Here we will assume that G is finitely generated without elliptics.

The corresponding manifold $\mathcal{M}(G)$ has the property that the inclusion from $S = \Omega/G, \pi_1(S) \hookrightarrow \pi_1(\mathcal{M}^3)$, is surjective. In particular, $\mathcal{M}(G)$ is a compression body.

Compression bodies are constructed by starting with a closed 3-ball B, cutting $n \geq 2$ holes in ∂B and attaching to the boundary of each of the these holes the boundary of a hole cut in: (i) a solid torus, and/or (ii) a closed surface bundle $S_g \times [0, 1]$, with S_g of genus $g \geq 2$.

In the opposite direction, on the compressible boundary component S there is a finite system of nontrivial simple loops that bound disks in $\mathcal{M}(G)$. Cut $\mathcal{M}(G)$ along these disks to get one or more pieces M_i. Here we are using Dehn's Lemma and the

Loop Theorem. If there is only one piece, then it is a ball and $\mathcal{M}(G)$ is a handlebody. Otherwise each M_i is either $\cong S_i \times [0, 1]$ where S_i is an incompressible boundary component of $\mathcal{M}(G)$, or it is a solid torus. Algebraically, G is the free product of surface groups and cyclic groups.

When there are parabolics and G is geometrically finite, "compactify" $\mathcal{M}(G)$ by removing solid cusp tori and solid pairing tubes. Then the analysis is essentially the same.

If G is an N-generator function group, find estimates for the number of pieces, and the genus and punctures of each [Marden 1974a].

Prove that a general geometrically finite manifold $\mathcal{M}(G)$ can be decomposed along incompressible surfaces into finitely many compression bodies and submanifolds with incompressible boundary. This is a result of Bonahon 1986, who also showed that the decomposition is unique up to isotopy. Two compact compression bodies with isomorphic fundamental groups are homeomorphic [McCullough and Miller 1986].

The simplest example is a (not necessarily classical) Schottky group representing a handlebody $\mathcal{M}(G)$ of genus $g \geq 1$. Equally important for a Schottky group are the simple loops which are not compressing. Show that there exist simple noncompressing loops that divide the surface into two parts; see Exercise (5-25).

In McCullough and Miller [1986] it is proved after a long argument that given a compressible boundary component S of a geometrically finite $\mathcal{M}(G)$ without parabolics, there is a submanifold $X \subset \mathcal{M}(G)$ with the following properties: (i) S is a boundary component of X, (ii) $\partial X \setminus S$ is incompressible in $\mathcal{M}(G)$, and (iii) the image of the inclusion $\pi_1(S) \hookrightarrow \pi_1(\mathcal{M}(G))$ is precisely $\pi_1(X)$, that is, X is a compression body with S its compressible boundary component.

Suppose $\mathcal{M}(G)$ is compact with an incompressible boundary component S. Show that $\pi_1(S)$ either has index at most two in G (Exercise (3-9)), or it has infinite index in G [Hempel 1976, Theorem 10.5].

3-12. *Diameters of ordinary set components.* Assume that $\infty \in \Omega(G)$ and Ω is a component of $\Omega(G)$ with $\infty \notin \Omega$. Suppose there exists a *relatively compact* fundamental set F for the action of $\mathrm{Stab}(\Omega)$ (recall that this is the group $\{g \in G : g(\Omega) = \Omega\}$). Prove that

$$\mathrm{Diam}(\Omega)^2 \leq \sum_{g \in \mathrm{Stab}(\Omega)} \mathrm{Diam}(g(F))^2.$$

Prove further that if $\{g_i(\mathrm{Stab}(\Omega))\}$ is the set of left cosets of $\mathrm{Stab}(\Omega)$ in G, then

$$\sum \mathrm{Diam}(g_i(\Omega))^2 < \infty.$$

Here Diam is the euclidean diameter of the set. There is a one-to-one correspondence between left or right cosets of $\mathrm{Stab}(\Omega)$ in G and components of the orbit $G(\Omega)$.

3-13. *Boundary fixed points* [Maskit 1974]. Suppose H is such that the quotient $\Omega(H)/H$ has a finite number of components each of which is a closed surface. Prove that if ∞ is not a limit point,

$$\sum_{h \in H} |c_h|^4 < \infty,$$

where $|c_h|^{-1}$ is the radius of the isometric circle of $h \neq$ id. *Hint:* The orbit of a fundamental region has finite spherical area since there is no overlap.

Now suppose G is a kleinian group, ∞ is not a fixed point, and $\Omega \subset \Omega(G)$ is a component of the regular set. Consider $\mathrm{Stab}(\Omega) = \{g \in G : g(\Omega) = \Omega\}$. Assume that the quotient $\Omega / \mathrm{Stab}(\Omega)$ is a closed surface. Let $\{\Omega_i\}$ denote the components of the G-orbit of Ω; that is, if $G = \bigcup g_i \mathrm{Stab}(\Omega)$ is the coset decomposition, then we can take $\Omega_i = g_i(\Omega)$. Prove for the spherical diameters that $\sum \mathrm{Diam}^4(\Omega_i) < \infty$. *Hint:* $\mathrm{Diam}(g_i(\Omega_i)) \leq |c_i|^{-2} d_i^{-1}$ where d_i is the spherical distance between $g_i^{-1}(\infty)$ and Ω.

Deduce that if a loxodromic $g \in G$ has a fixed point on $\partial\Omega$ then $g^k(\Omega) = \Omega$ for some k. *Hint:* If no power g^k preserves Ω then $\sum \mathrm{Diam}(g^k(\Omega)) = \infty$.

The same conclusion holds in the more general case that $\Omega / \mathrm{Stab}(\Omega)$ has in addition a finite number of punctures.

More generally, prove the following result from Anderson [1994]. Suppose G is a not necessarily finitely generated group but $G_1 \subset G$ is a finitely generated subgroup. Assume the loxodromic $\gamma \in G$ has a fixed point in $\Lambda(G_1)$. Prove that $\gamma^k \in G_1$ for some $k \geq 1$.

Analyze the following case. Suppose Ω is a component of the regular set of the nonelementary, torsion-free finitely generated kleinian group G. Assume the loxodromic $A \in \mathrm{Stab}(\Omega)$ represents a simple closed curve c on $\Omega / \mathrm{Stab}(\Omega)$. Suppose $A = B^n$, $n \geq 2$, where $B \in G$ preserves $\Omega_1 \neq \Omega$ and also represents a simple closed curve c'. Suppose c and c' are disjoint in the quotient. Now B^n preserves both Ω and Ω_1. In $\mathcal{M}(G)$, c is freely homotopic to c'^n. Since c and c' are disjoint curves, by the cylinder theorem there is a simple curve c^* near c' that is freely homotopic to c. When can you conclude that $n = \pm 1$?

For further analysis of this issue, see Leininger et al. [2011].

3-14. *Commensurability.* Two subgroups Γ_1, Γ_2 of a larger group G^* (which will usually be $\mathrm{PSL}(2, \mathbb{C})$ or $\mathrm{PSL}(2, \mathbb{R})$) are said to be *commensurable* ($\Gamma_2 \sim \Gamma_1$) if the subgroup of common elements $\Gamma_2 \cap \Gamma_1$ is of finite index in both Γ_1 and Γ_2. Prove that if Γ_1 is geometrically finite, Γ_2 is as well [Greenberg 1977].

In the context of kleinian groups, $\Gamma_2 \sim \Gamma_1$ if and only if $\mathcal{M}(\Gamma_1 \cap \Gamma_2)$ is a finite-sheeted cover of both $\mathcal{M}(\Gamma_1)$ and $\mathcal{M}(\Gamma_2)$.

The *commensurability group* or *commensurator* $C(\Gamma)$ of a kleinian group Γ is the group $C(\Gamma) = \{g \in \mathrm{PSL}(2, \mathbb{C}) : g\Gamma g^{-1} \sim \Gamma\}$.

Prove the following special case of Greenberg [1974, Theorem 2(4)]. If Γ is a finitely generated, nonelementary group whose limit set $\Lambda(\Gamma)$ is not a round circle on $\mathbb{S}^2$ nor is all $\mathbb{S}^2$, then the index $[C(\Gamma) : \Gamma]$ is finite.

To establish this, show that $C(\Gamma)$ is discrete. Then show it has the same limit set as Γ. Therefore if F is a fundamental region for $C(\Gamma)$, and $\{g_i\}$ is a set of coset representatives for Γ in $C(\Gamma)$, show that $\bigcup g_i(F)$ is a fundamental set for Γ on $\Omega(\Gamma)$.

Applying the Ahlfors Finiteness Theorem, conclude that $\{g_i\}$ is a finite set and therefore $[C(\Gamma) : \Gamma] < \infty$. Alternatively use the fact that the group K of elements that map $\Lambda(G)$ onto itself is discrete: As a closed subgroup of $\text{PSL}(2, \mathbb{C})$, the identity component of K is a connected Lie subgroup. Since it is not $\text{PSL}(2, \mathbb{C})$ or conjugate to $\text{PSL}(2, \mathbb{R})$ it either has a common fixed point in $\mathbb{H}^3 \cup \partial \mathbb{H}^3$, or it is the identity [Greenberg 1977], implying that K is discrete.

Prove that $C(\Gamma)$ contains any group H with the same limit set as Γ. That is $C(\Gamma)$ is the group of all Möbius transformations that map the ordinary set $\Omega(G)$ onto itself (or equivalently, map $\Lambda(G)$ onto itself). Therefore, if H contains Γ, Γ has finite index in H.

Prove that if h is an (orientation-preserving) isometry of a finite sheeted cover $\mathcal{M}(\Gamma_0)$ of $\mathcal{M}(\Gamma)$, then $h \in C(\Gamma)$. That is, $C(\Gamma)$ contains any orientation-preserving isometry of any finite cover.

If $\mathcal{M}(G_1)$, $\mathcal{M}(G_2)$ have finite volume, show that G_1, G_2 are commensurable if and only if there are *isomorphic* subgroups of finite index $H_1 \subset G_1$ and $H_2 \subset G_2$.

What do your results say about the normalizer $N(\Gamma)$ of Γ in $\text{PSL}(2, \mathbb{C})$?

In contrast to our analysis, if $\Gamma = \text{PSL}(2, \mathbb{Z})$ and $G^* = \text{PSL}(2, \mathbb{R})$, then $C(\Gamma)$ contains $\text{PSL}(2, \mathbb{Q})$ so that $[C(\Gamma) : \Gamma] = \infty$ and $C(\Gamma)$ is dense in G^*.

3-15. *Finiteness theorems.* Suppose that G is a finitely generated kleinian group. Prove:

(i) G has at most a finite number of conjugacy classes of rank one and rank two parabolic subgroups (Sullivan; see Feighn and McCullough [1987]).
(ii) G has at most a finite number of conjugacy classes of finite subgroups [Feighn and Mess 1991].

Hint: For (i), use the compact core or the relative compact core and the fact that, corresponding to the rank one and rank two cusps, there are mutually disjoint cusp cylinders and cusp tori with the property that a simple nontrivial loop on one is not freely homotopic to one on another [Kulkarni and Shalen 1989].

For (ii), first apply Selberg's Lemma (p. 73) to get a torsion-free, normal subgroup of finite index H. The finite group $F = G/H$ is isomorphic to a group of automorphisms of $\mathcal{M}(H)$. It is shown in Feighn and Mess [1991] that one can choose a compact core C of $\mathcal{M}(H)$ to be invariant under F; for this, C/F is compact.

3-16. *Retractions* [Epstein and Marden 1987]. Let K be a hyperbolically convex set in $\mathbb{H}^3$. The retraction map $r : \mathbb{H}^3 \setminus K \to K$ is defined as follows. For each $\vec{x} \in \mathbb{H}^3 \setminus K$, $r(\vec{x})$ is to be that point of K closest, in the hyperbolic metric, to $\vec{x}$. The closest point is uniquely attained. In the hyperbolic metric $d(\cdot, \cdot)$, show that the map r is Lipschitz: $d(r(\vec{x}), r(\vec{y})) < d(\vec{x}, \vec{y})$.

Hint: Normalize so that the geodesic from $r(\vec{x})$ to $r(\vec{y})$ lies on the vertical axis ℓ in the upper half-space model. Draw the planes orthogonal to ℓ through $r(\vec{x})$ and $r(\vec{y})$. The geodesic segment between $r(\vec{x})$ and $r(\vec{y})$ lies in K. The points $\vec{x}, \vec{y}$ cannot lie in the open set bounded by the two planes.

In Bridgeman and Canary [2010] a necessary and sufficient condition is given that the nearest point retraction of $\Omega(G)$ onto its convex hull boundary is Lipschitz in the hyperbolic metric on $\Omega(G)$. This condition is that Ω be uniformly perfect (Exercise (1-30)), [Marden and Markovic 2008].

3-17. *Cylindrical manifolds.* Suppose G is geometrically finite and $\partial \mathcal{M}(G)$ is incompressible. Let $C \subset \mathcal{M}(G)$ be an essential cylinder; $\mathcal{M}(G) \setminus C$ has one or two components M_1, M_2. Choose one of these, say M_1, and consider a lift $M_1^* \subset \mathbb{H}^3$. Set $G_1 = \text{Stab}(M_1^*)$. Describe $\Omega(G_1)$ in terms of $\Omega(G)$ and the topological type of $\partial \mathcal{M}(G_1)$ in terms of $\partial \mathcal{M}(G)$. In turn cut $\mathcal{M}(G_1)$ along an essential cylinder, if it has one. Show that this process must end after a finite number of steps. Classify the different possibilities you can end up with.

A parabolic $T \in G$ is called *accidental* if there is a component $\Omega \subset \Omega(G)$ such that $T(\Omega) = \Omega$ in which T has the three (equivalent) properties: (i) T has no horodisk in Ω; (ii) T corresponds to a loxodromic transformation in the Riemann map image of the universal cover $\mathbb{H}^2$ of Ω; (iii) the fixed point of T lies in the impression of two distinct prime ends of $\partial \Omega$. The simplest example is the transformation $z \mapsto z + 1$ acting in the strip $S = \{z : 0 < \text{Im } z < \pi\}$. The Riemann map $z \mapsto e^z$ maps S unto the upper half-plane. The parabolic T is transferred to the loxodromic $w \mapsto ew$. In truth the attribute "accidental" is singularly inappropriate, as there is nothing accidental about the appearance of an accidental parabolic.

From the three-dimensional point of view, it is possible that an "essential cylinder" just bounds a solid cusp cylinder for a rank one cusp. Then the result of cutting as proposed above does not change the group and indeed we have decided that such a cylinder is not officially called an essential cylinder. However it is entirely possible that one of the boundary components of an essential cylinder is retractable to a puncture, and the other not. This is exactly the situation of an "accidental" parabolic in a geometrically finite manifold. Suppose you only cut the manifold along essential cylinders associated with such parabolics. Show that after a finite number of steps you will end up with a group or groups that no longer have such any such parabolics: every cyclic parabolic group pairs exactly two punctures and is not represented by any homotopically different simple loop in the boundary; see Abikoff and Maskit [1977].

There is no reason to believe that a core curve c of an essential cylinder C is primitive: If $A \in G$ is determined by c, it is possible that $A = B^n$ for $B \in G$ and $n > 1$ (Exercise (4-20)).

3-18. *Conical limit points.* See also the proof of Theorem 3.13.7. If G is a nonelementary kleinian group, a point $\zeta \in \Lambda(G)$ is called a *conical limit point* if one of the following equivalent conditions are satisfied.

We are given the geodesic ray σ from $O \in \mathbb{H}^3$ to ζ, and $\varepsilon > 0$,

- The pointset

$$G(O) \cap C_\sigma(\varepsilon) \text{ accumulates to } \zeta.$$

Here $G(O)$ denotes the G orbit, and $C_\sigma(\varepsilon)$ the distance-ε neighborhood of σ.

- There is an infinite sequence of points $t_n \in \sigma$ with $\lim t_n = \zeta$, such that $\{\pi(t_n)\}$ is within distance ε of $\pi(O) \in \mathcal{M}(G)$ for all large indices. That is,

$$\pi(\sigma) \cap N_\varepsilon(\pi(O)) \subset \mathcal{M}(G)$$

has infinitely many components. Here $N_\sigma(\pi(O))$ is the ε-neighborhood.

In $\mathcal{M}(G)$, the condition means that the projection of the ray σ is *recurrent*

Any two geodesic rays ending at ζ have asymptotic distance 0 apart. Therefore $\rho \cap C_\sigma(\varepsilon)$ contains an infinite segment of ρ for any geodesic ray ρ ending at ζ. Therefore with given σ, O can be replaced by any other point in $\mathbb{H}^3$.

If $\mathcal{M}(G)$ is geometrically finite, and ζ is not a parabolic fixed point, then $\pi(\sigma)$ will lie in a compact set because it cannot asymptotically penetrate the universal horoballs.

Prove that G is geometrically finite if and only if all limit points except parabolic fixed points are conical limit points [Beardon and Maskit 1974]. *Hint:* All geodesics lie in the convex hull of $\Lambda(G)$.

3-19. *Quasiisometries.* A *quasiisometry* of $\mathbb{H}^3$ (or of any $\mathbb{H}^n$) is a map $f : \mathbb{H}^3 \to \mathbb{H}^3$ that satisfies

$$\frac{1}{L}d(x, y) - a \leq d(f(x), f(y)) \leq Ld(x, y) + a \qquad (3.17)$$

for some $L \geq 1$ and $a \geq 0$. The map f need not be a homeomorphism nor even continuous, just asymptotically Lipschitz. It is called a *Lipschitz map* if the right inequality holds for $a = 0$. The minimum factor L is called the *Lipschitz constant* for f. Initially Mostow used the term "pseudo-isometries", which satisfy (3.17) (except he takes $a = 0$ on the right side so that the map is Lipschitz). A homeomorphism f is *L-bilipschitz* if (3.17) holds with $a = 0$. An L-bilipschitz map on $\mathbb{H}^2$ or $\mathbb{H}^3$ is L^2-quasiconformal. The converse is not true in general (for an example, consider the radial stretch $\vec{x} \mapsto |\vec{x}|^\alpha \vec{x}$, $-1 < \alpha < 0$).

An equivalent definition is perhaps more illuminating: there exist constants $K \geq 1$ and $d_0 \geq 0$ such that

$$K^{-1}d(x, y) \leq d(f(x), f(y)) \quad \text{for all } x, y \in \mathbb{H}^3,$$
$$Kd(x, y) \geq d(f(x), f(y)) \quad \text{for all } x, y \in \mathbb{H}^3 \text{ and } d(x, y) > d_0$$

For example, $\mathbb{H}^2$ and $\mathbb{H}^3$ are quasiisometric to the Cayley graph dual to the tessellation by a fundamental polygon or polyhedron for a fuchsian group or a kleinian group whose respective quotients are closed [Cannon and Cooper 1992]: The graphs look like $\mathbb{H}^2$ and $\mathbb{H}^3$ if you look at them from afar. In fact, a hyperbolic group (Exercise (2-18)) is quasiisometric to $\mathbb{H}^2$ or $\mathbb{H}^3$ if and only if it is a fuchsian group representing a closed surface [Boileau et al. 2003, Theorem 6.18] or a kleinian group representing a closed manifold [Cannon and Cooper 1992].

Two Cayley graphs representing the same finitely generated group are quasiisometric.

Using Equation (8.27) of Exercise (8-10), prove that a quasiisometry f has the following properties [Efremovič and Tihomirova 1964; Thurston 1979a, p. 5.39]:

(i) If γ is a geodesic ray to a point $\zeta \in \partial \mathbb{H}^3$, then $f(\gamma)$ has a well-defined endpoint on $\partial \mathbb{H}^3$. Denote the endpoint by $f(\zeta)$.

(ii) There exists a constant $M < \infty$, such that for any $x \in \gamma$, $d(f(x), \gamma') < M$, where γ' denotes any other geodesic ray ending at $f(\zeta)$.

(iii) The extension of f to $\partial \mathbb{H}^3$ is a homeomorphism.

(iv) If $f(\zeta) = \zeta$ for all $\zeta \in \partial \mathbb{H}^3$, then $\sup_{x \in \mathbb{H}^3} d(x, f(x)) < \infty$.

(v) More generally, if f_1, f_2 are quasiisometries with the same boundary values, there exists a constant $B < \infty$ such that $d(f_1(x), f_2(x)) < B$ for all $x \in \mathbb{H}^3$.

An additional important property is that the extension to $\partial \mathbb{H}^3$ is quasiconformal [De-Spiller 1970]. To prove this it is necessary to show the *metric definition* of quasiconformality is satisfied: Let τ denote the spherical metric on $\partial \mathbb{H}^3$. Set

$$L(f, r)(\zeta) = \sup_{\tau(\zeta', \zeta) = r} \tau(f(\zeta'), f(\zeta)),$$

$$l(f, r)(\zeta) = \inf_{\tau(\zeta', \zeta) = r} \tau(f(\zeta'), f(\zeta)),$$

$$H(f, \zeta) = \overline{\lim}_{r \to 0} \frac{L(f, r)(\zeta)}{l(f, r)(\zeta)}.$$

The restriction of f to $\partial \mathbb{H}^3$ is quasiconformal if there exists $K^* < \infty$ such that $H(f, \zeta) < K^*$ for all $\zeta \in \partial \mathbb{H}^3$.

The use of this theory to prove Mostow's Rigidity Theorem is indicated at the end of Section 3.12. It is perhaps interesting to digress to summarize the history of Mostow's result. His original announcement and proof had the hypothesis that there is a quasiconformal mapping between closed manifolds $f : \mathcal{M}(G) \to \mathcal{M}(H)$ (actually in n-dimensions). Most of his 1968 paper was devoted to the proof, following earlier work of Gehring, that f, when lifted to $\mathbb{H}^n$, can be extended to $\partial \mathbb{H}^n$ and is there a quasiconformal mapping. He then used properties of quasiconformal mappings together with some ergodic theory to prove rigidity. Ahlfors immediately recognized the importance of Mostow's result and worked to make the proof more transparent to complex analysts. In a short unpublished manuscript, which assumed the boundary extension property which was known to him, Ahlfors simplified Mostow's proof, using fewer properties of quasiconformal mappings and less ergodic theory. The book [Mostow 1973] contains an entirely new proof of a more general theorem. There Mostow introduced the notion of a pseudo-isometry (now called a quasiisometry) and developed its properties. While Mostow was working on his generalization, G. A. Margulis independently published in 1970 a page-and-a-half "plan of a proof" of a less encompassing generalization of Mostow's original theorem. He too used a method akin to quasiisometries. The key feature that extension to the boundary is quasiconformal which was published in its own right in De-Spiller [1970] was thus apparently independently discovered by Mostow and Margulis in the course of their application.

3-20. *Hausdorff dimension.* The notion of Hausdorff dimension is used to measure the "size" of point sets. Smooth curves have dimension one and isolated points have dimension zero. The α-*dimensional Hausdorff dimension* of a closed set $X \subset \mathbb{C}$ (or more generally, of a Borel set) is defined in terms of

$$\Lambda_\alpha(X) = \lim_{\varepsilon \to 0} \left(\inf_{\{D_k\}} \sum_k \text{Diam}(D_k)^\alpha \right),$$

where the infimum is taken over all covers $\{D_k\}$ of X by euclidean disks of diameters at most ε. The *Hausdorff dimension* is defined as

$$\dim X = \inf\{\alpha : \Lambda_\alpha(X) = 0\}.$$

The inequality

$$\sum_k (\text{Diam } D_k)^\beta \le \varepsilon^{\beta - \alpha} \sum_k (\text{Diam } D_k)^\alpha,$$

which implies that $\Lambda_\beta(X) \le \varepsilon^{\beta-\alpha} \Lambda_\alpha(X)$, shows that $\Lambda_\alpha(X) = 0$ if $\alpha > \dim X$ while $\Lambda_\alpha(X) = \infty$ if $\alpha < \dim X$.

If f is an L-bilipschitz map of X, we have

$$L^{-\alpha} \Lambda_\alpha(X) \le \Lambda_\alpha(f(X)) \le L^\alpha \Lambda_\alpha(X).$$

For sets $X \subset \mathbb{C}$, $0 \le \dim X \le 2$. A connected closed set without interior which has Hausdorff dimension > 1 is called a *fractal*. Upper estimates of the Hausdorff dimension are often found by using a special covering and by the following estimate. Assume X is a bounded set. Let $N(\varepsilon)$ denote the minimum number of round disks of diameter ε needed to cover X. Then

$$\dim X \le \liminf_{\varepsilon \to 0} \frac{\log N(\varepsilon)}{-\log \varepsilon}.$$

In this estimate, a square grid of side length ε covering X and such that each square intersects X can replace the minimal cover by disks. For a careful development of the theory for plane point sets from the point of view of conformal mapping see Pommerenke [1992].

Thanks to the fundamental paper [Bishop and Jones 1997], added to earlier results (see the discussion in Canary and Taylor [1994]), we can assert:

Theorem 3.14.1. *Suppose G is a finitely generated kleinian group.*

(i) $\dim(\Lambda(G)) > 0$ *if and only if G is nonelementary.*
(ii) $\dim(\Lambda(G)) < 2$ *if and only if G is geometrically finite.*
(iii) $\dim(\Lambda(G)) > 1$ *if $\Lambda(G) \subset \mathbb{S}^2$ is connected but is not a circle (in which case $\dim(\Lambda(G)) = 1$).*
(iv) *If $\dim(\Lambda(G)) < 1$ then $\Lambda(G)$ is totally disconnected.*

For connections to eigenvalues of the hyperbolic laplacian see Exercises (5-5, 5-6).

In particular for a Schottky group G, $\Lambda(G)$ has zero area but positive Hausdorff dimension; in Mandelbrot's terminology, it is "fractal dust". There exists $\lambda > 0$ such that any nonelementary kleinian group with limit set of Hausdorff dimension $< \lambda$ is a classical Schottky group [Hou 2010]. On the other hand, Doyle [1988] proved there is a universal upper bound on the Hausdorff dimension of limit sets of finitely generated classical Schottky groups.

At the other extreme, for the singly degenerate groups of Section 5.8, in which $\Omega(G)$ is connected and simply connected, we have dim $\Lambda(G) = 2$. For further information see the excellent survey Matsuzaki and Taniguchi [1998].

Another measure for a kleinian group G acting on the ball model of $\mathbb{H}^3$ is its *critical exponent* (introduced by Beardon) which is defined as

$$\delta(G) = \inf\left\{s : \sum_{g \in G} e^{-sd(0, g(0))} < \infty\right\} = \inf\left\{s : \sum_{g \in G} \left(\frac{1 - |g(0)|}{1 + |g(0)|}\right)^s < \infty\right\},$$

where 0 is the center of the ball and $d(\cdot, \cdot)$ is hyperbolic distance. Use Exercise (1-14) to show that

$$\delta(G) = \inf\left\{s : \sum_{g \in G} (1 - |g(0)|)^s < \infty\right\} = \inf\{s : \sum_{g \in G} |g'(0)|^s < \infty\}.$$

We will see in Exercise (3-22) that $\delta(G) \le 2$. Moreover

$$\delta(G) = \dim \Lambda_c(G) \le \dim \Lambda(G),$$

where $\Lambda_c(G)$ denotes the set of conical limit points (see Exercise (3-18)).

It is an amazing fact that to know the critical exponent is to know the Hausdorff dimension. Combining the solution of Ahlfors' Conjecture with a result in Bishop and Jones [1997] yields:

Theorem 3.14.2. *If G is a finitely generated group with $\Omega(G) \ne \varnothing$, then*

$$\delta(G) = \dim(\Lambda(G)).$$

If $\Omega(G) = \varnothing$, both sides have the value 2.

A sharp formula for the change of Hausdorff dimension under quasiconformal mappings (the Gehring-Reich Conjecture) was established in Kari Astala's paper [Astala 1994]: For any compact set $E \in \mathbb{C}$ with Hausdorff dimension $0 < d < 1$, and any K-quasiconformal mapping $f : E \to \mathbb{C}$,

$$\frac{1}{K}\left(\frac{1}{d} - \frac{1}{2}\right) \le \frac{1}{\dim(f(E)) - \frac{1}{2}} \le K\left(\frac{1}{d} - \frac{1}{2}\right).$$

3-21. *Ergodic theory I.* This problem and the next are the first steps into the study of the ergodic theory of kleinian groups. See Nicholls [1989] for an introduction to the theory, especially the construction of invariant measures on the limit set. Also see Patterson [1987].

Let G be a discrete group acting in the ball model. Define the orbital counting function to be

$$N(r; \vec{x}, \vec{y}) = \text{card}\{g \in G : d(\vec{x}, g(\vec{y})) < r\}.$$

Prove that there is a constant C depending on G and $\vec{y}$, such that for any point $\vec{x}$,

$$N(r, \vec{x}, \vec{y}) < Ce^{2r}.$$

If in addition G has a fundamental polyhedron of finite volume, show that there is a constant $C_0 = C_0(G, \vec{x}, \vec{y})$ such that

$$N(r; \vec{x}, \vec{y}) > C_0 e^{2r} \quad \text{for all large } r.$$

Hint: Prove first that

$$\text{Vol}(\{\vec{x} : d(0, \vec{x}) < r\}) = 2\pi \int_0^r \sinh^2(t)dt,$$

where $t = |\vec{x}|$ and $d(0, \vec{x}) = \log \frac{1+t}{1-t}$. Find $\varepsilon > 0$ such that no two elements of the G-orbit of the ball $B_\varepsilon(\vec{y})$ of radius ε centered at $\vec{y}$ overlap. However if $\vec{y}$ is an elliptic fixed point of order m, then for each element of the orbit, m-images will coincide. Assuming $\vec{y}$ is not a fixed point, show that

$$\text{Vol}(B_\varepsilon(\vec{y}))N(r; \vec{x}, \vec{y}) < \text{Vol}(B_{r+\varepsilon}(\vec{x})) = 2\pi \int_0^{r+\varepsilon} \sinh^2(t)dt < \frac{1}{4} \int_0^{r+\varepsilon} e^{2t} dt.$$

Determine the corresponding statements for the action of a fuchsian group in the unit disk model of $\mathbb{H}^2$.

3-22. *Ergodic theory II.* Suppose G is a discrete group. Refer back to Exercise (1-14) and prove that

$$\sum_{g \in G} e^{-\alpha d(0, g(0))} < \infty \quad \text{for all } \alpha > 2.$$

Hint: Consider Exercise (3-21) and

$$\sum_{g \in G : d(0, g(0)) < r} e^{-\alpha d(0, g(0))} = \int_0^r e^{-\alpha t} dN(t; 0, 0)$$

$$= N(r; 0, 0)e^{-\alpha r} + \alpha \int_0^r N(t; 0, 0)e^{-\alpha t} dt.$$

The same expression for the disk model of $\mathbb{H}^2$ converges for $\alpha > 1$.

Let $\delta(G)$ be the critical exponent for G as defined in Exercise (3-20). We know from Exercise (3-20) that $\delta(G) \leq 2$ for groups acting in the ball model and that $\delta(G) \leq 1$ for fuchsian groups acting in the unit disk. The group is said to be of *convergence type* if

$$\sum_{g \in G} e^{-\delta(G)d(0, g(0))} < \infty;$$

otherwise G is said to be of *divergence type*.

Prove that if G has a fundamental polyhedron of finite volume in the ball model, or a fundamental polygon of finite area in the disk model, then G is of divergence type. If however the limit set $\Lambda(G) \neq \mathbb{S}^2$ in the ball model or $\neq \mathbb{S}^1$ in the disk model, then G is of convergence type.

Hint: Let D be a closed disk in the ordinary set $\Omega(G)$. We can assume that the elements of its G-orbit are mutually disjoint. Let $d\sigma$ denote the area form on $\mathbb{S}^2$: in spherical coordinates, $d\sigma = \sin\varphi \, d\varphi \, d\theta$. Then

$$4\pi > \sum_{g \in G} \text{Area}(g(D)) = \sum_{g \in G} \iint_D |g'(w)|^2 \, d\sigma.$$

To finish, deduce from Exercise (1-12) that when $|w| = |g(w)| = 1$,

$$|g'(w)||g(w) - g(0)|^2 = 1 - |g(0)|^2, \quad 1 - |g(0)| < 4|g'(w)|.$$

On the other hand, if G has a polyhedron with finite volume, write again

$$\sum_{g \in G : d(0, g(0)) < r} e^{-2d(0, g(0))} = \int_0^r e^{-2t} \, dN(t; 0, 0)$$

$$= N(r; 0, 0)e^{-2r} + 2\int_0^r N(t; 0, 0)e^{-2t} \, dt.$$

Apply Exercise (3-21) to finish the job.

The following result appears as Matsuzaki and Taniguchi [1998, Theorem 5.15] and incorporates results from Ahlfors [1981] and Sullivan [1981]. For the proof, see Nicholls [1989].

Theorem 3.14.3. *The following statements about a kleinian group G are equivalent:*

(i) *The conical limit set $\Lambda_c(G)$ has Lebesgue measure 4π on $\mathbb{S}^2$.*
(ii) *G is of divergence type.*
(iii) *$\mathcal{M}(G)$ does not support a hyperbolic Green's function.*
(iv) *G acts ergodically on $\mathbb{S}^2 \times \mathbb{S}^2$.*
(v) *The geodesic flow on the unit tangent bundle of $Int(\mathcal{M}(G))$ is ergodic.*

The group G is said to act *ergodically* on $\mathbb{S}^2$, or on $\mathbb{S}^2 \times \mathbb{S}^2$, if and only if the following holds: Given a measurable set X invariant under the action of G, the Lebesgue measure of either X or of the complement of X vanishes. Here the action of $g \in G$ on $\mathbb{S}^2 \times \mathbb{S}^2$ is $(x, y) \mapsto (g(x), g(y))$. Thus ergodic action on the product implies ergodic action on $\mathbb{S}^2$ itself. For a discussion of hyperbolically harmonic functions and their boundary values see Exercise (5-1).

3-23. *The Patterson-Sullivan measure on limit sets.* This measure has had many applications.

Theorem 3.14.4. *There exists a probability measure μ supported on $\Lambda(G)$ which satisfies in the ball model B:*

$$g_*\mu = j(g)^{-\delta(G)} \cdot \mu, \quad g \in G.$$

Here $g_* \mu$ is the g-pullback of μ:

$$(g_* \mu)(E) = \mu(g^{-1}(E)), \quad \text{and}$$

$$j(g) = j_B(g) = j(g; \zeta) = \frac{(1 - \|g(\zeta)\|^2)}{(1 - \|\zeta\|^2)}, \quad \text{for fixed } \zeta \in \mathbb{S}^2.$$

Here $\| \cdot \|$ is the euclidean norm. In $\mathbb{C}$ for the UHS model H, the expression for conformal distortion is instead

$$j_H(g) = j_H(g; \zeta) = \frac{\text{Im}(g(\zeta))}{\text{Im}(\zeta)}, \quad \zeta \in \mathbb{C} \, (\neq \infty).$$

For the proof and complete references see Sullivan [1979], Patterson [1987], Nicholls [1989].

In the case of geometrically finite groups G, μ is the Hausdorff density of dimension $\delta(G)$, the critical exponent.

3-24. *Poincaré series.* Suppose G is a fuchsian group of convergence type acting in the unit disk $\mathbb{D}$. Suppose $f(z)$ is a bounded analytic function in $\mathbb{D}$. Prove that the *Poincaré series*

$$\Phi(z) = \sum_{g \in G} f(g(z)) g'(z)$$

is an analytic function in $\mathbb{D}$ that satisfies the functional relation

$$\Phi(g(z)) g'(z) = \Phi(z) \quad \text{for all } g \in G, \, z \in \mathbb{D}.$$

That is, $\Phi(z) \, dz$ is an invariant form under G. It projects to a *holomorphic differential* on the quotient Riemann surface $\mathbb{D}/G$.

Now suppose instead that G is of divergence type. Prove that the Poincaré series

$$\Phi(z) = \sum_{g \in G} f(g(z)) (g')^2(z)$$

is analytic in $\mathbb{D}$ and satisfies

$$\Phi(g(z)) (g')^2(z) = \Phi(z) \quad \text{for all } g \in G, \, z \in \mathbb{D}.$$

The invariant form $\Phi(z) \, dz^2$ projects to an *holomorphic quadratic differential* on $\mathbb{D}/G$.

3-25. *Isotopy.* Two mappings between manifolds $f, g : M \to N$ are said to be *homotopic* if there is a continuous flow $F_t : M \to N$, $0 \leq t \leq 1$ such that $F_0 = f$, $F_1 = g$. In contrast f, g are *isotopic* if f, g are homeomorphisms and at every time t, $0 \leq t \leq 1$, F_t is a homeomorphism. Show that a homeomorphism of a 3-manifold onto itself can be homotopic but not isotopic to the identity. (*Hint:* $S \times [0, 1]$ flipped over. A homeomorphism which on the boundary is a Dehn twist (Example 5-12) about the boundary of a compressing disk.)

In contrast, on a surface, curves which are homotopic are also isotopic, and homeomorphisms which are homotopic are also isotopic [Epstein 1966]. A conformal map of a hyperbolic surface onto itself cannot be isotopic to the identity, unless it is the identity.

Two simple curves γ_1, γ_2 in a surface S are said to be *isotopic* if for every time t, $0 \leq t \leq$ the map $F_t : \mathbb{S}^1 \times \{t\} \rightarrow S$ is a homeomorphism with F_0 mapping to γ_1, and F_1 mapping to γ_2. Two simple curves that are freely homotopic in S are also isotopic.

On the other hand, the study of knots in $\mathbb{S}^3$ rests on the difference between isotopy and homotopy: Any knot is homotopic to an embedded circle, but is isotopic to one if and only if its complement is homeomorphic to the complement of an embedded circle.

3-26. *Homotopic isometries.* Prove that homotopic *isometries* of a hyperbolic manifold are identical, provided the fundamental group is nonabelian. *Hint:* Lift to $\mathbb{H}^3$.

3-27. *Voronoi diagrams, Delaunay triangulations, and polyhedra.*

Voronoi diagrams in $\mathbb{H}^2$. Given a discrete set of points $X \subset \mathbb{H}^2$, each point $x \in X$ has about it a polygon (or "cell")

$$C_x = \{p \in \mathbb{H}^2 : d(p, x) < d(p, y), y \neq x \in X\} = \{\cap_{y \neq x} P_y : x \in P_y\},$$

where P_y is the half-plane containing x determined by the perpendicular bisector of the line segment $[x, y]$.

Make this construction for all points of X; the result is a decomposition of $\mathbb{H}^2$ into polygons or cells. The Voronoi diagram is then

$$V(X) = \cup_{x \in X} C_x$$

The edges $\{e\}$ of $V(X)$ are geodesics equidistant to two points of X. The vertices $\{v\}$ of $V(X)$ are equidistant to three or more points of X. There is a circle σ_v about each vertex which passes through the centers $\{y\}$ of all the cells sharing v. The interior of $\delta(\sigma_v)$ contains no points of X.

The orbit $G(\mathcal{P})$ of a Dirichlet fundamental polygon $\mathcal{P}$ with center $\mathcal{O}$ is a special case of a Voronoi diagram for the set $X = G(\mathcal{O})$ (which is not a closed set!).

Delaunay triangulation in $\mathbb{H}^2$. Given the same closed set X, the Delaunay triangulation $D(X)$ is the graph dual to the Voronoi diagram.

Namely, each $x \in X$ is the endpoint of geodesic segments $\sigma_{x,y}$ from x to those points $y \in X$ such that the cell C_y shares an edge with C_x. These segments are the edges of $D(X)$. The vertices $\{v\}$ of $D(X)$ are the points of X. At each vertex v, each adjacent pair of edges determines a triangle in the graph. The interior of the circle circumscribing the three vertices of the triangle contains no points of X.

If $V(X) = G(\mathcal{P})$, then $D(X)$ is the dual graph with vertices $G(\mathcal{O})$ that represents the Cayley graph of G.

In $\mathbb{H}^3$ the definitions and properties are similar. There is an interesting limiting case. Suppose $\mathcal{M}(G)$ is a geometrically finite manifold of finite volume. In the upper half-space model, say, assume ∞ is a parabolic fixed point. Construct the Ford "polyhedron" $\mathcal{F}$ with "center" ∞. As we have seen on p. 139, $\mathcal{F}$ is invariant under the stabilizer Stab_∞ of ∞. Its orbit under the cosets of Stab_∞, is a Voronoi diagram. To obtain the Delaunay triangulation, draw the geodesics between ∞ and the centers of the polyhedra with share faces with $\mathcal{F}$, and so on. Show that there results a tessellation of $\mathbb{H}^3$ by ideal polyhedra centered on the interior vertices ζ of $\mathcal{F}$ and its orbit. Down below, there is a decomposition of $\mathcal{M}(G)$ into a finite number of ideal polyhedra. For more discussion see Weeks [1993]; Petronio and Weeks [2000].

Efficient ways of numerically finding Voronoi diagrams and Delaunay triangulations is an important issue in computer science.

3-28. What is the maximum and minimum number of sides that a Dirichlet region can have for a closed hyperbolic surface of genus $g \geq 0$ with $b \geq 0$ punctures?

The square once-punctured torus is defined by the property that there are two geodesics of equal length that are perpendicular. The hexagonal once-punctured torus has the property that there are three distinct geodesics of equal length that intersect at a point. Is it true that the hexagonal torus has the longest shortest geodesic in its deformation space? Construct the corresponding symmetric Dirichlet regions and determine the generating matrices.

3-29. *Generic Dirichlet regions: the Jørgenson-Marden Conjecture.* Given a geometrically finite, torsion-free kleinian group G, the conjecture is as follows: Is there a dense set of points (or even one point) in $\mathbb{H}^3$ such that the Dirichlet polyhedron $\mathcal{P}$ centered at one of these points has the following properties:

(i) Each edge ℓ which does not end at a parabolic fixed point has an edge cycle of length three (i.e., the edge is shared by three polyhedra in the G-orbit; in the G-orbit of ℓ, three are edges of $\mathcal{P}$).

(ii) If ℓ ends at a parabolic fixed point p it has an edge cycle of length three or four and every transformation in the cycle fixes p.

(iii) Three edges emanate from each vertex v. At most one of these edges ends at a parabolic fixed point. The order of v (number of G-images that are also vertices of $\mathcal{P}$) is either four or five (the count includes v itself). If the count is five, three of the four transformations $\neq \mathrm{id}$ are parabolic and fix the endpoint of an edge emanating from v.

(iv) Every boundary vertex $v^* \in \overline{\mathcal{P}} \cap \partial \mathbb{H}^3$ is an endpoint of exactly one edge ℓ of $\mathcal{P}$. The length of the vertex cycle at v^* is either three or four. In the latter case the four transformations fix v^*.

(v) No edges of $\mathcal{P}$ end at a rank one cusp ζ, but two faces are tangent at ζ with a face pairing transformation that fixes ζ. Every rank two cusp is the endpoint of four or six edges.

The authors (J-M) announced a proof of this conjecture in a forgettable paper in 1988; the mistake was found by Akira Ushijima in 2004. Recently Kapovich [2013] returned to this question. He discovered that triple intersections in the G-orbit of edges can be stable under deformations of the center, and the set of triple intersections is a semi-algebraic set, possibly with nonempty interior, not an algebraic set as previously claimed. In particular there is a Dirichlet region for a cyclic loxodromic group whose edges have a triple point of intersection which is stable under deformation of the center of $\mathcal{P}$. This triple intersection lies just outside the Dirichlet region selected.

3-30. (V. Markovic) In the ball model, suppose $\Omega \subset \mathbb{S}^2$ is a simply connected component of $\Omega(G)$, for some nonelementary group G. Take the Dirichlet region $\mathcal{P}_p$ centered at a point $p \in \text{Dome}(\Omega)$. Then let p approach a point on $\partial\text{Dome}(\Omega) \subset \Lambda(G)$. Show that the euclidean diameter of $\mathcal{P}_p$ tends to zero. In fact if the euclidean distance of p from $\mathbb{S}^2$ is ε, then the euclidean diameter of $\mathcal{P}_p$ is $\leq \sqrt{\varepsilon}$.

3-31. *A tower of covers.* Suppose $\mathcal{M}(G)$ is a closed manifold. Denote by L the minimum length of all closed geodesics in $\mathcal{M}(G)$. Show that the set of closed geodesics of length L is finite. Let ℓ_1 denote one of them.

Since G is residually finite (see Exercise (2-20)), there is a normal subgroup of finite index N_1 with $\ell_1 \notin N_1$. Denote the closed (hyperbolic) covering manifold determined by N_1 by $\mathcal{M}_1$; ℓ_1 does not lift to a closed geodesic. The minimum length L_1 of closed geodesics in $\mathcal{M}_1$ satisfies $L_1 \geq L$. There is at least one fewer geodesic in $\mathcal{M}_1$ that has length exactly L. Choose a minimal length geodesic ℓ_2 and corresponding construct a cover $\mathcal{M}_2$ that does not include it. Continue in this fashion. Find an infinite tower of closed manifolds $\{\mathcal{M}_k\}$ in which the minimal length closed geodesics $\rightarrow \infty$. Show (as remarked by Jeff Brock) that you can select infinite subsequences that converge geometrically to $\mathbb{H}^3$. Correspondingly $\cap_{k=1}^{\infty} N_k = \text{id}$.

Brock and Dunfield showed (to appear), using work of W. Lück, that the ratio of number of generators of the first homology group to the volume satisfies $\lim \frac{|b_1(M_k)|}{\text{Vol}(M_k)} = 0$.

3-32. *Isomorphisms that determine homeomorphisms* [Tukia 1985b]. Suppose that $\varphi : G \rightarrow H$ is an isomorphism between geometrically finite, nonelementary groups without elliptics such that $\varphi(G)$ is parabolic if and only if $g \in G$ is so. The problem is to determine when φ is induced by a homeomorphism.

The isomorphism determines a one-to-one map f_φ sending the attracting, repelling, or parabolic fixed point of $g \in G$ to the same for $\varphi(g) \in H$. According to Tukia [1985b, Lemma 3.4], there is a quasiisometry F of the convex hull $\mathcal{CH}(G)$ into $\mathcal{CH}(H)$ which induces φ and with the property that in the hyperbolic distance, $d(x, F(\mathcal{CH}(G)))$ is uniformly bounded for $x \in \mathcal{CH}(H)$. This is akin to one of the techniques used for Mostow's theorem.

There is a celebrated theorem of Fenchel and Nielsen concerning isomorphisms $\varphi : \Gamma \rightarrow \Gamma'$ between two fuchsian groups. Namely φ is induced by an orientation-preserving or-reversing homeomorphism $\mathbb{H}^2 \rightarrow \mathbb{H}^2$ if and only if the following property holds: The axes of loxodromics $g, h \in \Gamma$ intersect in $\mathbb{H}^2$ if and only if the images $\varphi(g), \varphi(h)$ are also loxodromic and have intersecting axes. Such an argument would work for quasifuchsian groups as well.

To generalize this we say that a loxodromic $g \in G$ and a quasifuchsian subgroup $\Gamma \subset G$ intersect provided the fixed points of g lie in $\Omega(\Gamma)$, one in each component. Suppose $\varphi : G \to H$ is an isomorphism between two geometrically finite groups without parabolics or elliptics, and φ preserves intersection in the sense that a loxodromic $g \in G$ and quasifuchsian $\Gamma \subset G$ intersect if and only if $\varphi(g)$ and $\varphi(\Gamma)$ intersect. Here $\varphi(\Gamma)$ is necessarily quasifuchsian because $f_\varphi(\Lambda(\Gamma))$ is a topological circle.

Suppose G is geometrically finite but not quasifuchsian. Prove that if Ω is a component of $\Omega(G)$ there is a uniquely determined component Ω' of $\Omega(H)$ such that $f_\varphi(\partial\Omega) = \partial(\Omega')$ and $\varphi(\mathrm{Stab}_\Omega) = \mathrm{Stab}_{\Omega'}$. The intersection property comes in to establish that if $\sigma \subset \Lambda(G)$ is the limit set of a quasifuchsian subgroup of G, then $x, y \in \Lambda(G) \setminus \sigma$ lie in different components of $\mathbb{S}^2 \setminus \sigma$ if and only if $f_\varphi(x), f_\varphi(y)$ are in different components of $\mathbb{S}^2 \setminus f_\varphi(\sigma)$. The bottom line is:

Theorem 3.14.5. [Tukia 1985b, Theorem 4.7]. *Suppose $\varphi : G \to H$ is an isomorphism between nonelementary, geometrically finite groups without elliptics such that $\varphi(g)$ is parabolic if and only if $g \in G$ is so. If $\Lambda(G)$ is connected, assume that φ preserves intersections. If $\Lambda(G)$ is not connected, certain orientability conditions must be satisfied for quasifuchsian subgroups and rank two parabolics. Then φ is induced by a quasiconformal homeomorphism Φ of $\mathbb{H}^3 \cup \mathbb{S}^2$.*

The orientability condition for a quasifuchsian subgroup $H \subset G$ is that the map f_φ restricted to $\Lambda(H)$ can be extended to an orientation-preserving map of $\mathbb{S}^2$ which sends attracting fixed points of loxodromics in H to attracting fixed points of their φ-images. For a rank two parabolic subgroup, φ needs to be induced by an orientation-preserving map of $\mathbb{S}^2$.

3-33. *Intersections.* If G_1, G_2 are finitely generated fuchsian groups, prove that the intersection $G_1 \cap G_2$ is also finitely generated.

If H is a finitely generated subgroup of the fuchsian group G and the limit sets are the same, then H has finite index in G.

Both these results can be found in Greenberg [1960].

If G_1 and G_2 are finitely generated subgroups of the not necessarily finitely generated H with $\Omega(H) \neq \varnothing$, then

$$\Lambda(G_1) \cap \Lambda(G_2) = \Lambda(G_1 \cap G_2).$$

This is proved in Anderson [1996].

3-34. [Greenberg 1977, Theorem 2.5.8] Suppose $\alpha : z \mapsto z + 1$ is a generator of a rank one parabolic subgroup of the finitely generated kleinian group G. Note that for any $g \in G$, $\mathcal{O} \in \mathbb{H}^3$, the perpendicular bisector of $[g(\mathcal{O}), \alpha g(\mathcal{O})]$ is a vertical plane. Show that the fundamental polyhedron $\mathcal{P}_\mathcal{O}$ lies in a slab $\{(z = x + iy, t) : a \leq x \leq a + 1\}$. There is a universal horoball at ∞. Suppose further that α has a horodisk $\mathcal{H} = \{z : y > b\} \subset \Omega(G)$. If for the euclidean closure $H = \overline{\mathcal{P}_\mathcal{O}} \cap \mathcal{H} \neq \varnothing$, show that

$H = \{z : a \le c_1 \le x \le c_2 \le a + 1\}$ for some c_1, c_2. From this prove MacMillan's theorem that $\overline{\mathcal{P}}_{\mathcal{O}} \cap \Omega(G)$ has a finite number of sides.

3-35. *Earthquakes.* This is to introduce Thurston's theory of earthquakes [1986a]. For this purpose let ℓ be the positive imaginary axis in the upper half-plane model UHP of $\mathbb{H}^2$. Denote the left and right quarter planes determined by ℓ by A and B; A and B have orientations inherited from $\mathbb{C}$. From the point of view of A, a *left earthquake* with fracture line ℓ is a discontinuous map which fixes A pointwise, and in B is an isometry moving B *to the left* with respect to A; that is, it moves B in the positive direction with respect to the positive orientation of ∂A. Therefore in B it has the form $z \mapsto kz$, for $k > 1$. It is uniquely determined once the displacement along ℓ is dictated.

If instead we require that B be fixed, the left earthquake along ℓ moves A *to the left* from the point of view of a person standing in B. In A it has the form $z \mapsto k^{-1}z$.

The inverse of a left earthquake is a right one.

Next suppose we have a finite lamination. Fix a gap σ as the base of operations. Suppose μ is a positive transverse measure—that is, each leaf of the lamination is assigned a positive number as atomic measure. The earthquake will be the identity on sigma. A transverse geodesic based in σ will cross a number of leaves. Carry out a sequence of left earthquakes in sequence along the various leaves, using the displacement assigned by μ.

Here is a more formal definition. Suppose $\Lambda \subset \mathbb{H}^2$ is a geodesic lamination. A *left earthquake* is a possibly discontinuous injective and surjective map $E : \mathbb{H}^2 \to \mathbb{H}^2$. It is an isometry on each leaf of Λ and on each complementary component. Given two gaps and/or leaves $X \ne Y$, a line ℓ is said to be weakly separating if any path from a point of X to a point of Y intersects ℓ. Let E_X, E_Y denote the respective isometric restrictions of E. We require that the *comparison isometry* $E_X^{-1} \circ E_Y$ be loxodromic, that its axis ℓ weakly separate X and Y, and that it translate to the left, when viewed from X. This last requirement means that the direction of translation along ℓ agrees with the orientation induced from $X \subset \mathbb{H}^2 \setminus \ell$. The case that one of X, Y is a line in the boundary of the other is exceptional in that the comparison map is the identity.

Thurston proves that finite left earthquakes are dense in all left earthquakes, in the topology of uniform convergence on compact subsets.

An earthquake maps a geodesic lamination Λ to another lamination Λ'.

A left earthquake between two Riemann surfaces R_1, R_2 is an injective, surjective map which sends a lamination $\Lambda_1 \subset R_1$ to $\Lambda_2 \subset R_2$. Does it lift to a left earthquake of $\mathbb{H}^2$ sending the lift Λ_1^* to Λ_2^*? Not if if one or more leaves of the Λ_i^* project to simple geodesics, for then lifts are determined only up to "twists" along the isolated geodesics. To avoid this ambiguity one can associate the earthquake with the homotopy type of a homeomorphism between the surfaces. A more common way, is to start with laminations in $\mathbb{H}^2$ invariant under the respective cover transformations.

Earthquake Theorem [Thurston 1986a]. *Every continuous orientation-preserving map $\partial \mathbb{H}^2 \to \partial \mathbb{H}^2$ is the boundary values of a left earthquake E of $\mathbb{H}^2$. The lamination Λ is uniquely determined. On Λ, E is uniquely determined except along those leaves ℓ on which it is discontinuous. For each such ℓ, there is a range of choices of translations ranging between the limiting values of E on the two sides; all the choices have the same image in $\mathbb{H}^2$.*

Suppose $R_i = \mathbb{H}^2 / G_i$, $i = 1, 2$, are arbitrary Riemann surfaces with possible boundary contours ∂R_i (coming from maximal open intervals of discontinuity on $\partial \mathbb{H}^2$). Assume $h : R_1 \to R_2$ is an (orientation-preserving) homeomorphism which extends to a continuous map $\partial R_1 \to \partial R_2$. Then the boundary values on $\partial \mathbb{H}^2$ of a lift of h are the boundary values of a left earthquake of $\mathbb{H}^2$ which projects back to a left earthquake $E : R_1 \to R_2$. Moreover, E has the same uniqueness indicated above.

This is a very general theorem. The second statement follows from the first as lifts of h extend to continuous maps of $\partial \mathbb{H}^2$.

Associated to any left earthquake is a nonnegative transverse Borel measure μ. The measure is constructed by a process akin to Riemann integration (see Epstein and Marden [1987]).

Normally one only works with the restricted class of *uniformly (locally) bounded* earthquakes. These are the class of earthquakes whose transverse measures have the property that for some $K < \infty$, $\mu(\tau) < K$ for all transverse geodesic segments τ of unit length.

The boundary values on $\partial \mathbb{H}^2$ of uniformly locally bounded earthquakes are quasisymmetric (that is, 1-quasiconformal) homeomorphisms, which means their boundary values have quasiconformal extensions to $\mathbb{H}^2$ (and which are equivariant if $(\Lambda; \mu)$ is invariant under cover transformations). In the other direction, the boundary values of a quasiconformal mapping $\mathbb{H}^2 \to \mathbb{H}^2$ (say the lift of a map between surfaces) are also the boundary values of a uniformly bounded left earthquake as described in the Earthquake Theorem. For the details, consult Thurston [1986a].

Given (Λ, μ), an *earthquake flow* is the earthquake E_t associated with $(\Lambda, t\mu)$ with $0 \le t$. For an application of this technique to the solution of the Nielsen Realization Problem, see Kerckhoff [1983].

In summary, in the dictionary entry relating geometry to complex analysis, earthquakes are the analogue of quasiconformal mappings used to deform conformal structure.

3-36. *The Nielsen kernel.* Suppose G is a fuchsian group in the unit disk $\mathbb{D}$ with $R = \mathbb{D}/G$ the interior of a compact, bordered Riemann surface $\overline{R}$ of genus $g \ge 0$, $n \ge 0$ punctures and $m \ge 1$ boundary contours. There is a set $C_1, \ldots, C_m$ of mutually disjoint simple geodesics such that C_i bounds an annular region A_i with the boundary contour γ_i. Each C_i is the projection of an axis of a loxodromic transformation. The convex core of R is $X = R \setminus \cup A_i$.

The convex core X is itself a compact bordered Riemann surface with the same genus, number of punctures, and number of boundary contours as R. Introduce on the interior of X its complete hyperbolic metric. Repeat the process; that is, let X_1 be the convex core of X. There results a nested sequence of subsets of R:

$$R \supset X_1 \supset X_2 \supset \cdots.$$

Set $Z = \bigcap_{i=1}^{\infty} X_i$. Bers first raised the problem: Describe Z. Following the insight provided by the special cases in Earle [1993], Jianguo Cao [1994] proved that Z has no interior, and that Z is the Hausdorff limit of souls $S(X_i)$.

Cao defines the *soul* $S(R)$ of R (or of any bordered surface) to be the set of points $z \in R$ such that there are at least two distinct shortest geodesic segments from z to $\bigcup C_i$. If there are no punctures, the soul is compact, without interior.

The soul is a union of geodesic arcs and is a deformation retract of R.

Explore this situation with the goal of gaining more precise information about Z, and finding a purely geometric proof of Cao's results. What about 3D?

3-37. *Extension from $\Omega(G)$ to $\mathbb{S}^2$.* Suppose as in Section 3.7.2 that F_2 is a quasiconformal map of $\mathbb{S}^2$ that restricts to $F_2 : \Omega(G) \to \Omega(H)$. It is assumed to induce an isomorphism $\varphi : G \to H$ between geometrically finite groups.

Suppose that $F : \Omega(G) \to \Omega(H)$ is quasiconformal, homotopic on $\Omega(G)$ to the restriction of F_2, and also induces φ.

Lemma 3.14.6. *F has a continuous extension to a homeomorphism of $\mathbb{S}^2$ that satisfies $F(\zeta) = F_2(\zeta)$ for all $\zeta \in \Lambda(G)$.*

Proof. Set $H = F_2^{-1} \circ F : \Omega(G) \to \Omega(G)$. The map H is homotopic to the identity on each component of $\Omega(G)$, and induces the identity automorphism of G. It also fixes every fixed point and is therefore equal to the identity on $\Lambda(G)$.

Denote spherical distance by $d(\cdot, \cdot)$ and hyperbolic distance in $\Omega(G)$ by $d_h(\cdot, \cdot)$.

Let γ_z be a shortest geodesic from z to $H(z)$. There is a constant $C_1 < \infty$ such that $L_h(z) = d_h(z, H(z)) < C_1$ for all $z \in \Omega(G)$ (lift from the quotient). Consequently if $z \to \zeta \in \Lambda(G)$ in the spherical metric, then $\lim \gamma_z = \zeta$ and $\lim H(z) = \zeta$ as well.   □

I am grateful to Vlad Markovic for allowing inclusion of his unpublished result as follows.

Proposition 3.14.7 (Markovic). *H is quasiconformal on $\mathbb{S}^2$; hence F itself is the restriction to $\Omega(G)$ of an equivariant quasiconformal map of $\mathbb{S}^2$.*

Proof. Markovic's proof is as follows. There exists a constant C_2 so that, using the notation introduced above,

- $d(\gamma_z, \Lambda(G)) < C_2, \quad z \in \Omega(G),$
- $d(w, \Lambda(G)) < C_2 d(z, \Lambda(G)), \quad w \in \gamma_z.$

Set $X = \Lambda(G)$. From Pommerenke [1984] we know that X has the property of uniform perfectness, see Equation (1.12), Exercise (1-30). That is, for the hyperbolic metric $\rho(w)|dw|$ in each component of $\Omega(G)$ and some constant $C_3 > 0$,

$$\frac{C_3|dw|}{d(w, X)} < \rho(w)|dw| < \frac{2|dw|}{d(w, X)}.$$

This holds for each component of $\Omega(G)$, whether or not it is simply connected.

Upon integrating over the shortest geodesic $\gamma_z = [z, H(z)]$ of hyperbolic length $L_h(z)$, we find that

$$C_3 d(z, H(z)) \le L_h(z) \sup_{w \in \gamma_z} d(w, X) < C_2 L_h(z) d(z, X), \quad \text{or}$$

$$d(z, H(z)) < C_4 d(z, \zeta) \text{ for any } \zeta \in X.$$

Now $d(H(z), \zeta) \le d(H(z), z) + d(z, \zeta)$. Consequently for some constant C_5, $d(H(z), \zeta) < C_5 d(z, \zeta)$. We conclude that

$$\frac{d(H(z), \zeta)}{C_5} \le d(z, \zeta) \le C_5 d(H(z), \zeta), \quad \frac{1}{C_5} < \frac{d(H(z), H(\zeta))}{d(z, \zeta)} < C_5$$

We have also used from the Lemma that $H(\zeta) = \zeta$. We are now in a position to apply the geometric definition (Section 2.8 footnote), of quasiconformality to show that $H(\zeta)$ is quasiconformal at ζ. For the ratio of distances of $H(z)$ to $H(\zeta)$ is uniformly bounded above 0 and below ∞. Since ζ was arbitrarily chosen this proves H is quasiconformal on $\Lambda(G)$. Therefore $F = F_2 \circ H$ is quasiconformal on $\mathbb{S}^2$ as required. $\qquad\square$

3-38. *Geometric intersection number estimates.* [Fathi et al. 1979, pp. 58–59]. Let R_0 denote the result of removing the universal horodisks from hyperbolic surface R. Prove:

- There is a constant $C > 0$ such that the intersection number of any two simple closed geodesics α, β satisfies

$$\iota(\alpha, \beta) < C \operatorname{Len}(\alpha) \operatorname{Len}(\beta).$$

- Let $\{\tau_i\}$ be a finite system of simple closed geodesics and simple arcs that cut R_0 into simply connected regions. There exists a constant $c = c(\cup \tau_i)$ such that for any simple closed geodesic α in R_0,

$$\sum \iota(\alpha, \tau_i) \ge c \operatorname{Len}(\alpha).$$

Hints: For the first item, cover α and β by ϵ-disks thereby dividing them into short geodesic segments. Each segment intersects another at most at one point. Show that

$$\iota(\alpha, \beta) < \left(\frac{\operatorname{Len}(\alpha)}{\epsilon} + 1\right)\left(\frac{\operatorname{Len}(\beta)}{\epsilon} + 1\right).$$

For the second item, let $c = 1/L$ where L is the length of the longest simple arc in the simply connected regions.

3-39. *Total angle measure of geodesic arcs.* This was introduced by Kerckhoff [1983]. Let τ be a geodesic segment transverse to the geodesic lamination $(\Lambda; \mu)$. Find the angle θ which τ makes with the leaves of Λ measured counterclockwise from τ to Λ, $0 \le \theta < \pi$. The *total angle* and *average angle* of τ with respect to Λ are defined, respectively, to be

$$\int_\tau \theta \, d\mu, \quad \frac{1}{i(\tau, \mu)} \int_\tau \theta \, d\mu.$$

The *total cosine* of τ with respect to Λ is defined as $\int_\tau \cos \theta \, d\mu$. This has many applications, see Kerckhoff's papers [1985], [1992]. In particular, in the former paper it is shown that if two measured laminations μ, ν on a surface R have the same total cosine over every closed geodesic on R, then $\mu = \nu$.

3-40. *Interval exchange transformations* [Bonahon 2001; Masur 1982]. Let $I \subset \mathbb{R}$ be the interval $(0,1]$. Write $I^\pm$ for its upper and lower edges. Suppose $\{I_i^+\}$ is a partition of I^+ into n half-closed intervals $\{[a_{i-1}, a_i)\}$ of various lengths, with $a_0 = 0$, $a_n = 1$. Take the same sequence of intervals on I^- but then permute them in any way. Label the result by the notation $\{I_i^-\}$ where I_i^- has the same length as I_i^+. The corresponding *interval exchange transformation* J is the piecewise euclidean isometry that maps I_i^+ onto I_i^-, $1 \le i \le n$. The map is one-to-one, except two-to-one at the endpoints of the closures of the intervals. We must chose the permutation so the interval exchange does not reduce to an exchange of fewer intervals, that is, so that J is not continuous at any interval endpoint.

There is a naturally associated closed Riemann surface R: View the complement of I in $\mathbb{S}^2$ as a polygon with n-pairs of edges $I_i^\pm$. Identify each pair of edges by the direction-preserving isometry; akin to what we did by "rolling up" fundamental regions by their edge identifications. The resulting surface will have singular points coming from the endpoints of the intervals. But there will be a natural complex structure at these points as well which maps the local neighborhoods into $\mathbb{C}$.

The vertical euclidean lines give rise to a *measured foliation* of R. Namely, except for a countable number of points, given $x \in I$ the forward and backward orbit $J^{\pm n}(x)$ will not hit an interval endpoint. These generic points will lie on a leaf of a foliation of R by vertical lines. The differential dx is the local vertical measure of the foliation.

The foliation is turned into a measured lamination by showing in the universal cover, the leaves have endpoints on $\partial \mathbb{H}^2$ and replacing each leaf by a geodesic.

By adjusting the interval lengths one can obtain minimal laminations, and uniquely ergodic ones as well. See Masur [1982] for more details and further references.

3-41. *Horocyclic foliations.* Assume we have a closed surface S, a hyperbolic metric g on S, and a maximal geodesic lamination Λ such that all complementary regions are ideal triangles. From these data we will construct a measured foliation called a *horocyclic foliation*.

First foliate each ideal triangle as follows. Model the triangle by an ideal triangle in the disk model $\mathbb{D}$ whose sides have equal euclidean lengths. Foliate a neighborhood of each vertex v_i by a family of arcs contained in concentric circles with center at v_i. Do this symmetrically about all three vertices. We are left with a small central curved triangle that is not foliated.

Using our model example, foliate $S \backslash \Lambda$. The leaves joint together to form a family of mutually disjoint open arcs of infinite length in S, each orthogonal to the leaves of Λ. Collapse the central triangles to points. This results in a singular foliation of a new surface S_0 equivalent to S with singular points at the collapsed triangles. Replace S by S_0.

The hyperbolic metric g determines a transverse measure μ_g by measuring vertical distances along the leaves of Λ. Thurston [1998] proved that the map $g \mapsto (\Lambda_g, \mu_g)$ is a homeomorphism of $\mathfrak{Teich}(S)$ onto its image in measured foliation space.

3-42. *Rigidity of points in* $\mathfrak{Teich}(R)$. Suppose R is a possibly punctured hyperbolic Riemann surface of finite topological type. Suppose $X, Y \in \mathfrak{Teich}(R)$ have the property that the respective hyperbolic lengths satisfy $L_X(\gamma) \leq L_Y(\gamma')$ for every pair γ, γ' of corresponding geodesics. Prove that $X = Y$ [McMullen 2009].

3-43. *Triangulations/cell decompositions of punctured surfaces.* Here we will outline the important paper [Bowditch and Epstein 1988]. Suppose R is a closed surface; denote by P the set of $n \geq 1$ punctures such that $R \setminus P$ is hyperbolic. To simplify the discussion assume horodisks are constructed about all the punctures. The size of the horodisk about $\zeta \in P$ is established as the euclidean length c of the horocycle taken when we pass to the universal cover, take $\zeta = +i\infty$, and the horodisk isometric to $\{z : \mathrm{Im}(z) \geq 1\}/[z \mapsto z+c]$; fix $c < 1$ small enough that it (or anything smaller) will serve for all points in P and that the resulting horodisks will be mutually disjoint.

Starting with these data, will will indicate the construction of a canonical triangulation of $R \setminus P$ depending only on the hyperbolic structure. It has the strong property of inducing a natural tessellation of the Teichmüller space, invariant under the modular group. In addition it determines not only a triangulation but also a compactification of moduli space.

For each $\zeta \in P$, there is a length $x(\zeta) \leq c < 1$ horocycle corresponding to the puncture ζ. Normalize these numbers $x(\zeta)$ so that

$$\sum_{\zeta \in P} x(\zeta) = 1.$$

Denote the union of all the horodisks by $\mathcal{B}$.

Next, let Σ denote the set of points $u \in R$ from which at least two distinct shortest geodesic rays end at points of $\mathcal{B}$.

The set Σ is best seen using the universal cover $\mathbb{D}$ of $S \setminus P$. The horodisks in $S \setminus P$ lift to horodisks at the fixed points of the parabolics in the covering group G; these are dense in $\partial \mathbb{D}$. Denote by $\mathcal{B}^*$ the totality of lifts of $\mathcal{B}$. Given a point $u \in S \setminus P$, and a point $u^* \in \mathbb{D}$ over it, let B_u^* denote the maximal disk in $\mathbb{D} \setminus \mathcal{B}^*$ centered at u^*. B_u^* is tangent to one or more of the horodisks in $\mathcal{B}^*$. The point u belongs to Σ if and only if B_u^* is tangent to two or more horodisks. The radius ℓ from u^* to a point of tangency, that is, a shortest length geodesic ray to $\mathcal{B}^*$, is orthogonal to a horocycle. It continues on inside the horodisk, ending up at the corresponding parabolic point, which we denote by $p(u^*)$.

Returning to $R \setminus P$, let $w(u)$ be the number $n \geq 1$ of distinct shortest rays from $u \in R \setminus \mathcal{B}$ to $\mathcal{B}$ determined in $\mathbb{D}$ as above. Define

$$V = \{u : w(u) \geq 3\}, \quad \text{while } \Sigma = \{u : w(u) \geq 2\}.$$

The points of V are vertices. Both V and Σ lie in $R \setminus P$.

Each ray constructed in $\mathbb{D}$ projects to a simple ray from u to a horocycle in R and ultimately to the corresponding puncture $p(u)$ (cf. Exercise (6-12)).

Prove with Bowditch and Epstein [1988]:

(i) V is a finite set of points,
(ii) $\Sigma \setminus V$ is a finite union of geodesic arcs, the endpoints of each are points in V,
(iii) Each $v \in V$ is the endpoint of at least three of the geodesic segments in Σ.
(iv) Σ is a deformation retract of $R \setminus P$ and is therefore a *spine* of the surface.

With respect to the last item, consider the case that $w(u) = 1$. If $u \in R \setminus \mathcal{B}$, there is a unique shortest ray ℓ from u, orthogonal to $\partial\mathcal{B}$. It continues on until it hits the parabolic point $p(u)$. Following ℓ in the opposite direction, it stops upon hitting a point of Σ. All points u' along the way have $w(u') = 1$. If $u \in \mathcal{B}$, denote by ℓ the geodesic from u to the cusp $p(u)$, likewise extend ℓ in the opposite direction until it hits Σ. The deformation retract to the spine Σ will be along the lines $\{\ell\}$, moving away from the punctures.

The geodesic segments $\{\ell\}$ from points $p(u)$ with $w(u) = 1$ to points of Σ are referred to as *retraction lines*.

Theorem 3.14.8 ([Bowditch and Epstein 1988]). *The vertices are the points of V. The edges E are the components of $\Sigma \setminus V$. These are geodesic segments, with endpoints in V. To these sets we add the retraction lines $\{\ell\}$. Together these three sets determine a finite triangulation of R, depending only on the hyperbolic structure and the length of the horocycles.*

- *Each triangle $\triangle$ has exactly one of the cusps $\zeta \in P$, and exactly one edge $e \in E$, namely the edge opposite ζ.*
- *The other two edges of $\triangle$ are retraction lines ℓ_1, ℓ_2, from the endpoints $u_1, u_2 \in V$ of e to $\zeta = p(u_1) = p(u_2)$.*
- *Reflection in the edge e, sends $\triangle$ onto the adjacent triangle $\triangle'$.*

Coordinates can be assigned for each triangle $\triangle$ as follows. The length $\alpha(e)$ of the horocyclic arc $\triangle \cap \partial\mathcal{B}$ at the cusp ζ of $\triangle$ facing the edge $e \in E$ uniquely determines $\triangle$. The two triangles sharing e have the same coordinate. Therefore define

$$\theta(e) = 2\frac{\alpha(e)}{c}.$$

Consequently,

$$\sum_{e \in E} \alpha(e) = c, \quad \sum_{e \in E} \theta(e) = 1, \quad \text{and}$$

$$2\sum_{e \sim \zeta} \theta(e) = 2x(\zeta).$$

where $e \sim \zeta$ indicates the sum is over all edges facing the cusp ζ. In these sums, only one of the two triangles determined by an edge appears.

If R is "generic", that is, if each vertex has degree three, the Euler characteristic is

$$\chi(R \setminus P) = 2 - 2g - \#P = -2\frac{\#E}{3} + \#E - 2\frac{\#E}{3} = -\#E/3.$$

In this case the Teichmüller space of $R \setminus P$ has real dimension $6g + 2\#P - 6$. Moreover the horocycle length map $x : P \to (0, 1]$ itself gives an additional $\#P - 1$ independent parameters giving a total dimension of $6g + 3\#P - 7$ for the dimension of the full deformation space varying both surface and triangulation.

The authors also prove a converse starting with a punctured topological surface (S, P), an "arc system" σ, and an assignment $\theta(e^*)$ to each arc of σ satisfying properties analogous to those above. They prove that there is a hyperbolic structure on $S \setminus P$, unique up to isotopy (with the points of P fixed), such that σ is the dual of the associated spine Σ with $\theta(e^*)$ assigned to each edge e of Σ.

For constructions of ideal triangulations, see Exercise (4-19).

3-44. *Ford polygon generalization.* Here we present a simpler approach to decomposing a punctured Riemann surface into polygons carried out in Epstein and Penner [1988]. We recall that on a once-punctured closed Riemann surface R, the Ford fundamental polygon can be interpreted as follows: Find the set Σ of points $u \in R$ such that there are two or more (distinct) geodesic rays from u to the puncture ζ. Equivalently ask for two or more geodesic segments from u to the boundary of a small horodisk D about ζ. Then $R \setminus \Sigma$ is a doubly connected region about ζ—the projection of the Ford domain or the Ford domain minus a horodisk.

In the case of multiple punctures, select small horodisks $\{D_i\}$ about the punctures. The horodisks may have different sizes. Take Σ to be the locus of points $u \subset R$ which have equal hyperbolic distance to two or more of the horodisks. Then $R \setminus \Sigma$ is a union of polygons; in general the interior is not connected. Each polygon contains exactly one of the punctures together with its horodisk. This "Ford region" depends on $(n-1)$ parameters, the projectivized vector of horodisk sizes, if there are n-punctures. A horodisk can be sized according to the hyperbolic length of its boundary.

3-45. *Lengths of intersecting geodesics.* Prove that for every integer $0 \le k < \infty$ there is an integer b_k such that on every hyperbolic Riemann surface, any closed geodesic of length $< b_k$ has at most k self-intersections. Basmajian [2011] proved that one can take $b_k = \frac{1}{4} \log 4k$. He also showed that as $k \to \infty$, necessarily $b_k \to \infty$. This generalizes Theorem 3.3.4.

Go on to prove that the figure-8 curve on a 3-punctured sphere is the minimal length geodesic with one self-intersection [Yamada 1982].

3-46. *Fuchsian horodisks.* Suppose G is a torsion-free fuchsian group acting in UHP, and assume G has a parabolic fixed point at ∞. If the parabolic subgroup is generated by $T(z) = z + 1$, then for any element $A \in G$, $A(\infty) \ne \infty$,

$$|c| \ge 2.$$

Here c is the lower left matrix entry. Note this is a better result than given by the universal horoball theorem. This interesting fact was discovered by Halpern [1981], see also Proposition 3.5.3. Can you generalize to 3D?

3-47. *More on horodisks.* For a fuchsian group in UHP with a parabolic subgroup at $+i\infty$ generated by $z \mapsto z + 1$, its horodisk at $+i\infty$ is described the equation

$$\{z : d(z, z+1) < d(i, i+1) = \log\left(\frac{\sqrt{5}+1}{\sqrt{5}-1}\right).$$

Instead suppose Ω is a region of discontinuity of Γ, and $+i\infty$ is fixed by a parabolic subgroup generated by $T(z) = z + 1$. Assume $\inf_\Omega d(z, z+1) = 0$ (that is, Ω is not, for example, a horozontal strip). There is then a conformal mapping of Ω to UHP conjugating T to itself.

Now the boundary $\partial\Omega$ is invariant under T. So we can find $h < \infty$, the maximum vertical height of $\partial\Omega$. Consider the horizontal line L of height h. The halfspace H above L necessarily lies in Ω, and so does the halfspace H' above the horizontal line L' of height $h + 1$. Show that H' is contained in the universal horodisk for Ω, that is,

$$d_\Omega(z, z+1) < d_H(z, z+1) < \log\left(\frac{\sqrt{5}+1}{\sqrt{5}-1}\right), \quad z \in H'.$$

See also Earle and Marden.

4

Algebraic and geometric convergence

The focus of this chapter is on sequences of kleinian groups, typically sequences that are becoming degenerate in some way. For these, it is necessary to distinguish between convergence of groups and convergence of quotient manifolds. The former has to do with sequences of groups whose generators converge, the latter with sequences of groups whose fundamental polyhedra converge.

We will continue on by introducing the operations called Dehn filling and surgery. This leads to the description of the set of volumes of finite volume hyperbolic 3-manifolds.

4.1 Algebraic convergence

In this section we will prove the two theorems which provide the basis for working with sequences of groups.

Let Γ be an abstract group and $\{\varphi_n : \Gamma \to G_n\}$ be a sequence of *homomorphisms* (also called *representations*) $\{\varphi_n\}$ of Γ to groups G_n of Möbius transformations. Suppose for each $\gamma \in \Gamma$, $\lim_{n\to\infty} \varphi_n(\gamma) = \varphi(\gamma)$ exists as a Möbius transformation. Then the sequence $\{\varphi_n\}$ is said to *converge algebraically* and its *algebraic limit* is the group of limits $G_\infty = \{\varphi(\gamma) : \gamma \in \Gamma\}$. When we say a sequence of groups converges algebraically, we are assuming that behind the statement is a sequence of homomorphisms or isomorphisms spawning the sequence.

In particular, a sequence of r-generator groups $G_n = \langle A_{1,n}, A_{2,n}, \ldots A_{r,n} \rangle$ is said to converge *algebraically* if $A_k = \lim_{n\to\infty} A_{k,n}$ exists as a Möbius transformation, $1 \leq k \leq r$. Its *algebraic limit* is the group $G = \langle A_1, A_2, \ldots A_r \rangle$. To make this terminology consistent with that used above, refer to the free group F_r on r-generators and express G_n as the sequence of representations $\varphi_n : F_r \to G_n$ determined by sending the k-th generator of F_r to $A_{k,n}$.

The convergence is controlled as spelled out by the following two fundamental results.

Theorem 4.1.1 [Jørgensen 1976; Jørgensen and Klein 1982]. *Let $\{G_n\}$ be a sequence of r-generator nonelementary kleinian groups converging algebraically to the group*

G. Then G is also a nonelementary kleinian group, and the map $A_k \to A_{k,n}$, $1 \le k \le r$, determines a homomorphism $\phi_n : G \to G_n$ for all large indices n.

In general ϕ_n will not be an isomorphism. For example, a sequence of elliptic transformations $\{A_{k,n}\}$ may converge to a parabolic transformation A_k. In the opposite direction, Theorem 4.1.1 implies that if some A_k is elliptic of order ρ, then $A_{k,n}$ remains elliptic of order ρ for all large n.

If the sequence $\{G_n\}$ consists of elementary groups, the algebraic limit may or may not be discrete. That is why we restrict our attention to sequences of nonelementary groups.

In contrast to Theorem 4.1.1, in applications we frequently work with isomorphisms from a fixed group:

Theorem 4.1.2 [Jørgensen 1976]. *Suppose G is a nonelementary kleinian group and $\{\theta_n : G \to G_n\}$ is a sequence of isomorphisms onto kleinian groups G_n. Assume that for each element $g \in G$, $\lim_{n\to\infty} \theta_n(g) = \theta(g)$ exists as a Möbius transformation. Then $G_\infty = \{\theta(g) : g \in G\}$ is a nonelementary kleinian group and $\theta : G \to G_\infty$ is an isomorphism.*

In Theorem 4.1.2, we do not need to require that G be finitely generated.

These two theorems are consequences of Jørgensen's inequality.

Proof of Theorem 4.1.2. First we will show $\theta : G \to \theta(G)$ is an isomorphism. If it is not, $\theta(g)$ is the identity for some $g \ne \mathrm{id} \in G$. If $g \in G$ has finite order, $\theta_n(g)$, in the limit $\theta(g)$ must have exactly the same order, since θ_n is an isomorphism. Therefore if $\theta(g) = \mathrm{id}$, g must have infinite order. Since G is not elementary, there is an element $h \in G$ also of infinite order but without a common fixed point with g. Each group $\langle \theta_n(g), \theta_n(h) \rangle$ is also nonelementary, since a nonelementary discrete group cannot be isomorphic to an elementary discrete group. But then we are in violation of Jørgensen's inequality (2.1) for all large n. Application of Jørgensen's inequality also shows that the sequence $\{\theta_n(g)\}$ cannot be loxodromic or parabolic.

Next we will show that $\theta(G)$ is discrete. If not, there is a sequence $\{g_k \ne \mathrm{id} \in G\}$ such that $\lim_{k\to\infty} \theta(g_k) = \mathrm{id}$. There is a sequence $n = n(k)$ so that $\lim_{k\to\infty} \theta_n(g_k) = \mathrm{id}$. We may assume that either all g_n lie in the same cyclic subgroup or else their fixed points are mutually disjoint. In either case we can find an element $h \in G$ of infinite order whose fixed points are distinct from those of all but a finite number of the elements g_n. The sequence of nonelementary groups $\langle \theta_n(h), \theta_n(g_k) \rangle$ for large k is in violation of Jørgensen's inequality.

Finally we have to show that $\theta(G)$ is nonelementary. That is now easy because an elementary discrete group is a finite extension of an abelian group. $\square$

Proof of Theorem 4.1.1. We start with a sequence of lemmas:

Lemma 4.1.3. *If each of the four transformations A, B, AB, $ABA^{-1}B^{-1}$ is elliptic or the identity, then they have a common fixed point in $\mathbb{H}^3$.*

Proof. If A and B commute, then according to Lemma 1.5.2, they either have the same axes, or each is of order two and their axes are orthogonal at a common point of intersection. In either case the conclusion is obvious. So assume they do not commute. Find U so that the conjugates UAU^{-1}, UBU^{-1} are such that the former has fixed points $0, \infty$. Then conjugate both by $V = \left(\begin{smallmatrix} \sqrt{c} & 0 \\ 0 & 1/\sqrt{c} \end{smallmatrix}\right)$, where c is the lower left entry of UBU^{-1}. Rename the results by A, B so as to end up with the following:

$$A = \begin{pmatrix} e^{i\theta} & 0 \\ 0 & e^{-i\theta} \end{pmatrix}, \quad B = \begin{pmatrix} a & b \\ 1 & d \end{pmatrix}, \quad 2\theta \not\equiv 0, \; ad - b = 1.$$

Now $\operatorname{tr}(B) = a + d = r_1$ and $\operatorname{tr}(AB) = e^{i\theta}a + e^{-i\theta}d = r_2$ are real while $e^{2i\theta} \neq 1$. Solving the two equations for a and d we find that $a = \bar{d}$. Since B is elliptic, its trace satisfies $-2 < \operatorname{tr}(B) < 2$. Hence the fixed points of B, namely $\frac{1}{2}\left(a - d \pm \sqrt{\operatorname{tr}^2(B) - 4}\right)$, are purely imaginary. Further the product of the fixed points is $1 - ad = -b$.

Next we find for the commutator that $\operatorname{tr}(ABA^{-1}B^{-1}) - 2 = 4b\sin^2\theta$. Since this is elliptic as well, we must have $b < 0$. Since the product of the fixed points of B is positive, they lie on opposite sides of $z = 0$. That is, in the upper half-space model of $\mathbb{H}^3$, the axes of A and B intersect. The point of intersection is fixed by both, and by the group they generate. □

Lemma 4.1.4. *Suppose that A, B are elliptic and with distinct axes properly intersecting in $\mathbb{H}^3$. Then the plane P containing their axes does not contain the axis of AB.*

Proof. Let x denote the spot that the rotation axes ℓ_A, ℓ_B intersect, $P \subset \mathbb{H}^3 \equiv \mathrm{UHP}$ the plane containing both ℓ_A, ℓ_B, and $\ell^\perp \subset \mathbb{H}^3$ the line orthogonal to P through x. We can conjugate A, B so that $\ell^\perp$ becomes the positive imaginary axis. In doing so, retain the nomenclature just introduced. In particular, P is now the plane (euclidean hemisphere) orthogonal to the newly named $\ell^\perp$ at x.

The fixed points (p_1, p_2), (q_1, q_2) of A and B are opposite on ∂P and so satisfy $p_2 = -p_1, q_2 = -q_1$. Introduce their normalized matrices $A = \left(\begin{smallmatrix} a & b \\ c & d \end{smallmatrix}\right)$, and $B = \left(\begin{smallmatrix} u & v \\ s & t \end{smallmatrix}\right)$. Because of the symmetry of the fixed points, we find that $a = d$ and $u = t$. Their fixed points are then the solution of the respective equations: $cz^2 = b$, and $sz^2 = v$.

Suppose the Lemma is false and the axis of AB also lies in P. Since its axis also goes through x, the fixed points of AB then have the same symmetry as A, B. Upon computing AB we find that $bs = cv$. But this implies that the fixed points of A and B are the same, in contradiction to the hypothesis. □

We will digress from the proof of Theorem 4.1.1 to draw the following important corollary:

Theorem 4.1.5. *A group G, discrete or not, composed entirely of elliptic elements either has a common axis or it has a unique common fixed point in $\mathbb{H}^3$.*

Proof. Suppose $A, B \in G$ have distinct axes ℓ_A, ℓ_B. Lemma 4.1.3 shows that they intersect in a point $x \in \mathbb{H}^3$. Let P be the plane they span. Consider a third element $C \in G$ with axis ℓ_C distinct from ℓ_A, ℓ_B. If $x \notin \ell_C$, then $\ell_C \subset P$ since it intersects both ℓ_A, ℓ_B. So all other axes from G either pass through x or lie in P. On the other hand by Lemma 4.1.4, the axis ℓ_{AB} of AB, which intersects ℓ_A at x, does not lie in P. So the plane P' spanned by ℓ_A, ℓ_{AB} does not coincide with P. But repetition of our argument shows that ℓ_C lies in P' as well, a contradiction. We have shown that either all elements of G have the same axis, or the set of axes of elements of G have a single common point of intersection in $\mathbb{H}^3$. $\qquad\square$

Lemma 4.1.6. *If $\langle A, B \rangle$ is nonelementary, then at most one of the three elements A, B, AB is elliptic of order two.*

Proof. An element of order two is conjugate to $z \mapsto -z$ whose axis in the upper half-space model is the half-line rising from $z = 0$. If both A and B are elliptic of order two and their axes do not coincide, there is a unique common perpendicular line ℓ to the two axes. Since A and B are rotations by π about their axes, each maps ℓ onto itself by rotating it by π about the crossing point with its axis. The cyclic subgroup $\langle AB \rangle$ which maps ℓ onto itself without reversing direction has index two. The bottom line is that the group $\langle A, B \rangle$ is elementary, a contradiction. $\qquad\square$

Lemma 4.1.7. *Suppose $A_1, \dots, A_r$ generate an infinite, discrete group G. The set*

$$\{A_i, \ A_i A_j, \ A_i A_j A_k, \ (A_i A_j) A_k (A_i A_j)^{-1} A_k^{-1} : \ i, j, k = 1, \dots, r\}$$

contains an element of infinite order.

Proof. The assertion is true for $r = 1$. For $r \geq 2$, assume the assertion is false so that all the listed elements have finite order. We will show that this implies they have a common fixed point in $\mathbb{H}^3$. This in turn will imply that G is a finite group, in contradiction to the hypothesis.

Use our current hypothesis and apply Lemma 4.1.3 to $A = A_i A_j$, $B = A_i$. We see that $A_i A_j$, A_i have a common fixed point x in $\mathbb{H}^3$ and therefore A_j fixes x as well. Consequently the axes of the generators $\{A_i\}$ pairwise intersect.

Choose a point $x \in \mathbb{H}^3$ at which a maximal number of generator axes intersect, say the axes of $A_1, \dots, A_m$, $2 \leq m \leq r$. If $m = r$ we are finished. If $m < r$, then the axis ℓ of A_{m+1} does not pass through the common point x of its predecessors. We may assume that the axes of A_1 and A_2 meet ℓ at different points. Let P be the plane spanned by the axes of A_1, A_2; necessarily P contains x and ℓ. According to Lemma 4.1.4, the axis of $A_1 A_2$, which goes through x, does not lie in P. So the axis of $A_1 A_2$ is disjoint from ℓ, in contradiction to Lemma 4.1.3. $\qquad\square$

Incidentally we have confirmed the following:

Corollary 4.1.8. *A discrete group in which every element is elliptic is a finite group.*

Lemma 4.1.9. *Suppose that $A_1, \ldots, A_r$ generate a nonelementary, discrete group G and that $g \in G$ is loxodromic or parabolic. Then $H_i = \langle A_i, g \rangle$ is nonelementary for at least one index i.*

Proof. Suppose to the contrary that for each index, H_i is elementary. If g is loxodromic, each A_i must fix or interchange the fixed points of g. Thus G itself must be elementary since it fixes the set of two fixed points of g. If g is parabolic, each A_i fixes the fixed point of g and again G is elementary. $\square$

Lemma 4.1.10. *Suppose that $A = \lim A_n$, $B = \lim B_n$ for two sequences of Möbius transformations where each group $G_n = \langle A_n, B_n \rangle$ is discrete. Then:*

(a) *If G_n is nonelementary for all indices, $A \neq \mathrm{id}$.*
(b) *If G_n is nonelementary for all indices and A is elliptic, its order is finite.*
(c) *If neither A nor B has order two, then A and B have a common fixed point on $\mathbb{S}^2$ if and only if A_n and B_n also do so for all large indices.*

Proof. We will first prove part (c) so assume that for all large n, A_n, B_n are not elliptic of order two. If A_n and B_n have a common fixed point on $\mathbb{S}^2$, the trace of their commutator is $+2$ by Lemma 1.5.1. By continuity the commutator of A and B also has trace $+2$. Consequently A and B have a common fixed point as well.

Conversely, suppose A and B have a common fixed point on $\mathbb{S}^2$. Then $K = \lim K_n = \lim A_n B_n A_n^{-1} B_n^{-1}$, and the trace of the commutator $K = ABA^{-1}B^{-1}$ is $+2$. For all large indices, either A_n and B_n have a common fixed point or K_n is not parabolic, since the trace $\neq -2$. If the former case occurs we are finished. So assume that for all large n, K_n is not parabolic.

Since A, B, K all share a fixed point,

$$\mathrm{tr}^2(K) - 4 = \mathrm{tr}(KAK^{-1}A^{-1}) - 2 = \mathrm{tr}(KBK^{-1}B^{-1}) - 2 = 0.$$

For all large indices then,

$$\left| \mathrm{tr}^2(K_n) - 4 \right| + \left| \mathrm{tr}(K_n A_n K_n^{-1} A_n^{-1}) - 2 \right| < 1,$$
$$\left| \mathrm{tr}^2(K_n) - 4 \right| + \left| \mathrm{tr}(K_n B_n K_n^{-1} B_n^{-1}) - 2 \right| < 1.$$

Since K_n is not parabolic and A_n and B_n are not of order two, according to Jørgensen's inequality Equation ((2.1)(i)), G_n is cyclic or a finite abelian extension of a cyclic group. Now K_n has two fixed points and neither A_n nor B_n exchange them. Therefore A_n and B_n share the fixed points of K_n. This completes the proof of (c).

Part (a) is a direct consequence of Jørgensen's inequality (p. 56). To prove (b), assume A is elliptic. If A is elliptic of infinite order then for some q, A^q is close to id. If $A_n^q \neq \mathrm{id}$, it is elliptic or loxodromic. For some q and all large indices,

$$\left| \mathrm{tr}^2(A_n^q) - 4 \right| + \left| \mathrm{tr}(A_n^q B_n A_n^{-q} B_n^{-1}) - 2 \right| < 1.$$

Either item (i) or (ii) of Jørgensen's inequality Equation ((2.1)) applies. If it is item (i), then G_n is elementary. If it is item (ii), then B_n is elliptic of order two and interchanges the fixed points of A_n^q, and hence of A_n. In either case $\langle A_n, B_n \rangle$ is elementary, a contradiction. $\square$

We can now continue with the proof of Theorem 4.1.1. The hardest part is to prove that G is not elementary. The case $r \geq 3$ can be reduced to the case $r = 2$. For if $r \geq 3$, according to Lemma 4.1.7, there exists a loxodromic or parabolic $g_n \in G_n$ whose length as a word in the generators $\{A_{i,n}\}$, $1 \leq i \leq r$, is uniformly bounded in n. Applying Lemma 4.1.9, for each n there is a nonelementary subgroup H_n of G_n generated by g_n and some generator $A_{i,n}$. Since the word length of g_n is uniformly bounded terms of the given generators, a subsequence of the two generator groups H_n converges algebraically to a subgroup H of G. If H is nonelementary, so is G.

So we may assume that $r = 2$ and $G_n = \langle A_{1,n}, A_{2,n} \rangle$ converges algebraically to $G = \langle A_1, A_2 \rangle$. In view of Lemma 4.1.6 after rearranging some more if necessary, we may assume that for all n, neither $A_{1,n}$ nor $A_{2,n}$ has order two. Then neither A_1 or A_2 can have order two, for if say $A_1^2 = $ id, replace G_n by $\langle A_{1,n}^2, A_{2,n} \rangle$ to get a contradiction to Lemma 4.1.10(a).

Since G_n is nonelementary, $A_{1,n}$ and $A_{2,n}$ have distinct fixed points. By Lemma 4.1.10(c), A_1 and A_2 have distinct fixed points as well.

Among the four elements $A_1, A_2, A_1 A_2, A_1 A_2 A_1^{-1} A_2^{-1}$ of G, there is an element X of infinite order. Otherwise these elements would be elliptic of finite order, and the same would hold for the corresponding elements of G_n for large indices. Lemma 4.1.3, G_n would then imply that G_n is a finite group. By Lemma 4.1.10(b), X cannot be elliptic.

If X is parabolic, then at least one of the two parabolic elements $A_1 X A_1^{-1}$, $A_2 X A_2^{-1}$ has no common fixed point with X. For otherwise A_1 and A_2 have a common fixed point. We confirm this by following the steps: (i) Apply Lemma 1.5.1(i) to $K = A_1 A_2 A_1^{-1} A_2^{-2}$, to $K A_i K^{-1} A_i^{-1}$, $i = 1, 2$ and then to their approximates. (ii) Show that for $i = 1, 2$ and all large indices,

$$| \operatorname{tr}^2(K_n) - 4| + | \operatorname{tr}(K_n A_{i,n} K_n^{-1} A_{i,n}^{-1})| < 1.$$

(iii) Finally apply Jørgensen's inequality, exception (i), followed by Lemma 1.5.2(i) to show that $A_{1,n}, A_{2,n}$ have the same fixed points thus forcing G_n to be elementary. We conclude that G is nonelementary.

If X is loxodromic, we claim that there exists $Y \in G$ such that X and $Y X Y^{-1}$ have no fixed points in common.

To establish this claim, we will first investigate what happens if for $Y \in G$, X and $Y X Y^{-1}$ do have a common fixed point. Applying again Lemma 4.1.10(c), we see that the corresponding elements in G_n also have a common fixed point, for all large indices. These approximates are loxodromic. Since G_n is discrete, two loxodromic elements cannot have exactly one fixed point in common by Lemma 2.3.1(ii),(iii). Therefore X and $Y X Y^{-1}$ have both fixed points in common. Unless Y has order two and interchanges the fixed points of X, X and Y have the same fixed points too.

Consequently by choosing Y as either A_1 or A_2 we obtain the desired result that X and YXY^{-1} have no fixed points in common. We conclude that G is not elementary.

Now return to the hypothesis of Theorem 4.1.1. We are given a sequence $G_n = \langle A_{1,n}, \ldots, A_{r,n} \rangle$ such that $\lim_{n \to \infty} A_{k,n} = A_k$ with $G = \langle A_1, \ldots, A_r \rangle$. We will use the correspondence $\phi : G \to G_n$ generated by $A_k \to A_{k,n}$.

We are ready to prove that G is discrete. Suppose otherwise. Then there exists a sequence of elements $B_k \in G$ with $\lim B_k = \text{id}$. We may assume that no B_n has order two. Since G is nonelementary, according to Exercise (2-1) there are two loxodromic elements $g_1, g_2 \in G$ without a common fixed point. By Lemma 4.1.10(c), g_i and B_k have a common fixed point if and only if $\phi_n(g_i)$ and $\phi_n(B_k)$ also do, for large n. Since G_n is discrete, this can occur only if the elements have the same fixed points, for neither is of order two. In this case g_i and B_k also have the same fixed points. The bottom line is that we can pick an infinite subsequence so that g_1, say, has no fixed point in common with $\{B_m\}$. Likewise $\phi_n(g_1)$ and $\phi_n(B_m)$ have no fixed points in common so generate a nonelementary subgroup. Now $\{\langle \phi_n(g_1), \phi_n(B_m) \rangle\}$ violates Jørgensen's inequality. Hence G is discrete.

The last step is to show that the correspondence $\phi_n : A_k \to A_{k,n}$ can be extended to a homomorphism $\phi_n : G = \langle A_1, \ldots, A_r \rangle \to G_n$. A necessary and sufficient condition that extension of ϕ_n to a homomorphism $G \to G_n$ is possible is that for each "relation" $R = \prod_k A_{i_k}^{m_k} = \text{id}$ in G also $\phi_n(R) = \prod_k A_{i_k,n}^{m_k} = \text{id}$. Now according to Theorem 2.6.1, there are only a finite number of relations in G: more precisely, there are a finite number of relations in G such that every other relation is a consequence of those. If to the contrary $\phi_n(R) \neq \text{id}$, application of Lemma 4.1.10(a) results in a contradiction. $\qquad \square$

There is another interesting corollary. Denote the space of ordered r-tuples of Möbius transformations which generate nonelementary groups by V_r. Let D_r be the subset consisting of discrete groups.

Corollary 4.1.11 [Jørgensen 1976]. *Each connected component of D_r consists of mutually isomorphic groups.*

Proof. Choose a component D and a group $G \in D$. Let X denote the set of all homomorphic images ϕ of G in D. By Theorem 4.1.1, X is relatively open in D. It is also closed in D. Therefore $X = D$. The same argument holds upon replacing G by any $G_1 \in D$. We conclude that G and G_1 are isomorphic (cf. [Sullivan 1985]) . See also Section 5.1. $\qquad \square$

4.2 Geometric convergence

Algebraic convergence deals not with geometry but with convergence of group generators. It is possible that in a sequence of groups $\{G_n\}$ there are words $W_n \in G_n$ in the generators, whose length increases without bound as $n \to \infty$, yet which converge to a Möbius transformation. Such phenomena are not detected by focusing on convergence of generators. Instead the phenomenon impacts the behavior of the sequence of quotient manifolds. From the point of view of a manifold, the generators of the fundamental group are rather arbitrarily chosen loops. What is fundamental are the geometrical quantities that determine its "shape". If we have a sequence of manifolds, we need a framework for discussing convergence to a limiting manifold.

If $\{G_n\}$ is a sequence of groups of Möbius transformations, define its *envelope* as

$$\text{Env}\{G_n\} = \{g \in \text{PSL}(2, \mathbb{C}) : g = \lim g_n, \ g_n \in G_n\}.$$

It follows that $\text{Env}\{G_n\}$ is itself a group.

Lemma 4.2.1. *If each G_n is discrete, then either $H = \text{Env}\{G_n\}$ is elementary, or it is a nonelementary, discrete group.*

Proof. According to Corollary 2.2.1 a group is discrete if and only if every two-generator subgroup is discrete. Assume that H is not elementary. Then given an element h_1 of infinite order there is another h_2 without a common fixed point. Suppose the nonelementary subgroup $\langle h_1, h_2 \rangle$ is not discrete. As in the final part of the proof of Theorem 4.1.1, there exist $h_1', h_2' \in \langle h_1, h_2 \rangle$ with the following property: h_2' is nearly the identity and $\langle h_1', h_2' \rangle$ is nonelementary yet in violation of Jørgensen's inequality (2.1). Now h_1', h_2' are each limits of elements in G_n. For large n, the pair of approximates in G_n will generate a nonelementary subgroup of G_n yet violate Jørgensen's inequality, a contradiction. This proves that any two-generator nonelementary subgroup of H is discrete. $\square$

A sequence of groups $\{G_n\}$ is said to *converge geometrically* (to $\text{Env}\{G_n\}$) if and only if for every subsequence $\{G_{n_j}\}$ of $\{G_n\}$, $\text{Env}\{G_{n_j}\} = \text{Env}\{G_n\}$. In other words, $\{G_n\}$ converges geometrically to H if and only if (i) each $h \in H$ is the limit $h = \lim g_n$, $g_n \in G_n$, and (ii) whenever $\lim g_{n_j} = g$ exists for a subsequence $\{n_j\}$ then $g \in H$. Necessarily $H = \text{Env}\{G_n\}$.

To justify use of the term "geometric convergence", and to give a precise meaning to the expression "convergent sequence of hyperbolic manifolds" we introduce the auxiliary concept of polyhedral convergence.

4.3 Polyhedral convergence

The sequence of discrete groups $\{G_n\}$ converges *polyhedrally* to the group H if H is discrete and for some point $\mathcal{O} \in \mathbb{H}^3$, the sequence of Dirichlet fundamental polyhedra $\{\mathcal{P}(G_n)\}$ centered at $\mathcal{O}$ converge to $\mathcal{P}(H)$ for H, also centered at $\mathcal{O}$, uniformly on compact subsets of $\mathbb{H}^3$.

We need to be more precise about the criterion for polyhedral convergence. Given $r > 0$, set

$$B_r = \{\vec{x} \in \mathbb{H}^3 : d(\mathcal{O}, \vec{x}) < r\},$$

where $d(\cdot, \cdot)$ denotes hyperbolic distance. We will work with the truncated polyhedra $\mathcal{P}_{n,r} = \mathcal{P}(G_n) \cap B_r$ and $\mathcal{P}_r = \mathcal{P}(H) \cap B_r$. A truncated polyhedron $\mathcal{P}_r$ has the property that its faces, that is the intersection with B_r of the faces of $\mathcal{P}$, are arranged in pairs, paired by the corresponding face pairing transformations of $\mathcal{P}$. Thus the projection of $\mathcal{P}_r$ into the quotient 3-manifold is a relatively compact submanifold, bounded by the projection of $\mathcal{P} \cap \partial B_r$ (Proposition 3.5.1).

The criterion for polyhedral convergence is as follows. Given any r sufficiently large, there exists $N = N(r) > 0$ such that (i) to each face pairing transformation h of $\mathcal{P}_r$, there corresponds a face pairing transformation g_n of $\mathcal{P}_{n,r}$ for all $n \geq N$ such that $\lim_{n \to \infty} g_n = h$, and (ii) if g_n is a face pairing transformation of $\mathcal{P}_{n,r}$ then the limit h of any convergent subsequence of $\{g_n\}$ is a face, edge or vertex pairing transformation of $\mathcal{P}_r$; in particular $h \neq \mathrm{id}$. In short, each pair of faces of $\mathcal{P}_r$ is the limit of a pair of faces of $\{\mathcal{P}_{n,r}\}$, and each convergent subsequence of a sequence of face pairs of $\{\mathcal{P}_{n,r}\}$, converges to a pair of faces, edges, or vertices of $\mathcal{P}_r$. We remark that it is possible that $\mathcal{P}_{n,r} = B_r$ for all large n. In this case the sequence of polyhedra converges to $\mathbb{H}^3$ itself.

If a given sequence of discrete groups is to converge polyhedrally, one must be allowed to conjugate the groups if necessary to find a point $\mathcal{O} \in \mathbb{H}^3$ that can effectively serve as center for all the polyhedra. We should be aware of the fact that a group can be conjugated so that for fixed $\mathcal{O}$, $\mathcal{P}_\mathcal{O}$ collapses. Namely conjugating $(az+b)/(cz+d)$ by $z \mapsto kz$ results in $k^{-1}(kaz + b)/(kcz + d)$. Its limit as $k \to \infty$ is 0, if $c \neq 0$.

The criterion needed is that there be a small ball about $\mathcal{O}$ that lies in the interior of the polyhedron for every group in the sequence. This is described in the following lemma.

Lemma 4.3.1. *To any infinite sequence of discrete groups $\{G_n\}$ corresponds a sequence of conjugates $\{A_n G_n A_n^{-1}\}$ which contains a polyhedrally convergent subsequence.*

Proof. Given $\mathcal{O} \in \mathbb{H}^3$, for each n choose a Möbius transformation A_n such that $G_n' = A_n G_n A_n^{-1}$ has the following property: Each truncated polyhedron $\mathcal{P}_{n,r}' = \mathcal{P}(G_n')_r$ centered at $\mathcal{O}$ contains the ball B_δ centered at $\mathcal{O}$ when $r > \delta$. Here $\delta > 0$ is a fixed number given by the universal ball property (Proposition 3.3.4). Thus the sequence of polyhedra centered at $\mathcal{O}$ of the conjugate groups cannot collapse to a convex object without interior... which is certainly possible in general.

We claim that for fixed $r > \delta$ the number of faces of the truncated polyhedra $\{\mathcal{P}_{n,r}'\}$ is uniformly bounded as $n \to \infty$. The reason for this is that there is an upper bound on the number of mutually disjoint balls of hyperbolic radius δ that fit inside B_{3r}. Therefore there is an upper bound, independent of n, on the number of points in the orbit $G_n'(\mathcal{O})$ that lie in B_{3r}. A face pairing transformation of $\mathcal{P}_{n,r}'$ satisfies $d(\mathcal{O}, g_n(\mathcal{O})) < 2r$, and the segment $[\mathcal{O}, g_n(\mathcal{O})]$ pierces a face. Hence there is also a uniform bound M independent of n on the number of faces.

Consequently given $s > 0$, there exists a large $r = r(s)$ such that the orbit of $\mathcal{P}_{n,r}'$ under its face pairing transformations covers the ball B_s for all n. This is because there is a uniform bound on the length of words W in the face pairing transformations of $\mathcal{P}(G_n')$, and the length of their segments $[\mathcal{O}, W(\mathcal{O})]$, required for the images of $\mathcal{P}(G_n')$ to cover B_s. For sufficiently large r all of the elements W are also words in the face pairing transformations of the truncated polyhedra. The number of polyhedra meeting B_s is uniformly bounded in n by some $N < \infty$.

For fixed r and each n make a list of the face pairing transformation $\{g_{i,n}\}$ of $\mathcal{P}'_{n,r}$, $1 \leq i \leq M$ (by repetition we may assume there are M faces for each n). Take a subsequence of $\{n\}$ and relabel so that for each i, $h_i = \lim_{n \to \infty} g_{i,n}$ exists; $h_i \neq \mathrm{id}$ because $d(\mathcal{O}, g_{i,n}(\mathcal{O})) > 2\delta$. Correspondingly construct the polyhedron

$$\mathcal{P}^*_r = \{\vec{x} \in \mathbb{H}^3 : d(\mathcal{O}, \vec{x}) \leq d(\vec{x}, h_i(\mathcal{O})), \ 1 \leq i \leq M\}.$$

Thus $B_\delta \subset \mathcal{P}^*_r \cap B_r = \lim \mathcal{P}'_{n,r}$.

Now take a sequence $r = r_k \to \infty$ and repeat the process for each r_k. We get a nested sequence of polyhedra $B_\delta \subset \mathcal{P}^*_{r_1} \subset \mathcal{P}^*_{r_2} \subset \cdots$. Set $\mathcal{P}_\infty = \bigcup_{i=1}^\infty \mathcal{P}^*_{r_i}$. The successive sets of side pairing transformations of the $\mathcal{P}^*_{r_k}$ are nested as well. Let $\{h_i\}$ denote the union. Let H denote the group they generate.

We claim that H is discrete, and $\mathcal{P}_\infty = \mathcal{P}_\mathcal{O}(H)$. Possibly $H = \{\mathrm{id}\}$ and $\mathcal{P}_\infty = \mathbb{H}^3$, the case that the groups $\{G_n'\}$ blow up completely.

First we claim that the orbit of $\mathcal{P}_\infty$ under H covers $\mathbb{H}^3$. As we have seen, given $s > 0$ there exists $r = r(s)$ such that the G_n'-orbit of $\mathcal{P}'_{n,r}$ covers the ball B_s for all n. For each n we can make a list of N transformations $W_{1,n}, W_{2,n}, \ldots, W_{N,n}$ such that $\bigcup_{i=1}^N W_{i,n}(\mathcal{P}'_{n,r}) \supset B_s$. Each $W_{i,n}$ is a word in the face pairing transformations of $\mathcal{P}'_{n,r}$, and the lengths are uniformly bounded as a function of s. Passing to a subsequence if necessary, each $W_i = \lim_{n \to \infty} W_{i,n}$ exists; necessarily $W_i \in H$. Therefore $(\cup_i W_i(\mathcal{P}_\infty)) \cap B_r$ covers B_s. Since s is arbitrarily chosen, our claim is established.

Next we claim that no two points in the interior of $\mathcal{P}_\infty$ are equivalent under H. For suppose that $W(x) = y$ for $x, y \in \mathrm{Int}(\mathcal{P}_\infty)$ and $W \in H$. The element W is a word in the generators $\{h_i\}$. For each n, let W_n denote the corresponding word in the approximates $\{g_{i,n}\}$ so that $W_n \in G_n'$ and $\lim W_n = W$, $\lim W_n(x) = y$. Choose $r > \max(d(\mathcal{O}, x), d(\mathcal{O}, y))$. Then for all large n, x and $W_n(x)$ lie in $\mathrm{Int}(\mathcal{P}'_{n,r})$. This is impossible unless $W_n = W = \mathrm{id}$.

We conclude that $\mathcal{P}_\infty$ is a fundamental polyhedron for H, and that H in turn is necessarily discrete. $\qquad\square$

We can now justify our use of the term "geometric convergence". But first note that it is possible for a sequence of nonelementary discrete groups to converge geometrically and polyhedrally to an elementary group: here are two examples:

$$\left\{\left\langle z \mapsto -\frac{3z-1}{z-3}, z \mapsto z+n \right\rangle\right\}, \qquad \left\{\left\langle z \mapsto -\frac{1}{n^2}\frac{3n^2 z - 1}{n^2 z - 3}, z \mapsto z + \frac{1}{n} \right\rangle\right\}.$$

This is why in the following fundamental result we have to explicitly assume that the limit groups are nonelementary.

Proposition 4.3.2. *A sequence $\{G_n\}$ of kleinian groups converges geometrically to a nonelementary kleinian group if and only if it converges polyhedrally to a nonelementary kleinian group. The geometric and polyhedral limits are the same.*

Proof. Suppose first the sequence converges polyhedrally to H. If $h \in H$ then h is a word W in the face pairing transformations of $\mathcal{P}(H) = \mathcal{P}_\infty$. As in the proof of Lemma 4.3.1, the word is the limit of a sequence of words $W_n \in G_n$. Next we have to show that if for a subsequence $h = \lim g_k$, $g_k \in G_k$, then $h \in H$. Again we return to the proof of Lemma 4.3.1. Let $s = 2d(\mathcal{O}, h(\mathcal{O}))$, where the polyhedra are centered at $\mathcal{O}$. We showed that $B_s \subset \bigcup_{i=1}^N W_i(\mathcal{P}_\infty \cap B_r)$ where $W_i = \lim_{k \to \infty} W_{i,k}$ and $W_{i,k} \in G_k$. Therefore $h = W_i$, for some i, and is the limit as $k \to \infty$ of the corresponding word $W_{i,k}$. We conclude that H is the geometric limit.

Conversely, suppose H is the geometric limit of $\{G_n\}$. By the universal ball property of Proposition 3.3.4, there exists $\mathcal{O} \subset \mathbb{H}^3$ such that the $h(B_\delta) \cap B_\delta = \varnothing$ for all $h \neq \mathrm{id} \in H$. We claim that the same property holds for G_n for all large n. Otherwise there would be a sequence $g_n \in G_n$ such that $g_n(B_\delta) \cap B_\delta \neq \varnothing$. A subsequence of $\{g_n\}$ converges to a Möbius transformation g_∞. If $g_\infty \neq \mathrm{id}$ then it would have to lie in H, a contradiction. If $g_\infty = \mathrm{id}$ we will find a contradiction to Jørgensen's inequality (2.1). Here we have to use the assumption that H is nonelementary. We can find two loxodromic transformations $h_1, h_2 \in H$ which have mutually disjoint fixed points (Exercise (2-1)). Each is the limit $h_i = \lim h_{i,n}$, $h_{i,n} \in G_n$. For large n at least one of the $h_{i,n}$, say $h_{1,n}$, does not share a fixed point with g_n. Now apply Jørgensen's inequality to $\langle g_n, h_{1,n} \rangle$. The conclusion is that $\{G_n\}$ converges polyhedrally. $\qquad\square$

We remark that the argument also applies in the following elementary situation. A sequence of cyclic loxodromic groups converges polyhedrally to a discrete parabolic group P if and only if it converges geometrically to P and no sequence of distinct elements converges to the identity. Here P may be of rank one or rank two. See Section 4.10.

Since geometric convergence makes no reference to a choice of center $\mathcal{O}$ for polyhedra, we can now remove any sign of dependence of polyhedral convergence on the choice of center $\mathcal{O}$.

Corollary 4.3.3. *If $\{G_n\}$ converges polyhedrally to the kleinian group H with one choice of center $\mathcal{O}$ for the polyhedra, it converges polyhedrally to H for any choice of center (which is not an elliptic fixed point of H).*

4.4 The geometric limit

We will need two lemmas. The first is a corollary of Theorem 4.1.1.

Lemma 4.4.1. *Suppose that $\{G_n\}$ is a sequence of nonelementary kleinian groups converging algebraically to G. There is no sequence of elements $g_k \in G_k$, $g_k \neq \mathrm{id}$, with either property:* $\lim g_k = \mathrm{id}$, *or* $\lim g_k = g$ *with g elliptic of infinite order.*

Proof. Present $G_n = \langle A_{1,n}, A_{2,n}, \ldots \rangle$, where $\lim A_{k,n} = A_k$ and no two generators have the same set of fixed points.

Case 1. g_n is elliptic for all large indices. For all large n, no generator $A_{k,n}$ can share exactly one fixed point with g_n. Otherwise $A_{k,n}$ would have to be parabolic and the order of g_n could not exceed six. Nor is it possible that every generator $A_{k,n}$ shares its fixed points with g_n or is of order two and interchanges the fixed points of g_n. For then G_n would be elementary. The conclusion is that for some k, $\langle g_n, A_{k,n} \rangle$ is nonelementary for all large indices, leading to a violation of Theorem 4.1.1.

Case 2. g_n is parabolic for all large indices. At most a finite number of elliptics can share its fixed point and at least one generator, say A_1 does not. This again leads to a violation of Jørgensen's inequality.

Case 3. g_n is loxodromic for all large indices. At most a bounded number of elliptics share a fixed point or interchange its fixed points, and at least one generator does neither. One is led to the usual contradiction. $\qquad \square$

Lemma 4.4.2. *Suppose the sequence of nonelementary kleinian groups G_n converges algebraically to G. There exists a point $\mathcal{O} \in \mathbb{H}^3$ and $\varepsilon > 0$ such that, for a subsequence G_k, no element of G_k has a fixed point in the ball $B_\varepsilon(\mathcal{O})$ of radius ε about $\mathcal{O}$.*

Furthermore, there exists $\delta < \varepsilon$ such that for all large indices, $T_k(B_\delta(\mathcal{O}))$ is disjoint from $B_\delta(\mathcal{O})$, for all $T_k \neq \mathrm{id} \in G_k$.

Proof. We begin by showing that given $x \in \mathbb{H}^3$, there exists ε with the following property. There exists a point $x_n \in B_\varepsilon(x)$ such that any axis of G_n which intersects $B_\varepsilon(x)$ passes through x_n.

If this assertion were false there would be a sequence $\varepsilon_n \to 0$ and rotation axes ℓ_n, ℓ'_n of $E_n, E'_n \in G_n$ which intersect $B_{\varepsilon_n}(x)$ but don't have common point of intersection in $B_\varepsilon(x)$. We may assume there is convergence $E_n \to E$, $E'_n \to E'$ where E, E' both fix x. Also, their rotation axes converge to lines ℓ, ℓ' through x. By Lemma 4.4.1 E, E' are elliptic of finite order with rotation axes ℓ, ℓ'.

According to Theorem 4.1.1, or the Universal Elementary Property, $\langle E_n, E'_n \rangle$ is necessarily elementary for all large indices.

Now E_n, E'_n cannot share a fixed point on $\mathbb{S}^2$, for their commutator would then be parabolic and by Lemma 4.4.1 would remain parabolic in the limit. Yet the limit would have to fix x.

Nor can E_n, E'_n both be elliptic of order two with disjoint axes, for $E_n E'_n$ would then be loxodromic and the limit, which fixes x, could only be the identity, which is impossible again by Lemma 4.4.1.

The remaining alternative is that for all large indices, $\langle E_n, E'_n \rangle$ is a finite, noncyclic group with a common fixed point $x_n \in \mathbb{H}^3$. As a noncyclic finite group, the number of elements (the order) of $\langle E_n, E'_n \rangle$ is uniformly bounded. Therefore it is isomorphic to $\langle E, E' \rangle$, and $x_n \to x$.

We conclude that there exists $\varepsilon > 0$ and $x_n \in B_\varepsilon(x)$ such that any rotation axis of G_n that intersects $B_\varepsilon(x)$ passes through x_n, for all large indices. Moreover, the finite subgroups $\mathrm{Stab}(x_n) \subset G_n$ are isomorphic to the limit group denoted by $\mathrm{Stab}(x)$ and $\lim x_n = x$.

There are only a finite number of possibilities for Stab(x), unless it is cyclic or a $\mathbb{Z}_2$ extension of a cyclic group. Find $\mathcal{O} \in B_\varepsilon(x)$ and $\varepsilon_1 < \varepsilon$ such that $T B_{\varepsilon_1}(\mathcal{O}) \cap B_{\varepsilon_1}(\mathcal{O}) = \emptyset$ for all $T \neq$ id $\in$ Stab(x). This property will persist for Stab(x_n), all large n.

Now consider the second assertion of Lemma 4.4.2. If it were false, corresponding to a sequence $\delta_n \to 0$ there would be a sequence $T_k \neq$ id $\in G_k$, $k = k(n)$, with

$$T_k(B_{\delta_n}(\mathcal{O})) \cap B_{\delta_n}(\mathcal{O}) \neq \emptyset.$$

Take a convergent subsequence, again labeled $\{T_k\}$. Its limit $T = \lim T_k$ fixes $\mathcal{O}$ but its approximates have no fixed point in $B_{\delta_n}(\mathcal{O})$. Therefore $T = $ id, again a violation of Lemma 4.4.1. $\qquad\square$

Theorem 4.4.3 [Jørgensen and Marden 1990]. *Suppose the nonelementary kleinian groups G_n converge algebraically to G. Then there is a geometrically convergent subsequence $\{G_k\}$. The limit H of any geometrically convergent subsequence contains G; consequently $\mathcal{M}(G)$ is a covering manifold of $\mathcal{M}(H)$.*

If the geometric limit H is finitely generated, there is a sequence of homomorphisms to its approximates $\psi_k : H \to G_k$, for all large k, such that $\lim \psi_k(h) = h$ for all $h \in H$. In addition if G is finitely generated, then $\psi_k(H) = G_k$.

Proof. Set $G_n = \langle g_{1,n}, g_{2,n}, \dots \rangle$ and $G = \langle g_1, g_2, \dots \rangle$ with $g_i = \lim_{n \to \infty} g_{i,n}$. According to Lemma 4.4.1 there is no subsequence $h_k \in G_k$, $h_k \neq$ id, with $\lim h_k = $ id. If the groups G_n contain no elliptic elements, then about any given point $\mathcal{O} \in \mathbb{H}^3$ there is a small ball B_ε which is contained in every polyhedron $\mathcal{P}(G_n)$ centered at $\mathcal{O}$. In this case we can find a polyhedrally convergent subsequence as in Lemma 4.3.1.

When the groups G_n contain elliptic elements, Lemma 4.4.2 tells us that the ball $B_\varepsilon(\mathcal{O})$, for some $\mathcal{O} \in \mathbb{H}^3$, is such that for a subsequence, no element of G_k has a fixed point in $B_\varepsilon(\mathcal{O})$. Then Lemma 4.4.2 tells us more strongly that for some $\delta < \varepsilon$, $T B_\delta(\mathcal{O}) \cap B_\delta(\mathcal{O}) = \emptyset$, for all $T \neq$ id $\in G_k$. So $B_\delta(\mathcal{O})$ will lie in the Dirichlet region for each G_k centered at $\mathcal{O}$.

Thus in all cases there is a subsequence $\{G_k\}$ that converges polyhedrally to a group H.

Given a compact subset $X \subset \mathbb{H}^3$, we claim that there exists $r > 0$ and N with the following property: X is covered by the images of the truncated polyhedron $\mathcal{P}(G_k)_r$ under all words of length $\leq N$ in the face pairing transformations of $\mathcal{P}(G_k)_r$, for all large k.

To see why, choose a larger compact set $X' \supset X$ containing X in its interior. For large enough r, N, the orbit $\mathcal{Q}_N$ of $\mathcal{P}(H)_r$ under words of length $\leq N$ in the face pairing transformations of $\mathcal{P}(H)_r$ covers X'. When k is large, $\mathcal{P}(G_k)_r$ is close to $\mathcal{P}(H)_r$ since the faces of $\mathcal{P}(G_k)_r$ converge to those of $\mathcal{P}(H)_r$. The corresponding orbit $\mathcal{Q}_{k,N}$ is close to $\mathcal{Q}_N$ and covers X.

From this we deduce that G is a subgroup of H as follows. Given $g \in G$, take X so that $\mathcal{O}$, $g(\mathcal{O})$ lie in its interior. We know $g = \lim g_k$, $g_k \in G_k$, and for large k,

$g_k(\mathcal{O}) \in X$. Therefore g_k is a word of length $\leq N$ in the face pairing transformations of $\mathcal{P}(G_k)_r$. In the limit, $g \in H$.

Now assume that H has a finite number of generators $\{h\}$. By Theorem 2.6.1, H is finitely presented. Fix a presentation. Each generator h of H is a word in the face pairing transformations of $\mathcal{P}(H)$ (centered at $\mathcal{O}$). For all sufficiently large k, say $k \geq k_0 = k_0(h)$, designate by $\psi_k(h)$ that element of G_k which is the same word in the corresponding face pairing transformations $\mathcal{P}(G_k)$. Then $\lim \psi_k(h) = h$.

The correspondence $h \mapsto \psi_k(h)$ determines a homomorphism $H \to G_k$, for large k. For if $R(h) = $ id is a relation in H, we have $\lim_{k\to\infty} \psi_k(R(h)) = $ id, and by Lemma 4.4.1, $\psi_k(R(h)) = $ id for $k \geq k_1$, where $k_1 > k_0$ is sufficiently large. Every relation in H is a consequence of the finite number in our presentation so it is only these we have to worry about. Therefore our argument shows that ψ_k determines a homomorphism as claimed.

If in addition $G = \langle g_1, \ldots, g_r \rangle$ is finitely generated, then for each index we have $g_i = \lim g_{i,k}$, with $g_{i,k} \in G_k$. By Theorem 4.1.1, the correspondence $\phi_k : g_i \to g_{i,k}$ determines a homomorphism for all large indices. Each generator g_i of G is also a word W_i in the generators $\{h\}$ of H. We know that $\lim_{k\to\infty} g_{i,k}^{-1} \psi_k(W_i) = $ id. Therefore for all large indices k, $g_{i,k} \psi_k(W_i) = $ id, that is $g_{i,k} = \psi_k(W_i)$ for all $1 \leq i \leq r$. So the homomorphism $\psi_k : H \to G_k$ is *onto* G_k; it restricts to the homomorphism $\phi_k : G \to G_k$ given by Theorem 4.1.1. $\square$

There are many examples, in particular examples of fuchsian groups, for which polyhedral convergence does not imply algebraic convergence. Taking this into account, we note that the existence of the homomorphism $\psi_k : H \to G_k$ a few lines above did not actually require that $\{G_k\}$ have an algebraic limit. Thus:

Corollary 4.4.4. *Suppose the sequence of kleinian groups $\{G_k\}$ converges polyhedrally to a finitely generated kleinian group H. Then there is a homomorphism ψ_k of H into G_k for all large k such that $\lim \psi_k(h) = h$ for all $h \in H$.*

4.5 Sequences of limit sets and regions of discontinuity

4.5.1 Hausdorff and Carathéodory convergence

In a discussion about convergence of sequences of kleinian groups, it is natural to ask about concomitant convergence of the regions of discontinuity, or of the limit sets. We begin by introducing the notion of Hausdorff convergence.

The *Hausdorff distance* between closed sets Λ and Λ_n in a metric space X (in our cases in $\mathbb{S}^2$, $\mathbb{H}^2$, or a hyperbolic surface) is defined as follows with respect to balls $B_r(x)$ of radius r about x measured by the metric in X:

$$d_H(\Lambda, \Lambda_n) = \inf\{r : \Lambda \subset \bigcup_{x \in \Lambda_n} B_r(x), \text{ and } \Lambda_n \subset \bigcup_{x \in \Lambda} B_r(x)\}.$$

Hausdorff distance defines a Hausdorff topology of closed sets.

We then say that there is *Hausdorff convergence* $\lim \Lambda_n = \Lambda$ if $d_H(\Lambda, \Lambda_n) \to 0$. In this case, an arbitrarily small collar neighborhood of Λ also contains all Λ_n for sufficiently large indices.

In short, given any $\varepsilon > 0$, there exists $N(\varepsilon)$ such that for all $n \geq N(\varepsilon)$, each point of Λ_n is within distance ε of a point on Λ, and conversely, each point on Λ is within distance ε of a point of Λ_n.

The following is a standard fact about Hausdorff distance:

Lemma 4.5.1. *If $\{\Lambda_n\}$ is a sequence of closed sets in a metric space, there is a subsequence $\{\Lambda_m\}$ which Hausdorff converges to a closed set $\Lambda \subset \mathbb{S}^2$.*

We will give two definitions of convergence of simply connected regions in $\mathbb{C}$, not the whole plane. The first assumes that the limiting region is known. In the second, the limiting region needs to be found as well. The latter is analogous to our criterion for geometric convergence of manifolds. For more details on this subject see Duren [1983] or Pommerenke [1992].

Situation 1. The sequence of regions $\{\Omega_n\}$ is said to converge to the region $\Omega \neq \mathbb{C}$ *in the sense of Carathéodory* if and only if every compact subset K of Ω lies in Ω_n for all large n *and* one of the following holds:

(i) Each $\zeta \in \partial\Omega$ is the limit $\zeta = \lim \zeta_n$, $\zeta_n \in \partial\Omega_n$.
(ii) Any open set U that lies in all elements of an infinite subsequence $\{\Omega_{i_j}\}$ also lies in Ω.

Carathéodory convergence of regions does not imply the Hausdorff convergence of the boundaries. For example, the sequence of boundaries may converge to a circle with an external ray while the regions themselves converge to the enclosed disk.

However, given a sequence of regions $\{\Omega_n\} \subset \mathbb{S}^2$, there is a subsequence such that $\{\mathbb{S}^2 \setminus \Omega_m\}$ Hausdorff converges to a closed set $\Lambda \subset \mathbb{S}^2$. Then $\{\Omega_m\}$ converges in the sense of Carathéodory to $\mathbb{S}^2 \setminus \Lambda$, if $\mathrm{Int}(\Lambda) \neq \varnothing$ and connected.

More generally, $\{\partial\Omega_n\}$ Hausdorff converges to $\partial\Omega$ if and only if both $\{\Omega_n\}$ converges to Ω, and $\{\mathbb{S}^2 \setminus \overline{\Omega_n}\}$ converges to $\mathbb{S}^2 \setminus \overline{\Omega}$, both in the sense of Carathéodory.

Situation 2. Suppose $\{\Omega_k\}$ is a sequence of regions on $\mathbb{S}^2$ all of which contain a point O serving as basepoint. To avoid shrinkage to O, we will assume that a small disk about O is contained in the members of the sequence. The *kernel* of the sequence is defined to be the largest region Y containing O with the property that $Y \subset \Omega_k$ for all k with at most a finite number of exceptions. More precisely, let Y_n denote the component of $\mathrm{Int}\left(\bigcap_{k \geq n} \Omega_k\right)$ that contains O. Then $Y = \cup_n Y_n$.

A sequence $\{\Omega_n\}$ converges in the sense of Carathéodory to its kernel Y if and only if every infinite subsequence also has Y as its kernel.

If Y has a hyperbolic metric, it is the limit of hyperbolic metrics on the approximating regions, uniformly on compact subsets of Y.

The kernel very much depends on the choice of basepoint O. For example, a sequence of simply connected regions may pinch in half, resulting in convergence, say, to the union of two disks. Depending on where the basepoint is chosen, the Carathéodory limit will be one or the other of the disks.

Carathéodory Convergence Theorem 4.5.2. *Suppose that* $\{\Omega_n\}$ *is a sequence of simply connected regions which converge in the sense of Carathéodory,* $\lim \Omega_n = \Omega$, *with respect to the basepoint* $O \in \cap \Omega_n$. *Assume that* $\partial \Omega \subset \mathbb{S}^2$ *contains at least two points. Let* $f_n : \mathbb{D} \to \Omega_n$ *be the Riemann map normalized by* $f(0) = O$, $f'(0) > 0$. *Then the sequence* $\{f_n\}$ *converges, uniformly on compact subsets of* $\mathbb{D}$, *to the normalized Riemann map* $f : \mathbb{D} \to \Omega$.

As a consequence, the sequence of hyperbolic metrics converges to the hyperbolic metric on the limiting region Ω.

Multiply connected regions can also be examined by normalizing the fuchsian covering groups with respect to O and then examining the groups with respect to geometric convergence. See Exercise (4-8).

More generally, the curvature of the metrics can be allowed to increase to 0 so the metric becomes euclidean as degeneration occurs. This results in other kinds of geometric limits.

4.5.2 Convergence of groups and regular sets

Proposition 4.5.3. *Suppose that* $\varphi_n : \Gamma \to G_n$ *is a sequence of homomorphisms from the finitely generated* Γ *onto kleinian groups* G_n *which converge algebraically to a kleinian group* G *with* $\Omega(G) \neq \varnothing$ *and geometrically to* H. *Suppose further that the convergence of* $\{\Omega(G_n)\}$ *to* $\Omega(G)$ *is in the sense that any given compact subset* $K \subset \Omega(G)$ *satisfies* $K \subset \Omega(G_n)$ *for all large indices. Then* $\Omega(G_n)$ *converges to* $\Omega(G)$ *in the sense of Carathéodory and* $\Lambda(G_n)$ *converges to* $\Lambda(G)$ *in the sense of Hausdorff. Moreover* G *has finite index in* H.

If in addition the $\{\varphi_n\}$ *are isomorphisms, then* $H = G$.

What is meant here is that each component Y of $\Omega(G)$ is the Carathéodory limit of components $\{Y_n\}$ of $\{\Omega(G_n)\}$, and conversely every sequence of components $\{Y_n\}$ of $\{\Omega(G_n)\}$ contains a subsequence $\{Y_k\}$ which converges to a component Y of $\Omega(G)$, in the sense of Carathéodory. Each of the components is governed by its stabilizer. There are only a finite number of conjugacy classes of component stabilizers in each of G_n, G (Ahlfors Finiteness Theorem).

Proof. (See Jørgensen and Marden [1990].) Suppose H properly contains G. Then there exists $h \in H$, $h \notin G$ such that $h = \lim g_n$, $g_n \in G_n$. Select compact sets K and K' such that $K \subset \text{Int}(K') \subset K' \subset \Omega(G)$. The sequence $\{g_n(K)\}$ converges to $h(K)$, and $\{g_n(\text{Int}(K'))\}$ to $\text{Int}(h(K'))$. We claim that $h(K) \subset \Omega(G)$.

If not, the interior $\text{Int}\, h(K')$ contains limit points of G, in particular fixed points of loxodromic elements of G. Choose a loxodromic $\gamma \in \Gamma$ such that $\varphi(\gamma) = \lim \varphi_n(\gamma)$ has a fixed point in $\text{Int}(K')$. For all large n, $\text{Int}\, g_n(K')$ contains a fixed point of $\varphi_n(\gamma)$, in contradiction to our hypothesis.

Consequently $h(K) \subset \Omega(G)$ for every compact subset K of $\Omega(G)$ and every $h \in H$. Therefore $h(\Omega(G)) \subset \Omega(G)$. The same argument can be applied to h^{-1}. We conclude that each $h \in H$ maps $\Omega(G)$ onto itself. In particular the fixed points of all loxodromic and parabolic elements of H lie in the limit set $\Lambda(G)$ showing that $\Lambda(H) = \Lambda(G)$.

In particular every fixed point of H is the limit of fixed points of G_n. Therefore every limit point of H is the limit of fixed points of G_n. We conclude that $\Omega(G_n)$ converges in the sense of Carathéodory to $\Omega(G)$ (situation 1(i) above) .

Furthermore, if an open set $U \supset \Lambda(G)$, then also $U \supset \Lambda(G_n)$ for large n. Conversely, if $U \supset \Lambda(G_n)$ for all large indices, then $U \supset \Lambda(G)$. Therefore the limit sets converge in the Hausdorff topology.

At this point we bring back Greenberg [1974] (see Exercise (3-14)) which implies that if $\Lambda(G)$ is not a round circle in $\mathbb{S}^2$, then G, which we know is contained in H, has finite index in H.

This holds even when $\Lambda(G)$ is a circle. For G is then a fuchsian group of finite area, or a $\mathbb{Z}^2$-extension of one (via an order two elliptic), so the larger discrete group H must contain G as a subgroup of finite index ($\mathbb{H}^2/G$ is necessarily a finite-sheeted covering surface of $\mathbb{H}^2/H$).

Now we come to the assumption that each φ_k is an isomorphism. In this case we claim that $H = G$. For suppose there were an element $h = \lim \varphi_n(\gamma_n)$, $\gamma_n \in \Gamma$, $h \notin G$.

Case 1: h is not elliptic. Since G has finite index in H, for some m, $h^m \neq \mathrm{id} \in G$ and $h^m = \varphi(\beta)$, $\beta \in \Gamma$. Therefore $\lim \varphi_n(\gamma_n^m \beta^{-1}) = \mathrm{id}$. By Lemma 4.4.1, $\beta = \gamma_n^m$ for all large n. In a discrete group an element of infinite order has fewer than m distinct m-th roots, by Lemma 1.5.3, unless $m - 2$. In either case, for a subsequence, we can assume all γ_n are the same and $h \in G$, a contradiction.

Case 2: h is elliptic. Choose a loxodromic element $g \in G$ whose fixed points are not interchanged by h. By a direct computation using a standard form for g, we see that for some integer m, $g^m h$ is not elliptic. Case 1 again applies to show that $g^m h \in G$ and hence $h \in G$, a contradiction. $\square$

Suppose $\{G_n\}$ converges geometrically to H, and H is geometrically finite. The fundamental polyhedron $\mathcal{P}$ for H at any suitable basepoint has a finite number of faces so that we can deform $\mathcal{P}$ backwards: The face pairing, edge pairing, and vertex pairing transformations associated with $\mathcal{P}$ are moved back to G_n. Just using this finite set of elements in G_n for large n form the corresponding Dirichlet region $\mathcal{P}_n^*$. One can show that $\mathcal{P}_n^* = \mathcal{P}_n$, the fundamental polyhedron for G_n. Using this idea, as in Jørgensen and Marden [1990], one concludes that $\{\Omega(G_n)\}$ converges to $\Omega(H)$ in the sense of Carathéodory. In view of polyhedral convergence Proposition 4.3.2, this argument leads to:

Theorem 4.5.4. *Suppose $\theta_n : \Gamma \to G_n$ is a sequence of isomorphisms of a group Γ onto kleinian groups G_n that converges algebraically to $\theta : \Gamma \to G$. Suppose G*

is geometrically finite with $\Omega(G) \neq \varnothing$. Then $\{G_n\}$ converges geometrically to G if and only if the regions of discontinuity converge, $\Omega(G_n) \to \Omega(G)$, in the sense of Carathéodory or equivalently, if and only if $\Lambda(G_n)$ converges to $\Lambda(G)$ in the sense of Hausdorff.

The definitive statement of limit set convergence follows from Bowditch [2013, Theorems 0, 1.1] and McMullen [1996, Prop. 2.4] (see Exercise (4-6)). It is particularly striking in that no assumption is made about parabolics, geometric finiteness, or boundary incompressibility.

Denote by $\mathcal{H}_r$ the set of hyperbolic 3-manifolds $\{\mathcal{M}(G)\}$ each of which has the properties

- Each is normalized so that the projection O of $O^* \in \mathbb{H}^3$ serves as a basepoint for π_1,
- For some $\varepsilon > 0$, the projection of the ε-ball with center $O^* \in \mathbb{H}^3$ is embedded in the interior of the convex core $\mathcal{CC}(G)$,
- For some $r < \infty$, every embedded ball in $\mathcal{M}(G)$ with center in $\mathcal{CC}(G)$ has radius at most r.

Lemma 4.5.5 [McMullen 1996]. *The set $\mathcal{H}_r$ is compact with respect to geometric limits. The limit set $\Lambda(G)$ varies continuously in the Hausdorff metric with $\mathcal{M}(G) \in \mathcal{H}_r$.*

Theorem 4.5.6 [Bowditch 2013; McMullen 1996]. *Suppose $\{\psi_n : G \to G_n\}$ is a sequence of isomorphisms to kleinian groups G_n from a finitely generated, torsion-free group G. Suppose $\{G_n\}$ converges algebraically to G_∞ and geometrically to H. Then $\Lambda(G_n)$ converges to $\Lambda(H)$ in the Hausdorff topology, and $\Omega(G_n)$ converges to $\Omega(H)$ in the sense of Carathéodory.*

In particular, if $G_\infty = H$ then $\lim \Lambda(G_n) = \Lambda(G_\infty)$.

Outline of proof. The radius r of balls is *not* the same as the Bowditch constant R as in Theorem 3.11.2. But applying a more refined statement of that Bowditch [2013, Theorem 1.1] one concludes that using the algebraic convergence of the sequence, the finiteness of the two numbers are equivalent. Form the space $\mathcal{H}_r$, and apply Proposition 4.5.5.

For boundary compressible manifolds one must beware of "long, thin, handles". On a compressible boundary component, there might be two short curves, each of which bounds an essential disk in the manifold, which, in the boundary, bound a very long cylinder. A generator of the fundamental group of the manifold must run through this "long handle" and thereby have great length. Up in $\mathbb{H}^3$, a lift of the generator would determine a large embedded ball, which would prevent algebraic convergence of the manifolds—think of carrying a suitcase. (Thanks to Brian Bowditch for the explanation.) □

In this context there is another interesting result of McMullen:

Proposition 4.5.7 ([McMullen 1996, Theorem 2.3]). *Suppose for an arbitrary sequence of kleinian manifolds $\{\mathcal{M}(G_n)\}$, the distance from a common basepoint $O \in \mathcal{CC}(G_n)$ to $\partial\mathcal{CC}(G_n)$ tends to ∞ with n. Then $\lim \Lambda(G_n) = \mathbb{S}^2$ in Hausdorff topology.*

This can easily happen under the conditions of Theorem 4.5.6, as for doubly degenerate groups, but Proposition 4.5.7 holds for arbitrary sequences.

Proof. (After McMullen.) In $\mathbb{S}^2 \setminus \Lambda(G_n)$, find a maximal spherical disk D_n. Erect the hyperbolic plane P_n on ∂D_n; P_n lies outside $\mathcal{CC}(G_i)$. The distance $d(O, P_n) \to \infty$. The spherical radius of D_n shrinks to 0. Consequently $\Lambda(G_n)$ comes arbitrarily close to every point in $\mathbb{S}^2$. $\square$

4.6 New parabolics

We can watch in the example of Section 4.9 how a sequence of cyclic loxodromic groups converges algebraically to a cyclic parabolic group and geometrically to a rank two parabolic group. It is also an example of how the algebraic limit acquires a "new" parabolic.

More generally, if $\theta_n : \Gamma \to G_n$ is a sequence of isomorphisms converging algebraically to the isomorphism $\theta : \Gamma \to G$, then we say $g \in G$ is a *new parabolic* if for all large indices, $\theta_n\theta^{-1}(g)$ is not parabolic. We may assume that the sequence also has a geometric limit $H \supset G$.

It was conjectured by Troels Jørgensen that if $\Omega(G) \neq \varnothing$ then $H = G$ provided G does not contain new parabolics (the converse is not true). When $\Omega(G) = \varnothing$, he conjectured that always $H = G$, since there is no "room" for new elements to appear. Both of these conjectures have been confirmed, as indicated below.

Here is the description of what happens in the geometrically finite cases.

Theorem 4.6.1 ([Jørgensen and Marden 1990]). *Suppose that Γ is a finitely generated abstract group without elements of finite order and $\{\theta_n : \Gamma \to G_n\}$ a sequence of isomorphisms onto kleinian groups that converges algebraically to $\theta : \Gamma \to G$. Assume that $\{G_n\}$ converges geometrically to a geometrically finite group H with $\Omega(H) \neq \varnothing$. Then:*

(i) *The limit sets converge $\lim \Lambda(G_n) \to \Lambda(H)$ in the Hausdorff topology and the sets of discontinuity converge, $\Omega(G_n) \to \Omega(H)$, in the sense of Carathéodory.*

(ii) *G is also geometrically finite.*

(iii) *For all large n, there is a homomorphism $\psi_n : H \to G_n$ with the properties that $\lim \psi_n(h) = h$ for all $h \in H$ and for $g \in G$, $\psi_n(g) = \theta_n\theta^{-1}(g)$.*

(iv) *Let $\{P_j\}$, $1 \leq j \leq N$, denote the rank two parabolic subgroups of H for which $\psi_n(P_j)$ is cyclic loxodromic, one representative from each conjugacy class in H. Let $T_{j,n} \in H$ denote a generator of the kernel of $\psi_n : P_j \to \psi_n(P_j)$. Then $\mathrm{Ker}(\psi_n)$ is the normal closure in H of the subgroup generated by $\{T_{j,n}\}$, $1 \leq j \leq N$.*

(v) *Assume each P_j contains an element of G. Then there exists $T_j \in P_j, T_j \notin G$ such that*

$$H = \langle G, T_1, T_2, \ldots, T_N \rangle.$$

(vi) *$H = G$ if and only if the class $\{P_j\}$ is empty.*

Outline of proof. As a finitely generated subgroup of the geometrically finite group H with $\Omega(H) \neq \varnothing$, G is also geometrically finite (Lemma 3.6.3). Item (i) follows from the remarks preceding Theorem 4.5.4. The first part of (iii) comes from Theorem 4.4.3. The second part of (iii) is a consequence of Lemma 4.4.1, namely $\theta_n \theta^{-1}(g) = \psi_n(g)$ for all large n, first for a set of generators of G and then for all G. Item (iv) is proved by working backward from a fundamental polyhedron for H to fundamental polyhedra for its approximates G_n. We will omit the detailed proof of this (the example of Section 4.9 shows how this happens). The proof of (v) begins with the fact that the common fixed point ζ_j of P_j is also a parabolic fixed point of G. Once again this is established by working backwards from a fundamental polyhedron for H; $\psi_n(P_j)$ represents a simple, short geodesic in $\mathcal{M}(G_n)$ which is associated with a word of uniformly bounded length in the face pairing transformations for G_n. So let $S_j \in G$ be a generator of the parabolic subgroup that fixes ζ_j. Then $\psi_n(S_j)$ is a generator of $\psi_n(P_j)$. Consequently $P_j = \langle S_j, T_{j,n} \rangle$. Take T_j to be any one of the $T_{j,n}$.

Item (vi) requires the elementary fact that the geometric limit of an algebraically convergent sequence of cyclic parabolic groups (which is again a cyclic parabolic group) is the same as the algebraic limit. The only way that H can differ from G is that there exist rank two groups $P_j \in H$ that are geometric limits of necessarily cyclic loxodromic subgroups of $\{G_k\}$ (while their algebraic limit is a cyclic parabolic group). □

Theorem 4.6.1 has been greatly generalized through the efforts of several authors, particularly Anderson, Brock, Bromberg, Canary, Evans, Ohshika, and Souto. Here is a statement of the final result, which incorporates the Tameness Theorem, that confirms Jørgensen's Conjecture.

Theorem 4.6.2. *Suppose $\{\theta_n : \Gamma \to G_n\}$ is a sequence of isomorphisms converging algebraically to $\theta : \Gamma \to G$. The sequence also converges geometrically to G under one of the following situations:*

(i) [Anderson and Canary 1996b; Evans 2004a] *If $\Omega(G) \neq \varnothing$ and G has "no new parabolics", that is, $g \in G$ is parabolic if and only if $\theta_n \theta^{-1}(g)$ is parabolic for all large indices n.*

(ii) [Canary 1996; Agol 2004; Calegari and Gabai 2004] *If $\Omega(G) = \varnothing$.*

Theorem 4.6.2 does not require that the approximating groups be geometrically finite (just finitely generated and torsion free). Of course the converse to (i) does not hold: G can be the geometric geometric limit even in the presence of new parabolics. Condition (ii) was initially established under additional assumptions, in particular when G is known to be tame. By incorporating the Tameness Theorem 5.6.6, we can make the general statement given here. In this case, whether or not there are new parabolics makes no difference.

A sequence is often said to be *strongly convergent* if it converges both algebraically and geometrically to the limiting group.

For algebraic and geometric convergence of sequences of discrete representations which are not faithful, see Biringer and Souto [2010]. This work shows that Theorem 4.6.2 then fails in general.

Here is another useful fact (especially in the context of the Density Theorem on p. 305):

Theorem 4.6.3 ([Brock et al. 2003; Brock and Souto 2006]). *If H is the algebraic limit of geometrically finite groups, then H is also the algebraic limit of geometrically finite groups $\{\theta'_n : \Gamma' \to G'_n\}$ with the property that $\theta'(g) = \lim \theta'_n(g)$ is parabolic if and only if $\theta'_n(g)$ is parabolic for all indices.*

By Theorem 4.6.2 H is also the geometric limit of $\{G'_n\}$. Of course in general, the groups Γ, Γ' will not lie in the same quasiconformal deformation space. What is remarkable about the theorem is that there is no requirement that H be geometrically finite.

4.7 Acylindrical manifolds

A compact 3-manifold with boundary M^3 and no parabolics is called *acylindrical* (or *anannular*) if M^3 contains no essential cylinders *and* is boundary incompressible. We recall from Section 3.7 that an *essential cylinder C* in M^3 is a cylinder C such that $C \cap \partial M^3 = \partial C$ and C is not homotopic into ∂M^3.

When there are parabolics the definition is as follows. As in Section 3.7, define an *essential cylinder* in $\mathcal{M}(G)$ as one which is *neither* homotopic into $\partial\mathcal{M}(G)$ *nor* retractable into a rank one cusp (or onto a pairing cylinder that pairs two rank one punctures). Then $\mathcal{M}(G)$ is said to be *acylindrical* if it is (i) boundary incompressible and (ii) contains no essential cylinders. In this case, every component of $\Omega(G)$ is simply connected, and a every loxodromic element preserves at most one component of $\Omega(G)$.

In particular in an acylindrical $\mathcal{M}(G)$ there can be no "accidental parabolics": a parabolic that acts in one component S of $\partial\mathcal{M}(G)$ as if it were loxodromic. For then there would be an essential cylinder one boundary component of which surrounds a puncture, while the other is a noncontractible simple loop $\gamma \subset S$.[1]

In any case, in $\mathcal{M}(G)$ a simple (nontrivial) loop on a cusp cylinder associated with a rank one cusp or on a cusp torus cannot be freely homotopic to a simple loop on a different cusp cylinder or cusp torus, for the corresponding parabolic subgroups belong to distinct conjugacy classes.

---

[1] Both endpoints of a lift of γ to $\Omega(G)$ end at the same parabolic fixed point.

This is a good place to point out that one way to exclude accidental parabolics in a geometrically finite manifold is to require that for each component $\Omega \subset \Omega(G)$, the subgroup Stab(Ω) is quasifuchsian (Exercise (3-10)), that is Ω is a quasidisk.

Acylindrical manifolds have compact deformation spaces:

Thurston Compactness Theorem [Thurston 1979b; Thurston 1986b] *Let G be a geometrically finite group such that $\mathcal{M}(G)$ is acylindrical (i.e., with nonempty, incompressible boundary). Then every sequence of parabolic-preserving isomorphisms to kleinian groups $\theta_n : G \to G_n$ has an algebraically convergent subsequence. That is, $\mathfrak{R}_{\mathrm{disc}}(G)$ is compact in the algebraic topology.*

The discrete deformation space $\mathfrak{R}_{\mathrm{disc}}(G)$ is defined on p. 279.

Suppose there were, in a geometrically finite manifold $\mathcal{M}(G)$, an essential cylinder C corresponding to the conjugacy class of a cyclic loxodromic subgroup. Suppose for example C divides $\mathcal{M}(G)$ into two components M, M'. Focus on M and fix a lift $M^* \subset \mathbb{H}^3 \cup \Omega(G)$. Normalize things so that a given point $O \in \mathbb{H}^3$ lies in a ball about O in M^*. Set $G_1 = \{g \in G : g(M^*) = M^*\}$. (The group $(\pi_1(\mathcal{M}(G)))$ is the free product of the fundamental groups of M, M' amalgamated over the common cyclic subgroup determined by C.)

Then we should expect that there is a sequence of degenerations of $\mathcal{M}(G_1) \cong M$ so that each cyclic loxodromic subgroup determined by C converges to a cyclic parabolic subgroup and C becomes a cusp cylinder in the limit. Here we keep the same normalization with respect to O.

If the lift (that is, a component of the preimage) M'^* of M' is adjacent to M^* and $G_2 = \mathrm{Stab}(M'^*)$, then except for the cyclic subgroup of G_2 that corresponds to the common boundary with M^*, the group G_2 will simply disappear in the degenerate limit—the Möbius transformations do not converge. This is why the acylindrical condition is necessary in the Compactness Theorem.

The same kind of situation could arise if M is not boundary incompressible.

For fuchsian groups Γ, such degenerations are always possible: The manifold $\mathrm{UHP} \cup \mathrm{LHP} \cup \mathbb{H}^3 / \Gamma$ is very cylindrical, see Exercise (4-8). One can start with a fuchsian group Γ and the lift of a simple geodesic from UHP/Γ, and "pinch" the geodesic so that in the limit it corresponds to a parabolic transformation.

To complement to the Compactness Theorem,

Theorem 4.7.1 ([[Johannson 1979]; [Matsuzaki and Taniguchi 1998, Theorem 3.29]]). *Suppose G is geometrically finite such that $\mathcal{M}(G)$ is acylindrical (i.e., boundary incompressible with nonempty boundary). Let $\theta : G \to G'$ be an isomorphism to a geometrically finite G' such that $\theta(g)$ is parabolic if and only if $g \in G$ is parabolic. Then there exists a possibly orientation-reversing quasiconformal mapping $F : \mathbb{S}^2 \to \mathbb{S}^2$ that satisfies $F \circ g \circ F^{-1}(z) = \theta(g)(z)$ for all $g \in G$, $z \in \mathbb{S}^2$. It can be chosen to project and extend to be a (quasiisometric) homeomorphism $F_* : \mathcal{M}(G) \to \mathcal{M}(G')$.*

Of special interest is the fact that θ dictates a bijection between components of $\Omega(G')$ and $\Omega(G)$. See Exercise (4-9).

4.8 Dehn filling and Dehn surgery

Dehn surgery and Dehn filling are both operations on tori embedded in some $\mathcal{M}(G)$. There is not a great distinction between the two terms so that authors are not always fastidious in doing so.

Dehn surgery is often identified as *drill and fill*, whereas Dehn filling is just *fill*. Consider four examples.

(i) A torus boundary component of a manifold.
(ii) The boundary of a solid cusp torus associated with a rank two parabolic subgroup, or
(iii) The boundary of a tubular neighborhood about a simple closed geodesic (or simple closed noncontractible loop).

Dehn filling usually refers to an operation of type (i) where the interior of the torus is filled in to create a solid torus

Dehn surgery usually refers to the operations of type (ii) or (iiii) where the interior of the given torus is first removed and then, using Dehn filling, replaced by a solid torus. The old has been "drilled out" and the new "filled in".

A `meridian` α on a torus $\mathbb{T}$ is a simple noncontractible loop that bounds an embedded disk in the filled torus. Its free homotopy class on $\mathbb{T}$ is uniquely defined.

The `longitude` β is a simple loop $\beta \subset \mathbb{T}$ which is transverse to the meridian α, with geometric intersection number one. It is also uniquely determined up to homotopy. If $\mathbb{T}$ is the boundary of a solid cusp torus, β is parallel to the core curve β^*. If $\mathbb{T}$ is embedded unknotted in $\mathbb{R}^3$, β bounds a disk in the exterior.

The meridian $\alpha \subset \mathbb{T}$, together with the longitude β generate both homology and homotopy of $\mathbb{T}$. Any simple loop $\gamma \subset \mathbb{T}$ can be expressed in the form $\gamma = m\alpha + n\beta$ where m, n are relatively prime integers. The ratio $0 \leq n/m \leq \infty$ is called the *slope* of γ. Thus α itself has slope 0 while β itself has slope ∞.

(m,n) Dehn filing (where (m,n) are relatively prime) is the operation of replacing the original solid torus with meridian α by a new solid torus with meridian $\gamma = m\alpha + n\beta$. Or, fill an empty torus with meridian α by a solid torus with meridian γ.

Another way to describe γ is as the image of α under an automorphism ϕ of $\mathbb{T}$. Fillings by two automorphisms ϕ, ϕ_1 determine homeomorphic manifolds if and only if $\phi_1 \circ \phi^{-1}$ extends to a homeomorphism of the solid torus.

(m,n)-Dehn surgery. Let f be a homeomorphism of the boundary $\mathbb{T}$ of a solid cusp torus that sends the meridian to γ with slope $\neq 0, \infty$ on $f(\mathbb{T})$. Refill the result with γ as the new meridian and, correspondingly, a new core curve.

The homotopy type of the ambient manifolds will change from the filling/surgery. The job is finished by putting a hyperbolic structure on the new manifold.

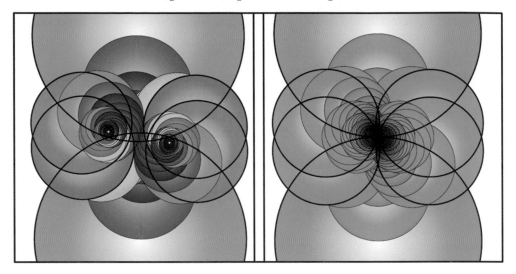

Fig. 4.1. A cyclic group generated by a loxodromic of approximate trace $1.919354+0.029772$ near its rank two parabolic geometric limit in the right frame. One can see how the 6-sided Ford polygon outside the outer circles is becoming a fundamental domain on $\mathbb{S}^2$ for the geometric limit, and the circles are becoming horocycles. See Jørgensen [1973] for a description of the combinatorics of the approximates.

4.9 The prototypical example

This is an explicit example both of Dehn filling in the simplest case and of differing algebraic and geometric limits. We will start with a solid cusp torus—a rank two parabolic group—and do $(1, n)$ Dehn filling on it. There results a cyclic loxodromic group. We will then watch what happens as $n \to \infty$. Figure 1.7 (p. 21) and Figure 4.1 (p. 242) show several generations of isometric circles of a cyclic loxodromic group.

Start with the parabolic group $\Gamma = \langle T_1(z) = z + \omega_1, \ T_2(z) = z + \omega_2 \rangle$. Set $\tau = \omega_2/\omega_1$, $\operatorname{Im} \tau > 0$. The quotient $\mathbb{C}/\Gamma = \mathcal{T}$ is a torus. The generating pair (ω_1, ω_2) corresponds to a pair of simple loops α, β on $\mathbb{T}$, crossing each other once.

Change the basis by the rule

$$\omega_{1,n} = \omega_1 + n\omega_2, \quad \omega_{2,n} = \omega_2, ; \quad \tau_n = \frac{\omega_{2,n}}{\omega_{1,n}} = \frac{\tau}{1 + n\tau},$$

so that $T_{1,n}(z) = z + \omega_{1,n}$, $T_{2,n}(z) = z + \omega_{2,n}$ also generate Γ. The pair $(\omega_{1,n}, \omega_{2,n})$ represents the simple loops $\alpha + n\beta$, β on $\mathbb{T}$.

Map $\mathbb{C}$ onto $\mathbb{C} \setminus \{0\}$ by $w_n(z) = e^{-2\pi i z/\omega_{1,n}}$.

Let U_n denote the loxodromic transformation

$$U_n(w) = e^{-2\pi i \tau_n} w = a_n w.$$

We have $(w_n \circ T_1)(z) = (U_n^{-n} \circ w_n)(z)$ and $(w_n \circ T_2)(z) = (U_n \circ w_n)(z)$, while $(w_n \circ T_{1,n})(z) = w_n(z)$ and $(w_n \circ T_{2,n})(z) = (U_n \circ w_n)(z)$.

The map w_n determines a conformal mapping

$$\mathbb{T} \to \mathbb{T}_n = (\mathbb{C} \setminus \{0\}) / \langle U_n \rangle$$

in which the image of $\alpha + n\beta$ is a meridian in the solid torus $(\mathbb{H}^3 \cup (\mathbb{C} \setminus \{0\})) / \langle U_n \rangle$. The image of straight lines with tangent vector $\omega_{1,n}$ is taken by w_n to concentric circles about $w = 0$ which in turn project to parallel meridians in $\mathbb{T}_n$. We have done Dehn filling on the original cusp torus $\mathcal{M}(\Gamma)$ by removing $\text{Int}(\mathcal{M}(\Gamma))$ and replacing it by a solid torus so the chosen simple loop $\alpha + n\beta$ becomes a meridian.

As $n \to \infty$, $\lim \tau_n = 0$, $\lim a_n = 1$, and $\lim U_n = \text{id}$. Renormalize U_n to have the fixed points $\omega_2 / (1 - a_n)$, ∞, thus

$$A_n(w) = w + \frac{\omega_2}{1 - a_n}, \qquad V_n(w) = A_n U_n A_n^{-1}(w) = a_n w + \omega_2.$$

Therefore $\lim V_n(w) = w + \omega_2$ and

$$V_n^k(w) = A_n U_n^k A_n^{-1}(w) = a_n^k w + \frac{a_n^k - 1}{a_n - 1} \omega_2.$$

Define

$$f_n(z) = \frac{\omega_2}{a_n - 1} (w_n(z) - 1).$$

Thus $f_n \circ T_1(z) = V_n^{-n} \circ f_n(z)$ and $f_n \circ T_2(z) = V_n \circ f_n(z)$ while $f_n \circ T_{1,n}(z) = f_n(z)$ and $f_n \circ T_{2,n}(z) = V_n \circ f_n(z)$. In short, f_n induces a conformal mapping to the renormalized solid torus

$$\mathbb{T} \to \left(\mathbb{C} \setminus \left\{\frac{\omega_2}{1 - a_n}\right\}\right) / \langle V_n \rangle.$$

in such a way that the image of $\alpha + n\beta$ remains a meridian.

Uniformly on compact subsets of $\mathbb{C}$ we have the following convergences:

(i) $\lim f_n(z) = z$.
(ii) $\lim V_n(w) = w + \omega_2$.
(iii) $\lim V_n^{-n}(w) = w + \omega_1$.

To prove (i), we use the estimate $e^x - 1 \sim x$ when x is small, and (iii) follows.

It is more complicated to show:

Claim. *The sequence of cyclic loxodromic groups $\{\langle V_n \rangle\}$ converges algebraically to the cyclic parabolic group $\langle T_2 \rangle$ and geometrically to the rank two parabolic group $\Gamma = \langle T_1, T_2 \rangle$.*

Proof. Suppose for a sequence $m \to \infty$ and $k = k(m) \to \infty$ that $\{V_m^k\}$ converges to a Möbius transformation. We must show the limit lies in Γ. For the limit to exist the ratio $(a_m^k - 1) / (a_m - 1)$ must remain bounded. Therefore $\lim a_m^k = 1$.

Write $k = pm + q$ where p, q are integral functions of m, as is k, and $0 \le q < m$. Since $a_m^k = \exp(-2\pi i k \tau_m)$ and $\text{Im}\, \tau_m = \text{Im}\, \tau / |1 + m\tau|^2$, we must have $k(m) = o(m^2)$. Therefore $p(m) = o(m)$, as $m \to \infty$.

Take the subsequence $\{m\}$ so that $\lim q(m)/m = c$ exists, $0 \leq c \leq 1$. We claim that either $c = 0$ or $c = 1$. For first of all $e^{-2\pi i k \tau_m} = e^{-2\pi i(k\tau_m - p)} = e^{-2\pi i(pm\tau_m - p + q\tau_m)}$. Also,

$$\lim \, p(m\tau_m - 1) = -\lim \frac{p}{1 + m\tau} = 0, \quad \lim q\tau_m = c.$$

So if c were not an integer, the ratio

$$\frac{a_m^k - 1}{a_m - 1} = \frac{e^{-2\pi i k \tau_m} - 1}{e^{-2\pi i \tau_m} - 1} \tag{4.1}$$

would become infinite.

We have to examine (4.1) in more detail. Write

$$e^{-2\pi i k \tau_m} = e^{-2\pi i(k\tau_m - p - c)}$$

so that the exponent approaches zero as $m \to \infty$. By Taylor's formula for e^x, the limit of the ratio (4.1) is the limit of

$$\frac{k\tau_m - p - c}{\tau_m} = \frac{p(m\tau_m - 1) + q\tau_m - c}{\tau_m} = \frac{-p + (q - cm)\tau - c}{\tau}.$$

Since $\operatorname{Im} \tau > 0$, if this is to have a finite limit then $\lim_{m \to \infty}(q - cm)$ must exist, necessarily as an integer. Then $\lim_{m \to \infty} p$ must exist as well, also as an integer. We conclude that

$$\lim_{m \to \infty} V_m^k(w) = w - \omega_1(c + \lim p) + \omega_2 \lim(q - cm). \qquad \square$$

Summing up, the solid tori $\mathcal{M}(\langle V_n \rangle)$ converge algebraically to $\mathcal{M}(\langle T_2 \rangle)$, which represents a solid cusp tube associated with a cyclic parabolic group. The boundary torus has become "pinched"—has become a doubly infinite cylinder: the hole in the bagel has coalesced to a single point since the length of the geodesic has gone to zero. In contrast, the manifolds $\mathcal{M}(\langle V_n \rangle)$ converge geometrically to the solid cusp torus $\mathcal{M}(\Gamma)$. The process of degeneration introduces so much twisting along the boundary torus, that in the limit the solid torus fractures, with the fracture line lying at its core. All this happens while the conformal type of the torus itself does not change, what changes is the presentation of its fundamental group.

For a generic choice of center $O \in \mathbb{H}^3$, the fundamental polyhedron $\mathcal{P}$ for Γ is a 6-sided chimney rising from $\mathbb{C}$. The approximates $\mathcal{P}_n$ acquire more and more faces as $n \to \infty$, but all of the faces, save six, collapse to the fixed point in the limit. The polyhedra truncated by intersection with a ball of radius r about their center $\mathcal{P}_{r,n}$ converge uniformly to $\mathcal{P}_r$.

Remark 4.9.1. We can now give an example of a sequence of cyclic loxodromic groups $\{\langle S_m \rangle\}$ that converge geometrically to a rank two parabolic group, yet which do not converge algebraically. In fact no subsequence of unbounded powers of the generators $\{S_m\}$ has a limit.

For an example, take from above $V_m(z) = a_m z + \omega_2$ with $\lim a_m = 1$. Pick any sequence of integers $n = n(m)$ which go to infinity with m. Set $d_m = \sqrt[n]{a_m}$ where the root is chosen in any way so long as *no* subsequence approaches 1. Set

$$S_m(z) = d_m z + \frac{d_m - 1}{d_m^n - 1} \omega_2.$$

Then $S_m^n(z) = V_m(z)$ and $S_m^{nk}(z) = V_m^k(z)$ but no infinite subsequence of $\{S_m\}$ converges since the constant term $\to \infty$.

At the other extreme a sequence of cyclic loxodromic groups $\langle z \mapsto c_n z \rangle$ with $c_n > 0$, $c_n \to 1$ can be conjugated to converge algebraically and geometrically to a cyclic parabolic group. Only loxodromics whose traces converge "tangentially" to ± 2 can have differing algebraic and geometric limits.

The space of cyclic loxodromic groups is completely described in terms of the combinatorics of the faces of the Ford polyhedron in Jørgensen [1973].

4.10 Manifolds of finite volume

Suppose $\{\mathcal{M}(G_n)\}$ is an infinite sequence of mutually nonisometric manifolds whose volumes $\{V_n\}$ do not exceed a number $V^* < \infty$. We can normalize the groups so that an ε-ball centered at a point $\mathcal{O} \in \mathbb{H}^3$, projects injectively into all of the manifolds. After passing to a subsequence, we may assume that the sequence $\{G_n\}$ converges geometrically to a group H.

Theorem 4.10.1. *Assume that the sequence $\{G_n\}$ with $\mathrm{Vol}(\mathcal{M}(G_n)) = V_n \leq V^* < \infty$ converges geometrically to H. Then $\mathrm{Vol}(\mathcal{M}(H)) = \lim V_n$. Consequently the set of volumes of finite volume manifolds is a closed subset of $\mathbb{R}$.*

Moreover, the number of solid cusp tori in $\mathcal{M}(H)$ is strictly greater than the number in its approximates, for large indices.

Proof. Take a maximal set of points in the ε-thick part $\mathcal{M}(G_n)^{\mathrm{thick}}$ such that the distance between any two of them is not less than $\varepsilon/2$. Then the ε-balls about these points cover $\mathcal{M}(G_n)^{\mathrm{thick}}$. The fact that $\mathrm{Vol}(\mathcal{M}(G_n) < V^*$ implies that the number of the covering balls is uniformly bounded in n. In particular there exists $d < \infty$ such that for all indices, the diameter of $\mathcal{M}(G_n)^{\mathrm{thick}}$ does not exceed d.

Now $\partial \mathcal{M}(G_n)^{\mathrm{thick}} = \partial \mathcal{M}(G_n)^{\mathrm{thin}}$ and $\lim_{\varepsilon \to 0} \mathrm{Vol}(\mathcal{M}(G_n)^{\mathrm{thin}}) = 0$, uniformly in n; the thin parts become successively thinner. Consequently

$$\lim_{\varepsilon \to 0} \mathrm{Area}(\partial \mathcal{M}(G_n)^{\mathrm{thin}}) = 0; \quad \lim_{\varepsilon \to 0} \mathrm{Vol}\,\mathcal{M}(G_n)^{\mathrm{thick}} = \mathrm{Vol}\,\mathcal{M}(G_n),$$

(see Exercise (2-11), Equation (2.9)).

Now as $n \to \infty$, $\text{Vol}(\mathcal{M}(G_n)^{\text{thick}})$ converges to $\text{Vol}(\mathcal{M}(H)^{\text{thick}})$, which therefore has finite volume $\leq V^*$, independent of ε. It follows that the volume of $\mathcal{M}(H)$ is bounded by V^* as well.

Now that we know $\mathcal{M}(H)$ has finite volume, the corresponding polyhedra for G_n must have a uniformly bounded number of faces and hence generators—see Lemma 3.6.4. The methods of the proof of Theorem 4.6.1 apply to describe the relation of the nearby polyhedra $\mathcal{P}_\odot(G_n)$ to the polyhedron $\mathcal{P}_\odot(H)$ for H.

For all large n there is a homomorphism $\psi_n : H \to G_n$. The ψ_n-image of each rank two parabolic subgroup of H is either a rank two subgroup of G_n or it represents the lift of a short geodesic in $\mathcal{M}(G_n)$.

If $\mathcal{M}(H)$ had the same number of solid cusp tori as $\mathcal{M}(G_n)$ for all large indices, then ψ_n would be an isomorphism and $\mathcal{M}(G_n)$ would be isometric to $\mathcal{M}(H)$, by Mostow's Rigidity Theorem, for the same large indices. Our assumption rules out this possibility. There are always strictly more solid cusp tori in the geometric limit than in the approximates. A sequence of cyclic loxodromic subgroups has become a rank two parabolic group in the geometric limit. See Theorem 4.6.1. $\qquad\square$

4.10.1 The Dehn Surgery Theorem

Suppose $\mathcal{M}(G)$ has finite volume and has $k \geq 1$ rank two cusps. Denote by $\mathcal{M}_c$ the compact manifold bounded by k tori $\{T_i\}$ resulting from removing the interior of the k solid cusp tori.

We start by presenting the argument in Thurston [1979a, Theorem 5.6] that *the dimension of the local deformation space of $\mathcal{M}(G)$ under the Dehn fillings is $\geq k$.* This includes deforming the rank two cusp groups to cyclic loxodromic groups. See Culler and Shalen [1983, Prop. 3.2.1] for an alternate treatment.

Choose a standard homology basis (γ_i, δ_i) on each T_i, $1 \leq i \leq k$, that is γ, δ_i is a pair of simple loops crossing each other once.

Let $\mathbb{Q} \subset \mathbb{S}^2$ denote the set of coprime vectors $\{d_i = (p_i, q_i)\}$. Given the vector $\text{vec}_k = (d_1, \ldots, d_k) \in \mathbb{Q}^k$, denote by M_{vec_k} the manifold resulting from (p_i, q_i)-Dehn filling on T_i, $1 \leq i \leq k$. Here the respective meridians are $\{p_i\gamma_i + q_i\delta_i\}$.

For each index, $1 \leq i \leq k$, choose in any way a simple, noncontractible loop $\{\alpha_i\}$ in the interior of $\mathcal{M}_c \subset \mathcal{M}(G)$ that corresponds to a loxodromic transformation in G. Fix a basepoint $O_i \in T_i$ and take that to be the basepoint for α_i. Remove a thin tube about each α_i and attach it to T_i so as to form a closed surface S_i of genus two. Do this for all indices, assuming the α_i are mutually disjoint, mutually not freely homotopic. We end up with a manifold $M' \subset \mathcal{M}_c$ bounded by k mutually disjoint closed surfaces $\{S_i\}$ of genus two.

On each S_i chose a simple loop β_i that bounds a compressing disk in $\mathcal{M}_c \setminus M'$. We can regard α_i to lie on S_i and also that the four simple loops $\gamma_i, \delta_i, \alpha_i, \beta_i$ have the common basepoint O_i. The four loops generate $\pi_1(S_i; O_i)$, and satisfy the commutator relation $[\delta_i, \gamma_i][\beta_i, \alpha_i] = \text{id}$.

The group $\pi_1(\mathcal{M}_c) \cong G$ is obtained from $\pi_1(M')$ by adding the relations $\{\beta_i = \mathrm{id}\}$; if we cut along the compressing disks bounded by the β_i, we obtain a manifold homeomorphic to $\mathcal{M}_c$.

The elements γ_i, δ_i, when lifted from a fixed $O_i^* \in \mathbb{H}^3$ over O_i determine generators γ_i^*, δ_i^* of a rank two parabolic subgroup of G. The element α_i when lifted from O_i^* determines a loxodromic element α_i^*. Under a small deformation, their traces change slightly, and also the location of the fixed points. Therefore when a homomorphism ψ is close to id, $\psi(\alpha_i^*)$ remains loxodromic with fixed points distinct from those of $\psi(\gamma_i^*)$, $\psi(\delta_i^*)$.

Lemma 4.10.2. *Suppose ψ is a homomorphism $\pi_1(M') \to \mathrm{PSL}(2, \mathbb{C})$ such that (a) $\psi(\langle \gamma_i, \delta_i \rangle) \neq \mathrm{id}$, (b) $\psi(\alpha_i)$ is loxodromic, and (c) $\psi(\langle \alpha_i, \gamma_i, \delta_i \rangle)$ is nonelementary. Then ψ extends to a homomorphism of $\pi_1(\mathcal{M}_c)$ if and only if for each index i the following two equations are satisfied:*

$$\mathrm{tr}\,\psi([\alpha_i, \beta_i]) = 2, \quad \mathrm{tr}\,\psi(\beta_i) = 2. \tag{4.2}$$

The necessity of the condition is obvious since a homomorphism of $\pi_1(M_c)$ must send both the commutator and the element β_i to the identity. The sufficiency is Exercise (4-5).

Now the compact 3-manifold M' can be triangulated in such a way that there is only one 0-simplex and the 1-simplices are generators of $\pi_1(\mathcal{M}_c)$. The 2-simplices then generate the relations among the chosen generators. Since the manifold has a nonempty boundary, the 2-skeleton of the triangulation is a deformation retract of M'. Its Euler characteristic is then

$$\chi(M') = +1 - h + r,$$

where h is the number of generators and r the number of relations.

Moreover $\chi(M') = \chi(\mathcal{M}_c) - k$ because $\mathcal{M}_c$ is obtained from M' by adding k relations. That is,

$$\chi(\mathcal{M}_c) = 1 - h + r + k.$$

The ψ-image of the h generators of G arising from our construction in $\mathcal{M}_c$ must satisfy the algebraic equations corresponding to each relation. Each Möbius transformation in turn depends on 3 complex parameters. In addition Equations (4.2) must be accounted for; that gives two more equations for each torus boundary. Thus the $3h$ parameters for the ψ-image of the generators are subject to constraints and the result is that ψ has the degree of freedom given by

$$3h - 3r - 2k = -3\chi(\mathcal{M}_c) + k + 3.$$

But if we rule out conjugations of the group G, we are left with the dimension

$$-3\chi(M_c) + k.$$

A closed 3-manifold has Euler characteristic zero. Therefore if $\widehat{\mathcal{M}_c}$ denotes the double of $\mathcal{M}_c$ across its boundary,

$$0 = \chi(\widehat{\mathcal{M}}_c) = 2\chi(\mathcal{M}_c) - \chi(\partial\mathcal{M}_c).$$

Since all the components of $\partial\mathcal{M}_c$ are tori, and $\chi(\partial\mathcal{M}_c) = 0$, necessarily $\chi(\mathcal{M}_c) = 0$ as well (for a general geometrically finite kleinian manifold $\mathcal{M}$ we would have instead $\chi(\partial\mathcal{M}) \leq 0$ and then $\chi(\mathcal{M}) \leq 0$). Thus ψ has k degrees of freedom; each rank two cusp contributes one degree. If $k = 0$, we have obtained a local version of Mostow's theorem

For a rigorous study of the deformation variety, see Kapovich [2001, Theorem 8.44].

The following result shows that there are lots of hyperbolic manifolds obtained independently of the criteria of the Hyperbolization Theorem (p. 385). The paper by Petronio and Porti [2000] is the current standard for a complete, rigorous proof of the first part of the following theorem. It is quite different from the one suggested in Thurston [1979a], and reflects the computational approach of SnapPy (the SnapPea sucessor) (p. 270). For another approach, see Hodgson and Kerckhoff [1998, §4].

Theorem 4.10.3 (Dehn Surgery Theorem). (Thurston 1979a, §5.8;Petronio and Porti 2000)

(i) *There exists a neighborhood U of $\infty^k = (\infty, \ldots, \infty) \in \mathbb{S}^2 \times \cdots \times \mathbb{S}^2$ such that for all vectors $v \in U \cap \mathbb{Q}^k$, the Dehn surgered manifold M_v has a complete hyperbolic metric.*

(ii) *More precisely, if a finite number of coprimes $\{(p_i, q_i)\}$ are excluded for each cusp torus $\{T_i\}$, $1 \leq i \leq k$, then all remaining Dehn surgeries on $\mathcal{M}(G)$ result in complete hyperbolic manifolds.*

(iii) *Suppose $\{v_n\}$ is a sequence of k-vectors with $\lim_{n\to\infty} v_n = \infty^k$ in U. The hyperbolic manifolds $\mathcal{M}(G_n) \equiv M_{v_n}$ converge geometrically back to $\mathcal{M}(G)$. The corresponding homomorphisms $\psi_n : G \to G_n$ converge to the identity.*

(iv) *Every cusped hyperbolic manifold of finite volume can be approximated in the geometric topology by closed manifolds.*

In particular there are arbitrarily small deformations $\{H\}$ of G which send any or all of the rank two parabolic subgroups to cyclic loxodromic groups. The result of removing from each such $\mathcal{M}(H)$ tubular neighborhoods of its new short geodesics is homeomorphic to $\mathcal{M}(G)$.

No finite volume hyperbolic manifold can have more than 12 exceptional Dehn fillings (or surgeries!)—fillings that don't result in hyperbolic manifolds. The figure-8 knot complement is known to have exactly 10 exceptional fillings. More generally, there are at most finitely many one cusped finite volume, orientable manifolds which have more than eight nonhyperbolic Dehn fillings; for the results cited see Agol [2000]; Agol [2010].

The simplest case of Dehn filling is when $\mathcal{M}(G)$ has finite volume but only one cusp—"torus boundary component". Denote by $M(s)$ the manifold obtained by doing Dehn filling of slope s on $\partial\mathcal{M}(G)$. Thurston had proved that there are at most a finite number of "exceptional" slopes s for which $M(s)$ does *not* have a hyperbolic structure As noted above, for the figure-8 knot complement, there are 10 exceptional slopes.

This gave rise to the conjecture that 10 is the maximum. Recently this conjecture was confirmed by a computer assisted proof:

Theorem 4.10.4 ([Lackenby and Meyerhoff 2013]). *In the set of all Dehn fillings $\{M(s)\}$ of any one-cusped finite volume manifold $\mathcal{M}(G)$, there are at most 10 exceptional slopes $\{s\}$.*

It is not known whether the figure-8 knot complement yields the unique maximum, and whether there are similar theorems for manifolds with multiple rank two cusps.

A process similar to Dehn filling allows the construction of orbifolds where the rank two parabolic groups are instead sent to cyclic elliptic groups with designated rotation angles.

4.10.2 Sequences of volumes

By the universal ball property, there is a uniform positive lower bound for all volumes. By the universal horoball property, there is a uniform upper bound on the number of solid cusp tori in manifolds of volume $\leq V$. Here the convergence theorems 4.1.1, 4.1.2 play a central role.

Theorem 4.10.5 [Thurston 1979a, §5.8; Gromov 1981b].

(i) *The set of hyperbolic 3-manifolds with a given volume V is finite.*

(ii) *Removing a simple geodesic α from a finite volume manifold **increases volume**: If $\mathcal{M}(H) \setminus \alpha$ is homeomorphic to $\mathcal{M}(G)$ of finite volume,*

$$\text{Vol}(\mathcal{M}(G)) > \text{Vol}(\mathcal{M}(H)).$$

(iii) *Dehn filling of a solid cusp torus of $\mathcal{M}(G)$, or a tubular neighborhood of a closed geodesic, **decreases volume**.*

(iv) *If $\{\mathcal{M}(G_k)\}$ is a sequence of manifolds whose volumes are nonincreasing,*

$$\cdots \geq \text{Vol}(\mathcal{M}(G_k)) \geq \text{Vol}(\mathcal{M}(G_{k+1})) \geq \cdots,$$

then $\text{Vol}\mathcal{M}(G_k) = \text{Vol}\mathcal{M}(G_m)$ for some m and all $k \geq m$.

(v) *For each constant V let $\mathfrak{M}_V$ denote the set of hyperbolic 3-manifolds with volume $\leq V$. There is a finite subset $\mathfrak{M}_{\text{moms}} \subset \mathfrak{M}_V$ with the following properties. (a) Each $\mathcal{M}(G) \in \mathfrak{M}_V \setminus \mathfrak{M}_{\text{moms}}$ contains a knot or link L such that $\mathcal{M}(G) \setminus L$ is homeomorphic to some $\mathcal{M}(H) \in \mathfrak{M}_{\text{moms}}$, and (b) In turn, $\mathcal{M}(G) \setminus L$ is homeomorphic to the result of Dehn filing on one or more rank two cusps of $\mathcal{M}(H)$.*

In fact according to Thurston [1979a, Theorem 5.11.2], there is a link $L_V \subset \mathbb{S}^3$ such that all manifolds in $\mathfrak{M}_V$ can be obtained by Dehn surgery along L_V (the limiting case of simply deleting components of L_V is allowed).

Heuristic discussion. The first item stems from the following argument. If there is an infinite sequence of nonisometric manifolds of volume exactly V there is a geometric limit of a subsequence. It must have at least one additional cusp which raises the volume by (iii).

We refer to Thurston 1979a, Chapter 6 for the proof of (ii); the volume $\mathrm{Vol}(\mathcal{M}(H))$ is less than $\mathrm{Vol}(\mathcal{M}(G))$ if $\mathcal{M}(H) \setminus \cup \gamma_i$ is homeomorphic to $\mathcal{M}(G)$ for a union of mutually disjoint simple geodesics γ_i. The proof is based on the analysis of the volumes of hyperbolic manifolds which are the images under degree $d \geq 1$ maps of a given finite volume manifold.

The most problematical issue in the background is to prove that the number of homeomorphism types for ε-thick parts M of hyperbolic manifolds of volume at most V is finite. Here is Thurston's argument. Take a maximal set of points of M with the property that the distance between every pair of the points exceeds $\varepsilon/2$. This insures that the $\varepsilon/2$-balls cover the thick part. The $\varepsilon/4$ balls about these points are mutually disjoint. The total volume of the $\varepsilon/4$-balls cannot exceed V so there are a finite number. The combinatorial pattern of intersections of the $\varepsilon/2$-balls determines the homeomorphism type of M; there are only a finite number of possibilities.

However, as pointed out in Benedetti and Petronio [1992, pp. 195–6] it is possible that a ε-tube may bore through an $\varepsilon/2$-ball, leaving one or more worm holes. This increases the possibilities for the topological type of M, beyond what is accounted for above. For this reason subsequent authors have to find lengthy alternate treatments to avoid this difficulty among others; see Petronio and Porti [2000].

The finiteness of topological types is coupled with the fact that two manifolds of finite volume which have homeomorphic ε-thick parts can be obtained from one another by Dehn surgery. As a consequence, given V, all manifolds of volume $\leq V$ are obtained by Dehn surgery on the cusp tori of a finite number of manifolds.

To analyze (iv), suppose a sequence of volumes is strictly decreasing. After passing to another subsequence if necessary, we may assume the groups G_n have a geometric limit H. Then $\mathrm{Vol}\mathcal{M}(H) = \lim \mathrm{Vol}\mathcal{M}(G_n)$. By Theorem 4.10.1 the geometric limit has at least one more rank two cusp than its close approximates. By Theorem 4.6.1, the close approximates arise from Dehn fillings on the rank two cusps of the geometric limit. By item (ii), the close approximates have lower volume.

On the one hand the volume of $\mathcal{M}(H)$ is greater than the volume of its close approximates, and on the other, the volume of its approximates is strictly decreasing. This contradiction proves that the volumes of the sequence must stabilize at a certain point, as claimed.

Item (v) holds because there are only a finite number of homeomorphism types of the solid cusp tori complements of elements of $\mathfrak{M}_V$. $\qquad\square$

4.10.3 Well ordering of volumes

Suppose there is at least one noncompact manifold of volume V. The finite set of cusped manifolds $\{\mathcal{M}(H)\}$ with volume V serves as the "mothers" of the manifolds $\{\mathcal{M}(G)\}$ of volume $< V$ with the following property. Each $\mathcal{M}(G)$ is obtained as the result of Dehn filling of one or more cusps of its mother $\mathcal{M}(H)$. There are a finite number of (unlinked) mutually disjoint nontrivial simple loops for which $\mathcal{M}(G) \setminus \bigcup_i \gamma_i$ is homeomorphic to a mother $\mathcal{M}(H)$. Each mother $\mathcal{M}(H)$ is the geometric limit of the manifolds $\mathcal{M}(G_n)$ obtained by Dehn fillings on its cusp tori.

If we start with the set of noncompact manifolds of lowest possible volume V, then their set of children comprise all closed hyperbolic manifolds of volume $< V$.

Theorem 4.10.1 leads to the conclusion that the set of volumes is well ordered (every subset has a least element), so there is a sequence of volumes indexed by the ordinals, ω being the ordinal of the positive integers.

$$v_1 < v_2 < \cdots \longrightarrow v_\omega < v_{\omega+1} < \cdots \longrightarrow v_{\omega+2} < \cdots \longrightarrow v_{2\omega} < \cdots \longrightarrow v_{3\omega} <$$

$$\cdots \longrightarrow v_{\omega^2} < \cdots \longrightarrow v_{\omega^3} < \cdots \longrightarrow v_{\omega^\omega}$$

Here v_1 is the lowest volume for closed hyperbolic manifolds, v_2 the second lowest, and so on, so that v_ω is the least accumulation point of volumes of closed manifolds obtained by Dehn filling on the least volume 1-cusped manifolds and is the lowest volume of 1-cusped manifold. Then $v_{2\omega}$ is the next lowest volume of 1-cusped manifolds and is the accumulation point of volumes $v_{\omega+1}, v_{\omega+2}, \ldots$ obtained by Dehn filling on these. And so on until reaching the first accumulation point v_{ω^2} of volumes $v_{k\omega}$ of 1-cusped manifolds; v_{ω^2} is the lowest volume for 2-cusped manifolds. This spawns the volume sequence $v_{2\omega^2}, v_{3\omega^2}, \ldots$ of 2-cusped manifolds which in turn accumulates at the least volume v_{ω^3} for 3-cusped manifolds. The last element ω^ω is the countable ordinal, the limiting ordinal as ω takes on larger and larger integers.

Thus the set of volumes form successive intervals on $\mathbb{R}$ of the form $[0, \omega]$, $[0, \omega^2], \ldots, [0, \omega^\omega)$.

The index t of a general element v_t of the volume sequence is an ordinal of the form

$$m_n \omega^n + m_{n-1}\omega^{n-1} + \cdots + m_0,$$

where m_j is a nonnegative integer. For example the index $2\omega^2 + 4\omega + 6$ corresponds to the volume of a manifold obtained first by Dehn filling on the 2nd lowest volume 2-cusped manifold resulting in a 1-cusped manifold of the 4th lowest volume followed by filling resulting in a closed manifold with 6th lowest volume.

For a report on the cusped hyperbolic manifolds composed of at most seven ideal tetrahedra and their Dehn surgery daughters, see Callahan et al. [1999].

4.10.4 Minimum volumes

In Cao and Meyerhoff [2001] it is shown that $v_\omega = 2v_\Delta \cong 2.03$, where v_Δ is the volume of the regular ideal tetrahedron. Among the cusped manifolds, only the complement of the figure-8 knot Figure 3.10 (p. 191) and its sibling in $\mathbb{S}^3$ achieve it (see Exercise (3-5)).

Least volume orientable hyperbolic manifold. Among all (orientable) manifolds, the minimum volume can be attained only by a closed manifold. By using the progam SnapPea, it was conjectured early on by Jeff Weeks that the minimum is $v_1 = 0.9427\ldots$, that value being attained by what is called the "Weeks manifold".

The Weeks manifold can be obtained by $(5, 1)$, $(5, 2)$ Dehn filling on the two components of the Whitehead link or $(5, 1)$ filling of the cusp of the minimum volume cusped manifolds. There are a chain of papers successively narrowing the seach, most recently 2004 and A. Przeworski, who proved $v_1 > 0.3325$.

That the Weeks manifold is indeed the unique (orientable) manifold of minimum volume $0.9427\ldots$ was finally established in deep work by D. Gabai, R. Meyerhoff, and P. Milley [Gabai et al. 2010], [Gabai et al. 2011], [Gabai et al. 2009].

Least volume orientable hyperbolic orbifold. Tim Marshall and Gaven Martin [2012] discovered the minimal volume orientable *orbifold*. It is generated by two elements, one of order 2, the other of order 3. It comes from an order two extension of the orientation-preserving subgroup of the reflection group of the following hyperbolic tetrahedron: Two faces form a $\pi/5$-dihedral angle and each of these faces form a $\pi/3$-dihedral angle with another; the remaining three dihedral angles are $\pi/2$. The minimum volume orbifold is uniquely determined and has volume $0.03905\ldots$. Its discovery allows the investigation of maximal automorphism groups of closed manifolds; see Conder et al. [2006].

Least volume one and two cusped manifolds. Earlier Meyerhoff [1987] had shown that the group H of orientation-preserving symmetries of the tessellation of $\mathbb{H}^3$ by regular ideal tetrahedra gives the smallest volume orientable orbifold with one cusp. In Agol [2010] the minimum volume orientable 2-cusped 3-manifolds are identified.

The well ordering of volumes of finite volume hyperbolic orbifolds is shown in Dunbar and Meyerhoff [1994].

Volumes of higher-dimensional manifolds

It is interesting to contrast the situation of three-dimensional finite volume manifolds with other dimensions. The areas of finite area two-dimensional hyperbolic manifolds are integral multiples of 2π (Exercise (3-1)). For a finite volume even-dimensional hyperbolic manifold M^{2n}, the formula also comes from the Gauss-Bonnet formula, for example see Kellerhals and Zehrt [2001].

$$\mathrm{Vol}(M^{2n}) = (-1)^n \frac{V_{2n}}{2} \chi(M^{2n}),$$

where V_{2n} is the surface area of the unit $(2n-1)$-dimensional sphere[2] in $\mathbb{R}^{2n}$ and χ denotes the Euler characteristic. The formula holds for orientable and nonorientable manifolds; in the former case the Euler characteristic is even for a closed manifold. The paper by Ratcliffe and Tschantz [2000] explicitly constructs the finite volume cusped (noncompact) hyperbolic 4-manifolds. Exactly 1171 of them have the minimum volume.

For odd-dimensional finite volume hyperbolic manifolds, the Euler characteristic is zero. However for *cusped* manifolds, using ball packing methods, good lower bounds can be found [Adams 1987; Kellerhals 1998].

---

[2] The surface area V_k of $\mathbb{S}^{k-1} \subset \mathbb{R}^k$ is $2\pi^{n/2}\Gamma(n/2)$.

It is known that in every dimension $n \geq 4$ the number of nonisometric manifolds with volume less than any prescribed number is finite [Wang 1972]; thus the set of volumes is a discrete set on $\mathbb{R}$. Furthermore, the number $N(V)$ of nonisometric manifolds of volume $\leq V$ grows to $+\infty$ with V; in fact, it is shown in Burger et al. [2002] that there are constants $a = a(n) > 0, b = b(n) > 0$ such that for all large V,

$$e^{aV \log V} \leq N(V) \leq e^{bV \log V}.$$

For an interesting discussion of what may or may not be possible for kleinian groups in dimensions $n \geq 4$ see Kapovich [2007].

4.11 Exercises and explorations

4-1. Suppose $\langle U, V \rangle$ is discrete and nonelementary. The subgroup $\langle U, [U, V] \rangle$ is also nonelementary provided U is not elliptic of order ≤ 60.

4-2. Suppose that the sequence of kleinian groups $\{G_k\}$ converges polyhedrally to a geometrically finite group H. Prove that there is a homomorphism ψ_k of H into G_k, for all large k, such that $\lim \psi_k(h) = h$, $h \in H$.

4-3. Prove with Mumford [1971] that the collection of all closed Riemann surfaces of genus $g \geq 2$ that have the property that the length of any closed geodesic exceeds some $\varepsilon > 0$ is compact: Every infinite sequence of such surfaces, or infinite sequence of normalized fuchsian covering groups, has a geometrically convergent subsequence to a group which represents a surface of the same type.

4-4. In contrast to the example of Section 4.9, prove the following. A sequence of cyclic loxodromic groups $\{\langle S_n \rangle\}$ with real traces which converges algebraically to the cyclic parabolic group $\langle S \rangle$ also converges to it geometrically.

Show that the conclusion remains the same if the hypothesis is weakened to the assumption that there exists $\delta > 0$ such that for all indices,

$$-\frac{\pi}{2} + \delta \leq \arg(\operatorname{tr} S_n) \leq \frac{\pi}{2} - \delta.$$

4-5. We will follow Thurston [1979a, Lemma 5.6.1] in outlining the sufficiency of Equation (4.2), p. 247, for the extension of ψ from a homomorphism of $\pi_1(M')$ to one of $\pi_1(M)$ (compare Lemma 4.10.2). The proof proceeds by considering each boundary torus separately. For simplicity of notation, we may therefore assume there is only one.

We have chosen generators of the genus two surface S so that $[\gamma, \delta][\alpha, \beta] = \mathrm{id}$. We are assuming that $\operatorname{tr}(\psi([\alpha, \beta])) = 2$, and $\operatorname{tr}\psi(\beta) = 2$. We are also assuming that $\psi(\langle \gamma, \delta \rangle) \neq \mathrm{id}$, $\psi(\alpha)$ is loxodromic (does elliptic and parabolic also work?), and $\psi(\langle \alpha, \gamma, \delta \rangle)$ is nonelementary. According to Lemma 1.5.2, $\psi(\alpha)$ and $\psi(\beta)$ have a common fixed point, say ∞.

Take $\psi(\alpha) = \left(\begin{smallmatrix} \lambda & 0 \\ 0 & \lambda^{-1} \end{smallmatrix} \right)$. Then $\psi(\beta) = \left(\begin{smallmatrix} a & b \\ 0 & a^{-1} \end{smallmatrix} \right)$. If $\psi([\alpha, \beta])$ is not the identity, then since $\psi([\gamma, \delta]) = \psi([\alpha, \beta]^{-1})$, all four of $\psi(\alpha), \psi(\beta), \psi(\gamma), \psi(\delta)$ have ∞ as a fixed point, a contradiction. Finally if $\psi(\beta) \neq$ id, then $\psi(\beta)$ is parabolic. This too is impossible for then it could not commute with $\psi(\alpha)$. So both $\psi([\alpha, \beta])$ and $\psi(\beta)$ are the identity.

4-6. *Convergence of limit sets.* [McMullen 1996, Prop. 2.4] Suppose $\{G_n\}$ is a sequence of kleinian groups normalized so that the respective convex hulls $\mathcal{CH}(G_n) \subset \mathbb{H}^3$ contain a fixed ball about a point $O \in \mathbb{H}^3$ that projects injectively to the quotients. We may then assume that the sequence converges geometrically to a kleinian group H.

A point z lies in $\liminf \Lambda(G_n)$ when every neighborhood U of z contains points of $\Lambda(G_n)$ for all n, with at most a finite number of exceptions. In contrast $z \in \limsup \Lambda(G_n)$ when every neighborhood U of z contains points of infinitely many $\Lambda(G_n)$. The sequence $\{\Lambda(G_n)\}$ converges when the two limits agree.

Because every loxodromic fixed point of H is the limit of loxodromic fixed points of G_n, conclude that $\Lambda(H) \subset \liminf \Lambda(G_n)$.

There is not always equality in the two limits. For example, there is a sequence of fuchsian groups of the first kind whose geometrical limit is just $\{$id$\}$—which has empty limit set. One such is the sequence of level-n subgroups $\{M_n\}$ of the modular group (Exercise (2-5): $M_n = \{g \in \text{Mod} : g \equiv I \bmod n\}$).

But suppose that there exists $\rho < \infty$ such that $\widetilde{\text{Inj}}_n(x) < \rho$ for all $x \in \tilde{\mathcal{C}}(G_n)$ and all n. Prove that $\lim \Lambda(G_n) = \Lambda(H)$ in the Hausdorff topology (cf. Theorem 3.11.2).

Hint: For a kleinian group G and $r < \infty$ set $M(r) = \{x \in \mathbb{H}^3 : \widetilde{\text{Inj}}(x) \leq r\}$. First show that $\overline{M}(r) \cap \mathbb{S}^2 \subset \Lambda(G)$, where $\overline{M}(r)$ denotes closure in the spherical metric. To see this descend to $\mathcal{M}(G)$. Suppose that $\text{Inj}(\pi(x)) \leq r$ for $\pi(x) \in \mathcal{M}(G)$. Then there is a noncontractible closed loop c of length $\leq 2r$ through $\pi(x)$. Shrink it so that it either becomes a geodesic $\gamma \subset \mathcal{C}(G)$ of length $\leq 2r$ or a simple loop γ on the boundary of a thin part at a finite distance from c. Back upstairs, in the former case, the G-orbit of x accumulates to $\Lambda(G)$. In the latter case, the orbit of x under a cyclic parabolic group accumulates to a parabolic fixed point.

Returning to our geometrically convergent sequence, show that $\{\widetilde{\text{Inj}}_n(x)\}$ for $\{G_n\}$ converges to $\widetilde{\text{Inj}}(x)$ for H, uniformly on compact subsets of $\mathbb{H}^3$. That is, $\limsup M_n(r) \subset M(r)$ with respect to Hausdorff convergence, for any $r > 0$. When $r = \rho$, $\tilde{\mathcal{C}}(G_n) \subset M_n(\rho)$, so $\limsup \tilde{\mathcal{C}}(G_n) \subset M(\rho)$. Since all rays from $O \in \tilde{\mathcal{C}}(G_n)$ to points on $\Lambda(G_n)$ lie in $\tilde{\mathcal{C}}(G_n)$, the limiting rays lie in $M(\rho)$. Conclude that $\limsup \Lambda(G_n) \subset \overline{M}(\rho) \subset \Lambda(H)$.

4-7. *Hyperbolic cone manifolds and orbifolds* We will start with the simplest of examples. Set $\alpha = 2\pi/n$ and consider the action of the elliptic $E_\alpha(z) = e^{i\alpha}z$ on $\mathbb{H}^3$. The quotient $\mathcal{M}(\langle E_\alpha \rangle)$ is an (oriented) orbifold. The *cone angle* at the set Σ of cone points, which is just the projection of the rotation axis, is $2\pi/n$. The elliptic E_α may be conjugated so that as $n \to \infty$, it converges to a parabolic (Exercise (2-4)). The quotient $M_{\mathcal{O}} = \mathbb{H}^3 \setminus \text{cone axis}/\langle E_\alpha \rangle$ has a hyperbolic metric but it is not complete. It does have a metric completion which is topologically a ball B and is called the *underlying space* of the orbifold. In B, the projection of the rotation axis is called the *singular locus* $\Sigma_{\mathcal{O}}$.

What about letting α be an irrational multiple of 2π? On the one hand it is perfectly reasonable for a cone to have an irrational cone angle. On the other hand the group is no longer discrete. Still we can use the model M_Θ for it. Again the underlying space is B and $M_\Theta \cong B \setminus \Sigma_\Theta$. The singular set Σ_Θ has cone angle α. We have no covering map π^{-1} to use, instead we have a more general map called a *developing map* that operates in the same way. The developing map d is a local isometry, unrolling M_Θ in $\mathbb{H}^3$: Given a closed loop γ encircling Σ_Θ, and a point O^* that projects to its initial point, the developing map "lifts" γ, starting at O^* and terminating at $E_\alpha^k(O^*)$ for some k. The developing map is coupled with a homomorphism φ, called the *holonomy map*. The holonomy map sends $\pi_1(M_\Theta)$ to the group of Möbius transformations generated in this case by E_α.

A 3-manifold M_C is called a *hyperbolic cone manifold* if there is a link $\Sigma \subset M_C$ (a union of simple loops not isotopic in $M_C \setminus \Sigma$ to a point) with the following properties.

(i) $M_C \setminus \Sigma$ has an incomplete hyperbolic structure.
(ii) Its metric completion in M_C is a singular metric with cone type singularities along Σ.
(iii) To each component σ of Σ corresponds a *cone angle* $0 < \alpha < \infty$. In a thin tubular neighborhood $N_r(\sigma)$ of radius r about σ the metric can be expressed in cylindrical coordinates (see Exercise (8-9)) as

$$dr^2 + \sinh^2 r \, d\theta + \cosh^2 r \, dh^2,$$

where h is the distance along σ and θ (mod α) is the angular measure about σ.

In other words there is a developing map $D : M_C \setminus \Sigma \to \mathbb{H}^3$ which is a local isometry that unrolls $M_C \setminus \Sigma$ in $\mathbb{H}^3$. The meridian on $\partial N_r(\sigma)$ lifts to an elliptic transformation E. Any longitude lifts to a loxodromic transformation with the same axis as E. The lift D^* to the universal cover of $M_C \setminus \Sigma$ induces a homomorphism from $\pi_1(M_C \setminus \Sigma)$ to $\mathrm{PSL}(2, \mathbb{C})$.

A cone manifold with rational cone angles $2\pi/n$ is an orbifold—but in general the singular set of an orbifold is not a link (Proposition 2.5.2). The limiting cases of cone angle zero corresponds to a rank one or two cusp while cone angle 2π signifies that there is no singularity at σ.

Assume (i) that $M_C \setminus \Sigma$ has finite volume and (ii) the cone angles along Σ are at most 2π. The fundamental local rigidity theorem of Hodgson and Kerckhoff [1998] asserts that the set of cone angles provides a local parameterization of the hyperbolic cone structures near the given one. In particular there are no infinitesimal deformations of the hyperbolic structure that keep the cone angles fixed. In fact $M_C \setminus \Sigma$ has a complete hyperbolic structure of finite volume [Kojima 1996]. If in addition the angles are at most π, there is a continuous, angle-decreasing family of deformations converging to the complete structure [Kojima 1998]. He also showed that if cone manifolds $M_C \setminus \Sigma$, $M_C' \setminus \Sigma'$ are homeomorphic with corresponding cone angles the same (and all at most π), the two manifolds are isometric.

One natural source of cone manifolds, used for example in Theorem 3.11.6, is the reflection in its boundary of a convex core for which the bending locus consists of simple geodesics. The bending lines become the singular locus and the cone angles are twice the interior bending angles. In this case the cone angles are less than 2π so the Hodgson-Kerckhoff deformation theory is operative.

For a recent study of cone manifolds with singular locus the union of circles and trivalent graphs (as with orbifolds), with cone angles $\leq \pi$, see Weiss [2004]. For more discussion see Exercise (6-3).

4-8. *Geometric limits of fuchsian groups.* In the surface $R = \mathbb{D}/G$ let γ be a simple loop cutting R into two components R_1, R_2. Let $R_1^* \subset \mathbb{D}$ be a lift of R_1 and let R_2^* be a lift of R_2 such that R_2^* is adjacent to R_1^* along a lift γ^* of γ. Let $G_i \subset G$ denote the stabilizer of R_i^*, $i = 1, 2$ and g^* a generator of the stabilizer of γ^*. Show that $G = \langle G_1, G_2 \rangle$ (in the language of combinatorial group theory, G is the free product of G_1 and G_2 with amalgamation over the cyclic subgroup that stabilizes γ^*). There is a general way to find a sequence of isomorphisms $\theta_n : G \to H_n$ to other fuchsian groups $\{H_n\}$ such that $\lim \theta_n(g^*)$ is parabolic and $\lim \theta_n(g)$ exists as a Möbius transformation for all $g \in G_1$. On the other hand for $g \in G_2$ with $g(\gamma^*) \neq \gamma^*$, no subsequence of $\theta_n(g)$ converges to a Möbius transformation, that is, the regions $\{R_{2,n}^*\}$ shrink to the fixed point of $\lim \theta_n(g^*)$.

Degeneration can be executed as follows. First note that a sequence of increasingly thick annuli can always be normalized to converge to a once punctured disk, for example the sequence $\{1/n < |z| < 1\}$. Cut R along γ and sew back in increasingly thick annuli thereby getting new surfaces $R_n = R_{1,n} \cup R_{2,n}$. Apply the Uniformization Theorem to R_n get a new fuchsian group H_n and an isomorphism $\theta_n : G \to H_n$. We are free to replace H_n by a conjugate; do so that for all n, the lift $R_{1,n}^*$ contains a fixed small disk about 0 that embeds in the quotient. The element $\theta_n(g^*) \in H_n$ and all its conjugates with respect to $H_{1,n}$ converge to parabolic transformations. The groups $\theta_n(G_1)$ converge algebraically to a fuchsian group representing a surface homeomorphic to R_1 but with a puncture in place of the boundary component γ. The subgroup of H_n that preserves any lift of R_2^n, or any lift of R_1^n other than R_1^{*n}, degenerates.

The geometric limit of $\{H_n\}$ is the algebraic limit of $\{H_{1,n}\}$. Note the role of the choice of basepoint as the focus of conjugation. One could have chosen it so that the convergent sequence was instead $\{H_{2,n}\}$. This is only one example in a complete description of all geometric limits of sequences of fuchsian groups. This process was first described by Bill Harvey [1977]. Can you formulate a general theorem describing all possible geometric limits?

Now consider the kleinian case. On the boundary of a geometrically finite $\mathcal{M}(G)$, suppose $\mathcal{S}$ is a set of mutually disjoint, noncompressing simple loops with the property that no two are freely homotopic within $\mathcal{M}(G)$.

Deformation theory allows the deformation of the group so that all the curves of the set $\mathcal{S}$ become "pinched"; the corresponding elements of G become parabolic. The example of the two surfaces resulting from a fuchsian group shows why the homotopy property is a necessary condition to carry this out. There results a new geometrically finite group H in various interesting combinatorial arrangements depending on $\mathcal{S}$. Show that a genus two Schottky group can be so pinched as to become a pair of once-punctured tori. See the Pinching Theorem (p. 337) and Exercise (5-5). Each once-punctured torus can be pinched at most once again in countably many ways so as to become a thrice-punctured sphere. For an elementary and detailed presentation of the two-generator Schottky case, see Mumford et al. [2002].

4-9. *Isomorphisms determining homeomorphisms.* Suppose G is a geometrically finite group without parabolics and $\mathcal{M}(G)$ is acylindrical. If Ω_1, Ω_2 are distinct components of $\Omega(G)$ and $G_i = \text{Stab}(\Omega_i)$, prove that $\Lambda(G_1) \cap \Lambda(G_2) = \varnothing$ (see Matsuzaki and Taniguchi [1998, §3.2.1, Theorem 3.29]). Conclude that if $\varphi : G \to H$ is an isomorphism to another geometrically finite group without parabolics then $\mathcal{M}(H)$ is also acylindrical and there is a bijection between the components of $\Omega(G)$ and $\Omega(H)$ (see Exercise (3-32)).

Hint: Assume to the contrary that $\Lambda(G_1) \cap \Lambda(G_2) \neq \varnothing$. Take a ray ℓ in the convex hull $\mathcal{CH}(G_2)$ of $\Lambda(G_2)$ ending at $\zeta \in \Lambda(G_1) \cap \Lambda(G_2)$. Given $O \in \mathbb{H}^3$ there exists $\{g_n \in G_1\}$ such that $\{g_n(O)\}$ lies in a conical neighborhood of ℓ. Project the sequence into $\mathcal{M}(G_2)$. It is contained in a finite distance neighborhood of the convex core $\mathcal{CC}(G_2)$ of $\mathcal{M}(G_2)$. Therefore there is a sequence $h_n \in G_2$ such that (for a subsequence) $\lim h_n g_n(O) = O' \in \mathbb{H}^3$. Discreteness requires that $h_n g_n = \text{id}$ for all large indices. Therefore $h_n \in G_1 \cap G_2$, a contradiction.

4-10. *Alternate definitions of geometric convergence.* A quantitative version of the definition of Section 4.2 goes as follows. Suppose we have a sequence $\{G_n\}$ and a point $O \in \mathbb{H}^3$ such that the corresponding fundamental polyhedra $\{\mathcal{P}_n\}$ with origins at O all contain small ball B_ϵ about O with $g_n(B_\epsilon) \cap B_\epsilon = \varnothing$ for all $g_n \neq \text{id} \in G_n$ and all indices n.

The sequence converges geometrically to the group H if and only if the following holds: There exists a sequence of K_n-bilipschitz maps $F_n : \mathbb{H}^3 \to \mathbb{H}^3$, $F_n(O) = O$, with $\lim K_n = 1$, which have the following two properties. (i) On every compact subset of $\mathbb{H}^3$, $\lim F_n(x) = x$. (ii) Choose a sequence $r_n > 0$, $\lim r_n = \infty$. Set $M_n = (\mathcal{P}_n \cap B_{r_n})/G_n$, where B_{r_n} is the ball of radius r_n about O. Then F_n projects to a K_n-bilipschitz map f_n of M_n into $\mathbb{H}^3/H$, see Canary and Minsky [1996, Lem. 3.1].

From a more general perspective, Gromov presented the following definition of geometric convergence [Gromov 1981b].

We are given two metric spaces X, Y and a map $f : X \to Y$. Using the respective metrics set

$$L(f) = \sup_{x_1 \neq x_2 \in X} \left| \log \frac{d(x_1, x_2)}{d(f(x_1), f(x_2))} \right|.$$

A sequence of metric spaces $\{(X_n; O_n)\}$ with basepoints O_n is said to *converge* to the metric space with basepoint $(Y; O)$ if the following holds. Given any $r > 0$

and $\varepsilon > 0$, there exists N such that for each $n \geq N$, there exists a map f_n from the radius-r ball $B_r(O_n) \subset X_n$ into Y such that

(i) $f_n(O_n) = O$,
(ii) $f_n(B_r(O_n)) \subset Y$ contains the ball $B_{r-\varepsilon}(O) \subset Y$, and
(iii) $L(f_n) \leq \varepsilon$, computed on $B_r(O_n)$.

For an application to our situation, set $X_n = [\mathcal{P}_O(G_n) \cap B_r(O)]/G_n$ and $Y = \mathbb{H}^3/H$.

4-11. $\mathbb{R}$-*trees*. Another way of representing the degeneration of hyperbolic manifolds is by $\mathbb{R}$-trees. This point of view was pioneered by Morgan and Shalen [1984; 1988a; 1988b]. (See also references in Bestvina [1988, 2008]; Ohshika [1998b]; Otal [1996]). Here we will define $\mathbb{R}$-trees and in the next exercise show how they arise in degenerations of kleinian groups.

A metric space $\mathcal{T} = (X, d)$ is called a *real tree* or $\mathbb{R}$-*tree* if there is a *unique arc* (up to parameter change) connecting any two points $x, y \in X$ and that arc has length $d(x, y)$.

Thus if $[a, b, c] \subset X$ is a triangle, each side is contained in the union of the other two. This is called the *tripod property*. The center of the tripod is the unique point $[a, b] \cap [b, c] \cap [c, a]$. Conversely suppose a metric space (Y, d) has the properties that (i) there is an arc of length $d(x, y)$ between any two points $x, y \in X$, and (ii) the tripod property holds for any geodesic triangle. Then there is a unique arc between any two points, and (Y, d) is an $\mathbb{R}$-tree.

For example, let $X \subset \mathbb{C}$ be the union of three rays from the origin and let d be the metric on X induced from $\mathbb{C}$. Then X is an $\mathbb{R}$-tree.

Another example comes from taking the "dual graph" of a measured lamination (Λ, μ) in $\mathbb{H}^2$, the lift of one on a closed surface S of genus exceeding one [Otal 1996, §2.3]. For simplicity we will assume that Λ is minimal without closed leaves.

The points of $\mathcal{T}$ will be of two types: (i) the closure of a component (gap) of $\mathbb{H}^2 \setminus \Lambda$, and (ii) a leaf $\lambda \subset \Lambda$ that is not in such a closure. Define the distance between the points as follows. Suppose $x, y \in \mathcal{T}$ are points that correspond to two gaps. Choose points, also denoted x, y, in each gap and consider the geodesic segment $[x, y]$ between them. Then the positive number $\mu[x, y]$ is the transverse measure of the segment. "Integration" of this measure determines a distance between the closed

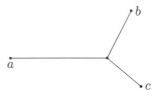

Fig. 4.2. A tripod.

subsets that intersect $[x, y]$. This definition is independent of the choice of x, y in their gaps. The distance between two points that lie on leaves is defined in terms of the gaps that separate them, and correspondingly for a point on a leaf and one in a gap. Thus the distance $d(\cdot, \cdot)$ can be defined between any pair of points of $\mathcal{T}$.

With this distance, $\mathcal{T}$ is an $\mathbb{R}$-tree. If (Λ, μ) is invariant under the action of a fuchsian group G, the action of G determines a fixed-point-free isometry of $\mathcal{T}$. The action is also minimal. Conversely, we have the following basic theorem of Skóra (see Otal [1996, Theorem 2.3.5]):

Theorem 4.11.1. *Suppose $G \times Y \to Y$ is a nontrivial isometric, minimal action of the fuchsian group G on the $\mathbb{R}$-tree Y. Assume that (i) every subgroup of G that fixes an arc of Y has a finite index abelian subgroup, and (ii) the action of parabolic elements of G have translation distance zero in Y. Then the action of G on Y is isometric to the action of G on the tree $\mathcal{T}$ determined by a measured geodesic lamination in $\mathbb{H}^2$.*

4-12. *$\mathbb{R}$-trees and the degeneration of manifolds.* Because of its significance in the general theory of group deformations, we will provide a somewhat lengthy outline of the approach in Bestvina [1988] to show how degeneration arises in an $\mathbb{R}$-tree. This approach has proved useful in establishing hyperbolization for fibered 3-manifolds [Otal 1996; Kapovich 2001]. First we will describe Gromov-convergence of a sequence of metric spaces.

A sequence $\{Y_n\}$ of compact, connected metric spaces is said to converge in the *Gromov-Hausdorff sense* to the metric space Y if the following holds. There is a compact metric space Z and isometric embeddings $Y_n \hookrightarrow Z$ and $Y \hookrightarrow Z$ so that as subsets of Z, $\{Y_n\}$ converges to Y in the Hausdorff topology. That is, in terms of the images in Z, given any $\varepsilon > 0$ there exists $n(\varepsilon)$ so that when $n > n(\varepsilon)$, Y is contained in the ε-neighborhood of Y_n and Y_n is contained in the ε-neighborhood of Y. Gromov showed that the limit is uniquely determined up to isometry.

According to Gromov [1981a, §7], a necessary condition on the metric spaces for there to exist a Gromov-Hausdorff convergent subsequence is as follows: To every $\varepsilon > 0$, there is an integer $N(\varepsilon)$ such that *every Y_n can be covered by $N(\varepsilon)$ ε-balls.*

It is interesting to see Gromov's construction of Z. Set $\varepsilon_i = 2^{-i}$. For each i there exists an integer N_i so that *every Y_n is covered by N_i ε_i-balls.* For each i, consider the i-tuple of integers $A_i = \{(n_1, n_2, \ldots, n_i) : 1 \leq n_j \leq N_j, \ 1 \leq j \leq i\}$. As $i \to \infty$, $\varepsilon \to 0$ and $N_i \to \infty$.

For each index n, define inductively a sequence of maps of $A_1, A_2, \ldots, A_k, \ldots$ into Y_n as follows:

(1) Cover Y_n by N_1 ε_1-balls and then choose any one-to-one map I_n^1 from the set of N_1-integers $A_1 = \{n_1\}$ to the N_1 centers of the N_1 ε_1-balls that cover Y_n.
(2) From the N_2 ε_2-balls that cover Y_2, choose subsets that cover each ε_1-ball. Denote by I_n^2 the map of the set A_2 of integer pairs to the set of centers of the ε_2-balls such that (n_1, n_2) goes to the center of an ε_2-ball used to cover the ε_1-ball centered at $I_n^1(n_1)$.

(3) In turn, from the N_3 ε_3-balls that cover Y_3, choose subsets that cover each ε_2-ball. Denote by I_n^3 the map of A_3 onto the centers of the ε_3-balls such that (n_1, n_2, n_3) goes to the center of an ε_3-ball used to cover the ε_2-ball centered at $I_n^2(n_1, n_2)$. And so on. As $i \to \infty$, the image sets $\{I_n^i(A_i)\}$ become more and more dense in each Y_n.

Set $A = \bigcup_1^\infty A_k$ and let $I_n : A \to Y_n$ denote the corresponding map. Consider the metric space of maps $B = \{f : A \to \mathbb{R} : f \text{ is bounded}\}$ with metric determined by the norm $\|f\| = \sup_{a \in A} |f(a)|$. Let $Z \subset B$ be the compact metric subspace of those functions satisfying

$$\text{if} \quad a \in A_1, \qquad \text{then} \quad 0 \le f(a) \le \sup_n\{\text{Diam} Y_n\};$$
$$\text{if} \quad a \in A_k, \ k > 1, \quad \text{then} \quad |f(a) - f(p_{k-1}(a))| \le 2\varepsilon_{k-1},$$

where $p_k : A_{k+1} \to A_k$ is the natural projection. The construction is such that for each $a \in A_k$, $I_n^k(a)$ is contained in the $2\varepsilon_{k-1}$-ball centered at $I_n^{k-1}(p_{k-1}(a))$.

Define $h_n : Y_n \to B$ by

$$(h_n(x))(a) = \text{dist}(x, I_n(a)), \quad x \in Y_n, \ a \in A;$$

here "dist" is the metric in Y_n. Verify that the image of h_n is contained in Z and that h_n gives an isometric embedding of Y_n.

Finally, the space of all compact subsets of a compact set is itself compact with respect to the Hausdorff topology, so there is a convergent subsequence of $\{Y_n\}$.

Now we apply Gromov's construction to our situation.

Suppose G is a (nonelementary, finitely generated) kleinian group. Fix a set of generators $\{g_1, g_2, \ldots, g_r\}$ of G. Let $\{\theta_n : G \to G_n\}$ be a sequence of isomorphisms to discrete groups. Renormalize if necessary so that a given basepoint point $O \in \mathbb{H}^3$ is moved least by these generators in the sense that

$$d_n = d(G_n) = \max_{1 \le i \le r}\{d(O, \theta_n(g_i)(O))\} \le \max_{1 \le i \le r}\{d(\vec{x}, \theta_n(g_i)(\vec{x}))\}$$

for all $\vec{x} \in \mathbb{H}^3$. Here $d(\cdot, \cdot)$ denotes hyperbolic distance. Show that if the sequence $\{d_n\}$ is uniformly bounded, a subsequence can be chosen so that $\{\theta_k\}$ converges to an isomorphism θ.

Since we want to study degenerations, assume that $\lim d_n = \infty$.

Let $\mathcal{W}^k$ denote the set of words in the given generators of G which have length $\le k$. Let $\mathcal{F}_n^k$ denote the convex hull in $\mathbb{H}^3$ of the point set $\{\theta_n(g)(O) : g \in \mathcal{W}^k\}$.

Now rescale $\mathcal{F}_n^k$: let $\mathcal{X}_n^k$ denote the abstract metric space whose set of points is $\mathcal{F}_n^k$ but the distance between two points is rescaled as $\rho_n(x, y) = d(x, y)/d_n$, for $x, y \in \mathcal{F}_n^k$. Thus $\rho_n(x, y) \le 2$.

Theorem 4.11.2. *For each k, there is a subsequence of $\{\mathcal{X}_n^k\}$ which in the Gromov sense converges to a compact metric space T^k.*

Proof. Given $\varepsilon > 0$ we will verify that there exists $N(\varepsilon)$ such that each $\mathcal{X}_n^k$ can be covered by $N(\varepsilon)$ ε-balls, or equivalently, $\mathcal{F}_n^k$ can be covered by $N(\varepsilon)$ $d_n\varepsilon$-balls. If

$\mathcal{W}^k$ has $W(k)$ elements, then $\mathcal{F}_n^k$ has $W(k)$ vertices and $\frac{1}{2}W(k)(W(k)-1)$ diagonals (geodesic segments between distinct vertices). The length of each diagonal cannot exceed $2kd_n$. Cover the diagonals by εd_n-balls centered at points on the diagonals spaced at distance $\leq \varepsilon d_n$. This requires at most

$$\frac{W(k)(W(k)-1)}{2}\left(\left[\frac{2kd_n}{\varepsilon d_n}\right]+1\right) = \frac{W(k)(W(k)-1)}{2}\left(\left[\frac{2k}{\varepsilon}\right]+1\right)$$

balls.

Now, by Exercise (1-24), there exists a constant $C > 0$ such that for each $\vec{x}$ in any triangle or in any tetrahedron in $\mathbb{H}^3$, the shortest distance of $\vec{x}$ to the edges does not exceed C. Therefore the collection of balls covers $\mathcal{F}_n^k$ if $\varepsilon d_n \geq 2C$, since each point of $\mathcal{F}_n^k$ lies in a tetrahedron whose vertices are among the vertices of $\mathcal{F}_n^k$.

Therefore, after passing to a subsequence if necessary and taking a diagonal subsequence, we can arrange matters so that (1) $\mathcal{X}_n^k \to T^k$ in the Gromov sense, (2) $\cdots \subset T^{k-1} \subset T^k \subset T^{k+1} \subset \cdots$, and (3) for $g \in \mathcal{W}^k$, $\lim_{n\to\infty} \theta_n(g) = x^*(g) \in T^k$ exists. $\qquad\square$

In short, the rescaled diagonals of $\mathcal{F}_n^k$ converge to segments or points in an ambient compact metric space and the rescaled convex hulls collapse upon them. More precisely:

Theorem 4.11.3. T^k *is a finite $\mathbb{R}$-tree, meaning that*:

(1) *Any two points of T^k can be joined by a segment, that is, a subspace isometric to a closed interval or a point. The segments can be chosen with endpoints in the set $\{x^*(g) : g \in \mathcal{W}^k\}$.*
(2) *The intersection of any two nondisjoint segments in T^k is a segment or point.*
(3) *The union of two segments with a common endpoint is again a segment.*

That T^k is finite means it is the union of finitely many segments.

The diagonals in $\mathcal{F}_n^k$ give rise to segments in $\mathcal{X}_n^k$ that converge to segments or points in T^k; the endpoints converge to points $x^*(e) \in T^k$, $e \in \mathcal{W}^k$. We claim that the limiting segments and points cover T^k. If not, for some $x \in T^k$ and some ε, the ε-ball about x does not intersect any of the limiting segments or points. Suppose $x_n \in \mathcal{X}_n^k$ converges to x. For large n the $\mathcal{X}_n^k$-distance between x_n and the rescaled diagonals exceeds $> \varepsilon/2$. In $\mathcal{F}_n^k$, this distance exceeds $\varepsilon d_n/2$. Since we have taken ε so that $\varepsilon d_n \geq 2C$, the corresponding point $x_n \in \mathcal{F}_n^k$ lies in an εd_n-ball centered at a point on the diagonal, a contradiction.

For the remainder of the proof see Bestvina [1988].

Now set $T = \bigcup_k T^k$. Then T is a metric space with the three properties of the theorem above. Therefore it is an $\mathbb{R}$-tree. The theory continues by studying the action of G on T. The basic result is that it acts by isometries so that $gz(h) = z(gh)$ for $z \in T$ and $g, h \in G$, where $z(g) = \lim \theta_n(g)(z_n)$ and $z = \lim z_n$. Deeper study leads to a compactification of the space of algebraic limits and a proof of Thurston's compactness theorem (p. 240).

4-13. *The isoparametric inequality for* $\mathbb{H}^3$. As stated in Chavel [1993, §6.4], and referring back to the formulas on p. 17,

Theorem 4.11.4. *Suppose* $X \subset \mathbb{H}^3$ *is a compact set with piecewise smooth boundary. Then measured in the hyperbolic metric,*

$$\frac{\mathrm{Area}(\partial X)}{\mathrm{Volume}(X)} > 2.$$

The smallest ratio is attained when X *is a ball. The ratio 2 is reached only in the limiting case that* X *is a horoball.*

A variation (suggested by Cooper) of an argument of Cooper [1999] yields the following interesting fact.

Proposition 4.11.5. *Suppose* $\mathcal{M}(G)$ *is a closed manifold. There exists a presentation of* G *consisting of* m *generators and* $\frac{2}{5}(m+1)$ *relations for which* $\mathrm{Vol}(\mathcal{M}(G)) < \frac{2}{5}\pi(m+1)$.

The *presentation length* of a group G is defined by Cooper to be $\ell(G) := \pi(L - 2n)$, where L is the sum of the word lengths of the relations in G, and n is the number of relations of word length at least 3 (relations of less word length are exceptional). For a closed manifold $\mathcal{M}(G)$ we have [Cooper 1999],

$$\mathrm{Vol}(\mathcal{M}(G)) < \ell(G).$$

In Agol and Liu [2012] it is shown there exists a universal constant C such that for any closed manifold,

$$\mathrm{Diam}(\mathcal{M}(G)) < C\ell(G).$$

Proof of Proposition 4.11.5. Choose a generic polyhedron $\mathcal{P}_\mathcal{O}$ such that each vertex is shared by three edges and each edge relation has length three, that is, three polyhedra in the orbit share the edge.

Construct the dual graph Γ to the orbit of $\mathcal{P}_\mathcal{O}$. Recall this is done by connecting the center $\mathcal{O}$ to the centers of the polyhedra of the orbit that share a face with $\mathcal{P}_\mathcal{O}$, and so on. The vertices of Γ are the points of the orbit $G(\mathcal{O})$.

Each piecewise geodesic loop in Γ that surrounds a single edge is a geodesic triangle transverse to faces of the orbit of $\mathcal{P}_\mathcal{O}$. If two edges e, e' of $\mathcal{P}_\mathcal{O}$ are related $e' = g(e)$ for $g \in G$ then the triangle about e' is the g-image of the triangle about e.

The number of triangles in Γ with vertex $\mathcal{O}$ is equal to the number s of edges of $\mathcal{P}_\mathcal{O}$. The number of triangles which are inequivalent under G is $s/3$. Corresponding to each vertex of $\mathcal{P}_\mathcal{O}$ is a cone in $\mathcal{P}_\mathcal{O}$ with vertex $\mathcal{O}$. Its three faces with vertex $\mathcal{O}$ geodesic triangles.

The union of all these triangles $\{\Delta\}$ separates $\mathbb{H}^3$ into simply connected polyhedra $\{X\}$ each of which contains exactly one vertex in the orbit of the vertices of $\mathcal{P}_\mathcal{O}$. Each component X projects injectively into $\mathcal{M}(G)$.

Since $\mathcal{P}_{\circlearrowleft}$ is a compact convex polyhedron, its boundary is a topological sphere. Euler's formula for the boundary is $E - F + V = 2$. The number of vertices V is related to the number of edges E by $V = 2E/3$. The face pairing transformations generate the group there are are $m = F/2$ of these. Therefore $E = 6(m + 1)/5$. There are E edge relations, but the edge cycles all have length three leaving $E/3 = 2(m + 1)/5$ independent relations.

Now we bring in the fact that the surface area of each component X exceeds twice its volume. Each geodesic triangle in ∂X lies in the boundary of exactly two of the polyhedra $\{X\}$. Down in $\mathcal{M}(G)$ count the distinct projections $\{\pi(X)\}$, which fill up $\mathcal{M}(G)$ without overlap. The area of ∂X exceeds twice the volume of X by Theorem 4.11.4. There are $N = E/3$ distinct triangles $\{\pi(\Delta)\}$. The sum of the area of the distinct $\{\pi(\Delta)\}$ exceeds the volume of $\mathcal{M}(G)$, since each $\pi(\Delta)$ borders two of the components $\{X\}$ and is counted twice when totaling up the volume. The area of each triangle is less than π. The sum of the areas of the distinct triangles is therefore bounded above by πN, where $N = \frac{2}{5}(m + 1)$ is the number of independent edge relations of $\mathcal{P}_0$. $\qquad\square$

4-14. *The Gromov norm.* Let $\mathcal{M}$ be a closed hyperbolic 3-manifold. Gromov showed that its volume can be obtained as the limit of a process of approximation by 3-chains:

We consider all continuous maps $\{\sigma\}$ of a standard euclidean regular tetrahedron Δ into $\mathcal{M}$. The theory allows us to assume that $\sigma(\Delta)$ is a hyperbolic tetrahedron. These maps are a basis of the vector space of singular real chain complexes $\{\sum a_i\sigma_i\}$ in $\mathcal{M}$, $a_i \in \mathbb{R}$. The most natural source of 3-chains are triangulations of M^3 by tetrahedra (these are easily constructed from a fundamental polyhedron).

Recall from Exercise (1-23) that among all tetrahedra, the regular ideal tetrahedron has the maximum volume.

Gromov's norm is a seminorm on the real singular homology $H_3(\mathcal{M}; \mathbb{R})$ (here we are only considering the 3-homology). Let $[\mathcal{M}]$ denote the third homology class of the whole manifold (the fundamental class). Consider all singular 3-cycles c that represent this class, for example, the cycles coming from triangulations. We can express c as $c = \sum a_i\sigma_i$. Define $|c| = \sum |a_i|$. Then define the *Gromov norm* of $\mathcal{M}$

$$\|\mathcal{M}\| = \inf\{|c| : c \text{ is a singular cycle representing } [\mathcal{M}]\}.$$

(See Thurston [1979a, §6.1] or Benedetti and Petronio [1992, §C.3].) The Gromov norm has the property that for any continuous map $f : \mathcal{M} \to \mathcal{M}_1$,

$$\|\mathcal{M}\| \geq |\deg f| \|\mathcal{M}_1\|.$$

Thus if $\mathcal{M}_1 = \mathcal{M}$ and $\|\mathcal{M}\| \neq 0$, $|\deg f| = 0$, or 1.

Gromov's Theorem says that

$$\|\mathcal{M}\| = \frac{\mathrm{Vol}(\mathcal{M})}{V_3},$$

where $V_3 = 1.01294\cdots$ is the volume of the regular ideal tetrahedron.

This is an remarkable formula. One can view it as showing that the volume of $\mathcal{M}$ a topological invariant, or as giving a topological interpretation of the volume of closed hyperbolic manifolds. The same formula holds for hyperbolic n-manifolds, $n \geq 2$. (Prove it for $n = 2$, when the volume is $4\pi(g - 1)$ and $V_2 = \pi$.)

For an orientable, closed manifold M^3, not necessarily hyperbolic, the number $V_3 \Vert M^3 \Vert$ is called the *simplicial volume* of M^3.

Can the theory be extended to cusped manifolds of finite volume?

4-15. *The space of geodesics.* Show that the space of geodesics in $\mathbb{H}^2$, in the topology coming from the Hausdorff metric, is homeomorphic to the quotient $(\mathbb{S}^1 \times \mathbb{S}^1 \setminus \delta)/\langle J \rangle$, where J is the involution $(\zeta_1, \zeta_2) \to (\zeta_2, \zeta_1)$ and $\delta = \{(\zeta, \zeta), \ \zeta \in \mathbb{S}^1\}$ is the diagonal.

Show that the quotient space is in turn homeomorphic to an *open* Möbius band. *Hint*: Represent $\mathbb{S}^1 \times \mathbb{S}^1 \setminus \delta$ as a square torus less a simple loop representing the diagonal: take the torus to be the quotient of a square lattice, and in a fundamental square represent δ as a diagonal. Let Δ_1, Δ_2 denote the resulting triangles and J the reflection $J : \Delta_1 \leftrightarrow \Delta_2$ in δ. Note that Δ_1 is a fundamental domain for the torus group augmented by J. Label the sides s_1, s_2 of Δ_1 and s_1', s_2' of Δ_2 where $J : s_i \leftrightarrow s_i'$. Under the augmented group a pair of points on side s_1 is equivalent to a pair on s_2, in the opposite order. Using this correspondence, glue the side s_1 to s_2. This forms a Möbius band.

4-16. *Circle packings I.* This is to report on Bob Brooks' important papers 1985; 1986. We start by noting that three circles each externally tangent to the other two, bound a circular triangle—an ideal triangle (actually two of them, one containing ∞). In $\mathbb{S}^2$ such a configuration is uniquely determined up to Möbius equivalence.

Next consider four circles, each externally tangent to exactly two others. They bound a circular quadrilateral Q. Two such configurations are generally not Möbius equivalent. Recall that two rectangles of widths a, a' and heights b, b' are Möbius equivalent in such a way that the horizontal sides and vertical sides correspond if

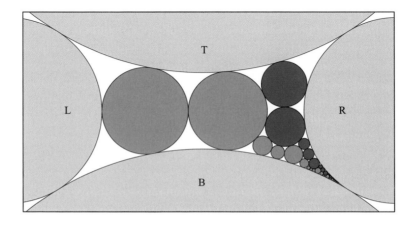

Fig. 4.3. Computation of the modulus of a marked quadrilateral.

and only if $a/b = a'''/b'$. Brooks discovered an analogous modulus for circular quadrilaterals:

Designate one pair of opposite sides of Q as *horizontal* and the other as *vertical*. Order the sides as top, bottom, right and left. This makes Q into a *marked quadrilateral*. A circle bounding a disk in Q will be called *horizontal* if it is tangent to the left vertical side and the two horizontal sides; it will be called *vertical* if it is tangent to the top horizontal side and to the right and left vertical sides.

Suppose, for example, we can insert a horizontal circle. Its exterior in Q consists of two circular triangles and perhaps another quadrilateral Q_1. Insert another horizontal circle in Q_1, if possible. Continue the process of inserting horizontal circles until after $n_1 \geq 1$ steps we can no longer do so (this process must stop after a finite number of times). Then with the remaining quadrilateral Q_{n_1}, start inserting vertical circles until that is no longer possible. Say the number of vertical circles is $n_2 \geq 1$. Then again start inserting horizontal ones. And so on. Form the continued fraction

$$r(Q) = n_1 + \cfrac{1}{n_2 + \cfrac{1}{n_3 + \cdots}};$$

it converges to a positive real number $r(Q)$ which is rational if and only if the sequence $\{n_i\}$ terminates after a finite number of terms.

Theorem 4.11.6. *Two marked quadrilaterals Q, Q' are Möbius equivalent consistent with the marking if and only if $r(Q) = r(Q')$.*

The set of marked Q with finitely many circles is dense in the configuration space of all marked quadrilaterals Q.

Next consider the case of a geometrically finite group G without parabolics. By a circle packing in $\Omega(G)$ we will mean a G-invariant collection of round circles with mutually disjoint interiors which projects to $\partial\mathcal{M}(G) = \Omega(G)/G$ to give a finite packing by simple curves, which we will again call circles, bounding mutually disjoint regions. Add additional circles as required so as to arrive at a G-invariant packing in $\Omega(G)$, finite in the union of closed surfaces comprising $\partial\mathcal{M}(G)$, with the property that all *interstices* are either triangles or quadrilaterals. This will now be our definition of circle packings.

Each circle packing has a *nerve* which is the graph obtained by associating each circle with a vertex and each pair of tangent circles with an edge.

From the point of view of $\partial\mathcal{M}(G)$ there are a finite number of quadrilaterals. Mark the quadrilaterals and compute their moduli $\{r(Q)\}$.

Theorem 4.11.7 (Brooks). *Suppose G, G' are geometrically finite (torsion free) groups. Assume there is an isomorphism θ between nerves of circle packings on $\partial\mathcal{M}(G)$ and $\partial\mathcal{M}(G')$ such that corresponding marked quadrilaterals have the same moduli. Then G and G' are conjugate; $\mathcal{M}(G)$ and $\mathcal{M}(G')$ are isometric.*

The group G may be quasiconformally deformed an arbitrarily small amount to a new group G_ϵ which has an isomorphic circle packing such that the moduli of *the corresponding quadrilaterals on $\partial \mathcal{M}(G_\epsilon)$ are rational. Therefore G_ϵ has a circle packing such that all interstices are triangles, and $\partial \mathcal{M}(G)$ has a finite circle packing.*

Theorem 4.11.8 (Brooks). *Given $\varepsilon > 0$, the ε-quasiconformal deformation G_ε is a subgroup of a finite volume group Γ_ϵ^*.*

Also,

$$G_\varepsilon \subset \Gamma^* \subset \Gamma^{**}$$

where Γ^{**} is a closed manifold.

Here are some remarks concerning the proofs. First assume that G has no parabolics. Given a circle packing of $\Omega(G)$, the group Γ generated by G plus the reflections in the hyperbolic planes supported by the circles is geometrically finite. (It suffices to consider the orientation-preserving index two subgroup.) The deformation space of Γ depends on one real parameter for each quadrilateral. Small changes in the parameter can be realized by a quasiconformal map with small dilatation that maps the circle packing onto a combinatorially identical one close by.

Once we get a circle packing such that the moduli of all quadrilaterals is rational, we can add circles to get a larger packing such that all interstices are triangles. Denote by Γ_ϵ the group generated by G_ϵ and reflection in all the planes supported by circles. Γ_ϵ too is geometrically finite.

In Ω_ϵ, put a new circle through the three vertices of each interstitial triangle. It is orthogonal to the three forming the triangle. Then form the even larger group Γ_ϵ^* generated by Γ_ϵ and reflections in the hyperbolic planes supported by the new circles. Again we may consider the orientation-preserving subgroup, which is of index two. It too is geometrically finite. There are only a finite number of points which, mod Γ_ϵ^*, are not contained in the interior of any disk; these become rank two parabolic fixed points. The corresponding manifold has finite volume.

So far the proof can be adapted to apply when there are parabolics in G. The circle packing of $\Omega(G)$ is to include horodisks about the punctures.

We now have the finite volume group Γ_ϵ^*, where we have replaced the original by its orientation-preserving subgroup, now of index four in the group with all the circle reflections. The only parabolics are those associated with the rank two subgroups which are the artifacts of the construction plus the rank two subgroups of G_ε which are not visible in $\Omega(G)$. Next do Dehn filling on the rank two groups arising in the construction. We obtain Γ_ϵ^{**} whose quotient manifold has finite volume and has only those rank two groups which belong to G. We cannot get rid of these without changing the isomorphism class of G. Our original deformation G_ϵ is a subgroup of Γ_ϵ^{**} without any new parabolics. Because Γ_ϵ^{**} can be taken arbitrarily close to Γ_ϵ^*, a small quasiconformal deformation G_ϵ' of G_ϵ appears as a subgroup of Γ_ϵ^{**}. This completes the argument.

4-17. *Circle packings II.* There is an important technique based on circle packings, motivated by results of Koebe, Andreev, and Thurston, and by a conjecture of Thurston (confirmed in Rodin and Sullivan [1987]). The theory has been extensively

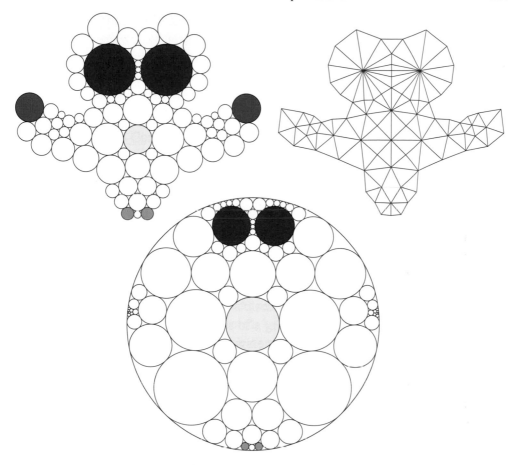

Fig. 4.4. Top: a triangulation of $\mathbb{S}^2$ (right) taken from an owl (left) by drawing edges between the centers of tangent circles and lines to ∞ from boundary circles. Bottom: its realization as a circle packing in the unit disk.

developed into a tool in pure and applied mathematics, especially by Ken Stephenson and colleagues [Stephenson 2005]. We will base our report here on Beardon and Stephenson [1990]; see also the expositions [Stephenson 1999; 2003].

Suppose S is an abstract orientable, closed polyhedral surface, composed of triangles (actually one can begin with a topologically triangulated surface or even an abstract 2-complex). The prototypical example is a topological 2-sphere. There are likely to be vertices of both positive and negative discrete curvature. Denote the graph of edges and vertices by Γ. The basic theorem asserts that there exists a homeomorphic closed Riemann surface R—a sphere, a torus, or a hyperbolic surface—such that Γ is combinatorially isomorphic to the nerve of a circle packing of R. That is, there is a circle packing of R (in the spherical, euclidean or hyperbolic metric) such that the graph (nerve) formed by taking vertices to be the centers of the circles, and edges the geodesic segments between centers of circles, is isomorphic to Γ. Furthermore, R is uniquely determined by the combinatorics of the packing, up to conformal equivalence. For the case $g \geq 2$, the set of Riemann surfaces that can be circle packed is dense in the Teichmüller space [Bowers and Stephenson 1992].

If the original surface S is a region in $\mathbb{C}$, say with fractal boundary, one typically proceeds as follows. Restrict the regular hexagonal packing of $\mathbb{C}$ by circles of radii δ to S. Select the connected component containing a prescribed basepoint O. Form the nerve of the resulting configuration in S and join the boundary vertices to ∞ by arcs, thereby creating a graph on $\mathbb{S}^2$. Construct the corresponding circle packing of $\mathbb{S}^2$. Matters can be normalized so that ∞ corresponds to ∞ and the disk in the packing with center ∞ is the exterior of the unit disk. In effect, the packing of S becomes a packing of the unit disk $\mathbb{D}$. The euclidean circles are hyperbolic circles, or, if tangent to $\partial\mathbb{D}$, they are horocycles. It is proved in Rodin and Sullivan [1987] that as the radius $\delta \to 0$, the quasiconformal simplicial map set up by mapping the nerve of the packing of S to that in $\mathbb{D}$ converges uniformly on compact subsets of S to a Riemann mapping (there needs to be another normalization to account for rotation about O). Many other investigations have followed from this pioneering result.

If S is a surface with boundary, there are two intrinsic methods of assigning boundary values. Start a finite triangulation of S. (i) Assign the value one to each boundary edge, and require that the nerve of the circle packing have the same property. This gives a different shape to the circle packing of a plane region than the process above. (ii) Assign a "radius" to each boundary vertex and require the boundary circles in the resulting packing to have the assigned radii. This can be done in hyperbolic geometry where the packing is done in $\mathbb{D}$ and radius ∞ corresponds to a horocycle. In either case the circle packing is uniquely determined up to Möbius equivalence.

The circle packing method can be described as a process to change combinatorics into geometry! For example, experimental work in neurology uses this method in an attempt to set up a universal coordinate system for the cortex surface of the human brain. The surface of the cerebral cortex is highly convoluted and varies from person to person. Using MRI scans it can be triangulated into many tiny triangles. The circle packing method can be used to replace the triangulated cortex by a circle packing of a plane region. The hope is that this "flattening" of the cortex will make those of different individuals more easily comparable.

There is a theory of packings where the circles have prescribed intersection angles given by a function on the edges of the triangulation [Bobenko and Springborn 2004; Rivin 1996; Thurston 1979a, Chapter 13]; [Bowers and Stephenson 1996]; [Roeder et al. 2007], and see Section 1.2.2. It is closely tied up with the study of hyperbolic polyhedra in $\mathbb{H}^3$, see Section 1.2.2.

4-18. *Right–angled hyperbolic polyhedra.* Start with the fact that given three mutually tangent circles, there is a 3rd circle that intersects the other three at right angles.

Construct a circle packing of $\mathbb{S}^2$ so that all circles have the same radius and all interstices are triangles; there are many choices. Then construct the dual packing of circles orthogonal to the triples in the packing. In turn, erect hemispheres, that is hyperbolic planes, upon the dual circles. Show that the result is a finite volume hyperbolic polyhedron P. Then show there is a kleinian group with P a fundamental polygon.

4-19. *Ideal triangulations; spinning.* Start with a closed hyperbolic surface S of genus $g \geq 2$. Take a family P of $3g - 3$ mutually disjoint simple geodesics forming a *pants* decomposition of $2g - 2$ triply connected regions $\{\mathcal{P}\}$ (see Section 5.3). On each boundary geodesic of P fix one point. Draw 3 mutually disjoint geodesic arcs between the pairs of points, dividing each pants into two simply connected regions. Now "spin" the "triangulation" by applying Dehn twists (Exercise (5-10)) of higher and higher order about the simple geodesics of P. At each stage realize the edges of the auxiliary arcs by geodesic segments. In the limit there will result a geodesic lamination Λ of S. The leaves of Λ will consist of the geodesics of P, plus a finite number of infinite length leaves spiraling about these, three in each pants $\mathcal{P}$. Up in the universal cover, say $\mathbb{D}$, the lifts will give a tessellation by ideal triangles. Each ideal triangle will project to a complementary component of Λ; a finite number will cover S, except for a set of measure zero. In fact there will be exactly $2(2g - 2)$ triangles.

Return again to S with the geodesics giving a pants decomposition P. Suppose $\theta : \pi_1(S) \to \pi_1(\mathcal{M})$ is an injection to some hyperbolic manifold. Each geodesic of P corresponds to a uniquely determined geodesic of $\mathcal{M}$. Even more, each geodesic of Λ corresponds to a unique geodesic of $\mathcal{M}$. Now fill in the spaces with ideal triangles. This results in a pleated surface in $\mathcal{M}$. It is easiest to carry out this construction in $\mathbb{H}^3$ directly—there the collection of ideal triangles is embedded. In visualizing the result in $\mathcal{M}$, recall the uniform injectivity property (p. 173).

Designate a point v on S. We can divide S into $4g-2$ triangles where all the vertices are at v. The triangulation has $6g - 3$ edges. Now puncture S at v. In the hyperbolic metric on the punctured surface, in the same combinatorics we can take all the edges to be geodesics, dividing the surface into $4g - 3$ ideal triangles. (Correspondingly, the universal cover will be tessellated by ideal triangles.)

For example a construction when $g = 1$ is as follows. Represent the torus as a rectangle, with all four vertices corresponding to v. The pair of horizontal and vertical boundary arcs correspond to arcs α, β on S, transverse at v but otherwise disjoint. Now draw the diagonal C in rectangle. It corresponds to a third arc $\gamma \subset S$ disjoint from the other two except at v where it is transverse to the other two. The general surface of genus g with one puncture at v is constructed by starting with a torus and successively introducing additional handles. After each additional handle, carry out the construction as above, with arcs disjoint from the prior ones, except for the common endpoint at v.

Triangulations (not necessarily ideal triangulations) of punctured surfaces are described in Exercises (3-43, 3-44).

Once an ideal triangulation is specified on S it is specified at all points of $\mathfrak{Teich}(S)$. Denote by $\mathcal{C}(S)$ the space of all possible ideal triangulations based on v.

We will now describe how an ideal triangulation on the once-punctured S is determined by a real vector $\vec{x} \in \mathbb{R}_+^{6g-3}$. To each edge e_k associate a positive number as follows: There are two ideal triangles that share the edge e_k. Take the orthogonal projection of the third ideal vertex of each onto e_k. The k-th component of $\vec{x}$ is taken to be e^{x_k} where $x_k \geq 0$ is the distance between the two projections. In fact, once we fix a basepoint, there is a unique left earthquake (Exercise (3-35)) that realizes the described motion.

The vectors $\{\vec{x}\}$ give a coordinate system for $\mathcal{C}(S)$ over $\mathfrak{Teich}(S)$. It is no accident that $(6g - 3)$ is the real dimension of $\mathfrak{Teich}(R)$ up to normalization. It is shown in Bonahon [1996] how to get real analytic coordinates of $\mathfrak{Teich}(R)$ by using these numbers.

According to Jeff Weeks (personal communication) it is "almost surely true" that every finite volume hyperbolic manifold with cusps can be decomposed into positively oriented ideal tetrahedra, but at this writing this remains a conjecture. However it is true that every such manifold has a decomposition into ideal polyhedra. (Exercise (3-5) works this out for the figure-8 knot complement.) Even so, there is no proof that one can subdivide the polyhedra into ideal tetrahedra so that the subdivisions agree on the common faces of adjacent polyhedra [Petronio and Weeks 2000; Weeks 1993; 2005a].

On the other hand, the decomposition can be achieved if the "positively oriented" requirement is dropped to allow some flattened tetrahedra; the flattened tetrahedra appear where the polyhedral subdivisions do not agree on a common face. Here a flattened tetrahedron is a planar quadrilateral whose four ideal vertices lie on a circle in $\mathbb{S}^2$.

SnapPea [Weeks 2012] and its enhancement SnapPy [Culler and Dunfield 2012] (see also Section 6.7) uses such decompositions to numerically approximate the volumes of cusped manifolds, using the formulas recorded in Exercise (1-23). More generally, the program computes hyperbolic structures on manifolds, if such exist. It is a fundamental and productive tool in experimental work on the subject. The process SnapPea uses to search for a hyperbolic structure is as follows. Start with a topological ideal triangulation τ of n simplices of a cusped manifold, for example, a knot complement. Using the combinatorial data implicit in τ, write down a system of equations expressing the fact that a solution allows τ to be homotoped to a (geodesic) ideal triangulation that is compatible with the given hyperbolic structure.

In more detail, a (geodesic) ideal tetrahedron is uniquely determined up to isometry by the cross ratio of its ideal vertices (Exercise (1-22)). To form the manifold, the ideal tetrahedra need to be put together by a sequence of face identifications. The total angle about each edge of the complex needs to be 2π. At the cusps there must be horosphere cross sections so as to become rank two cusps (in more general cases, there is a condition that in effect says that a Dehn surgery must be undertaken). The bottom line is that there results a finite number of algebraic conditions to be satisfied by the tetrahedra if the required identification can be accomplished in $\mathbb{H}^3$. If SnapPea finds a solution, it is shown to be mathematically correct using Snap [Goodman 2006], another software tool that uses exact arithmetic. See [Neumann 1999; Coulson et al. 2000; Neumann and Reid 1992] and Section 6.6 for more details. As indicated above, there may be some complications in the construction. See also Exercise (3-5).

There is a more complicated procedure used for closed hyperbolic manifolds.

4-20. Prove that $\text{Int}(\mathcal{M}(G))$ has finite volume if and only if the injectivity radius $\text{Inj}(x) \to 0$ as any $x \in \text{Int}(\mathcal{M}(G))$ approaches the ideal boundary. See also Theorem 3.11.2.

4-21. *Simple loops in $\mathcal{M}(G)$; primitive curves.* What should it mean that $\gamma \subset \mathcal{M}(G)$ is a simple loop, assuming γ is not retractable to a point or a cusp? On one level the answer is obvious: Every closed curve can be deformed in space so it becomes a simple loop. But on a deeper level we may want to use the criterion that arises as follows: Given a closed curve γ find the corresponding geodesic γ'. Let $g \in G$ be the loxodromic element determined by a lift of γ' to $\mathbb{H}^3$. The stabilizers of the pair of fixed points of g determine a cyclic subgroup. Is g a generator of this cyclic group? Or, is there another element g_1 of it such that $g = g_1^n$ with $n \geq 2$? If g is indeed a generator, the curve γ is called *primitive*. Even if γ is freely homotopic to a simple loop on $\partial\mathcal{M}(G)$, it may not be primitive.

Here is an example. In Exercise (4-28) we will show that there is a $\mathcal{M}(G)$ with a single rank two cusp and an essential cylinder C with one boundary component on a cusp torus and the other in $\partial\mathcal{M}(G)$. Let $c \subset C$ be a central curve. Let c' denote the freely homotopic curve on the solid torus. Let d be a simple loop on the cusp torus with is transverse to c'. Do $(1, 2)$ Dehn filling on the cusp so that in the new manifold, which now has no rank two cusps, $c' \sim d^2$. In the new manifold, c remains freely homotopic to a simple loop on the boundary, but it is not primitive.

In the original manifold there is also an essential cylinder C^* with both boundary components on $\partial\mathcal{M}(G)$ and whose central curve is freely homotopic to c. In the new manifold, C^* remains an essential cylinder, and its core curve is not primitive.

On the other hand, a simple closed *geodesic* in $\mathcal{M}(G)$ is automatically primitive! See also Example 5-17.

4-22. *Geometric limits by renormalization.* Suppose $\{\gamma_n\}$ is a collection of closed geodesics with the property that only a finite number meet any given compact set $K \in \mathcal{M}(G)$ and that their lengths are uniformly bounded by $L < \infty$. Fix origins $\{O_n \in \gamma_n\}$. Consider the sequence $\{T_n G T_n^{-1}\}$ where T_n maps O_n to the origin O in say the ball model. Prove that there is a geometrically convergent subsequence to a limit H. Show that in $\mathcal{M}(H)$, the renormalized geodesics $\{\gamma_m'\}$ converge to a closed geodesic γ.

4-23. *Benjamini-Schramm convergence* This is a weak form of geometric convergence called BS convergence that applies to special kinds of manifolds.

Suppose $\{M_n\}$ is a sequence of closed hyperbolic manifolds. Given a number $r > 0$, which may be very large, the *r-thin part* of a hyperbolic manifold M is region $\text{thin}_r M$ consisting of all points $x \in M$ whose injectivity radius does *not exceed* r.

We say $\lim_{BS} M_n = \mathbb{H}^2$ if

$$\lim_{n \to \infty} \frac{\text{Vol} thin_r M_n)}{\text{Vol}(M_n)} = 0, \quad \text{for every } r > 0,$$

That is, no matter how large each r-thin part is, the total volume of the manifold increases much more quickly; the vast majority of the manifold consists of points with larger injectivity radii.

4-24. *Every surface has a decomposition by pants of medium size.* Consider the Teich-müller space $\mathfrak{Teich}_g$ of closed surfaces of genus g. Prove with Bers [1985] that there exists a number L_g, depending only on g, such that every surface $S \in \mathfrak{Teich}_g$ has a pants decomposition (see Exercise (4-24) and Section 5.3) in which no boundary curve of a pants has length exceeding L_g.

4-25. Prove that the limit set $\Lambda(G)$ is not uniformly perfect (Exercise (1-30)), if and only if there is a sequence of Möbius transformations $\{A_n\}$ such that $\{A_n(\Omega(G))\}$ converges in the to a hyperbolic region Ω^* whose boundary $\partial\Omega^*$ contains an isolated point [Marden and Markovic 2008, Prop.3.1].

4-26. *Rank two subgroups.* Suppose M^3 is a hyperbolic manifold and $K \subset \pi_1(M^3)$ is a rank two subgroup of *infinite index*. Then K is either an abelian group or a free group of rank two [Jaco and Shalen 1979, Theorem VI.4.1]. Compare with Exercise (4-27).

4-27. *The* $\log(2k - 1)$ *theorem.* Suppose the kleinian group G is a free group on k-generators $\{A_i\}$, $1 \le i \le k$. For $x \in \mathbb{H}^3$, set $d_i = d(x, A_i x)$ (hyperbolic distance).

Theorem 4.11.9 ([Anderson et al. 1996]).

$$\sum \frac{1}{1 + e^{d_i}} \le \frac{1}{2}.$$

In particular, for at least one index j,

$$d_j \ge \log(2k - 1).$$

The original statement has been strengthened here by application of the Tameness Theorem 5.6.6.

If $k = 2$, the inequaity is that at least one of d_1, d_2 satisfies $d_j \ge \log 3$.

Compare with McShane's identity Exercise (7-15).

4-28. *Joining unpaired or paired punctures; Plumbing.* Here are two related constructions.

Construction 1. Suppose Δ_1^*, Δ_2^* are horodisks at different parabolic fixed points corresponding to two rank one parabolic subgroups of a group G (or of two different groups G_1, G_2). For definiteness suppose they are both in the same component $\Omega \subset \Omega(G)$ over the component S of $\partial \mathcal{M}(G)$. And suppose their partners $\Delta_{1'}^*, \Delta_{2'}^*$ are on different components. Correspondingly, in $\partial \mathcal{M}(G)$, there are disjoint solid pairing tubes $C_i \times [0, 1)$ joining the projections Δ_i and Δ_i', $i = 1, 2$.

Take any Möbius transformation T that sends the exterior of Δ_2^* onto the interior of Δ_1^* and conjugates the parabolic subgroups. Form the augmented group $G^* = \langle G, T \rangle$.

Down in the quotient $\mathcal{M}(G^*)$, $\partial \Delta_2$ is identified with $\partial \Delta_1$ forming a handle on S which now has two fewer punctures. The pairing tube C_2 joins with C_1 to form the boundary of a solid pairing tube between Δ_2' and Δ_1'. See Figure 4.5.

Construction 2. Suppose Δ^*, Δ'^* are disjoint horodisks at the same rank one parabolic fixed point. In $\mathcal{M}(G)$ there is a solid pairing tube $C \times [0, 1)$ joining Δ and Δ'.

Take any Möbius transformation T that sends the exterior of Δ'^* onto the interior of Δ^* and conjugates the rank one parabolic group to itself. Form the augmented group $G^* = \langle G, T \rangle$.

The effect of adjoining T is to paste together the two punctured disks and the horocycles bounding them. This forms a solid cusp torus $C \times \{0 \le z < 1\}$ in $\mathcal{M}(G^*)$ corresponding to the rank two parabolic group that has been created. And S has lost a puncture.

In both cases the new manifold is geometrically finite if and only if this is true of the original.

The simplest application of Construction 2 is to the modular group M_2 of Exercise (2-5), which acts in the upper and lower half-planes. The new parabolic has the form $Tz = z + \tau$ where $\mathrm{Im}\, \tau \ne 0$. The process gives us a 4-punctured sphere with complex parameter τ. For this case we can take τ so that $|\mathrm{Im}\, \tau| \ge 1 + 2\epsilon$. The resulting group has infinitely many regions of discontinuity, each of which is simply connected. The manifold $\mathcal{M}(G^*)$ is bounded by a 4-punctured sphere with the punctures arranged in two pairs. In addition $\mathcal{M}(G^*)$ has a rank two cusp which is hidden from the viewer

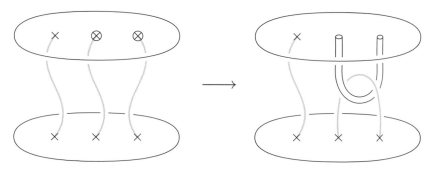

Fig. 4.5. Two unpaired punctures determine a handle. The gray curves represent solid pairing tubes.

looking only at $\mathbb{S}^2$. A simple loop that separates the two punctures of each pair on $\partial\mathcal{M}(G^*)$ is freely homotopic within $\mathcal{M}(G^*)$ to a simple loop on the cusp torus.

Construction 2 can be used to give an explicit example of the following situation referred to in Exercise (4-21): Suppose $\mathcal{M}(G)$ has a rank two cusp. Denote e the result of removing a corresponding solid cusp torus from $\mathcal{M}(G)$ by M_0. It is possible that there are two nontrivial, nonparallel simple loops α_1, α_2 in $\partial\mathcal{M}(G)$ and a simple loop α_c in the cusp torus of M_0, such that the pairs (α_c, α_1), and (α_c, α_2) both bound essential cylinders in M_0. Of course, (α_1, α_2) then bound an essential cylinder in $\mathcal{M}(G)$.

To obtain such an example, start with the algebraic limit of a quasifuchsian group, say closed surface group, that has a surface of genus g on the bottom, and a pinched surface—say a singly pinched surface—on top; the top surface gains two punctures. There is a simple loop α_2 in the bottom surface that bounds an essential cylinder with a loop about either of the punctures on top. Now apply Construction 2 to replace the top pinched surface by a closed surface S of genus g and a rank two cusp. The simple loops about the punctures determine a simple loop α_1 on S, which is freely homotopic to α_2. The two simple loops about the punctures also determine the meridian α_c on a cusp torus corresponding to the new rank two cusp. These three simple loops fulfill the requirements.

Figure 4.6 is an implementation of Construction 2. The starting point is the thrice-punctured sphere fuchsian group. The 6 punctures are arranged in 3 pairs, each pair supports two horodisks. The first picture shows the result of joining one pair of horodisks resulting in the quotient manifold $\mathcal{M}_1$ with one rank two cusp and $\partial\mathcal{M}_1$ a 4-punctured sphere. For the second picture, another pair of horodisks are joined resulting in $\mathcal{M}_2$ with two rank two cusps and $\partial\mathcal{M}_2$ a 2-punctured torus. The final step gives Figure 4.7. Here the final pair of horodisks are joined resulting in $\mathcal{M}_3$ with

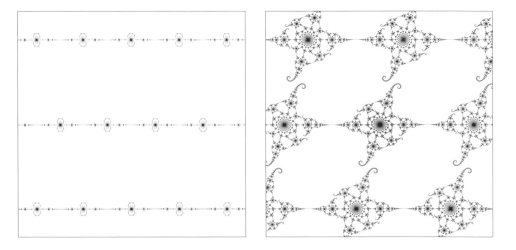

Fig. 4.6. Earle–Marden coordinates for 4-punctured spheres and 2-punctured tori.

Fig. 4.7. Earle–Marden coordinates for genus-2 surfaces.

three rank two cusps and $\partial \mathcal{M}_3$ a genus-two closed surface. There is one free complex parameter for each rank two cusp. These extend to become holomorphic coordinates of the Teichmüller spaces of the respective surfaces and are called the *Earle–Marden coordinates*; see Earle and Marden [2012]. In this sequence of figures, the parameters are chosen so that each group is a subgroup of the next.

5

Deformation spaces and the ends of manifolds

Our work in the earlier chapters, especially our understanding of the structure of geometrically finite manifolds, has prepared the ground for understanding the results that will be presented here, without most proofs. At center stage are the three fundamental conjectures, now theorems, concerning the structure of hyperbolic manifolds with finitely generated fundamental groups but which are not geometrically finite: the Tameness Theorem, Ending Lamination Theorem and Density Theorem.

The chapter begins with a study of the representation variety. We go on to discuss deformation spaces of geometrically finite groups and degeneration of their boundaries. This requires the understanding of the *topology* of geometrically infinite ends provided by the Tameness Theorem. Once we know the topology we are in the position to understand their geometry. This is what is explained by the Ending Lamination Theorem: In place of the Riemann surface boundaries of geometrically finite groups, the geometrically infinite ends turn out to be degenerated Riemann surfaces, the surfaces being crunched up into "ending laminations". To complete the picture, the Density Theorem tells us that geometrically finite groups are dense in the set of all finitely generated kleinian groups.

The chapter closes with a discussion of the new results on Cannon-Thurston mappings which lead to the affirmation of the vexing old question whether connected limit sets are also locally connected.

5.1 The representation variety

Suppose $G = \langle g_1, g_2, \ldots, g_r \rangle$ is a finitely generated kleinian group (without elliptics), that is G is isomorphic to the fundamental group of $\mathcal{M}(G)$. By the Scott–Shalen theorem, there are a finite number of relations $\{R_k(g_1, \ldots, g_r) = \mathrm{id}\}$, each of which is a word in the generators, such that any relation in the group is a consequence of these.

If we vary the entries in the generating matrices $\{g_i\}$ we will get a new set of Möbius transformations. The group generated by the new set will be a homomorphic image of G if and only if the new set of generators $\{g_i'\}$ satisfy the same relations $\{R_k(g_1', \ldots, g_r') = \mathrm{id}\}$. These are algebraic equations in the matrix entries. In this book, a *representation* of G is a homomorphism into $\mathrm{PSL}(2, \mathbb{C})$ or into $\mathrm{PSL}(2, \mathbb{R})$.

A representation is called *elementary* if the image groupis elementary. It is called *reducible* if the image group has a common fixed point (Exercise (2-1)).

We will not want to distinguish between conjugate representations. Therefore, for our purposes, we define the *representation variety* $\mathfrak{R}(G)$ as the quotient of the set of *nonelementary* representations modulo conjugation: two representations are equivalent $\varphi_1 \equiv \varphi_2$ if and only if there is a Möbius transformation T such that $\varphi_2(g) = T \circ \varphi_1(g) \circ T^{-1}$, for all $g \in G$. We will work with the quotient space

$$\mathfrak{R}(G) =$$

$$\{\varphi \mid \varphi : G \to H, \text{ a type-preserving homomorphism to a nonelementary } H\}/\equiv.$$

By *type preserving* we will mean that if $g \in G$ is parabolic, so is $\varphi(g)$. This means that rank one and rank two parabolic groups are preserved, but it does not prevent new parabolics (or elliptics) appearing in the target groups.

For the representation variety specifically of fuchsian groups see Exercise (5-28).

We have taken a certain liberty in our use of the term "representation variety". Actually $\mathfrak{R}(G)$ is the quotient of an open subset of the affine algebraic variety of type-preserving representations into $\mathrm{PSL}(2, \mathbb{C})$. It is not itself a "variety", but rather a Zariski open subset of the *character variety* described in Culler and Shalen [1983], Kapovich [2001], or Exercise (5-29).

We emphasize that $\mathfrak{R}(G)$ is not just a space of groups, but a space of *marked groups*, marked by the images of the chosen generators of G. A group H which the target of nonconjugate homomorphisms $\varphi_1, \varphi_2, G \to H$, is represented by *distinct* points of $\mathfrak{R}(G)$.

Each (normalized) matrix depends on 3 complex parameters; the set of r-generators depends on $3r$ complex parameters. That s-relations must be satisfied costs $3s$ conditions giving a dimension of at most $(3r - 3s)$. But then we do not distinguish between two groups that are conjugate. Conjugacy depends again on 3 complex parameters, so the $\mathbb{C}$-dimension of the quotient is now at most $(3r - 3s - 3)$.

The parabolicity condition further reduces the dimension of $\mathfrak{R}(G)$. Each cyclic parabolic conjugacy class gives rise to a relation of the form

$$\mathrm{tr}^2(R'_k(g_1, \dots, g_r)) = 4.$$

If there are b_1 rank one classes, and b_2 rank two classes, $\mathfrak{R}(G)$ has complex dimension $(3r - 3s - b_1 - 2b_2 - 3)$. So if G is a fuchsian group representing a closed surface of genus g, the $\mathbb{C}$-dimension of $\mathfrak{R}(G)$ is $3 \cdot 2g - 3 - 3 = 6g - 6$. If G represents a finite area surface of genus g and n punctures the dimension count is $3(2g+b-1)-b-3 = 6g+2n-6$. Also see Exercise (4-13). For rigorous computations see Culler and Shalen [1983], or the discussion of $\mathfrak{T}(G)$ below.

The groups represented by points of $\mathfrak{R}(G)$ are in general not discrete and perhaps not even finitely presented. Furthermore, $\mathfrak{R}(G)$ is typically not connected. For example if X is an orientation-reversing Möbius transformation there is no flow $\{A_t G A_t^{-1}\}$ by Möbius transformations A_t connecting G to XGX^{-1} if $YGY^{-1} \neq G$ for all orientation-reversing transformations Y.

We have removed the elementary groups. Of these, the most significant are subgroups conjugate into SO(3), rotations of the ball. For example there is an isomorphism of a closed surface group of any genus ≥ 2 to a subgroup of SO(3) [Greenberg 1981]. Besides fitting better into our theory, a big advantage of removing them is that the quotient is then Hausdorff: disjoint points have disjoint neighborhoods. Not only for nonelementary groups, but more generally for irreducible groups, the Hausdorff property holds, as shown by the following Lemma. Denote the space of irreducible representations by $\mathrm{Hom}_{ir}(G; \mathrm{PSL}(2, \mathbb{C}))$.

In all these spaces, the topology used is the topology of algebraic convergence.

Lemma 5.1.1. *Assume that for $\varphi_1, \varphi_2 \in \mathrm{Hom}_{ir}(G; \mathrm{PSL}(2, \mathbb{C}))$ the target groups $H_1 = \varphi_1(G)$ and $H_2 = \varphi_2(G)$ are not conjugate. There exist neighborhoods N_1 of H_1 and N_2 of H_2 in $\mathrm{Hom}_{ir}(G; \mathrm{PSL}(2, \mathbb{C}))$ such that $T N_1 T^{-1} \cap N_2 = \varnothing$ for all $T \in \mathrm{PSL}(2, \mathbb{C})$.*

Proof. Given a neighborhood $N(H_1) \subset \mathrm{Hom}_{ir}(G; \mathrm{PSL}(2, \mathbb{C}))$ of H_1 with $H_2 \notin N(H_1)$, we claim that there exists a neighborhood $N(H_2)$ of H_2 so small that no representation $\theta_1(G) \in N(H_1)$ is conjugate to a representation $\theta_2(G) \in N(H_2)$.

Assume our claim is false. Then for a sequence $\{\theta_{1,n}\} \in \mathrm{Hom}_{ir}(G; \mathrm{PSL}(2, \mathbb{C})$ in $N(H_1)$ and a corresponding sequence of Möbius transformations $\{T_n\}$ with the following property. Setting $\theta_{2,n}(G) := T_n \circ \theta_{1,n}(G) \circ T_n^{-1}$, in algebraic convergence,

$$\lim \theta_{1,n}(G) = H_1, \quad \lim \theta_{2,n}(G) = H_2.$$

By Proposition 2.1.1 we may assume the fixed point(s) (p_n, q_n) of T_n converge to $(p, q) \in \mathbb{S}^2$, and that for all indices T_n is loxodromic with p_n the repelling fixed point, or parabolic $p_n = q_n$, or elliptic. By hypothesis, no subsequence of $\{T_n\}$ can converge to a proper Möbius transformation.

For definiteness, consider the case that $\{T_n\}$ is a loxodromic sequence. By Exercise (2-1) there exists $g, g' \in G$, such that $h = \varphi_1(g)$, $h' = \varphi_1(g') \in H_1$ have distinct fixed points. The same will be true of $\theta_{1,n}(g), \theta_{1,n}(g')$ for large indices. We may assume that it is h and $\theta_{1,n}(g)$ that does not fix the limit p of the repelling fixed points of $\{T_n\}$.

Choose a small enough neighborhood N_p of p so that $\theta_{1,n}(g)(N_p) \cap N_p = \varnothing$ for all large indices. Choose $z \notin N_p$. Then choose a small neighborhood N_q of q. If $p = q$ take $N_q \subset N_p$.

For all large indices, $T_n^{-1}(z) \in N_p$. So $\theta_{1,n}(g)(T_n^{-1}(z)) \notin N_p$. In fact for all large indices $T_n \theta_{1,n}(g) T_n^{-1}(z) \in N_q$. Since N_q can be taken arbitrarily small, we conclude that $\lim T_n \theta_{1,n}(g_1) T_n^{-1}(z) = q$, and then that this holds for all $z \neq p$. We have reached a contradiction to our assumption that $\lim \theta_{2,n}(G) = H_2$.

Similarly parabolic and elliptic sequences are excluded as well. □

5.1.1 The discreteness locus

The *discreteness locus* of a geometrically finite G is the following *closed* subset:

$$\mathfrak{R}_{\text{disc}}(G)X = \{\theta \in \mathfrak{R}(G) \mid \theta : G \to H$$

is a parabolic-preserving isomorphism to a discrete group H}.

In other terms, setting $M^3 = \mathfrak{M}(G)$, $\mathfrak{R}_{\text{disc}}(G)$ is the *space of hyperbolic structures on the interior of M^3*, modulo conjugation, with the topology of algebraic convergence.

It follows from the Thurston Compactness Theorem, p. 240, and Theorem 5.2.4 that if $\mathfrak{M}(G)$ is geometrically finite, acylindrical, and boundary incompressible then $\mathfrak{R}_{\text{disc}}(G)$ is compact in the topology of algebraic convergence (see also Thurston [1979b, Theorem 7]).

The target group H may be a quasiconformal deformation of G, but the isomorphism θ itself might *not* be induced by a quasiconformal map of $\mathbb{S}^2$, as θ may so distort the presentation of G (see Section 5.2 below). The image H may have additional parabolics and in general will not be geometrically finite. In fact the interior of the discreteness locus may have many components. These correspond to the marked homeomorphism types (orientation-preserving or -reversing) of isomorphisms of G that send parabolics to parabolics; two isomorphisms $\theta_1, \theta_2 : \pi_1(\mathfrak{M}(G)) \to \pi_1(\mathfrak{M}_i)$ correspond to the same component of $\text{Int}(\mathfrak{R}_{\text{disc}}(G))$ if and only if there is a quasiconformal mapping $h : \mathfrak{M}_1 \to \mathfrak{M}_2$ which induces the isomorphism $\theta_2\theta_1^{-1} : \pi_1(\mathfrak{M}_1) \to \pi_1(\mathfrak{M}_2)$.

In the literature the commonly used notation for $\mathfrak{R}_{\text{disc}}(G)$ is $\mathcal{AH}(G)$, where H stands for homotopy equivalence and Λ reminds us that the topology is that of algebraic convergence.

Curiously, the convex cores of groups in $\mathfrak{R}_{\text{disc}}(G)$ are of a uniform "size", as indicated below. Incorporating the Tameness Theorem in the original papers, this was first shown with restrictions [Canary 1996], and in the unpublished preprints [Evans 2006], [Evans 2005]. Repeating here part of Theorem 3.11.2:

Theorem 5.1.2 ([Bowditch 2013]). *There exists a uniform bound $r > 0$ for the injectivity radii about points in the convex core of any manifold $\mathfrak{M}(H)$ with $H \in \mathfrak{R}_{\text{disc}}(G)$.*

Theorem 5.1.2 was not unexpected since there are pleated surfaces of fixed finite genus that exit each end. That the embedded balls are uniformly bounded prevents the manifolds from flying apart at the ends. For a totally degenerated group, the convex core coincides with the whole manifold.

Whether there exist maximal balls of uniform size centered in the convex hull was first raised by McMullen. A still unsolved conjecture of McMullen is the following: The rank k of a group is the minimal number of generators. Given k, does there exist

$R = R(k)$ such that for any *closed* manifold $\mathcal{M}(G)$ of rank k, every point has injectivity radius $\leq R(k)$? If this were to fail, there would be a sequence whose geometric limit is $\mathbb{H}^3$; see Biringer and Souto [≥ 2015].

5.1.2 The quasiconformal deformation space $\mathfrak{T}(G)$

The group H is a *quasiconformal deformation* of the geometrically finite group G if there is a quasiconformal map $F : \mathbb{S}^2 \to \mathbb{S}^2$ that induces an isomorphism $\theta : G \to H$ for which $F \circ g(z) = \theta(g) \circ F(z)$, for all $g \in G$, $z \in \mathbb{S}^2$. Such a group H is necessarily discrete and nonelementary (as G is assumed to be). In particular, $g \in G$ is parabolic if and only if $\theta(g) \in H$ is parabolic. The map F in turn is uniquely determined up to normalization by its Beltrami differential defined on $\Omega(G)$ with the invariance property given by Equation (2.7). When we explicitly *normalize*, we will normally arrange things so that $(0, 1, \infty) \subset \Lambda(G)$ and F fixes these points.

The *quasiconformal deformation space* of G is defined as the following open, connected subset of the interior of $\mathfrak{R}_{\text{disc}}(G)$:

$$\mathfrak{T}(G) = \{\theta \in \mathfrak{R}(G) \mid \theta \text{ is induced by a quasiconformal deformation of } G\}.$$

Thus two normalized deformations $F_1 \sim F_2$ are taken to be equivalent if they induce the same isomorphism θ. Another way of putting this is that if f_1, f_2 denote their projections to $\partial\mathcal{M}(G) \to \partial\mathcal{M}(\varphi(G))$, then $f_2^{-1} \circ f_1$ extends to $\mathcal{M}(G) \to \mathcal{M}(G)$ and is homotopic to the identity on $\text{Int}\,\mathcal{M}(G)$—see Theorem 3.7.4. It is not necessarily true that $f_2^{-1} \circ f_1$ is then homotopic to the identity on $\partial\mathcal{M}(G)$; see Exercise (5-35).

The deformations of a geometrically finite group G can be followed from the changes in a fundamental polyhedron $\mathcal{P}_\Theta(G)$ [Marden 1974a]. Along with any small deformation of G (or of any other point of $\mathfrak{T}(G)$) $\mathcal{P}_\Theta(G)$ is correspondingly deformed. When parabolics are preserved, no essential change occurs at the cusps. It follows that parabolic-preserving homomorphisms of G close to the identity are actually isomorphisms induced by quasiconformal deformations of maximal dilatation close to one (a notion known as "strong stability") [Marden 1974a]). For this reason $\mathfrak{T}(G)$ is an open subset of $\text{Int}\,\mathfrak{R}_{\text{disc}}(G)$. It is a connected subset containing the identity because if μ is a Beltrami differential so is $t\mu$ for any $t \in \mathbb{C}$, $|t| < 1$. (If G is geometrically infinite, $\mathfrak{T}(G)$ is not open as nearby groups are nondiscrete.)

As a consequence of the Thurston Compactness Theorem, p. 240, we know that if $\mathcal{M}(G)$ is both acylindrical and boundary incompressible, the closure of $\mathfrak{T}(G)$ is compact in $\mathfrak{R}_{disc}(G)$.

Moreover, if all components of $\Omega(G)$ are simply connected, then $\mathfrak{T}(G)$ is a complex manifold biholomorphically equivalent to the product of the Teichmüller spaces of the individual components of $\partial\mathcal{M}(G)$. For then each component of $\Omega(G)$ serves as the universal cover of the associated quotient surface.

On the other hand, if the components $\{\Omega_i\}$ over some S_i are not simply connected, there is a little problem because a conformal change of S_i may not cause a deformation in G. A simple loop γ around a handle of S_i may be compressible so a deformation in itself of the element of $\pi_1(S_i)$ corresponding to γ may result in no change to G. For an example, apply a Dehn twist (Exercise (5-10)) to a compressing loop on the boundary of a handlebody.

The way out of this conundrum is given by Theorem 5.1.3. It was originally proved at the level of Beltrami differentials (using "strong stability" [Marden 1974a]) by Bers [1970b], Maskit [1971], and Kra [1972], with a three-dimensional interpretation in Marden [2007]. Details of the proof are in Exercise (5-35).

Theorem 5.1.3. *Suppose G is geometrically finite. Denote the components of $\partial\mathcal{M}(G)$ by $\{S_i\}$. Then there is biholomorphic equivalence:*

$$\mathfrak{T}(G) = \mathfrak{Teich}(S_1)/\mathrm{Mod}_0(S_1) \times \cdots \times \mathfrak{Teich}(S_k)/\mathrm{Mod}_0(S_k).$$

Here $\mathrm{Mod}_0(S_i)$ is the fixed-point-free subgroup of biholomorphic automorphisms of $\mathfrak{Teich}(S_i)$ generated by automorphisms of S_i that have an extension homotopic to the identity in the interior of $\mathcal{M}(G)$. If S_i is incompressible, $\mathrm{Mod}_0(S_i) = \mathrm{id}$.

The product $\mathfrak{Teich}(S_1) \times \cdots \times \mathrm{Teich}(S_k)$ represents the universal cover of $\mathfrak{T}(G)$; the product equals $\mathfrak{T}(G)$ if and only if $\partial\mathcal{M}(G)$ is incompressible. The spaces $\mathfrak{T}(G)$ and $\mathfrak{Teich}(S_i)/\mathrm{Mod}_0(S_i)$ are complex analytic manifolds; $\mathfrak{T}(G)$ has dimension

$$\sum_{i=1}^{m}(3g_i + n_i - 3),$$

where g_i is the genus of the i-th component of $\partial\mathcal{M}(G)$ and n_i is the number of its punctures.

In the case $\mathcal{M}(G)$ is geometrically infinite Theorem 5.1.3 still holds for any conformal boundries that exist (Theorem 3.13.5). For example, if $\Omega(G)$ is connected and simply connected (a singly degenerate group (see Section 6.1)), then $\mathfrak{T}(G)$ has dimension $(3g + n - 3)$. In contrast if G_1 is fuchsian then $\mathfrak{T}(G_1)$ has dimension $(6g+2n-6)$. If all components S_i are triply punctured spheres, G is quasiconformally rigid, geometrically finite or not.

5.2 Homotopy equivalence

Understanding of isomorphisms to discrete groups in $\mathfrak{R}(G)$ involves the notion of *homotopy equivalence*. A homotopy equivalence between two manifolds M_1, M_2 is a pair of continuous mappings $f_1 : M_1 \to M_2$ and $f_2 : M_2 \to M_1$ such that $f_2 \circ f_1 : M_1 \to M_1$ is homotopic to the identity and $f_1 \circ f_2 : M_2 \to M_2$ is homotopic to the identity. In particular M_1 and M_2 have isomorphic fundamental groups. For example, the 3-manifold $S \times (0, 1)$ is homotopy equivalent to the surface S; the one-holed torus is homotopy equivalent to the three-holed sphere. A hyperbolic manifold is homotopy equivalent to its compact or relative compact core. For a general development see Johannson [1979].

Here are two basic facts about homotopy equivalences. The first statement is well known in topology [Whitehead 1978, Theorems 3.5, 7.1]; for the second, known as the Baer-Nielsen theorem, see Canary and McCullough [2004, Theorem 2.5.5]; 5.2.2 is a generalization.

- *Two manifolds whose higher homotopy groups vanish (as is the case for hyperbolic manifolds) are homotopy equivalent if and only if they have isomorphic fundamental groups.*

    In certain cases, an isomorphism is induced by a homeomorphism.

- *If f is a homotopy equivalence between closed surfaces S, S' then f is homotopic to an homeomorphism g. Instead if S, S' are compact with $f(\partial S) = \partial S'$ then f is properly homotopic[1] to a homeomorphism.*

Theorem 5.2.1 [Swarup 1980]. *Suppose $\mathcal{M}(G)$ with $\partial \mathcal{M}(G) \neq \varnothing$ is geometrically finite without parabolics. There are at most a finite number of compact, orientable, irreducible, nonhomeomorphic manifolds M^3 with boundary whose fundamental group is isomorphic to G.*

In particular there are at most a finite number of such manifolds whose peripheral structure is different from that of $\partial \mathcal{M}(G)$.

Theorem 5.2.2 [Waldhausen 1968]. *Suppose M, N are compact, boundary incompressible hyperbolic 3-manifolds and $\psi : \pi_1(M) \rightarrow \pi_1(N)$ is an isomorphism (a homotopy equivalence) which respects the peripheral structure; that is, for each component $S \subset \partial M$, there is a component $S' \subset \partial N$ such that*

$$\psi(i_*(\pi_1(S)) \text{ is conjugte to } i_*(\pi_1(S')),$$

where i_* is the inclusion isomorphism. Then there exists a diffeomorphism $F : M \rightarrow N$ which induces ψ

If ψ is induced by a homotopy equivalence $f : M \rightarrow N$, then f is homotopic to F.

If instead ∂M is compressible, so long as f restricts to a homeomorphism $\partial M \rightarrow \partial N$, there exists a diffeomorphism $F : M \rightarrow N$ homotopic to f.

For references see Waldhausen [1968], Hempel [1976, Theorem 13.6], Bonahon [2002, Theorem 3.11].

In the simplest case that $M_1, M_2 \cong S \times [0, 1]$ where S is a closed surface of genus ≥ 2, not every homotopy equivalence $f : M_1 \rightarrow M_2$ is homotopic to a orientation-preserving or -reversing homeomorphism. To do so it would have to send nonintersecting geodesics to nonintersecting geodesics, or more specifically, it would have to preserve the order of their fixed points on $\partial \mathbb{D}$, taking $\mathbb{D}$ as universal cover (by a classical theorem of Nielsen). But see Theorem 5.2.4(ii).

Rigidity in homotopy equivalences

Here we give more details of the discussion begun on p. 182 that homotopy equivalences between closed manifolds have a certain topological rigidity. We will restate Theorem 3.13.2:

---

[1] A proper map is one such that the preimage of every compact set is compact. A proper homotopy between f and g is a proper map $F : S \times [0, 1] \rightarrow S'$ from f to g. See Exercise (5-15) for an improper homotopy.

Theorem 5.2.3 ([Gabai 1997; Gabai et al. 2003]). *Assume* $\mathcal{M}(G)$ *is a closed hyperbolic manifold.*

(i) *If* $f : M^3 \to \mathcal{M}(G)$ *is a homotopy equivalence from a closed, irreducible 3-manifold* M^3, *then* f *is homotopic to a homeomorphism.*

(ii) *If* $f, f_1 : M^3 \to \mathcal{M}(G)$ *are homotopic homeomorphisms, then* f_1 *is isotopic to* f,

(iii) *The space of hyperbolic metrics on* $\mathcal{M}(G)$ *is path connected.*

In short, while Mostow's Rigidity Theorem (p. 182) implies that hyperbolic structures are unique up to homotopy, Theorem 5.2.3 implies that hyperbolic structures are unique up to isotopy.

If M^3 is itself hyperbolic, Property (i) follows from Mostow's Rigidity Theorem: Mostow's theorem implies that a homotopy equivalence $f : \mathcal{M}(G) \to \mathcal{M}(H)$ between two finite volume manifolds is homotopic to an isometry.

In the case that M^3 is a Haken manifold (see Section 6.3.1) then Waldhausen [1968] proves that two homotopic homeomorphisms f_1, f_2 from M^3 to $\mathcal{M}(G)$ are isotopic. From this Property (ii) follows.

In more detail, think of a closed manifold M^3 with two hyperbolic metrics ρ_1, ρ_2. By Mostow's theorem, there is a homeomorphism $F : M^3 \to M^3$, *homotopic* to the identity, taking the geodesic α in a free homotopy class in metric ρ_1 to the geodesic α' in the same free homotopy class in metric ρ_2. Property (ii) says that α' is not just homotopic but is isotopic to α, a subtle but nontrivial distinction! It is this that implies that the space of hyperbolic metrics on M^3 is path connected.

Property (ii) also implies that homotopy classes of automorphisms of a closed $\mathcal{M}(G)$ are the same as isotopy classes. Consequently the group of automorphisms of $\mathcal{M}(G)$, modulo the subgroup of those isotopic to the identity, is isomorphic to the outer automorphism group[2] of $\pi_1(\mathcal{M}(G)) = G$. By Mostow's theorem, the outer automorphisms of $\pi_1(\mathcal{M}(G))$ are isometries.

An automorphism of a surface S should not be confused with an automorphism of $\pi_1(S)$. While any self homeomorphism f of S belongs to $\mathrm{Out}(G)$, f determines an element of $\mathrm{Aut}(\pi_1(S))$ only if it fixes the basepoint. Instead, we can assert that f preserves $\mathrm{Aut}(\pi_1(S))$ up to conjugacy necessitated by moving the basepoint.

Unless stated otherwise, "automorphism" will mean "orientation-preserving automorphism".

Turning to Property (iii), consider a diffeomorphism $F : \mathcal{M}(G) \to \mathcal{M}(G)$, homotopic to the identity. Let M denote the manifold underlying $\mathcal{M}(G)$. Denote by η the hyperbolic metric on M given by $\mathcal{M}(G)$. The pull back $F_*\eta$ is a hyperbolic metric on M (the local structure at $p \in M$ is defined to be the local structure given by η at $F(p)$). There is a path of riemannian metrics on M connecting the two. If M were Haken, F would be isotopic to the identity, and (iii) would follow.

---

[2] For a group H with automorphism group $\mathrm{Aut}(H)$, the *inner automorphism group* $\mathrm{Inn}(H)$ of H is its normal subgroup generated by the elements $\{ghg^{-1} : h \in H\}$ for each $g \in H$. The *outer automorphism group* $\mathrm{Out}(H)$ is the quotient $\mathrm{Out}(H) = \mathrm{Aut}(H)/\mathrm{Inn}(H)$.

However there are many closed hyperbolic manifolds which are not Haken. To establish Property (iii) for non-Haken manifolds, a deep analysis of the geometry of $\mathcal{M}(G)$ is required, involving computer calculations using up to one gigabyte of data. Every simple geodesic has an embedded tubular neighborhood about it. The central issue is identifying closed manifolds containing short geodesics whose embedded tubular neighborhoods have radius $< \log(3)/2$.

5.2.1 Components of the discreteness locus

The rank one parabolic subgroups in every geometrically finite group G can be "opened up" to a geometrically finite group G^* without rank one parabolic subgroups: $\mathcal{M}(G^*)$ is homeomorphic to the result of removing the interior of the solid pairing tubes from $\mathcal{M}(G)$. Jørgensen first studied this operation [1974a], but today we can apply the Hyperbolization Theorem (p. 385). In other words, G^* has the property that $g \in G^*$ is parabolic if and only if g lies in a rank two parabolic subgroup.

Geometrically finite groups without rank one parabolics have been called *minimally parabolic* since their parabolics cannot be removed by small deformations. The boundary components of the corresponding manifolds are closed surfaces.

(Contrast minimally parabolic with "maximally parabolic" as in Theorem 4.6.3.)

As the basepoint for our study, we will fix a minimally parabolic group G^*. Then (see Anderson et al. [2000] and Theorem 5.15.14)

$$\text{Int } \mathfrak{R}_{\text{disc}}(G^*) = \{H \in \mathfrak{R}_{\text{disc}}(G^*) : H \text{ is minimally parabolic}\}.$$

The Density Theorem (p. 305) implies that $\mathfrak{R}_{\text{disc}}(G^*) = \overline{\text{Int } \mathfrak{R}_{\text{disc}}(G^*)}$.

Denote by $\mathcal{M}_0(G^*)$ the compact manifold resulting from removing the interiors of the cusp tori; if G^* has no rank two parabolics, $\mathcal{M}_0(G^*) = \mathcal{M}(G^*)$.

Via the compact cores, and the Hyperbolization Theorem (p. 385), the elements of $\mathfrak{R}_{\text{disc}}(G^*)$ are "marked" by the equivalence classes of homotopy equivalences $\mathcal{HE}(\mathcal{M}_0)$ of $\mathcal{M}_0 = \mathcal{M}_0(G^*)$,

$$\mathcal{HE}(\mathcal{M}_0) = \{(M, h) : h \text{ is a homotopy equivalence } h : \mathcal{M}_0 \to M/ \equiv\}.$$

Here M is another compact manifold with hyperbolizable interior. The equivalence is as follows: Two homotopy equivalences of $\mathcal{M}_0$ are identified, $(M_1, h_1) \equiv (M_2, h_2)$, if and only if there is an *orientation-preserving* homeomorphism $f : M_1 \to M_2$ such that $f \circ h_1$ is homotopic to h_2.

Each component of Int $\mathfrak{R}_{\text{disc}}(G^*)$ is determined by member of $\mathcal{HE}(\mathcal{M}_0(G^*))$. Within each component, the marked groups are quasiconformally equivalent. Two marked groups are the same if the quasiconformal map f can be taken to be conformal on the boundary.

The case of a fuchsian closed surface group G is special. There is a reflection, for example $J : z \mapsto \bar{z}$, that induces the *identity* automorphism of G. An orientation-reversing map $h : \mathcal{M}(G) \to M$ that induces an isomorphism $\theta : \pi_1(\mathcal{M}(G)) \to \pi_1(M)$ can be replaced by the orientation-preserving hJ that also induces θ. In short, $\mathcal{HE}(\mathcal{M}(G))$ has two components, which differ by a reflection.

An interesting example of homotopy equivalence is the *shuffle* of a rolodex or pages of a book first recognized in this context by Jim Anderson and Dick Canary [1996a]. Start with a solid torus T and its core curve c. Fix a finite system of mutually disjoint, parallel simple loops $\{\gamma_k\}$ on ∂T which are not contractible in T (longitudes). Correspondingly fix a collection of surfaces $\{S_k\}$, each of some genus $g_k \geq 1$ and with a single boundary component. For greater effect assume the genera g_k are all different. Slightly thicken each S_k to obtain the compact manifolds $\{S_k \times [-\epsilon, +\epsilon]\}$. The boundary of each contains the annulus $\partial S_k \times [-\epsilon, \epsilon]$.

Attach $S_k \times [-\epsilon, \epsilon]$ by gluing $\partial S_k \times [-\epsilon, \epsilon]$ to a thin neighborhood of γ_k. The resulting manifold M is orientable and compact. By "rearranging the pages"—taking a noncyclic permutation of $\{S_k\}$—we get another manifold M_τ which is homotopy equivalent but not homeomorphic to M. The manifolds M_τ have a hyperbolic structure. When the core curve of T is removed, the corresponding hyperbolic manifolds gain a rank two cusp for which ∂T becomes a cusp torus. For more details see Exercise (5-15).

Operations akin to the shuffles just described operate in a general compact manifold M^3 with incompressible boundary. Choose one or more mutually disjoint, nonparallel simple loops $\{\gamma_i\}$ on ∂M^3. Attach embedded solid tori $\{T_i\}$ tori in M^3 in such a way that $\partial M^3 \cap T_i$ is an annulus with central curve γ_i; γ_i generates the homotopy of T_i (a longitude). Such a T_i is called "primitive". The above example has such a structure. A homotopy equivalence $h : M_1^3 \to M_2^3$ is called a *primitive shuffle* if there exists finite sets of mutually disjoint primitive tori $\{T_i\} \subset M_1^3$ and $\{T_i'\} \subset M_2^3$ such that h restricts to an orientation-preserving homeomorphism $h : M_1^3 \setminus \overline{\cup T_i} \to M_2^3 \setminus \overline{\cup T_i'}$. For details see Anderson et al. [2000, §2].

The following compilation of results reveals much about the structure of $\mathfrak{R}_{\mathrm{disc}}(G^*)$. Compare with the more general Theorem (5.2.1) above. As before, $\mathcal{M}_0(G^*)$ denotes the result of removing from a minimally parabolic manifold $\mathcal{M}(G^*)$ its solid cusp tori, if any.

Theorem 5.2.4 ([Anderson et al. 2000; Canary and McCullough 2004]). *Assume that G^* is minimally parabolic, and $\mathcal{M}_0 = \mathcal{M}_0(G^*)$ is boundary incompressible with $\partial \mathcal{M}_0 \neq \varnothing$.*

(i) *$\mathfrak{R}_{\mathrm{disc}}(G^*)$ is the union of the closures of the components of $\mathrm{Int}\mathfrak{R}_{\mathrm{disc}}(G^*)$).*

(ii) *If G^* contains no parabolics at all, then $\mathfrak{R}_{\mathrm{disc}}(G^*)$ has a finite number of components, that is, $\mathcal{HE}(\mathcal{M}_0)$ is finite.*

(iii) *$\mathfrak{R}_{\mathrm{disc}}(G^*)$ has infinitely many components if and only if $\mathcal{M}_0$ has the following structure: $\partial \mathcal{M}_0$ contains a cusp torus T and there exist simple, mutually disjoint loops $\alpha_t, \alpha_1, \alpha_2 \in \partial \mathcal{M}_0$ such that (a) $\alpha_t \subset T$, (b) $\alpha_1, \alpha_2 \subset \partial \mathcal{M}_0 \setminus T$ and are not freely homotopic in $\partial \mathcal{M}_0$, but (c) α_t, α_1 and α_t, α_2 bound essential cylinders in $\mathcal{M}_0$.*

(iv) *The components of $\mathfrak{R}_{\mathrm{disc}}(G^*)$ are in one-to-one correspondence with the finite quotient $\widehat{\mathcal{HE}}(\mathcal{M}_0) = \mathcal{HE}(\mathcal{M}_0)/\{primitive\ shuffles\}$ that identifies two members differing by a primitive shuffle.*

(v) *The closures of two components X_1, X_2 of $\mathrm{Int}\,\mathfrak{R}_{\mathrm{disc}}(G^*)$ have nonempty intersection if and only if they correspond to two members of $\mathcal{HE}(\mathcal{M}_0)$ which are primitive shuffle equivalent, that is, if and only if they correspond to the same element in $\widehat{\mathcal{HE}}(\mathcal{M}_0)$.*

Manifolds with the property (iii) of Theorem 5.2.4 are constructed in Exercise (4-21).

Theorem 5.2.5 ([Canary and McCullough 2004]). *If the boundary $\partial\mathcal{M}_0(G^*)$ is compressible, then $\mathcal{HE}(\mathcal{M}_0(G^*))$ is infinite, with the following exceptions: G^* is a free group, the free product of two closed surface groups, or of a closed surface group with a cyclic group, or of a cyclic group with rank two parabolic group.*

Using the upcoming Ending Lamination Theorem (p. 300), Theorem 5.2.4 can be extended to manifolds which are not geometrically finite; see the new foreword to Canary et al. [1987] in Canary et al. [2006].

To see why in most cases a compressible boundary leads to an infinite number of homotopy equivalences consider the following example [Canary and McCullough 2004, p. 7]. Suppose $\mathcal{M}_0$ has compressible boundary but is not a compression body. Then it contains a geodesic α which is not homotopic into the boundary. Let $D \subset \mathcal{M}_0$ be a compressing disk and let $N \cong D \times [-\varepsilon, \varepsilon]$ be a neighborhood. Take α' with origin in $D_0 = D \times \{0\}$ to be freely homotopic to α. Cut N along D_0; drag the right side of D_0 once around α' and glue the two sides back together. This will change the homotopy type of any loop having essential intersections with D_0. A homotopy equivalence h can be constructed which is the identity outside N and inside N is the map resulting from the dragging. A loop in $\partial\mathcal{M}_0$ with nonzero intersection number with ∂D will be sent to a loop which can no longer be homotoped into $\partial\mathcal{M}_0(G^*)$. Therefore h is not homotopic to a homeomorphism. If there are infinitely many free homotopy classes that contain such a curve α, then $\mathcal{HE}(\mathcal{M}_0)$ is infinite.

The memoir of Canary and McCullough [2004] contains a comprehensive investigation of the following topological question for compact, irreducible, orientable 3-manifolds M with $\partial M \neq \varnothing$: Denote by $R(M)$ the subgroup of the outer automorphism group $\mathrm{Out}(\pi_1((M))$ that is realized by automorphisms of M. When does $R(M)$ have finite index in the full group $\mathrm{Out}(\pi_1(M))$? When are there a finite number of self-homotopy equivalences? (See also pp. 353, 369.)

Local connectivity and the discreteness locus

In preparation for our next theorem and the results of Section 5.12.2 we remind the reader of the following classic definition:

Local connectivity. A closed set $X \in \mathbb{C}$ is locally connected if it has the following property. Given $\varepsilon > 0$ there exists $\delta > 0$ such that for any two points, $x, y \in X$ with $|x - y| < \delta$, there exists a continuum $\gamma \subset X$ such that

$$x, y \in \gamma, \text{ and } \operatorname{diam}(\gamma) < \varepsilon.$$

More generally, a topological space X is said to be *locally connected* if the following holds at each point $x \in X$: For every open subset $V \subset X$ containing x, there is a *connected* open subset $U \subset V$ containing x.

According to Theorem 5.2.4(ii), for a fuchsian group G representing a closed surface R of genus ≥ 2, $\mathfrak{R}_{\mathrm{disc}}(G) = \mathcal{AH}(G)$ has a finite number of components. Any automorphism α of $\pi_1(R) \cong G$ that is not realizable by an orientation-preserving homeomorphism of R spawns a new component of $\mathfrak{R}_{\mathrm{disc}}(G)$. Two automorphisms α_1, α_2 determine distinct components if there is no orientation-preserving homeomorphsm h with $\alpha_2 = h \circ \alpha_1$.

Aaron Magid proved the following deep result:

Theorem 5.2.6 ([Magid 2012]). *Let G be a fuchsian group representing a closed surface of genus ≥ 2. Then $\mathfrak{R}_{\mathrm{disc}}(G)$ is* **not** *locally connected.*

Earlier Bromberg [2011] proved the result for once-punctured torus groups.

5.3 The quasiconformal deformation space boundary

It is an interesting fact (Theorem 5.15.14 (p. 363)) that $\mathfrak{T}(G)$ is the interior of its closure $\overline{\mathfrak{T}(G)} \subset \mathfrak{R}_{\mathrm{disc}}(G)$—it is not like an open ball with a slit to the boundary removed.

The study of the boundary $\partial \mathfrak{T}(G) = \overline{\mathfrak{T}(G)} \setminus \mathfrak{T}(G)$, which is contained in $\mathfrak{R}_{\mathrm{disc}}(G)$, is one of the most fascinating aspects of the subject. Every group $H \in \partial \mathfrak{T}(G)$ is discrete, the limit of an algebraically convergent sequence from $\mathfrak{T}(G)$, and the target of an isomorphism $\varphi : G \to H$, by Theorem 4.1.2. Yet a boundary group is no longer a quasiconformal deformation of G; some kind of degeneration must occur as we approach it from inside $\mathfrak{T}(G)$.

A boundary group $H = \varphi(G) \in \partial \mathfrak{T}(G)$ is called a *cusp* if it is geometrically finite. It is called a *maximal cusp* if in addition all components of $\partial \mathcal{M}(H)$ are triply punctured spheres. Thus, for a maximal cusp H, all components of $\Omega(H)$ are round disks (Exercise (2-6)). The limit sets of such groups are quite visually attractive, as seen in the pictures in Mumford et al. [2002] and in Figures 5.1 and 5.9.

Here is how maximal cusps arise by *pinching* in the case that $\mathcal{M}(G)$ is a geometrically finite, *acylindrical* manifold. (For more general cases see the Pinching Theorem in Exercise (5-5).) Choose a maximal system $\mathcal{S}$ of mutually disjoint, simple geodesics in the hyperbolic metric on $\partial \mathcal{M}(G)$. The geodesics in $\mathcal{S}$ have the property that no two are parallel in $\partial \mathcal{M}(G)$, and of course none can be homotoped to a puncture or a point. The system $\mathcal{S}$ divides $\partial \mathcal{M}(G)$ into a union of triply connected regions—Bers coined the term *pants decomposition* because each component is homeomorphic to a pair of

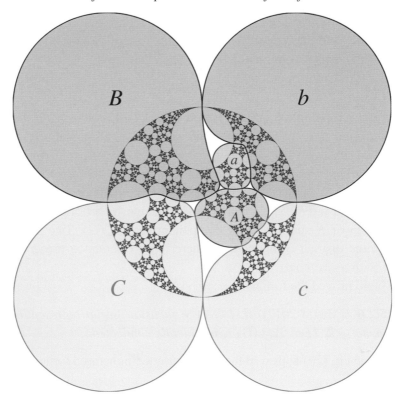

Fig. 5.1. A maximal cusp on the boundary of three-generator Schottky space. The limiting Schottky curves are indicated.

pants. On a surface of genus g with n punctures a pants decomposition $\mathcal{S}$ consists of $3g + n - 3$ simple loops; these divide the surface into $2g + n - 2$ pants. There are countably many homotopically different pants decompositions.

Fix an annular neighborhood A_i about each $\alpha_i \in \mathcal{S}$; choose these to be mutually disjoint. Set $B_n = \{z \in \mathbb{C} : 1/n < |z| < 1\}$. There is a quasiconformal map $f_{i,n} : A_i \to B_n$. On $\partial \mathcal{M}(G)$ take the Beltrami differential which is $(f_{i,n})_{\bar{z}}/(f_{i,n})_z$ in A_i and zero in the complement of $\bigcup A_i$. This defines a quasiconformal deformation $F_n : G \to G_n$. By the Thurston Compactness Theorem, p. 240, there is a subsequence $\{G_m\}$ which converges algebraically. The limit group H is a kleinian group isomorphic to G which necessarily lies on $\partial \mathfrak{T}(G)$. Furthermore H is a geometrically finite group with $\partial \mathcal{M}(H)$ homeomorphic to $\partial \mathcal{M}(G) \setminus \mathcal{S}$.

The proof involves showing that the length of the geodesics on $\{\partial \mathcal{M}(G_m)\}$ in the free homotopy classes of $\mathcal{S}$ go to zero by the pinching estimate (5.14). In the complementary components $\partial \mathcal{M}(G) \setminus \bigcup A_i$, lifted into $\Omega(G) \to \Omega(G_m)$, the conformal maps $\{F_m\}$ converge to conformal maps. Concurrently, the elements of $\{G_m\}$ corresponding to the curves of $\mathcal{S}$ converge to parabolic transformations of H.

The following result makes use of McMullen's technique [1991].

Theorem 5.3.1 (**Cusps are dense** [Canary et al. 2003; Canary and Hersonsky 2004]). *Cusps are dense on $\partial \mathfrak{T}(G)$ for any geometrically finite group G with $\partial \mathcal{M}(G) \neq \varnothing$.*

If $\partial \mathcal{M}(G)$ is connected, maximal cusps are dense on $\partial \mathfrak{T}(G)$.

In particular, maximal cusps are dense on the boundary of the deformation space of a Schottky group. However in general maximal cusps are not dense (Exercise (5-7)).

5.3.1 Bumping and self-bumping

Two components X_1, X_2 of Int $\Re(H)$ are said to *bump* at ζ if (i) $\zeta \in \overline{X}_1 \cap \overline{X}_2$, and (ii) every small neighborhood $U \in \Re(H)$ of ζ intersects both X_1 and X_2.

Theorem 5.2.4(v) with $H = G^*$ gives such an example.

Holt [2003] has shown, more generally, that if $X_1, \ldots, X_n$ are components of Int $\Re_{\mathrm{disc}}(G^*)$ such that each pair X_i, X_j bumps, then for any $K \geq 1$, there exists a geometrically finite member of $\Re_{\mathrm{disc}}(G^*)$ such that any K-quasiconformal deformation of it belongs to $\bigcap_i \overline{X}_i$. In particular, whenever a collection of components $\{X_i\}$ is "primitive shuffle equivalent", there exists a geometrically finite point $\zeta \in \bigcap \overline{X}_i$.

The Teichmüller space $\mathfrak{T}(G)$ is said to *self-bump* at $\zeta \in \mathfrak{T}(G)$ if for all small neighborhoods $U \in \Re(G)$ of ζ, $U \cap \mathfrak{T}(G)$ is not connected.

If there is such self-bumping, $\overline{\mathfrak{T}(G)}$ is not a manifold.

The following generalizes Theorem 6.9.11, p. 416:

Theorem 5.3.2 [Bromberg and Holt 2001]. *Suppose G is geometrically finite without rank one cusps. Assume the result $\mathcal{M}_0$ of cutting out the solid cusp tori contains an essential cylinder C such that a central loop c on C is primitive and not freely homotopic to a loop in a torus boundary component. Then there exists a point $\zeta \in \partial \mathfrak{T}(G)$ at which $\mathfrak{T}(G)$ self-bumps. Thus $\overline{\mathfrak{T}(G)}$ is not a manifold.*

For information about the primitiveness hypothesis see Exercise (4-21). The boundary point ζ will be a cusp resulting from pinching a component of ∂C. The same statement holds at the boundary of any component of Int $\Re_{\mathrm{disc}}(G)$. The condition on loops in C assures us that they do not determine parabolics in G. The prototype of theorems of this type is presented in Exercise (6-10), Theorem 6.9.11. Compare with Theorems 5.2.4 and 5.2.5.

For more infomation about bumping and the local connectivity, or lack of same, on $\partial \Re_{\mathrm{disc}}(G)$ see Exercise (6-9).

5.4 The three conjectures for geometrically infinite manifolds

Once geometrically finite manifolds were understood and the existence of geometrically infinite manifolds established, there emerged the daunting task of classifying the structure of these manifolds. Where are the "missing" components of $\partial \mathcal{M}(G)$? Have they left any geometric trace behind? Are the ends "wild" without a local product structure? Once the hyperbolization theorem was firmly nailed down, the following conjectures came to the fore.

The Tameness Conjecture. *The interior of every hyperbolic manifold $\mathcal{M}(G)$ with finitely generated G is homeomorphic to the interior of a compact 3-manifold.*

The tameness question was raised by Marden very early in the 3D theory.

The Density Conjecture. *Every finitely generated kleinian G group is the algebraic limit of geometrically finite groups isomorphic to G. Every group on ∂𝔗eich(G) is the algebraic limit of groups lying within 𝔗eich(G).*

Groups on $\partial\mathfrak{T}(G)$ have this property. But how do we know that every group with the appropriate structure actually appears on the boundary? This enigma appeared in the context of the Bers–Maskit discovery of geometrically infinite groups on the Bers boundary. For this reason it is called the *Bers Conjecture*.

The Ending Lamination Conjecture. *Hyperbolic manifolds* $\mathcal{M}(G)$ *with finitely generated fundamental group are completely determined by the conformal structure of* $\partial\mathcal{M}(G)$ *and the "ending laminations" of the geometrically infinite ends of* $\mathcal{M}(G)$.

This was proposed by Thurston when he discovered the phenomenon of ending laminations. Its proof requires the deepest analysis of the three conjectures.

For full generality, both Density and Ending Lamination depend on the resolution of Tameness.

It is a tribute to the power of the researchers who have entered the field within nearly 40 years after Thurston's proof of the Hyperbolization Theorem for Haken manifolds that all three conjectures have now been solved. We can now say that the main structural features of geometrically infinite hyperbolic manifolds with finitely generated fundamental group are understood. The formal statements begin in Section 5.6.

5.5 Ends of hyperbolic manifolds

Here is a precise definition of an end, or "ideal boundary component", of an open manifold M. Exhaust M by a strictly increasing sequence of connected, compact submanifolds $\{U_i\}$ with boundary:

$$\cdots \subset U_{i-1} \subset U_i \subset U_{i+1} \subset \cdots, \quad \bigcup_i U_i = M.$$

Choose the exhaustion so that no component of $M \setminus U_i$ is compact in M (this is called a *regular exhaustion*). Correspondingly consider a nested sequence $\{V_i\}$ of open subsets of M,

$$\cdots \supset V_{i-1} \supset V_i \supset V_{i+1} \supset \cdots,$$

where each V_i is a connected component of $M \setminus U_i$. Then $\bigcap V_i = \varnothing$. The sequence $\{V_i\}$ defines an *end* or *ideal boundary component* of M.

Another such sequence $\{V_k'\}$, perhaps coming from a different exhaustion of M, defines the same end if for each large j, V_j lies in some V_k' and for each large k, V_k' lies in some V_j; an *end* is formally an equivalence class of such nested sequences. The simplest example is that the end of the manifold $\mathbb{C}$ is defined by the equivalence class of the nested sequence $V_n = \{z : |z| > n\}$.

For a hyperbolic manifold $\mathcal{M}(G)$ without parabolics, by an "end" we will mean an end of the interior $\mathbb{H}^3/G$. The ends are in one-to-one correspondence with the boundary components of the *compact* core C—this is independent of the particular choice of core C. Consequently there are only a finite number of ends (we are always assuming G is finitely generated). The appropriate complement of the compact core serves as a *neighborhood* of the end, as does any V_j that lies in it.

In the presence of parabolics the above definition of end must be modified because it does not take account of the fact that rank one cusps correspond to what in effect are extra parts of the boundary, namely cusp cylinders. Instead we will use the refined notion of relative compact core $C = C_{\mathrm{rel}}$ introduced in Section 3.12. There we started by choosing $\mathcal{P}$ to be a G-invariant union of (open) horoballs at the cusps. The relative boundary in $\mathbb{H}^3/G$ of $(\mathbb{H}^3 \setminus \mathcal{P})/G$ consists of cusp tori and doubly infinite cusp cylinders. The boundary of the relative compact core C_{rel} contains the cusp tori, and it also intersects each doubly infinite cusp cylinder in a closed essential annulus. Thus $\partial C_{\mathrm{rel}}$ consists of closed surfaces and cusp tori.

Let $A = \cup A_j$ denote the union of the annuli resulting from the intersection of $\partial C_{\mathrm{rel}}$ with the cusp cylinders. A *relative end* E_{rel} of $\mathcal{M}(G)$ corresponds to a component of $\partial C_{\mathrm{rel}} \setminus A$.

The simplest case is a fuchsian group G representing two surfaces with $n \geq 1$ punctures. Remove the n-solid pairing tubes. The relative compact core is bounded in part by n-annuli, one about each pairing tube. If these annuli are removed from the boundary of the relative core, what is left is two surfaces. Each of these is parallel one of the boundary components of $\mathcal{M}(G)$ with n-horodisks about its punctures removed. Suppose the two components of $\partial \mathcal{M}(G)$ are degenerated,

While the relative core C_{rel} has one boundary component, the degenerated $\mathcal{M}(G)$ has two relative ends, each modelled by a component of $\partial C_{\mathrm{rel}} \setminus A$.

In general, the boundary of the relative core of $\mathcal{M}(G)$ can be quite different than what is expected of $\partial \mathcal{M}(G)$. It is the boundary of the relative core *minus* its annular intersections with the cusp cylinders, that model the boundary components (degenerated or not) of $\mathcal{M}(G)$.

An end E of $\mathcal{M}(G)$ or of $\mathcal{M}_{nc}$, the noncuspidal part of $\mathcal{M}(G)$ (see Section 3.12), is called *geometrically finite* if it has a neighborhood which does not intersect the *convex core* of $\mathcal{M}(G)$. Geometrically finite ends correspond to the components of $\partial \mathcal{M}(G)$. It is appropriate to declare ends that correspond to cusp tori to be geometrically finite as well but then are generally ignored as they are simple special cases.

Our concern now lies with *geometrically infinite ends*: those which are not geometrically finite.

5.6 Tame manifolds

The interior of a hyperbolic manifold is called *(topologically)* tame if it is homeomorphic to the interior of a compact manifold.

An end E or relative end E_{rel} is called *tame* if it has a neighborhood V in $(\mathbb{H}^3 \setminus \mathcal{P})/G$ with relative boundary $\partial V = S$ such that V is homeomorphic to $S \times [0, \infty)$.

It should be noted that there is not a unique way to express a neighborhood of a compressible end as a product $S \times [0, \infty)$: Suppose $f : S \to S$ is an orientation-preserving homeomorphism, not homotopic to the identity, but which extends to a homeomorphism of V which is homotopic to the identity in V. Then V is also homeomorphic to $f(S) \times [0, \infty)$.

The manifold $\mathcal{M}(G)$ itself is said to be *tame* if each of its ends or relative ends is tame. Actually before the fundamental paper [Canary 1993], Thurston had worked with two notions: topologically tame (as we have defined it) and geometrically tame which requires in addition that each end be exhaustible by pleated surfaces Section 3.10.2. Canary proved the two notions equivalent:

Theorem 5.6.1 ([Canary 1993]). *Every geometrically infinite end or relative end is exhaustible by pleated surfaces.*

From the point of view of projections from covering manifolds, geometrically infinite ends behave as if the missing surface were actually present. This is the import of the main case of the important Covering Theorem, which we have here modified by incorporating the Tameness Theorem:

Theorem 5.6.2 (The Covering Theorem [Thurston 1979a, Theorem 9.2.2; Canary 1996; Corollary B]). *Suppose $\mathcal{M}(H)$ has infinite volume and $G \subset H$ is a finitely generated subgroup such that the covering $\mathcal{M}(G)$ of $\mathcal{M}(H)$ has a geometrically infinite relative end $\widehat{E}_{\mathrm{rel}} \subset \mathcal{M}(G)$. Then there exists a neighborhood $\widehat{V} \cong S \times [0, 1)$ of $\widehat{E}_{\mathrm{rel}}$ such that the projection $\pi : \widehat{V} \to \mathcal{M}(H)$ is k-to-one for some $1 \leq k < \infty$.*

This would apply for example to a geometric limit H at an algebraic limit G. The geometric limit $\mathcal{M}(H)$ has a geometrically infinite end if $\mathcal{M}(G)$ does. Even though G may have infinite index in H, each infinite end of $\mathcal{M}(G)$ behaves as a finite-sheeted cover over an end of $\mathcal{M}(H)$. This is consistent with Lemma 3.6.3.

The Tameness Theorem

Each of the following sufficient conditions of Bonahon has been a fundamental tool in dealing with the tameness question:

Bonahon's Tameness Criteria ([Bonahon 1986]). *Either of the following conditions implies that $\mathcal{M}(G)$ is tame.*

A. *G cannot be split as a free product $G = A * B$ with $A, B \neq \{\mathrm{id}\}$.*
B. *In any splitting $G = A * B$ of G with $A, B \neq \{\mathrm{id}\}$ there is a parabolic $g \in G$ none of whose conjugates is contained in A or in B.*

In short, no simple loop on $\partial \mathcal{M}(G)$ bounds an essential disk within $\mathrm{Int}(\mathcal{M}(G))$.

A group G satisfying either hypothesis is called *indecomposable*. Free product decompositions of the fundamental group G arise in manifolds from compression disks that divide the manifold $\mathcal{M}(G)$ into two components; if the compression disk D does not divide $\mathcal{M}(G)$ $\mathcal{M}(G)$ is a *HNN extension* of $\mathcal{M}(G) \setminus D$, and correspondingly for the respective fundamental groups. A group G without any free product decomposition is called *freely indecomposable* Condition B says that any compression disk must cut through at least one cusp cylinder. For example if G is fuchsian, then after removing the interior of the pairing tubes from $\mathcal{M}(G)$ the resulting manifold has a compressible boundary. Equivalently, the boundary of every compressing disk contains an arc along a pairing tube. Thus Condition B holds for fuchsian groups.

Thus an end E or relative end E_{rel} (cf. 290) is said to be *compressible* if $S = \partial V$ is compressible. Otherwise E or E_{rel} is *incompressible*—there is no compressing disk based in the corresponding end of $\mathcal{M}^*$ The whole manifold $\mathcal{M}(G)$ is called boundary incompressible if ∂M^* supports no compressing disk.

One can also speak of an end E being indecomposable if the fundamental group of a neighborhood is indecomposable.

There had been a steady advance in understanding tameness before the complete answer was found. Partial results had been obtained by Thurston, Brock, Bromberg, Canary, Evans, Minsky, Ohshika, individually and in collaboration; for example see Brock et al. [2003]; Ohshika [2005]. Here we record two of the notable results (compare with Theorem 4.6.2).

Theorem 5.6.3 [Canary and Minsky 1996; Evans 2004b]. *Suppose* $\{\theta_n : G \to G_n\}$ *is a type-preserving sequence of isomorphisms where each* $\mathcal{M}(G_n)$ *is known to be tame. Suppose the sequence converges algebraically and geometrically to a group H which has no new parabolics. Then H is tame as well.*

Theorem 5.6.4 [Brock and Souto 2006]. *The algebraic limit of any sequence of geometrically finite groups is tame.*

This implies:

Corollary 5.6.5. *If the density conjecture is true, all hyperbolic manifolds* $\mathcal{M}(G)$ *with finitely generated G are tame.*

The resolution of the tameness conjecture, also called the *Marden Conjecture*, was announced by Ian Agol, and a different proof by Danny Calegari and David Gabai. Both sets of authors credit discussions with Mike Freedman.

Theorem 5.6.6 (The Tameness Theorem [Agol 2004; Calegari and Gabai 2004]). *The interior of every hyperbolic manifold* $\mathcal{M}(G)$ *with finitely generated G is homeomorphic to the interior of a compact manifold; it is tame.*

It is appropriately designated the **Fundamental theorem of geometrically infinite hyperbolic manifolds**.

An immediate consequence is that $\pi_1(\mathcal{M}(G))$ is finitely presented, giving another proof of this fact.

Remark 5.6.7. *Relative cores; the "noncuspidal" part.* On p. 181 we introduced the notion of relative compact cores and the notion of relative ends on p. 290. Removing from a geometrically infinite $\mathcal{M}(G)$ a complete set of doubly infinite solid cusp cylinders and solid cusp tori, with mutually disjoint closures, as in Section 3.12, results in a manifold $\mathcal{M}_{nc}(G)$ which is also tame [Bowditch 2011, Theorem 1.4].

Consequently there is a compact core C or C_{rel}, depending on the situation, with each boundary component either a torus, or a closed surface of genus ≥ 2. There is then a homeomorphism

$$\Phi : C_{rel} \setminus \partial C_{rel} \to \mathcal{M}_{nc}(G), \tag{5.1}$$

and correspondingly for C.

In view of Bonahon's Conditions A and B, to prove the Tameness Theorem, it suffices to restrict consideration to the compressible ends.

Agol's proof makes heavy use of manifolds of pinched negative curvature. In particular this allows him to remove the rank one and rank two cusps. He then uses Canary's trick of finding a "diskbusting" curve to construct a two-sheeted cover which also has finitely generated fundamental group but a given end is now incompressible. More of the details are outlined in Exercises (5-20, 5-26).

The Calegari–Gabai proof is centered on the existence of "shrinkwrapped surfaces". Namely, suppose S is a finite collection of mutually disjoint, simple closed geodesics in $\mathcal{M}(G)$ where G has no parabolics. Let S be a closed, incompressible surface in $M = \mathcal{M}(G) \setminus S$. Then S can be *shrinkwrapped* in $\overline{M}$: There is an isotopy $F : S \times [0, 1] \to \overline{M}$ with $F(\cdot, 0) = S$, $F(\cdot, t)$ is an embedding of S in M, $0 \leq t \leq 1$, and $T = F(\cdot, 1)$ is a CAT(-1) surface (this has a hyperbolic-like metric property [Bridson and Haefliger 1999], which is a minimum for hyperbolic area among all surfaces in the homotopy class. The minimizers are likely to abut upon S so the actual structure may be more complicated, although it will remain a CAT(-1). Using a sequence of geodesics which exit a geometrically infinite end, the authors find a sequence of shrinkwrapped surfaces trapped between the successive geodesics. By establishing a uniform bound on their diameters, they show the shrinkwrapped surfaces also exit the end. Recently their proof has been simplified in Soma [2006]. See also Bowditch [2010].

Earlier it was shown in Canary [1993] that for an end to be tame there must exist a neighborhood V of the end, and a sequence of (not necessarily embedded) pleated surfaces in V, each homotopic within V to $\partial V = S$, that exit the end or relative end. This has turned out to be an important ingredient in the tameness proofs.

Souto [2005] had proved that $\mathcal{M}(G)$ is tame if its interior can be exhausted by a nested union of compact cores. It is not enough to find a sequence of mutually disjoint surfaces $\{S_n\}$ of the same topological type exiting each end. It is necessary to know that each pair (S_n, S_{n+1}) bounds a region homeomorphic to $S_n \times [0, 1]$; that is, S_{n+1} is homotopic to S_n in such a way that given any compact set, the homotopy does not meet it for all large indices. Yet, it would be nice to be able to study the ends using only a fundamental polyhedron!

As an example, if G is a free group of rank two, Int $\mathcal{M}(G)$ is homeomorphic to the interior of a handlebody, even if $\partial \mathcal{M}(G) = \varnothing$!

Building on earlier results of Thurston and Bonahon, Canary [1993] proved the forty-year old Ahlfors' Conjecture for tame manifolds. Canary's result, coupled with Theorem 5.6.4, guarantees that any algebraic limit of geometrically finite groups satisfies Ahlfors' Conjecture. Ahlfors himself had proved it for geometrically finite manifolds. A prior special case was treated in Ohshika [2005]. Adding in the Tameness Theorem, we now have:

Theorem 5.6.8 (**Ahlfors' Conjecture/Theorem**). *For any finitely generated group* G, *either* $\Lambda(G) = \mathbb{S}^2$, *or* $\Lambda(G)$ *has two-dimensional Lebesgue measure zero. Moreover if* $\Lambda(G) = \mathbb{S}^2$ *the action of* G *is ergodic: There does not exist a pair of disjoint* G-*invariant sets in* $\mathbb{S}^2$ *each of positive measure.*

Untame (wild) manifolds

Lest one think that tameness is self-evident, it is worth pondering an example of Fox and Artin [1948]: There exists a wild embedding of $\mathbb{S}^2$ into $\mathbb{S}^3$ such that both complementary components are simply connected, and neither component is homeomorphic to the (open) 3-ball $\mathbb{B}^3$. An embedding σ of $\mathbb{S}^2$ is called *wild* if there is no homomorphism of S^3 that carries σ to S^2. In contrast, wild embeddings of $\mathbb{S}^1$ in $\mathbb{S}^2$ do not exist; the closure of each complementary component of a simple closed curve is homeomorphic to the closed ball.

The same example of Fox and Artin shows that there exists a polyhedral plane in $\mathbb{R}^3$ such that the closure of neither complementary region is homeomorphic to a closed half-space $\mathbb{R}^2 \times [0, \infty)$. We are grateful to Tom Tucker for the reassurance that this awful possibility does not arise in our analysis:

Theorem 5.6.9 [Tucker 1975]. *Suppose* $S \subset \mathcal{M}(G)$ *is an incompressible surface embedded in Int* $\mathcal{M}(G)$ *with nontrivial fundamental group. Let* S^* *be a component of* $\{\pi^{-1}(S)\}$ *in* $\mathbb{H}^3$. *Then the closure of each complementary component of* S^* *in* $\mathbb{H}^3$ *is homeomorphic to a closed half-space.*

A cornucopia of other weird examples are presented in Scott and Tucker [1989], including one of Peter Scott already mentioned in Marden [1974b]: Given a closed surface S of genus ≥ 1 there is a 3-manifold M^3 with the properties (i) $\partial M^3 = S$, (ii) the injection $\pi_1(S) \hookrightarrow \pi_1(M^3)$ is an isomorphism, (iii) the universal cover of M^3 is the closed upper half 3-space H^+, (iv) for any cover transformation T, $H^+/\langle T \rangle \cong (\mathbb{S}^1 \times \mathbb{R}) \times [0, 1)$, but (v) $M^3 \neq S \times [0, 1)$! Singly degenerate groups have the first four properties.

Tucker [1974] gave the following example. Let T_0 be a solid torus $\mathbb{D} \times \mathbb{S}^1$. Embed T_0 in a larger solid torus $T_0 \subset T_1$ so that T_0 is knotted, and homotopy equivalent to T_1. Take an infinite sequence of nested solid tori $T_2 \subset T_3 \subset T_4 \subset \ldots$ so that there is a homeomorphism of $T_{k+1} \setminus T_k$ onto $T_1 \setminus T_0$. Set $M = \bigcup_k T_k$. Then $\pi_1(M)$ is infinite cyclic and covered by $\mathbb{R}^3$, but M cannot be embedded in any compact manifold. Also $\pi_1(M \setminus T_0)$ is not finitely generated.

A necessary condition that a noncompact irreducible manifold M^3 be homeomorphic to the interior of a compact manifold is that for every compact submanifold $K \subset M^3$, $\pi_1(M^3 \setminus K)$ is finitely generated. It is not so far from being sufficient; see Tucker [1974].

It is also worth pondering the following example of Freedman and Gabai [2007]. There is a noncompact 3-manifold M^3, with finitely generated fundamental group and universal cover $\mathbb{R}^3$ with the following property: There exists a simple closed curve $\gamma \subset M^3$ such that the fundamental group of $M^3 \setminus \{\gamma\}$ is *infinitely* generated and the set of preimages of γ in $\mathbb{R}^3$ is equivalent to a locally finite set Γ of vertical lines. That is, there is a homeomorphism of $\mathbb{R}^3$ taking Γ to a subset of $(\mathbb{Z}, 0) \times \mathbb{R} \subset \mathbb{R}^2 \times \mathbb{R}$.

5.7 The Ending Lamination Theorem

In this section we will discover how the essence of the "missing" boundary component of a geometrically infinite end is still retained in the manifold.

The simplest situation is the case of an incompressible end E of $\mathcal{M}(G)$ without parabolics. Since infinite ends are tame, it has a neighborhood $V \subset \mathcal{M}(G)$ homeomorphic to $S \times [0, 1)$ with injective inclusion $\phi : \pi_1(S) \to G$. Here $S = \partial V \subset \mathcal{M}(G)$.

In this section we will use the notation E_{rel} so as to refer not just to a relative ends but also to an ordinary ends E when no parabolic is involved.

A geodesic lamination $\Lambda \subset \mathbb{H}^2/\Gamma = S$ is said to be *realizable* in $\mathcal{M}(G)$ if there is a pleated surface $P : S \to S_{\mathrm{pl}} \subset \mathcal{M}(G)$ whose bending lamination contains $P(\Lambda)$ [Canary et al. 1987, Theorem 5.3.9]. In particular, each leaf corresponds to a geodesic in $\mathcal{M}(G)$. A measured lamination is said to be realizable in $\mathcal{M}(G)$ if and only if its support is realizable.

For compressible ends, lots of geodesics on S are not realizable—the compressible ones for example. We will see later that, in general, nonrealizable laminations correspond to ending laminations. Only in the case that the targeted end E_{rel} of $\mathcal{M}(G)$ is geometrically finite and boundary incompressible are all laminations realizable for the end.

Suppose $\Lambda = \lim \gamma_k$ where $\gamma_k \subset S$ are simple geodesics. Then $\phi(\Lambda) = \lim \phi(\gamma_k)$ exists as a lamination in $\mathcal{M}(G)$ if and only if the sequence $\{\phi(\gamma_k)\}$ lies in a compact subset of $\mathcal{M}(G)$.

The simplest example of a nonrealizable lamination occurs when V is a neighborhood of a pinched surface in $\partial \mathcal{M}(G)$. The parabolic is not represented by a geodesic in $\mathrm{Int}(\mathcal{M}(G))$ hence not a bending line in a pleated surface. The corresponding geodesic in the fuchsian Γ is therefore not realizable in $\mathcal{M}(G)$.

Here is a useful estimate.

Lemma 5.7.1. *Suppose $\gamma \subset \mathcal{M}(G)$ is a closed geodesic, and $\gamma^* \subset \mathcal{M}(G)$ is a simple loop freely homotopic to γ such that the closest distance of γ^* to γ is r. Then the length (Len) of γ satisfies,*

$$\text{Len}(\gamma) \leq \frac{\text{Len}(\gamma^*)}{\cosh r}. \tag{5.2}$$

Proof. Consider the tubular neighborhood C_r of radius r about γ. The shortest simple loop on ∂C_r freely homotopic to γ has length $\cosh r \, \text{Len}(\gamma)$ (see Exercise (1-4)). □

Suppose $V \cong S \times [0, 1)$ is a neighborhood of a geometrically infinite, incompressible end E_{rel}. Here S is embedded in $\mathcal{M}(G)$, we can also take it to be a finite area hyperbolic surface, for example a pleated surface as in Section 3.10.2, by adding cross sections of solid cusp cylinders as needed (see Section 3.12, p. 290).

A sequence of geodesics in E_{rel} is said to *exit* E_{rel} if, given any compact subset $K = S \times [0, r] \subset V$, at most a finite number of elements of the sequence have nonempty intersection with K.

Lemma 5.7.2 [Thurston 1979a, §9.3; Bonahon 1986].

(i) *Assume $\{\gamma_n\}$ is a sequence of closed geodesics exiting E_{rel}. Each γ_n is freely homotopic within V to a simple loop $\gamma_n^* \subset S$ which is a geodesic in the hyperbolic metric on S. Also $\text{Len}_S(\gamma_n^*) \to \infty$.*

(ii) *Suppose $\alpha, \beta \subset V$ are closed geodesics each of distance exceeding r from S; they are freely homotopic within V to simple geodesics $\alpha^*, \beta^* \subset S$. Given $\varepsilon > 0$, assume that each of α, β is either disjoint from the ε-thin part of $\mathcal{M}(G)$ or is the core of an maximal ε-tubular neighborhood about itself. Then on S, there exists a constant $C = C(\varepsilon)$ such that*

$$\iota(\alpha^*, \beta^*) \leq Ce^{-r} \, \text{Len}_S(\alpha^*) \, \text{Len}_S(\beta^*) + 2. \tag{5.3}$$

(iii) *Suppose $\{\alpha_n\}, \{\beta_n\}$ are exiting sequences such that their realizations in S projectively converge to measured laminations: $\alpha_n^*/c_n \to (\Lambda_1, \mu)$, $\beta_n^*/d_n \to (\Lambda_2, \nu)$. Then $\iota(\mu, \nu) = 0$, so that no leaf of Λ_1 crosses a leaf of Λ_2 and $\Lambda_1 \cup \Lambda_2$ is also a lamination; possibly $\Lambda_1 = \Lambda_2$.*

Outline of proof. Statement (i) has an elementary proof depending only on the fact that S has finite topological type. Suppose for an infinite subsequence $\text{Len}_S(\gamma_k^*) < M < \infty$. Passing to another subsequence if necessary, for all except at most a finite number of indices, either the geodesics $\{\gamma_k^*\}$ coincide with a fixed closed geodesic γ^* or they collapse to a puncture. In the former case, the corresponding $\{\gamma_k\}$ would also coincide. The second case cannot arise by the construction of relative ends in Section 3.12.

The proof of (ii) is based on a detailed study of the interaction of the free homotopy cylinders between each pair of geodesics (α, α^*), (β, β^*). The cylinders can be assumed to be transverse to each other (if $\iota(\alpha^*, \beta^*) = 0$, (ii) is vacuous). Also used is Equation (8.32).

Property (iii) follows from Equation (5.3). Equation (5.3) plays a key role in the existence theory. Compare with Exercise (3-38). □

The following theorem was initiated by Thurston [1979a], then filled out by Bona-hon [1986] and Canary [1993]. A good overview can be found in Minsky [1994a]. It was originally proved under the assumption of tameness; here we complete their theo-rem by taking account of that. Canary's paper dealt with compressible ends, reducing them to incompressible ends as described in Exercise (5-26), and clearing the path to their analysis.

Theorem 5.7.3 (**Existence of ending laminations**). *Suppose E_{rel} is an incompress-ible relative end with given neighborhood $V \cong S \times [0, 1) \subset M(G)$, where $S = \partial V$ is a finite area pleated surface. Set $\kappa(S) = 3g + n - 3$, which can be viewed as a measure of the complexity of S.*

- (i) *The end E_{rel} is geometrically infinite if and only if there is a sequence of closed geodesics $\{\gamma_n\}$ which exit E_{rel} with each γ_n freely homotopic in V to a simple geodesic $\gamma_n^* \subset S$.*
  There exists $L = L(\kappa(S))$ such that each γ_n may be chosen so that its length is $< L$ ([Bowditch 2011, Prop. 2.2]).
- (ii) *There is a uniquely determined measurable lamination $\Lambda(E_{\text{rel}}) \subset S$ such that $\Lambda(E_{\text{rel}}) = \lim \gamma_n^*$, for any such exiting sequence.*
- (iii) *Either a simply degenerate end E_{rel} has bounded geometry, or there exists $\eta > 0$ such that all closed geodesics in the end whose lengths are $\leq \eta$ are unlinked ([Bowditch 2011, Prop. 23.1]).*
- (iv) *Suppose instead that E_{rel} is a compressible end. The end E_{rel} is geometrically infinite if and only if there is a sequence of closed geodesics $\{\gamma_n\}$ which exit E_{rel} with each γ_n freely homotopic in V to a simple geodesic $\gamma_n^* \subset S$ which is incompressible.*
- (v) *Once the product structure of the neighborhood $V \cong S \times [0, 1)$ is fixed, there is a uniquely determined measurable lamination $\Lambda(E_{\text{rel}}) \subset S$ such that $\Lambda(E_{\text{rel}}) = \lim \gamma_n^*$ for any such exiting sequence.*

It needs to be emphasized that in the compressible case, there are countably many possibilities for expressing the end V as a product. This issue is analyzed at some length in terms of the Masur domain in Exercise (5-25). Without a specific specifica-tion, $\Lambda(E_{\text{rel}})$ is determined only up to the action by the group of automorphisms of S which are homotopic to the identity in $\mathcal{M}(G)$—but not in S itself. Here S is taken as the boundary component of a compact core that faces E_{rel}.

The lamination $\Lambda(E_{\text{rel}})$ is called the *ending lamination* of the end E_{rel}. It was first described in Thurston [1979a, Theorem 9.3.2]. The key point is that if Λ_1, Λ_2 are realizations of ending laminations on S, as a consequence of Lemma 5.7.2 their leaves do not cross. Therefore $\Lambda(E_{\text{rel}})$ can be defined as the union of all the limits. For compressible ends see Theorem 5.15.5, p. 356, for more details.

An ending lamination does not come with any particular measure and may support projectively a number of measures, or may be uniquely ergodic and support only one up to positive multiple.

The ending lamination $\Lambda(E_{\mathrm{rel}})$ of a geometrically infinite, incompressible relative end is a *filling/arational*[3] measurable geodesic lamination. In particular:

- Each simple geodesic on S is transverse to $\Lambda(E_{\mathrm{rel}})$. In fact if ν is any measured lamination on S with support different than $\Lambda(E_{\mathrm{rel}})$, then the geometric intersection number (Section 3.9) satisfies $\iota(\mu, \nu) \neq 0$, where μ is a measure on $\Lambda(E_{\mathrm{rel}})$.
- Each half-leaf in $\Lambda(E_{\mathrm{rel}})$ is dense in $\Lambda(E_{\mathrm{rel}})$.
- $\Lambda(E_{\mathrm{rel}})$ is not a proper sublamination of any measurable lamination; each component of $S \setminus \Lambda(E_{\mathrm{rel}})$ is an ideal polygon, possibly containing a puncture. (A geodesic that divides an ideal polygon is isolated and its endpoints are cusps so it cannot support a measure.)
- Ending laminations do not penetrate small neighborhoods of cusps (punctures).

The blockbuster proof of the Ending Lamination Conjecture is built on the pioneering work of Yair Minsky [2010] in constructing an ingenious Lipschitz model of $\mathcal{M}(G)$. The model is related to the curve complex (Exercise (5-20)), studying the combinatorics of what corresponds in the curve complex to short geodesics and the graph distances between them (Exercise (5-17)). The basis for the analysis in the curve complex was worked out jointly with Masur [1999; 2000]. The proof for the incompressible case is in Minsky [2010]. There, the problem is reduced to one of surface groups. The proof for the case of surface groups by Brock, Canary, and Minsky is in Brock et al. [2012]. For a new proof of the full result see Bowditch [2011]. Their work shows that the lipschitz map from the curve complex is actually bilipschitz. The *coup de grâce* was delivered by applying the Tameness Theorem [Agol 2004; Calegari and Gabai 2004].

The magic of the Minsky model is that it *depends only on the topology of the interior, and the ending lamination/conformal structure* of each end of $\mathcal{M}(G)$. Therefore, two manifolds $\mathcal{M}(G), \mathcal{M}(G_1)$ with homeomorphic interiors and such that corresponding ends have the *same* ending laminations are related by a bilipschitz homeomorphism. Such a homeomorphism is in particular quasiconformal. Therefore when lifted to $\mathbb{H}^3$ it extends to be a quasiconformal map $\mathbb{S}^2 \to \mathbb{S}^2$ that conjugates G to G_1, and is conformal $\Omega(G) \to \Omega(G_1)$. Such a map can only be a Möbius transformation (Theorem 3.13.5).

To indicate the starting point of the model, here is Minsky's basic construction [Minsky 2010, p. 4]: Suppose $\mathcal{M}(G) \cong S \times \mathbb{R}$ is a doubly degenerate group without parabolics based on a closed surface S. Denote the ending laminations by $\nu(= \nu^+ \cup \nu^-)$. A model manifold M_ν with a homeomorphism ρ onto $\mathcal{M}(G)$, and a piecewise riemannian metric is constructed to have the following properties: There is a subset $\mathcal{U} \subset M_\nu$ of mutually disjoint open solid tori $\{U\}$ the boundaries of which are euclidean tori in the metric. Each boundary ∂U has a modulus $\omega(U)$. Each element

---

[3] The two terms mean the same thing and both are widely used.

U is isometric to a tubular neighborhood of a hyperbolic geodesic γ_U in $\mathcal{M}(G)$ so that $|\omega(U)| \to \infty$ as the geodesic length of $\gamma_U \to 0$.

Denote by $\mathcal{U}[k]$ the union of components of $\mathcal{U}$ with $|\omega(U)| \geq k$, and set $M_\nu[k] = M_\nu \setminus \mathcal{U}[k]$. Here $M_\nu[0] = M_\nu \setminus \mathcal{U}$, with U, is akin to a thick/thin decomposition. Correspondingly there is a decomposition of $\mathcal{M}(G)$ by mutually disjoint tubular neighborhoods of geodesics of length $< \varepsilon$, for some ε. Denote by $\mathcal{T}[k] \subset \mathcal{M}(G)$ the union of the tubular neighborhoods corresponding to $\mathcal{U}[k]$ under the homeomorphism $\rho : M_\nu \to \mathcal{M}(G)$.

LIPSCHITZ MODEL THEOREM [Minsky 2010] *There exists $K, k > 0$ and a homeomorphism $f : M_\nu \to \mathcal{M}(G)$ such that,*

(i) *f induces ρ on $\pi_1(M_\nu)$, is proper, and maps the ends of M_ν to $\nu^\pm$ of $\mathcal{M}(G)$,*
(ii) *f is K-Lipschitz on $M_\nu[k]$ in the induced path metric,*
(iii) *$f : \mathcal{U}[k] \to \mathcal{T}[k]$, $f : M_\nu[k] \to \mathcal{M}(G) \setminus \mathcal{T}[k]$,*
(iv) *For $U \in \mathcal{U} \subset \mathcal{U}[k]$, the restriction f_U is λ-Lipschitz, $\lambda = \lambda(K, |\omega(U)|)$.*

The key idea of the proof is to use the geometry of the curve complex (Example 5-17) to obtain bounds on the length of curves in $\mathcal{M}(G)$.

The universal tubular neighborhoods property (p. 127) gives some insight into why it is possible to build a model based on short geodesics.

The Minsky model is not only the basis of the proof of the Ending Lamination Conjecture, but also the basis of the proof of local connectivity of limit sets, see Section 5.14 and the classification of geometric limits at the Bers boundary Section 5.13.

A somewhat simpler, more combinatorial, construction of a model manifold is described in Bowditch [2011].

Minsky describes the combinatorial basis of the proof in [2003b]. Earlier [1999] he had solved the Ending Lamination Conjecture for the once-punctured torus case—which satisfies Bonahon's Condition B.

A more recent, comprehensive proof of the entire theorem was developed by Brian Bowditch [2011]. It also is based on the Minsky model, although it is constructed differently. For the doubly degenerate case there are two (mutually transverse) ending laminations. These are two boundary points of the curve complex [Klarreich 1999b]. The starting point of Bowditch' model is a bi-infinite geodesic between the two boundary points. This corresponds to a sequence of successively disjoint simple curves in the reference surface.

Theorem 5.7.4 (The Ending Lamination Theorem). *Suppose $\phi : G_1 \to G_2$ is isomorphism between finitely generated groups so that $\phi(g)$ is parabolic if and only if g is so. Assume ϕ is induced by a homeomorphism $\Phi : \mathcal{M}(G_1) \to \mathcal{M}(G_2)$ such that*

* *$\Phi : \partial \mathcal{M}(G_1) \to \partial \mathcal{M}(G_2)$ is homotopic to a conformal mapping, and*
* *Corresponding geometrically infinite ends E_{rel} and $\Phi(E_{\mathrm{rel}})$ have the same ending laminations.*

Then $\phi : G_1 \to G_2$ is realized by an conjugation and Φ by an isometry.

The ending invariants are Riemann surfaces or filling laminations, which in compressible cases support Masur domain measured laminations.

In short,

A hyperbolic manifold is uniquely determined up to isometry by its topological structure and its ending invariants.

An end designated by E_{rel} is a true relative end only if it has parabolics. It is shown in the course of the proofs that the ending lamination at an end or relative end e is precisely the unrealized lamination in reference to e.

Complementing the Ending Lamination Theorem is the following.

Theorem 5.7.5 (All ending possibilities occur [Namazi and Souto 2012], [Ohshika 2011]). *On the boundary of a given geometrically finite $\mathcal{M}(G)$, place the relative ends, that is the components of $\partial\mathcal{M}(G)$, into three disjoint sets as follows:*

(i) *$f \geq 0$ marked Riemann surfaces $R_1 \dots, R_f$,*
(ii) *$m_{incomp} \geq 0$ incompressible ends $e_1, \dots, e_{m_{incomp}}$,*
(iii) *$n_{comp} \geq 0$ compressible ends $E_1, \dots, E_{n_{comp}}$.*

Prescribe a

(i) *Possibly new conformal structure on each $R_1, \dots, R_m$,*
(ii) *Filling/aratrional measurable lamination for each of $e_1, \dots, e_{m_{incomp}}$,*
(iii) *Filling/arational measurable lamination in the Masur domain for each of $E_1, \dots, E_{n_{comp}}$.*

There exists a manifold $\mathcal{M}(H) \subset \mathcal{R}_{disk}(G)$ which has the stated laminations as ending laminations.

By the Ending Lamination Theorem, $\mathcal{M}(H)$ is uniquely determined up to isometry by the given data and the topology of $\mathcal{M}(G)$.

There is an interesting special case:

Corollary 5.7.6 ([Namazi and Souto 2012]). *Suppose $\mathcal{M}(H)$ homeomorphic to the interior of a handlebody X (of genus ≥ 2) and $\lambda \subset \partial X$ is a filling Masur domain lamination which is not realized in $\mathcal{M}(H)$. Then there is a homeomorphism $\phi : X \to \mathcal{M}(H)$ such that $\phi(\lambda)$ is the ending lamination of $\mathcal{M}(H)$.*

The space of ending laminations

The space $\mathcal{EL}(S)$ is defined as those geodesic laminations $\{\Lambda\}$ on a finite volume hyperbolic surface S which are *minimal and filling*. In brief this means that every leaf λ is dense in Λ and every simple closed geodesic crosses Λ, see p. 166. Ending laminations carry one or more inequivalent bounded measures, but here we do not need them.

The topology on $\mathcal{EL}(S)$ is the Hausdorff topology, see p. 232. In this topology $\mathcal{EL}(S)$ is totally disconnected [Thurston 1998], and has Hausdorff dimension zero [Zhu and Bonahon 2004].

The space $\mathcal{EL}(S)$ defined above is exactly the space of ending laminations of geometrically infinite ends as described in Section 5.7.

Gabai [2014; 2009] found interesting properties of $\mathcal{EL}(S)$ instead using its *coarse topology*. This is weaker than the Hausdorff topology in that $\{\Lambda_n\}$ converges to Λ if there is a subsequence $\{\Lambda_m\}$ that converges in the Hausdorff topology to the possible extension Λ^+ of Λ by adding a finite number of extra leaves. In this topology he proved that $\mathcal{EL}(S)$ becomes path connected and locally path connected.[4]

Denote the space of ending laminations on a hyperbolic surface S bounding a neighborhood of an end by $\mathcal{EL}(S)$. This means that we are looking at the space of measured laminations $\mathcal{ML}(S)$ on S, and then ignoring the measure. It consists of all filling/arational measured geodesic laminations with the topology induced from $\mathcal{ML}$ (see Section 3.9), after ignoring the measures. Dave Gabai proved [2009] that $\mathcal{EL}(S)$ is path connected and locally path connected, provided S is not a 3- or 4-punctured sphere, or a 1-punctured torus.

However, the end invariants are not continuous in $\mathfrak{R}(G)$. For a simple example consider a cusp on $\partial\mathfrak{R}_{\mathrm{disc}}(G)$. As we make a "tangential" approach to the cusp from within the deformation space, or approach the cusp along its boundary, the geometric limit (as we may assume) is larger than the algebraic; the limit sets of the approximates will look more and more like the limit set of the geometric limit. For a full discussion see Brock [2000; 2001a].

The following assertion follows from Theorem 5.7.3 and Formula (5.2), p. 297; see also Exercise (5-31).

Corollary 5.7.7. *Suppose $\{G_n\}$ is a sequence of quasifuchsian groups converging algebraically and geometrically to a singly degenerate group H. Let (Λ_n, β_n) denote the bending lamination of the boundary component C_n of the convex core $\mathcal{CC}(G_n)$ of $\mathcal{M}(G_n)$ that is approaching the infinite end E of H. Then $\{(\Lambda_n, \beta_n)\}$ converges to a measured lamination (Λ, β) in the reference surface S such that Λ is not realizable in $\mathcal{M}(H)$. That is, Λ is the ending lamination of E.*

When applied more generally to algebraic and geometric limits of geometrically finite manifolds, Corollary 5.7.7 gives a "natural" way of finding the ending laminations.

Alternately, instead of bringing in the bending laminations, the process can be described in terms of the hyperbolic metrics g_n on the degenerating component or components of $\partial\mathcal{M}(G_n)$. These metrics are obtained from pulling over the metric in $\mathbb{H}^2$ to the degenerating component or components of $\Omega(G_n)$ by the Riemann maps. The one or two sequences $\{g_n\}$ converge to measured laminations whose support(s)

---

[4] Unless S is the 3-punctured sphere, when $\mathcal{EL}(S) = \varnothing$, or S is the 1-punctured torus or 4-punctured sphere, when $\mathcal{EL}(S) = \mathbb{R} \setminus \mathbb{Q}$.

are the ending lamination(s) in the sense of Thurston (p. 296); see the Double Limit Theorem (p. 303). Theorem 3.11.5 displays how the geometry of the convex core boundary and the geometry of the surface facing it "at infinity" are related.

The Ending Lamination Theorem has the following consequence proved in Brock et al. [2012] under the assumption of tameness; a prior version (including compressible ends) was given by Ohshika [1998b]. The second statement is an application of Sullivan's Theorem (p. 183). Once again we have applied tameness to complete the original statements.

Theorem 5.7.8 (**Quasiconformal and conformal rigidity**). *Suppose $\mathcal{M}(G)$ and $\mathcal{M}(H)$ have incompressible ends. Assume there is an orientation-preserving homeomorphism $\Psi : \mathbb{S}^2 \to \mathbb{S}^2$ which induces an isomorphism $\psi : G \to H$ so that $\Psi \circ g(z) = \psi(g) \circ \Psi(z)$ for all $z \in \mathbb{S}^2$ and $g \in G$. Then there is a quasiconformal mapping $F : \mathbb{S}^2 \to \mathbb{S}^2$ that likewise satisfies $F \circ g(z) = \psi(g) \circ F(z)$ for all $z \in \mathbb{S}^2$ and all $g \in G$.*

If Ψ is conformal on $\Omega(G)$, or if $\Omega(G) = \varnothing$, then F is a Möbius transformation.

The essence of the proof is to show $\mathcal{M}(G)$ and $\mathcal{M}(H)$ have the same ending laminations.

Before the Ending Lamination Theorem was announced, the Ending Lamination Conjecture was proved in Minsky [2001] under the following assumption: There exists a positive lower bound $\delta > 0$ for the length of all closed geodesics in the interior of the manifold. This condition forces a certain uniformity in how pleated surfaces converge to the ends. Yet in general for geometrically infinite manifolds, in fact at a dense set of points on the boundary of the deformation space of any geometrically finite group, the uniform lower bound condition is not satisfied [McMullen 1991, Corollary 1.6; Canary et al. 2003; Canary and Hersonsky 2004].

A manifold $\mathcal{M}(G)$ satisfying the uniform lower bound condition is said to have *bounded geometry*. Geometrically finite manifolds automatically have bounded geometry. For a geometrically infinite example, see Section 5.12.

5.8 The Double Limit Theorem

The following result is a precursor both to the Density Theorem 5.9.1 and to the fibered case of the Hyperbolization Theorem 6.1.1. It is a general limit theorem for sequences of quasifuchsian groups, the general case of simultaneous uniformization. By its means we can find singly and doubly degenerate groups.

We will work with a finite area hyperbolic surface R. See Section 3.9 for the basics of the theory of geodesic laminations.

The Double Limit Theorem [Thurston 1986c]. *Set $\overline{\mathfrak{Teich}}(R) = \mathfrak{Teich}(R) \cup \partial_{\mathrm{th}}\mathfrak{Teich}(R)$. Choose a binding pair of laminations with compact support Λ_-, Λ_+ in $\overline{\mathfrak{Teich}}(R)$. Choose any pair of convergent sequences of hyperbolic metrics $\{g_n\}$ and $\{h_n\}$ in $\mathfrak{Teich}(R)$ which converge respectively to Λ_- and Λ_+ in $\overline{\mathfrak{Teich}}(R)$. Let H_n be a normalized quasifuchsian group where* (i) *the bottom surface R_{n-} of $\partial\mathcal{M}(H_n)$ is* *determined by* g_n, *the top* R_{n+} *by* h_n, *and* (ii) *the natural involution between them respects the markings.*

   Then the quasifuchsian groups $\{H_n\}$ *converge algebraically to a group* H, *where* $\mathcal{M}(H) \in \overline{\mathfrak{Teich}}(G)$ *has ending laminations* Λ_- *and* Λ_+, *respectively.*

   In fact from Section 5.7, there are sequences $\{\mu_n, \ \nu_n\}$ of measured laminations converging in $\mathcal{ML}$ to Λ_- Λ_+ respectively such that for the length in each metric

$$\lim_{n\to\infty} \mathrm{Len}_{g_n}(\mu_n) \ = \ \lim_{n\to\infty} \mathrm{Len}_{h_n}(\nu_n) \ = 0.$$

This reinforces the picture of "pinching" the approximating surfaces along the ending laminations.

   Convergence to a point on the Thurston boundary $\partial_{\mathrm{th}}\mathfrak{Teich}(R)$ is discussed starting p. 314. If R has punctures the laminations are of course confined to a compact submanifold. We have modified the original statement by bringing in the Ending Lamination Theorem 5.7.4, which releases us from the obligation of passing to subsequences. The limit group H will be doubly degenerate without *new* parabolics only if both laminations are arational. If instead $\Lambda_{\mathrm{bot}}, \Lambda_{\mathrm{top}}$ are transverse pants decompositions, the limit group H is a maximal cusp; the two boundary components correspond to the result of pinching the top and the bottom along the respective pants loops.

   Doubly degenerate groups appear as subgroups of hyperbolic 3-manifolds that fiber over the circle, as we will see later. The first doubly degenerate group appears in Jørgensen [1977a], it was given by explicit generating matrices. For other explicitly constructed degenerate groups see p. 372 and Jørgensen and Marden [1979].

   In Kleineidam and Souto [2002] it is shown how to degenerate a Schottky group or more generally a compression body. They show that it suffices to take (i) for compressing boundary components, a sequence of points in the Teichmüller space of each compressing component that converge to a filling lamination on the Thurston boundary which lies in the Masur domain, and (ii) for incompressible components, a sequence in the Teichmüller space of each boundary component that converges to a filling lamination in the corresponding Thurston boundary.

   As for all infinite ends or relative ends e there is a surface S embedded in the interior of the manifold and "parallel" to e. The surface S carries the ending lamination or the Masur domain lamination as the case may be. The laminations also can be realized in the universal covering surface, say $\mathbb{D}$ of S where the endpoints of the leaves are dense in $\partial\mathbb{D}$.

5.9 The Density Theorem

There is a long list of mathematicians who have contributed to the complete solution of the forty-year old density conjecture under a decreasing number of special assumptions. The succession of proofs began with Bromberg [2007] and Brock-Bromberg [Brock and Bromberg 2004], and finally, independently, the general case was established recently by Ohshika [2011], Bromberg-Souto [≥2015], Namazi-Souto [2012].

The Ohshika and Namazi-Souto proofs depend on the Tameness and Ending Lamination Theorem, while Bromberg-Souto proceed without ELC by using the deformation theory of cone manifolds. The case of freely decomposible ends is the toughest part of the proof.

Theorem 5.9.1 (The Density Theorem). *Every finitely generated, torsion-free kleinian group H is the algebraic limit of geometrically finite groups. In more detail, suppose G is geometrically finite and $\theta : G \to H$ is a parabolic-preserving isomorphism to the group $H \in \mathfrak{R}_{\mathrm{disc}}(G)$. There exists a sequence of isomorphisms $\{\theta_n : G \to H_n \in \mathrm{Int}\,\mathfrak{R}_{\mathrm{disc}}(G)\}$ to geometrically finite groups which converges algebraically to θ. That is, the closure $\mathrm{Int}\,\mathfrak{R}_{\mathrm{disc}}$ equals $\mathfrak{R}_{\mathrm{disc}}(G)$.*

In fact one can choose algebraic approximates $\{H_n^*\}$ to H to be minimally parabolic.

The last assertion is proved in Ohshika [2011].

Recall from Theorem 4.6.3 that the sequence can be chosen so that $\theta_n \theta^{-1}$ also preserves parabolics, but then $\theta_n(G)$ will not be in the same deformation space if H has new parabolics. In the opposite direction, the theorem can be formulated [Ohshika 2011] so that it is assumed that G has no rank one parabolic subgroups. Then H is also approximated by geometrically finite groups without rank one parabolics; all of the rank one parabolic subgroups of H are "new".

Density implies tameness by Corollary 5.6.5. That tameness implies density is a consequence of the Ending Lamination Theorem, via Ohshika's work using the Double Limit Theorem Section 5.8.

Thus we have a complete answer to what was suspected earlier:

Corollary 5.9.2. *The Tameness Conjecture holds if and only if the Density Conjecture holds.*

The Double Limit Theorem is a particular case of Theorem 5.9.1; the doubly degenerate case is special because of the extra requirement that a pair of potential ending laminations must be chosen to be mutually transverse (they have a positive intersection number).

5.10 Bers slices

This section begins with a fuchsian group G acting in the upper and lower half-planes such that $R = \mathrm{LHP}/G$ is a surface of genus g with $b \geq 0$ punctures satisfying $3g + b - 3 > 0$. The reflected surface $R' = \mathrm{UHP}/G$ is anticonformally equivalent to R under reflection $J_0(z) = \bar{z}$ in $\mathbb{R}$.

The simplest deformation spaces are the quasifuchsian spaces $\mathfrak{T}(G)$. By the principle of simultaneous uniformization (Section 3.8), the points of this space can be described by the triples

$$\mathfrak{T}(G) = \{(S_{\mathrm{bot}}, S^{\mathrm{top}}; J), F\}.$$

Here $S_{\mathrm{bot}}, S^{\mathrm{top}}$ are Riemann surfaces quasiconformally equivalent to R, R' respectively and J is an orientation-reversing involution $S_{\mathrm{bot}} \leftrightarrow S^{\mathrm{top}}$. The marking

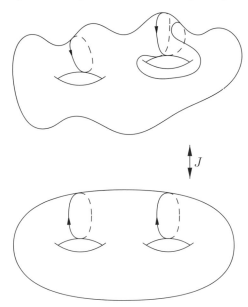

Fig. 5.2. A genus-2 quasifuchsian group. J is an orientation-reversing homeomorphism pairing the curves.

is fixed by the homotopy class $[F]$ of a quasiconformal mapping $R \to S_{\text{bot}}$. Each triple is associated with a quasifuchsian group H with $\mathcal{M}(H) \cong S_{\text{bot}} \times [0, 1]$ in the following manner:

(i) H is uniquely determined by the triple $(S_{\text{bot}}, S^{\text{top}}; J)$, up to conjugation.
(ii) $\partial \mathcal{M}(H) = S_{\text{bot}} \cup S^{\text{top}}$ and J extends to an orientation-reversing, fiber-preserving involution $J : \mathcal{M}(H) \to \mathcal{M}(H)$.
(iii) There exists a quasiconformal map $F : \mathcal{M}(G) \to \mathcal{M}(H)$ such that $F(R) = S_{\text{bot}}$, $F(R') = S^{\text{top}}$ and $F J_0 F^{-1}$ is homotopic to J.
(iv) F is the projection of the restriction to $\Omega(G) = \text{LHP} \cup \text{UHP}$ of a quasiconformal map $F^* : \mathbb{S}^2 \to \mathbb{S}^2$; F^* determines an isomorphism $\varphi : G \to H$.

We will write $\partial_{\text{bot}} \mathcal{M}(H) = S_{\text{bot}}$ and $\partial^{\text{top}} \mathcal{M}(H) = S^{\text{top}}$.

The *Bers slice* $\mathcal{B}(R) \equiv \mathcal{B}(G) \subset \mathfrak{T}(G)$ determined by the Riemann surface $R = \text{LHP}/G$ is defined as the subset of triples

$$\mathcal{B}(R) = \{((S_{\text{bot}}, S^{\text{top}}; J), F) \in \mathfrak{T}(G) \,|\, F : R \to S_{\text{bot}} \text{ is conformal}\}.$$

In the deformations, the bottom surface remains conformally equivalent to R; the mapping F is quasiconformal on R' and *conformal* on R. A lifted map F^* is a Riemann map of the LHP onto the component $F^*(\text{LHP}) = \Omega_{\text{bot}}$ of $\Omega(H)$. Of course F^* satisfies the relations $F^* \circ g(z) = \varphi(g) \circ F^*(z)$ for all $g \in G$, $z \in \text{LHP}$. So the image group H is determined, up to normalization, by the quasiconformal mapping F^* on UHP.

Fig. 5.3. A once-punctured torus quasifuchsian group near the boundary of the quasifuchsian space.

Fig. 5.4. Slightly opening the cusp of Figure 5.3 results in this Schottky group. The limit set is totally disconnected but very close to a quasicircle.

A Bers slice is perhaps the most useful realization of the Teichmüller space (Section 2.8) of the Riemann surface R'; S^{top} runs through all possible quasiconformal deformations of R', all the while S_{bot} remaining conformally fixed as R. The initial discovery by Bers thrilled all in the field because in its realization as a slice within a space of complex matrices, the complex structure on $\mathfrak{Teich}(R)$ becomes more accessible [Ahlfors 1966].

The slice $\mathcal{B}(R)$ is a complex analytic manifold of $\mathbb{C}$-dimension $3g + b - 3$. It is also a metric space in the Teichmüller metric (p. 89). The full quasifuchsian space (p. 280) $\mathfrak{T}(G)$ as a holomorphic submanifold of the representation variety has twice the dimension of a slice.

At least in the case that R is a *closed* surface, there is an extension $\mathcal{B}^*(R)$ of the Bers slice, called the *extended Bers slice*, which is a *properly embedded* submanifold of $\mathfrak{R}(G)$ [Gallo et al. 2000, Theorem 11.4.1]. The extension is the space of projective structures on the Riemann surface R (see Exercise (6-9)).

Each choice of complex structure on R determines a different Bers slice.

Masaaki Wada's Mac program OPTi [Wada 2011; Wada 2006] visualizes quasi-conformal deformations of the once-punctured torus groups. It interactively draws in two and three dimensions isometric circles, Ford regions, limit sets, etc. The groups are parametrized by Jørgensen complex probabilities Exercise (1-34).

The team of Y. Komori, T. Sugawa, M. Wada, and Y. Yamashita was the first to succeed in visualizing the complex one-dimensional Bers slice of once-punctured torus space—based on the square torus. The method is explained in Komori and Sugawa [2004]. Their work was augmented by studies of David Dumas [2009], who analyzed the slice as an island in the archipelago which is the discreteness locus of $\mathcal{B}^*(R)$, lying in the sea of indiscreteness, see Figure 6-9 with a closeup of the Bers slice in Figure 5.5 embedded in the once-punctured torus space. Figure 5.5 is based on the hexagonal torus. The combinatorics of the Ford polyhedron give a tiling of the quasifuchsian space (Jørgensen). This tiling, restricted to the Bers slice, gives the tiling by triangular regions shown in Figure 5.5—this is a special property of the hexagonal slice.

The boundary of this slice is a Jordan curve [Minsky 1999] and the cusps you see are dense on it [McMullen 1991]. These are indeed geometric cusps [Miyachi 2003].

The Teichmüller modular group/mapping class group (see p. 89 and Exercise (5-12)) acts by isometries (and biholomorphic mappings) on $\mathcal{B}(R)$, akin to the action of a fuchsian group on $\mathbb{H}^2$. If τ is an orientation-preserving automorphism of the surface $R = S_{\text{bot}}$, realized as a quasiconformal automorphism if there are punctures, then τ induces the following action:

$$\tau : (S_{\text{bot}}, S^{\text{top}}, J) \mapsto (S_{\text{bot}}, S^{\text{top}}, J \circ \tau).$$

In this action the conformal types of the surfaces do not change, what changes is the *topological* relationship between the bottom and the top as suggested in Figure 5.2. The action of the iterates $\{\tau^n\}$ is described in Theorem 5.12.1 and Exercise (5-14).

Bers slices are not the only $(3g + b - 3)$-dimensional slices of quasifuchsian space. One can take a slice based on any boundary group of quasifuchsian space, except a doubly degenerate group. The bottom component S_{bot} might be a cusp, especially a maximal cusp (composed of triply punctured spheres). Such a slice is called a *Maskit slice*. One could as well take S_{bot} to be a singly degenerate end (in view of Sullivan's Theorem). Or one could require that $S_{\text{bot}} = S^{\text{top}}$ but that J match each $\gamma \in \pi_1(S^{\text{top}})$

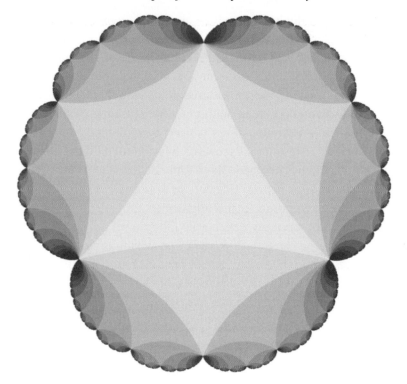

Fig. 5.5. Bers slice based on the hexagonal torus, inside the once-punctured torus space.

with $\alpha(\gamma) \in \pi_1(S_{\text{bot}})$ where α is a fixed automorphism taken on all surfaces in the Teichmüller space. For example one could start with a reflection in the top surface of a fuchsian group arranged so that the positive imaginary axis is fixed, followed by the reflection in the real axis. Keeping the symmetry under deformations results in a slice that depends only on the complex structure of one surface. Such submanifolds are called *Earle slices*.

5.11 The quasifuchsian space boundary

We start by considering the closure $\overline{\mathfrak{T}(G)}$ of a quasifuchsian space $\mathfrak{T}(G)$. This is contained in $\mathfrak{R}_{\text{disc}}(G)$. Its boundary is denoted by $\partial\mathfrak{T}(G)$. In the two subsections to follow, we will specialize to Bers slices which represent Teichmüller spaces. A Bers slice has two important boundaries: the analytic (Bers) boundary, and the geometric (Thurston) boundary.

The groups on $\partial\mathfrak{T}(G)$, referred to as boundary groups or *B-groups*, are classified as cusps (geometrically finite groups), singly and doubly degenerate groups, or partially degenerate groups.

We can approach a boundary cusp from $\mathfrak{T}(G)$ most directly by a process of pinching, as in Section 5.3 and Exercise (5-5).

In contrast, a singly or doubly degenerate group is a group H isomorphic to G such that either $\Omega(H)$ is connected or $\Omega(H) = \varnothing$. By Bonahon's criteria (p. 292), $\mathcal{M}(H)$ is homeomorphic to $R \times [0, \infty)$ in the singly degenerate case, or to $R \times (-\infty, +\infty)$ in the doubly degenerate case. Either one component or the whole boundary has become degenerated. By the Density Theorem, all manifolds with this topological structure lie on $\partial \mathfrak{T}(G)$. By Sullivan's Theorem, a doubly generate group is quasiconformally rigid (other than Möbius conjugations).

Doubly degenerate groups were discovered by Jørgensen on the boundary of once-punctured torus quasifuchsian space. His doubly degenerate groups H are periodic: There is a Möbius transformation $T \notin H$ satisfying $THT^{-1} = H$ such that the manifold $\mathcal{M}(H^*)$ corresponding to the augmented group $H^* = \langle H, T \rangle$ has finite volume. The limit set of a doubly degenerate group is all $\mathbb{S}^2$. Yet any fiber $R \times \{s\}$, $0 < s < 1$, lifts to a planar object P in $\mathbb{H}^3$ on which H acts as a surface group. Even so, ∂P is dense in $\mathbb{S}^2$. For more details see the Double Limit Theorem (p. 303), the discussion following it, and our discussion of Cannon-Thurston maps. A doubly degenerate group might well contain new parabolics.

The cases of singly and partially degenerate groups will be discussed on p. 317, again in the context of the Bers boundary.

The boundary has a lot of self-bumping as described in Theorem 5.3.2 because there are a lot of independent essential cylinders. The self-bumping occurs at most cusps. However there can be no self-bumping at maximal cusps (see Exercise (6-10)). Nor (per Bromberg and Holt) is there any bumping at singly or doubly degenerate groups, with or without parabolics—geometric limits agree with algebraic at such points.

5.11.1 The Bers (analytic) boundary

As before, suppose G is a fuchsian group acting un $\text{UHP} \cup \text{LHP}$ with $R = \text{LHP}/G$ and $R' = \text{UHP}/G$. In the future we will simply refer to R instead of R', relying on the context to indicate whether the reference is to the "bottom" or "top" of $\mathcal{M}(G)$. The Bers slice

$$\mathcal{B}(G) \equiv \mathcal{B}(R) \equiv \mathfrak{Teich}(R)$$

has a boundary

$$\partial \mathcal{B}(G) \equiv \partial \mathcal{B}(R) \subset \partial \mathfrak{Teich}(R) \subset \mathfrak{R}(R),$$

called the *Bers boundary* or *analytic boundary*. Unlike the boundary of the full quasifuchsian space, $\mathcal{B}(R) \cup \partial \mathcal{B}(R)$ is compact; its topology is the topology of algebraic convergence of kleinian groups. The compactness follows an once from the fact that the family of normalized conformal maps of LHP that conjugate G to another group H is compact. Therefore for every boundary group H exactly one component $\Omega_{\text{bot}}(H)$ of $\Omega(H)$ is invariant under the full group H. The theory of the boundary was first worked out in the pioneering papers [Bers 1970a] and [Maskit 1970].

Fig. 5.6. The limit set of a two-loxodromic generator group with elliptic commutator of order-3. It corresponds to a point in the quasifuchsian orbifold space of manifolds homeomorphic to $P \times [0, 1]$ where P is a torus T with an elliptic cone point of order-3.

A Bers slice coincides with the interior of its closure (see Exercise (5-30)); this fact is also a consequence of the proof of Density Theorem p. 305).

In the case that G represents a once-punctured torus so that $\mathcal{B}(R)$ has complex dimension one, it is known [Minsky 1999] that $\partial \mathcal{B}(R)$ is a Jordan curve (in a planar embedding), and hence locally connected. Also see McMullen [1998].

In general, as a subset of $\mathfrak{R}_{\mathrm{disc}}(G)$, it is not known whether or not the compact closure $\overline{\mathcal{B}(R)}$ is locally connected (Bromberg conjectures that it is not in general, like the full discreteness locus itself).

If a boundary group H is geometrically finite, that is if H is a cusp, H must contain new parabolics. Corresponding to these are a finite number of mutually disjoint simple closed geodesics $\{\alpha_i\}$ on $S_{\mathrm{bot}} = \partial_{\mathrm{bot}}\mathcal{M}(H)$, which determine parabolic transformations in H. These divide S_{bot} into one or more components $\{S_i\}$. Each S_i is parallel in $\mathcal{M}(H)$ to a component S_i' of $\partial^{\mathrm{top}}\mathcal{M}(H)$ which is a finitely punctured closed

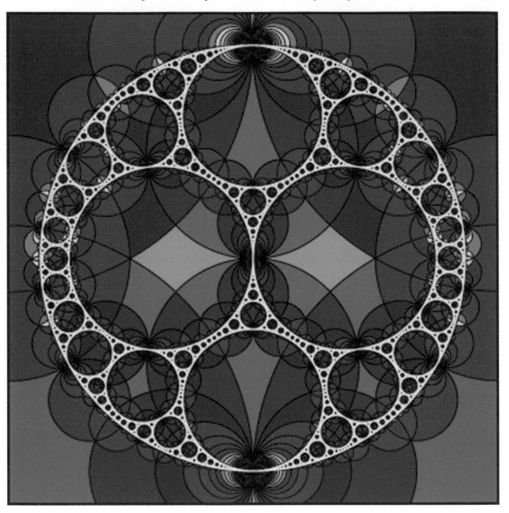

Fig. 5.7. A double cusp on the boundary of the quasifuchsian orbifold space of Figure 5.6. Namely on a surface P, transverse simple geodesics are pinched, one at the top boundary component and the other at the bottom. This gives rise to a rigid manifold whose top and bottom boundaries are 2-punctured spheres with a cone point of order-3: each disk shown is stabilized by a $(2, \infty, \infty)$ triangle group. Compare with Figure 5.9.

surface homeomorphic to S_i. The stabilizing subgroup of a component Ω_i' over S_i' is quasifuchsian. The configuration $\bigcup S_i'$ results from pinching the top surface in the free homotopy classes of the geodesics in S_{bot}.

As the images $\theta_n(g)$ of a loxodromic element $g \in G$ become parabolic $\theta(g)$ in H, the two fixed points coalesce into one, pinching $\Omega^{\mathrm{top}}(H)$ into two simply connected pieces. This happens simultaneously in the conjugacy class of $\theta(g)$ so that Ω^{top} becomes pinched into countably many simply connected regions.

Moving down to the quotient manifolds, when this happens for a number of conjugacy classes, we end up with the newly punctured surfaces $\{S_i'\} \subset \partial \mathcal{M}^{\mathrm{top}}(H)$ lying over the subsurfaces $\{S_i\} \subset S_{\mathrm{bot}}$. If we focus on the geodesics in the original manifold $\mathcal{M}(G)$ that are in the free homotopy classes of $\{\alpha_i\}$ then as we approach $\mathcal{M}(H)$ these

Fig. 5.8. This beautifully crafted limit set of Hausdorff dimension two belongs to a singly degenerate once-punctured torus group on the boundary of a Bers slice. The complement is connected (can you find your way out of the maze?) The "holes" in the picture contain horodisks at the parabolic fixed points and are an artifact of the algorithm: extremely long words in the generators would be required to fill them in.

and only these "exit" $\partial^{\text{top}}\mathcal{M}(H)$ in the sense that they are "becoming" parabolic fixed points.

For details see Maskit [1970; 1988]; Marden [1974a; 1977] and Exercise (5-5).

A *maximal cusp* on the Bers boundary is a cusp for which $\partial\mathcal{M}^{\text{top}}(H)$ is a union of triply punctured spheres. It corresponds to a set of pinching loops which form a pants decomposition of S_{bot}—each complementary component is a 3-holed sphere.

One or all of the top surfaces S_i' can themselves be degenerated so long as they are not triply punctured spheres. When not all are degenerated, the group is referred to as a *partially degenerate group*. Even if all become degenerated, one will be left with parabolic transformations which on Ω_{bot} act as hyperbolic transformations. If S^{top} is entirely degenerated with (or without) such pinchings, the resulting boundary group is called *singly degenerate*. Singly degenerate groups H are characterized by the property that $\Omega(H)$ is connected and simply connected. Such groups are constructed in Section 5.12.

Suppose G has no parabolics. There exist boundary groups H without any parabolics either, in fact there are lots of boundary groups which have no parabolics. For as pointed out by Bers [1970a], since G has a countable number of elements, the set of boundary groups $H \in \partial\mathcal{B}(G)$ which contain a parabolic element h has positive codimension in $\partial\mathcal{B}(G)$. Therefore most boundary groups H are without parabolics, and therefore are necessarily singly degenerate.

By Bonahon's criteria (p. 292), the interior of the manifold coming from a boundary group is, like all other manifolds of the deformation space, homeomorphic to $S_{\text{bot}} \times (0, 1)$. The longstanding question as to whether, conversely, every kleinian group with the structure of a singly degenerate group is a boundary group of some Bers slice was first answered affirmatively as a special case of the Density Conjecture:

Bromberg's Theorem. *Suppose* Γ *has no parabolics, is isomorphic to the fundamental group of a closed surface, and* $\mathcal{M}(\Gamma)$ *has exactly one geometrically finite end. Then* Γ *is a boundary point of the Bers slice determined by its geometrically finite end. Moreover,* Γ *is the algebraic limit of a sequence of quasifuchsian groups lying in this Bers slice.*

This was extended to all quasifuchsian spaces by Brock and Bromberg, and finally incorporated into the framework of the Density Theorem (p. 305).

McMullen [1991] proved the strongest form of a longstanding conjecture of Bers and introduced ideas that have been used for the more general Theorem 5.3.1.

Theorem 5.11.1 (Maximal cusps are dense). *In the topology of algebraic convergence, maximal cusps are dense on the boundary of any Bers slice* $\mathcal{B}(R)$ *in quasifuchsian space, where* R *is a Riemann surface of finite hyperbolic area.*

It was established in Kerckhoff and Thurston [1990] that when the genus of R exceeds one, the natural (biholomorphic) map from one Bers slice to another does *not* have a continuous extension to a map between the Bers boundaries, at least for closed surfaces of even genus. On the other hand, for the once-punctured torus case, in fact for all dimension-one Teichmüller spaces, the map does extend continuously to the slices based on different surfaces R [Bers 1981].

5.11.2 The Thurston (geometric) boundary

In about 1976, Thurston announced [Thurston 1988] that the closed, projectivized set of measured geodesic laminations $\mathcal{PML}(R)$, (see Section 3.9) which is homeomorphic to the sphere $\mathbb{S}^{6g+2b-7}$, serves as a boundary of $\mathfrak{Teich}(R)$. That is, the closure of $\mathfrak{Teich}(R)$ is $\mathfrak{Teich}(R) \cup \mathcal{PML}(R)$. The subset of $\mathcal{PML}(R)$ composed of finite disjoint unions of simple geodesics serves as the set of "rational points". The closure $\mathfrak{Teich}(R) \cup \mathcal{PML}(R)$ is homeomorphic to the closed ball $\mathbb{B}^{6g+2b-6}$.

Recall that a measured geodesic lamination on R, like a simple closed geodesic, transfers to a measured geodesic lamination at every point in $\mathfrak{Teich}(R)$. That is why its closure can be realized as above.

The main references for this section are Thurston [1986c]; Fathi et al. [1979, §8]; Bonahon [1988], and also Section 3.9. Fix a hyperbolic surface R as our reference surface.

Thurston identifies $\mathfrak{Teich}(R)$ with the space of hyperbolic metrics $\{\rho\}$ on R of curvature -1, modulo isotopy.[5] The topology of $\mathfrak{Teich}(R)$ is taken to be the minimum topology that the geodesic length function $\ell_\rho(\gamma)$ is a continuous function of ρ for all geodesics γ.

A geodesic lamination ν on R is the limit of simple geodesics in the hyperbolic metric on R. In a different hyperbolic metric ρ the corresponding limit exists as well (see Section 3.9).

With each hyperbolic metric ρ we can find the ρ-length $\ell_\rho(\nu)$.

---

[5] Two metrics ρ_1, ρ_2 are isotopic if there is a diffeomorphism f of R, isotopic to id, taking one metric to the other; the ρ_1-length of a loop a equals the ρ_2-length of $f(a)$.

A sequence of points $x_n \in \mathfrak{Teich}(R)$ has a subsequence which either converges to a point in $\mathfrak{Teich}(R)$, or to a measured lamination.

A sequence of hyperbolic structures $\{\rho_n\} \in \mathfrak{Teich}(R)$ is said to *converge* to $(\Lambda; \mu) \in \mathcal{PML}(R)$ if and only if there is a sequence of positive numbers $\{c_n \to +\infty\}$ such that for all $\nu \in \mathcal{ML}(R)$ with $\iota(\nu, \mu) \neq 0$,

$$\lim_{n \to \infty} \frac{\ell_{\rho_n}(\nu)}{c_n} = \iota(\nu, \mu). \tag{5.4}$$

Now as in Section 3.9 a measure μ is automatically associated with a geodesic lamination $|\mu|$, namely its support. Formula (5.4) reflects the fact that if $\{\rho_n\}$ converges the metrics become more and more concentrated on a geodesic lamination $\Lambda = |\mu|$. At the same time the measure $\mu(\nu)$ of a transverse measured lamination ν is just the intersection number. This in turn is the limit of the normalized ρ_n-length of ν.

It suffices to take measures ν to be supported on simple closed geodesics. We can then express the convergence criterion as either of

$$\lim_{n \to \infty} \frac{\ell_{\rho_n}(a)}{c_n} = \iota(a, \mu), \quad \lim_{n \to \infty} \frac{\ell_{\rho_n}(a)}{\ell_{\rho_n}(b)} = \frac{\iota(a, \mu)}{\iota(b, \mu)}, \tag{5.5}$$

for any simple loop a, or pair a, b of simple loops on R with $\iota(b, \mu) \neq 0$. Note that μ is determined only up to a multiplicative constant.

In classical language, the characterization of convergence is this. Let G be a fuchsian group G and $\mathfrak{T}(G) = \mathfrak{T}(R)$, $R = \mathbb{H}^2/G$, its associated Teichmüller space. A sequence of points $\{(S_n, f_n)\} \subset \mathfrak{T}(R)$, $f_n : R \to S_n$, is said to converge to the projective measured geodesic lamination μ realized on R if there exists a sequence $\{c_n > 0, c_n \to +\infty\}$ such that

$$\lim \frac{\ell_{S_n}(f_n(\gamma))}{c_n} = \iota(\gamma, \mu), \quad \text{for all simple loops } \gamma \subset R, \ \iota(\gamma, \mu) \neq 0. \tag{5.6}$$

As a somewhat different way of characterizing the convergence, Thurston [1986c, Theorem 2.2] proves that $\{\rho_n\}$ converges on R to the lamination $(\Lambda; \mu) \in \mathcal{PML}(R)$ if and only if there is a sequence of measured laminations $\{(\Lambda_n, \mu_n)\}$ converging projectively to $(\Lambda; \mu)$. When $\nu \in \mathcal{PML}(R)$ satisfies $\iota(\nu, \mu) \neq 0$ then:

$$\lim \frac{\ell_{\rho_n}(\nu)}{\iota(\mu_n, \nu)} = 1, \quad \ell_\rho(\mu_n) \to \infty, \quad \ell_{\rho_n}(\mu_n) < C < \infty, \tag{5.7}$$

for some constant C and all indices. Here ρ is a fixed metric on R. Moreover, there is a constant C' for which

$$\iota(\nu, \mu_n) \leq \ell_{\rho_n}(\nu) \leq \iota(\nu, \mu_n) + C'\ell_\rho(\nu).$$

This shows that near $\partial_{\mathrm{th}}\mathfrak{Teich}(R)$, $\iota(\nu, \mu_n) \sim \ell_{\rho_n}(\nu)$.

In particular, $\ell_{\rho_n}(\widehat{\mu}_n) \to 0$, where $\widehat{\mu}_n = \mu_n/\ell_\rho(\mu_n)$ lies in the projective class of μ_n. If μ_n is supported on a simple loop γ_n, $\ell_{\rho_n}(\gamma_n)/\ell_\rho(\gamma_n) \to 0$.

The significance of the Thurston boundary lies in the fact that the mapping class group $\mathcal{MCG}(R)$ extends continuously to it. The orbit of any point $x \in \partial_{\mathrm{th}}\mathfrak{Teich}(R)$ is dense. Furthermore the action is *ergodic*; any invariant subset of $\partial_{\mathrm{th}}\mathfrak{Teich}(R)$ either has full or zero measure [Fathi et al. 1979; Masur 1985].

The compactified $\mathfrak{Teich}(R) \cup \partial_{th}\mathfrak{Teich}(R)$ for a surface R with $g \geq 0$ and $n \geq 1$ punctures is homeomorphic via a map F to a finite-sided convex polyhedron $\mathbf{P} \subset \mathbb{R}^{6g+2n-6}$, the intersection of finitely many closed half-spaces, *modulo* the projective quotient [Hamenstädt 2003]. The $(6g + 2n - 5)$-dimensional image of F is the projection $\pi\mathbf{P}^{6g+2m-6}$ by factoring out constant multiples of the k-tuple.

The result is obtained by finding the length of $(6g + 2n - 6)$ mutually disjoint closed geodesics which together are filling on R, each with double self intersections, $(\ell_1, \ldots, \ell_{6g+2n-6})$. The F-image of this set of lengths defined over $\mathfrak{Teich}(R)$ fills out the interior of $\mathbf{P}$. The map F extends to the Thurston boundary upon the extension of its domain to the (2g+2n-6)-tuples $\{(\iota(\mu, \ell_1), \ldots, \iota(\mu, \ell_{6g+2n-6}))\}$, given any measured lamination μ. Passing to the projective quotient gives the final result.

Comparison of the Bers and Thurston boundaries

A boundary point of a Bers slice $\mathcal{B}(G)$ corresponds to a quasifuchsian manifold $\mathcal{M}(G)$ whose top surface has become degenerated in some way. The manifold $\mathcal{M}(H)$ is uniquely determined up to conjugacy by its ending lamination (the bottom surface of the degenerating manifolds remaining fixed)

The ending geodesic lamination λ corresponds to a single Thurston boundary point provided the space of projective measures with support on λ consists only of positive multiples of λ. Thus a maximal cusp on the Bers boundary determined by $3g - 3$ pinching curves corresponds to a $(3g - 4)$-dimensional subspace on the Thurston boundary $\partial_{th}\mathfrak{Teich}(R)$. On the other hand the result of pinching a single curve gives rise to a $(3g - 4)$-dimensional boundary space of $\partial\mathcal{B}(G)$ but a single point of $\partial_{th}\mathfrak{Teich}(R)$.

However the two boundaries agree for a dense set of points on each: A uniquely ergodic ending lamination is a single point on both boundaries, as all measured laminations with the same support are projectively equivalent. In particular the fixed points of pseudo-Anosov mappings are dense both on the analytic and the geometric boundaries [Fathi et al. 1979, §12.3]. These points are geodesic laminations (see Section 6.1).

As is put in Brock [2001a], a point on the Bers boundary of $\mathcal{B}(G)$ corresponds to a geodesic lamination, namely the ending lamination, the end result of pinching or degenerating of some manifold. On the other hand, a point on the Thurston boundary corresponds consists the projective class of a measured geodesic lamination on $\partial\mathcal{M}(G)$, a measure of "stretching" of a divergent sequence of hyperbolic metrics, as described above. While the map from the Bers boundary to the projective quotient of the Thurston boundary is an injection, it is not continuous. For the proof, see Brock [2001a].

In contrast to the Bers boundary, the Teichmüller modular group, i.e. $\mathcal{MCG}$, extends so as to become a group of automorphisms (self-homeomorphisms) of $\mathfrak{T}(G) \cup \partial_{th}\mathfrak{T}(G)$. It is because of this property that ∂_{th} is such an important structure in Teichmüller theory.

As cited earlier, the orbit under the modular group of each point of ∂_{th} is dense in ∂_{th}.

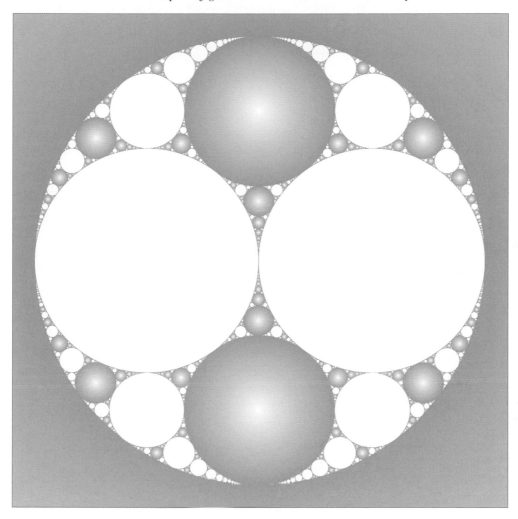

Fig. 5.9. A pinched, rigid, 2-generated Schottky group G is constructed by choosing two pairs of specially chosen generating disks: Select (i) the lavender disk D^* centered at ∞, (ii) the points of tangency u, w of ∂D^* with the two large white disks, (iii) the tangency x of δD^* with the bottom lavender disk, (iv) the point of tangency v of the two large white disks. The pairs of disks are: (i) the disks D_2 and D_3 tangent at x and orthogonal to D^* at u, w; (ii) (a) the half-plane/disk above the line through u, v, w, and (b) the disk tangent to it at v and to D_2 and D_3. G is generated by 2 parabolics with fixed points at tangencies of the pairs. The handlebody boundary is pinched so as to become two 3-punctured spheres, while the interior remains an open handlebody. The sets of lavender disks, and white disks, are invariant under the group. Alternately, the gasket can be built by placing chains of maximal disks in triangular interstices.

For the torus, both the Bers and the Thurston boundaries coincide with $\mathbb{R} \cup \{\infty\}$, with its usual topology.

5.12 Examples of geometric limits at the Bers boundary

We ask, if a sequence converges algebraically to a boundary point H of a quasifuchsian deformation space, what are its possible geometric limits?

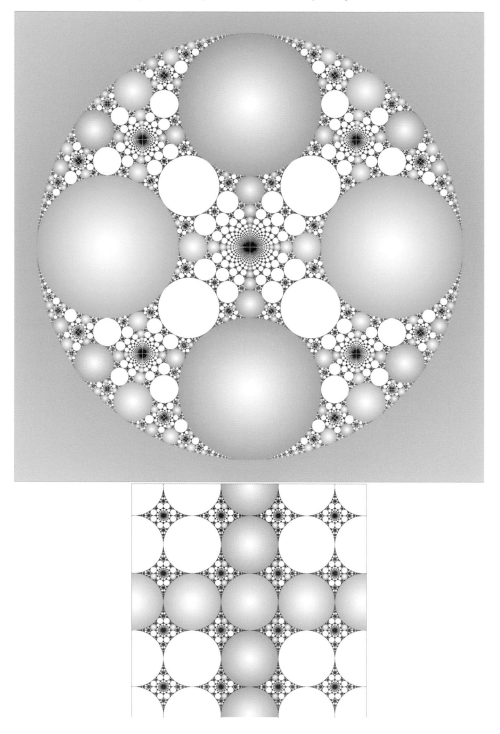

Fig. 5.10. The Apollonian Gasket group G of Figure 5.9 is here augmented by a rank two parabolic group G_v with fixed point at the center v. The picture below results when v is moved to ∞, with the disk at ∞ now becoming the disk centered at the origin v. From this the action of the two orthogonal translations generating G_0 is clear. Each disk is the universal cover of a 3-punctured sphere.

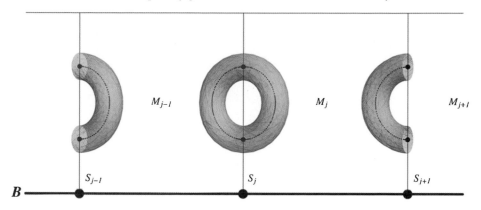

Fig. 5.11. An infinitely generated geometric limit at the boundary of quasifuchsian space of once-punctured tori.

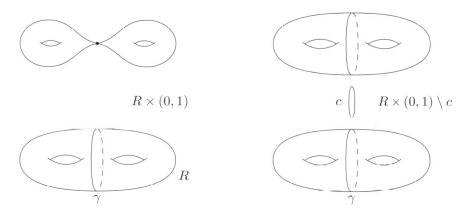

Fig. 5.12. The algebraic (left) and geometric (right) limits at a cusp on the boundary of the quasifuchsian space of a closed, genus-2 surface obtained as the limit of the iteration of a point O in a Bers slice by powers of the Dehn twist about γ. The top surface of the algebraic limit is the result of pinching $\gamma \subset R^{\text{top}}$ of O while the top surface of the geometric is conformally equivalent to R^{top}. See Exercises (4-21, 5-12, 5-14).

We know that if $\Omega(H) = \varnothing$, any geometric limit coincides with the algebraic. If the algebraic has no new parabolics, the same is true for $\Omega(H) \neq \varnothing$.

At a maximal cusp $X \in \partial \mathfrak{T}(G)$, the convex core $\mathcal{CC}(X)$ of $\mathcal{M}(X)$ is embedded in $\mathcal{M}(X^*)$ for any geometric limit X^* at X; see [Anderson et al. 1996; Prop. 32].

The once-punctured torus case

It is illuminating to consider consider the case of the once-punctured torus quasifuchsian space, $\mathfrak{T}(G)$ where the possibilities for geometric limits were enumerated by Troels Jørgensen in unpublished work. There, a manifold on $\partial \mathfrak{T}(R)$ is either singly or doubly degenerate, or one or two ends are thrice-punctured spheres. Only in the latter cases can a geometric limit be strictly larger than the algebraic.

It is easiest to understand Jørgensen's description if we start with a group $H_0 = \theta(G_0)$ such that $\partial \mathcal{M}(H_0)$ consists of two thrice-punctured spheres. Of the three

Fig. 5.13. The limit set of an algebraic limit corresponding to Figure 5.12. The cusp is chosen so that the pinched components of the manifold are covered by round disks.

Fig. 5.14. The limit set of the geometric limit at the cusp of Figure 5.14. It can be constructed by reflecting the algebraic limit in the circles.

punctures on each end, one puncture on each is represented by the same conjugacy class of parabolic transformations. Then we will degenerate both ends getting a manifold $\mathcal{M}(H)$. It is pictured as a doubly infinite strip with a "backbone" B along the bottom, signifying the locus of the common puncture (conjugacy class) of our groups. Refer to Figure 5.11.

The strip is divided into an infinite number of "chunks" by the presence of a discrete sequence of singular fibers $\{S_j\}$ exiting both ends.

Each singular fiber S_j is a triply punctured sphere while each chunk M_j is an open submanifold homeomorphic to $T_j \times (-1, +1)$, where T_j is a once-punctured torus with its puncture on the backbone.

Each S_j results from pinching a simple geodesic γ_j in M_j. For $k \neq j$, the geodesics γ_k, γ_j are not freely homotopic within $\mathcal{M}(G)$

Each fiber S_j corresponds to a distinct generator pair A_j, B_j of H with the following properties. (1) The commutator $[A_j, B_j]$ generates a rank one parabolic subgroup whose conjugacy class forms the backbone B. (2) The element A_j is parabolic, corresponding to the pair of new parabolics on S_j. (3) The element B_j is loxodromic. Finally (4) $\langle A_j, B_j \rangle = \pi_1(M_j)$.

The pair (A_j, B_j) can be found by inserting the in M_j the solid pairing tube associated with the two new punctures on S_j forming again a once-punctured torus bounding M_j thereby giving rise to A_j, B_j.

In M_{j-1} there is also a solid pairing tube pairing the two punctures on S_j. By joining them together we obtain a solid cusp torus in $M_{j-1} \cup S_j \cup M_j$. We thereby obtain a rank two parabolic group. No two of these rank two groups are conjugate within H.

The manifold corresponding to $\langle A_j, B_j \rangle = \pi_1(M_j)$ is a double cusp group.

The generator pairs associated with S_j and S_{j+1} are related by a Nielsen transformation as in Exercise (5-4).

For an explicit construction of groups of the type considered see Exercise (5-14). The analogue for general quasifuchsian groups is presented in Thurston 1986c, Theorem 7.2.

Brock's partially degenarate examples

We have already indicated in the context of the Jørgensen picture one class of geometric limits. More generally, suppose the group H represents a cusp on a $\partial\mathcal{B}(G)$. The Jørgensen picture suggests that if H is approached "radially" from within $\mathcal{B}(G)$, the geometric and algebraic limits coincide. To get a larger geometric limit, it is necessary to approach the cusp "tangentially" or even "ultratangentially". By analogy, in the modular group, contrast the radially approach $z_n \to \zeta$ to the fixed point of a parabolic T with the tangential approach $z_n = T^n(z) \to \zeta$.

Brock [2001a] has given a complete description of the geometric limits resulting from the iteration $\{\tau^n(O)\}$ of a point $O \in \mathcal{B}(G)$ by a reducible pseudo-Anosov automorphism τ. This generalizes the case that τ is a Dehn twist (Exercise (5-10)), which was examined in Marden [1980]; Kerckhoff and Thurston [1990] and is presented in Exercise (5-14).

A *reducible* automorphism (see Exercise (5-12)) is an automorphism $\tau : R \to R$ that fixes a set of free homotopy classes represented by mutually disjoint, nontrivial, simple closed curves $\{\gamma_i\}$ on the surface R, none of which can be homotoped to a puncture. After replacing τ by a power τ^n we may assume that the curves themselves, and furthermore each complementary component $R \setminus \bigcup \gamma_i$, is fixed. These are neither simply nor doubly connected. Then in each complementary component τ^n is either (homotopic to) the identity, or is a pseudo-Anosov—fixing the boundary curves. If τ_n is the identity on both sides of a loop γ_j, then τ is a Dehn twist (Exercise (5-10)) about γ_j. A pseudo-Anosov automorphism τ^n is called a *reducible pseudo-Anosov* if it is pseudo-Anosov in at least one complementary component. The once-punctured torus is special in that there are no reducible pseudo-Anosovs.

Now let τ act on the Bers slice $\mathcal{B}(R)$. Choose a point $O \in \mathcal{B}(R)$ and denote its top surface by S and bottom, which is constant throughout the slice, by R. Let $\{R_{pA}\}$ denote the components of $R \setminus \bigcup \gamma_i$ on which τ is pseudo-Anosov and $\{R_{id}\}$ the remaining components, if any, on which τ acts as the identity. Denote the subsurfaces of S parallel to them by $\{S_{pA}\}$ and $\{S_{id}\}$. In the algebraic limit, the components of $\{S_{pA}\}$ are headed for degeneration and those of $\{S_{id}\}$ are due to have their boundary components pinched. In fact all of the loops $\{\gamma_j\}$ will be pinched.

For simplicity of description, assume there is one component of each set, S_{pA} and S_{id}. Also assume that their common boundary is formed by a single simple loop γ. The parallel subsurfaces of R, namely R_{pA} and R_{id}, are bounded by a loop parallel to γ. Suppose the basepoint O corresponds to the quasifuchsian group G.

Theorem 5.12.1 [Brock 2001a]. *The sequence of iterates $\{\tau^n(O)\}$ described above converges algebraically to a group $\varphi : G \to H \in \partial \mathcal{B}(G)$ and geometrically to a group H^* properly containing H. These have the following properties:*

(i) *The isomorphism φ sends the cyclic loxodromic subgroup of G corresponding to each lift of γ to a rank one parabolic subgroup of H.*

(ii) *φ is associated with a conformal map Φ between the bottom surfaces R of $\partial \mathcal{M}(G)$ and R of $\partial \mathcal{M}(H)$. and a homeomorphism $\Phi : S_{id} \to \partial \mathcal{M}(H) \setminus R$; γ corresponds to the puncture on $\Phi(S_{id})$.*

(iii) *The H-stabilizer of each lift of R_{id} to $\mathbb{H}^3$ is a quasifuchsian subgroup of H; the H-stabilizer of each lift of $\{R_{pA}\}$ is a singly degenerate subgroup of H.*

(iv) *$\partial \mathcal{M}(H^*)$ has two boundary components, one conformally equivalent to R the other conformally equivalent to S.*

(v) *The H^*-stabilizer of each lift of R_{id} and S_{id} is a quasifuchsian subgroup of H^*. The H^*-stabilizer of each lift of R_{pA} and of S_{pA} is a singly degenerate subgroup of H^*.*

(vi) *The interior $\mathcal{M}(H^*)$ is homeomorphic to $R \times [0, 1] \setminus [R_{pA} \times \{1/2\}]$. Thus $\mathcal{M}(H^*)$ has two degenerate ends, corresponding to the top and bottom sides of $R_{pA} \times \{1/2\}$, and two geometically finite ends.*

There is an analogous description for the general case. The only difference occurs

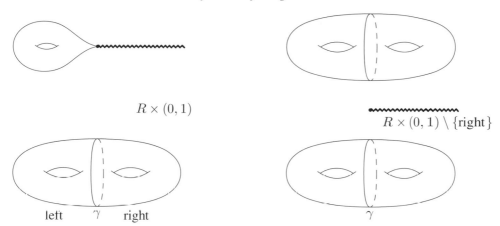

$R \times (0, 1)$

$R \times (0, 1) \setminus \{\text{right}\}$

left γ right

γ

Fig. 5.15. The algebraic and geometric limit at a boundary point of a genus-2 Bers slice. In the algebraic limit, half the top surface has degenerated leaving a once-punctured torus. The geometric limit is homeomorphic to the result of removing a once-punctured torus from $R \times (0, 1)$. It is the result of iterating the right side by a pseudo-Anosov that fixes the left side pointwise, as in Theorem 5.12.1.

when both sides of a loop γ_j belong to S_{id}, that is either γ_j is the common boundary of two components, or is a pinching loop of a single component. In either case γ_j determines a rank two parabolic subgroup of the geometric limit H^*, as described in Exercise (5-14).

Brock showed that by applying the techniques of the Skinning Lemma (Section 6.2), groups with the properties of Theorem 5.12.1 can be directly constructed as follows: Find groups $H_1, H_2 \in \partial \mathfrak{T}(G)$ with $\partial \mathcal{M}(H_1) = R \cup S_{\text{id}}$, $\partial \mathcal{M}(H_2) = S \cup R_{\text{id}}$, that is degenerate $S_{\text{pA}}, R_{\text{pA}}$, respectively. Then use the skinning lemma to identify R_{id} on the bottom of $\partial \mathcal{M}(H_2)$ with S_{id} on the top of $\partial \mathcal{M}(H_1)$. In the case that τ is a Dehn twist, this is explained in Exercises (5-12, 5-4). See Figures 5.15, 5.16, 5.17.

5.13 Classification of the geometric limits

This is to report on an important paper of Terukhiko Soma [2007] followed up by the more extensive work Ken'ichi Ohshika and Teruhiko Soma [2010].

(1) *Classification of the topological possibilities for geometric limits at boundary points of quasifuchsian space of a closed hyperbolic surface,*
(2) *Construction showing that all of the possibilities can occur, and*
(3) *A geometric limit is uniquely determined (up to isometry) by its ending data.*

Thus they have announced a far-reaching generalization of the Ending Lamination Theorem.

Here we will give an account of Soma [2007] and refer the reader to the original announcement for the many technical details. I am grateful to Ken'ichi Ohshika for his help in understanding their work.

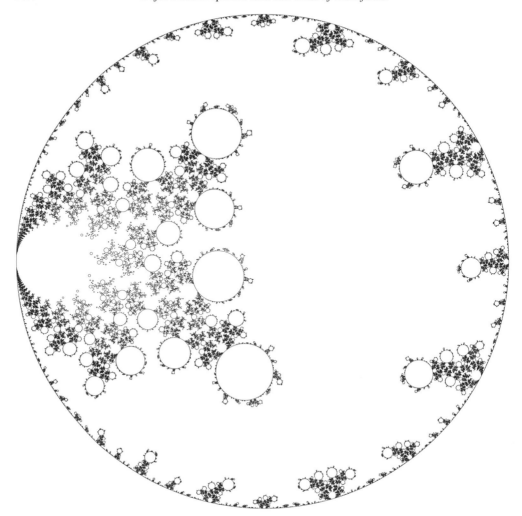

Fig. 5.16. The limit set of an algebraic limit corresponding to Figure 5.15. The boundary point is chosen so that the good half of the top is covered by a round disk.

Fix a closed hyperbolic surface S and an algebraic limit H on the boundary of its quasifuchsian space $\mathfrak{T}(R)$. We know there are geometric limits $\Gamma \supset H$ only if H has new parabolics, Theorem 4.6.2. Assume that this is the case. We will provide a picture of the quotient manifold $\mathcal{M}(\Gamma)$ of a geometric limit Γ in terms of $\mathrm{Int}\,\mathcal{M}(H) \cong S \times (-1, 1)$.

In particular Soma shows that the geometric limit manifold $\mathcal{M}(\Gamma)$ is homeomorphic ($\cong$) to an open subset of $S \times (0, 1)$ (this fact is also pointed out in Brock et al. [2012]).

The totality of possible geometric limits contains an amazing richness of structure, yet is obtained in an organized process. As we have seen earlier, new parabolics can appear in the form of either new rank one or new rank two parabolic subgroups. The Soma and Ohshika-Soma proofs build on the Minsky bilipschitz models of hyperbolic manifolds; in fact the proof involves taking geometric limits of sequences of the Minsky models.

Fig. 5.17. The limit set of the geometric limit corresponding to Figure 5.15. It can be obtained by reflecting the algebraic limit set in the round circles.

We have already seen some relatively elementary examples of geometric limits: (a) iterated Dehn twists acting on a curve γ in a manifold $S \times I$ in a Bers slice resulting in a manifold $S \times I \setminus \{\gamma\}$ (Marden-Earle, Exercise (5-14)), (b) Brock's example of an iterated pseudo-Anosov acting in a subsurface S_0 of S resulting in a manifold $S \times I \setminus S_0$, and (c) Jorgensen's original example of an infinitely generated geometric limit generated by Dehn twists of increasing orders about a succession of homotopically independent simple loops yielding a geometric limit with infinitely many rank two cusps.

Notations for Soma's theorem

(i) $I = [0, 1]$; and the horizontal and vertical projections $P_h : S \times I \to S$, $P_v : S \times I \to I$.

(ii) $S_y = P_v^{-1}(y)$; S_y is given a hyperbolic structure so that $P_h(S_y) \to S$ is an isometry for all $y \in I$.

(iii) Given a closed subset $\mathbf{C} \subset S \times I$ (called a *crevasse*), define

- $\mathcal{Y} = P_v(\mathbf{C})$;
- $\mathcal{X}_y = S_y \cap \mathbf{C}, \; y \in \mathcal{Y}$.
- $\Lambda_y^+ = S_y \cap \overline{\{(S \times \{(y, 1]\} \cap \mathbf{C}\}}$, if $y < 1$; $\quad \Lambda_y^- = S_y \cap \overline{\{(S \times [0, y)\} \cap \mathbf{C}\}}$, if $y > 0$.

For each $0 \leq y \leq 1$, Soma shows the existence of a disjoint union $\lambda(\Lambda_y^{\pm})$ of simple closed geodesics with the properties:

- There is a uniquely determined, maximal open geodesic subsurface $Z(\Lambda_y^{\pm}) \subset \mathcal{X}_y$, not necessarily connected, but (i) disjoint from $\Lambda_y^{\pm}$, and (ii) no component is homeomorphic to a pants.
- $\lambda(\Lambda_y^{\pm})$ is the union of simple closed geodesics $\{\ell\} \subset \mathcal{X}_y \setminus Z(\Lambda_y^{\pm})$ such that at least one side of ℓ is isolated from all other geodesics in the set $\Lambda_y^{\pm}$, and
- $\lambda(\Lambda_y^{\pm})$ contains (in particular) the boundary components of $\mathcal{X}_y \subset S_y$.

Theorem 5.13.1 (Soma). *Let Γ denote the geometric limit of an algebraically convergent sequence to the boundary of a quasifuchsian space of a closed hyperbolic surface S. Then $\mathcal{M}(\Gamma) \cong (S \times I) \setminus \mathbf{C}$ where the crevasse $\mathbf{C} \subset S \times I$ is closed and has the following properties:*

(i) *$(S \times I) \setminus \mathbf{C}$ is connected and contains $S_{1/2}$.*
(ii) *For each $y \in \mathcal{Y}$, $\mathcal{X}_y$ is a disjoint union of geodesic subsurfaces and simple closed geodesics in S_y,*
(iii) *No component of $Z(\Lambda_y^{\pm})$ is homeomorphic to an open pair of pants.*
(iv) *If $y < y' \in \mathcal{Y}$, a geodesic $\ell_y \in \lambda(\Lambda_y^+)$ is parallel to a geodesic $\ell_{y'} \in \lambda(\Lambda_y^-)$ in $(S \times I) \setminus \mathbf{C}$ only when they are horizontally parallel in $\mathbf{C}$.*

Conversely, given any crevasse $\mathbf{C}$, there exists a geometric limit Γ such that $\mathcal{M}(\Gamma) \cong (S \times I) \setminus \mathbf{C}$.

A component of of $Z(\Lambda_y^{\pm})$ is called a "*nonperipheral front*". In contrast $(S_0 \cup S_1) \setminus \mathbf{C} \cong \partial\mathcal{M}(\Gamma)$ is a "*peripheral front*". The end of $\mathcal{M}(\Gamma)$ corresponding to a non-peripheral front is geometrically infinite. Ohshika and Soma discovered widespread new phenomena occurring in geometric limits called *wild ends*. Unlike a tame end, a wild end is an end without a product neighborhood in $\mathcal{M}(\Gamma)$. The infinite ends in Jørgensen's picture are wild. The Ohshika-Soma groups contain a very large class of infinitely generated kleinian groups.

Most of the technical details are necessarily omitted here; still it seems safe to declare that the authors have shown that anything resulting from a natural maximal extension of the earlier examples can exist as a geometric limit.

5.14 Cannon-Thurston mappings

5.14.1 The Cannon-Thurston Theorem

The story begins with a theorem of Jim Cannon and Bill Thurston appearing in a preprint in 1989 and only recently in print.

Let Γ be a doubly degenerate quasifuchsian group, without any parabolics (or elliptics), isomorphic to a closed surface group G. Its limit set $\Lambda(\Gamma)$ is $\mathbb{S}^2$ itself. Choose any embedding $F : \mathbb{H}^2 \to F(\mathbb{H}^2) \subset \mathbb{H}^3$, which conjujates G to Γ.

Assume in addition that $\Gamma \subset \Gamma^*$ where Γ^* is an extension of Γ so that $\mathbb{H}^3/\Gamma^*$ is a closed hyperbolic 3-manifold fibered over $\mathbb{S}^1$ with fibers homeomorphic to S. We refer to Section 6.3.1 for a description of such manifolds.

With these assumptions the transformational theorem in question is this:

Theorem 5.14.1 (Doubly degenerate case; [Cannon and Thurston 2007]). *The mapping $F : \mathbb{H}^2 \hookrightarrow \mathbb{H}^3$ has a continuous extension to become an equivariant surjective map*

$$F : \mathbb{S}^1 \equiv \partial\mathbb{H}^2 \to \mathbb{S}^2.$$

That is, $F(\mathbb{S}^1)$ is a Γ-invariant Peano curve.

Outline of proof. First consider the case that both ends of $\mathcal{M}(\Gamma)$ are infinite. We have assumed that the ending laminations are those associated with the stable and unstable laminations of the pseudo-Anosov automorphism α of LHP/G, which are also the attracting and repelling fixed points of α in the Thurston boundary (see Exercise (5-12) and p. 303)). Realize the ending laminations (with their endpoints) in the respective half-planes $\Lambda_- \subset \overline{\text{LHP}}$ and $\Lambda_+ \in \overline{\text{UHP}}$. Construct a new space $\mathcal{S}$ from $\mathbb{S}^2$ as follows.

The elements of $\mathcal{S}$ are (i) the components of $\overline{\text{UHP}} \setminus \Lambda_+$ and the components of $\overline{\text{LHP}} \setminus \Lambda_-$, (ii) the leaves of Λ_+ and the leaves of Λ_-, which do not appear in (i), and (iii) the points on $\mathbb{S}^2$ that are not in (i) nor in (ii). The collections of elements in $\mathcal{S}$ is a "cellular semicontinuous decomposition" of $\mathbb{S}^2$ because of the special nature of the ending laminations.

The operative requirements, which can be proved to be satisfied here, are that the elements of $\mathcal{S}$ are mutually disjoint, proper, nonseparating continua with union $\mathbb{S}^2$, invariant under G, and have the "upper semicontinuity property": Given an element $s \subset \mathcal{S}$, and an open subset $s \in U \subset \mathbb{S}^2$, the union of all the elements of $\mathcal{S}$ that lie in U is open in $\mathbb{S}^2$.

So the decomposition consisting of a closed disk and its complement does not qualify.

Form the identification space $\mathbb{S}^2/\mathcal{S}$. The celebrated cellular decomposition theorem of R.L. Moore [Moore 1927; Cannon and Thurston 2007] says that $\mathbb{S}^2/\mathcal{S}$ is *homeomorphic* to $\mathbb{S}^2$ itself!

Choose a continuous equivariant embedding f of each of the two half-planes UHP, LHP into $\mathbb{H}^3$ so the projection of the image bounds a compact submanifold of $\mathcal{M}^{\text{int}}(\Gamma)$. The f-images serve as a Γ-invariant duplicate $Q_U \cup Q_L$ of the G-invariant UHP $\cup$ LHP; the common boundry is now $\Lambda(\Gamma)$. Using now $Q_U \cup Q_L \cup \Lambda(\Gamma)$ one can find a similar semicontinuous cellular decomposition of $\mathbb{S}^2$, this time invariant under Γ. The corresponding identification space is homeomorphic to $\mathbb{S}^2$.

The following sequence of mappings result in a an equivariant continuous mapping $F : \mathbb{S}^2 \to \mathbb{S}^2$ that has the effect, via passing through the decomposition space, of sending S^1 onto $\Lambda(\Gamma)$:

$$\mathbb{S}^2 \cong \mathrm{UHP} \cup \mathrm{LHP} \cup \mathbb{S}^1 \to Q_U \cup Q_L \cup \Lambda(\Gamma) \to (Q_U \cup Q_L \cup \Lambda(\Gamma))/\mathcal{S} \cong \mathbb{S}^2.$$

So $F(\mathbb{S}^1)$ is an Γ-invariant Peano curve. We think of this curve as resulting from the collapse of the two laminations in UHP and LHP. (Think of pinching S^1 by collapsing a leaf. Then pinch again, and again, etc., from both the inside and the outside of $\mathbb{S}^1$, remembering that the set of endpoints of the leaves in UHP is disjoint from the set for LHP.) □

Limit sets of singly degenerate groups. The paper [Cannon and Thurston 2007] includes an analogue for some singly degenerate groups without parabolics. Let α be a pseudo-Anosov automorphism of the closed surface group G (and correspondingly of the surface $\mathbb{H}^2/G$). The iterates $\{\alpha^n(G)\}$ acting on the Bers slice $\mathcal{B}(G)$ converge algebraically and geometrically to a singly degenerate group Γ on the boundary. The ending lamination of $\mathcal{M}(\Gamma)$ corresponds to the stable measured lamination of α, which is also the attracting fixed point of α in its action on the Thurston boundary (see Section 6.1.1).

Now suppose that the bottom end of $\mathcal{M}(\Gamma)$ is homeomorphic to $R = \mathrm{LHP}/G$ while Λ_+, the ending lamination of the top, corresponds to the attracting fixed point for the pseudo-Anosov map α. Form the space $\mathcal{S}$ using (i) and (ii) in $\overline{\mathrm{UHP}}$ but replace (iii) by (iii') points in $\overline{\mathrm{LHP}}$ which are not in (i) or (ii). Again the identification space $\mathcal{S}$ is homeomorphic to $\mathbb{S}^2$ resulting in a continuous map $f : \mathbb{S}^2 \to \mathbb{S}^2$ that conjugates G to Γ and is a homeomorphism on LHP. In this case $f(\mathbb{S}^1)$ is the limit set of Γ; the limit set results from collapsing the ending lamination so it has a tree structure. See also Exercise (5-13).

Theorem 5.14.2 (Singly degenerate case [Cannon and Thurston 2007]). *The mapping F has a continuous extension to become an equivariant surjective map*

$$\mathbb{S}^1 \equiv \partial \mathbb{H}^2 \to \Lambda(\Gamma).$$

Our account is based on Minsky [1994a; 1994b].

Start again with a fuchsian group G group acting in UHP and LHP and an isomorphism $\theta : G \to \Gamma$ to a discrete group Γ.

Consider first the simplest case. Take a simple closed geodesic $\gamma \in R' = \mathrm{UHP}/G$, and the set of lifts $\{\gamma_n^*\} \subset \mathrm{UHP}$. Suppose H is the cusp group arising from R' by pinching γ. There is a continuous equivariant surjection $h : \mathbb{S}^1 \equiv \mathbb{R} \cup \infty \to \Lambda(H)$ which sends the two endpoints of each γ_n^* to a single point but is otherwise one-one.

Now consider a general manifold in the Bers boundary $\mathcal{M}(\Gamma) \in \partial \mathcal{B}(G)$. Denote the ending laminations of $\mathcal{M}(\Gamma)$ by Λ_+ and Λ_- realized in UHP and LHP. One or both may correspond to geometrically finite ends, but not both laminations are empty. The leaves of Λ_- have no endpoints in common with the leaves of Λ_+ (Double Limit Theorem, p. 303).

Theorem 5.14.3 [Minsky 1994a]. *Assume that G is a closed surface group and Γ has bounded geometry. There exists a continuous map $h : \mathbb{S}^2 \to \mathbb{S}^2$ which is equivariant: $h \circ g(z) = \theta(g) \circ h(z)$ for all $g \in G$ and $z \in \mathbb{S}^2$. It collapses the leaves to points: $h(u) = h(v)$ with $u \neq v$ if and only if u and v lie on the closure of the same leaf of Λ_+, or of Λ_-. If say $\Lambda_- = \varnothing$ then h is an equivariant homeomorphism* LHP$\to \Omega(\Gamma)$.

Thus if a component P of UHP$\backslash \Lambda_+$ is an ideal polygon, then $h(P)$ is a single point. See also Exercise (5-9).

For simply or doubly degenerate groups without new parabolics, bounded geometry[6] is determined by a property of the ending laminations [Minsky 2001].

In the case that $\mathcal{M}(\Gamma)$ is singly degenerate without new parabolics, the continuous image of $\mathbb{S}^1$ is the boundary of $\Omega(\Gamma)$. Therefore the application of Pommerenke [1992, Theorem 2.1], results in:

Corollary 5.14.4 [Minsky 1994a]. *Under the same hypotheses, if Γ is singly degenerate, then $\Lambda(\Gamma)$ is locally connected.*

Minsky conjectured that Theorem 5.14.3 holds without the requirement of bounded geometry. McMullen [2001a] proved that indeed, this is true for once-punctured tori. His argument makes heavy use of Minsky's model manifolds.

A less specific collapsing theorem that parallels the geometrically finite case was proved by Erica Klarreich (her original statement required tameness):

Theorem 5.14.5 [Klarreich 1999]. *Assume G is freely indecomposable[7] and $\mathcal{M}(G)$ is geometrically finite. Suppose there is a homeomorphism $f : \mathbb{H}^3/G \to \mathbb{H}^3/\Gamma$, where $\mathbb{H}^3/\Gamma$ has bounded geometry. Then there is a homotopic homeomorphism $h \sim f$ whose lift to $\mathbb{H}^3$ extends to be a continuous, surjective, equivariant map $\mathbb{S}^2 \to \mathbb{S}^2$.*

5.14.2 Cannon-Thurston mappings and local connectivity

Fundamental question 1: Are the limit sets of all finitely generated kleinian groups with connected limit sets ($\neq \mathbb{S}^2$) locally connected (see p. 286)?

For years this has been seen as a notoriously hard problem (just like the Ahlfors Conjecture before it too was solved). It is certainly true for geometrically finite groups [Anderson and Maskit 1996], which can be pieced together from Schottky and quasifuchsian groups, and for doubly degenerate groups. The situations at a number of special limit set points have been considered in Brock et al. [2011].

Fundamental question 2: Do Cannon-Thurston maps exist for finitely generated kleinian groups in full generality? Generalizations of Theorems 5.14.1, 5.14.2 are called Cannon-Thurston mappings. The most important cases involve incompressible infinite ends.

Suppose S is a finite area incompressible hyperbolic surface, and $\widehat{S}$ is its universal cover, $\partial \widehat{S} = \mathbb{S}^1$. A *Cannon-Thurston mapping* refers to extending an embedding of

---

[6] The length of all closed geodesics in $\mathcal{M}(\Gamma)$ has a positive lower bound.

[7] G is not a nontrivial free product of subgroups.

$\widehat{S}$ into $\mathbb{H}^3$ to a continuous mapping $\partial \widehat{S} \to \mathbb{S}^2$, and the generalization to compressible surfaces.

Mahan Mj (Mahan Mitra before adopting the new name) has announced a complete solution of the Cannon-Thurston problem. We will present his results in a series of steps, leaving aside discussion of the final result which involves more cases.

The Cannon-Thurston Theorem I [Mj 2014; Mj 2011]. *Suppose* $\mathcal{M}(G)$ *is compact and* G' *is isomorphic to* G. *There is a continuous map of limit sets* $\phi : \Lambda(G) \to \Lambda(G')$ *taking a fixed point of* $g \in G$ *to the corresponding fixed point of* $\phi(g)$.

This answers a question of Thurston. We will expand this theorem in two cases.

The Cannon-Thurston Theorem II [Mj 2014; Mj 2014]. *Let* Γ *be a singly or doubly degenerate kleinian surface group without accidental parabolics.*

(i) *There exist one or two equivariant Cannon-Thurston mappings* $f_\pm : \mathbb{S}^1 \to \Lambda(\Gamma)$ *corresponding to the one or two infinite ends of* $\mathcal{M}(\Gamma)$, *with ending laminations* $e_\pm$,

(ii) *If there are two infinite ends, the preimage of a point in* $\Lambda(\Gamma)$ *under* f_+^{-1} *is either the two endpoints of a leaf of* e_+, *or it is the ideal boundary points of an ideal polygon. Likewise the preimage* f_-^{-1} *is either the pair of endpoints of a leaf of* e_-, *or it is the ideal boundary points of an ideal polygon,*

(iii) *If* $\mathcal{M}(\Gamma)$ *has only one infinite end with ending lamination* e_+, f_- *is an equivariant homeomorphism of the unit disk onto* $\Omega(\Gamma)$.

Thus $\Lambda(\Gamma)$ is formed either by collapsing both the interior and exterior of $\mathbb{S}^1$ along the leaves of the ending laminations for $e_\pm$, or collapsing just the exterior of $\mathbb{S}^1$.

Next we consider the general case that G is a finitely generated, geometrically infinite kleinian group without compressible or relative compressible ends. Each infinite or relative infinite end e can be realized in the interior of $\mathcal{M}(G)$ as a surface $S(e)$ parallel to e. The ending lamination of e can be realized on S as a geodesic lamination $\lambda(e)$. Pass to the universal cover of S, say $\mathbb{D}$. The endpoints of the leaves of the lift $\widehat{\lambda}(e)$ are dense on the unit circle. The complement of $\widehat{\lambda}(e)$ is a union of ideal polygons from S.

The Cannon-Thurston Theorem III. *There is a continuous mapping of the unit circle:*

$$\phi : \partial \mathbb{D} \to \Lambda(G) \subset \partial \mathbb{H}^3.$$

Two points of $\partial \mathbb{D}$ *have the same image,* $\phi(a) = \phi(b)$, *if and only if either*

(i) a, b *are the endpoints of a leaf of* $\widehat{\lambda}(e)$, *or*

(ii) *there is a finite-sided ideal polygon* $P \subset \mathbb{D} \setminus \widehat{\lambda}(e)$ *so that* a *and* b, *along with the full set of ideal points of* P, *are mapped to the common point* $\phi(a) = \phi(b)$.

Mj has announced the solution in all cases including the case of compressible ends which are more complicated to deal with, and perhaps less important. We will discuss the case of handlebodies below.

General CT Theorem [Mj 2014; Mj 2011]. *Cannon-Thurston maps exist for any finitely generated, geometrically infinite kleinian group.*

Mj's work on Cannon-Thurston mappings is related to the Ending Lamination Theorem [Mj 2014]. In fact the Minski model and "electrification" (see p. 108) of the cusps are used in his proofs.

Local connectivity in turn follows from Theorem III above since $\Lambda(G)$ connected implies all components of $\Omega(G)$ are simply connected, all ends of $\mathcal{M}(G)$ are incompressible. Then apply the fact that the continuous extension of the Riemann map to the closure of each simply connected component implies local connectivity [Pommerenke 1992, Theorem 2.1].

Local Connectivity of Limit Sets [Mj 2014; Mj 2011]. *For any finitely generated kleinian group G with connected limit set $\Lambda(G)$, $\Lambda(G)$ is locally connected.*

There are two special cases that illuminate the general theorem for Cannon-Thurston maps.

1. Γ *is a singly or doubly degenerated quasifuchsian group with new parabolics.* In this case the limit set $\Lambda(\Gamma)$ is connected, but the existence of new parabolics may change the geometry of the ends. The components of the ordinary set $\Omega(\Gamma)$ are still simply connected. But Γ itself can be singly, doubly, or partially degenerated quasifuchsian. Corresponding to each component Ω_i of $\Omega(\Gamma)$ with stabilizer Γ_i is a Cannon-Thurston map arising from a fuchsian equivalent of Ω_i/Γ_i.

2. Γ *is a degenerated Schottky group.* This is the more intriguing situation. A Schottky group can be degenerated, $G \to \Gamma$ by iterating a pseudo-Anosov transformation on $\partial\mathcal{M}(G)$.

Let G be a classical Schottky group (determined by circles) such that there is a homeomorphism $\mathrm{Int}(\mathcal{M}(G)) \to \mathcal{M}(\Gamma)$. The limit set $\Lambda(G) \subset \mathbb{S}^2$ is homeomorphic to a Cantor set. The convex hull $\mathcal{CH}(\Lambda) \subset \mathbb{H}^3$ of $\Lambda(G)$ is embedded in $\mathbb{H}^3$ as a thickened tree with the very tips of its branches forming a dense subset of $\Lambda(G)$.

Set $S^* = \partial\mathcal{CH}(\Lambda)$. Because there are no isolated bending lines, the nearest point retraction of S^* is a homeomorphism to $\mathbb{S}^2 \setminus \Lambda(G)$.

Choose an embedding $f : S = S^*/G \hookrightarrow \mathcal{M}(\Gamma)$ so that $f(S) \times [0, 1)$ is a neighborhood of the end e. Correspondingly consider the lift $f^* : S^* \hookrightarrow \mathbb{H}^3$ that conjugates G to Γ. The image $f^*(S^*)$ is likewise a thickened tree and the tips of its branches are a dense subset of $\Lambda(\Gamma) = \mathbb{S}^2$. Again, $f^*(S^*) \cup \Lambda(\Gamma)$ is homeomorphic to $\mathbb{S}^2$.

We can realize the ending lamination of the end e of $\mathcal{M}(\Gamma)$ as a geodesic lamination $\lambda \subset f^*(S^*)$ (corresponding to its representation in in the Masur domain of S—see Exercise (5-19)). Their endpoints are dense in $\Lambda(\Gamma)$.

Akin to the construction in the proof of Theorem 5.14.1, let $\mathcal{S}$ denote the elements of the subdivision of $S^* = \partial\mathcal{CH}(\Lambda)$ into three classes of subsets: (i) the closure of the complements in S^* of the leaves of λ, (ii) the leaves of λ which do not appear in the set (i), and (iii) the points of S^* which do not lie in (i) or (ii).

Project the decomposition $\mathcal{S}$ to a decomposition $f^*(\mathcal{S}) \cup \Lambda(\Gamma) = \mathbb{S}^2$ by adding to (iii) the points of $\Lambda(\Gamma)$ not represented in the f^*-image of (i) and (ii). This is an admissible subdivision of $\mathbb{S}^2$ and the identification space $\mathbb{S}^2/\mathcal{S}$ is homeomorphic to $\mathbb{S}^2$.

The bottom line is that there exists a continuous mapping $f : \mathbb{S}^2 \to \mathbb{S}^2$ that conjugates G to Γ. It collapses the leaves of λ to points and is a Cannon-Thurston map of the Cantor set $\Lambda(G)$ to a dense subset of the space filling curve formed by the going through the identification space.

A more direct construction is suggested by Mj (personal communication). Let S be the boundary of a handlebody within $\mathcal{M}(\Gamma)$ parallel to the degenerated end e. Denote by K the kernel of the inclusion $\pi_1(S) \hookrightarrow \pi_1(\mathcal{M}(\Gamma))$. Form the corresponding covering surface $\widehat{S}_K$. This is a planar surface with limit set a Cantor set. The Masur domain ending lamination $\lambda(e)$ lifts to $\widehat{\lambda}(e)$ in $\widehat{S}_K$, and any complementary polygons lift as well. A set homeomorphic to $\Lambda(G)$ is obtained by identifying the pairs of endpoints of the leaves of $\widehat{\lambda}(e)$ and collapsing each ideal polygon to a point.

5.15 Exercises and explorations

5-1. *Hyperbolic Poisson integral formula* [Ahlfors 1981, §5.7]. A hyperbolically harmonic function $u : \mathbb{H}^3 \to \mathbb{R}$ is a function that vanishes under the Laplace–Beltrami operator $\Delta_h u = 0$. The Laplace–Beltrami operator in the ball model with $|\vec{x}| = r < 1$ is

$$\Delta_h u = \frac{(1 - r^2)^2}{4}\left(\Delta u + \frac{2r}{1 - r^2}\frac{\partial u}{\partial r}\right), \tag{5.8}$$

and in the upper half-space model $\{(z, t), \ t > 0\}$ is

$$\Delta_h u = t^2\left(\Delta u - \frac{1}{t}\frac{\partial u}{\partial t}\right). \tag{5.9}$$

Here Δ denotes the euclidean laplacian. The Laplace–Beltrami operator has the property that for any isometry g,

$$\Delta_h(u \circ g) = (\Delta_h u) \circ g.$$

Thus u and $u \circ g$ are simultaneously hyperbolically harmonic.

We will use the ball model of $\mathbb{H}^3$ and denote spherical measure on $\mathbb{S}^2 \equiv \partial\mathbb{H}^3$ by $d\omega(\zeta)$. Suppose $f(\zeta)$ is a measurable function on $\mathbb{S}^2$ with $\iint_{\mathbb{S}^2} |f(\zeta)| \, d\omega(\zeta) < \infty$. The function

$$u(x) = u_f(x) = \frac{1}{4\pi}\iint_{\mathbb{S}^2}\left(\frac{1 - |x|^2}{|\zeta - x|^2}\right)^2 f(\zeta) \, d\omega(\zeta), \quad x \in \mathbb{H}^3, \tag{5.10}$$

has the following properties [Ahlfors 1981, Chapter V].

(i) $u(x)$ is hyperbolically harmonic, and in particular real analytic, for $x \in \mathbb{H}^3$.
(ii) $u(x)$ has radial limits $f(\zeta)$ a.e.
(iii) If g is a Möbius transformation, then $u_f \circ g(x) = u_{f \circ g}(x)$.
(iv) If $f \circ g(\zeta) = f(\zeta)$ for a Möbius transformation g and almost all $\zeta \in \mathbb{S}^2$ then $u \circ g(x) = u(x)$.

Note that the expression $(1 - |z|^2)/|\zeta - z|^2$ is the Poisson kernel in the unit disk.

For the following theorem, assume that G is not quasifuchsian or a $\mathbb{Z}/2$ extension of a quasifuchsian group. Assume further that components Ω_1, Ω_2 of $\Omega(G)$ are quasidisks with $\mathrm{Stab}(\Omega_i) = G_i$ and the Möbius transformation $T \in G$ sends the exterior of Ω_2 onto the interior of Ω_1.

Theorem 5.15.1 (Existence of invariant embedded surfaces). *Let $\chi(\zeta)$ be the characteristic function of Ω_1, namely with value 1 for $\zeta \in \Omega_1$ and zero elsewhere. Denote by $u(x)$ the "harmonic measure" defined by Equation (5.10) above with $f(\zeta) = \chi(\zeta)$. Choose $r > \frac{1}{2}$ such that the level surface $S = \{x \in \mathbb{H}^3 : u(x) = r\}$ is smooth; it is also embedded. Then $S \cap h(S) = \varnothing$ for all $h \in \langle G, T G T^{-1} \rangle$, $h \notin G_1$.*

Proof. See Kapovich [2001, Lemma 4.102]. The surface S separates points of $\mathbb{H}^3$ with values $u(x) > r$ from points with $u(x) < r$. It is embedded because the gradient flow is orthogonal to S at all points. We know already that S is invariant under G_1.

Suppose for some $g \in G$, $g \notin G_1$ we had $y \in S \cap g(S)$. Recall that $u_\chi \circ g^{-1}(x) = u_{\chi \circ g^{-1}}(x)$ so $g(S)$ is the level surface $u_{\chi \circ g^{-1}}(x) = r$ for the characteristic function $\chi \circ g^{-1}$ of $g(\Omega_1)$. Adding the two Poisson integrals evaluated at y,

$$1 < 2r = \frac{1}{4\pi} \int\!\!\int_{\Omega_1 \cup g(\Omega_1)} \left(\frac{1 - |y|^2}{|\zeta - y|^2} \right)^2 d\omega(\zeta). \tag{5.11}$$

But this is impossible since the regions $\Omega_1, g(\Omega_1)$ are disjoint so the Poisson integral on the right represents the harmonic measure of their union. Its values must be strictly between 0 and 1 in $\mathbb{H}^2$.

The argument shows that g maps the level-r surface over Ω_1 onto the level-r surface over $g(\Omega_1)$. We have shown the totality of all such surfaces are mutually disjoint. Likewise the map T maps the level-r surface S_2 over Ω_2 to the level r-surface $T(S_2)$ over the complement Ω_1' of Ω_1. Now $T(S_2)$ is the level-$(1 - r)$ surface over Ω_1. This can have no points in common with the level-r surface over Ω_1. Note that T maps the side of S_2 facing Ω_2 to the side of $T(S_2)$ facing away from Ω_1. We could take $r = \frac{1}{2}$ unless we wanted the surfaces to be smooth, then we can take r arbitrarily close to $\frac{1}{2}$.

There is one more case to check. Take a component $\Omega_3 \subset \Omega_1'$. Can its level-$r$ surface intersect $T(S_2)$? If y were an intersection point then the integral on the right side of Equation (5.11) would have the value $1 = r + (1 - r)$. This could happen only if $\Omega_3 = \Omega_1'$ and G were quasifuchsian or an extension of a quasifuchsian by an element that interchanged the two components. We have assumed that this is not the case.

It follows that the orbit of S under the group $\langle G, T G T^{-1} \rangle$ is the union of mutually disjoint surfaces. □

5-2. *The bottom of the spectrum of eigenvalues.* The bottom of the L^2-spectrum of the hyperbolic laplacian $-\Delta_h$ on the hyperbolic manifold $\mathcal{M} = \mathcal{M}(G)$ is given by

$$\lambda_0(\mathcal{M}) = \inf_{f \in C_c^\infty(\mathcal{M})} \frac{\int_\mathcal{M} |\nabla f|^2 \, dV}{\int_\mathcal{M} |f|^2 \, dV}, \tag{5.12}$$

where the infimum is taken over all C^∞ functions with compact support, and dV is the volume element. Thus $\lambda_0(\mathcal{M}) = 0$ if $\mathcal{M}$ has finite volume (for then the constants are in the competition).

For a geometrically finite, infinite volume $\mathcal{M} = \mathcal{M}(G)$ with $\text{Area}\,\partial\mathcal{C}(\mathcal{M}) = 2\pi|\chi(\partial\mathcal{C}(\mathcal{M}))|$,

$$\frac{K}{(\text{Vol}\,\mathcal{C}_1(\mathcal{M}))^2} \le \lambda_0(\mathcal{M}) \le 4\pi\,\frac{\text{Area}\,\partial\mathcal{C}(\mathcal{M})}{\text{Vol}\,\mathcal{C}(\mathcal{M})}. \tag{5.13}$$

Here $K > 0$ is a universal constant and $\mathcal{C}_1(\mathcal{M})$ denotes the distance-1 neighborhood of the convex core $\mathcal{C}(\mathcal{M})$. The volume $\text{Vol}\,\mathcal{C}_1(\mathcal{M})$ is finite if $\text{Vol}\,\mathcal{C}(\mathcal{M}) < \infty$ and if for some $\delta > 0$, $\partial\mathcal{C}(\mathcal{M})$ contains no compressible curves of length $< \delta$—this condition prevents long thin waists with compressible cross section for which the 1-neighborhood will have large volume [Thurston 1979a, Proposition 8.12.1]. Also $\text{Vol}\,\mathcal{C}_1(\mathcal{M}) > 0$ if G is fuchsian. The right inequality is proved in Canary [1992] and the left in Burger and Canary [1994].

It is amazing that the lowest eigenvalue is precisely determined by the Hausdorff dimension $d(\mathcal{M})$ of the limit set of G. Making use of the information in Exercise (3-20), and bringing in the Tameness Theorem, we can state the situation as follows:

Theorem 5.15.2 [Sullivan 1987; Bishop and Jones 1997; Canary 1992]. *Suppose $\mathcal{M} = \mathcal{M}(G)$ has infinite volume and G is nonelementary.*

(i) $\lambda_0(\mathcal{M}) = 0$ *and* $d(\mathcal{M}) = 2$ *if and only if G is geometrically infinite, in particular, singly degenerate.*

(ii) $\lambda_0(\mathcal{M}) = 1$ *if and only if G is geometrically finite with* $d(\mathcal{M}) \le 1$.

(iii) $0 < \lambda_0(\mathcal{M}) = d(\mathcal{M})(2 - d(\mathcal{M})) < 1$ *if and only if G is geometrically finite with* $1 < d(\mathcal{M}) < 2$.

In case (iii) there is an L^2 eigenfunction f_0 corresponding to $\lambda_0(\mathcal{M})$:

$$-\Delta_h f_0 = \lambda_0(\mathcal{M}) f_0.$$

There is a global version of (iii). Suppose G is geometrically finite and nonelementary. Set $\lambda_0^*(G) = \sup\lambda_0(\mathcal{M})$, $d^*(G) = \inf d(\mathcal{M})$ as $\mathcal{M} = \mathcal{M}(H)$ ranges over $\mathfrak{R}_{\text{disc}}(G)$. Then $\lambda_0^*(G) = d^*(G)(2 - d^*(G))$, provided $\mathcal{M}(G)$ is not a handlebody [Canary et al. 1999]. In addition the cases that $\lambda_0^*(G) = 1$ or $d^*(G) = 1$ are identified.

5-3. *The eigenvalue* $\lambda_1(\mathcal{M})$. This is to report on the interesting paper of Biringer and Souto [2011].

As pointed out above, Equation (5.12) implies that $\lambda_0(\mathcal{M}) = 0$. The spectrum of eigenvalues of the hyperbolic laplacian then is

$$0 = \lambda_0(\mathcal{M}) < \lambda_1(\mathcal{M}) \le \lambda_2(\mathcal{M}) \le \cdots.$$

Concerning λ_1, we record from their paper the inequality due to Cheeger and Buser:

$$\frac{1}{4}h(\mathcal{M})^2 \le \lambda_1(\mathcal{M}) \le 4h(\mathcal{M})^2 + 10h(\mathcal{M}).$$

Here $h(\mathcal{M})$ is the *Cheeger constant* of $\mathcal{M}$:

$$h(\mathcal{M}) = \inf_{U \subset \mathcal{M}} \frac{\text{Area}(\partial U)}{\min[\text{Vol}(U), \ \text{Vol}(\mathcal{M} \setminus U)]},$$

with the infimum taken over submanifolds U.

The authors derive two very interesting consequences:

Theorem 5.15.3 (Biringer-Souto).

(i) *Suppose we are given $2 \leq k < \infty$, and $\varepsilon > 0$, $\delta > 0$. There are at most finitely many isometry classes of closed hyperbolic manifolds $\mathcal{M}$ with rank at most k, such that $\text{Inj}(\mathcal{M}) \geq \varepsilon$, and $\lambda_1(\mathcal{M}) \geq \delta$.*

(ii) *Suppose $\{\mathcal{M}_j\}$ is a sequence of distinct closed manifolds satisfying the above bounds except now with $\lim_{j \to \infty} \lambda_1(\mathcal{M}_j) = 0$. Normalize each $\mathcal{M}_j$ so that it contains a given point O and small ball $B_\rho(O)$ about it. Then a subsequence $\{\mathcal{M}_k\}$ converges geometrically to a doubly degenerate group $\cong S \times \mathbb{R}$ where S is a closed surface with genus $\leq k$.*

Here the term "rank" refers to the least number of generators in the fundamental group.

5-4. *Nielsen transformations.* Suppose G is a free group on two generators X, Y. Nielsen proved that every automorphism of G is the composition of a finite number of automorphisms of the following elementary types.

(i) Interchange the two generators: $(X, Y) \rightarrow (Y, X)$.
(ii) Replace one generator by its inverse: $(X, Y) \rightarrow (X, Y^{-1})$.
(iii) Replace one generator by its product with the other: $(X, Y) \rightarrow (X, XY)$.

Thus if we start with a particular generating pair, we can systematically find every other generating pair.

This is applied to once-punctured torus groups, where there is another relation that has to be maintained. By a generator pair of a once-punctured torus group G we mean elements $X, Y \in G$ such that $G = \langle X, Y \rangle$ and $[X, Y]$ generates a parabolic subgroup.

The commutator requirement is preserved by the Nielsen transformations: Application of the first Nielsen transformation sends the commutator $[X, Y] = XYX^{-1}Y^{-1}$ to its inverse $[Y, X]$, the second sends it to $[X, Y^{-1}] = Y^{-1}[Y, X]Y$ and the third sends it to $X[X, Y]X^{-1}$. In any case the trace of the commutator remains -2.

Show using a cancellation argument that if both (X, Y) and (X, Z) are generator pairs in this sense then $Z = X^n Y$, modulo conjugation by some X^k.

For a quasifuchsian once-punctured torus group G, set as usual $S^{\text{top}} = \partial^{\text{top}}\mathcal{M}(G)$, and $S_{\text{bot}} = \partial_{\text{bot}}\mathcal{M}(G)$. Suppose $\langle X^{\text{top}}, Y^{\text{top}} \rangle$ is a generator pair for $\pi_1(S^{\text{top}})$ and the elements $X_{\text{bot}}, Y_{\text{bot}}$ are a generator pair for $\pi_1(S_{\text{bot}})$. Assume these ordered generating pairs are not conjugate within G. Then we can pinch S^{top} and S_{bot} by requiring X^{top} and X_{bot} to become parabolic independently of each other. This results in a manifold whose boundary is two triply punctured spheres. Still, the two generator pairs are Nielsen transforms of each other.

Show that the Nielsen transformations generate the mapping class group for a once-punctured torus.

5-5. *The pinching estimate* [Bers 1970a; McMullen 1999]. The following estimate is frequently used in situations of pinching.

Theorem 5.15.4. *Suppose Ω is a simply connected component of the ordinary set $\Omega(H)$ corresponding to an incompressible component S of $\partial \mathcal{M}(H)$. Suppose $h \in \mathrm{Stab}(\Omega)$ is loxodromic and corresponds to the geodesics α in the hyperbolic metric on S and $\alpha_* \subset \mathcal{M}(H)$. Then*

$$\mathrm{Len}_{\mathcal{M}(H)}(\alpha_*) \leq 2\,\mathrm{Len}_S(\alpha). \tag{5.14}$$

Suppose instead Ω_1, Ω_2 are the components of a quasifuchsian group H corresponding to the surfaces $S^t = \partial^{\mathrm{top}}\mathcal{M}(H)$, and $S_b = \partial_{\mathrm{bot}}\mathcal{M}(H)$. Denote by $\alpha_*, \alpha^t, \alpha_b$ the geodesics in the corresponding hyperbolic metrics. Then

$$\mathrm{Len}_{\mathcal{M}(H)}(\alpha_*) \leq 2 \min \left(\mathrm{Len}_{S_b}(\alpha_b), \mathrm{Len}_{S^t}(\alpha^t) \right). \tag{5.15}$$

Proof. The modulus of the annulus $A = \{z : r < |z| < R\}$ is defined as

$$M_A = \frac{\log(R/r)}{2\pi} = \frac{1}{\lambda(c_A)},$$

where $\lambda(c_A)$ is the extremal length of the free homotopy class of curves c_A separating the boundary components; see Ahlfors [1973, Chapter 4]. The length of the shortest geodesic in c_A with respect to the hyperbolic metric of A is $\pi\lambda(c_A)$ (Exercise (2-2)).

Suppose first we have a quasifuchsian group G with invariant components Ω_1, Ω_2. Assume that the loxodromic g is in G. Then $A_i = \Omega_i / \langle g \rangle$ is conformally equivalent to an annulus, $i = 1, 2$. Let L_i denote the length of the geodesic in c_{A_i} in the hyperbolic metric in A_i. As above, $\lambda(c_{A_i}) = L_i/\pi$. These are the same as the lengths of the geodesics corresponding to g in each of the two surfaces $\partial \mathcal{M}(G)$.

On the other hand suppose g is conjugate to $z \mapsto kz$, $|k| > 1$, and consider the torus $T = (\mathbb{C} \setminus \{0\})/\langle g \rangle$. The annular regions A_i are embedded in T and are disjoint there. The central curves of both A_1 and A_2 belong to the free homotopy class of curves c_T in T. A well-known inequality, in particular in Ahlfors [1973, Theorem 4.2], says that

$$\frac{1}{\lambda(c_T)} \geq M_{A_1} + M_{A_2} = \frac{1}{\lambda(c_{A_1})} + \frac{1}{\lambda(c_{A_2})} = \frac{\pi}{L_1} + \frac{\pi}{L_2}.$$

In particular, $\pi\lambda(c_T) \leq \min(L_1, L_2)$.

Calculate $\lambda(c_T)$ as follows. Set $\varphi = \arg k$, $0 \leq \varphi < 2\pi$, and $\tau = \log k = \log|k| + \varphi i$. Consider the group $X = \langle z \mapsto z + 2\pi i, z \mapsto z + \tau \rangle$. A fundamental polygon is spanned by the vectors $(2\pi i, \tau)$. Its area is $2\pi|\tau|\cos\phi$, where $0 \leq \phi < \pi/2$ is a vertex angle; $\cos\phi = \pm\cos\varphi$.

The map $z = e^w$ sends $\mathbb{C}/X$ onto the torus T in such a way that $[0, 2\pi]$ is mapped to a circle and the line segments parallel to $[0, \tau]$ are mapped into the class c_T. The translation $w \mapsto w + \tau$ is sent to $z \mapsto kz$.

Now $\pm \cos\varphi = \operatorname{Re}\tau/|\tau|$. The rectangle of length $|\tau|$ and height $2\pi \operatorname{Re}\tau/|\tau|$ is foliated by the line segments parallel to $[0,\tau]$; it serves as a fundamental region for T. Conclude that $\lambda(c_T) = |\tau|^2/(2\pi \operatorname{Re}\tau)$.

For the quasifuchsian group we end up with

$$|\log k| \leq 2 \min(L_1, L_2).$$

Now return to the hypothesis of Theorem 5.15.4. Here we use the annulus $A = \Omega/\langle h \rangle$ with $\operatorname{Len}_{\Omega/H}(h) = \pi\lambda(c_A)$. If h is conjugate to $z \mapsto kz$ (we can assume $|k| > 1$) then

$$\log|k| \leq |\log k| \leq \frac{2\pi}{M_A} = 2\pi\lambda(c_A) = 2\operatorname{Len}_{\Omega/\operatorname{Stab}(\Omega)}(h). \qquad \square$$

In particular, suppose $\mathcal{M}(H)$ has bounded geometry, that is, $\operatorname{Len}_{\mathcal{M}(H)}(\alpha_*) \geq \epsilon > 0$, for all geodesics $\alpha_* \subset \mathcal{M}(H)$. In the quasifuchsian case, the lengths of geodesics in both the top and bottom component are uniformly bounded below. If we can find a path out to the Bers boundary in a Bers slice, and if all manifolds along this path have uniformly bounded geometry with geodesic lengths $\geq \epsilon > 0$, then no pinching can occur. The projection of the path to the moduli space lies in a compact set.

Suppose instead that Ω is an infinitely connected component of $\Omega(H)$ invariant under a function group G. Assume there is a positive lower bound for the length of all closed curves in Ω in the hyperbolic metric in Ω. According to Canary [1991], there is a constant $\kappa > 0$ with the following property. If $c \subset S = \Omega/G$ is a closed geodesic in the hyperbolic metric on S, and c^* is the geodesic or a point representing c in $\mathcal{M}(H)$, then in the respective hyperbolic metrics,

$$\operatorname{Len}_{\mathcal{M}(H)}(c^*) \leq \kappa \operatorname{Len}_S(c).$$

5-6. *Pinching.* Given a geometrically finite, boundary incompressible manifold $\mathcal{M}(G)$ we will say that disjoint simple loops $\alpha, \beta \subset \partial\mathcal{M}(G)$ are *parallel* if neither one can be homotoped to a puncture (or to a point), yet the loops are freely homotopic in $\mathcal{M}(G)$. That is, they bound an annular region on $\partial\mathcal{M}(G)$, or they bound an essential cylinder in $\mathcal{M}(G)$. Another way of saying this is that if $A \in G$ is a primitive loxodromic associated with α, and $B \in G$ one associated with β then the cyclic groups $\langle A \rangle$ and $\langle B \rangle$ are conjugate in G. By a primitive loxodromic A we mean that, for some lift $\alpha^* \in \Omega(G)$, A is a generator of cyclic subgroup that maps α^* onto itself. We have often said more simply that $A \in G$ is associated with α.

The best general result about pinching is as follows: The setting is a geometrically finite group G without elliptics. Let $\alpha_1, \ldots, \alpha_n$ be mutually disjoint and nonparallel simple loops on $\partial\mathcal{M}(G)$ that are represented by the *loxodromics* $A_1, \ldots, A_N$ and their conjugacy classes in G.

Pinching Theorem [Ohshika 1998a]. *The manifold $\mathcal{M}(G)$ can be pinched along the loops $\{\alpha_i\}$ resulting in a geometrically finite manifold $\mathcal{M}(H)$.*

More precisely there is a sequence of points $\{\theta_n : G \to G_n\}$ *in the deformation space* $\mathfrak{T}(G)$ *such that*

(i) $\{G_n\}$ *converges algebraically and geometrically to* $H = \lim \theta_n(G)$;
(ii) $\lim \theta_n(A_i) = A_i^*$ *is parabolic for* $1 \leq i \leq N$, *and every new parabolic in* H *is in the conjugacy class of some* $\langle A_i^* \rangle$;
(iii) *the interior* $\mathbb{H}^3/G$ *is homeomorphic to* $\mathbb{H}^3/H$; *and*
(iv) *if* $\alpha_{i_1}, \ldots, \alpha_{i_k}$ *lie in the component* R_i *of* $\partial \mathcal{M}(G)$, *then* $R_i \setminus \bigcup_{1 \leq j \leq k} \alpha_{i_j}$ *is homeomorphic to a union of components of* $\partial \mathcal{M}(H)$ *such that each* α_{i_j} *determines a pair of punctures.*

The statement has been slightly modified from Ohshika's. His proof depends on Thurston's stronger version of his Compactness Theorem (p. 240) and brings in techniques used in Maskit [1983] to prove a weaker result. In any case the existence of the limit group H follows from the Hyperbolization Theorem.

Prove this for quasifuchsian groups by replacing the α_i by ever thicker annuli as in Exercise (4-8). The estimate (5.14) is needed to show that the sequences $\{\theta_n(A_i)\}$ converge to parabolics.

Hint: Remove from the original manifold $\mathcal{M}(G)$ the geodesics which are parallel to the initial loops α_i. Apply the Hyperbolization Theorem to get a manifold where these geodesics correspond to rank two cusps. Do Dehn surgery on solid cusp tori.

For a different approach, apply Theorem 3.11.6.

5-7. *Nondensity of maximal cusps: an example of Curt McMullen.* Consider the quasifuchsian space of a geometrically finite group without parabolics. Choose a boundary group H without parabolics where the bottom end of $\mathcal{M}(H)$ is geometrically infinite and the top end is geometrically finite (a closed surface). This group cannot be approximated (algebraically) by maximal cusps. For suppose otherwise so that $H = \lim H_n$. Each maximal cusp H_n has the property that all the components of $\partial \mathcal{M}(H_n)$ are triply punctured spheres. Therefore the convex core $\mathcal{CC}(H_n)$ of $\mathcal{M}(H_n)$ is bounded by a finite union of totally geodesic 3-punctured spheres, the number of components being independent of n. In view of Theorem 4.6.2 we know that all sequences that converge algebraically to H also converge geometrically. In particular this is true of $\{H_n\}$. On the other hand geometric convergence implies convergence of the convex hulls. This is impossible at the bottom end.

5-8. *Limiting laminations.* Consider now the set of lifts of the simple loop $\alpha \subset R$ to $\mathbb{H}^2$ and the set of lifts of a transverse ℓ. What is the action in $\mathbb{H}^2$ of a lift of the twist τ about α?

What happens when τ^{-1} is applied?. What happens if $\tau^{\pm n}$ is applied, $n \to \infty$?

Next suppose the projection Δ to R of an ideal triangle in $\mathbb{H}^2$ is injective. Assume that Δ is transverse to α. What happens to the sequence $\tau^n(\Delta)$ as $n \to \infty$?.

Based on this idea, can you show that every geodesic lamination on R is the limit in the Hausdorff topology of a sequence of ideal triangles (Saul Schleimer)?

5-9. *More on collapsing mappings* (Thurston; see Minsky [1994b]). Let $f : \mathbb{S}^2 \to \mathbb{S}^2$ be any continuous map whose restriction to the lower half-plane LHP is a homeomorphism. Set $\sigma = f(\mathbb{R} \cup \{\infty\})$. $y \in \sigma$ corresponds the closed set $C_y = \{f^{-1}(y)\} \subset \mathbb{S}^1 \equiv \mathbb{R} \cup \{\infty\}$. Let $\mathcal{C}_y$ denote the hyperbolic convex hull in LHP of C_y.

Verify that the collection of closed sets $\{\mathcal{C}_y : y \in \sigma\}$ is a partition of LHP into disjoint closed sets. (*Hint:* If $\mathcal{C}_y \cap \mathcal{C}_z \neq \varnothing$ there would be two pairs of points in C_y and C_z that separate each other on $\mathbb{S}^1$.)

Each $\mathcal{C}_y$ is a polygon with possibly infinitely many edges. The totality of edges forms a geodesic lamination Λ in LHP.

Now suppose in addition there is a fuchsian group G such that $f \circ g = \theta(g) \circ f$ for all $g \in G$ and θ is an isomorphism to a kleinian group $\theta(G)$. Then Λ is invariant under G and projects to a geodesic lamination on LHP$/G$.

5-10. *Dehn twists.* A Dehn twist in the round annulus $A_0 = \{z : 1 < |z| < u\}$ is the following mapping. Positively orient the boundary components a, b of A_0. Hold one fixed and twist A_0 by rotating the other by 2π in its positive direction. For example we can take (Figure 5.18)

$$\tau : re^{i\theta} \mapsto re^{i\theta s(r)}, \quad \text{where } 1 \leq r \leq u \text{ and } s(r) = \frac{(u - 2\pi) + (2\pi - 1)r}{u - 1}.$$

Now if $A \subset R$ is instead a neighborhood of a simple loop $\alpha \not\sim 1$ on some surface, we can bring over the twist from the round annulus to A. Extend the Dehn twist in A to all R by setting it equal to the identity on $R \setminus A$. This is a *Dehn twist* of R about α—actually we are only interested in its homotopy class. More precisely this is a Dehn twist of degree $+1$. Similarly we can define a Dehn twist of any positive or negative degree about α. A Dehn twist has no effect on the free homotopy class of a simple loop that does not make essential crossings with A.

Suppose γ is a simple loop which has essential crossings with α, that is the number of crossings cannot be reduced within the free homotopy class of γ—this minimal number of crossings is the geometric intersection number $\iota(\gamma, \alpha)$. In the annular region A about α, we may assume $\gamma \cap A$ consists of a finite number of mutually

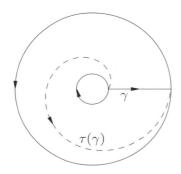

Fig. 5.18. A positive Dehn twist in an annulus.

disjoint segments running between the two boundary components. After A is twisted, each of the segments winds once around A but keeps the same endpoints. The collection remains mutually disjoint. The twisted segments can then be reattached to the part of γ outside A to become a new simple loop.

For example, consider the following construction on a genus 2 handlebody $\mathcal{M}(G)$ with boundary surface S. Let $\gamma \subset S$ be a simple dividing loop which in addition bounds a disk in the handlebody. Let $\alpha \subset S$ be simple nondividing loop which crosses γ twice and is not compressible in $\mathcal{M}(G)$. Let γ' be the result of applying a Dehn twist τ about α to γ. Show that γ' still divides S into two tori each with one boundary component but γ' is no longer compressible. In one of the components choose two simple loops A and B crossing each other exactly once. Show that the simple loop $[A, B]$ is freely homotopic to the boundary γ', and neither A or B is compressible.

Suppose α, β are simple geodesics which cross once. Describe the action of $\tau(\beta) \circ \tau(\alpha)$ on a simple geodesic γ which has a single crossing with each.

Extension of twists to $\mathcal{M}(G)$. A Dehn twist about a compressing curve α on $\partial\mathcal{M}(G)$ can be extended to a *twist map* of $\mathcal{M}(G)$, which is isotopic to the identity. First carry this out in the round model $\mathbb{D} \times [0, 1]$, via $(re^{2\pi i\theta}, t) \mapsto (re^{2\pi i(\theta+t)}, t)$. Then thicken to D^* an essential disk D bounded by α, so that $D^* \cap \partial\mathcal{M}(G) = A$. Bring over the twist to D^* and set it equal to the identity outside.

A Dehn twist about an essential cylinder C in $\mathcal{M}(G)$ is constructed as follows. Thicken C a bit to get the homeomorphic image C^* of the round $A_0^* = A_0 \times [-\delta, \delta]$. The Dehn twist in A_0 can be extended to A_0^* then C^*. Set it equal to the identity outside C^*.

5-11. *Anosov mappings of a torus.* Show that the automorphisms A of a torus with $\mathrm{tr}^2(A) > 4$ preserve no simple closed geodesic while if $\mathrm{tr}^2 = 4$ a simple geodesic is preserved. In the former case, the automorphisms are called *Anosov mappings.* Prototypical examples of such A are $\left(\begin{smallmatrix}1 & 1\\ 0 & 1\end{smallmatrix}\right)$ and $\left(\begin{smallmatrix}2 & 1\\ 1 & 1\end{smallmatrix}\right)^8$.

Hint: Work with the square lattice and the quotient torus

$$\mathbb{T}^2 = \mathbb{C}/\langle z \mapsto z + 1, \ z \mapsto z + i\rangle.$$

Referring back to Exercise (2-6), show that an affine map that sends the square lattice onto itself has the form $A : (x, y) \mapsto (u, v)$ where

$$u = ax + by, \ y = cx + dy, \quad a, b, c, d \in \mathbb{Z}, \ ad - bc = 1.$$

Simple loops on the square torus come from lines through the origin of rational slope. Show that A preserves a simple loop on the quotient if and only if $\mathrm{tr}^2(A) = 4$, that is if there is only one eigenvalue $\lambda = \pm 1$, or if and only if the fixed point of A, if $\neq \infty$, is a rational number. Otherwise the projection to $\mathbb{T}^2$ of each of the lines of given irrational slope is dense and uniformly distributed in the torus by a famous classical

---

[8] The linear map $\left(\begin{smallmatrix}2 & 1\\ 1 & 1\end{smallmatrix}\right)$ maps the square torus onto itself. Being symmetric, its eigenvectors L_1, L_2 are orthogonal. There is expansion along lines parallel to L_1 and contraction along lines parallel to L_2.

theorem of Weyl. The projection of the family of parallel lines is called a *foliation* of $\mathbb{T}^2$. Find the eigenvalues and the eigenvectors. Identify the set of orthogonal lines that are preserved by A. On one set of lines, A is expanding by a factor exceeding one; these are called the *unstable leaves*. On the orthogonal lines, A is contracting; these are called the *stable leaves*. An Anosov map is characterized by having such a pair of transverse, invariant foliations.

On tori, the Anosov maps have no singularities, but on hyperbolic surfaces, the topology forces singularities.

5-12. *Surface automorphisms.* Suppose R is a surface of finite topological type. Let τ be an automorphism of R, that is, an orientation-preserving homeomorphism of R onto itself. We are actually interested not so much in τ itself but in its equivalence class, also denoted by τ, under the homotopy relation $\sim$. These homotopy classes of automorphisms, as we saw on p. 92, form the mapping class group $\mathcal{MCG}(R)$. (One could use the isotopy equivalence relation instead, with the same results.)

Assume $\tau \not\sim$ id. Thurston's classification (and earlier Nielsen's classification) of the automorphisms parallels the classification of Möbius transformations:

τ **has finite order** if for some integer, $\tau^m \sim$ id;

τ **is reducible** if there is a set of mutually disjoint, unoriented, nontrivial simple loops $\mathcal{S} \subset R$, no two of which are parallel and no one of which is parallel to a boundary component or puncture, such that τ preserves the set $\mathcal{S}$ up to free homotopy of its elements;

τ **is pseudo-Anosov** if it has neither property (i) nor (ii).

Finite order elements. An element of finite order is analogous to an elliptic Möbius transformation. As mentioned on p. 92, there was a longstanding conjecture of Nielsen that given a finite group (up to homotopy) of automorphisms τ there is a hyperbolic structure that can be put on R so that the group becomes a group of isometries of R onto itself. This was first proved in Kerckhoff [1983], with different proofs in Gabai [1992]; Casson and Jungreis [1994].

It is a useful fact, established by Serre, that there is a torsion-free, normal subgroup of finite index in the mapping class group. It is the kernel of the map of the mapping class group of a genus $g > 1$ closed surface R to its group of automorphisms mod 3 of homology $H_1(R, \mathbb{Z}/3Z)$. In other words, the orbifold $\mathcal{M}_g$ has a finite-sheeted manifold cover.

Reducible mappings. The term "reducible" is used to suggest that if one cuts R along the curves of $\mathcal{S}$, some power τ^m ($m \geq 1$) maps each of the complementary regions onto itself and therefore can be analyzed in terms of its action on simpler surfaces. A Dehn twist is clearly reducible (with $m = 1$). Dehn twists are analogous to parabolic Möbius transformations. The "fixed point" of a Dehn twist (as well as of its inverse) is the geodesic lamination consisting of the geodesic representative of the simple loop. A set of Dehn twists about mutually disjoint loops generates a free abelian subgroup of $\mathcal{MCG}(R)$. (The mapping class group itself is generated by $2g+1$ Dehn twists—see Birman [1974].)

Pseudo-Anosov mappings. This is the generic case. For example, on a once-punctured torus choose simple closed geodesics a, b crossing each other exactly once. The composition of the Dehn twist about a with the Dehn twist about b is pseudo-Anosov.

More generally, suppose α, β are multicurves on R with components $\alpha = (a_1, \ldots, a_m)$, $\beta = (b_1, \ldots, b_n)$. Assume that $\alpha \cup \beta$ fills R. (It is a lot easier to find such multicurves, than to find a pair of simple loops which fill!) Denote the positive Dehn twists by $\{T_{a_i}\}$ and $\{T_{b_j}\}$. Let $T_\alpha = \Pi_1^m T_{a_i}$, and $T_\beta = \Pi_1^n T_{b_i}$. Then $T_\alpha T_\beta^{-1}$ is pseudo-Anosov [Thurston 1982].

A pseudo-Anosov transformation τ acting in $\mathfrak{Teich}(R)$ has a unique *axis* preserved by τ, similar to the axis of a loxodromic transformation acting in $\mathbb{H}^3$. It is the line L through (S, f) in the direction determined by the quadratic differential φ associated with the Teichmüller map homotopic to $f\tau f^{-1} : S \to S$, see Marden and Strebel [1993]. Namely L is the set $\{(S_t, f_t f)\}$ where $f_t : S \to S_t$, $-1 < t < 1$, is the Teichmüller map satisfying

$$(f_t)_{\bar{z}}/(f_t)_z = t\bar{\varphi}/|\varphi|,$$

A pseudo-Anosov τ does not preserve the homotopy class of any simple closed curve on the surface. Rather it preserves a pair of geodesic laminations which, like all geodesic laminations, are limits of simple closed curves. The situation is as follows:

Given the Riemann surface R on which τ is acting, there is a uniquely determined holomorphic quadratic differential with the following property: The quadratic differential determines a singular euclidean structure on R in terms of which τ is an affine map on neighborhoods of nonsingular points. In analogy to the torus case when one has eigenvectors, in the geometry of quadratic differentials there is a horizontal (unstable) and vertical (stable) foliation. These are determined by the horizontal and vertical trajectories—the pullback of the horizontal and vertical lines in $\mathbb{R}^2$ by the locally conformal map determined by the quadratic differential—which are invariant under τ. The quadratic differential is associated with the extremal Teichmüller map in the homotopy class of τ. See Exercise (5-33).

From this one obtains projective measured laminations $(\Lambda_\pm, \mu_\pm)$, uniquely associated with fixed points of τ, in analogy with the fixed points of a loxodromic. They are called the *attracting* and *repelling* (or *stable* and *unstable*) laminations associated with the pseudo-Anosov τ. The fixed points $\Lambda_\pm$ are elements of $\mathcal{EL}$.

The laminations arise as Λ_+ when $\lim \tau^{+n}(\gamma)$ and Λ_- when $\lim \tau^{-n}(\gamma)$, for any simple closed geodesic γ. The measures $\mu_\pm$ are obtained as the projectivized limit of the counting measures. A pseudo-Anosov τ is associated with a pair of "stretch factors" or "eigenvalues" $\lambda^{\pm 1}$, such that $\lambda > 1$, and

$$\tau : (\Lambda^+, \mu^+) \to (\Lambda^+, \lambda\mu^+), \quad \tau : (\Lambda^-, \mu^-) \to (\Lambda^-, \lambda^{-1}\mu^-).$$

For the intersection numbers $\iota(\tau(\gamma), \Lambda_\pm) = \lambda^{\mp 1}\iota(\gamma, \Lambda_\pm)$ for any simple closed geodesic γ.

$\log \lambda$ is also the translation length of τ as an isometry of $\mathfrak{Teich}(R)$, see Farb [2006].

For a closed surface of genus $g \geq 2$ the $\lambda = \lambda(g)$ is an algebraic integer whose degree does not exceed $6g - 6$. Moreover there are at most finitely many conjugacy classes in $\mathcal{MCG}(R)$ with $\lambda < N$ for any N [Arnuox and Yoccoz 1981], [Ivanov 2002]. On the other hand, $\lim_{g \to \infty} \lambda(g) = 1$ [Penner 1991].

The fixed point laminations $\Lambda_\pm$ are *uniquely ergodic* in that, up to projective equivalence, there is only one transverse measure μ on $\Lambda_\pm$. The corresponding projective measured laminations are dense in all projective measured laminations.

Just like loxodromic Möbius transformations, two pseudo-Anosov automorphisms τ_1, τ_2 either have the same pair of fixed points or no common fixed points. In the latter case the subgroup $\langle \tau_1^m, \tau_2^n \rangle \subset \mathcal{MCG}(R)$ is a free group, for sufficiently large m, n; see Mosher [2007].

It has recently been proved that on any finite area surface R, $\mathcal{MCG}(R)$ has a normal subgroup, which is also a free group, in which every (nontrivial) element is pseudo-Anosov [Dahmani et al $\geq$ 2015, Theorem 2.28].

For a comprehensive introduction to the mapping class group see Farb and Margalit [2012] or Ivanov [2002], where the reader will find details of the statements made above. For the interpretation in terms of quadratic differentials see Exercise (5-33) and Section 6.1.1.

For the details of this theory in the topological case see Thurston [1988]; Fathi et al. [1979]; Mosher [2007]. For a discussion of the analytic interpretation presented above, see Marden and Strebel [1984; 1986].

Finally, show that a 5-punctured sphere has the property that if you choose any simple closed geodesic γ, the orbit of γ under the mapping class group comprises the totality of simple closed geodesics on the surface. No other surface has this property.

5-13. *Twists and traces.* Suppose G is a fuchsian group without elliptics. Let T be a loxodromic element, and $A, B \in G$ any two other elements with fixed points distinct from those of T. Prove that $\text{tr}^2(AT^n B)$ goes to ∞ when n goes to $\pm\infty$. Conclude that the length of the geodesics on the quotient surface determined by the members of the corresponding sequence become infinite. (*Hint:* Work in UHP, take T to have fixed points $0, \infty$.)

Prove that $\text{tr}^2(AT^n BT^{-n}) \to \infty$ as well.

Suppose on the surface $R = \mathbb{H}^2/G$, γ is a simple nontrivial loop. Let τ denote the Dehn twist about γ. First assume that γ separates R. Let δ be a simple loop with geometric intersection number $\iota(\delta, \gamma) = 2$. The loop δ is homotopic to a composition $\alpha\beta$ where α, β lie in the two complementary components except for a common origin on γ. The twist τ^n sends $\alpha\beta$ to, say, $\alpha\gamma^n\beta\gamma^{-n}$. If instead γ does not separate, let δ be a simple loop with $\iota(\delta, \gamma) = 1$ and crossing γ once, at its origin. The result of applying τ^n is then to send δ to $\delta\gamma^n$. In either case, applied to the corresponding elements of G, show that the length of the corresponding geodesics becomes infinite as $n \to \pm\infty$.

5-14. *The orbit of a point of* $\mathfrak{Teich}(R)$ *under iterated Dehn twists: Constructing geometric limits on the Bers boundary.* Recall that on the surface R, a Dehn twist about a geodesic. The group of all such twists is the mapping class group of R. By mapping by quasiconformal homotopy classes, $\mathfrak{Teich}(R)$ is generated. The mapping class group of R then carries over to form a discrete group in $\mathfrak{Teich}(R)$. In the following we will view $\mathfrak{Teich}(R)$ as a Bers slice $\mathcal{B}(R)$. Fix a basepoint $O \in \mathcal{B}(R)$, that is, fix a hyperbolic manifold, a point in $\mathcal{B}(R)$.

The goal of this exercise is to identify the limit of a sequence $\{\tau^n(O)\}$ where τ is now a Dehn twist acting in $\mathcal{B}(R)$. The construction was first explored by Earle-Marden see 1980 in their study of the boundary of Teichmüller space, and independently discovered by Kerckhoff and Thurston [1990] in their study of continuity on the boundary.

The point $O \in \mathcal{B}(R)$ corresponds to a manifold $\mathcal{M}(G)$ which is described in the Bers slice as $(R, S^{\text{top}}; J)$ where $R \equiv S_{\text{bot}}$ is the bottom surface S^{top} the top surface of $\mathcal{M}(G)$ and J is a fiber-preserving involution.

The Dehn twist τ is applied as follows. Choose a simple loop $\gamma \subset R$. For definiteness, we will assume that γ divides R into two subsurfaces R_{bot1}, R_{bot2}. A parallel curve $\gamma^{\text{top}} \subset S^{\text{top}}$ divides S^{top} into subsurfaces S^{top1}, S^{top2} parallel to R_{bot1}, R_{bot2}.

The action of τ^n on O is

$$\tau^n : (R, S^{\text{top}}; J) \mapsto (R, S^{\text{top}}; J \circ \tau^n).$$

The point O is sent to $\tau^n(O) \in \mathcal{B}(R)$. Neither the conformal type of the bottom or the top surface changes, they remain conformally equivalent to R and S^{top}. Rather, the topological relationship determined by $J \circ \tau^n$ causes increasing distortion between the bottom and top surfaces which in the end will cause a fracture in the manifold; S^{top2} is being more and more twisted about S^{top1} and in reference to the bottom surface.

Suppose $\delta \in S^{\text{top}}$ is a simple loop crossing γ^{top} exactly twice (geometric intersection number two). We can write $\delta = \alpha\beta$ with α a closed loop in S^{top1} and β in S^{top2} with a common point on γ^{top}. Then τ^n acts as

$$\tau^n : \alpha\beta \mapsto \gamma_n = \alpha\gamma_t^{\,n}\beta.$$

According to Exercise (5-13) the length $\text{Len}(\gamma_n)$ approaches ∞ conclude that the sequence $\{\tau^n(O)\}$ and corresponding manifolds $\{\mathcal{M}(H_n)\}$ converges algebraically to the result of pinching S^{top} along γ^{top}. This occurs in both cases $n \to \pm\infty$. Denote the algebraic limit by H and the corresponding manifold $\mathcal{M}(H)$. The top boundary of $\mathcal{M}(H)$ has two components which we label S^{H_1}, S^{H_2}.

Yet there remains a conformal map f_n of S^{top} onto $\partial\mathcal{M}^{\text{top}}(H_n)$ (and also a conformal map onto the bottom). If we choose a point P in, say, S^{top1} we see that a subsequence $\{f_n\}$ converge to a conformal map f. In fact the sequence will converge to a conformal map of the whole surface S^{top} onto a surface S^*.

We will find a geometric limit $\mathcal{M}(H^*)$, with $H^* \supset H$, with the following properties:

(i) $\partial \mathcal{M}(H^*)$ has two boundary components, one conformally equivalent to $R \equiv S_{\text{bot}}$, the other conformally equivalent to S^{top},

(ii) $\mathcal{M}(H^*)$ is homeomorphic to $(R \times [0, 1]) \setminus \{c\}$, where c is a simple loop parallel to γ and γ_{top}; here c represents a solid cusp torus in $\mathcal{M}(H^*)$.

Consider the situation in more detail. The regular set $\Omega(G)$ has two components. To simplify notation, assume G is fuchsian so that $\Omega^{\text{top}} = \text{UHP}$, $\Omega_{\text{bot}} = \text{LHP}$. The totality of lifts $\{\gamma^{\text{top}^*}\}$ of γ^{top} divides Ω^{top} into countably many regions. Consider two adjacent regions $\Omega^{\text{top}1}$, $\Omega^{\text{top}2}$ where the first listed covers $S^{\text{top}1}$ and the second $S^{\text{top}2}$. Each is preserved by a subgroup G_1, G_2 of G, and G itself is the free product of these two subgroups, amalgamated over the common cyclic subgroup generated by the element preserving their common boundary, a lift of γ^{top}.

Each point $\tau^n(O)$ in the Bers slice is uniquely associated with a manifold $\mathcal{M}(H_n)$, up to conjugacy. Denote by F_n the conformal map of LHP onto the component $\Omega_-(H_n)$, corresponding to the bottom boundary of $\mathcal{M}(H_n)$, normalized by requiring $(0, 1, \infty)$ to be fixed. On LHP, the map F_n determines an isomorphism $\theta_n : G \to H_n$. There, the sequence $\{F_n\}$ converges to a conformal map $F : \text{LHP} \to \Omega_{\text{bot}}(H)$. In particular $\{\theta_n(\gamma)\}$ coverges to a parabolic element $g^* \in H$.

The two regions $\Omega^{\text{top}1}$ and $\Omega^{\text{top}2}$ are adjacent in UHP and share a geodesic γ^{top}. There is a sequence of conformal maps, denoted by $\{F_n'\}$, of UHP onto the complement of $F_n(\text{LHP})$ in $\mathbb{S}^2$. Focus on the $\{F_n'\}$-image of $\Omega^{\text{top}1}$, and $\Omega^{\text{top}2}$.

The group element $g \in G$ fixes $\gamma^{\text{top}} \subset \Omega^{\text{top}}(G)$ as well as $\gamma \subset \Omega_{\text{bot}}(G)$. The sequence $\{\theta_n(\gamma)\}$ converges to a parabolic g^* simultaneously with its conjugates.

The sequence $\{H_n\}$ also has a geometric limit H^*.

With respect to $\Omega^{\text{top}1}(G)$, a conformal map F_n' induces the isomorphism θ_n of G_1 onto the subgroup of H_n that stabilizes $F_n'(\Omega^{\text{top}1}(G))$. In contrast, on $\Omega^{\text{top}2}$, F_n' induces the isomorphism

$$G_2 \to \theta_n(g^{-n} G_2 g^n).$$

But $\{F_n'\}$ converges on $\Omega^{\text{top}2}$ as well. This forces $\{\theta_n(g)\}$ to converge via the *geometric limit* to a necessarily parabolic element $h^* \in H^*$. This in turn forces the relation $h^{*-1} g^* h^* = g^*$ to hold. This relation implies that both g^* and h^* are parabolic with the same fixed point since they cannot both lie in a cyclic subgroup of H^*. So $\langle g^*, h^* \rangle$ is a rank two parabolic subgroup thereby determining a solid cusp torus in $\mathcal{M}(H^*)$.

In fact, $H^* = \langle H, h^* \rangle$. See also Exercise (4-28).

To see the picture more clearly, pass to the quotient manifold $\mathcal{M}(H)$, specifically to the top surface which has broken into two parts S^{H_1}, S^{H_2}, attached via the fixed point p of g^*. At p, in each component, we can choose horodisks Δ_1, and Δ_2 with the following property.

In $\Omega(H)$, choose a lift p^* of p and lifts $\widehat{S}_1, \widehat{S}_2$ that share the fixed point p^*. Inside these regions, find lifts Δ_1^*, Δ_2^* that are respective horodisks at p^*. Furthermore, choose the lifted horodisks so that the parabolic h^* sends the exterior of Δ_2^* (with respect to $\mathbb{S}^2$) onto the interior of Δ_1^*. And make this construction equivariant under H.

The result is that the geometrically finite $\mathcal{M}(H^*)$ has the structure announced earlier. We will deal further with manifolds of this type in the next exercise.

Once we have the new groups H, H^* we can repeat the process with a new simple loop, and keep going until the loops selected determine a pants decomposition on $\partial\mathcal{M}(G)$.

Finally, the algebraic and geometric limits are independent of the subsequences used to attain them. This is a consequence of the Rigidity (or Isomorphism) Theorem 3.13.4.

For another exposition with application to Riemann surface theory, see Marden [1980]. Published details appear in the independent development in Kerckhoff and Thurston [1990], where the result is applied to show that the extension of the mapping class group to the Bers boundary is not necessarily continuous.

What happens to $\mathcal{M}(G)$ if you apply $\alpha\gamma^n\beta$ to the top surface of $\mathcal{M}(G)$ and $\alpha\gamma^n\beta\gamma^{-n}$ to the bottom surface?

5-15. *Piling up double cusps.* This construction is simplest for boundary cusps of a Bers slice in the once-punctured torus quasifuchsian space. For in this quasifuchsian space, a boundary cusp group H is such that $\partial\mathcal{M}(H)$ is the union either of a triply punctured sphere and a once-punctured torus (which we will here call a single cusp group), or of two triply punctured spheres (a double cusp group).

We can make an arbitrarily high pile of double cusp groups, where the top and/or bottom of the pile, if the pile is finite in that direction, is either a double cusp group or a single cusp group. It should suffice to illustrate the method in the simplest case.

Suppose H_1, H_2 are single cusp groups. We may arrange things so that the triply punctured sphere is the top component $S^{\text{top}1}$ of $\partial\mathcal{M}(H_1)$ and the bottom $S_{\text{bot}2}$ of $\partial\mathcal{M}(H_2)$. This means if we start with a fuchsian once-punctured torus group G, then there are orientation-preserving homeomorphisms Φ_1, Φ_2 of the interiors $\mathbb{H}^3/G \to \mathbb{H}^3/H_i$ that take the top and bottom ends of $\mathbb{H}^3/G$ to the respectively labeled ends of $\mathbb{H}^3/H_i$.

Now $S^{\text{top}1}$ arises by pinching the top punctured torus along a simple closed curve and likewise pinching $S_{\text{bot}2}$. The pinching curves are represented on the other boundary components by simple loops $\gamma_1 \subset S_{\text{bot}1}$ and $\gamma_2 \subset S^{\text{top}2}$ respectively, parallel to the pinching curves. For our construction we must require that γ_1 and γ_2 are in the Φ_1, Φ_2 images of the *same* free homotopy class of $\mathbb{H}^3/G$.

Now of the three punctures on $S^{\text{top}1}$ and $S_{\text{bot}2}$, one is spoken for as the puncture coming from the paired punctures on $\partial\mathcal{M}(G)$. We will refer to these as the basic punctures. The other two punctures will be called new punctures. If we draw a small circle about each of them, there is a solid pairing tube T_1, T_2 in each of $\mathcal{M}(H_1)$, $\mathcal{M}(H_2)$ that pairs them—its boundary is bounded by the two small circles. When we stack $\mathcal{M}(H_2)$ on top of $\mathcal{M}(H_1)$ so that the basic punctures are matched, the two solid pairing tubes will join up to form a solid cusp torus—and to determine a rank two cusp of the new manifold $\mathcal{M}(H^*)$. The group H^* is not quasifuchsian, still, the top and bottom are once-punctured tori.

How do we do the construction so that the new manifold is hyperbolic? Represent S^{top1}, S_{bot2} as totally geodesic surfaces S^{top1*}, $S_{bot2}*$ within the corresponding manifolds—just replace a disk in $\partial\mathbb{H}^3$ by the hyperbolic plane supported by its boundary and project. Let $P_1 \subset \mathbb{H}^3$ be a lift of S^{top1*} and $\vec{n}_1$ the lift of an inner pointing normal. We may conjugate H_2 so that a lift P_2 of $S_{bot2}*$ coincides with P_1 but $\vec{n}_2$ points to the opposite side of $P_1 = P_2$ as $\vec{n}_1$. We must also arrange things so that the respective subgroups preserving $P_1 = P_2$ are identical and the parabolic conjugacy class associated with the basic punctures coincide. Now $\langle H_1, H_2 \rangle$ is discrete. How does the new rank two cusp arise?

Up in $\mathbb{H}^3$, given a lift P_1 of S^{top1*}, and a new parabolic α acting on P_1, there is another lift P_1' of S^{top1*} which uniquely determined by the property that α preserves both—that the fixed point of α is the point of tangency of the boundaries of P_1 and P_1'. Choose an inner normal vector $\vec{n}_1'$ to P_1'. Correspondingly for $S_{bot2}*$ there is another lift P_2' that is also preserved by α; here $P_2' \neq P_1'$. Choose an inner normal vector $\vec{n}_2'$. When H_1 and H_2 are joined across $P_1 = P_2$, consider the associated planes P_1' and P_2' which also share the fixed point of α on their boundaries. To complete the combination of the two groups, we must add another parabolic β that maps the side of P_1' containing $\vec{n}_1'$ onto the side of P_2' opposite that determined by $\vec{n}_2'$, and satisfying

$$\beta\alpha\beta^{-1} = \alpha, \quad \beta(P_1') = P_2', \quad \beta\,\mathrm{Stab}(P_1')\beta^{-1} = \mathrm{Stab}(P_2').$$

These conditions uniquely determine β. In essence we are making a construction as in Exercise (4-28). We end up with the group $H = \langle H_1, H_2, \beta \rangle$.

Verify that the hyperbolic construction works, and that the resulting manifold $\mathcal{M}('H')$ has the following properties:

(i) $\partial\mathcal{M}(H)$ has two components, one conformally equivalent to the bottom component of $\partial\mathcal{M}(H_1)$, the other to the top component of $\partial\mathcal{M}(H_2)$.
(ii) In the interior of $\mathcal{M}(H)$ there is a "singular" totally geodesic surface representing the thrice-punctured sphere.
(iii) $\mathcal{M}(H)$ is homeomorphic to $(S \times [0, 1]) \setminus c$ where c is a circle homotopic to the representative of the pinching curve on each of the two boundary components, and S is a once-punctured torus.

Actually we made no essential use of the fact that the bottom component of $\partial\mathcal{M}(H_1)$ and top of $\partial\mathcal{M}(H_2)$ remained a once-punctured torus. H_1 and H_2 could as well have been double cusp groups. In this case we can continue the process and build an arbitrarily large pile of manifolds. This is the process required to directly construct the groups representing geometric limits in quasifuchsian space, as described in Section 5.12.

5-16. *Shuffling a rolodex.* This and the next exercise is a report on Anderson and Canary [1996a]. First we will construct the basic manifold.

Start with a solid torus T and its core curve c. Fix a finite system of mutually disjoint, parallel simple loops $\{\gamma_k\}$ on ∂T which are not contractible in T. Correspondingly fix a collection of surfaces $\{S_k\}$, each of some genus $g_k \geq 1$ and with a single boundary component. For greater effect assume the genera g_i are all different. Slightly thicken each S_k to obtain the compact manifolds $\{S_k \times [-\epsilon, +\epsilon]\}$. The boundary of each contains the annulus $\partial S_k \times [-\epsilon, \epsilon]$.

Attach $S_k \times [-\epsilon, \epsilon]$ by gluing $\partial S_k \times [-\epsilon, \epsilon]$ to a thin neighborhood of γ_k. The resulting manifold $M = T \cup_k S_k \times [-\epsilon, \epsilon]$ is orientable and compact. By "rearranging the pages"—taking a noncyclic permutation τ of $\{S_k\}$, we get another manifold M_τ which is homotopy equivalent but not homeomorphic to M. The manifolds M_τ have a hyperbolic structure. For more details see Exercise (5-15).

By the Hyperbolization Theorem (p. 385) we can write $M = \mathcal{M}(G)$, for a kleinian group G and likewise $M_\tau = \mathcal{M}(G_\tau)$. The original solid torus becomes a tubular neighborhood about the core geodesic c. So this construction results in a multitude (depending on the number of pages chosen) of hyperbolic manifolds homotopy equivalent but not homeomorphic to $\mathcal{M}(G)$ or to each other.

Let $\widehat{M}$ denote the result of removing from M the core curve c of T. Now $T \setminus \{c\} \subset \widehat{M}$ has the structure of a solid cusp torus. By applying the Hyperbolization Theorem we can assume that $\widehat{M} \cong \mathcal{M}(H)$ for a geometrically finite H with now a rank two cusp.

To get bumping, do Dehn surgery (Section 4.9) on the rank two cusp of $\mathcal{M}(H)$. Take a cusp torus parallel to ∂T, a meridian α which is contractible in T, and a longitude β which is parallel to c. Set $\delta_n = \alpha + n\beta$, $0 \leq n$. After Dehn surgery δ_n becomes a meridian on the solid torus that replaces the solid cusp torus. The resulting manifold M_n also has a hyperbolic structure $\mathcal{M}(H_n)$ and is homeomorphic to $\mathcal{M}(G)$. We claim that $\{H_n\}$ converges algebraically to a geometrically finite group G^* with a rank one parabolic and geometrically to $H \supset G^*$ in analogy with Section 4.9.

We will explicitly construct G^* and H in the next exercise. It will turn out that in H the $\{S_k\}$ are once-punctured surfaces and the system $T \setminus (\bigcup \partial S_k \times (-\epsilon, \epsilon))$ appears as pairing tubes, pairing successive punctures.

Next, choose a noncyclic permutation τ of $(1, 2, \ldots, k)$. Using the permutation of indices given by τ, build M_τ on T as M was built. As pointed out earlier, the manifolds M, M_τ have isomorphic fundamental groups and are are homotopy equivalent, but they are not homeomorphic.

The hardest part is to construct an immersion $f_\tau : M_\tau \to \widehat{M}$ and equally $\mathcal{M}(G_\tau) \to \mathcal{M}(H)$ such that on the level of fundamental groups f_τ has the properties (i) it determines an isomorphism of G_τ onto a geometrically finite subgroup of H, and (ii) it determines an isomorphism $\theta_{n,\tau} : G_\tau \to H_n$.

The sequence of isomorphisms $\{\theta_{n,\tau}\}$ converges algebraically to the isomorphism $\theta_\tau : G_\tau \to G^* \subset H$ and the sequence of groups geometrically to H itself. Thus the group G^* is a boundary group of the deformation space of G and of the deformation space of G_τ. These groups represent the nonhomeomorphic manifolds $\mathcal{M}(G)$ and $\mathcal{M}(G_\tau)$. The two deformation spaces bump at G^* (see 5.3.1).

Manifolds of the type constructed above exhibit other interesting properties (lest one believes such phenomena cannot occur!).

A hyperbolic manifold which has a simple geodesic γ which is not freely homotopic to curve in the boundary, yet γ^n is freely homotopic to a simple loop in the boundary.

Instead of doing $(1, n)$-Dehn surgery on $\mathcal{M}(H)$ do $(n, 1)$-Dehn surgery. That results in a manifold $\mathcal{M}(H'_n)$ in which $\beta\alpha^n$ is homotopic to a point. That is, $\beta \sim \alpha^{-n}$. Now in $\mathcal{M}(H'_n)$, α cannot be homotoped into $\partial\mathcal{M}(H'_n)$ (or to a point). Yet β can be homotoped into the boundary—it is parallel to the central curves of the annuli we used.

An essential cylinder whose central curve is not the generator of a maximal cyclic subgroup.

An essential cylinder/annulus is said to be a *primitive* if its fundamental group is a maximal cyclic subgroup. This is how one normally thinks about such things, like an essential cylinder in a fuchsian manifold. An example of a nonprimitive cylinder is the following. Start again with a solid torus T and a simple loop γ on its boundary that wraps around $n \geq 2$-times. If c denotes the core curve of the solid torus, we take γ to be homotopic in $T \setminus c$ to c^n.

Put a thin annular neighborhood N about γ in ∂T. Then choose a compact surface S with one boundary component and genus ≥ 1. Take the product manifold $S \times [-\epsilon, \epsilon]$ and glue $\partial S \times [-\epsilon, \epsilon]$ to the annular neighborhood about γ. Form the compact hyperbolizable manifold $M = T \cup (S \times [-\epsilon, \epsilon])$. Then N is an essential cylinder within M, and $\pi_1(N)$ is a proper subgroup of $\langle c \rangle$. If we remove from M a small tubular neighborhood about c then c is represented by a simple boundary curve.

5-17. *Constructing a rolodex.* In this exercise we will explicitly construct the rolodex used in Exercise (5-15). Again we closely follow Anderson and Canary [1996a].

We used above $k \geq 3$ surfaces $\{S_k\}$ with one boundary component and distinct genera $\{g_i \geq 1\}$. We will now assume the surfaces are closed Riemann surfaces each with one puncture.

Uniformize each of the surfaces by a fuchsian group $\{\Gamma_k\}$ acting in the upper and lower half-plane so normalized so that ∞ is a parabolic fixed point and $z \mapsto z + 1$ generates the rank one parabolic group at that point. By the universal disk property, the horizontal strip $\sigma = \{z \in \mathbb{C} : -1 - \varepsilon < \operatorname{Im} z < 1 + \varepsilon\}$, $\varepsilon > 0$, has the following property. For any element $g \neq \operatorname{id}$ of any group Γ_k, $g(\mathbb{C} \setminus \sigma) \subset \sigma$, unless g is in the rank one parabolic group at ∞. Moreover, the vertical slab σ^* in upper half-space over σ has the property that $\sigma^*/\Gamma_k \cong S_k \times (-1, 1)$. Let σ' be the result of truncating σ^* at height $1 + \varepsilon$. By the Universal Horoball Theorem, we have that $g(\mathbb{C} \cup \mathbb{H}^3 \setminus \sigma') \subset \sigma^*$ for any element g of any Γ_k, provided g does not fix ∞. Instead of lining the surfaces up on a solid torus as above, we will line them up in vertical translates of σ.

Next choose $\mu > 2k(1 + \varepsilon)$. Conjugate each Γ_k by a vertical translation over $\mathbb{C} : z \mapsto z + a_k i$ so that the horizontal strips σ_k for the resulting groups all lie in $\{z : 0 < \operatorname{Im} z < \mu\}$, with mutually disjoint closures. Choose them in the order $\sigma_1, \sigma_2, \ldots, \sigma_k$ as $\operatorname{Im} z$ increases from $\operatorname{Im} z = 0$ to $\operatorname{Im} z = \mu$. Denote the conjugated groups by $\{\Gamma'_k\}$.

We claim that the group $G^* = \langle \Gamma'_1, \Gamma'_2, \ldots, \Gamma'_n \rangle$ is a kleinian group such that $\mathcal{M}(G^*)$, with the interior of a solid pairing tube removed, is homeomorphic to the complement of the interior of T in the manifold $M = \mathcal{M}(G)$ constructed in the previous Exercise (5-16). In G^* there is only one parabolic conjugacy class, namely that generated by $z \mapsto z + 1$, while if $g(\infty) \neq \infty$, $g \in \Gamma_j$ maps the exterior of σ'_j into σ'_j.

Let $U(z) = z + \mu i$ and set $H = \langle G^*, U \rangle$. We claim that $\mathcal{M}(H)$ is homeomorphic to $\widehat{M}$ constructed above. H is geometrically finite with a rank two parabolic group at ∞.

Finally we construct a hyperbolic structure for M_τ. To do that we have to rearrange the strips $\{S_i\}$ to have the new ordering dictated by τ. A simple way of doing that is as follows:

$$H_\tau = \langle U\Gamma_{\tau(1)}U^{-1}, U^2\Gamma_{\tau(2)}U^{-2}, \ldots, U^n\Gamma_{\tau(n)}U^{-n} \rangle.$$

5-18. *The innards of an interesting manifold.* Describe the internal structure of the manifold presented in Figure 5.9.

5-19. *Boundary area* The hyperbolic area Area(S) of a component S of $\partial\mathcal{M}(G)$ is given by the formula Exercise (3-1). The total area of $\partial\mathcal{M}(R)$ is the sum of these expressions. Consider an arbitrary kleinian group G with rank $N < \infty$ (minimal number of generators). Prove Bers' inequality

$$\text{Area}(\partial\mathcal{M}(G)) \leq 4\pi(N - 1).$$

Prove in addition [Abikoff and Harvey 2012] that equality occurs if and only if G is a torsion-free, free group (Schottky or pinched Schottky). (*Hint*: Use Remark 3.7.2 and Equation (2.5).)

5-20. *The curve complex.* In the hands of Howard Masur and Yair Minsky [1999; 2000] this has proved to be an essential tool in analyzing the short geodesics near the ends of hyperbolic manifolds. The definition of the complex was originally given by Bill Harvey [1981]. Let Σ denote a compact surface of genus $g \geq 0$ and $n \geq 0$ punctures.

The simplicial complex $\mathcal{C}(\Sigma)$ is defined as follows:

(i) The vertices V are the simple closed curves on Σ that cannot be homotoped into a boundary component or to a point.

(ii) Two vertices $v_1, v_2 \in V$ are joined by an edge e if v_1 and v_2 correspond to disjoint simple loops, not in the same free homotopy class.

(iii) More generally, a finite set $\sigma \in V$ will bound a simplex in $\mathcal{C}(\Sigma)$ if the elements of σ can be realized by mutually disjoint, nonparallel, simple loops.

Special cases are:

- $g = 0, \ n \leq 3$. Then $V = \varnothing$; there are no simple closed curves, not retractable to a puncture.
- $g = 0, n = 4$, or $g = 1, n = 0, 1$. Then $\mathcal{C}(\Sigma) = V$; there are no edges.

In the cases $g = 0, n = 4$ or $g = 1, n = 1$ the edges must be defined slightly differently: Vertices v_1, v_2 are to be connected by an edge if the simple loops representing the vertices have intersection number 2 in the first case and 1 in the second case. Then $\mathcal{C}(\Sigma)$ is isomorphic to the Farey graph in $\mathbb{H}^2$ (Exercise (2-10)). For the remainder of the discussion we will exclude these cases.

The complex $\mathcal{C}(\Sigma)$ is connected and of dimension $(3g + n - 4)$ meaning the largest simplices have $(3g + n - 3)$ vertices. For example, if $g = 2, n = 0$ there are at most three simple, mutually disjoint, nonparallel, nontrivial simple loops. Thus the largest simplices in $\mathcal{C}(\Sigma)$ are triangles.

The curve complex is connected. However it is not locally finite. Suppose α, β are simple loops which cross once. They cannot be the endpoints of an edge. Rather, there are infinitely many distinct two-edge paths connecting them. $\mathcal{C}(\Sigma)$ is however δ-hyperbolic.

The 1-skeleton of $\mathcal{C}(\Sigma)$ is a connected graph. With the assignment of length one to each edge, it becomes a metric space. In fact [Masur and Minsky 1999] it is a Gromov hyperbolic space, as well as the Farey graph! See Exercise (2-18). The mapping class group of Σ, namely the group of homotopy classes of homeomorphisms of Σ onto itself, acts on $\mathcal{C}(\Sigma)$ and is its full group of isometries.

Except for the special cases listed above plus $(g, n) = (1, 2), (2, 0)$, mostly due to the presence of hyperelliptic involutions, the automorphisms of the curve complex have the following property: In Ivanov [1997] followed by Korkmaz [1999] and Luo [1999] it is shown that the group of automorphisms of the curve complex is isomorphic to the *extended* mapping class group (includes orientation-reversing mappings). The group of quasiisometries is also isomorphic to it [Rafi and Schleimer 2011].

Moreover, in the unpublished work [Klarreich 1999b], and in Hamenstädt [2006] it is proved that the Gromov boundary of the curve complex can be identified with the space of ending laminations associated with Σ. That is, the boundary consists of filling (irrational) geodesic laminations such that each halfleaf is dense in the whole lamination.

The arc complex. Suppose S is the compact, oriented surface of genus g and $n \geq 1$ boundary curves and/or punctures. In addition to closed curves we consider arcs with both endpoints on ∂S. As with closed curves we only consider "essential arcs": those which, together with a seqment of ∂S, do not cut off a disk.

From the collection of arcs and curves we can form two graphs, the arc complex, $\mathcal{A}(S)$, and the arc-curve complex $\mathcal{AC}(S)$. In each case the vertices are proper isotopy classes of essential arcs or arcs and curves, and the edges pairs of disjoint representatives. $\mathcal{AC}(S)$ is quasiisometric to the curve complex, while $\mathcal{A}(S)$, is, like $\mathcal{C}(S)$, Gromov-hyperbolic.

The boundary of $\mathcal{A}(S)$ [Schleimer and Wickens 2013] consists of ending laminations

$$\mathcal{E}\mathcal{L}_{\mathrm{arc}}(S) = \{\text{minimal laminations } \lambda \subset S : \lambda \cap \alpha \neq \varnothing, \text{ all arcs } \alpha\}.$$

The disk complex. Correspondingly, for handlebodies H, there is the *disk complex*. Here, the vertices are essential disks D in H (boundaries on ∂H), up to proper isotopy. The (k-1)-simplices $\{D_1, \ldots, D_k\}$ are mutually disjoint (isotopy classes) of disks. Here $k \leq 3g - 3$ if ∂H has genus g. We refer to Masur and Schleimer [2013] for its geometry including the proof that it is Gromov hyperbolic.

5-21. *The pants complex/graph.* The pants complex (or graph) $\mathcal{P}(S)$ for a surface S with $3g + n - 3 \geq 2$ is the metric graph whose vertices are the pants decompositions of S. Two vertices P, P' are connected with an edge of unit length if P' arises from P as the result of an *elementary move*.

In an elementary move, all edges of P are fixed, except for one edge α. Removing α from the pants decomposition leaves one component V which is not a pants, rather it is either a 4-holed sphere or a 1-holed torus. If V is a 4-holed sphere, replace α by any curve $\beta \subset V$ that crosses α twice (nontrivially). If V is a one-holed torus, replace α by β chosen to cross α once. In either case, β is uniquely determined up to free homotopy. Place the resulting pants decomposition P' is at the other end of an edge emanating from P. There are countably possible elementary moves P' of P and thus countably many distinct edges emanating from P.

The distance between two vertices of $\mathcal{P}(S)$ is the minimal number of elementary moves from one to the other.

In Brock [2003] it is shown that $\mathcal{P}(S)$ is quasiisometric to $\mathfrak{Teich}(R)$ in its Weil-Petersson metric. Even more is shown: Let G be any quasifuchsian group and X, Y the top and bottom surfaces of $\mathcal{M}(G)$ taken as Riemann surfaces lying in the same Teichmüller space (so for a fuchsian group they are the same surface). Let $\mathcal{CC}(G)$ denote the convex core of $\mathcal{M}(G)$. Then the Weil-Petersson distance $d_{\mathrm{WP}}(X, Y)$ is *comparable* to the volume $V(\mathcal{CC}(G))$! For definitions and proof see Jeff Brock's paper.

The simplest pants complex is the Farey graph (see Example 2-10) which is the complex for once-punctured tori.

It is proved in [Masur and Schleimer 2006] that the pants complex has only one end. This means take any ball centered at any vertex of the complex. The complement of the ball has exactly one unbounded component. Also proved is that there are constants $K(S)$ and $L(S)$ so that for $r > M(S)$ the following holds. If P and Q are pants decompositions both of distance $> K(S)r$ from some vertex V, then P and Q may be connected by a path in the complex which remains at distance $\geq r$ from V.

Summary of facts about Gromov hyperbolicity

- Teichmüller space in the Teichmüller metric is *not* Gromov hyperbolic [Masur and Wolf 1995], [McCarthy and Papadopoulos 1996], [McCarthy and Papadopoulos 1999].

- Teichmüller space in the Weil-Petersson metric is *not* Gromov hyperbolic when $3g + n - 3 \geq 3$ [Brock and Farb 2006].
- The curve complex with its path metric *is* Gromov hyperbolic [Masur and Minsky 1999].
- The pants complex is *not* Gromov hyperbolic: When R is a closed surface, there is a quasiisometry from $\mathfrak{T}(R)$ in the Weil-Petersson metric to the pants complex of R in the path metric [Brock 2003]; each point of $\mathfrak{T}(R)$ is sent to a Bers-shortest pants decomposition (see Exercise (4-24)).
- The extended mapping class group $\mathcal{MCG}^{\pm}(R)$ is the full group of isometries of $\mathfrak{T}(R)$ in the Weil-Petersson metric ($\mathcal{MCG}(R)$ itself is the full group for the Teichmüller metric); $\mathcal{MCG}^{\pm}(R)$ is the full group of automorphisms of the pants graph [Margalit 2004], [Masur and Wolf 2002], [Brock and Margalit 2007].

The Weil-Petersson (WP) metric is an incomplete Kahler metric on $\mathfrak{T}(R)$ with negative curvature which however is both unbounded below, and unbounded above below zero. Its good properties include the facts that its geodesics are unique, and its completion in augmented Teichmüller space is similar to the Bers boundary.

The W-P metric arises from the tangent space of $\mathfrak{Teich}(R)$. On a surface X the tangent space is represented by (equivalence classes of) Beltrami differentials; these have the invariant form $\{\mu(z)\frac{\overline{dz}}{dz}\}$ (see Section 2.9). There is a pairing with the cotangent space of quadratic differentials $\{\phi(z)\,dz^2\}$: $\langle \mu, \phi \rangle_X = \int_X \mu \, \phi \, dx dy$. The W-P metric is based on the following Hermitian inner product which combines quadratic differentials with the hyperbolic metric $\rho|dz|$ on X:

$$\frac{i}{2} \int_X \frac{\phi \, \overline{\psi}}{\rho} \, dx dy.$$

For an exposition and further references see Wolpert [2009].

5-22. *Outer space.* Recall from p. 284 that the outer automorphism group of a group G, typically $G = \pi_1(S)$, is defined as

$$\mathrm{Out}(G) = \mathrm{Aut}(G)/\mathrm{Inn}(G).$$

The *Dehn-Nielsen-Baer Theorem*, see [Farb and Margalit 2012, Theorem 8.1], establishes an *isomorphism*:

$$\mathrm{Out}(\pi_1(S)) \leftrightarrow \mathcal{MCG}^{\pm}(S),$$

with $\mathcal{MCG}(S)$ a subgroup of index two.

Denote the free group of rank $n \geq 3$ by F_n. *Outer space* was introduced by Culler and Voghtmann [2008], [1986] to study the group $\mathrm{Out}(F_n)$. It is comparable to Teichmüller space in that it consists of projective classes of marked metric graphs with fundamental group F_n.

Construct the graph $\mathcal{R}$ consisting of n circles joined at a common point but otherwise mutually disjoint. These circles are the n edges of $\mathcal{R}$.

Outer space is defined to be the set of pairs

$$\mathcal{O}_n = \{(X, g)\},$$

where X is a finite connected graph, normally without vertices of valence one or two, whose edges are isometric to intervals of $\mathbb{R}$, and $g : \mathcal{R} \to X$ is a homotopy equivalence (rather than just a homeomorphism or marking). Two pairs are equivalent, $(X, g) \equiv (X', g')$, if X and X' are isometric while g is homotopic to g' by means of an isometry.

The goal is to obtain an action of $\mathrm{Out}(F_n)$ on outer space $\mathcal{O}_n$. As $\mathcal{MCG}$ acts on Teichmüller space by changing the marking, so an element of $\mathrm{Out}(F_n)$, represented by a homotopy equivalence of $\mathcal{R}$, acts on $\mathcal{O}_n$ by changing the homotopy equivalence ("marking") $\mathcal{R} \to X$, without changing the graph X itself. The stabilizer of a point $(X, g) \in \mathcal{O}_n$ under the action of $\mathrm{Out}(F_n)$ is isomorphic to the (finite) group of isometries of X.

Recently constructed [Bestvina and Feighn 2010] is a δ-hyperbolic complex $\mathcal{F}$, analogous to the curve complex (see Exercise (5-20)), acted on by $\mathrm{Out}(F_n)$. The vertices of $\mathcal{F}$ are conjugacy classes of free factors of F_n. Simplices are chains of free factors.

5-23. *Coset graph* [Lyndon and Schupp 1977]. Here G is a group and H is a subgroup. The vertices are to be the cosets $\{Hg : g \in G\}$. If H has finite index in G, the set of vertices is finite as well. The edges $\{(Hx, Hgx) : g \in G\}$ eminate from the vertex Hx. The inverse of the edge (Hx, Hgx) is the edge (Hgx, Hx^{-1}).

We can take H itself as the basepoint of the graph Γ. For any element $w \in G$, given a word $w = g_1 \cdots g_n$ in terms of a fixed set of generators of G, there is a unique path from H to Hw. If $Hw = H$, the path is a closed loop.

If H has finite index N in G, there is a finite basis for the loops, and $\pi_1(\Gamma)$ is freely generated by these loops. In fact $\pi_1(\Gamma)$ has rank

$$r = \#\text{unoriented edges} - \#\text{vertices} - 1.$$

If there are n unoriented edges, there are $2n$ oriented edges that share each vertex. Hence,

$$\#(\text{vertices}) = N, \quad \#(\text{unoriented edges}) = nN,$$

$$r = nN - N + 1.$$

5-24. *Graph of 3-manifolds.* The vertices are the finite volume hyperbolic 3-manifolds. There is an edge between M and N if N is obtained by Dehn filling, or surgery, on M. Does this graph have an interesting structure (Thurston)?

5-25. *The Masur domain.* Compressible ends have been the most difficult to analyze. The basic structure in the analysis is the Masur domain (see Canary [1993]; Kleineidam and Souto [2003]) first defined for compression bodies. For the extension to all geometrically finite manifolds see Lecuire [2006]. For this exercise, all kleinian groups are assumed to have *no parabolics* (although rank two parabolics could be allowed if the interior of solid cusp tori were removed from the manifolds to make them compact).

First we make some comments about compressible boundary components.

Suppose $\mathcal{M}(G)$ is compact (no parabolics). Assume that f is a quasiconformal automorphism $\mathcal{M}(G) \to \mathcal{M}(G)$ that fixes a basepoint $O \in \text{Int}\,\mathcal{M}(G)$. Then the restriction of f to $\text{Int}\,\mathcal{M}(G)$ induces the identity automorphism of $\pi_1(\text{Int}\,\mathcal{M}(G))$ if and only if f is isotopic in $\text{Int}\,\mathcal{M}(G)$ to a composition of Dehn twists about compressing curves [McCullough and Miller 1986, Theorem 6.2.1] (recall from Exercise (5-10) that such a twist can be extended to $\mathcal{M}(G)$ and is isotopic to the identity there). Presumably some form of this result can be applied if rank one cusps are present as well (rank two cusps are not involved).

Suppose now $F : \Omega(G) \to \Omega(H)$ is a quasiconformal map that induces an isomorphism $\varphi : G \to H$. By Theorem 3.7.3, we may assume that F extends to a quasiconformal mapping of $\mathbb{S}^2$ and projects to a quasiconformal map $f : \mathcal{M}(G) \to \mathcal{M}(H)$ that also induces the isomorphism $\varphi : \pi_1(\mathcal{M}(G)) \to \pi_1(\mathcal{M}(H))$. If $\varphi = \text{id}$, necessarily F pointwise fixes $\Lambda(G)$.

We will now restrict our attention to a compact compression body $\mathcal{M}(G)$ and to its one compressible boundary component $S \subset \partial \mathcal{M}(G)$.

In general, there are infinitely many free homotopy classes of compressible loops— the exceptions occur when there is essentially only one compressing disk in $\mathcal{M}(H)$. A quasiconformal $f : \partial \mathcal{M}(G) \to \partial \mathcal{M}(G)$, extends to a quasiconformal map of $\mathcal{M}(G)$ if and only if f preserves the set of free homotopy classes of compression curves [McCullough and Miller 1986]. As mentioned above, an automorphism f of $\mathcal{M}(G)$ that fixes a basepoint O in its interior induces the identity automorphism of the group $\pi_1(\mathcal{M}(G); O)$ if and only if it is isotopic (with O fixed) to a composition of Dehn twists about compression loops. (If f were orientation reversing, $\mathcal{M}(G)$ would be a handlebody, [McCullough and Miller 1986, Theorem 5.3.1].)

The compression body is called *small* if there is only one compression loop up to free homotopy, that is if for a compression disk D, $\mathcal{M}(G) \setminus D$ is one or two manifolds of the form $S_0 \times [0, 1]$ where S_0 is an incompressible component of $\partial \mathcal{M}(G)$.

In Section 3.9 we introduced the measured lamination space $\mathcal{ML}(S)$ and projective measured lamination space $\mathcal{PML}(S)$, using the hyperbolic metric on S. A finite leaved geodesic lamination is called compressible if each of its leaves is compressible. Let $C(S)$ be the set of projective classes of *compressible* finite leaved measured laminations with atomic measures on the leaves. Let $\mathcal{C}(S)$ denote the closure of $C(S)$ in $\mathcal{PML}(S)$.

Suppose the compact compression body $\mathcal{M}(G)$ is not small. In addition to the compressing leaves of $C(S)$, the support Λ of a lamination in $\mathcal{C}(S)$ also may contain the following (Otal's thesis; see Kleineidam and Souto [2003]):

(i) Λ is the union of two disjoint, simple loops which are not parallel on S but which bound an essential cylinder C within $\mathcal{M}(G)$, or

(ii) Λ is a simple loop which homotopic to α^k, for a loop $\alpha \in \pi_1(\mathcal{M}(G))$ and $|k| \geq 2$, or

(iii) Λ is a simple loop freely homotopic to a simple loop on $\partial \mathcal{M}(G) \setminus S$.

(iv) [Kleineidam and Souto 2003, Lemma 3.6] Λ is a minimal lamination for which $S \setminus \Lambda$ is compressible.

The first statement follows from the fact that since $\mathcal{M}(G)$ is not "small", there is a compressing loop $\gamma \subset S$ which is transverse to both components of $C \cap S$ [Kleineidam and Souto 2003]. Let $\tau : \mathcal{M}(G) \to \mathcal{M}(G)$ be the Dehn twist about C (see Exercise (5-12)). The sequence of compressing loops $\{\tau^k(\gamma)\}$ converge to a measured lamination with support in Λ.

If $\mathcal{M}(G)$ is not small the *Masur domain* of the compressible boundary component S is defined to be

$$\mathcal{O}(S) = \{\mu \in \mathcal{PML} : \iota(\lambda, \mu) > 0 \text{ for all } \lambda \in \mathcal{C}(S)\}.$$

That is it consists of projective measured laminations which are "complementary" to the set of compressible ones.

If instead $\mathcal{M}(G)$ is small then $\mathcal{O}(S)$ is defined to consist of $\lambda \in \mathcal{PML}$ for which $\iota(\lambda, \mu) > 0$ for all those $\mu \in \mathcal{PML}(S)$ for which there exists $\nu \in \mathcal{C}(S)$ with $\iota(\mu, \nu) = 0$.

The Masur domain is open in $\mathcal{PML}$. If the support of $\mu \in \mathcal{O}(S)$ is a finite number of simple geodesics, then every component of $S \setminus \mu$ is *incompressible* and even *acylindrical*. When the support of μ is a simple geodesic γ, γ is transverse to every simple compressing geodesic. Note that two simple compressing geodesics may well cross each other.

Denote by $\text{Mod}(S)$ the group of orientation-preserving automorphisms of S that extend to diffeomorphisms of $\mathcal{M}(G)$. Denote the subgroup whose extensions are homotopic to the identity by $\text{Mod}_0(S)$.

The group $\text{Mod}(S)$ acts on $\mathcal{PML}(S)$ and $\mathcal{C}(S)$ is its limit set. $\text{Mod}(S)$ also acts properly discontinuously on the Masur domain. Suppose the supports of both μ_1, $\mu_2 \in \mathcal{O}(S)$ are collections of simple geodesics on S. Then if the components of the supports are respectively freely homotopic within the compression body M, there is a element $h \in \text{Mod}(S)$ with $h(\mu_1) = \mu_2$ (Otal; see Canary [1993]). For further details see Masur [1986]; Kerckhoff [1990]; Canary [1993]; Kleineidam and Souto [2003].

Referring back to Section 5.6 we can now state:

Theorem 5.15.5 ([Canary 1993]). *Suppose S is a boundary component of the compact core facing a geometrically infinite compressible end e of some $\mathcal{M}(G)$. The ending lamination λ determined by e lies in the Masur domain of S.*

Regarding compressible ends, we have the following older result; see also Kleineidam and Souto [2002]; Ohshika [2005, Theorem 4.1].

Theorem 5.15.6 ([Kleineidam and Souto 2003, Corollary 1.2]). *Suppose H has no parabolics and is not a free group. Then $\mathcal{M}(H)$ is tame if and only if for every geometrically infinite, compressible end E of $\mathcal{M}(H)$, there is a Masur domain lamination on ∂E that is not realized in $\mathcal{M}(H)$.*

Of course we now know every $\mathcal{M}(H)$ is tame so this theorem follows from the previous Theorem 5.15.5. Recall that an end can be represented by a boundary component of the compact core. The case of incompressible ends is covered by Bonahon's criteria (p. 292).

5-26. *Diskbusting curves; Canary's trick.* See also Theorem 6.9.3. Here we are only concerned with manifolds $\mathcal{M}(G)$ that have nontrivial splittings into free products $\pi_1(\mathcal{M}(G)) = A * B$. A *diskbusting curve* $\sigma \in \pi_1(\mathcal{M}(G))$ is one that is not contained in either factor of *any* nontrivial free product decomposition of $\pi_1(\mathcal{M}(G)) \cong G$. Thus if $\pi_1(\mathcal{M}(G)) = \langle g_1, g_2 \ldots g_N \rangle$ is a free group on N generators, then the element $g = g_1^2 g_2^2 \cdots g_N^2$ is diskbusting as it is the relator of a closed, nonorientable surface with an N-generator fundamental group.

In the case of compact, hyperbolizable manifolds M^3 the algebraic definition we have just given is equivalent to the geometric definition: A curve $\sigma \in \pi_1(M^3)$ is *diskbusting* if any curve $\sigma' \subset M^3$ freely homotopic to σ intersects every compressing disk in M^3. For according to Proposition 3.7.1 every free product decomposition of M^3 is generated by a compressing disk. Thus a simple loop on the boundary of a compression body M^3 such that it, with its counting measure, lies in the Masur domain is diskbusting [Canary 1993, Proposition 3.4]. In fact, there is a countable collection of them, no two of which are freely homotopic in M^3.

A curve β that is freely homotopic to a diskbusting curve is also diskbusting.

There exists an infinite collection of diskbusting curves on the compressible boundary component S of a compression body $\mathcal{M}(G)$, no pair being freely homotopic in $\mathcal{M}(G)$. Moreover, if the support of $\sigma \in \mathcal{O}(S)$ (Exercise (5-25)) is a simple geodesic on S, then σ is diskbusting [Canary 1993, Proposition 3.4 and Corollary 3.5].

Now every diskbusting curve β has the property that if D is any compressing disk based on the boundary, then β intersects D. Thus if $\mathcal{M}(G)$ is geometrically finite with compressible boundary, or if we are in a relative compact core C, then $\mathcal{M}(G) \setminus \beta$ or $C \setminus \beta$, is incompressible. We can take diskbusting curves to be geodesics, but we cannot be sure they are simple.

A union of mutually disjoint simple closed curves is called a *diskbusting link* if for every free product decomposition, at least one of the components of the link is not contained in either factor of the decomposition. We can add simple curves to the link and its diskbusting role will not change. Such a diskbusting link can be taken to be composed on geodesics; they will be mutually disjoint but we cannot be sure they are simple.

Now suppose σ is diskbusting link in the interior of $\mathcal{M}(G)$ which is homologous to zero. For example σ might lie on the boundary of a compact core as in Exercise (5-25). We can take σ to consist of closed geodesics. Suppose they are all simple. According to Canary [1993, Lemma 3.1], σ then bounds an embedded oriented surface $\Sigma \subset \mathcal{M}(G)$.

Next we will construct a new 3-manifold $\widehat{M}$ which is a two-sheeted branched cover of $\mathcal{M}(G)$, branched over σ. This is easily effected by cutting open $\mathcal{M}(G)$ along the surface Σ, bounded by σ, and designating the two sides of the cut by $\Sigma_\pm$. Set $M = \mathcal{M}(G) \setminus \Sigma$. Take two copies of M and identify the $\pm$ sides of Σ on one copy to the $\mp$ sides of Σ on the other.

Lemma 5.15.7 [Canary 1993, 5.2]. *$\partial \widehat{M}$ has incompressible boundary.*

This is a great advantage for analyzing ends. The bad news is that we have introduced cone axes. In Exercise (5-27), it is shown how to smooth these out, and also how to deal with the situation that one or more of the geodesics in the link is not simple.

Diskbusting links are used to deal with compressible ends. In studying the corresponding branched cover, the *engulfing property* of Brin and Thickstun is used (see Agol [2004], Myers [2005]): Suppose X is a compact, connected submanifold of an orientable, irreducible 3-manifold M^3 with no compact complementary components (for example, complementary components of a compact core with a geometrically infinite ends). There exists an open (not necessarily properly embedded[9]) submanifold Y containing X, uniquely determined in M^3 up to isotopy fixing X, by the following properties:

 (i) No complementary components $M^3 \setminus Y$ are compact,
 (ii) Y has a regular exhaustion $\{Y_n\}$ such that ∂Y_n is incompressible in the complement of X,
(iii) Given compact submanifold Z with $X \subset \mathrm{Int}\, Z \subset Z \subset M^3$ and ∂Z incompressible in the complement of X, then Z can be isotoped into Y with the isotopy fixing X.

The submanifold Y is called an *end reduction* at X.

It is directly applicable in the following:

Theorem 5.15.8. [Myers 2005]. *Let $\alpha \subset M^3$ be an algebraically diskbusting link with $X = N(\alpha)$ a thin tubular closed neighborhood of α. If Y is an end reduction at X, then the inclusion $\iota : \pi_1(Y) \hookrightarrow \pi_1(M^3)$ induces an isomorphism $\pi_1(Y) \to \pi_1(M^3)$, and the inclusion $\pi_1(Y \setminus \alpha) \hookrightarrow \pi_1(M^3 \setminus \alpha)$ is an injection.*

5-27. *Pinched negative curvature manifolds.* It has been quite useful to eliminate pesky cusps, inconvenient cone singularities, and self-intersections of geodesics by locally changing the constant negative curvature hyperbolic metric (singular on the cone axes) to a complete riemannian metric of *pinched negative curvature* (PNC): the sectional curvature $\sigma(x)$ lies between two constants $-\infty < -b^2 < \sigma(x) < -a^2 < 0$. For PNC-manifolds, many of the qualitative properties of hyperbolic manifolds remain true—see Canary [1993] and additional references listed there. A general reference is Ballmann et al. [1985].

The following result is attributed to Gromov and Thurston and proved in Bleiler and Hodgson [1996].

---

[9] The embedding $F : Y \to M^3$ is proper if $F(Y) \subset M^3$ is closed.

Theorem 5.15.9 (The "2π Lemma" or "Theorem A"). *Let V be a solid torus with a hyperbolic metric near ∂V so that ∂V is the quotient of a horosphere. The hyperbolic metric can be extended to a PNC metric in V provided that the length of the euclidean geodesic on ∂V serving as meridian has length $> 2\pi$.*

The necessity of 2π follows from the Gauss–Bonnet formula Equation (1.4) applied to a geodesic meridian, which inherits curvature $+1$ from the horosphere Exercise (3-1), and bounds a disk of negative curvature in V.

Agol [2004] recognized a very useful generalization to be used in the frequent cases when the 2π condition is not satisfied. I thank Juan Souto for pointing this out.

Theorem 5.15.10 (The "$2\pi/k$ Lemma" or "Theorem A$^+$"). *Let V be a solid torus or an infinite cylinder with a hyperbolic metric near ∂V so that ∂V is the quotient of a horosphere. Find $k \geq 1$ such that the length of the euclidean geodesic on ∂V serving as meridian has length $\geq 2\pi/k$. Then the hyperbolic metric near ∂V can be extended so that the solid torus or solid cylinder V becomes a PNC $2\pi/k$ orbifold (when $k \geq 2$).*

Consequently rank one or rank two cusps can always be eliminated by the use of Theorem A or A$^+$ at the cost of locally pinching the hyperbolic metric. For a rank two cusp, one can instead first do Dehn surgery of high enough order so that the length of the chosen meridian satisfies Theorem A.

For Agol's tameness proof, in view of Bonahon's theorem, it suffices to assume that the relative core C of $\mathcal{M}(G)$ is a compression body. Extend C to a compact core C' of the PNC manifold M' by replacing the interior of each pairing cylinder associated with a rank one cusp by an orbifold $\mathbb{D} \times [0, 1]$. Agol then applied the Orbifold Theorem to get a hyperbolic orbifold structure on M'. Then he used Selberg's Lemma to get a finite cover of M'. If the cover is tame, M' and hence the original $\mathcal{M}(G)$ will be tame as well. See Agol [2004], Bleiler and Hodgson [1996] for details.

Now we consider somewhat different situations which also lead to PNC manifolds. Suppose the compact core C of a some $\mathcal{M}(G)$ is a compression body with compressible boundary component S. There is a diskbusting link $\{\alpha_i\} \subset S$. We may assume that $\sum \alpha_i$ is homologous to zero. Denote the mutually disjoint geodesics in $\mathcal{M}(G)$ that are freely homotopic to the elements $\{\alpha_i\}$ by $\{\gamma_i\}$. Unfortunately there is no way of knowing whether or not these geodesics are simple. This problem is eliminated as follows:

Theorem 5.15.11 ("Theorem B" [Canary 1993]). *Fix a small tubular neighborhood U_i about each of the closed geodesics γ_i. The hyperbolic metric of $\mathcal{M}(G)$ can be changed within each U_i so as to obtain a complete PNC manifold M^3 in which each α_i is freely homotopic to a simple geodesic $\gamma_i' \subset U_i$. The PNC metric may be chosen so that its restriction to a thin tubular subneighborhood $U_i' \subset U_i$ about γ_i' is hyperbolic.*

As the $\sum \gamma_i'$ of simple geodesics is homologous to zero, an embedded surface $\Sigma \subset M^3$ can be constructed so that $\partial \Sigma = \bigcup \gamma_i'$ [Canary 1993, Lemma 3.1]. The surface

Σ determines a 2-fold cover $\widehat{M}$ of M^3 branched over $\bigcup \gamma_i'$: there is a cyclic group X of order two such that $\widehat{M}/X = M^3$. This is constructed by excising Σ, taking two copies of $\mathcal{M}(G) \setminus \Sigma$ and cross identifying them over Σ. For a simple example consider upper half-space UHS as a 2-sheeted cover of itself under the map $(z, t) \mapsto (z^2, t)$.

By Lemma 5.15.7, $\widehat{M}$ has incompressible boundary! Although the lifted metric in $\widehat{M}$ agrees with the original hyperbolic metric near the ends, it is now singular over the branch lines $\{\gamma_i'\}$. The job is completed by removing the branch locus by some more local pinching as follows:

Theorem 5.15.12 ("Theorem C" [Gromov and Thurston 1987; Canary 1993]). *Given the small tubular neighborhood $\bigcup U_i'^* \subset \widehat{M}$ about the branch locus of simple geodesics $\bigcup \gamma_i'^*$, $\widehat{M}$ can be given a complete PNC metric that agrees outside $\bigcup U_i'^*$ with the lifted metric from M^3.*

The bottom line is that by the process outlined, and Lemma 5.15.7, compressible ends can be eliminated at the cost of obtaining a PNC manifold $\widehat{M}$.

Finally, the tame ends of $\widehat{M}$ are projected back to M^3 to deduce that the ends of M^3 are tame as well. The original goal, the tameness of the ends of $\mathcal{M}(G)$, follows, one compressible end at a time.

Finally we cite an immediate consequence of the Tameness Theorem coupled with the Hyperbolization Theorem (p. 385):

Corollary 5.15.13 *If M^3 is a noncompact, complete PNC manifold of finite or infinite volume, then M^3 has a complete hyperbolic metric of finite or infinite volume, respectively.*

For closed PNC manifolds M^3, this is a consequence of Perelman's confirmation of the Geometrization Conjecture (Section 6.4), since M^3 is known to be irreducible, atoroidal, and not Seifert fibered [Cooper and Lackenby 1998].

5-28. *Representation varieties of fuchsian groups.* Fix a fuchsian closed surface group $G = \langle A_1, B_1, \ldots, A_g, B_g \rangle$ with $\prod [A_i, B_i] = 1$. Normalize so that $A_1 = \left(\begin{smallmatrix} k & 0 \\ 0 & k^{-1} \end{smallmatrix} \right)$ and $B_1 = \left(\begin{smallmatrix} a & c \\ c & d \end{smallmatrix} \right)$, where $k > 1$, $ad - c^2 = 1$, $c \neq 0$. Set

$$R_g = \{\varphi : G \to \varphi(G) \in \mathrm{PSL}(2, \mathbb{R}) \text{ is a normalized homomorphism}\}.$$

Normalized means that $\varphi(A_1)$, $\varphi(B_1)$ are normalized as A_1, B_1. R_g is an irreducible variety of real dimension $6g - 6$. Prove that R_g is a real analytic manifold.

Hint: Consider $H = \varphi(G) \in R_g$. For $2 \leq j \leq g$, choose any X_j, Y_j close to $\varphi(A_j)$, $\varphi(B_j)$. Show that there exist normalized X_1, Y_1 obtained by solving the matrix equation

$$[X_1, Y_1] = Z = \prod_{j=g}^{2} [X_j, Y_j]^{-1}.$$

5-29. *Character variety.* Earlier we had defined the representation variety $\mathfrak{R}(G)$ as as the $\mathrm{PSL}(2, \mathbb{C})$ quotient of nonelementary representations of G, restricting our attention to nonelementary, torsion-free kleinian groups $G \cong \pi_1(\mathcal{M}(G))$. As it stands

$\mathfrak{R}(G)$ is a Hausdorff space, but it is not strictly a variety. But if we had not removed the elementary or reducible groups, it would not be Hausdorff.

Often, researchers instead work with the *character variety* which is a true variety. Let $H(G;\mathcal{P})$ denote the space of representations (homomorphisms) of G into $\mathcal{P} = \text{PSL}(2,\mathbb{C})$. $\mathcal{P}$ acts on $H(G;\mathcal{P})$ by sending a representation ρ to the collection of representations

$$\mathcal{P} \cdot \rho = \{T \circ \rho \circ T^{-1} : T \in \mathcal{P}\}.$$

The *character variety* can be interpreted as the set of equivalence classes $[\rho]$ of representations $\{[\rho]\} in \mathcal{P}$:

$$X(G) := H(G;\mathcal{P})//\mathcal{P} = \{\rho\}, \text{ where}$$

$$\rho_1 \equiv \rho_2 \text{ iff } \overline{\mathcal{P} \cdot \rho_1} \bigcap \overline{\mathcal{P} \cdot \rho_2} \neq \varnothing.$$

The principal difference is that if $\rho(G)$ is a rank one parabolic group, by means of a sequence of conjugations, ρ can be conjugated to the trivial representation that sends every element of G to the identity. Thus in the character variety, $\rho \equiv \text{id}$. We have eliminated such a representation, and more, from our definition of $\mathfrak{R}(G)$.

The character variety[10] $X(G)$ is an affine algebraic variety (of dimension 6g-6 for a closed, genus $g \geq 2$ surface group G), although this is not seen from our formulation.

The algebraic approach to the character variety starts by the association of each representation with its trace function $\text{tr}^2(\rho) : g \in G \to \text{tr}^2(\rho(g))$. Two representations are called equivalent if and only if they are they have the same trace function. These traces are referred to as the *characters* of the representation ρ.

Corresponding to a generating set $\{g_i\}_{i=1}^n$ of G, and a representation ρ, is the map $\rho \to \mathbb{C}^{9n}$ obtained by uniquely representing each matrix of a pair $\bot \rho(g_j) \in \text{PSL}(2,\mathbb{C})$ as a 3×3 matrix. Then the relations in G determine a complex affine variety $R(G)$ in $\mathbb{C}^{9n}$. There is a bijection between the points of $R(G)$ and the representations ρ so that $R(G)$ can equally be regarded as the representation space; $R(G)$ is invariant under conjugation of G.

Each $g \in G$ determines a function $\tau_g : R(G) \to \mathbb{C}$ by

$$\tau_g = \{\text{tr}^2 \rho(g) : \rho \in R(G)\}.$$

Let T denote the ring generated by all functions $\{\tau_g, \ g \in G\}$. It is finitely generated (cf. Exercise (7-7)). Fix a generating set $\{\tau_j\}_{j=1}^k$. Consider the map $R(G) \to \mathbb{C}^k$ defined by

$$\rho \mapsto (\tau_1, \ldots, \tau_k) \in \mathbb{C}^k.$$

The *character variety* $X(G)$ of G is defined to be the *image* of this map in $\mathbb{C}^k$; it is an affine algebraic variety.

The Culler-Shalen theory finds a compactification of $X(G)$ consisting of equivalence classes of group actions on trees. When G corresponds to a closed surface

[10] Thanks to M. Kapovich for the given interpretation.

group S, points "at ∞" of the compactification are related to measured laminations on S. The C-S theory finds important application in hyperbolization.

The bottom line is that what we called the representation variety $\mathfrak{R}(G)$ is a actually a smooth open subset of a Zariski open subset of $X(G)$, the subset of "nonsingular" points. Thus the term "variety" is used in our context with only a mild abuse of nomenclature.

5-30. *Holomorphic motions.* Suppose $B \subset \mathbb{S}^2$ is an arbitrary set containing at least three points. Let $\{f_\lambda(z) : B \to \mathbb{S}^2\}$ be a family of functions with parameter λ in the open unit disk $\mathbb{D}$. The family is called a *holomorphic motion* of B if the following hold:

- For each fixed $\lambda \in \mathbb{D}$, the map $f_\lambda : z \in B \mapsto f_\lambda(z) \in \mathbb{S}^2$ is one-to-one.
- For each fixed $z \in B$, the map $\lambda \in \mathbb{D} \mapsto f_\lambda(z) \in \mathbb{S}^2$ is holomorphic.
- $f_0(z) = z$ for each $z \in B$.

The great utility of this notion in complex analysis comes from the λ-*Lemma* discovered by Ricardo Mané, Paulo Sad, and Dennis Sullivan. Its ultimate expression is due to Slodkowski [1991], to which we refer for history and development; also see Earle et al. [1994].

The λ-Lemma. *Let G be a Möbius group which maps the set $B \subset \mathbb{S}^2$ onto itself. Suppose $\{f_\lambda\}$ is a holomorphic motion of B. Assume further that, for each $\lambda \in \mathbb{D}$, there is an isomorphism to another Möbius group $\phi_\lambda : G \to G_\lambda$ such that $f_\lambda \circ g(z) = \phi_\lambda(g) \circ f_\lambda(z)$ for all $z \in B$, $g \in G$. Then*

(i) *$f_\lambda(z)$ is jointly continuous in λ and z.*
(ii) *For fixed λ, $f_\lambda(z)$ is the restriction to B of a K_λ quasiconformal mapping $f_\lambda^* : \mathbb{S}^2 \to \mathbb{S}^2$ which satisfies $f_\lambda^* \circ g(z) = \phi_\lambda(g) \circ f_\lambda^*(z)$ for all $z \in \mathbb{S}^2$, $g \in G$.*
(iii) *$\{f_\lambda^*\}$ is a holomorphic motion of $\mathbb{S}^2$.*

Explicitly, $K_\lambda = (1 + |\lambda|)/(1 - |\lambda|)$. Note that continuity in z is not assumed, it is a conclusion. A special case is $G = \text{id}$. On reflection, one finds the conclusions remarkably strong from what at first sight appears as rather weak hypotheses.

Here is the essence of the proof of the λ-Lemma kindly provided by Vlad Markovic. Assume that $f_\lambda(z)$ is an orientation-preserving differentiable mapping of a region $B = \Omega$, for each $\lambda \in \mathbb{D}$. Then the complex dilatation $\mu_\lambda(z)$ is defined for all $z \in \Omega$, and it is holomorphic in λ for each fixed z. Moreover $\mu_\lambda(0) = 0$ and $\sup_{z \in \Omega} |\mu_\lambda(z)| \leq 1$. Therefore by the Schwarz Lemma applied as a function of λ, we find that $|\mu_\lambda(z)| \leq |\lambda|$ for all $\lambda \in \mathbb{D}$ and $z \in \Omega$. In particular $f_\lambda(z)$ is quasiconformal in Ω for each $\lambda \in \mathbb{D}$.

An applicable situation might arise as follows. Suppose the generators of the kleinian group G depend analytically on a parameter $\lambda \in \mathbb{D}$ and that there is a corresponding family of isomorphisms $\{\phi_\lambda : G \to G_\lambda\}$ onto other kleinian groups such that $\phi_\lambda(g)$ is parabolic if and only if $g \in G$ is so. For each λ, ϕ_λ then determines an injection f_λ of the set B of loxodromic and parabolic fixed points of G to those of G_λ.

We see that $\{f_\lambda\}$ satisfies the hypothesis of the λ-Lemma. Therefore the family $\{G_\lambda\}$ is continuously quasiconformally conjugate to G.

A striking consequence of the λ-lemma is the following result illustrating the *structural stability* of kleinian groups in their representation variety: Sufficiently close representations of kleinian groups into PSL(2, $\mathbb{C}$) are quasiconformally conjugate. Compare with Corollary 4.1.11.

Theorem 5.15.14 ([Sullivan 1985]). *Suppose $h : X \to \mathfrak{R}(G)$ is a holomorphic map of a complex manifold X into the representation variety of a nonelementary kleinian group G such that each $x \in X$ is sent to a group $\rho_x(G)$ isomorphic to G. Then for any $x, y \in X$, the groups $\rho_x(G), \rho_y(G)$ are quasiconformally conjugate on $\mathbb{S}^2$.*

That $\rho_x(G)$ is holomorphic in x means that $\rho_x(g)$ is holomorphic in x for every $g \in G$.

Another striking theorem concerns finitely generated, nonelementary subgroups $H \subset \mathrm{PSL}(2, \mathbb{C})$ without parabolics. The subgroup H is called *structurally stable* if all sufficiently close representations into PSL(2, $\mathbb{C}$) are injective.

Theorem 5.15.15 ([Sullivan 1985]). *A group $H \subset \mathrm{PSL}(2, \mathbb{C})$ which is structurally stable is either geometrically finite or all nearby representations $H \to \mathrm{PSL}(2, \mathbb{C})$ are "rigid", that is, are conjugates of H.*

The theorem also applies to groups with parabolics if the rank one parabolic subgroups arise from punctures on $\partial \mathcal{M}(G)$, and the nearby representations preserve the parabolics and are injective. While we know geometrically finite groups are structurally stable (and not "rigid"), whether general groups in PSL(2, $\mathbb{C}$) are is another matter entirely.

For the quasiconformal deformation space of a geometrically finite group G we ask, is $\mathfrak{T}(G) = \mathrm{Int}\,\overline{\mathfrak{T}(G)}$? Or, is $\mathfrak{T}(G)$ like a ball with a radius removed?

According to Kapovich [2001, Theorem 8.44], there is a complex manifold V with $\overline{\mathfrak{T}(G)} \subset V \subset \mathfrak{R}(G)$. Apply Theorem 5.15.14 to the inclusion map of the submanifold $\mathrm{Int}(\overline{\mathfrak{T}(G)})$ into $V \subset \mathfrak{R}(G)$. We deduce that all representations in $\mathrm{Int}\,\overline{\mathfrak{T}(G)}$ are quasiconformal conjugate. Therefore the interior is just $\mathfrak{T}(G)$ itself (see also McMullen [1998, appendix]).

5-31. Confirm that the Tameness Theorem implies the Ahlfors Finiteness Theorem and the finiteness of the conjugacy classes of parabolics and of finite (elliptic) subgroups as well. To do this make use of the convex core boundary and the fact that there is a sequence of pleated surfaces of uniformly bounded area exiting each geometrically infinite end and relative end; see Canary [1993, Theorem 8.1].

5-32. *Two-generator groups.* Prove by the Tameness Theorem that a two-generator nonelementary kleinian group G (without elliptics) for which $\mathcal{M}(G)$ has infinite volume is a free group (Jaco, Shalen, Agol).

5-33. *Quadratic differentials and measured laminations.* Choose as basepoint $O = (R, \mathrm{id})$ in $\mathfrak{Teich}(R)$, where R is a closed Riemann surface of genus $g \geq 2$. A *Teichmüller ray* from O is determined by a holomorphic quadratic differential $\varphi(z)\, dz^2$ on R. The ray consists of the targets of the solutions of the Beltrami equation on R, $F_{\bar{z}} = t(\overline{\varphi(z)}/|\varphi(z)|) F_z$ for $0 \leq t < 1$ with $F(z; 0) = z$ (see Section 2.10.1). The solution $F_t : R \to R_t$ determines a quadratic differential ψ on R_t. The inverse $F_t^{-1} : R_t \to R$ is a Teichmüller map associated with the quadratic differential $-\psi$ on R_t and $-\varphi$ on R.

A quadratic differential defines a local euclidean metric on R except at its $4g-4$ zeros: $w = \int^z \sqrt{\varphi}\, dz$ is a locally univalent map into $\mathbb{C}$. The preimage of horizontal line segments can be extended globally. We get as a result the *horizontal foliation* Λ_h of R determined by φ. Likewise the inverse images of vertical line segments determine the *vertical foliation* Λ_v. It is customary to *normalize* quadratic differentials $\{\varphi\}$ so that $\|\varphi\| = \iint_R |\varphi|\, dx dy = 1$.

As we have seen in Section 2.8.1, a Teichmüller map $F : R \to S$ is associated with uniquely determined normalized quadratic differentials φ on R and ψ on S. Each determines a locally euclidean coordinate system, away from its zeros. In terms of these pairs of orthogonal coordinate systems, F is a local affine mapping of the form $w = z \mapsto z + t\bar{z}$.

The horizontal foliation Λ_h comes with an associated transverse measure, namely the vertical measure $|dv| = |\mathrm{Im}\, dw|$. Likewise $|du| = |\mathrm{Re}\, dw|$ gives the transverse measure for Λ_v. The Teichmüller map expands along the horizontal foliation in the form $\mathrm{Re}(z) \mapsto K^{\frac{1}{2}}\, \mathrm{Re}(z)$ and contracts along the vertical foliation in the form $\mathrm{Im}(z) \mapsto K^{-\frac{1}{2}}\, \mathrm{Im}(z)$, $K > 1$. The length of a transverse segment to Λ_h is decreased by the factor $K^{-\frac{1}{2}}$ while the length of a transverse segment to Λ_v is increased by $K^{\frac{1}{2}}$. As a result the vertical foliation is called the *stable foliation* while the horizontal foliation is the *unstable foliation*. Since foliations can be replaced by geodesic laminations, the words apply to laminations as well.

The association of quadratic differentials with measured laminations is as follows. The leaves of Λ_h which do not have an endpoint at a zero (the noncritical leaves) have two well-determined endpoints when lifted to the universal cover $\mathbb{H}^2$. For each pair of endpoints draw the geodesic with the same endpoints. Doing this for all noncritical leaves results in a geodesic lamination we again denote by Λ_h. It is equivariant under the group of cover transformations and therefore can also be viewed on R; it has a transverse measure determined by $|dv|$. Likewise the vertical foliation is associated with the measured lamination supported on Λ_v. For an account of the theory of quadratic differentials, see for example Strebel [1984]; Masur [1975]; Marden and Strebel [1984; 1986; 1993].

We know that measured laminations with support consisting of simple closed geodesics are dense in $\mathcal{ML}$. What corresponds to closed geodesics are the *simple Jenkins–Strebel differentials*: Given the free homotopy class $[\gamma]$ of a simple closed geodesic $\gamma \subset R$ we can ask, what is the thickest annulus A that can be embedded in the Riemann surface R whose central curves are in $[\gamma]$? In terms of a conformal map of A onto a proper annulus $\{1 < |w| < M\} \subset \mathbb{C}$, the problem is to maximize M over all such embedded A. There exists a unique solution: There is a uniquely determined (normalized) quadratic differential $\varphi[\gamma](z)\,dz^2$ on S such that all of its noncritical horizontal trajectories are simple loops in $[\gamma]$ covering R to a set of measure zero.

These horizontal trajectories sweep out an annulus A^*. The complement of A^* is the "critical graph" whose edges are critical trajectories of finite length running between critical points. If A^* is cut along a vertical trajectory segment of $\varphi[\gamma]$, then $w = \int^z \sqrt{\varphi[\gamma]}\,dz$ maps the result onto a rectangle in $\mathbb{C}$. Denote the length of the rectangle by $L[\gamma]$ and the height by $H[\gamma]$; its area is $\|\varphi[\gamma]\| = L[\gamma]H[\gamma] = 1$. The transverse measure to $\Lambda_h = \gamma$ associated with $\varphi[\gamma]$ is $H[\gamma] = 1/L[\gamma]$.

The Teichmüller ray determined by $-\varphi[\gamma]$ consists of surfaces R_t resulting from successively thickening A^* so that in the limit, the result is that R becomes pinched in the class $[\gamma]$.

Extremal length theory in complex analysis implies that for all pairs of simple geodesics on R,

$$L[\alpha] \geq \iota(\alpha, \gamma)H[\gamma], \quad L[\alpha]L[\gamma] \geq \iota(\alpha, \gamma),$$

using the fact that the extremal length of the class $[\alpha]$ is $L[\alpha]^2$ [Strebel 1984].

Now the space of normalized differentials on R is compact, and the simple differentials are dense. From Kerckhoff [1980] we learn that the functions $L[\gamma]$ extend continuously to all $\mathcal{ML}$, using $L[a\gamma] = aL[\gamma]$, $a > 0$.

There exists $c > 0$ such that in comparison with hyperbolic length $\ell(\cdot)$

$$0 < c < \frac{\ell(\gamma)}{L[\gamma]} < \sqrt{4\pi(g-1)} \tag{5.16}$$

for all simple geodesics and hence measured laminations $\gamma \in R$; see Minsky [1992, §8]; Minsky [1993, Lemma 2.1]. The ratio is positive and invariant under scaling by positive constants. As a positive function on the compact set $\mathcal{PML}(R)$, $\frac{\ell(\gamma)}{L[\gamma]}$ has a positive maximum and minimum. The constant on the right comes from an extremal length comparison. Equation (5.16) implies in particular that noncritical horizontal φ-trajectories are quasigeodesics in the hyperbolic metric on S—each lift to $\mathbb{H}^2$ has bounded distance from the hyperbolic geodesic with the same endpoints [Marden and Strebel 1985].

From Kerckhoff [1980] or Marden and Strebel [1984, Theorem 5.9] for example we learn that $\lim \varphi[\gamma_n] = \varphi_g$ if and only if $\lim(\gamma_n/c_n) = (\Lambda_g, \mu_g)$ exists. In fact, if we have both $\lim \varphi[\alpha_n] = \varphi_a$ and $\lim(\alpha_n/a_n) = (\Lambda_a, \mu_a)$, then in the limit

$$L_{\varphi_a}(\Lambda_a)L_{\varphi_g}(\Lambda_g) \geq \iota(\mu_a, \mu_g);$$

see Kerckhoff [1980]; Minsky [1993]. Here

$$L_{\varphi_a}(\Lambda_a) = \lim \frac{L[\alpha_n]}{a_n},$$

and correspondingly for $L_{\varphi_g}(\Lambda_g)$.

Moreover, the horizontal laminations corresponding to φ_a and φ_g are Λ_a and Λ_g [Minsky 1994a].

The upshot of these considerations is that a given (projective) measured lamination on R is associated with a uniquely determined normalized quadratic differential.

Now suppose we are dealing with an sequence of simple geodesics, or a sequence of pleated surfaces $\{f_n : R \to P_n\}$, exiting an infinite end E of the quasifuchsian manifold based on R. (Or a sequence of hyperbolic metrics $\{\rho_n\}$ converging to a Thurston boundary point.) Represented on R, the sequence of simple geodesics $\{\gamma_n\}$, or sequence of bending laminations, is converging to the ending lamination (Λ_E, μ). Correspondingly the normalized quadratic differentials converge $\lim \varphi[\gamma_m] = \varphi[\Lambda_E]$. As seen below, we do not need to pass to subsequences.

Under the assumption of bounded geometry and incompressible ends, Minsky [1993; 1994a] proved the following. Suppose we have a sequence of pleated surfaces $f_n : R \to P_n$ exiting E. Then the hyperbolic structures on $\{P_m\}$ are of bounded distance from the Teichmüller rays determined by $\{-\varphi[\gamma_m]\}$. The horizontal trajectories of $\varphi[\Lambda_E]$ are equivalent to the ending lamination Λ_E. By way of analogy, if $\gamma \subset R$ is a simple loop then the Teichmüller ray determined by $-\varphi[\gamma]$ thickens the annular region about γ so that the corresponding sequence of Riemann surfaces R_t pinches along γ. In the quasifuchsian manifold, the corresponding ending lamination is just the geodesic represented by $\gamma \in R$.

The Teichmüller ray determined by $-\varphi[\Lambda_E]$ in $\mathfrak{Teich}(R)$ on the other hand projects to a compact subset of moduli space $\mathfrak{Teich}(R)/\mathfrak{M}(R)$, because of the bounded injectivity radius hypothesis. This implies that Λ_E is uniquely ergodic, according to Masur [1992, Theorem 1.1]. Therefore the quadratic differential $\varphi[\Lambda_E]$ is uniquely determined by Λ_E.

Thus the sequence of pleated surfaces are being "pinched" to the pleating loci—the horizontal trajectories of $\varphi[\gamma_n]$. Along the sequence, the measure of a given transverse segment is increasing without bound. For more discussion see Section 6.1.1.

We have touched on the "dictionary" between measured foliations in topology, measured laminations in geometry, quadratic differentials in complex analysis, and there are also train tracks in combinatorics, as we will see next.

5-34. *Train tracks.* Suppose S is a closed hyperbolic surface. A *train track* $\tau \in S$ is a finite one-dimensional graph such that all vertices are trivalent. The relation of vertices to edges is to be like a switching point for a train. The three edges e_1, e_2, e_3 at a vertex v are placed so that a train coming in on either track e_1 or e_2 must exit on e_3, and conversely, a train coming in on track e_3 can exit on either e_1 or e_2.

Formally, the edges are C^1-arcs and the tangent lines have one-sided limits at their endpoints. At each vertex, the tangent lines of the three edges coincide (thus there is one line ℓ_v at each vertex so that ℓ is the limit of the tangent lines to all three tracks at v).

It is also assumed that each component of $S \setminus \tau$ is a triangle.

Fig. 5.19. The local structure of a train track.

A train track with *weights* has numbers $c > 0$ assigned to each edge. The *switch condition* is that at a vertex v, the numbers c_1, c_2, c_3 assigned to the three edges must satisfy $c_1 + c_2 = c_3$, using the labeling introduced above. If these are integers we can interpret the assumption to be that e_1 and e_2 each carry c_1 and c_2 parallel tracks coming in to v, and e_3 carries $c_1 + c_2$ parallel tracks leaving v. Thus if all the assigned numbers are integers, a particular train can take many possible journeys over the set of tracks. The journey will be of finite length before the trip repeats itself. If the weights are not integers, the journey of a train may be infinitely long.

A train track with weights uniquely determines a measured foliation on S: the leaves run along the branches with transverse measure given by the weights. Conversely, by pinching together nearly parallel leaves, every measured foliation is represented by some track τ. More precisely, a foliation F is mapped onto τ if there is a map ϕ of $S \setminus \{$singularities of $F\}$ such that ϕ is homotopic to the identity in such a way that tangent lines to leaves of F are sent to tangent lines of τ.

Likewise each measured geodesic lamination can be represented by a weighted train track, and conversely, each weighted train track uniquely determines a measured lamination.

The theory of train tracks was created by Thurston [1979a, §§8.9, 9.7]. For an extended exposition of the theory see Penner and Harer [1992].

5-35. *Extension of boundary deformations to* $\mathcal{M}(G)$. The purpose of this exercise is to sketch the proof of Theorem 5.1.3. We have to explain the relation of the quasiconformal deformation space $\mathfrak{T}(G)$ to the product of the classical Teichmüller spaces of the components $\{S_i\}$ of $\partial \mathcal{M}(G)$: $\mathfrak{Teich}(S_1) \times \cdots \times \mathfrak{Teich}(S_k)$.

We recall that two normalized quasiconformal deformations of G are equivalent $(F_1 \sim F_2)$ if they induce the same isomorphism $\varphi : G \to H$. This means in terms of their projections $f_1, f_2 : \partial \mathcal{M}(G) \to \mathcal{M}(H)$, that $f_2^{-1} \circ f_1$ extends to $\mathcal{M}(G)$ and is homotopic to the identity on Int $\mathcal{M}(G)$; see Section 3.7.2 and Exercise (5-25).

We also have to consider the stronger equivalence, namely

$$F_1 \simeq F_2 \iff f_2^{-1} \circ f_1 : \partial \mathcal{M}(G) \to \partial \mathcal{M}(G) \text{ is homotopic to id}.$$

It follows that $F_1 \sim F_2$ in the earlier definition. These two equivalences differ only when $\partial \mathcal{M}(G)$ is compressible.

To mirror the difference in the two equivalence relations we introduce the group $X(G)$ consisting of normalized quasiconformal deformations that preserve each component of $\Omega(G)$ *and* induce the identity automorphism of G. The latter property implies that the quasiconformal mappings fix all the fixed points of G. Here we refer to Theorem 3.7.3.

From the point of view of the manifolds, $X(G)$ consists of equivalence classes of quasiconformal automorphisms $h : \partial \mathcal{M}(G) \to \partial \mathcal{M}(G)$ that extend to the manifolds $\mathcal{M}(G) \to \mathcal{M}(G)$ where they are homotopic to the identity on the interior $\operatorname{Int} \mathcal{M}(G)$. Two such maps h_1, h_2 are to be identified if and only if $h_2^{-1} \circ h_1$ is homotopic to the identity on $\partial \mathcal{M}(G)$ too; specifically, $h_2^{-1} \circ h_2$ maps each component S_i onto itself and is homotopic on S_i to the identity.

If $\mathcal{M}(G)$ is boundary incompressible, $X(G) = \operatorname{id}$.

Denote by $\operatorname{Mod}_0(S_i)$ the group of homotopy classes of quasiconformal mappings $h : S_i \to S_i$ which extend to $\mathcal{M}(G)$ to be homotopic in $\operatorname{Int} \mathcal{M}(G)$ to the identity— Theorem 3.7.3 again. To be more precise, such a map h in particular fixes the punctures on S_i, and the set of compressing loops. Extend h from S_i to all $\partial \mathcal{M}(G)$ by setting it equal to the identity on S_m, $m \neq i$. Then h extends to $\mathcal{M}(G)$ and is homotopic in $\operatorname{Int} \mathcal{M}(G)$ to the identity. In other terms, h is the projection of a quasiconformal automorphism h^* of each component $\Omega_{i,j}$ over S_i with the property that h^* induces the identity automorphism of $\operatorname{Stab}(\Omega_{i,j})$ and extends continuously to the identity map of $\partial \Omega_{i,j}$. The group $\operatorname{Mod}_0(S_i)$ is a subgroup of the mapping class group $\mathcal{MCG}(S_i)$.

Therefore the group $X(G)$ splits into a direct product

$$X(G) = \operatorname{Mod}_0(S_1) \times \operatorname{Mod}_0(S_2) \times \cdots \times \operatorname{Mod}_0(S_k).$$

The group $\operatorname{Mod}_0(S_i)$ acts without fixed points on $\mathfrak{Teich}(S_i)$. For suppose, for example, that $h \in \operatorname{Mod}_0(S_i)$ fixes the origin (S_i, id) in $\mathfrak{Teich}(S_i)$. Then h is homotopic to a conformal map $h_0 : S_i \to S_i$. Now h and then h_0 lift to automorphisms h^* and h_0^* of $\Omega_{i,j}$ over R_i; we can choose h^* to be homotopic in $\Omega_{i,j}$ to h_0^*. We know that h^* extends continuously to $\partial \Omega_{i,j}$ and fixes every point, Exercise (3-37). So the same is true of h_0^* which therefore must be the identity since it is a conformal automorphism. Consequently h^* is homotopic in $\Omega_{i,j}$ to the identity and h is homotopic in S_i to the identity.

The classical results obtained by projection from the space of Beltrami differentials with respect to G on $\Omega(G)$ that imply $\mathfrak{T}(G)$ is a complex analytic manifold [Ahlfors 1966].

Examine now the quotient $\mathfrak{Teich}(S_i)/\operatorname{Mod}_0(S_i)$. Here we are identifying those elements of the Teichmüller space of S_i that are related by a mapping that is the identity with respect to the interior of the 3-manifold.

Since we are taking the quotient of an analytic manifold by a discrete group of fixed-point-free biholomorphic automorphisms, $\mathfrak{Teich}(S_i)/\operatorname{M}_0(S_i)$ is an analytic manifold of the same dimension as $\mathfrak{Teich}(S_i)$.

This completes the proof of Theorem 5.1.3.

Remark 5.15.16. Suppose all components $\Omega_{i,j}$ of $\Omega(G)$ are simply connected but that there may be torsion in their stabilizers $G_{i,j}$. Then in addition to the punctures on each component $S_i \subset \partial \mathcal{M}(G)$ there will be $b_i \geq 0$ cone points. In this case the dimension count will be

$$\sum (3g_i + b_i + n_i - 3).$$

For it is an interesting fact that $\mathfrak{Teich}(S_i)$ is biholomorphically equivalent to $\mathfrak{Teich}(S_i')$ where S_i' is the result of removing the cone points. That is, the dimension is the same whether you have $b_i + n_i$ punctures, or n_i punctures and b_i cone points [Marden 1969], [Bers and Greenberg 1971].

The basis for the equivalence is the following fact: A homeomorphism $f : S_i \to S_i$ lifts to a homeomorphism f^* of $\Omega_{i,j}$ which induces the identity automorphism of $G_{i,j}$ if and only if f is homotopic in S_i' to the identity map. This follows from the fact that γ is freely homotopic in S_i' to $f(\gamma)$ for all simple loops $\gamma \subset S_i'$.

Consequently Theorem 5.1.3 remains true at least when the components of $\Omega(G)$ are simply connected but $\{G_{i,j}\}$ contains elliptics and the deformations preserve elliptics and their orders. The original papers [Bers 1970b; Maskit 1971; Kra 1972] include the general case and contain the formal proofs.

5-36. *Representations into* $\mathrm{SL}(3, \mathbb{R})$ [Long et al. 2011] Prove that the following family of representations of the triangle group Δ are discrete and faithful for every parameter value $t \in \mathbb{R}$.

$$\Delta = \langle a, b : a^3 = b^3 = (ab)^3 = 1 \rangle :$$

$$a \mapsto \begin{pmatrix} 0 & 0 & 1 \\ 1 & 0 & 0 \\ 0 & 1 & 0 \end{pmatrix}, \quad b \mapsto \begin{pmatrix} 1 & 2 - t + t^2 & 3 + t^2 \\ 0 & -2 + 2t - t^2 & -1 + t - t^2 \\ 0 & 3 - 3t + t^2 & (-1 + t)^2 \end{pmatrix}.$$

This is an example from *higher Teichmüller theory*. Instead of the fuchsian representations of a surface R:

$$\mathrm{Hom}(\pi_1(S), \mathrm{PSL}(2, \mathbb{R}))/\mathrm{PSL}(2.\mathbb{R}),$$

one investigates representations into a Lie group, in particular

$$\mathrm{Hom}(\pi_1(S), \mathrm{PSL}(n, \mathbb{R}))/\mathrm{PSL}(n.\mathbb{R}).$$

In this theory, the group $\mathrm{PSL}(2, \mathbb{R})$ acts on the vector space of homogeneous polynomials $P(x, y) = \sum_{i=1}^n a_i x^{n-i} y^{i-1}$ of degree $(n-1)$ in two real variables. If $\gamma = \begin{pmatrix} a & b \\ c & d \end{pmatrix}$, its action on each term is $\gamma \cdot x^{n-i} y^{i-1} = (ax + by)^{n-i}(cx + dy)^{i-1}$.

The component of the image in $\mathrm{PSL}(n, \mathbb{R})$ of the real $(6g - 6)$-dimensional Teichmüller space $\mathfrak{Teich}(R)$ is called the *Hitchin component*. There is now an extensive algebraic theory on this subject, for references see Pollicott and Sharp [2014].

5-37. *Moduli space of kleinian manifolds.* [Canary and Storm 2013; Canary and Storm 2012] For a group $G \cong \pi_1(\mathcal{M}(G))$, the inner automorphisms $\mathrm{Inn}(G)$ are the maps $G \to gGg^{-1}$, $g \in G$. The outer automorphisms are then defined as the following quotient of all automorphisms

$$\mathrm{Out}(G) = \mathrm{Aut}(G)/\mathrm{Inn}(G).$$

The action of Out(G) on $\mathfrak{R}_{disc}(G)$ is given by

$$\alpha : \rho \in \mathfrak{R}_{disc}(G) \mapsto \rho \circ \alpha, \quad \alpha \in \text{Out}(G).$$

The action of OUT(G) would be properly discontinuous if it were restricted to the interior. Acting on the closure causes weird phenomena.

The moduli space of $\mathcal{M}(G)$, the space of "unmarked manifolds", can be defined either in parallel with the Teichmüller case as the quotient

$$\text{Int}(\mathfrak{R}_{disc}(G))/\text{Out}(G),$$

or simply the quotient of the closed space

$$\mathfrak{R}_{disc}(G))/\text{Out}(G).$$

The latter definition results in rather strange phenomena as a not properly discontinuous action resulting in non-Hausdorff moduli space.

5-38. Choose a simple geodesic γ, not retractible to the boundary, in some $\mathcal{M}(G)$. Let $\widehat{\mathcal{M}(G)}$ be the cyclic cover corresponding to γ. Therefore $\widehat{\mathcal{M}(G)}$ is hyperbolic, without boundary, and with infinite cyclic fundamental group. Describe the geometry of $\widehat{\mathcal{M}(G)}$. Can there be an automorphism of $\mathcal{M}(G)$ that preserves the homotopy class of γ? Then examine the case where the covering is determined by a two-generator free subgroup of G.

5-39. *Sufficient condition for Cannon-Thurston Maps.* Mahan Mj [Mj 2014, §1.1] asserted that existence of a Cannon-Thurston map would follow upon verifying the following for a manifold $\mathcal{M}(G)$ homotopy equivalent to a closed surface S. The surface S may be assumed to be an incompressible surface embedded in $\mathcal{M}(G)$. Consider a lift $\widehat{S} \subset \mathbb{H}^3$, which is a universal cover of S. Let $\ell \subset \widehat{S}$ be a geodesic segment in the hyperbolic geometry of $\widehat{S}$ and lying outside a large ball $B \subset \widehat{S}$ centered at a given point. Show that the geodesic in $\mathbb{H}^3$ connecting the endpoints of ℓ lies outside a large ball $B^* \subset \mathbb{H}^3$. Can you find a direct proof of this statement? Mj's proof is indirect and long.

6

Hyperbolization

The focus of this chapter is the Hyperbolization Theorem for essentially compact 3-manifolds, one of the most influential mathematical discoveries of the twentieth century, and one for the ages. This theorem shows that the interiors of "most" of these 3-manifolds, in particular most knot and link complements, can be realized as kleinian manifolds. As a consequence, such 3-manifolds can be described and classified not just in terms of their topology, but more powerfully, in terms of their geometrical properties—shape and volume.

We will continue on with a presentation of the recent profound discoveries concerning the structure of closed hyperbolic manifolds. In particular, each contains infinitely many essential immersed surfaces, and has a finite-sheeted cover which fibers over the circle.

6.1 Hyperbolic manifolds that fiber over a circle

6.1.1 Automorphisms of surfaces

We begin by reviewing some facts about automorphisms (orientation-preserving self-homeomorphisms) of hyperbolic surfaces, but not the 3-punctured sphere. We will continue using as basepoint a fuchsian group G and associated hyperbolic Riemann surfaces $R = \text{LHP}/G$, $R' = \text{UHP}/G$, closed with at most a finite number of punctures. Suppose $\alpha : R \to R$ is an orientation-preserving automorphism which is not homotopic to the identity. As we learned in Section 5.5.1, the automorphism α, or rather its homotopy/isotopy class, induces an automorphism of the Bers slice $\mathcal{B}(R) \equiv \mathfrak{Teich}(R)$ based on R, which we also denote by α, by the action

$$\alpha : (S_{\text{bot}} = R, S^{\text{top}}; \text{id}) \mapsto (S_{\text{bot}} = R, S^{\text{top}}; J \circ \alpha \circ J).$$

Here J, $J^2 \sim 1$ is a fiber-preserving reflection of $\mathcal{M}(R)$ with $J : R \leftrightarrow R'$. The identity automorphism of R is sent to the automorphism $J \circ \alpha \circ J$ of R' which does not change conformal type of R'. Instead the *relationship* between R and R' changes (since α is not homotopic to the identity). The group of homotopy classes of orientation-preserving automorphisms α is called the *mapping class group* or

Teichmüller modular group. It is the group of all isometries of Teichmüller space $\mathfrak{Teich}(R)$ in the Teichmüller metric of Section 2.8.

6.1.2 Pseudo-Anosov mappings

Recall from Exercise (5-12) that a *pseudo-Anosov automorphism* is a homeomorphism α of a surface R onto itself with these properties: (i) No power α^n is homotopic to the identity; and (ii) α does not preserve the set of free homotopy classes of any system of mutually disjoint, simple geodesics on R. Such automorphisms of R are the "generic" automorphisms.

The automorphism α sends simple geodesics to simple geodesics. As described in Section 3.9, p. 158, or Fathi et al. [1979], by passing from measured simple geodesics to measured geodesic laminations $\mathcal{ML}$ in the Hausdorff topology, α becomes an automorphism of $\mathcal{ML}(R)$.

For any simple geodesic γ and hyperbolic length $\ell(\gamma)$ on R, in the topology of $\mathcal{ML}$,

$$(\Lambda_{\text{attr}}, \mu_{\text{attr}}) = \lim_{n \to +\infty} \frac{\alpha^n(\gamma)}{\ell(\alpha^n(\gamma))}, \quad (\Lambda_{\text{rep}}, \mu_{\text{rep}}) = \lim_{n \to -\infty} \frac{\alpha^n(\gamma)}{\ell(\alpha^n(\gamma))}. \qquad (6.1)$$

This notation encapsulates both the laminations and the measures, as for a segment τ transverse to γ,

$$\mu(\tau) = \lim \frac{\iota(\tau, \alpha^n(\gamma))}{\ell(\alpha^n(\gamma))}.$$

There exists $K > 1$ such that for any simple closed geodesic γ, the generalized intersection numbers satisfy [Otal 1996, §1.5]

$$\iota(\alpha(\gamma), \mu_{\text{attr}}) = K^{-1}\iota(\gamma, \mu_{\text{attr}}) \quad \text{and} \quad \iota(\alpha(\gamma), \mu_{\text{rep}}) = K\iota(\gamma, \mu_{\text{rep}}). \qquad (6.2)$$

Since $\iota(\gamma, \mu_{\text{attr}}) \neq 0$ for all γ, Equation (6.2) shows unequivocally that for no $n \neq 0$ does α^n fix a simple geodesic.

In analogy to the case of loxodromic Möbius transformations, (i) the fixed points of two pseudo-Anosovs are either distinct or identical, and (ii) if α_1, α_2 have distinct fixed points on ∂_{th}, then $\langle \alpha_1^m, \alpha_2^n \rangle$ is a free abelian group for sufficiently large $m, n > 0$ [Ivanov 1992].

The pseudo-Anosov automorphism α (Exercise (5-12)) acts not only on R but also on $\mathfrak{Teich}(R)$, as $\alpha : (S, f) \mapsto (S, f \circ \alpha)$. It extends to an automorphism of the Thurston compactification $\partial_{\text{th}}\mathfrak{Teich}(R) = \mathcal{PML}(R)$. The points $(\Lambda_{\text{attr}}, \mu_{\text{attr}})$ and $(\Lambda_{\text{rep}}, \mu_{\text{rep}})$ on $\partial_{\text{th}}\mathfrak{Teich}(R)$, are the attracting (stable), and repelling (unstable) fixed points of α respectively.

Now α is homotopic to a uniquely determined Teichmüller map Ψ. There is a uniquely determined Teichmüller geodesic A_α, called the *axis* of α invariant under Ψ. Like a loxodromic, the forward iteration of a point converges to the attractive fixed point of α on $\partial_{\text{th}}\mathfrak{Teich}(R)$, and the backwards iteration converges to the repelling fixed point, [Bers 1978, 1985; Marden and Strebel 1993]. This axis A_α is found by minimizing the Teichmüller distance $d(p, \alpha(p))$ over points of $\mathfrak{Teich}(R)$.

The map Ψ is a solution of an associated Beltrami Equation (2.6) with $\mu = k\overline{\psi}/|\psi|$ for a uniquely detemined normalized quadratic differential ψ, and suitable constant k. The inverse map Ψ^{-1} is associated with $-\psi$; see Marden and Strebel [1986] for details.

Iteration in $\mathfrak{Teich}(R)$. For every point x in $\mathfrak{Teich}(R)$, typically represented as a Bers slice $\mathcal{B}(R)$, the orbit of x under a pseudo-Anosov α converge to its fixed points on $\partial_{th}\mathfrak{Teich}(R)$,

$$(\Lambda_{attr}, \mu_{attr}) = \lim_{n \to +\infty} \alpha^n(x), \quad (\Lambda_{rep}, \mu_{rep}) = \lim_{n \to -\infty} \alpha^n(x).$$

This situation is analogous to Anosov maps of the torus (Exercise (5-11)), especially when expressed in the context of the theory of quadratic differentials.

In the Bers slice $\mathcal{B}(G)$, consider the sequence of triples $\{(R, R'; J \circ \alpha^n \circ J)\}$. It converges algebraically to a singly degenerate group $H_{attr} \in \partial\mathcal{B}(R)$ as $n \to +\infty$. The limit Λ_{attr} of the sequence $\{J\alpha^n J(R')\}$ is the ending lamination of the top (geometrically infinite) end of $\mathcal{M}(H_{attr})$. There are no new parabolics in H_{attr} so the convergence is not only algebraic but is also geometric by Theorem 4.6.2.

If instead $n \to -\infty$, the corresponding points of $\mathcal{B}(R)$ converge algebraically to a different singly degenerate group $H_{rep} \in \partial\mathcal{B}(R)$ with ending lamination Λ_{rep}. The two laminations $\Lambda_{attr}, \Lambda_{rep}$ form a *filling or binding pair*: In the reference surface R they have no leaves in common and each complementary component $R \setminus \Lambda_{attr} \cup \Lambda_{rep}$ is a polygon possibly containing a single puncture. An alternate description is that $\iota(\nu, \mu_{attr}) + \iota(\nu, \mu_{rep}) > 0$ for any measured lamination $(\Lambda, \nu) \neq 0$.

In the full quasifuchsian deformation space $\mathfrak{T}(G)$, consider the sequence of quasifuchsian groups given by $\{[\alpha^{m-n}(R_{bot}), J \circ \alpha^m \circ J(R^{top})]\}$, with $n, (m-n) \to +\infty$. We are independently applying $J\alpha^m J$ to degenerate the top surface R' and α^{m-n} to degenerate the bottom R, with the "reflection" $J(R) = R'$. As a consequence of the Double Limit Theorem (p. 303), the sequence of groups converges algebraically (and also geometrically) to a doubly degenerate group $H \in \partial\mathfrak{T}(G)$. The ending laminations for the top and bottom ends of $\mathcal{M}(H)$ are $\Lambda_{attr}, \Lambda_{rep}$ [Thurston 1986c §4; McMullen 1996 §§3.3-5]. The Ending Lamination Theorem 5.7.4 obviates the necessity of taking subsequences.

There is an interesting special case. Following, suppose α is a pseudo-Anosov acting on a closed hyperbolic surface S. Consider the sequence of paired automorphisms $\mathbb{Q}_n = \mathbb{Q}(\alpha^{-n}(S), \alpha^{+n}(S))$. Thurston showed that both the algebraic and the geometric limit exist:

$$\lim_{n \to +\infty} \mathbb{Q}_n = \mathbb{Q}_\infty$$

where $\mathbb{Q}_\infty$ is a doubly degenerate product manifold invariant under the action of α. In fact $\mathbb{Q}_\infty$ is the infinite cyclic cover of the mapping torus $\mathbb{Q}_\infty/\langle\alpha\rangle$ with the fiber S and isometry α. The sequence $\lim_{n \to \infty}(S, \mathbb{Q}_{+n}(S))$ alone converges algebraically and geometrically to a point on the Bers slice of S. The above cited work contains estimates the growth rate of the convex core diameter in terms of the curve complex.

6.1.3 The space of hyperbolic metrics

In the interpretation of Teichmüller space as the space of hyperbolic metrics on R, modulo isotopy—see Section 2.10.1, start with the hyperbolic structure ρ_0 on R. Shift ρ_0 from the point z to $\alpha^n(z)$. We thereby determine a sequence of new hyperbolic structures $\{\rho_n(z) = \rho_0(\alpha^n(z))\}$. Let γ_n denote the geodesic on R freely homotopic to $\alpha^n(\gamma)$. In the respective lengths,

$$\ell_{\rho_n}(\gamma) = \ell_{\rho_0}(\gamma_n), \quad \lim \ell_{\rho_n}(\gamma) = \infty.$$

Restating Equation (6.1), in the $\mathcal{ML}$ topology [Otal 1996, §1.5],

$$\lim_{n \to +\infty} \frac{\gamma_n}{\ell_{\rho_0}(\gamma_n)} = (\Lambda_{\text{attr}}, \mu_{\text{attr}}), \quad \lim_{n \to -\infty} \frac{\gamma_n}{\ell_{\rho_0}(\gamma_n)} = (\Lambda_{\text{rep}}, \mu_{\text{rep}}). \qquad (6.3)$$

6.1.4 Fibering

Suppose R is a oriented surface and M^3 is the product manifold $R \times [0, 1]$. Choose a homeomorphism or diffeomorphism $\alpha : R \to R$. The corresponding *mapping torus* is the oriented 3-manifold

$$\widehat{M}^3 = R \times [0, 1]/\langle (x, 0) \equiv (\alpha(x), 1)\rangle,$$

obtained by gluing the bottom to the top. Thus if R is instead a line segment l, $l \times [0, 1]$ is a rectangle, and $l \times [0, 1]/\langle (x, 0) \equiv (x, 1)\rangle$ is an ordinary cylinder.

The first case of the Hyperbolization Theorem has the following beautifully succinct statement for a quasifuchsian manifold $\mathcal{M}(G)$.

Manifolds Fibered over the Circle 6.1.1 ([Thurston 1986c; Otal 1996; Kapovich 2001]). *Necessary and sufficient for the manifold $\mathcal{M}(G)/ \sim \tau$ to have A hyperbolic structure $\mathcal{M}(X)$ is that $\alpha : R \to R$ be pseudo-Anosov. In this case, $\mathcal{M}(X)$ has finite volume.*

In contrast, consider what happens if α is not pseudo-Anosov. Suppose it fixes the free homotopy class of a simple geodesic $c \in R$. Then c and $\tau(c)$ bound a cylinder C in $\mathcal{M}(G)$. The cylinder C rolls up to form an incompressible torus in M^3. This is possible in a hyperbolic manifold only if the torus comes from a rank two parabolic, which is the case only when C is a pairing cylinder for two parabolics.

The lifts of the fibers to $\mathbb{H}^3$ are fascinating. Suppose $\mathcal{M}(X)$ is fibered over a circle with fibers homeomorphic, say, to a finitely punctured, closed surface R. Choose a fiber Y. The lifts—components of the preimage—$\{Y^*\}$ of Y in $\mathbb{H}^3$ form a discrete set of mutually disjoint simply connected surfaces. There is a Möbius transformation $T \in X$ such that if Y^* is one lift, the orbit of Y^* under $\langle T \rangle$ comprises the complete set. Setting $H = \text{Stab}_X(Y^*)$, $THT^{-1} = H$. And $X = \langle H, T \rangle$, where T represents the pseudo-Anosov α that determines $\mathcal{M}(X)$.

We claim that $\Omega(H) = \varnothing$. Otherwise, choose $K \subset \Omega(H)$ compact. Then $K \cap \overline{T^n(Y^*)} = \varnothing$ for all n. In particular, no fixed point of T lies in K. Therefore $K \subset \Omega(X)$, which is impossible. Consequently H is a periodic doubly degenerate group, without new parabolics, isomorphic to G.

The Möbius transformation T projects to an automorphism Φ of $\mathcal{M}(H)$. The sequence of planes $\{T^n(Y^*)\} \subset \mathbb{H}^3$ projects to a discrete Φ-invariant sequence of surfaces $\{Y_n\} \subset \mathcal{M}(H)$ exiting its two ends. If $\gamma \subset Y$ is a simple loop, and γ_g is its geodesic representative in $\mathcal{M}(H)$, then the two ending laminations are determined by the exiting sequences of equal length geodesics $\lim_{n \to +\infty} \Phi^n(\gamma_g)$ and $\lim_{n \to -\infty} \Phi^n(\gamma_g)$ [Minsky 2003b]. These are the "fixed points" of the pseudo-Anosov α, as described earlier.

Such a manifold $\mathcal{M}(H)$ has the property of bounded geometry.

It is shown in Section 5.14 (see also Minsky [1994a]; Mj(Mitra) [1998a; 1998b]) that there exists a quasiisometric map $f : \mathbb{H}^2 \to Y^*$, which induces an isomorphism $G \to H$ and extends continuously to a map $\mathbb{S}^1 \to \partial Y^* \subset \mathbb{S}^2$, which is therefore a space-filling (Peano) curve. It is the image of a collapsing map of $\partial \mathbb{H}^2$ with respect to the two laminations associated with the pseudo-Anosov—placing one in the upper half-plane, for example, and the other in the lower.

It was originally believed that manifolds fibered over the circle could not be hyperbolic because of the strange properties their coverings would have. Thus Jørgensen's example of a periodic doubly degenerate groups with fiber the once-punctured torus was instrumental in inspiring the early development of the subject (see Thurston [1986c, §0]). An oft cited, closely related example is the hyperbolic manifold, also fibered over the circle with once-punctured torus fibers, which is homeomorphic to the complement of the figure-8 knot (p. 191). See also Jørgensen and Marden [1979]. Earlier, Robert Riley [1975] showed that the figure-8 knot complement itself has a hyperbolic structure.

So, starting with the fuchsian G and pseudo-Anosov α, to find H and T we have to move through the deformation space $\mathfrak{T}(G)$ until we find the doubly degenerate H on its boundary with ending laminations associated with α. H will be "periodic" with respect to a loxodromic T representing α.

6.2 Hyperbolic gluing boundary components

Manifolds can be assembled by putting together simpler manifolds. The key to finding hyperbolic structures on manifolds not fibered over the circle is being able to carry out the assembly by isometries—Möbius transformations. The skinning technique is the key element that makes this and related results possible.

6.2.1 Skinning a bordered manifold

Suppose $\mathcal{M}(G)$ is a geometrically finite, acylindrical (and hence boundary incompressible) manifold. Because of the hypothesis, if Ω is a component of $\Omega(G)$, the subgroup $G_\Omega = \mathrm{Stab}(\Omega)$ is a quasifuchsian group—see Exercise (3-10). Now the ordinary set of G_Ω has two components one of which is Ω. Denote the other by $\Omega' = \mathbb{S}^2 \setminus \overline{\Omega}$. Since G itself cannot be quasifuchsian, Ω' is not a component of $\Omega(G)$. There is an orientation-reversing quasiconformal involution $\sigma : \Omega' \leftrightarrow \Omega$ that induces the identity automorphism of G_Ω. Its extension pointwise fixes the common boundary. It projects to an orientation-reversing, fiber-preserving involution

$\sigma : S = \Omega/G_\Omega \leftrightarrow S' = \Omega'/G_\Omega$. In the case of a fuchsian group acting in UHP and LHP, σ is the map $z \mapsto \bar{z}$ and is an isometry.

Choose a fundamental set of components $\Omega_1, \ldots, \Omega_r$ of $\Omega(G)$ in the sense that no two are equivalent under G yet their G-orbits cover $\Omega(G)$. Thus we have the surfaces $\{S_i\}$, their partners $\{S_i'\}$, together with the simply connected lifts $\{\Omega_i\}$ and their partners $\{\Omega_i'\}$. Each partner set is associated with a quasifuchsian subgroup of G.

We have an associated set of Riemann surfaces, their covers, and the corresponding orientation-reversing quasiconformal involutions:

$$\sigma : (S_1, \ldots, S_r) \mapsto (S_1', \ldots, S_r'), \text{ and } \sigma : (\Omega_1, \ldots, \Omega_r) \mapsto (\Omega_1', \ldots, \Omega_r').$$

Lemma 6.2.1 (The Skinning Lemma). *The map σ:*

$$\sigma : \mathfrak{Teich}(\partial \mathcal{M}(G)) \to \mathfrak{Teich}(\overline{\partial \mathcal{M}(G)})$$

is holomorphic and uniformly contracting, if $\mathcal{M}(G)$ is acylindrical.

The map σ is called the *skinning map* since it removes the "skin", the internal topology, that hides the structures $\{S_i'\}$, as skinning an apple exposes the yummy stuff underneath. The i'th factor of σ commutes with the elements of the corresponding quasifuchsian group G_{Ω_i}, and pointwise fixes its limit set.

Recently Jonah Gaster [2012] showed that σ is not necessarily a diffeomorphism onto its image, nor is it necessarily an immersion. By working with genus-2 pared surfaces he discovered a skinning map that has a critical point. It is the kleinian group that represents this point that is used in the construction of the frontispiece of this book.

The simplest example is the case that Γ is a fuchsian group acting in LHP and UHP. Then the skinning map is reflection in $\mathbb{R}$ and is an isometry.

6.2.2 Totally geodesic boundary

Next, denote by ρ the map that operates on each surface by reversing its orientation without additional distortion of its complex structure. The composition is an orientation-preserving quasiconformal map

$$\rho \circ \sigma : (S_1, \ldots, S_r) \xrightarrow{\sigma, \text{ o.rev}} (S_1', \ldots, S_r') \xrightarrow{\rho, \text{ o.rev}} (\widehat{S_1'}, \ldots \widehat{S_r'}); \text{ and}$$

$$\rho \circ \sigma : (\Omega_1, \ldots, \Omega_r) \xrightarrow{\sigma, \text{ o.rev}} (\Omega_1', \ldots, \Omega_r') \xrightarrow{\rho, \text{ o.rev}} (\widehat{\Omega_1'}, \ldots \widehat{\Omega_r'}).$$

A lift of $\rho \circ \sigma$ to $\Omega(G)$ extends to a quasiconformal map of $\mathbb{S}^2$. Each map $S_i \to S_i'$ lifts and extends to a quasiconformal mapping of $\mathbb{S}^2$ since $S_i \cup S_i'$ covers $\mathbb{S}^2$ up to a set of zero measure.

We can view $\rho \circ \sigma$ as inducing an automorphism of the quasifuchsian space

$$\mathfrak{T}(G) \cong \mathfrak{Teich}(S_1) \times \cdots \times \mathfrak{Teich}(S_r).$$

It is a homeomorphism of the manifold $\mathcal{M}(G)$ with conformal boundary components $S_1, \ldots S_r$ to the manifold with conformal boundary $S_1', \ldots, S_r'$.

The question is: Is there a quasiconformal deformation of $\mathcal{M}(G)$ resulting in a manifold $\mathcal{M}(G^*) \in \mathfrak{T}(G)$ on which $\rho \circ \sigma$ acts as the identity? For this, we have to find a fixed point for the action of $\rho \circ \sigma$ on $\mathfrak{T}(G)$. If we can find one, at the fixed point there is anticonformal equivalence,

$$((S'_1, \ldots, S'_r) \equiv_{\text{anticonf}} (S_1, \ldots, S_r).$$

For simplicity, look at the case $r = 1$ when $\mathcal{M}(G)$ has only one boundary component Ω which is simply connected. We have

$$\Omega \xrightarrow{\;\sigma\;} \mathbb{S}^2 \setminus \overline{\Omega} = \Omega' \xrightarrow{\;\rho\;} \widehat{\Omega}'.$$

Suppose in moving through $\mathfrak{Teich}(G)$ we find a point such that

$$\rho \circ \sigma : \Omega^{\heartsuit} \to \widehat{\Omega}'^{\heartsuit}$$

is conformal. Then $\sigma : \Omega^{\heartsuit} \to \Omega'^{\heartsuit}$ is anticonformal and pointwise fixes the common boundary. This is only possible if the common boundary is a circle and $\partial \mathcal{CC}(G^{\heartsuit})$ reduces to a plane.

A geometrically finite manifold $\mathcal{M}(G^*)$ is said to be *totally geodesic* if the each boundary component of its convex core $\mathcal{CC}(G^*)$ is the projection of a geodesic plane in $\mathbb{H}^3$. That is, each convex core boundary component contains the geodesic segment between any two of its points. Equivalently, each component of $\Omega(G^*) \subset \mathbb{S}^2$ is a round disk.

The conclusion in the general case is that if we can solve the fixed point problem, namely find a fixed point $\mathcal{M}(G^*)$ of the automorphism $\rho \circ \sigma$ of $\mathfrak{T}(G)$, then $\mathcal{M}(G^*)$ is totally geodesic. We will say more about this fixed point problem in the next section.

Theorem 6.2.2 [Thurston 1979b; McMullen 1990]. *If $\mathcal{M}(G)$ is geometrically finite and acylindrical, there is a unique manifold $\mathcal{M}(G^*)$ in its quasiconformal deformation space $\mathfrak{T}(G)$ which has totally geodesic boundary (more precisely, the convex hull of $\mathcal{M}(G^*)$ has totally geodesic boundary).*

Uniqueness is a consequence of Theorem 3.13.4, or of Theorem 3.13.1 when applied to the double.

Without the acylindrical requirement the theorem would be false For example uniqueness is false in a quasifuchsian space, for there are many fuchsian deformations of a fuchsian group. Existence is also false: if a boundary component is compressible, it cannot be represented by a totally geodesic surface.

The geodesic boundary of $\mathcal{M}(G^*)$ makes it possible to directly construct its hyperbolic double when there is one boundary component. Look at a typical component Ω of $\Omega(G^*)$. Let J denote the reflection in the hyperbolic plane P rising from its bounding circle $\partial \Omega$. Consider the new group $\langle G, JGJ \rangle$. Here J conjugates the action of G in the exterior of P to a new action in the space between P and Ω, under the dome over Ω. But also $JgJ = g$ for each element $g \in \text{Stab}(\Omega)$. Topologically we have attached two copies of $\mathcal{M}(G^*)$ along the common boundary $S = \Omega/\text{Stab}(\Omega)$. Simultaneously we have built a kleinian group representing the double, which is a manifold of finite volume. Finding a manifold with totally geodesic boundary is equivalent to finding one whose double has a hyperbolic structure—in particular one which can be represented by a kleinian group.

The corresponding construction when there are multiple totally geodesic boundary components is described in Exercise (6-2).

Mostow's Rigidity Theorem applied to the double also shows that $\mathcal{M}(G^*)$ is unique. The first proof by Thurston [1979b] applied the Hyperbolization Theorem (see p. 385 below) to the topological double of $\mathcal{M}(G)$. See also Exercise (6-2).

Peter Storm answered a related conjecture of Bonahon about convex core volumes when he proved:

Theorem 6.2.3 ([Storm 2002; 2007]). *Suppose $\mathcal{M}(G)$ is geometrically finite, acylindrical and without rank two cusps. In the quasiconformal deformation space of $\mathcal{M}(G)$, the volume of the convex core $\mathcal{CC}(G)$ is uniquely minimal for the manifold $\mathcal{M}(G^*)$ with totally geodesic boundary.*

In the same vein, Bridgeman and Kahn [2010] establish a lower bound for the volume of a convex core $\mathcal{CC}(G)$ which has totally geodesic boundary in terms of the hyperbolic area of $\partial \mathcal{C}(G)$.

An analogous result for surfaces was discussed in Exercise (4-13).

David Wright's Figure 6.1 is a wonderful illustration of this section. The limit set is a *Sierpiński gasket*, that is, it is a closed subset of $\mathbb{S}^2$ with empty interior whose complement is the union of round disks with mutually disjoint closures. The complementary set is called a *Sierpiński carpet*. The group G has no parabolics and is constructed by identifying the faces of two adjacent truncated ideal tetrahedra. Details are in Thurston [1997, p. 133].

It is known that there are exactly eight nonisometric manifolds with genus two geodesic boundaries formed from two ideal tetrahedra [Fujii 1990]. Their convex cores all have the same volume. More generally, there are 151 manifolds with geodesic boundary constructed from three tetrahedra, and 5,033 ones constructed from four [Frigerio et al. 2004].

6.2.3 Gluing boundary components

By a *gluing map* we mean an orientation-reversing involution between boundary surfaces. A homeomorphism becomes orientation reversing simply by reversing the orientation of local neighborhoods in the target manifold akin to the result of reflectiing a fuchsian group in $\mathbb{R}$.

The Gluing Theorem 6.2.4 [Thurston 1980; McMullen 1990]. (See also Kapovich [2001]; Otal [1998].) *Suppose $M = \bigcup \mathcal{M}(G_i)$ is a finite union of geometrically finite, boundary incompressible manifolds. We are given a gluing map $\tau : \partial M \leftrightarrow \partial M$ which is an orientation-reversing involution. Assume that the identification manifold $M/\langle \sim \tau \rangle$ is connected. Then $M/\langle \sim \tau \rangle$ has a hyperbolic structure if and only if it is atoroidal. Its hyperbolic structure is then unique.*

Remarks on the proof. The proof of the Gluing Theorem is accomplished by solving a certain fixed point theorem, as follows.

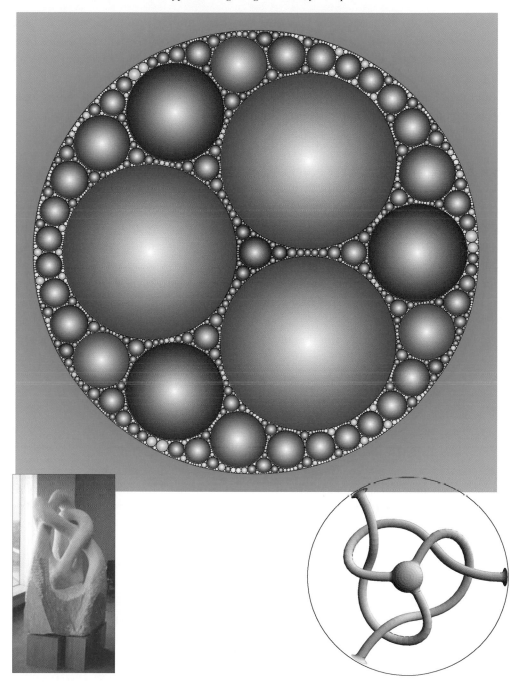

Fig. 6.1. This limit set is a Sierpiński gasket. The boundary of $\mathcal{M}(G)$ is a totally geodesic closed surface of genus two. $\mathcal{M}(G)$ itself, explained in Thurston [1997, p. 133–138], can be embedded in $\mathbb{S}^3$, with complement a handlebody: it is what remains of an apple once a three-legged wormhole—a knotted Y shape—is eaten out (bottom right). Helaman Ferguson's marble *Knotted Wye*, standing six feet tall, was inspired by this example of Thurston. The sculpture was commissioned for the Geometry Center at the University of Minnesota, and now adorns the University's mathematics library.

We will distinguish between between two Teichmüller spaces $\mathfrak{Teich}(\partial M)$ and $\mathfrak{Teich}(\overline{\partial M})$, the Teichmüller spaces of ∂M, with the opposite orientation. The only difference is that the orientations are reversed in the latter. We can represent the gluing map τ as an orientation-reversing map of the surface(s) themselves $\tau : S \to \overline{S}$, [Thurston 1980; Otal 1998], or, upon moving around in $\mathfrak{Teich}(\partial M)$, the gluing map can be taken to be an isometry [McMullen 1990]:

$$\tau : \mathfrak{Teich}(\overline{\partial M}) \to \mathfrak{Teich}(\partial M).$$

The fixed point of $\tau \circ \sigma$ in $\mathfrak{Teich}(\partial M)$ solves the gluing problem.

To simplify, assume that $\mathcal{M}(G)$ has two boundary components S_1, S_2 which are incompressible and homeomorphic. For the method to work, $\mathcal{M}(G)$ cannot be quasifuchsian. Choose an orientation-reversing involution gluing map $\tau : S_1 \to S_2$. Moving in $\mathfrak{Teich}(\overline{\partial M})$ we can assume the gluing map is an (orientation-reversing) isometry.

In the previous section we introduced the orientation-reversing *skinning map* $\sigma : S \to S'$, sending each boundary component to its quasifuchsian partner.

For the quasifuchsian groups (S_1, S_1'), $S_2, S_2')$ choose lifts $(\Omega_1, \Omega_2) \subset \Omega(G)$, and then their partners (Ω_1', Ω_2'). On this level the action of $F = \tau \circ \sigma$ is:

$$(\Omega_1, \Omega_2) \xrightarrow{\sigma, \text{ o.rev}} (\Omega_1', \Omega_2') \xrightarrow{\tau, \text{ o.rev}} (\Omega_2, \Omega_1).$$

In terms of the closures $\overline{\Omega_1}, \overline{\Omega_2}$,

$$F : \Omega_1 \to \mathbb{S}^2 \setminus \overline{\Omega_1}; \quad F : \Omega_2 \to \mathbb{S}^2 \setminus \overline{\Omega_2}. \tag{6.4}$$

Thus F has an extension so as to become a quasiconformal map $\mathbb{S}^2 \to \mathbb{S}^2$.

If $\tau \circ \sigma : \mathfrak{Teich}(G) \to \mathfrak{Teich}(G)$ fixes the point $\mathcal{M}(G^*)$, the map $F = \tau \circ \sigma$ becomes a conformal automorphism of $\mathbb{S}^2$, a loxodromic Möbius transformation A. Equation (6.4) shows its action.

We claim that the HNN extension $H = \langle G^*, A \rangle$ is discrete and $\mathbb{H}^3/H$ has the called for structure $\cong M_\tau$. In our case, this is a closed manifold.

Here we could either refer to Maskit's combination theorem II as in Exercise (3-8), or Maskit [1988, p. 160], or an argument from first principles. The essential facts are: (i) G_1 is the full stabilizer of Ω_1, and G_2 is of Ω_2; (ii) the loxodromic A maps the interior of Ω_1 to the exterior of Ω_2, which is Ω_2', and the exterior of Ω_1 to the interior of Ω_2; (iii) A conjugates the stabilizer G_1 of Ω_1 to that G_2 of Ω_2.

The situation may be a bit clearer if we make the following construction. There is a nonsingular surface $C_1^* \subset \mathbb{H}^3$ facing Ω_1 which is invariant under G_1. Together they bound a "halfspace" X_1^*. The projection $C_1 = C_1^*/G_1$ into $\mathcal{M}(G^*)$ is embedded, parallel to $S_1 = \Omega_1/G_1$; together S_1 and C_1 bound the submanifold $X_1 = X_1^*/G_1$.

Correspondingly the image $C_2'^* = A(C_1^*)$ faces Ω_2'; its projection $C_2' = C_2^*/G_2$ faces $S_2' = S_2^*/G_2$. Together S_2' and C_2' bound a submanifold $X_2' = X_2'^*/G_2$.

Now in $\mathbb{H}^3$, A maps X_1^* onto the complement $\mathbb{H}^3 \setminus X_2'^*$, and maps C_1^* onto $C_2'^*$. In $\mathcal{M}(G^*)$, C_1 and C_2' are glued together as dictated by A. There is a corresponding construction for $C_1'^*$ and C_2^*.

The bottom line is that the group $H = \langle G^*, A \rangle$ is such that $\mathcal{M}(H) \cong M_\tau$. The manifold $\mathcal{M}(H)$ has finite volume. Inside is the surface $S_1 \equiv S_2$.

To prove The Gluing Theorem it is necessary to find a fixed point of the action of $\tau \circ \sigma$:

$$\tau \circ \sigma : \mathfrak{Teich}(\partial M) \to \mathfrak{Teich}(\partial M).$$

This is the deep part of the proof; it is a consequence of The Skinning Lemma. Uniqueness then follows from Mostow's theorem.

Existence follows from Lemma 6.2.1. $\tau \circ \sigma$ and the gluing problem has a solution only when when the identification manifold M_τ is atoroidal. Presence or absence of rank two cusps is immaterial. Thurston's original proof uses the Bounded Image Theorem 6.2.4. McMullen's complete proof [1990] uses quadratic differentials and complex analysis. □

When $\tau \circ \sigma$ is replaced by $\rho \circ \sigma$, the corresponding fixed point in $\mathfrak{T}(G)$ solves the problem of finding manifold with totally geodesic boundary, as discussed in Section 6.2.2.

In practice, a still more general construction is needed to prove the Hyperbolization Theorem. We have to allow an incompressible subsurface of one component of $\partial \mathcal{M}(G)$ to be identified with an incompressible subsurface of another. This is carried out using a technique, or "orbifold trick", involving the reflection of $\mathcal{M}(G)$ over subsurfaces of $\partial \mathcal{M}(G)$. It is explained in Exercise (6-14).

Concerning the problem of finding a fixed point, we have the recent discovery:

Theorem 6.2.5 ([Antonakoudis 2015]). *If $F : \mathfrak{Teich}(R) \to \mathfrak{Teich}(R)$ is a holomorphic map such that there exists a point $p \in \mathfrak{Teich}(R)$ with bounded orbit $\{F^n(p)\}_{n=1}^\infty$, then F has a fixed point in $\mathfrak{Teich}(R)$.*

6.2.4 The Bounded Image Theorem

We recall that the *discreteness locus* $\mathfrak{R}_{\mathrm{disc}}(R)$ of an *acylindrical,*[1] and hence *boundary incompressible,* surface R is compact in the topology of algebraic convergence (Thurston's Compactness Theorem, p. 240). When G has no parabolics and $\partial \mathcal{M}(G)$ has a single, incompressible component S then $\mathfrak{R}_{\mathrm{disc}}(G)$ is connected and equal to the closure $\overline{\mathfrak{Teich}(S)}$.

Theorem 6.2.6 (The Bounded Image Theorem). [Thurston 1979b, Prop. 8], [Kent 2010, Ch. 7]. *Assume $\mathcal{M}(G)$ is geometrically finite and acylindrical. The skinning map*

$$\sigma : \mathfrak{Teich}(\partial \mathcal{M}(G)) \to \mathfrak{Teich}(\overline{\partial \mathcal{M}(G)})$$

has a continuous extension to a map

$$\sigma^+ : \mathfrak{R}_{\mathrm{disc}}(G) \to \mathfrak{Teich}(\overline{\partial \mathcal{M}(G)}).$$

Therefore, the image of σ has bounded closure.

---

[1] Quasifuchsian groups do not qualify.

Moreover for each point $X \in \mathfrak{Teich}(\overline{\partial \mathcal{M}(G)})$, *there are a finite number of preimages* $\sigma^{-1}(X) \in \mathfrak{Teich}(\partial \mathcal{M}(G))$, [Dumas $\geq$ 2015, Theorem A].

A complete proof is provided in the above cited paper of Kent.

For an extensive discussion of skinning see Dumas $\geq$ [2015]; Kent [2010], Dumas and Kent [2009].

6.3 Hyperbolization of 3-manifolds

We will begin by reviewing our list of definitions and at the same time adding some new ones. For technical details see Hempel [1976]; Jaco [1980]. Suppose M^3 is a compact, orientable 3-manifold, possibly with boundary. The technical disclaimer is that we are implicitly working in the piecewise linear or equivalently in the differentiable category (see Hempel [1976, p. 4]). The point is that we do not want to deal with "wild" embeddings. The following definitions are made with this understanding.

The Sphere Theorem is the statement that any embedded sphere in a hyperbolic manifold bounds a ball; in turn the statement that M^3 is irreducible is equivalent to the condition that $\pi_2(M^3) = 0$, since the Poincaré Conjecture is now known to be true.

As elsewhere in this book, we are assuming that embedded surfaces S are two-sided, hence orientable. An embedding has a collar neighborhood $\mathcal{E} : S \times [-1, 1] \hookrightarrow \mathcal{M}(G)$ with $\mathcal{E}(x, 0) = x$, $x \in S$, and $\mathcal{E}((\partial \mathcal{M}(G) \cap S) \times [-1, 1]) = \mathcal{E}(\partial S \times [-1, 1])$, if $\partial S \neq \varnothing$. See Hempel [1976, Chapter 6]; Jaco [1980, §III.12].)

6.3.1 Review of definitions in 3-manifold topology

Irreducible: Every embedded 2-sphere in M^3 bounds a closed 3-ball.[2]

Boundary incompressible: If $\gamma \subset \partial M^3$ is homotopic to a point in M^3, it is already homotopic to a point in ∂M^3.

Incompressible surface: A compact, embedded surface $S \neq \mathbb{S}^2 \subset M^3$ with $S \cap \partial M^3 = \partial S$, if $\partial S \neq \varnothing$, with the property that if $\gamma \subset S$ is homotopic to a point in $M^3 \setminus S$, it is already homotopic within S to a point.

Atoroidal: Every map of a torus into M^3 that is injective on its fundamental group is homotopic to a map to a torus boundary component of M^3.[3]

Acylindrical: $(M^3; P)$ has the property that $\partial M^3 \setminus P$ is incompressible and every incompressible cylinder $C \subset M^3$ bounded by simple loops $C \cap (\partial M^3 \setminus P) = \partial C$ can be homotoped (relative to ∂M^3) into ∂M^3.

Pared manifold: $(M^3; P)$ is pared if M is irreducible and atoroidal, $P \subset \partial M^3$ is the union of a finite number of mutually nonconjugate, disjoint, incompressible tori and cylinders $\equiv$ annuli such that (i) every incompressible cylinder with both boundary components in P can be homotoped (relative to its boundary) into P, and (ii) every torus component of ∂M^3 is incompressible and already included in P.

---

[2] The term "sphere" and "ball" refers to the homeomorphic image of a euclidean round sphere and ball.

[3] This definition can be called *homotopically atoroidal* as compared with *geometrically atoroidal* which is that every incompressible torus is homotopic to a boundary component. The latter is weaker: the complement of an open tubular neighborhood of a torus knot Section 6.5.1 satisfies the former but not the latter.

Haken: M^3 is a Haken manifold if it is compact, orientable,[4] irreducible and contains a (two-sided) incompressible surface $S \neq \mathbb{S}^2$ with $\partial S \subset \partial M^3$, if $\partial S \neq \varnothing$.

In an atoroidal manifold M^3, every embedded, incompressible torus is parallel to a boundary component. The following example shows that this condition alone is not sufficient for M^3 to be atoroidal: R is a closed surface of genus two or the 3-punctured sphere and α is a figure-8 loop in R. In the 3-manifold $M^3 = R \times \mathbb{S}^1$, the immersed torus $\alpha \times \mathbb{S}^1$ has fundamental group $\mathbb{Z} \oplus \mathbb{Z}$.

The concept of "pared" manifold arises to characterize the "parabolic loci" in those hyperbolic manifolds made compact by removing solid cusp tori and solid pairing tubes. As by paring an apple we remove blemishes on its skin, so by paring a manifold we mark the places occupied by pesky parabolics.

Less often, a paring P can come from an embedded annulus $A \subset \partial M$. It is incompressible if a central curve does not bound a disk in M. The situation can be seen more easily by pushing A into the interior of M while leaving $\partial A \subset \partial (M \setminus A)$ fixed.

Suppose M^3 is a compact, orientable, irreducible manifold.

If $\partial M^3 = \varnothing$, then M^3 is Haken if and only if (i) the cardinality of the first homology group is infinite and/or (ii) $\pi_1(M^3)$ is a free group with amalgamation over a closed surface group or is an HNN-extension [Waldhausen 1968, Lemma 1.1.6].

Otherwise, M^3 is Haken if $\partial M^3 \neq \varnothing$. Condition (ii) then implies that there is an embedded incompressible surface S which is not parallel to a boundary component.[5] This includes handlebodies; the incompressible surface is an essential disk. The converse is true by van Kampen's theorem.

Haken's original term for the class of Haken manifolds, "sufficiently large", is suggestive of what is required of the fundamental group, although no precise characterization in terms of this group is known. A criterion in distinguishing between Haken and non-Haken closed manifolds that has been useful is the following curious theorem of Hyman Bass:

Theorem 6.3.1 [Bass 1980; Maclachlan and Reid 2003, Corollary 5.2.3]. *Suppose* $\mathcal{M}(G)$ *has finite volume. Either* $\mathcal{M}(G)$ *contains a closed incompressible surface not homotopic to a cusp torus, or* $G \subset \mathrm{SL}(2, \mathbb{C})$ *is conjugate to a subgroup of* $\mathrm{SL}(2, \mathbb{A})$, *where* $\mathbb{A}$ *is the ring of algebraic integers in the algebraic closure* $\overline{\mathbb{Q}}$.

The importance of the class of Haken manifolds arises from the fact that if M^3 is Haken, it has a "hierarchy"

$$M^3 = M_1 \supset M_2 \supset \cdots \supset M_n = \mathbb{B}^3\left(\bigcup \mathbb{B}^3\right).$$

Here $S_k \subset M_k$ is an orientable, incompressible, nonseparating (for $k > 1$) surface, and

$$M_{k+1} = M_k \setminus S_k.$$

The surface S_k is not boundary parallel and it is properly embedded. Possibly S_k has a boundary in which case $S_k \cap \partial M_k = \partial S_k$ a union of incompressible simple loops.

---

[4] Non Haken manifolds are automatically orientable otherwise they would have a 2-sheeted orientable cover.

[5] Such surfaces with $\partial S \neq \varnothing$ are always assumed to be "properly embedded", that is $S \cap \partial M^3 = \partial S$.

If M^3 is closed, then the first surface may disconnect M^3 resulting at the end of the decomposition in two balls. For details see Hempel [1976]; Jaco [1980].

Since a Haken M^3 can be systematically decomposed by a sequence of incompressible surfaces; equally it can be systematically composed from one or

Fig. 6.2. The limit set pictured is that of Figure 7.2. Here it is colored and plotted using high accuracy computations with Wright's automorphic function. The spots result from its zeros and poles and its argument determines the coloring.

two 3-balls by successively forming M_k from M_{k+1} by identifying a pair of disjoint incompressible surfaces or subsurfaces on the boundary of M_{k+1} and gluing them together to form M_k. If M_{k+1} is hyperbolic, the resulting M_k can be made hyperbolic as well by an application of the Skinning Lemma (p. 378)—if full boundary components are to be joined; otherwise an elaboration is needed (see Exercise (6-14)).

6.3.2 Hyperbolization

The Hyperbolization Theorem 6.3.2 {Thurston, Perelman}. *Assume that M^3 is compact, orientable, irreducible, (homotopically) atoroidal, and pared $(M^3; P)$.*

(1) *Suppose $\partial M^3 \neq \varnothing$. Then $M^3 \setminus P$ is homeomorphic to $\mathcal{M}(G)$ for some geometrically finite group G. $\mathcal{M}(G)$ is compact if and only if $P = \varnothing$; $\mathcal{M}(G)$ is noncompact with finite volume if and only if $\partial M^3 = P$ consists of incompressible tori.*[6]

(2) *Suppose M^3 is a closed manifold. Then M^3 is homeomophic to some $\mathcal{M}(G)$ if and only if M^3 has an infinite fundamental group and satisfies one, hence both (i) and (ii), of the following conditions:*

  (i) *$\pi_1(M^3)$ contains no noncyclic abelian subgroup, in other words, is homotopically atoroidal,*
  (ii) *$\pi_1(M^3)$ contains no cyclic normal subgroup,*
  (iii) *M^3 is Haken.*

In short,

Every closed, irreducible, atoroidal, 3-manifold with infinite fundamental group is hyperbolic.

The interior of every compact, irreducible, atoroidal 3-manifold with boundary is hyperbolic.

Thurston proved items (1) and (2iii) in the mid 1970s. He proved (1) by reassembling hyperbolically the Haken decomposition of M^3 using the Skinning Lemma, and proved (2iii) again using the Skinning Lemma. Perelman proved parts (2i,ii) in (2003) in the larger context of the Geometrization Theorem.

For an overview of Thurston's work see Thurston [1980; 1982]; Morgan [1984]; Bonahon [2002]; Scott [1983]. For proofs of (1) see Kapovich [2001]; Otal [1996; 1998].

Before, it had been a conjecture that hyperbolization is implied by the conditions (2i) or (2ii) coupled with the requirement of an infinite fundamental group, all of which are obviously necessary conditions.

A bridge between conditions (2i) and (2ii) is Scott's Strong Torus Theorem 1980. This says, for an orientable, irreducible, compact M^3 for which $\pi_1(M^3)$ has a rank two abelian subgroup, that either M^3 contains an embedded incompressible torus or

---

[6] $T^2 \times [0, 1]$ is the only exception to this rule; the hyperbolic structure on the interior has infinite volume. The interior of the twisted I-bundle over the Klein bottle $N^3 = T^2 \times [0, 1]/(\mathbb{Z}/2)$ is the only exception to the rule that the interior of an atoroidal manifold has a hyperbolic structure. Both manifolds are also exceptional in that their interiors have euclidean structures. See Exercise 6-1.

$\pi_1(M^3)$ contains a cyclic normal subgroup K. The Seifert Conjecture, confirmed in Gabai [1992, §8.6] and Casson and Jungreis [1994], says that if $\pi_1(M^3)$ is infinite, the latter case occurs if and only if M^3 is a Seifert fibered space and then $\pi_1(M^3)/K$ is either the fundamental group of a 2-orbifold, or is fuchsian (with elliptics).

The proof of tameness also establishes that noncompact, complete, orientable riemannian 3-manifolds with pinched negative sectional curvatures $-\infty < -L < \kappa < -l < 0$ and finitely generated fundamental groups have tame ends (Agol, personal communication). Therefore they too carry hyperbolic metrics, settling a longstanding question.

6.4 The three big conjectures, now theorems, for closed manifolds

In his survey on hyperbolization, Thurston posed 24 fundamental questions or conjectures which ever since have focused research in the subject. Now after 40 years of continual development and partial results by many mathematicians, the last of these, the notorious #16, #18 plus the related Surface Subgroup Conjecture, are finally solved. The three blockbuster discoveries are described below. For further background see Thurston [1982]; Bergeron [2012]; Gabai [1986].

The preeminent conjecture was the Virtual Haken Conjecture; it had been raised earlier [Waldhausen 1968] but it was brought to the fore by Thurston. Its confirmation was announced in 2012 by Ian Agol [(2012)], integrating the deep contributions to geometric group theory by Dani Wise [2011] and hyperbolic geometry by Jeremy Kahn and Vlad Markovic. Others making important contributions to the final result include Nicolas Bergeron, Daniel Groves, Frédéric Haglund, Jason Manning, and especially the pioneering work of Micah Sageev. An exposition of Agol's proof can be seen in the video of Dunfield's talk [(2012)] and read in Bestvina [2014].

We already know that hyperbolic manifolds with boundary and/or cusps are automatically Haken so the manifolds involved here can be assumed to be closed, without boundary. The term "virtual" has come to mean that a finite-sheeted cover has the required property, if not the initial manifold. On the other hand the "virtual manifold" may have a high number of sheets (trillions?) over the initial; this seems to be the case for VHT and VFT below.

Virtual Haken Theorem (VHT) 6.4.1. *Suppose M^3 is an orientable, irreducible, closed, aspherical 3-manifold[7] with infinite fundamental group. Then M^3 has a finite-sheeted cover which is a Haken manifold.*

Setting $G = \pi_1(M^3)$, the Haken cover corresponds to a finite index subgroup N of G. It is a general fact in algebra that there is a finite index normal subgroup N^* of G contained in N; N^* is the kernel of the homomorphism $G \to G/N$. The M^3 cover determined by N^* is also Haken.

It may seem that this Theorem implies the full Hyperbolization Theorem for closed, irreducible, atoroidal 3-manifolds with infinite fundamental groups. However what is

---

[7] Otherwise known as a $K(\pi, 1)$ manifold: the two- and three-dimensional homotopy groups vanish.

actually proved is VHT for *hyperbolic* 3-manifolds [Agol (2012), §9]. The reduction
to the hyperbolic case is a consequence of the Geometrization Theorem. There are
still major hurdles to overcome before having a purely algebraic/topological proof;
one possible approach is via the Cannon Conjecture, see Exercise (2-18).

This comment also holds for VFT and SST presented below for the case of
hyperbolic manifolds.

For hyperbolic manifolds, Agol [(2012)] in conjunction with Dani Wise [2011] also
proved the even stronger result, asked in Thurston [1982, Ques. 18].

Virtual Fibering Theorem (VFT) 6.4.2. *Suppose $\mathcal{M}(G)$ is a closed hyperbolic
manifold. There is a finite-sheeted cover $\mathcal{M}(H)$ which is a closed surface bundle
(genus ≥ 2) over the circle.*

Such a fibered covering manifold is in particular Haken. But the assumption that
the Virtual Haken Theorem holds is required for the proof of Theorem VFT. This was
shown earlier by Dani Wise [2011] as a consequence of his deep investigations in geometric group theory. His work also validates the assumption made in an earlier paper
of Agol [2008]. This application of Agol's work then completes the proof of VFT.

Recall that closed hyperbolic manifolds fibered over a circle and their construction
were described in Section 6.1.2.

For finite volume noncompact manifolds $\mathcal{M}(G)$, as well as for closed manifolds,
the following properties are established, making use of [Cooper et al. 1997]:

(i) $\mathcal{M}(G)$ is virtually Haken. Moreover there exists a finite index normal subgroup
H of G such that $\mathcal{M}(H)$ fibers over the circle.[8]

(ii) $\mathcal{M}(H)$ is "large": There is a homomorphism of H onto the free group of rank
two; $\mathcal{M}(H)$ has positive first Betti number $b_1(\mathcal{M}(H))$.[9]

(iii) G is LERF: Given any finitely generated subgroup H and any element $g \notin H$,
there is a finite index, normal subgroup $H^* \supseteq H$ in G such that $g \notin H^*$. See
Exercise (2-21) for details.

6.4.1 Surface subgroups of $\pi_1(\mathcal{M}(G)) = G$

Being not only an essential component in the solution of VHT, but having broader
implications as well is the solution of the *Surface Subgroup Conjecture*. This well-
known conjecture (said to be due to Waldhausen) was confirmed in 2009, before the
VHT was solved, by Jeremy Kahn and Vlad Markovic [Kahn and Markovic 2012a;
2012b] in a much stronger form than that being asked—the original question asked
for just one surface—and results in the following theorem:

Surface Subgroup Theorem (SST) 6.4.3. *Suppose $\mathcal{M}(G)$ is a closed hyperbolic manifold. Then G has infinitely many nonconjugate subgroups isomorphic
to fundamental groups of surfaces. For each case there exists a closed surface
S_g of genus $g \geq 2$ and a continuous immersion $F : S_g \hookrightarrow \mathcal{M}(G)$ such that*

---

[8] Of course finite volume, noncompact hyperbolic manifolds are automatically Haken, but the fibering result is new.
[9] $b_1(M^3)$ is the rank—the smallest number of generators—of the first homology group $H_1(M^3)$.

the immersed surface $F(S_g)$ is nearly "flat" (nearly totally geodesic) and hence incompressible—the inclusion $F_* : \pi_1(S_g) \to G$ is injective.

More precisely, the authors show [Kahn and Markovic 2012a, Theorem 1.1] that given a closed manifold $\mathcal{M}(G)$, and $\varepsilon > 0$ the following holds: There is a surface $S_\varepsilon = \mathbb{H}^2/F_\varepsilon$, where F_ε is fuchsian, and a $K = (1 + \varepsilon)$ equivariant quasiconformal map $\chi : \partial\mathbb{H}^3 \to \partial\mathbb{H}^3$ such that the quasifuchsian group $H^* = \chi \circ F_\varepsilon \circ \chi^{-1}$ is a subgroup $H_\varepsilon \subset G$. The immersed surface $\chi(S_\varepsilon)$ is nearly totally geodesic in $\mathcal{M}(G)$ and its lifts to $\mathbb{H}^3$ are bounded by near-circles on $\partial\mathbb{H}^3$. In fact there exists a closed surface S with the property that for *any* closed hyperbolic two- or three-dimensional manifold M and $\varepsilon > 0$, there exists a finite cover $\widehat{S}$ and map $f : \widehat{S} \to M$ that is, locally, within ε of being an isometric immersion. (The two-dimensional case implies the Ehrenpreis Conjecture, p. 89.)

The LERF condition to be established for $\mathcal{M}(G)$ [Agol 2012] implies that if we set $H_\varepsilon = \pi_1(S_\varepsilon) \subset G$ and choose any element $p \in G$, $p \notin H_\varepsilon$, there is a finite index normal subgroup $H_\varepsilon^* \supseteq H_\varepsilon$ with $p \notin H_\varepsilon^*$ as well. The surface S_ε has a lift that is embedded in the finite-sheeted, quasifuchsian covering $\mathcal{M}(H^*) = \mathbb{H}^3/H_\varepsilon^*$, and it is compact.

It is quite surprising that given *any* round circle $C \subset \mathbb{H}^3$, and $\epsilon > 0$, there is a Kahn-Markovic surface $S_\epsilon \subset \mathcal{M}(G)$ that lifts to a surface $\widetilde{S}_\epsilon$ whose boundary is in the ϵ-neighborhood of C. This is found by choosing a unit vector $v \in \mathbb{H}^3$ that is orthogonal to the plane bounded by C. According to the K-M construction, there is a K-M surface $\widetilde{S}_\epsilon \subset \mathbb{H}^3$ nearly orthogonal to v. It projects to surface in $\mathcal{M}(G)$.

Furthermore Kahn-Markovic count the number of essentially different immersed surfaces in a manifold $\mathcal{M}(G)$. These immersed surfaces are counted in terms of both conjugacy classes and commensurability classes of their fundamental groups.[10]

Theorem 6.4.4 [Kahn and Markovic 2012b]. *Given a closed manifold $\mathcal{M}(G)$, define*

(i) $s_1(\mathcal{M}(G); g) :=$*number of conjugacy classes of essential*[11] *surface subgroups of genus at most g in $\mathcal{M}(G)$.*

(ii) $s_2(\mathcal{M}(G); g) :=$*number of commensurability classes of essential surface subgroups of G of genus at most g.*

There are constants $0 < c \le d$ such that

$$(c\,g)^{2g} \le s_2(\mathcal{M}(G); g) \le s_1(\mathcal{M}(G); g) \le (d\,g)^{2g}, \quad \text{all large } g, \text{ and} \tag{6.5}$$

$$\lim_{g \to +\infty} \frac{\log s_1(\mathcal{M}(G); g)}{2g \log g} = \lim_{g \to +\infty} \frac{\log s_2(\mathcal{M}(G); g)}{2g \log g} = 1. \tag{6.6}$$

The constant d depends only on the injectivity radius of $\mathcal{M}(G)$.

---

[10] Groups G_1, G_2 are *commensurable* if $G_1 \cap G_2$ has finite index in both G_1 and G_2.

[11] That is, there is a continuous immersion $S_g \hookrightarrow \mathcal{M}(G)$ such that the induced homomorphism $\pi_1(S_g) \to G$ is injective.

The authors compare Equation (6.5) with classical results that count the number of closed geodesics $b(r; M^3)$ of length $\leq r$ in a 3-manifold M^3 and the number of primes $\pi(N)$ less than N. That is,

$$b(r; M^3) \sim \frac{e^r}{2r} = \frac{X}{\log X}, \quad X = e^{2r}, \quad \text{(Margulis),}$$

$$c\frac{N}{\log N} \leq \pi(N) \leq d\frac{N}{\log N}, \quad 0 < c < d, \quad \text{(Chebyshev Prime Number Theorem).}$$

Thus there are infinitely many homotopy types of immersed surfaces in a given closed manifold $\mathcal{M}(G)$. The number grows exponentially with the genus.

The authors construct surfaces by assembling pairs of pants with geodesic boundaries (Exercise (8-7)). Exponential mixing of the pants in the geodesic flow of the unit tangent bundle of $\mathcal{M}(G)$ is required to match their boundaries with small bending angles. They can then be assembled to make the nearly geodesic K-M surfaces. If the cuff lengths are approximately L, then the genus of the surfaces constructed from them is bounded by e^{3L}. The dimension of the cubulation Y formed from them in Section 6.4.2 will have the form e^{kL} for some constant k; see Kahn and Markovic [2012a, §2.4].

Here are three additional consequences of SST,

Theorem 6.4.5. *In a closed hyperbolic manifold $\mathcal{M}(G)$:*

(i) [Liu and Markovic 2013]. *Suppose γ is a closed, oriented 1-submanifold composed of (at most) finitely many mutually disjoint simple loops immersed in a closed manifold $\mathcal{M}(G)$. Assume that all components of γ are homotopically nontrivial and γ itself is rationally homologous to zero. Then γ has a finite cover γ^*, the same degree over all components, which bounds an oriented, compact, nearly flat, π_1-injective, immersed image of a quasifuchsian surface.*

(ii) [Liu and Markovic 2013]. *Every rational 2nd homology class of a closed hyperbolic 3-manifold has a positive integral multiple represented by an oriented, connected, closed, nearly flat surface $S_c \subset \mathcal{M}(G)$, of genus ≥ 2, which is π_1-injective. Moreover, $H = \pi_1(S_c) \subset G$ is a quasifuchsian subgroup.*

(iii) (I. Agol, V. Markovic). *There exists a sequence of coverings $\{M_n\}$ of $M(G)$, with $n =$ degree of the cover, such that*

$$\underline{\lim}_{n \to \infty} \frac{\log |H_1^{\text{tor}}(M_n; \mathbb{Z})|}{n} > 0.^{12}$$

Item (iii) was announced by Markovic at his Ordway lectures (UMN, 2013). Compare it with the following result announced by Thang Le [$\geq$ 2014] at the Columbia conference 2013:

---

[12] The elements of finite order in the first homology group $H_1(M_n; \mathbb{Z})$ form a finite subgroup H_1^{tor}, the *torsion subgroup*. By Poincaré duality $H_1(\mathcal{M}(G); \mathbb{Z}))$ is isomorphic to $H_2(\mathcal{M}(G); \mathbb{Z})$

If $\{M_n\}$ is a sequence of coverings of $\mathcal{M}(G)$ of degree n then

$$\overline{\lim}_{n\to\infty} \frac{\log |H_1^{\mathrm{tor}}(M_n; \mathbb{Z})|}{n} \leq \frac{\mathrm{Vol}(\mathcal{M}(G))}{6\pi},$$

provided $\lim_{BS} M_n = \mathbb{H}^3$ *as defined on p. 271.*

Currently, there is a search for better understanding of the quantitative impact of the growth in torsion on manifold geometry. Brock and Dunfield [2012] find examples where the torsion grows while the first betti number says constant or zero.[13]

Theorem 6.4.6 (Brock-Dunfield). *Given a finitely generated abelian group A, and positive constants r, ε, there exists a closed hyperbolic manifold $\mathcal{M}(G)$ for which*

$$H_1(\mathcal{M}(G); \mathbb{Z}) = A, \quad \text{and} \quad \frac{\mathrm{Vol}\{x \in M(G) : \mathrm{Inj}_x \mathcal{M}(G) < r\}}{\mathrm{Vol}(\mathcal{M}(G))} < \varepsilon.$$

For these examples, most of $\mathcal{M}(G)$ has large injectivity radii.

There is a surprising recent discovery of Hongbin Sen [2015] which confirms a conjecture of Ian Agol. Its proof is based on SST and the techniques used by Agol, Wise and collaborators applied to the virtual theorems cited above. It shows that closed hyperbolic manifolds dominate all others.

Virtual Domination Theorem (VDT) 6.4.7. *Given any closed, hyperbolic manifold $\mathcal{M}$ and any closed, oriented 3-manifold N, the following holds. There is a finite-sheeted cover $\mathcal{M}^*$ of $\mathcal{M}$ and a degree two proper map[14] $f : \mathcal{M}^* \to N$.*

6.4.2 Remarks on the proof of VHT and VFT: Cubulation

This exposition is based on Sageev [$\geq$ 2015]; Bestvina [2014] to which the reader should turn for more details. It is rather cryptic and can only present some of the flavor of a deep new development in geometric group theory.

A *cube* is an isometric image of the unit euclidean cube in k-dimensions: $[-1, 1]^k$, for $k \geq 1$. A *cube complex* X is built as follows. Take an assemblage C of mutually disjoint cubes of various dimensions. A *face* of a cube is what is obtained by setting some of its coordinates $= \pm 1$. For example a 2-cube has four faces of dimension one (i.e., edges) and four faces of dimension zero (i.e., vertices). Correspondingly there is given a set of isometries F between faces. The resulting cube complex is the quotient space $X = q(C) = C/F$; thus the "cubes" of X are the $q-$images of the various faces of the cubes of C. We will assume the complex X is "nonpositively curved" (NPC), that is, the link of every vertex is a simplicial complex.

---

[13] Consider a knot in $\mathbb{S}^3$ and do p/q-Dehn surgery. If $p = 1$ and q is large enough the result is a hyperbolic manifold M with $H_1(M; \mathbb{Z}) = \mathbb{Z}/\mathbb{Z} = 0$, for example the Poincaré dodecahedral space is an integer homology sphere (see Section 1.2.2); in fact it is the only homology 3-sphere $\neq \mathbb{S}^3$ with a finite fundamental group. When $p > 1$, M is a rational homology sphere with $H_1(M; \mathbb{Q}) = \mathbb{Z}/p\mathbb{Z}$.

[14] A degree two map means that $\mathcal{M}^*$ "winds twice" around N, as a degree two map of circles c^* to c is a map that winds c^* twice around c. Formally, this is expressed in terms of the respective third homology groups.

For example a closed surface of genus g can be represented by a $4g$ sided regular polygon with center O and side identifications. Draw a line from the midpoint of each edge to O. This divides the polygon into "squares". The complex so formed is NPC.

One takes an appropriate configuration of cubes so as to form a simply connected CAT(0) space[15] with isometric pairs of parallel faces. It turns out that $G = \pi_1(X)$ is a Gromov hyperbolic group; it acts properly (inverses of compact sets are compact) on the CAT(0) cube complex X with compact quotient.

A choice of a finite number of round circles on $\partial \mathbb{H}^3$ is made as follows. In $\mathbb{H}^3$, choose a fundamental polyhedron F for G. Find ε so that the distance on $\partial \mathbb{H}^3$ between the two endpoints of every geodesic in $\mathbb{H}^3$ that intersects F exceeds ε; there are an infinite number of such geodesics. We can find a finite number of not necessarily disjoint round circles so that every pair of endpoints is separated by at least one of the circles. Choose $0 < \epsilon \leq \varepsilon$. Corresponding to each circle, construct a K-M surface for which the boundary of a lift to $\mathbb{H}^3$ is contained in its ϵ-neighborhood. The compact cube complex X referred to below is built from these surfaces which may have numerous intersections and self-intersections. For an overview, see Bestvina [2014].

Bergeron and Wise [2012] build the cubulation from the division of $\mathbb{H}^3$ by the multiply intersecting, embedded, simply connected lifts of the large number of K-M surfaces $\{S_j\}$ mentioned earlier. Each lift S_i^* separates $\mathbb{H}^3$ into two halfspaces H_+. This is akin to the picture in $\mathbb{H}^2$ of lifts of a finite collection of closed geodesics on a surface S. As it is for lifts of geodesics, at most a finite number of the lifted K-M surfaces intersect any given compact subset of $\mathbb{H}^3$,

A cube complex can be associated with a graph (a "wall space") constructed as follows: A vertex, that is a 0-cell, is a choice of half-space $S_{i\pm}^*$ for each lifted S_i^*. The collection of half-spaces is to be chosen to have the property that there is a point $p \subset \mathbb{H}^3$ that lies in all but a finite number of them. It is a bit like forming the Dirichlet region with center p.

Two vertices v_1, v_2 are connected by an edge if their collections of half-spaces agree with the exception of exactly one halfspace. The exceptional halfspace may contain p_2 but not p_1. The pair $[v_1, v_2]$ then are said to form a 1-cube, that is a 1-simplex—a 1-skeleton.

A 2-cube is formed by two 1-cubes that differ by exactly two halfspaces, giving rise to a 2-skeleton. The construction continues until there is no $(n-1)$-skeleton; the n-cube contains half-spaces for all except a finite number of lifts $\{S_i^*\}$

Hyperplanes in X are constructed by taking $(k-1)$-dimensional slices halfway between two opposite faces in a k-cube and then using the identifications at the edges

---

[15] A triangle Δ in a geodesic space Y consists of three vertices (a, b, c) and three geodesics connecting the pairs of vertices. A comparison triangle $\Delta' \subset \mathbb{R}^2$ consists of 3 vertices (a', b', c') and three edges between them of the same d_{euc}-length as the d-length corresponding edges of Δ. A point p' in an edge, say $[a', b']$, is called the companion to $p \in [a, b]$ if $d(a, p) = d_{euc}(a', p')$. The CAT(0) property of Y is that $d(p, q) \leq d_{euc}(p', q')$ for all companion points $p'q' \in \Delta'$ and for all geodesic triangles $\Delta \subset Y$. In short, triangles are thinner in Y than in $\mathbb{R}^2$ [Bridson and Haefliger 1999]. Y is also said to have "nonpositive curvature".

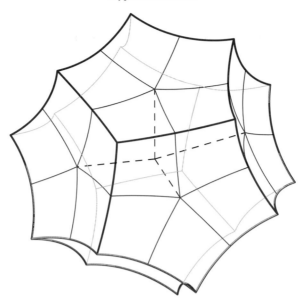

Fig. 6.3. Cubulation of a dodecahedron.

of the slices to continue it over the whole of X. A special type of CAT(0) cube complex X is required so that the hyperplanes are embedded. Hyperplanes are used to encode the pattern of mutual intersections of the lifts of K-M surfaces to $\mathbb{H}^3$.

The dimension n of X can be exponentially large. According to Markovic (personal communication), the dimension reflects the way X is constructed from the nearly geodesic K-M surfaces [Sageev $\geq$ 2015; Sageev 1995]. The dimension is proportional to the number of sheets of the K-M surfaces with uniformly bounded distance apart, for example the number of sheets intersecting a ball in $\mathbb{H}^3$ of unit radius. The more the intersections of the lifted surfaces, the higher the dimension. A special case is the cubulation of a Riemann surface determined by a closed, not necessarily simple geodesic.

A closed manifold M^3 is said to be cubulated if $\pi_1(M^3)$ acts freely with compact quotient on a CAT(0) complex. The end result of the construction is a *cubulation Y* of $\mathcal{M}(G)$ based on the K-M surfaces. It is a consequence of SST, coupled with results of Bergeron-Wise [2012]. It has the property that $\pi_1(Y)$ is isomorphic to G; $\mathcal{M}(G)$ is homotopy equivalent to Y. Furthermore Y is a "right-angled Artin group" (RAAG).[16]

A consequence (cf. [Bestvina 2014]) of the RAAG property is that

A kleinian group G with $\mathcal{M}(G)$ a closed manifold embeds as a group of matrices with integer entries: $G \hookrightarrow SL_n(\mathbb{Z})$ for some n.

From the cubulation Y, Agol constructs a finite cover of $\mathcal{M}(G)$ in which has an embedded closed surface of possibly large genus. This uses his earlier result [Agol 2008], and contributions of Groves and Manning.

---

[16] A RAAG has the presentation [Charney 2007]: $\langle x_1, \ldots, x_m | x_i x_j x_i^{-1} x_j^{-1} = 1 \rangle$, *for all $i \neq j$.*

Agol's proof of VHT coupled with an earlier result of Wise, proving that closed Haken manifolds are virtually fibered, result in the proof of VFT.

The details of cubulation are too lengthy to go into here. A full account of cubulation is presented in Wise [2012]. It has become a powerful new tool for investigations in geometric group theory.

Topological coverings of closed manifolds. In the context of coverings of closed manifolds, the following result is interesting.

Theorem 6.4.8 ([Edmonds 1979b]). *Suppose M^3, N^3 are closed, orientable 3-manifolds and the map $f : M^3 \to N^3$ has the properties* (i) *the induced map between fundamental groups is surjective, and* (ii) *the degree*[17] *of f is at least 3. Then f is homotopic to a branched covering of N^3 with the branch locus in N^3 a simple closed curve.*

Versions applicable to compact surfaces and manifolds with boundary are presented in the Edmonds papers [1979b; 1979a].

6.4.3 Prior computational evidence

We restrict our attention to (orientable), finite volume hyperbolic manifolds. For a finite cover to be Haken, or to be fibered, it must have infinite homology, which is automatically true for cusped manifolds.

In Dunfield and Thurston [2003] the Virtual Haken Conjecture was tested on the Hodgson–Weeks census of the 10,986 smallest volume closed hyperbolic manifolds, most but not all of which have finite homology. The fundamental groups all are two- or three-generator. In every case the conjecture was confirmed by showing each had a cover with infinite homology. The authors also show that every nontrivial Dehn surgery on the figure-8 knot complement results in a Virtual Haken manifold. Whether the Haken cover is hyperbolic depends on whether it is atoroidal.

To explore the situation further, Button [2005] checked the census of more than 5,000 cusped (Callahan-Hildebrand-Weeks census) together with the 10, 986 closed hyperbolic manifolds .

Button finds more than 100 each of closed and cusped hyperbolic manifolds, including the Weeks lowest volume manifold, that are themselves not fibered but which have finite covers which are fibered. Dunfield (2003) earlier had shown that more than 87% of the cusped manifolds in the list are already fibered, so it was surprising to find so many nonfibered ones. Only one of those Button found had itself infinite homology.

---

[17] The third homology groups of M^3, N^3 are cyclic, choose generators α, β. Degree is defined as that number such that $f_*(\alpha) = (degf)(\beta)$.

6.5 Geometrization

In 1977 William Thurston proclaimed a general conjecture of grand sweep called the *Geometrization Conjecture*. It has been the focus of 3-manifold topology ever since. When he proposed the conjecture, Thurston announced the solution of a substantial part of it.[18] Today, more than 35 years later, it is completely proved. The world now has a complete classification of compact 3-manifolds, not just a topological classification but a *geometric* classification. For many explicit examples of three-dimensional manifolds and orbifolds see Montesinos [1987].

It had been known that there is a process of canonically cutting down a compact, orientable 3-manifold into compact pieces by embedded spheres and two-sided incompressible tori. First cut along an embedded sphere that does not bound a ball and cap off the resulting two pieces. After a finite number of steps the process will stop. The summands in the decomposition are unique up to homeomorphisms; see Hempel [1976]; Jaco [1980]. Then for each irreducible piece there is the Johannson–Jaco–Shalen decomposition by a finite set of mutually disjoint, embedded, incompressible tori with the following properties: none of the tori is parallel to a boundary component, and each component resulting from cutting the manifold along the tori is either a *Seifert fiber space* or it contains no incompressible torus. A minimal set of tori is uniquely determined up to isotopy; see Jaco [1980].

Seifert fiber spaces (Seifert manifolds). One example is the manifold $M^3 = R \times \mathbb{S}^1$, with R a compact surface; the boundary components of R, if any, become incompressible boundary tori for M^3. Other examples are obtained by replacing a finite number of circles in M^3 with "singular fibers"; a singular fiber has a neighborhood homeomorphic to the quotient $\mathbb{D} \times \mathbb{S}^1$ of $\mathbb{D} \times \mathbb{R}$ under the action $(z, t) \mapsto (\omega z, t + 1/q)$, where ω is a primitive q-th root of unity and $\mathbb{D}$ is the open unit disk centered at $z = 0$. Each nonsingular fiber wraps q-times around the singular one. An orientable 3-manifold M^3 is called Seifert fibered if it is a union of pairwise disjoint simple loops each with a closed neighborhood, and a finite number of singular fibers with neighborhood as above. Each neighborhood is fiber-homeomorphic to a fibered solid torus $\mathbb{D} \times \mathbb{S}^1$; see [Jaco 1980; Scott 1983]. The only Seifert fibered manifolds that appear inside hyperbolic manifolds are interiors of solid tori or solid cusp tori.

The Geometrization Conjecture is the statement that the interior of each resulting submanifold has a uniquely determined geometric structure. Here "geometric structure" means the following: Assume that X is a simply connected, complete riemannian 3-manifold which is homogeneous—there is an isometry taking any point to any other. The manifold X is diffeomorphic to $\mathbb{S}^3$ or $\mathbb{R}^3$, unless it is modeled on $\mathbb{S}^2 \times \mathbb{R}$. A complete riemannian manifold M is said to be *modeled* on (X, g) if $M = X/G$, where G is a group of fixed-point-free isometries of X in the riemannian metric g.

Thurston conjectured that the interior of each (compact) piece is modeled on one of eight kinds of geometries (see Thurston [1997, Theorem 3.8.4]). The most familiar are the constant sectional-curvature geometries: spherical, euclidean and hyperbolic.

---

[18] In a 2001 Harvard talk, he speculated that a proof for all manifolds "could happen in maybe 50 years".

There are five other geometries possible as well: $\mathbb{S}^2 \times \mathbb{R}$, $\mathbb{H}^2 \times \mathbb{R}$ (both of which have product metrics), $\widetilde{SL(2, \mathbb{R})}$ (the universal covering group of SL(2, $\mathbb{R}$), and Nil and Sol [Thurston 1982; Scott 1983; Boileau et al. 2003], three-dimensional Lie groups like $\widetilde{SL(2, \mathbb{R})}$. The formerly unresolved cases in the Geometrization Conjecture fall into two types according to the nature of the fundamental group: (i) $\pi_1(M^3)$ is finite; in this case the conjecture is that $M^3 \cong \mathbb{S}^3 / \Gamma$ where Γ is a finite orthogonal group acting in $\mathbb{R}^4$ (the Poincaré Conjecture is the case $\Gamma = $ id); (ii) $\pi_1(M^3)$ does not contain a cyclic normal subgroup; this is the Hyperbolization Conjecture.

The confirmation of the Seifert Conjecture by Gabai [1992] and independently by Casson and Jungreis [1994] settled the issue for manifolds for which $\pi_1(M^3)$ does contain a cyclic normal subgroup. Namely if M^3 is compact, orientable, irreducible with infinite $\pi_1(M^3)$, then M^3 is a Seifert fibered space if and only if $\pi_1(M^3)$ contains a cyclic normal subgroup. A Seifert fibered space cannot have hyperbolic interior, unless it is the elementary case of a solid torus or thickened torus (solid torus minus its core).

The proofs depend on showing that a discrete group G of homeomorphisms acting on $\mathbb{S}^1$ with the convergence properties of a group of Möbius transformations (a convergence group) is topologically conjugate to a fuchsian group. Recently Markovic [Markovic 2006] complemented this result by showing that if G is a group of quasisymmetric maps, the conjugacy can be taken to be quasisymmetric as well. (See also [Epstein and Markovic 2007] where it is proved: if $q : QC \to QS$ denotes the restriction (a homeomorphism) of quasiconformal maps of the unit disk to the quasisymmetric maps of the circle, there is no reverse homeomorphism $e : QS \to QC$ such that $q \circ e = $ id.)

Of the eight model geometries, only closed manifolds based on $\mathbb{H}^3$ are not fibered. (Sol is fibered over a line in E^2, with fiber a plane. Manifolds with one of the six other model geometries are Seifert fiber spaces [Boileau et al. 2003, Prop. 1.11].)

The mathematical world has confirmed many times over Grigori Perelman's astounding initial announcement of a complete proof of the Geometrization Conjecture. This includes both the completion of the proof of the hyperbolization theorem for closed manifolds, and the proof of the Poincaré Conjecture [Perelman 2003a; 2003b]. Complete proofs, from differing points of view, are available in Kleiner and Lott [2008]; Morgan and Tian [2007; 2008] and Bessières et al. [2010]; the authors of the latter work present a proof "more attractive to topologists". These accounts also give extensive references.

For expositions of Perelman's work see Chow and Knopf [2004]; Milnor [2003]; Morgan [2005]. The strategy of the ultimately successful proof was set out by Richard Hamilton some years ago: Allow the Ricci flow to act on a given 3-manifold and deal with the singularities as they arise, pinching off pieces of the manifold.

Perleman's result can be stated as follows:

Theorem 6.5.1. *An orientable, closed, irreducible manifold M^3 has exactly one of the following properties:*

(i) $\pi_1(M^3)$ *is finite if and only if* $M^3 \cong \mathbb{S}^3 / \Gamma$, *where* Γ *is a finite orthogonal group acting on* $\mathbb{R}^4$. $\pi_1 M^3 = \mathrm{id}$ *if and only if* $\mathcal{M}^3 \cong \mathbb{S}^3$.

(ii) $\pi_1(M^3)$ *is infinite without noncyclic abelian subgroups (or, atoroidal) if and only if* M^3 *has a hyperbolic structure (modelled on* $\mathbb{H}^3$*).*

(iii) M^3 *is a Seifert fibered manifold with infinite* $\pi_1(M^3)$ *if and only if* $\pi_1(M^3)$ *has a cyclic normal subgroup.*

A sufficient condition for hyperbolicity of a closed manifold in terms of Heegaard splitting is presented in Exercise (6-17). The approach to hyperbolicity via Cayley graphs and the Cannon Conjecture is discussed in Exercise (2-17).

In the context of the solution of the Geometrization Conjecture, one can proclaim that the "vast majority" of the interiors of compact 3-manifolds support hyperbolic structures, although there is not yet a formal theorem making matters precise. Compare the manifold case with the case of word-hyperbolic groups (Exercise (2-18); see also Exercise (6-18) and Dunfield and Thurston [2003]).

For interesting comparisons of hyperbolic volumes and volumes of riemannian 3-manifolds, assuming Perelman's work, see Agol et al. [2005].

6.6 Hyperbolic knots and links

We now cite Thurston's famous theorems for knot and link complements. These are special cases of his Hyperbolization Theorem and clearly display the power of his theory.

6.6.1 *Knot complements*

Hyperbolic Structures on Knot Complements 6.6.1. *If* $K \subset \mathbb{S}^3$ *is a knot,* $\mathbb{S}^3 \setminus K$ *has a hyperbolic structure if and only if* K *is not a torus knot or a satellite knot.*

A torus knot is a knot that lies on the boundary of a torus neighborhood of an unknot in $\mathbb{S}^3$ (and not shrinkable on the boundary to a point). A satellite (also called a companion) knot K is one that is embedded in a small solid torus neighborhood of some knot K_0, not the unknot, and K is not isotopic to K_0 nor is contained in a ball inside the solid torus. The torus about K_0 is incompressible in $\mathbb{S}^3 \setminus K$, which makes it impossible for K to have a hyperbolic complement.

A knot or link whose complement in $\mathbb{S}^3$ has a hyperbolic metric is called a *hyperbolic knot or link.*

By a celebrated theorem of Gordon and Luecke [1989], knots are determined by their complements: If K_1, K_2 are two knots, there is a homeomorphism (orientation-preserving or -reversing) of $\mathbb{S}^3$ taking K_1 to K_2 if and only if $\mathbb{S}^3 \setminus K_1$ is homeomorphic to $\mathbb{S}^3 \setminus K_2$. (This is not true for link complements).

Theorem 6.6.2 ([Gordon and Luecke 1989]). *A knot is uniquely determined by its complement up to ambient isotopy and mirror image. The volume of its complement is an invariant of a hyperbolic knot.*

One might think it very rare that a closed manifold with infinite fundamental group has zero first homology. Yet given a hyperbolic knot K, there are infinitely many

Dehn surgeries on K which result in a closed manifold with exactly this property. To construct examples, start with tubular neighborhood $\mathcal{T}$ of K. On $\partial\mathcal{T}$ there is a uniquely defined (up to free homotopy) pair of simple closed curves which cross each other once: The *meridian* μ bounds a disk within $\mathcal{T}$ and generates the homology of $\mathbb{S}^3 \setminus \mathcal{T}$. The *longitude* λ is parallel to K. The longitude is homologous to zero in $\mathbb{S}^3 \setminus \mathcal{T}$. This is an immediate consequence of the fact that there exists a *Seifert surface* for K—an orientable surface $S \subset \mathbb{S}^3 \setminus K$ with $\partial S = K$ [Lickorish 1997]. See also the Heegaard construction of Exercise (6-16).

According to surveys of Hoste, Thistlethwaite, and Weeks reported in Cannon [2011], all but 32 of $1,701,936$ prime knots of 16 or fewer crossings are hyperbolic. A prime knot is an "indecomposable" knot, one that is not the connected sum of two other knots (which are not unknots).

In a finite volume manifold with a single cusp, the solid cusp torus $\cong \{0 < |z| < 1\} \times \mathbb{S}^1$ constitutes a neighborhood of what can be regarded as a knot in the closed manifold obtained by Dehn filling the interior. This manifold is however not necessarily hyperbolic. In particular, there is no way known to determine whether it is $\mathbb{S}^3$.

Return now to the solid torus neighborhood $\mathcal{T}$ of K, and choose a meridian μ and longitude λ on $\partial\mathcal{T}$. Do $(n, 1)$-Dehn filling: replace the inside of $\mathcal{T}$ by a new solid torus in which the simple loop $\lambda^n\mu$ becomes the meridian. This yields a new closed manifold M^*. In M^*, both λ and λ^n are homologous to 0; hence so is μ. Consequently the first homology group of M^* is zero. Thurston shows that for all except a finite number of integers n, M^* has a hyperbolic structure.

One might ask, what determines when a compact manifold can be embedded in $\mathbb{S}^3$? Toward an answer, the following fact discovered by Purcell and Souto is interesting:

Theorem 6.6.3 ([Purcell and Souto 2010]). *Suppose M^3 is a compact compression body whose interior is hyperbolic and has exactly one end. If M^3 is homeomorphic to a submanifold of $\mathbb{S}^3$, then M^3 is the geometric limit of a sequence of hyperbolic knot complements in $\mathbb{S}^3$.*

From this, the authors deduce (1) given *any* $r > 0$, there exists a hyperbolic knot complement containing a point with injectivity radius $> r$, but (2) if the interior of M has more than one end, M cannot not be the geometric limit of hyperbolic knot complements in $\mathbb{S}^2$.

On the other hand, in Koundouros [2004] the following interesting conjecture is proposed and explored: If the injectivity radius of the closed manifold $\mathcal{M}(G)$ is sufficiently large, then $\mathcal{M}(G)$ cannot be obtained by Dehn filling on a knot in $\mathbb{S}^3$. The *injectivity radius* of a closed manifold is the largest number r such that every point in the manifold is the center of an embedded ball of radius r.

6.6.2 Link complements

A link $L \subset \mathbb{S}^3$ is called *indecomposable* if it cannot be separated into two parts which can be isotoped into disjoint balls. A satellite (or companion) link L is a satellite to

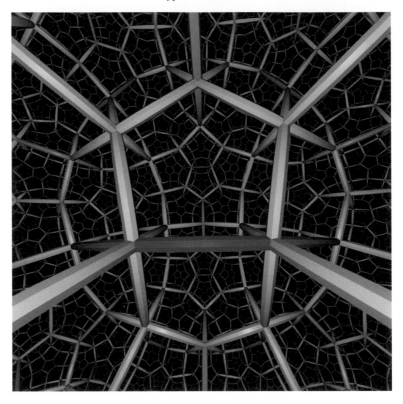

Fig. 6.4. The Borromean ring (BR) complement C_∞ is the algebraic limit of closed hyperbolic orbifolds $\{C_n\}$. Here C_4 is represented in $\mathbb{H}^3$ in true hyperbolic perspective; this is what we would *see* if we lived inside C_4. Its fundamental domain $F_4 \subset \mathbb{H}^3$ is a right-angled dodecahedron whose colored (red, blue, green) edges lie on order-4 elliptic axes (here thickened). A partial orbit of F_4 is displayed. The 6-colored edge segments, which together represent the BR, shrink forming 3 rank two cusps as their orders $n \to \infty$. What remains is a right-angled rhombic dodecahedron F_∞ of 12 4-sided faces, 24 edges, and 14 vertices. F_∞ is a fundamental domain for C_∞ [Epstein and Gunn 1991]. Graphics created at The Geometry Center, UMN.

another link L_0 if one or more components of L is satellite to a component of L_0. Thurston proved the following two theorems. The first is recorded in Epstein and Gunn [1991, p. 41]; the second appears in Thurston [1982; 1979a, p. 5.38].

Hyperbolic Structures on Link complements 6.6.4. *Suppose $L \subset \mathbb{S}^3$ is an indecomposable link of $m \geq 2$ components. Suppose no component is a torus knot and L is not a satellite link. Then $\mathbb{S}^3 \setminus L$ has a hyperbolic structure.*

The hyperbolic structure for the Borromean rings complement is visualized in Figure 6.4 and in Gunn and Maxwell [1991].

Dehn Surgeries on Hyperbolic Link Complements 6.6.5. *Suppose $L \subset M^3$ is a link in the 3-manifold M^3, in particular in $\mathbb{S}^3$, such that $M^3 \setminus L \cong \mathcal{M}(G)$ has a hyperbolic structure. For each cusp of $\mathcal{M}(G)$ there are a finite number of Dehn fillings* *that must be excluded. The manifolds resulting from all Dehn fillings on* $\mathcal{M}(G)$, *except for those excluded, have a hyperbolic structure.*

For an example, consider the Borromean rings complement $\mathcal{M}(G) \subset \mathbb{S}^3$ as in Figure 6.4. The $(1, 0)$ filling on one of the links results in a manifold homeomorphic to $\mathbb{S}^3$ minus two unlinked circles. This is not hyperbolic, nor is the result of any further filling—there are noncontractible embedded spheres in the complement that separate the components.

For a different slant on this matter see the Dehn Surgery Theorem, p. 248.

Every closed, orientable 3-manifold is obtained by Dehn filling along some link $L \subset \mathbb{S}^3$ whose complement is hyperbolic [Thurston 1982, p. 362], [Lickorish 1962]. Most of these Dehn fillings also give rise to non-Haken manifolds [Thurston 1982]. See also Exercise (6-3).

It is also true that every closed orientable 3-manifold is a branched cover of $\mathbb{S}^3$, branched over the Borromean rings [Hilden et al. 1985]. For the construction of k-fold unbranched covers of $\mathbb{S}^3 \setminus L$, see Rolfsen [1976, §10.F].

In particular Thurston [1979a] showed there exists a hyperbolic orbifold $\mathcal{M}(B)$ branched of degree four over the Brorromean rings in $\mathbb{S}^3$. From this the following universal fact is drawn.

Theorem 6.6.6 ([Hilden et al. 1987]). *Given any closed, oriented 3-manifold N, there exists a finite index subgroup $B_0 \subset B$ such that N is homeomorphic to the hyperbolic orbifold $\mathcal{M}(B_0)$.*

6.7 Computation of hyperbolic manifolds

It turns out that, contrary to prior assumption, computing the hyperbolic structure is easier in the conformal model than in the hyperboloid model [Floyd et al. 2002].

Marc Culler's and Nathan Dunfield's *SnapPy* [2012] is now the preferred user interface to the older *SnapPea* kernel written by Jeff Weeks [Weeks 2012; Weeks 2005a]. SnapPy is used for the study of the topology and geometry of 3-manifolds, with the focus primarily on hyperbolic structures. It computes hyperbolic manifolds including knot complements. The functionality of the original SnapPea is continued and extended as an essential research tool (see also Exercise (4-19)). SnapPy runs on Mac OS X, Linux, and Windows.

As described in the extensive documentation of SnapPy, a manifold is presented as an ideal triangulation of the interior of a compact 3-manifold with torus boundary, where each tetrahedron has been assigned the geometry of an ideal tetrahedron in hyperbolic 3-space. A Dehn-filling can be specified for each boundary component, allowing the description of closed 3-manifolds and some orbifolds. The manifold can then be visualized in 3D graphics as a Dirichlet region. In particular, a knot or link can be drawn with the mouse and from there, if the knot is hyperbolic, the volume, etc., of the hyperbolic structure of its complement computed. The knot complement can then be visualized as a Dirichlet domain.

Verification that the algorithm underlying SnapPea in principle results in the correct structure is contained in Oliver Goodman's enhancement to SnapPea called *Snap* [Coulson et al. 2000; Goodman 2006], which can compute arithmetic invariants, such as volumes and trace fields to very high precision. Using Snap one can verify the hyperbolic structure as discovered by SnapPy by finding exact solutions of the equations satisfied by the tetrahedral parameters needed to construct the cusped manifold from ideal tetrahedra (if the degree is not too high). Thus it is possible, in principle, to decide whether or not a given knot, presented say by over and under crossings, is the unknot. Likewise the question of whether two hyperbolic knots are the same or not can be answered in principle when SnapPy can find a hyperbolic structure for each. See also Exercise (1-23) on computing volumes.

Another augmentation to SnapPea is the program *Orb* written by Heard [2007]. Orb can start with a projection of a graph embedded in the 3-sphere, and produce and simplify a triangulation with some prescribed subgraph as part of the 1-skeleton, and the remainder of the graph drilled out. It enables computation of hyperbolic structures on knot complements, graph complements and orbifolds whose underlying space is the 3-sphere minus a finite number of points.

Orb has been applied to computing commensurators of certain finite volume cusped manifolds without too many cusps. Margulis [1991] showed (for dimension ≥ 3) that a group G is arithmetic if and only if its commensurator (see Section 2.5.1) is *not* discrete. Therefore if G has finite volume and is not arithmetic then $C(G)$ consists of conjugates of its finite index subgroups. This situation is amenable to computation [Goodman et al. 2008].

For rigorous verification that a given ideal triangulation of a cusped manifold $\mathcal{M}(G)$, or a Dehn filling of $\mathcal{M}(G)$ admits a complete hyperbolic structure see the program of Neil Hoffman et al. [Hoffman et al. 2013].

There is a program for computing limit sets of kleinian groups created by Curt McMullen called *lim*, see [2001b].

Throughout the early 1970s Robert Riley did extensive research looking for $PSL(2, \mathbb{C})$ representations of knot and link groups. He had the help of a computer program he developed to test a group for discreteness, given the generators, by checking whether a fundamental polyhedron existed. In particular his paper [1975] (see also Wielenberg [1978]) records a variety of kleinian groups among the Bianchi groups $\Gamma_d = PSL(2, \mathcal{O}_d)$ and their finite index subgroups. Here $\mathcal{O}_d$ denotes the ring of integers in the quadratic imaginary number field $\mathbb{Q}(\sqrt{-d})$, where d is a positive integer. Many of these model familiar knots and links. The ring $\mathcal{O}_d$ has $\mathbb{Z}$-basis $\{1, \omega\}$ where $\omega = \sqrt{-d}$, unless $d + 1$ is divisible by 4 in which case $\omega = \frac{1}{2}(1 + \sqrt{-d})$. So the normalized matrices in the group have entries of this form. All of the Bianchi groups, together with their finite index subgroups, give rise to manifolds of finite volume. In many cases, an exact formula can be given for the volume; see Milnor [1994, p. 257].

In particular Riley showed that the group representing the figure-8 knot complement is generated by

$$\begin{pmatrix} 1 & 1 \\ 0 & 1 \end{pmatrix}, \text{ and } \begin{pmatrix} 1 & 0 \\ -\omega & 1 \end{pmatrix}.$$

Here ω is a solution of the representation polynomial $1 + x + x^2 = 0$, a primitive cube root of one.

In Jørgensen and Marden [1979] a doubly degenerated once-punctured torus quasi-fuchsian group is described based on earlier studies of Jorgensen. The group has an additional property: it admits an infinite cyclic extension T generated by an isometry. The quotient manifold associated with the extended group is the figure-8 knot complement, and is therefore conjugate to Riley's prior discovery.

Moreover the Picard group of Exercise (2-12) is $PSL(2, \mathcal{O}_1)$. The Borromean rings come from the one torsion-free normal subgroup of index 24 in $PSL(2, \mathcal{O}_1)$ [Brunner et al. 1984; Wielenberg 1978]. The figure-8 knot complement (Exercise (3-5)) has index 12 in $PSL(2, \mathcal{O}_3)$.

The field $\mathbb{Q}(x)$, where $x^4 - x^2 + 3x - 2 = 0$, spawns a closed manifold all of whose geodesics are simple—in fact an infinite family of such manifolds has been found! The definitive reference on this subject of *arithmetic kleinian groups* is Maclachlan and Reid [2003]. There the reader will also find an exhaustive list of known examples.

The Weeks manifold is the lowest volume (orientable) manifold, among *all* hyperbolic 3-manifolds, see Section 4.10.4. It is a closed, arithmetic manifold of approximate volume 0.94 obtained from $\mathbb{Q}(x)$ with $x^3 - x + 1 = 0$.

There is another approach to knot invariants, namely via the several known knot polynomials, for example the Alexander and the Jones polynomials It is not known how to find these polynomials directly from the hyperbolic structure.

6.8 The Orbifold Theorem

As discussed in Section 2.5, if the kleinian group G has elliptic elements, its quotient $\mathcal{M}(G)$ is called an (orientable) *orbifold*. Assume that $\mathcal{M}(G)$ is geometrically finite.

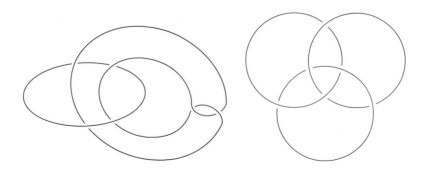

Fig. 6.5. The Whitehead link and the Borromean rings.

We will collect the properties of its singular set $\sigma(G)$, which is a graph as described in Corollary 2.5.2:

(i) Each edge e is labeled by a positive integer $k \geq 2$ which is the order of a primitive elliptic element that pointwise fixes a lift $e^* \subset \mathbb{H}^3$.

(ii) Each vertex of $\sigma(G)$ is the endpoint of three edges. These edges must have orders $(2, 3, 3)$, $(2, 3, 4)$, $(2, 3, 5)$ or $(2, 2, n)$, $n \geq 2$.

(iii) A component of $\sigma(G)$ is either a simple loop, a full geodesic with each endpoint in $\partial \mathcal{M}(G)$ or at a cusp, or it contains vertices. An edge from a vertex ends either at another vertex, a point on $\partial \mathcal{M}(G)$, or at a cusp.

(iv) If an edge ends at a rank one cusp, there are exactly two such edges and each has the label 2.

(v) If an edge ends at a rank two cusp, there are three or four such edges with the labels $(3, 3, 3)$, $(2, 3, 6)$, $(2, 4, 4)$, or $(2, 2, 2, 2)$.

(vi) There are no euclidean or spherical suborbifolds except as arise automatically from a cusp or a fixed point of a finite subgroup.

A spherical submanifold means that there is a smooth topological sphere S that is transverse to $\sigma(G)$, cuts each edge at most once, and doesn't intersect the vertices. Set $S \cap \sigma(G) = \{p_i\}$. Each point p_i has a label r_i coming from the edge it lies on. In view of Equation (3.16), we must have $\sum(1 - 1/r_i) > 2$.

A definition of orbifold and references to the literature are presented in Section 2.5.2, and further discussion is in Exercise (4-7). Thurston proclaimed a complete geometrization theory for compact, irreducible, atoroidal orbifolds. Proofs have been developed by Boileau, Leeb, and Porti [Boileau et al. 2005], and independently by Cooper, Hodgson, and Kerckhoff [Cooper et al. 2000]. A proof when the singular set is a union of circles is in Boileau and Porti [2001]. The proofs require analysis of cone manifolds with cone angles $\leq \pi$ when trivalent vertices are present.

Instead of plunging into the general theory we will be more than content to present a beautiful application to hyperbolic manifolds:

Designated Hyperbolic Orbifolds. *Suppose $\mathcal{M}(G)$ is a geometrically finite hyperbolic manifold, and $\{\gamma_i\}$ is a finite set of mutually disjoint, simple closed geodesics with assigned integers $\{2 \leq r_i \leq \infty\}$. There exists a hyperbolic orbifold $\mathcal{M}(H)$ such that the labeled graph $\sigma(H)$ is isotopic to $\bigcup \gamma_i$ with the given labeling and $\mathcal{M}(H) \setminus \sigma(H)$ is homeomorphic to $\mathcal{M}(G) \setminus \bigcup \gamma_i$.*

Proof. This is a direct consequence of Boileau et al. [2005, Theorem 2.3]. The presence of the graph determines orbifold structure $\mathcal{O}$ with "base space" $\mathcal{M}(G)$. It is topologically atoroidal, not a Seifert fibered orbifold, nor is there a spherical or euclidean suborbifold. For more details see Exercise (6-3). $\square$

There are many other possibilities for a hyperbolic orbifold besides the ones stated above, but the analysis is more complicated. One problem is that a singular set cannot consist of a "knotted" geodesic. Another is that the underlying manifold may or may not have a hyperbolic structure. The example of $\mathbb{H}^2$ covering $\mathbb{S}^2$ with branch points satisfying Equation (3.14) or (3.15) shows why. It requires the extensive analysis of Boileau et al. [2005]; Cooper et al. [2000] to explore all the possibilities.

Suppose $\mathcal{M}(G)$ is a closed manifold. We can ask, of all the hyperbolic orbifolds whose underlying space is $\mathcal{M}(G)$, which has the least volume? Peter Storm gave the following answer, which is consistent with the Dehn Surgery Theorem (p. 248) (see also Exercise 6-3):

Theorem 6.8.1 [Storm 2002]. *The volume of the closed manifold $\mathcal{M}(G)$ is strictly less than the volume of any hyperbolic orbifold $\mathcal{O}$ (or, more generally, any cone manifold $\mathcal{O}$ with cone angles $\leq 2\pi$) with the property that the underlying space of $\mathcal{O}$ is homeomorphic to $\mathcal{M}(G)$.*

Thurston's general Orbifold Theorem forms part of Geometrization Theory. We conclude by stating part of its general form following Boileau et al. [2005, Cor. 1.2]:

A compact, orientable, irreducible, atoroidal orbifold $\mathcal{O}$ either carries one of Thurston's 8-geometries, or is the quotient of the 3-ball by a finite orthogonal action.

6.9 Exercises and explorations

6-1. The interior of $M^3 = \mathbb{T}^2 \times [0, 1]$ resulting from thickening the torus $\mathbb{T}^2$ with $P = \mathbb{T}^2 \times \{0\}$ would fall under the aegis of the Hyperbolization Theorem (p. 385), except that it is not pared since there are essential cylinders with one boundary component on P. Its interior is homeomorphic to $M^3 = \mathbb{R}^3 / \Gamma_0$, where Γ_0 is the group generated by $\langle (x, y, t) \mapsto (x+1, y, t), (x, y, t) \mapsto (x, y+1, t) \rangle$. M^3 is the only manifold whose boundary components are tori whose interior does not have a complete hyperbolic structure of *finite volume*.

Consider instead group Γ of euclidean isometries of $\mathbb{R}^3$ generated by the two maps $(x, y, t) \mapsto (x+1, y, t), (x, y, t) \mapsto (-x, y+1, -t)$. The group Γ preserves $\mathbb{C}$ and $\mathbb{C}/\Gamma$ is a Klein bottle. The torus $\mathbb{C}/\Gamma_0$ is its two-sheeted orientable cover and its flip is the cover transformation. The corresponding 3-manifold $N^3 = \mathbb{R}^3/\Gamma$, which is called the *twisted I-bundle* over the Klein bottle, is the only homotopically atoroidal 3-manifold whose interior does not have a hyperbolic structure (see Hempel [1976, Theorem 10.3]).

6-2. *Doubling a manifold.* The following construction leads to another proof of Theorem 6.2.2. Suppose $M = \mathcal{M}(X)$ is geometrically finite, acylindrical, without cusps, and with $b \geq 1$ boundary components.

Here is a process for constructing a *closed* manifold $\widehat{M}$ that is 2^b-sheeted over M and is orientable, atoroidal, and Haken. In the case $b = 1$ it is just the double of M.

Order the boundary components of $M : (S_1, S_2, \ldots)$. Take 2^b copies of M. On each copy, assign to each of its boundary components one of the two symbols $+$ and $-$. We do this so that the ordered boundary of each copy is associated with a symbol sequence and no two copies are assigned the same ordered symbol sequence. Index the copies as M_i.

Glue M_i to M_j along the surface S_k if and only if the symbol sequence for M_i differs from that of M_j only in the k-th position. Think of this operation as M_j being the reflection of M_i across S_k. If we start with a manifold with totally geodesic boundaries, we can carry out this operation in the universal cover $\mathbb{H}^3$.

There is a finite group G of order 2^b of automorphisms acting on $\widehat{M}$. We can see this by considering the connected b-valent graph $\mathcal{G}$ constructed as follows. The vertices correspond to the copies $\{M_i\}$. Two vertices are joined by an edge if one manifold is glued to the other. Denote by J_k the involution of $\mathcal{G}$ determined by replacing the sign attached to the k-th boundary component of each M_i by the opposite sign. Each J_k acts on $\widehat{M}$ and is an orientation-reversing involution: "reflection" in each S_k. Denote the group generated by the $\{J_k\}$ by G. No element of G (other than the identity) fixes a vertex of $\mathcal{G}$, and for each index i, there is an element w_i taking the vertex corresponding to M_1 to that corresponding to M_i. The expression for w_i as a word in the generators is not uniquely determined. Rather what is determined is the parity of its length, even or odd, orientation-preserving or -reversing.

The original M appears as the quotient space $\widehat{M}/G$. Hyperbolize $\widehat{M}$ to become the closed manifold $\mathcal{M}(Y)$. For example by Theorem 3.13.2, there is a homeomorphism $f : \widehat{M} \to \mathcal{M}(Y)$. Moreover each element of fGf^{-1} is isotopic to an isometry by Mostow's theorem and G is isomorphic to a group H of isometries. The boundary components of $\mathcal{M}(Y)/H \cong M$ are totally geodesic.

6-3. *Drilling out simple geodesics.* Start with a geometrically finite hyperbolic manifold $\mathcal{M}(G)$ with boundary. The result of removing a simple loop which is *not* primitive cannot have a hyperbolic structure (Exercise (4-21)) but simple geodesics are automatically primitive. The result of removing a finite number of mutually disjoint simple closed geodesics is also atoroidal and Haken and consequently has a hyperbolic structure. On the other hand, if $\mathcal{M}(G)$ has finite volume with cusps, a closed hyperbolic manifold can be obtained by Dehn surgeries. The original manifold can in turn be recovered by removing the resulting new geodesics and introducing the original hyperbolic structure on the complement. The general theorem covering this matter is as follows.

Theorem 6.9.1 (Kerckhoff, Kojima, Sakai) . *If $\mathcal{M}(G)$ is geometrically finite and $L \subset \mathcal{M}(G)$ is a disjoint union of a finite number of simple closed geodesics, then $\mathcal{M}(G) \setminus L$ has a complete hyperbolic structure.*

As a consequence of the Dehn Surgery Theorem (p. 248), the volume of $\mathcal{M}(G)$, if it does have finite volume, is less than that of the hyperbolic structure on $\mathcal{M}(G) \setminus L$.

We will assume that $\mathcal{M}(G)$ is closed and indicate the proof as in Kojima [1988, Prop. 4]. First, $\mathcal{M}(G) \setminus L$ is irreducible—every embedded 2-sphere bounds a ball. (*Hint*: Lift to $\mathbb{H}^2$.) Nor does it contain an essential embedded torus T which cannot be homotoped into L. (*Hint*: Suppose there is an essential embedded torus T. With respect to $\mathcal{M}(G)$, T is compressible so there is a compressing disk D. Choose a lift $\widehat{T}$ of T to $\mathbb{H}^3$. Then $\widehat{T}$ is either a torus or an open cylinder. In the former case, the region bounded by $\widehat{T}$ could not contain a lift of any of the geodesic components of L. Therefore $\widehat{T}$ would be contractible in $\mathbb{H}^3 \setminus L^*$, where L^* is the totality of lifts of L. As such T would be contractible in $\mathcal{M}(G) \setminus L$, a contradiction. If the second case occurred, $\widehat{T}$ would be invariant under a cover transformation $g \in G$, and hence its closure would have two ideal points on $\partial \mathbb{H}^3$, the fixed points of g. $\widehat{T}$ encloses a cigar-shaped region which would contain exactly one component of L^*, the lift of some component L_0 of L. In this case T is the boundary of a tubular neighborhood of L_0.)

It remains to show that $M' = \mathcal{M}(G) \setminus L$ cannot be a Seifert fiber space (SFS)— see Section 6.4 or Scott [1983] for the formal definition. If to the contrary it were a SFS, M' would have the following structure (mandated by the Torus Theorem— [Jaco 1980]). There would be a fuchsian group H with possible elliptic elements such that $S = \mathbb{H}^2/H$ is a closed surface and $\pi_1(M')/K \cong H$, where K is an infinite cyclic normal subgroup. The singular fibers come from the elliptic fixed points. Since $\mathcal{M}(G)$ is formed from M' by adding back L, there would be at most one singular fiber and that would have to be a meridian of T, because $\mathcal{M}(G)$ is not a SFS. Therefore we would have $G \cong H$. Now G is torsion free so H must be as well and hence is a surface group. This is impossible as the third homology of S is zero while that of $\mathcal{M}(G)$ is not.

The argument presented was kindly provided by Peter Scott.

Can you hyperbolize $\mathcal{M}(G) \setminus \{\gamma\}$ where γ is a simple loop in the interior which does not bound an embedded disk in its complement?

Under some circumstances it may be advantageous not to globally change the hyperbolic structure because there is no control of how the geometry changes. Instead one can replace the hyperbolic metric near γ by a complete PNC metric.[19] What is needed is the following result, formulated by Souto (see also Hodgson and Kerckhoff [1998]):

Theorem 6.9.2. *Given $\varepsilon > 0$ there exists $C > 0$ with the following property. If $\gamma \subset \mathcal{M}(G)$ is a simple closed geodesic such that its radius-r tube V about it is embedded, then the hyperbolic metric on $\mathcal{M}(G) \setminus V$ can be extended to a complete PNC metric on $\mathcal{M}(G) \setminus \gamma$ which in V has curvature κ satisfying $-1-\varepsilon < \kappa < -1+\varepsilon$.*

It is shown in [Agol 2002] that if a closed geodesic γ is removed from a hyperbolic 3-manifold $\mathcal{M}(G)$ of finite volume, and if γ is contained in an embedded tube of radius r, then the volume of the hyperbolic structure M_γ on $\mathcal{M}(G) \setminus \gamma$ satisfies

$$\mathrm{Vol}(M_\gamma) \leq (\coth r)^{5/2}(\coth 2r)^{1/2}\mathrm{Vol}(\mathcal{M}(G)).$$

Compare Theorem 6.9.1 with the following striking result of Robert Myers [1982]; it is the most general formulation of "diskbusting", see Exercise (5-26).

---

[19] PNC stands for "pinched negative curvature".

Theorem 6.9.3. *Let M be a compact, orientable 3-manifold for which ∂M, if non-empty, does not contain a 2-sphere. Then M contains a simple closed curve K whose open tubular neighborhood N_K has the following property. The complement $M \setminus N_K$ is irreducible, boundary incompressible, and atoroidal.*

In particular $\text{Int}(M) \setminus K$ has a complete hyperbolic structure. For example if $M \cong S \times [0, 1]$ for a closed surface S, there is a knot K so that $M \setminus K$ is atoroidal. For this to happen the projection of K to S must fill up S in the sense that each complementary component is simply connected.

Meyers goes on to prove that M is completely determined as a 3-manifold by the countably infinite set of subgroups $\{\pi_1(M \setminus J)\}$ as J runs over all simple loops in M.

Instead of drilling out a geodesic we turn to the following, related situation. Suppose instead of a manifold we have a hyperbolic cone manifold M (Exercise (4-7)) such that the singular locus c is a simple closed curve, or disjoint union of them, with common cone angle. Still M will have a geometrically finite structure and in particular a conformal boundary as befits geometrically finite hyperbolic manifolds. The following results are based on the pioneering prior paper of Craig Hodgson and Steve Kerckhoff [1998] on the deformation of cone manifolds.

For the following two results, M_α denotes a geometrically finite, hyperbolic cone manifold, without rank one cusps (this condition may not be necessary), and with singular locus $\mathcal{S}$. The components $\{c_i\}$ of $\mathcal{S}$ are mutually disjoint, simple closed curves with a common cone angle α.

Theorem 6.9.4 ([Bromberg 2004]). *Assume* (i) $\mathcal{S}$ *has one component c,* (ii) c *is contained an a tubular neighborhood of radius at least* $\text{arcsinh} \sqrt{2}$, (iii) $0 \leq \alpha \leq 4\pi$, *and* (iv) *there exists* $\varepsilon = \varepsilon(\alpha)$ *such that the length* $L_\alpha(c) < \varepsilon$ *in* M_α. *Then there is a 1-parameter family of geometrically finite cone manifolds M_t with cone angle $0 \leq t \leq \alpha$ and conformal boundary fixed.*

More generally suppose $\mathcal{S}$ has a finite number of components consisting of mutually disjoint simple loops in M_α. We are given $\alpha > 0$. Assume property (ii) *holds for each component and the neighborhoods are mutually disjoint and like property* (iv), *there exists $\ell > 0$ such that the total length of the components $\{c_i\}$ of $\mathcal{S}$ satisfies $L(\mathcal{S}) < \ell$. Then there is a 1-parameter family $\{M_t\}$ of geometrically finite manifolds with fixed conformal boundary and with all cone angles $t \in [0, \alpha]$.*

A component of the singular set arises from the rotation axis of an elliptic, from which its length can be determined.

The Drilling Theorem 6.9.5 [Brock and Bromberg 2004]. *Given $\alpha > 0$, and M_α as above, and $K > 1$, there exists $\ell^* > 0$ with the following property. If the total length $L(\mathcal{S}) < \ell^*$ there is, for each $t \in [0, \alpha]$,* (i) *a collection of mutually disjoint tubular neighborhoods $\{V_i\}$ of the components of $\mathcal{S}$, and* (ii) *a K-bilipschitz diffeomorphism (quasiconformal map) h_t of pairs,*

$$h_t : (M_\alpha \setminus V_\alpha(\mathcal{S}), \partial V_\alpha(\mathcal{S})) \mapsto (M_t \setminus V_t(\mathcal{S}), \partial V_t(\mathcal{S})),$$

which extends to a homeomorphism $h_t{}^* : M_\alpha \to M_t$ for each $t \in [0, \alpha]$.

Here the $\{V_*\}$ are tubular neighborhoods about the components of S in M_*. They are contained in the universal tubular neighborhoods of tube radius $\leq \varepsilon$ embedded and mutually disjoint in M_*. The tube radii are functions of ℓ^* and α. There is a simplified version applicable to drilling out a geodesic as follows.

The Drilling Theorem I 6.9.6 [Brock and Bromberg 2004]. *Suppose* $\mathcal{M}(G)$ *is geometrically finite. Given* $L > 1$, *there exists* $\ell > 0$ *with the following property. If* $c \subset \mathcal{M}(G)$ *is a geodesic with length less than* ℓ, *there is an L-bilipschitz diffeomorphism*

$$h : (\mathcal{M}(G) \setminus V(c), \partial V(c)) \mapsto (\mathcal{M}(G^*) \setminus \mathcal{T}(c), \partial \mathcal{T}(c)).$$

Here as before $V(c)$ *is a tubular neighborhood of radius* $\leq \varepsilon$ *about* c, $\mathcal{M}(G^*)$ *is the complete hyperbolic structure on* $\mathcal{M}(G) \setminus c$, *and* $\mathcal{T}(c) \subset \mathcal{M}(G^*)$ *is a solid cusp torus corresponding to* c.

Theorem 6.9.4 allows a continuously decreasing cone angle from 4π to 2π, at which point $M_{2\pi}$ is a complete hyperbolic manifold, and on to 0, when S becomes a union of rank two cusps. This latter is the case of Theorem 6.9.1. Cone angles of 4π arise naturally from the construction of Exercise (6-7).

The Drilling Theorem is quite strong: It describes the change in the geometry of the manifolds. It says in quantitative terms that the manifolds $\{M_t\}$ remain a bounded distance apart as $t \to 0$. When a simple geodesic γ is drilled out, away from the geodesic the hyperbolic structure does not change much. As $t \to 0$, the tubular neighborhoods approach horoballs. As $K \to 1, t \to \alpha$.

The Drilling Theorem has been applied in the study of the density of geometrically finite groups (Section 5.9), the density of cusps on boundaries of deformation spaces, the Ending Lamination Conjecture, and tameness of manifolds.

Another application is to the proof of Theorem 4.6.3. In an approximating sequence one drills out the short geodesics coming from pinching which are destined to become the rank one parabolics in the algebraic limit. This does not change the hyperbolic structure away from these geodesics very much. Consider the covering manifolds determined by the marked fundamental groups of the sequence of drilled out cores. These covers converge algebraically to the expected algebraic limit.

Still another application is in the uniqueness proof of Bonahon and Otal's Theorem 3.11.6. If we have a convex core whose bending lamination is finite, we can construct the double of the convex core across its boundary components. If a particular bending line ℓ has bending angle α so that the internal bending angle is $\pi - \alpha$, then the result of doubling gives a cone manifold with cone axis ℓ and cone angle $2(\pi - \theta)$ and similarly the other bending lines become cone axes as well. The manifold can be continuously deformed until all the cone angles become zero. Then Mostow rigidity can be applied.

As pointed out in Kent [2010, p. 3], Drilling Theorem I can be applied in reverse to analyze certain skinning maps. Suppose $\mathcal{M}(G)$ is geometrically finite and has a connected, closed boundary component S and in addition a rank two cusp. We can find a solid cusp torus $\mathcal{T}$ which lies outside the convex core $\mathcal{C}$ of $\mathcal{M}(G)$. In this case Stab$(S) \subset G$ is a quasifuchsian group G^{qf} in which the skinning map σ pairs S with S'.

Do higher and higher Dehn fillings on $\mathcal{T}$. The new geodesics c_n in $\mathcal{M}(G_n^{qf})$ become shorter and shorter while lying outside $\mathcal{C}$. The L_n-bilipschitz diffeomorphisms mapping from $\mathcal{T}$ to neighborhoods V_n of c_n can be taken with L_n closer and closer to one. The bottom line is that the sequence of skinning maps $\sigma_n : S \to S_n'$ converge back to σ, uniformly over all surfaces $\{S\}$ in $\mathfrak{T}(G)$.

6-4. *A nongeometrically finite limit on* $\partial\mathfrak{T}(G)$. Consider a (finitely generated) quasifuchsian group G without parabolics. Given a compact submanifold K in the interior of $\mathcal{M}(G)$, show there exists $L > 0$ such that if γ is a closed geodesic of length not exceeding L, then γ lies in the complement of K and is parallel to a simple loop in a boundary component of $\mathcal{M}(G)$.

Prove that the manifold obtained by putting a hyperbolic structure on the result of drilling out a Myers curve (see Theorem 6.9.3) cannot appear as a geometric limit on $\partial\mathfrak{T}(G)$ [Soma 2007].

6-5. *Geometrically finite manifolds with a single, incompressible boundary component.* We will sketch a construction of Jonas Gaster [2012]. Let M be, say, a genus-2 handlebody. On ∂M, construct two disjoint simple loops such that their complement on ∂M is a 4-holed sphere; the two curves are "disk-busting" (Exercise (5-26)). Let P be the union of two thin, disjoint annuli about the curves. Move the annuli a little inside M while fixing their boundaries on ∂M. We can interpret the result as a manifold M^* pared by these two cylinders. M^* is now boundary incompressible and acylindrical.

Hyperbolize M^* by applying Theorem 6.3.2. The result is a geometrically finite manifold $\mathcal{M}(G)$ with connected incompressible boundary that is topologically a 4-punctured sphere; the two curves on ∂M become pinched.

The region of discontinuity $\Omega(G)$ has infinitely many simply connected components.

One can now "skin" the resulting manifold. For an example with a geodesic boundary, see Gaster [2012].

Example 2. This construction was motivated by the explicit constructions of Hidalgo and Maskit [2006]. Let S denote the boundary of, say, a genus three handlebody $\mathcal{M}(G)$. We will find 6 incompressible in $\mathcal{M}(G)$, mutually disjoint, simple, homotopically independent loops on S that divide S into 4 pants.

Choose 3 of the loops around the 3 "holes". Choose 2 loops around the first two holes, one on either side of S. Choose the last loop around all three holes.

Now pinch S along these 6 loops to wind up with 4-triply punctured spheres, mutually connected through the punctures. The interior of the manifold will remain homeomorphic to the interior of a handlebody, even though its boundary is now incompressible. The regular set $\Omega(G)$ will be the union of infinitely many round disks.

6-6. *Knottedness.* What should it mean that a simple closed geodesic γ in $\mathcal{M}(G)$ be *unknotted*? In $\mathbb{S}^3$, a simple loop γ is unknotted if it is isotopic to circle, or, if it is isotopic to a simple loop on a sphere.

The solid cusp tubes associated with rank one parabolics are "unlinked" in a similar sense. For they are mutually disjoint, and any one of them can be shrunk towards a cusp (by using smaller and smaller horoballs) without bumping any other. Likewise the core curves in the solid tori obtained by Dehn filling on rank two cusps are unknotted, at least when the resulting manifold is close enough to the cusp.

Thus one definition might be that a simple loop $\gamma \in \mathcal{M}(G)$ is unknotted if a thin tubular neighborhood is compressible in its exterior.

Another definition might be as follows. Consider the countable set in $\mathbb{H}^3$ of all lifts of γ. Call γ *unknotted* if any geodesic γ^* over γ can be isotoped into $\mathbb{S}^2$ without intersecting any other lift. This property is used in Gabai et al. [2003].

Finally Otal defined unknottedness by the following property: γ is *unknotted* if the simple loop γ lies in, or is isotopic to, a simple loop that lies in an incompressible closed surface S properly embedded in the interior of $\mathcal{M}(G)$. If γ satisfies Otal's definition, it also satisfies the previous definition. For each lift S^* of S to $\mathbb{H}^3$ is a topological plane separating $\mathbb{H}^3$ and different lifts are mutually disjoint. Therefore in principle, γ^* can moved slightly off S^* and then homotoped into $\mathbb{S}^2$ without hitting any other lift of S.

In any case Otal's has been a fruitful definition, as illustrated by the following theorem:

Unknottedness Theorem I 6.9.7 [Otal 1995]. *Suppose there is a diffeomorphism* $\Phi : \mathcal{M}(G)^{\mathrm{int}} \to S \times \mathbb{R}$. *There exists a constant* $0 < c = c(S)$ *such that any simple closed geodesic* $\gamma \subset \mathcal{M}(G)$ *of length* $<\ c$ *is isotopic to a curve* γ^* *contained in* $S^* = \Phi^{-1}(S \times \{0\})$. *Moreover the union of all simple closed geodesics* $\{\gamma_i\}$ *in* $\mathcal{M}(G)$ *of length* $<\ c$ *is isotopic to a union of simple loops in disjoint surfaces in* $\Phi^{-1}(S \times \mathbb{Z})$.

Therefore the collection of short curves is not only unknotted, but also unlinked. If $\mathcal{M}(G)$ is geometrically infinite, an infinite number of distinct short geodesics may exit one or both ends. (Such a sequence will exist in $\mathcal{M}(G)$ if and only if there is no positive lower bound for the length of the geodesics.)

Continuing further with this definition, a family of mutually disjoint simple loops, or geodesics $\{\gamma_i\}$, in a manifold $\cong S \times \mathbb{R}$, where S is a surface, can be said to be *unlinked* if there is a homeomorphism $\phi : S \times \mathbb{R} \to S \times \mathbb{R}$ such that $\psi(\cup_i \gamma_i) \subset S \times \mathbb{Z}$. For an example of unlinked geodesics, see Theorem 5.7.3(iii).

Souto found the following generalization for unlinked geodesics in compact compression bodies M^3 (here handlebodies are included) with hyperbolic interiors: As a compression body, there is one boundary component S of M^3 for which the inclusion $\pi_1(S) \hookrightarrow \pi_1(M^3)$ is surjective. A set of simple geodesics $\{\gamma_i\}_{i=1}^n$ in $\mathrm{Int}(M^3)$ is called *unlinked* if there exists a collection of disjoint embedded surface $\{S_i\}$ which (i) are parallel to S, and (ii) $\gamma_i \subset S_i$, $1 \leq i \leq n$ (consistent with the definition above).

Unknottedness Theorem II 6.9.8 [Souto 2008]. *With M^3 as above, there exists $\varepsilon_g > 0$, depending* **only** *on the genus g of S, such that any finite collection of disjoint geodesics $\{\gamma_i\}$ with the property that the length of each γ_i is $< \varepsilon_g$, is unlinked.*

At this point the prudent reader may think that Souto's result is vacuous for handlebodies because there is a positive lower bound for the length of a geodesic in a handlebody which depends only on the length of the shortest geodesic on its boundary. Such a belief is entirely wrong! Take a simple geodesic in the interior, and do higher and higher orders of Dehn surgery about it so that the manifolds nearly have rank two cusps. The length of the core geodesics will become arbitrarily small, independent of the length of the shortest geodesic on $\partial \mathcal{M}(G)$.

In addition, Souto (personal communication) showed that if $\mathcal{M}(G)$ is a closed manifold with a Heegaard splitting surface Σ_g of genus g and $\gamma \subset \mathcal{M}(G)$ is a geodesic of length $< \varepsilon_g$, there is an embedded surface S isotopic to Σ_g such that $\gamma \subset S$. Short geodesics not only are simple, but at least in some cases and perhaps in all, are also unknotted and unlinked.

6-7. *Constructing a cone manifold on an unknotted geodesic: Bromberg's construction.* This process is analogous to Exercise (6-8) but it is harder to implement. Let $\mathcal{M}(G)$ be a quasifuchsian manifold whose bottom end $S = \partial_{\text{bot}}\mathcal{M}(G)$ is a closed surface and whose top is geometrically infinite (and no parabolics). Minsky's paper [2001] covers the case that there is a positive lower bound for the length of all geodesics. So assume there is a sequence of simple loops exiting the infinite end for which the geodesic realizations $\{\gamma_i \in \mathcal{M}(G)\}$ are Otal-unknotted with lengths approaching 0.

Fix an index i. We can assume that $\gamma = \Phi(\sigma)$ lies in $S^* = \Phi(S \times \{t_0\})$, here $\Phi(S \times \{0\}) = \partial \text{bot}(M(G)$ and $t_0 > 0$. Assume for definiteness that γ does not divide S^*. A lift $\hat{\gamma}$ of γ to $\mathbb{H}^3$ determines a cyclic loxodromic subgroup $\langle g \rangle$ of G. The quotient $T = \mathbb{H}^3/\langle g \rangle$ is the interior of a solid torus, and we can designate its core loop again by γ.

The set $C = \Phi(\sigma \times [t, 1))$, $t_0 \leq t < 1$ is a half-open cylinder in the interior of $\mathcal{M}(G)$ which is bounded at one end by $\gamma \subset \partial \text{bot}\mathcal{M}(G)$ while its other end exits the top of $\mathcal{M}(G)$. The cylinder C lifts homeomorphically to give a half-open cylinder $\widetilde{C}$ in the (open) solid torus T. Cut $\mathcal{M}(G)$ along C and T along $\widetilde{C}$. Isometrically glue $T \setminus \widetilde{C}$ to $\mathcal{M}(G) \setminus C$; $\widetilde{C}$ has two sides in T and likewise C in $\mathcal{M}(G)$. After the gluing, as you travel in $\mathcal{M}(G)$ and hit the left side of C you enter T through the right side of $\widetilde{C}$. Then travel around until until you hit the left side of $\widetilde{C}$ and then pass back into $\mathcal{M}(G)$. This will give a hyperbolic cone manifold $\mathcal{M}^*$ with cone axis γ and cone angle 4π, like a two-sheeted branched cover of an arc in the plane. Still $\mathcal{M}^*$ is homeomorphic to $\mathcal{M}(G)$, but now the hyperbolic structure is singular along γ.

Bromberg made the striking discovery that the hyperbolic cone manifold $\mathcal{M}^*$ is a quasifuchsian cone manifold: the bottom end is conformally the same as that of

$\mathcal{M}(G)$ but the top end is now geometrically finite as well. An additional interesting fact that the subgroup $G_0 \cong \pi_1(\mathcal{M}(G) \setminus C)$, being a free loxodromic group, is a Schottky group. Cutting along C has the profound effect on the limit set of totally disconnecting it. Bromberg [2004], and 6.9.5 then proved that the cone angle can be deformed to 2π without changing the conformal structures on the two ends. We end up with a geometrically finite, hyperbolic quasifuchsian manifold.

By applying the result to the sequence of simple loops $\{\sigma_i \subset S \times \{t_i\}\}$ for which the geodesic representatives of $\Phi(\sigma_i)$ shrink to zero, one can complete the proof of Bromberg's Theorem, stated on p. 313.

6-8. *Grafting.* This construction was used by Bill Goldman [Goldman 1987] to describe all real projective structures—those with fuchsian holonomy—over the Teichmüller space $\mathfrak{Teich}(R)$ of a closed Riemann surface R (see Exercise (6-9)). Let G be a fuchsian group acting on the upper (UHP) and lower (LHP) half-planes and representing $R = \text{LHP}/G$. Suppose the negative imaginary axis is the lift $\hat{c}$ of a simple closed geodesic $c \subset R$ of length L.

The element $g \in G$ determined by lifting c into $\hat{c}$ is $g : z \mapsto e^L z$. The quotient $T = (\mathbb{C} \setminus \{0\})/\langle g \rangle$ is a torus. There is a homeomorphic lift of c into T that we will also denote by c. Cut T along c to get a cylinder; the size of the cylinder—ratio of height to circumference—is determined by L. Likewise cut R along c to get one or two subsurfaces. Attach the cylinder $T \setminus c$ to $R \setminus c$; that is pull open the cut of R and insert the cylinder $T \setminus c$ in the cut. This is possible because the ends of the cylinder have length L. We get a new Riemann surface R_c homeomorphic to R with a different conformal structure. Uniformize to get a representation $R_c = \text{LHP}/G_c$.

Such a construction is called 2π-*grafting*.

We will now carry out this operation in $\mathbb{C}$: Slit $\mathbb{S}^2$ along $\hat{c}$ and denote the result by $\mathbb{S}^2_{\text{cut}}$; the cut has $+$ and $-$ edges. Then slit LHP along $\hat{c}$ and denote the result by LHP_{cut}, it has corresponding $+$ and $-$ edges. Attach $\mathbb{S}^2_{\text{cut}}$ to LHP_{cut} by sewing the $-$ and $+$ edge of $\mathbb{S}^2_{\text{cut}}$ to the $+$ and $-$ edge of LHP_{cut}. Likewise sew $\mathbb{S}^2_{\text{cut}}$ equivariantly along all lifts of c to LHP. We get an abstract simply connected surface $\widehat{\mathbb{H}}$ lying over $\mathbb{S}^2$ and G acts on $\widehat{\mathbb{H}}$. The quotient $\widehat{\mathbb{H}}/G$ is conformally equivalent to the Riemann surface R_c we obtained by sewing into R a particular cylinder. Let π denote the projection of $\widehat{\mathbb{H}}$ onto $\mathbb{S}^2$. Under π, the action of G on $\widehat{\mathbb{H}}$ projects to the action of the original fuchsian group G on $\mathbb{S}^2$.

Now the abstract $\widehat{\mathbb{H}}/G$ is conformally equivalent to R_c so there is a conformal map $F : \text{LHP} \equiv \mathbb{H}^2 \to \widehat{\mathbb{H}}$ which conjugates G_c to G. The composition $f = \pi \circ F : \mathbb{H}^2 \to \mathbb{S}^2$ is a locally univalent meromorphic function; G is referred to as its monodromy group. This is an example of a *real projective structure* on R_c or on LHP with respect to G_c. It is called "real" because the monodromy group is a fuchsian group. The components of $f^{-1}(\mathbb{R})$ consist of mutually disjoint arcs which project to simple loops on R_c bounding an annulus (earlier called a cylinder) in the homotopy class of c.

The process of cutting a simple geodesic $c \subset R$ and inserting an annulus is called *grafting*. More particularly we have just done 2π-grafting along c as up in the universal covering we have wrapped the slit sphere once around.

A *multicurve* is a finite geodesic lamination λ, that is, it is a mutually disjoint set of simple geodesics on R. With the assignment of an integral weight $2\pi m_j$ to each component ℓ_j, where m_j is a positive integer, it becomes a measured lamination. Such measured laminations are denoted by $\mathcal{ML}_{\mathbb{Z}}$ or when considered projectively, $\mathcal{PML}_{\mathbb{Z}}$. In the example we have just described c was assigned the weight 2π and we have accordingly grafted R. If the weight were instead $2\pi m$ we would have attached m copies of the cylinder $T \setminus c$ to $c \in R$ and m copies of the slit sphere to $\hat{c}$. Correspondingly, *integral grafting* can be effected by any element of $\mathcal{ML}_{\mathbb{Z}}$. In all cases the construction results in a structure $\widehat{\mathbb{H}}$ acted on by G, a new Riemann surface $R_\lambda = \mathrm{LHP}/G_\lambda$, and a locally injective meromorphic function conjugating G_λ to G.

Once the lamination $\lambda \in \mathcal{ML}_{\mathbb{Z}}$ has been fixed, the integral grafting map can be interpreted as acting on the full deformation (Teichmüller) space, $\mathrm{gr}_\lambda : \mathfrak{Teich}(R) \to \mathfrak{Teich}(R)$. Here are three important theorems about grafting on a closed surface R:

Theorem 6.9.9.

   (i) [Tanigawa 1997] *Fix a simple closed curve γ defined on $\mathfrak{Teich}(R)$. Integral grafting on λ is a real analytic homeomorphism of $\mathfrak{Teich}(R)$ onto itself.*

  (ii) [Scannell and Wolf 2002] *On a fixed surface R, grafting is a homeomorphism of $\mathcal{ML}(R)$ onto $\mathfrak{Teich}(R)$.*

 (iii) [Dumas and Wolf 2008] *Fix a measured lamination λ defined on $\mathfrak{Teich}(R)$. Grafting on λ is then a real analytic diffeomorphism of $\mathfrak{Teich}(R)$ onto itself. This generalizes (i), and replaces the fixed surface in (ii), with a fixed lamination λ in $\mathfrak{Teich}(R)$.*

In particuar for each $\lambda \in \mathcal{ML}_{\mathbb{Z}}$, there is a Riemann surface R_λ such that the λ-grafting on $R_\lambda = \mathbb{H}^2/G_\lambda$ is realized by a locally univalent function f_λ for $R = \mathbb{H}^2/G$. We will pursue this in Exercise (6-9).

Grafting can be defined for nonintegral weights on multicurves, and by continuity for any measured lamination $(\Lambda; \mu)$. As a consequence of Goldman [1987], the more general graftings do not result in fuchsian holonomy.

The construction of each convex core boundary can be interpreted as grafting on the component of $\partial \mathcal{M}(G)$ that it faces (see Epstein et al. [2006] for a discussion).

6-9. *Complex projective structures.* Instead of grafting in Exercise (6-7), complex analytic mappings can be used, as explained below.

Let $R = \mathrm{LHP}/G$ be a closed Riemann surface, and G a fuchsian group. We recall that a Bers slice $\mathcal{B}(R)$ is the collection of *conformal* mappings $f : \mathrm{LHP} \to \mathbb{S}^2$ that have an equivariant *quasiconformal extension* to $\mathbb{S}^2$. The extension, which will also be denoted by f, induces an isomorphism θ to a quasifuchsian group by $f(g(z)) = \theta(g)(f(z))$, for all $z \in \mathbb{S}^2$, $g \in G$. The closure of the Bers slice is the closure of this space of conformal maps on LHP (modulo conjugation).

Add to the mix of conformal maps as follows. Take the much larger class of *locally injective meromorphic functions* $f : \mathrm{LHP} \to \mathbb{S}^2$ for which there exists a *homomorphism* $\varphi : G \to \mathrm{PSL}(2, \mathbb{C})$ satisfying $f(g(z)) = \varphi(g)(f(z))$ for all $z \in \mathrm{LHP}$, $g \in G$. Locally injective meromorphic functions form the solution class of *schwarzian differential equations* on LHP over R (see Exercise (1-37)),

$$S_\phi(f_\phi)(z) = \left(\frac{f_\phi''}{f_\phi'}\right)' - \frac{1}{2}\left(\frac{f_\phi''}{f_\phi'}\right)^2 = \phi(z), \quad z \in \text{LHP}.$$

Here ϕ is a lift of a holomorphic quadratic differential on R, namely it satisfies

$$\phi(g(z))g'(z)^2 = \phi(z), \quad \forall z \in \text{LHP}, \ g \in G.$$

Solutions are uniquely determined by ϕ up to postcomposition with Möbius transformations so solutions can be appropriately normalized. The schwarzian derivative (the differential operator) is zero if and only if f is Möbius. A solution will in general map LHP *onto* $\mathbb{S}^2$.

On a *fixed* Riemann surface R there is the correspondence,

$$[f_\phi] \leftrightarrow \phi \leftrightarrow [\varphi_\phi],$$

where the brackets indicate $\text{PSL}(2, \mathbb{C})$ equivalence.

In current terminology, $f_\phi : R \to \mathbb{S}^2$ is called the *developing map*; it "unrolls" R over $\mathbb{S}^2$. The homomorphism φ associated with the differential ϕ or f_ϕ is called the *holonomy representation*. The *holonomy groups* or *monodromy groups* $\{\varphi(G)\}$ are in general not discrete or even finite presentable, but they are nonelementary. For a general introduction see for example Gallo et al. [2000]. For another slant on projective structures relating to circle packings in $\mathbb{S}^2$, see Kojima et al. [2006].

The collection of all projective structures on the *fixed* Riemann surface $R = \mathbb{H}^2/G$ (for definiteness we continue to think of $\mathbb{H}^2$ as LHP) is called the *extended Bers slice* $\mathcal{B}^*(R)$. It is parameterized by the $(3g-3)$-complex dimensional vector space of quadratic differentials on R. The extended slice $\mathcal{B}^*(R)$ is properly embedded in the representation variety $\mathfrak{R}(G)$ [Gallo et al. 2000]. We can ask about the components of its *quasifuchsian locus*

$$\mathcal{B}^*_{\text{qf}}(R) = \{\varphi_\phi : \varphi_\phi \text{ is an isomorphism to a quasifuchsian group}\} \subset \mathcal{B}^*(R)$$

This is an open set in $\mathcal{B}^*(R)$. The component containing the basepoint (R, id) is the Bers slice.

Suppose we are given an element $(\lambda, \mu) \in \mathcal{ML}_{\mathbb{Z}}$, that is a multicurve λ with integral weights. There exists a uniquely determined Riemann surface $S^* = \mathbb{H}^2/G^*$ such that integral grafting on S^* results in the initial Riemann surface R: The grafting map sends $\mathbb{H}^2$ over S^* onto a simply connected surface $\widehat{\mathbb{H}}$ lying over $\mathbb{S}^2$ on which G^* acts resulting in $\widehat{\mathbb{H}}/G^*$ being conformally equivalent to R, as we saw in Exercise (6-8). The bottom line is that for each multicurve λ and assignment of integral weights giving an element of $\mathcal{ML}_{\mathbb{Z}}$, there is a projective structure f_ϕ on R such that the homomorphism φ corresponding to $f_\phi = \pi \circ F_\phi$ is an isomorphism $G \to G^*$ while F_ϕ is a conformal map of $\mathbb{H}^2$ onto $\widehat{\mathbb{H}}$, conjugating the action of G to G^*. The picture is filled out by the following important result, the second part of which has been studied and visualized by David Dumas as in Figure 6.6.

Fig. 6.6. A red Bers slice for the once-punctured hexagonal torus in the archipelago of blue components of $\mathfrak{Q}_{\mathrm{disc}}(G)$ with their fuchsian centers indicated by x. See Fig. 5.5 for a picture of the Bers slice itself.

Theorem 6.9.10 [Shiga and Tanigawa 1999; Dumas 2009]. *Each component of the discreteness locus of $\mathcal{B}^*(R)$ consists of quasifuchsian groups, and is biholomorphically equivalent to the Bers slice $\mathcal{B}(R)$. The component $\mathcal{B}_\lambda$ is indexed by its fuchsian center $\{c_\lambda = (R, \varphi_\phi)\}$. This point is determined by the quadratic differential ϕ on R with fuchsian monodromy group $\varphi_\phi(G) = G^*$ with G^* determined by that surface $S^* = \mathbb{H}^2/G^*$ on which λ-grafting yields R.*

Each component of $\mathcal{B}^*(R) \cap \mathfrak{T}(G)$ is a "generalized Bers slice" consisting of certain deformations of a projective structure on R.

The theorem applies to finite area surfaces R more generally. When there are punctures the projective structures must be taken so that $\varphi(g)$ is parabolic whenever $g \in G$ is so.

In the case of a once-punctured torus, the one-dimensional space of quadratic differentials for which the holonomy map preserves the parabolics, can be given explicitly. The first computation and visualization of the resulting Bers slice was carried out by the Japan team (Komori, Sugawa, Wada, Yamashita) in terms of complex probabilities. The result was dramatic. More recently they have computed the extended Bers slice. Using his own software, David Dumas has computed slices in addition for the hexagonal torus (Figure 6.6) and more generally has broadened their explorations. See Komori and Sugawa [2004]; Wada [2006]; Dumas [2009], and Figure 6.6.

We can also consider the totality of all projective structures on *all* Riemann surfaces in the deformation space $\mathfrak{Teich}(R) \equiv \mathfrak{Teich}(G)$, $R = \mathbb{H}^2/G$. This is given by the $(6g - 6)$-complex dimensional bundle of quadratic differentials $\mathfrak{Q}(G)$ over $\mathfrak{Teich}(G)$ (equivalently, we may write $\mathfrak{Q}(R)$ over $\mathfrak{Teich}(R)$). The solution of the schwarzian equation for ϕ on $R' = \mathbb{H}^2/G'$ gives rise to the holonomy representation $\varphi_\phi \in \mathfrak{R}(G)$ onto the holonomy group $\varphi_\phi(G)$ (as usual, quotienting out by conjugations). According to Gallo et al. [2000] the totality of monodromy groups $\{\varphi_\phi(G)\}$, $\phi \in \mathfrak{Q}(G)$ comprise the component $\mathfrak{R}_+(G)$ of $\mathfrak{R}(G)$ consisting of those representations that lift to $SL(2, \mathbb{C})$: $\varphi(G)$ lifts if each generator can be assigned a matrix so that the designated matrices satisfy the surface relation satisfied by $\pi_1(R)$. Representations that so lift comprise one of the two components of $\mathfrak{R}(G)$.

The surjective holomorphic map $\mathrm{Hol} : \mathfrak{Q}(G) \to \mathfrak{R}_+(G)$ is a local homeomorphism, but it is not a covering mapping—closed arcs do not necessarily lift in their entirety [Hejhal 1975]. In Hejhal's examples this occurs because continuation over the path leads to a pinching of the underlying Riemann surfaces, which occurs before continuation is complete.

In the present context, the *discreteness locus* is defined as the closed set

$$\mathfrak{Q}_{\mathrm{disc}}(G) = \{\phi \in \mathfrak{Q}(G) : \varphi_\phi \text{ is an isomorphism to a discrete group}\}.$$

It is analogous to the discreteness locus $\mathfrak{R}_{\mathrm{disc}}(G)$ considered in Section 5.2.2. Its interior $\mathrm{Int}(\mathfrak{Q}_{\mathrm{disc}}(G))$ consists of quasifuchsian groups. Hejhal [1975] proved that Hol is injective on the components of $\mathrm{Int}(\mathfrak{Q}_{\mathrm{disc}}(G))$

The extended Bers slice directly involves (in our setup) only LHP. The action of nonzero integral grafting sends the extended Bers slice based on $R = \mathrm{LHP}/G$ to the extended slice based on the surface $S_\lambda = \mathrm{LHP}/G_\lambda$.

Shinpei Baba showed that the method developed in Gallo et al. [2000] can be extended so that it can be used to construct *any* complex projective structure on a surface S (without solving all possible schwarzian equations): Suppose S is a closed, orientable genus $g \geq 2$ surface and C is a complex projective structure on S. Then there exists a decomposition of S into pants and cylinders such that the restriction of C to each component has an injective developing map and a discrete, faithful holonomy map. By sewing back together all the pieces, one recovers both S and the projective structure C on S [Baba 2010].

The recent paper [Baba $\geq$ 2015] includes the proof of the following, responding to an old question for a closed surface S. Suppose two generic (purely loxodromic) complex projective structures on S have the *same* holonomy ρ. Then one can be transformed to the other by 2π-graftings, or by their inverses.

6-10. *Local connectedness of* $\overline{\mathfrak{T}(G)}$. For the definition see p. 286. An associated question is *self-bumping*: Suppose X is a component of $\mathfrak{R}_{\mathrm{disc}}(G)$. For X to bump itself at $\zeta \subset \partial X$ means that for all small enough neighborhoods $U \subset \mathfrak{R}(G)$ of ζ, $U \cap \mathrm{Int}(X)$ is not connected. For X to *bump* at ζ means that there is some component $Y \neq X$ of $\mathfrak{R}_{\mathrm{disc}}(G)$ such that $\zeta \in \overline{X} \cap \overline{Y} \in \partial \mathfrak{R}_{\mathrm{disc}}(G)$

Making use of the technique of Anderson and Canary [1996a], the prototypical case of a fuchsian group G representing a closed surface $R = \mathbb{H}^2/G$ was analyzed by Curt McMullen. (For a more general case see Theorem 5.3.2.)

Theorem 6.9.11 [McMullen 1998, Appendix]. *There exists a cusp ζ on the boundary of quasifuchsian space* $\mathfrak{T}(G)$ *such that for all small neighborhoods $U \subset \mathfrak{R}(G)$ of ζ,* $U \cap \mathfrak{T}(G)$ *is not connected.*

The closure $\overline{\mathfrak{T}(G)} \subset \mathfrak{R}(G)$ *is not a manifold with boundary.*

The theorem does *not* say that the closure $\overline{\mathfrak{T}(G)}$ is not locally connected. However, Bromberg [2011] has proved that $\overline{\mathfrak{T}(G)}$ is *not* locally connected in the once-punctured torus quasifuchsian case. In contrast, the boundary of a Bers slice of the once-punctured space is known to be locally connected (Minsky).

Because of its application in future developments and its intrinsic interest, we will outline Bromberg's argument in the next exercise. But first, here are the recent results of Brock et al. [2011] concerning this issue. (For item (iv) see the Density Theorem p. 305.)

(i) *If $\mathcal{M}(G)$ is compact, and $\zeta = \mathcal{M}(G^*) \in \partial \mathfrak{R}_{\mathrm{disc}}(G)$ is such that $\partial \mathcal{M}(G^*)$ is a union of triply punctured spheres, then no two components of $\mathfrak{R}_{\mathrm{disc}}(G)$ bump at ζ.*

(ii) *If in addition $\mathcal{M}(G)$ is either acylindrical or homeomorphic to $S \times [0, 1]$ for a closed hyperbolic surface S, then there is no self-bumping at ζ. In fact, $\mathfrak{R}_{\mathrm{disc}}(G)$ is locally connected at ζ.*

(iii) *If more generally, $\mathcal{M}(G)$ is compact with incompressible boundary, and $\zeta = \mathcal{M}(G^*) \in \partial \mathfrak{R}_{\mathrm{disc}}(G)$ has no rank one cusps, then $\partial \mathfrak{R}_{\mathrm{disc}}$ is locally connected at ζ.*

(iv) *If neither self-bumping nor bumping occurs at $\zeta \in \partial \mathfrak{R}_{\mathrm{disc}}(G)$, then the boundary $\partial \mathfrak{R}_{\mathrm{disc}}(G)$ is locally connected at ζ.*

6-11. *Bromberg's proof of nonlocal connectedness of* $\overline{\mathfrak{T}(G)}$. The proof involves two interesting, more general constructions.

The first construction, called "wrapping", results in an immersion of the Riemann surface R which is not homotopic to an embedding. The construction originates in Anderson and Canary [1996a]—see Exercise (5-16).

We will start by constructing the target manifold. Let d be a simple geodesic on say $\partial_{\mathrm{bot}}\mathcal{M}(G)$. Inside $\mathcal{M}(G) \cong R \times [0, 1]$, set $\delta = d \times \{\frac{1}{2}\}$. The manifold $\mathcal{M}(G) \setminus \{\delta\}$ is represented by a geometrically finite $\mathcal{M}(H)$. For an explicit construction see

Exercise (4-28). We can take $\partial\mathcal{M}(H)$ to be conformally equivalent to the bottom and top components of $\partial\mathcal{M}(G)$.

Suppose for purposes of explaining the wrapping, $R = \mathbb{H}^2/G$ is a closed surface of genus two. Let $a_1, b_2; a_2, b_2$ be simple loops about each of the two handles so that,

$$\pi_1(R) \cong \langle a_1, b_2, a_2, b_2 \; : \; [a_1, b_1] = [a_2, b_2]\rangle.$$

Choose $d \subset R$ to be a simple loop that divides R into two subsurfaces of genus one. Set $\delta = d \times \{\frac{1}{2}\}$, a parallel simple loop inside $\mathcal{M}(G)$. $\mathcal{M}(G) \setminus \delta$ has a hyperbolic structure $\mathcal{M}(H)$—see Exercise (4-28). Referring to a thin torus T_0 about δ with simple loops d' parallel to d and meridian c, we find that H has the presentation

$$H \cong \langle a_1, b_2, a_2, b_2, c, d : d = [a_1, b_1] = [a_2, b_2], \; [c, d] = 1\rangle.$$

"Wrap" a closed surface homeomorphic to R about δ as follows. Start with the surface $R_1 = R \times \{3/4\}$ containing $\delta' = d \times \{\frac{3}{4}\}$, parallel to δ. Choose a solid torus in $\mathcal{M}(H)$ with core curve δ bounded by a torus T_0 of such a size that T_0 is "tangent" to R_1 along δ' but is otherwise disjoint from R_1. Join together $R_1 \setminus \delta'$ and $T_0 \setminus \delta'$ by cross identification over δ'. Note that $T_0 \setminus \delta'$ is a cylinder. The result is an immersed surface S in $\mathcal{M}(H)$, homeomorphic to R_1, and wrapping once around δ. The subgroup $\pi_1(S) \subset H$ is generated by $\langle a_1, b_1, ca_2c^{-1}, cb_2c^{-1}\rangle$.

The immersion is not homotopic in $\mathcal{M}(H)$ to an embedding, for $\pi_1(S)$ is not conjugate in H to the fundamental group of either component of $\partial\mathcal{M}(H)$.

We digress to make two remarks.

(i) The wrapping of R_1 about δ can be done any integral number of times, and can be done for any collection of mutually disjoint simple loops $\{d\}$ on R.

(ii) There are intrinsic restrictions on wrapping, which are hidden under our assumption that the top and bottom of $\mathcal{M}(G)$ are closed surfaces. Suppose more generally they are finitely punctured surfaces. First of all, $d_\pm$ on the top and bottom of $\mathcal{M}(G)$ cannot be homotoped on $\partial\mathcal{M}(G)$ to a puncture. Secondly, if $\mathcal{M}(G) \setminus \delta$ is to have a hyperbolic structure, no loop which has nonzero geometric intersection with one of $d_\pm$ can determine a parabolic transformation. Thus if $\mathcal{M}(G)$ were, for example, a maximal cusp group on $\mathfrak{T}(G)$, the construction would be impossible.

Continuing, let $\varphi : G \to G^* \subset H$ be the isomorphism induced by the immersion S. We deduce

(i) $\mathcal{M}(G^*)$ covers $\mathcal{M}(H)$.

(ii) $\mathrm{Int}\mathcal{M}(G^*) \cong R \times (0, 1)$.

(iii) One component of $\partial\mathcal{M}(G^*)$ is a closed surface homeomorphic to R and the other end is the union of two surfaces sharing a parabolic resulting from pinching.

(iv) $\varphi : G \to G^* \in \partial\mathfrak{T}(G)$.

The group $G^* \in \partial\mathfrak{T}(G)$ is a cusp. As such it is the limit of quasifuchsian groups in the slice. Specifically, there is a sequence of isomorphisms $\{\varphi_n : G \to G_n\} \in \mathfrak{T}(G)$ which converge algebraically and geometrically to $\varphi : G \to G^*$. In particular,

$\lim \Omega(G_n) = \Omega(G^*)$. The group H also arises as the geometric limit of a sequence of Dehn twists about δ, while G^* arises by pinching δ.

The second construction is the application of $(1, n)$ Dehn filling on the rank two cusp of $\mathcal{M}(H)$, where 1 corresponds to the meridian c, and n to the longitude d. Take the cusp torus, remove its interior, and replace it by a solid torus so that $\gamma = cd^n$ becomes the meridian, that is, $\gamma = cd^n \sim 1$. For all large n there results a hyperbolic manifold $\mathcal{M}(H_n)$ which is a quasifuchsian manifold homeomorphic to $\mathcal{M}(G)$. Carried along is the inclusion

$$S \subset \mathcal{M}(H) \xrightarrow{\text{immersion}} \mathcal{M}(H_n),$$

which is homotopic to an embedding now that δ is no longer there. It induces the algebraically converging isomorphisms

$$\{\rho_n : G \to H_n\} \xrightarrow{n \to \infty} \varphi : G \to G^*\}.$$

On the other hand, $\{\mathcal{M}(H_n)\}$ converges geometrically back to $\mathcal{M}(H)$ (Exercise (5-14)). According to Theorem 4.5.4, $\lim \Omega(H_n) = \Omega(H)$.

To complete the argument we must draw on the theory of projective structures $\mathfrak{Q}(R)$ as outlined in Exercise (6-9).

The representation $\varphi : G \to G^*$ in $\partial \mathfrak{T}(G)$ is the holonomy representation of some $\phi \in \mathfrak{Q}_{\text{disc}}(G)$. Let U be any small enough neighborhood of ϕ in $\mathfrak{Q}(G)$. For U small, the holonomy representation

$$\text{Hol} : U \mapsto V = \text{Hol}(U) \subset \mathfrak{R}(G)$$

is a homeomorphism onto a neighborhood V of $(\varphi, G^*) \subset \mathfrak{R}(G)$.

Set $U_d = U \cap \mathfrak{Q}_{\text{disc}}(G)$. There exists $\{\phi_n\} \subset U_d$ such that its holonomy is $\varphi_n : G \to G_n$. There also exists $\{\phi'_n\} \subset U_d$ with holonomy $\rho_n : G \to H_n$. Both sequences converge to ϕ. Now $\lim \Omega(G_n) = \Omega(G^*)$ while $\lim \Omega(H_n) = \Omega(H)$. Because $\Omega(G^*)$ and $\Omega(H)$ have no component in common it follows from looking at the corresponding developing mappings that $\{\phi_n\}$ and $\{\phi'_n\}$ cannot lie in the same component of U_d. Therefore $\{(\varphi_n, G_n)\}$ and $\{(\rho_n, H_n)\}$ do not lie in the same component of $\text{Hol}(U_d) = V \cap \mathfrak{T}(G)$. Yet both sequences converge to $(\varphi, G^*) \in \partial \mathfrak{T}(G)$. It follows as a consequence of Theorem 5.15.14 that the closure $\overline{\mathfrak{T}(G)}$ is not a manifold.

6-12. *Ito's expansion of Theorem 6.9.11.* Kentaro Ito [2000] made a detailed study of the situation and proved (compare with Theorem 6.9.10):

Theorem 6.9.12. *The components of* $\text{Int}(\mathfrak{Q}_{\text{disc}}(G))$ *are in one-to-one correspondence with the elements of* $\mathcal{ML}_{\mathbb{Z}}$. *Denote the component of the interior corresponding to* $\lambda \in \mathcal{ML}_{\mathbb{Z}}$ *by* $\mathfrak{Q}_\lambda$; $\mathfrak{Q}_0 = \mathfrak{T}(G)$. *The component* $\mathfrak{Q}_\lambda$ *has a uniquely determined fuchsian center* c_λ: *There is a surface* $S_\lambda = \mathbb{H}^2/G_\lambda$ *on which* λ-*grafting determines a projective structure on* R *with monodromy group* G_λ. $\mathfrak{Q}_\lambda$ *is biholomorphically equivalent to quasifuchsian space* $\mathfrak{T}(G) = \mathfrak{Q}_0$.

Choose any $\lambda, \mu \in \mathcal{ML}_{\mathbb{Z}}$.

(i) While $\text{Hol} : \mathfrak{Q}_\lambda \to \mathfrak{R}(G)$ is injective [Hejhal 1975], it is not injective on the closure $\overline{\mathfrak{Q}_\lambda}$ unless $\lambda = 0$.

(ii) For the closures in $\mathfrak{Q}(G)$, $\overline{\mathfrak{Q}}_\lambda \cap \overline{\mathfrak{Q}}_\mu \neq \varnothing$; in particular $\overline{\mathfrak{Q}}_\lambda \cap \overline{\mathfrak{Q}}_0 \neq \varnothing$.

(iii) There exists $\zeta \in \partial\mathfrak{Q}_0 \cap \partial\mathfrak{Q}_\lambda$ such that for all small neighborhoods U of ζ, $U \cap \mathfrak{Q}_\lambda$ is not connected; $\overline{\mathfrak{Q}}_\lambda$ is not a manifold.

(iv) The closed set $\mathfrak{Q}_{\text{disc}}(G)$ is connected.

Another way of identifying $\mathfrak{Q}_\lambda$ is that it contains the fuchsian center c_λ.

6-13. *Meromorphic functions and laminations.* This construction is due to Thurston; a proof appears in Kamishima and Tan [1992]. Suppose $f : \mathbb{D} \to \mathbb{S}^2$ is a locally injective meromorphic function in the unit disk $\mathbb{D} \equiv \mathbb{H}^2$. Consider round disks $\{D \subset \mathbb{S}^2\}$ with the property that there is a single valued branch of $f^{-1} : D \to \mathbb{D}$. We may assume all such disks $\mathcal{D} = \{D\}$ are maximal in the sense none is contained in a larger disk on which f^{-1} has a branch.

Consider the set $\mathcal{U} = \{U = f^{-1}(D) : D \in \mathcal{D}\}$. Set $U^\infty = \overline{U} \cap \partial\mathbb{D}$. Each U^∞ contains at least two points. Construct the hyperbolic convex hull $\mathcal{CH}(U^\infty)$ in $\mathbb{H}^2$; it is either a geodesic or a hyperbolic polygon.

The following properties hold for $\mathcal{U}$:

(i) Corresponding to each point $z \in \mathbb{D}$ is a unique element $U_z \in \mathcal{U}$ with $\mathcal{CH}(U_z^\infty)$.

(ii) Two hulls $\mathcal{CH}(U_1^\infty)$, $\mathcal{CH}(U_2^\infty)$ are either disjoint, or they share a common edge and/or vertex on $\partial\mathbb{D}$.

(iii) The collection of hulls $\{\mathcal{CH}(U^\infty)\}$ covers $\mathbb{D}$ without overlapping interiors.

(iv) Suppose $f \circ g(z) = \varphi(g) \circ f(z)$, (i) for a homomorphism $\varphi : G \to \text{PSL}(2, \mathbb{C})$, (ii) for all elements g of a fuchsian group G, and (iii) for all $z \in \mathbb{D}$. Then the action of G permutes the elements of $\{\mathcal{CH}(U^\infty)\}$ while the action of $\varphi(G)$ permutes the maximal disks in $\mathcal{D}$.

In this analysis, it is helpful to examine for a given $z \in \mathbb{D}$, the set $W_z = \bigcup_{z \in U} U$, the union of those elements of $\mathcal{U}$ that contain z. Then set $W_z^\infty = \overline{W}_z \cap \partial\mathbb{D}$. The image set $f(W_z^\infty) \subset \mathbb{S}^2$ is well defined and we can pass to its convex hull $\mathcal{CH}(fW_z^\infty)$, now taken with respect to $\mathbb{H}^3$. There is a "closest" point $r(f(z)) \in \mathcal{CH}(fW_z^\infty)$ to $f(z)$, where r denotes the nearest point retraction. Construct the hyperbolic plane $P_{f(z)}$ which is orthogonal to the segment $[f(z), r(f(z))]$ at the point $r(f(z))$. The boundary of $P_{f(z)}$ on $\partial\mathbb{S}^2$ is a circle, and one of the disks $D_{f(z)}$ that it bounds on $\mathbb{S}^2$ contains $f(z)$. In fact $D_{f(z)} \in \mathcal{D}$ and $f^{-1}(D_{f(z)}) \in \mathcal{U}$. The map $\Psi : \mathcal{CH}(U_z^\infty) \subset \mathbb{H}^2 \to \mathcal{CH}(fW_z^\infty)$ sending u to $r(f(u))$ is an isometry; more generally the map $z \in \mathbb{D} \to r(f(z)) \in \mathbb{H}^3$ determines an isometry to a pleated surface in $\mathbb{H}^3$.

The set of all bending lines of all the convex hulls in $\mathbb{H}^2$ form a geodesic lamination Λ. It will be invariant under the group G, if there is one. There is a naturally determined bending measure on this lamination. Namely if for $U_1, U_2 \in \mathcal{U}$, $U_1 \cap U_2$ intersect with exterior angle $0 < \alpha < \pi$, then so do the image disks in $\mathcal{D}$, since f is locally a conformal mapping. The corresponding $\mathcal{CH}(U_1^\infty)$, $\mathcal{CH}(U_2^\infty)$ share an edge. We assign the bending angle α to that. However the geodesics in Λ are unlikely to be isolated and then a process akin to Riemann integration is used to obtain the bending measure.

Thurston's insight was that this construction, disseminated to the world by Bill Goldman and proved in Kamishima and Tan [1992], results in a coordinate system for the projective space $\mathfrak{Q}(G)$, where the fuchsian group G can represent either a closed or a finite area surface:

Theorem 6.9.13. *The construction described above results in a homeomorphism* Θ : $\mathfrak{Q}(G) \rightarrow \mathfrak{T}(G) \times \mathcal{ML}(G)$.

That the map is surjective is a consequence of the fact that grafting on a Riemann surface R' gives rise to a projective structure on another Riemann surface S', as described in Exercise (6-8).

Theorem 6.9.13 shows that corresponding to each Riemann surface $R' \in \mathfrak{Teich}(R)$ and projective structure on it is a uniquely determined Riemann surface S_λ and measured lamination λ on S_λ with the following property: Grafting on S_λ determines the given structure on R'. However we can no longer be restricted to integral grafting. That is, $\mathfrak{Q}(R) \cong \mathfrak{Teich}(R) \times \mathcal{ML}(R)$. These are called the *Thurston coordinates*. Using the Thurston coordinates, David Dumas has proved that one can compactify the space so that $\overline{\mathfrak{Q}(R)} \cong \mathcal{PML}(R) \times \mathcal{PML}(R)$. See his exposition [Dumas 2009] and the important paper [Dumas 2007] showing the relationship between quadratic differentials an measured laminations.

6-14. *Hyperbolic manifolds with corners.* This exercise was inspired by Otal [1998, §§7,8], Kapovich [2001, Ch. 13, 19], to which the reader is referred for more detail and for application to the proof of the Hyperbolization Theorem. The process is often referred to as Thurston's *orbifold or reflection trick*.

Start with a compact, acylindrical manifold $M = \mathcal{M}(X)$ with nonempty boundary. Let $\{R_i\}$ be a decomposition of ∂M into compact, (connected) subsurfaces with mutually disjoint interiors. That is, each R_i is a compact bordered surface, or an entire boundary component of M. But we require that R_i is not a topological annulus or disk.

Associated with the decomposition is the decomposition graph $\mathcal{G}$: Each vertex corresponds to an R_i, and two vertices are joined by an edge if the subsurfaces are adjacent along a common border. We will call the graph an *admissible decomposition graph* and the corresponding $\{R_i\}$ an *admissible decomposition*, if we can give each R_i a label "+" or "−" in such a way that if R_i and R_j share a boundary component, they have different labels. Thus every simple loop in $\mathcal{G}$ is composed of an even number of edges.

Suppose then we have an admissible decomposition of ∂M. Denote the subsurfaces with the label "+" by $(\Sigma_1, \Sigma_2, \ldots)$, and the subsurfaces with the "−" label, namely the complementary subsurfaces, by $(\Sigma_1', \Sigma_2', \ldots)$.

The idea is to "reflect" M over the b subsurfaces $\{\Sigma_i\}$. Thus if $b = 1$ we "reflect" M across Σ and get a manifold whose boundary consists of two copies of Σ' with opposite orientations attached along their common boundary.

To carry out this scheme in general we will follow the construction of Exercise (6-2). Order the b-subsurfaces $\Sigma_1, \ldots, \Sigma_b$. Make a pile of 2^b-copies of $\cup_{i=1}^b \Sigma_i$, each layer with the attached manifold M. In each layer assign either "+" or "−" to each of $\{\Sigma_i\}$ in such a way that different layers have different assigments. If two layers have assignments differing only in Σ_k, then join the two layers by "reflecting" M in Σ_k. When this is completed we will end up with a connected manifold $\widehat{M}_{\Sigma'}$ that is 2^b sheeted over M. The boundary components of $\widehat{M}_{\Sigma'}$ will consist of 2^b copies of each of the surfaces Σ_i'. Again $\widehat{M}_{\Sigma'}$ is an orientable, irreducible, acylindrical, atoroidal, Haken manifold.

There will be a group of orientation-reversing automorphisms G of order 2^b acting on $\widehat{M}_{\Sigma'}$ generated by the "reflections" in the elements of $\{\Sigma_i\}$.

Realize $\widehat{M}_{\Sigma'}$ as a hyperbolic manifold $\mathcal{M}(Y)$; we may assume that its boundary is totally geodesic. The "reflections" we introduced in the surfaces $\{\Sigma_i\}$ now become orientation-reversing isometries of $\mathcal{M}(Y)$; each pointwise fixes a totally geodesic subsurface, a copy of some Σ_i. Each boundary contour of this copy of Σ_i is contained in $\partial \mathcal{M}(Y)$. Along this contour Σ_i is orthogonal to $\partial \mathcal{M}(Y)$.

The quotient $\mathcal{M}(Z) = \mathcal{M}(Y)/G$ is homeomorphic to the result of cutting along the surfaces $\{\Sigma_i\}$ which are orthogonal to $\partial \mathcal{M}(Y)$. Thus $\mathcal{M}(Z)$ is homeomorphic to the original M.

Our construction can be summarized as follows.

Hyperbolic Manifolds with Corners 6.9.14. *Given M as above, there is a uniquely determined kleinian group Z with the following properties:*

(i) *$\mathcal{M}(Z)$ is compact and $\partial \mathcal{M}(Z)$ is the union of two systems of subsurfaces $\{\Sigma_i\}$, and $\{\Sigma_i'\}$. None of the subsurfaces are simply or doubly connected. The totality of interiors are mutually disjoint. The closures of the members of each system are mutually disjoint.*

(ii) *The ordinary set $\Omega(Z)$ is the union of round disks $\{D_i\}$. Either $D_i \cap D_j = \varnothing$ for all $j \neq i$, or D_i is orthogonal to D_j for some $j \neq i$. In the former case, D_i covers a boundary component of $\mathcal{M}(Z)$ which is a member of one of the systems. In the latter case, D_i and D_j contain lifts of subsurfaces of different systems, and the stabilizer of $D_i \cap D_j$ is the cyclic group determined by a common boundary component.*

Note the symmetry in properties between the two systems of subsurfaces. An interesting special case is when the elements of the systems comprise a pants decomposition.

This construction is then used for the hyperbolic assembly of pieces required to build a Haken manifold. A better approach is illustrated in Example 4-17.

6-15. *Infinitely generated kleinian groups.* Riemann surfaces of infinite genus and/or an infinite number of ends can be represented by fuchsian groups. Each such fuchsian group has in turn a quasiconformal deformation space. The elementary combination theorems can be used to paste together such groups over round disks or horodisks to construct a range of infinitely generated (non-finitely-generated) groups, just as in the finitely generated case. Likewise infinitely generated Schottky groups can be constructed.

Is there a classification of infinitely generated kleinian groups?

Here is an interesting example of Bromberg and Souto (private communication).

Start with a closed hyperbolic manifold M with a nondividing incompressible surface $S \subset M$. Set $N = M \setminus S$. Assume that N does not have the form $S \times (0, 1)$. Then $\pi_1(S)$ is a proper subgroup of $\pi_1(N)$. We may choose the hyperbolic structure on N so as to have totally geodesic boundary components S_-, $S_+ \cong S$ (it will not be fuchsian).

Set $N_0 = N$. Form the hyperbolic manifold N_1 by reflecting N_0 across S_+. In other words take two copies of N_0 and glue the top boundary component S_+ of N_0 to the bottom S_- of the copy. Correspondingly we can form N_{-1}. Let N_k denote the hyperbolic manifold formed from N_0 by successively gluing together the $2k+1$ copies of N_0: glue to N_0 k copies in the positive and in the negative direction.

N_k is a hyperbolic manifold whose two boundary components are conformally equivalent to S_-, S_+. We have that $\pi_1(N_k)$ is a proper subgroup of $\pi_1(N_{k+1})$. Normalize the representations $N_k = \mathcal{M}(G_k)$ so that $G_k \subset G_{k+1}$, $k = 1, 2, \ldots$. Then $\{G_k\}$ converges algebraically and geometrically to a infinitely generated group H with $\mathcal{M}(H) \ncong S \times \mathbb{R}$. Its limit set is all $\mathbb{S}^2$.

Bromberg and Souto show as a consequence of the finite area of the boundary components of the approximates $\{\mathcal{M}(G_k)\}$ that $\mathcal{M}(H)$ is quasiconformally rigid: any quasiconformal conjugation of H to another group is Möbius. They then show that if M^* is another hyperbolic manifold that is homeomorphic to $M(H)$, there is a bilipschitz map between them. By quasiconformal rigidity the two manifolds are in fact isometric.

More generally, they conjecture that any manifold $\mathcal{M}(H)$ which can be exhausted by tame manifolds whose boundaries have uniformly bounded areas is either rigid or it has a tame end. Soma's and Ohshika's work (page 323) gives a multitude of infinitely generated examples.

Long ago, Agol conjectured that any irreducible, atoroidal, orientable 3-manifold with infinitely generated fundamental group is hyperbolic provided the covering corresponding to any nonelementary finitely generated subgroup is tame. Now prove it.

Thurston (MathOverflow 11) explained why there exists an infinitely generated kleinian group whose limit has Hausdorff dimension two, yet every finitely generated subgroup has Hausdorff dimension less than a preassigned $\varepsilon > 0$. Work out the details.

6-16. Suppose R is a closed surface of genus $g \geq 2$ and $\gamma \subset R$ is a closed geodesic with multiple self-intersections. Show that there is a regular, finite-sheeted covering surface $\widetilde{R}$ for which some lift of γ is a simple closed curve. This is analogous to the goal of cubulation in Section 6.4.2; see also the summary of properties on p. 114. Compare with Exercise (3-45).

6-17. *Splitting distance.* Suppose M^3 is an orientable, closed manifold, and $S \subset M^3$ is a Heegaard splitting of M^3 (see Section 2.7). The surface S divides M^3 into two handlebodies H_1, H_2. Correspondingly there are two sets C_1, C_2, each of $(k + 1)$ mutually disjoint simple curves on S, with those in C_i compressible (that is, bounding mutually disjoint disks) in H_i, $i = 1, 2$. The *splitting distance* $d(C_1, C_2)$ of S is the distance in the curve complex $\mathcal{C}(S)$ (see Exercise (5-17)) between the between the k-dimensional simplexes determined by C_1 and C_2. Alternately, the *splitting distance* $d(x, y)$ can be taken as the minimum distance in $\mathcal{C}(S)$ between points $x \in C_1$ and $y \in C_2$.

Theorem 6.9.15 ([Hempel 2001], Perelman). *If for some splitting surface $S \subset M^3$ the splitting distance satisfies $d(C_1, C_2) > 2$, then M^3 carries a hyperbolic metric.*

Theorem 6.9.16 ([Hartshorn 2002]). *If a closed manifold $\mathcal{M}(G)$ contains an incompressible surface of genus g, any Heegaard splitting of $\mathcal{M}(G)$ has distance at most 2g.*

6-18. *Random Heegaard splitting.* This is to report on interesting papers of Joseph Maher [2010] and Dunfield and Thurston [2006].

Given two handlebodies of the same genus, let $\varphi : S \to S'$ denote a homeomorphism between the two bounding surfaces. Glue the two surfaces together, using φ as a gluing map, so as to form a closed manifold $M^3(\varphi)$, which depends only on the homotopy type of φ.

Now random closed 3-manifolds can be generated as follows.

Consider the Cayley graph $\mathcal{G}$ of the mapping class group $\mathcal{MCG}(S)$ of S. This is a metric space; it is not δ-hyperbolic. Fix a set of generators $\{g_i\}_{i=1}^k$ of $\mathcal{MCG}(S)$. Do a random walk on this Cayley graph, starting at the origin. At each vertex randomly choose the next vertex, and so on. At the n-th vertex the walk can be described as a word of n-letters in the generators: $w_n = w_{n-1} \cdots w_0$.

Corresponding to w_n is an automorphism of S, the composition of $(n - 1)$ prior automorphisms. We can as well interpret w_n as a homeomorphism of S onto S'. Denote the corresponding 3-manifold by $M^3(w_n)$.

It is shown in Maher [2010] that the splitting distances of the random Heegaard splittings grow linearly as $n \to \infty$. This implies that the probability that a random Heegaard splitting yields a hyperbolic manifold tends to one as $n \to \infty$. It also follows that the volume of the hyperbolic manifolds created grows linearly with the length of the walk.

There is a much studied subgroup of $\mathfrak{M}(S)$ called the *Torelli group* characterized by the property that it acts as the identity on the first homology group: every cycle is homologous to its image. One can equally take a random walk in the Cayley graph of this group thereby getting a sequence of homeomorphisms all of which fix the elements of the homology group. If these elements are applied to the standard Heegaard splitting of $\mathbb{S}^3$ (think $\mathbb{R}^3$) there results a sequence of homology spheres, which are also hyperbolic with asymptotic probability one [Maher 2010].

The typical splitting by a closed surface S of genus g has the properties: (i) there is a homology basis of simple loops $\{\alpha_i, \beta_i\}_{i=1}^g$ such that α_i crosses β_i, once and the i-pair is disjoint from the j-pairs, $j \neq i$, and (ii) α_i bounds a disk on one side of S, while β_j bounds a disk on the other side. So all of $\{\alpha_i, \beta_j\}$ are homologous to zero in the 3-manifold.

An earlier application of the same construction of random manifolds was carried out by Dunfield and Thurston [2006]. They studied the Virtual Haken Conjecture for closed, irreducible 3-manifolds with infinite fundamental groups. In this, they searched for finite covering manifolds with particular properties. In particular they show that for a certain type of abelian covers of manifolds of Heegaard genus 2, the probability of their having infinite first homology group is 0.

6-19. *Twisted face parings* We will briefly describe the surprising method of Cannon et al. [2002] to build "random" 3-manifolds.

The analogy is to a simple two-dimensional topological construction. Start with a polygon with an even number of sides. Pair these sides in any manner. Then identify the sides of each pair by an orientation-reversing map so as to build an orientable, closed two-dimensional surface. We must be free to adjust the vertex angles so that after identification, the angles at each vertex add to 2π.

The same method can be attempted for face pairing of polyhedra, where each face paring identifies polygons the same number of edges by an orientation-reversing map. Gluing together the edges yields a closed three-dimensional "pseudomanifold". The problem is that after gluing some or all of the vertices may not be manifold points. The authors develop a systematic method to correct this problem by edge subdivision and modifying the face pairing transformations by "twisting". The end result is that, by a specific, relatively simple process, an infinite family of closed, orientable 3-manifolds is constructed.

7

Line geometry

This chapter describes an elegant method that makes it easier to quantitatively analyze geometric situations in hyperbolic space that involve lines and planes. The approach below here was developed by Troels Jørgensen.

7.1 Half-rotations

We will identify each line $\ell \in \mathbb{H}^3$ with the *half-rotation* about ℓ, that is, the elliptic transformation of order 2 having ℓ as its axis.

The Cayley–Hamilton identity satisfied by normalized matrices A is

$$A + A^{-1} = \tau_A I, \quad \text{or} \quad A^2 = \tau_A A - I. \tag{7.1}$$

As always, τ_A is the trace and I is the 2×2 identity matrix. In particular, $\tau_{(A^2)} = (\tau_A)^2 - 2$.

Lemma 7.1.1. *Half-rotations correspond to normalized matrices A of trace $\tau_A = 0$ (eigenvalues $\pm i$), or equivalently, matrices A that satisfy*

$$A^2 = -I. \tag{7.2}$$

Consider the half-rotation $M = \left(\begin{smallmatrix} a & b \\ c & -a \end{smallmatrix} \right)$, with $-a^2 - bc = 1$ and $c \neq 0$. Its fixed points are $P = (a + i)/c$ and $Q = (a - i)/c$. Thus we can write

$$M = \frac{i}{(P - Q)} \begin{pmatrix} P + Q & -2PQ \\ 2 & -(P + Q) \end{pmatrix}. \tag{7.3}$$

If P or Q is ∞, take the corresponding limit of the expression.

The determinant of the matrix, $-(P - Q)^2$, does not have a uniquely determined root. Correspondingly, there is no way to distinguish between the two fixed points P, Q.

We are reminded of the useful Lemma 2.1.3:

Lemma 7.1.2. *A Möbius transformation T interchanges two distinct points x, y on $\mathbb{S}^2$ if and only if it is a half-rotation. If so, T is the half-rotation about a line ℓ orthogonal to the line τ between x and y.*

Proof. We may assume τ is the vertical axis in the upper half-space model. The map T maps τ onto itself, interchanging its endpoints. Therefore T has a fixed point $p \in \tau$. T^2 fixes the two endpoints and also fixes p so it must be the identity. Let P be the plane through p orthogonal to τ. T also maps P onto itself, but interchanges its two sides in $\mathbb{H}^3$. Therefore the rotation axis ℓ of T lies in P and is necessarily orthogonal to τ at p.

Conversely, suppose T is the half-rotation about ℓ. Taking ℓ to be the vertical conjugates T to the map $z \mapsto -z$. It interchanges the opposite points on every concentric circle about $z = 0$. $\qquad\square$

7.2 The Lie product

The *Lie product* φ of two nonsingular 2×2 matrices A, B is defined as

$$\varphi = AB - BA := \{A, B\}.$$

Interchanging A and B changes φ to $-\varphi$ but leave the corresponding Möbius transformation unchanged. Likewise the Möbius transformation corresponding to the matrix φ is independent of the sign chosen for A or B.

We have the relations

$$U\{A, B\}U^{-1} = \{UAU^{-1}, UBU^{-1}\} \tag{7.4}$$

and

$$\det \varphi = 2 - \operatorname{tr}(ABA^{-1}B^{-1}). \tag{7.5}$$

The second one follows from the identity $\varphi = (ABA^{-1}B^{-1} - I)BA$ and the formula $\det(X - I) = 2 - \tau_X$. If $\det \varphi \neq 0$, since φ has zero trace,

$$\varphi^2 = -(\det \varphi)^2 I.$$

Of course, interpreted as a Möbius transformation, $\varphi^2 = \operatorname{id}$.

In preparation for the next result, note that two distinct lines in $\mathbb{H}^3$ not having a common endpoint always have a common perpendicular. This can be seen in the upper half-space model by taking one of the lines as the vertical axis. If the two lines intersect in $\mathbb{H}^3$, the common perpendicular is the line through the point of intersection and orthogonal to the plane containing the two lines.

Proposition 7.2.1. *The Lie product $\varphi = \{A, B\}$ has the following properties:*

(1) $\varphi = 0$ *if and only if either*

- *A and/or B is kI for some $k \in \mathbb{C}$, or*
- *A and B correspond to transformations with the same set of fixed points on $\partial\mathbb{H}^3$.*

More generally, φ is singular ($\det \varphi = 0$) if and only if $\operatorname{tr}(ABA^{-1}B^{-1}) = +2$. In other words, φ is singular if and only if A and B have at least one fixed point in common on $\partial\mathbb{H}^3$.

(2)
$$\varphi A^{-1} = A\varphi, \quad \varphi B^{-1} = B\varphi. \tag{7.6}$$

(3) *If φ is nonsingular, it corresponds to a half-rotation. Its axis is:*

- *the common perpendicular to the axes of A and B, if neither is parabolic;*
- *the line between the fixed points if both A and B are parabolic, or, if only one is parabolic, the line from the parabolic fixed point which is orthogonal to the axis of the other;*
- *the line between the diametrically opposite points $\mathcal{C} \pm i/c$ on the isometric circle of φ with center $\mathcal{C}$ and radius $|c|^{-1}$, where c is the lower left term in the normalized matrix for φ.*

(4) *If $\tau_A \neq \pm 2$, then $\psi = A - A^{-1}$ is the half-rotation about the axis of A.*

(5) *Suppose A and B are half-rotations. Then AB has the same axis as φ. If the axes of A and B intersect at a point $x \in \mathbb{H}^3$, then AB is necessarily elliptic and fixes x as well. If all three A, B, AB are half-rotations, then the three rotation axes have a common point of intersection in $\mathbb{H}^3$ and are mutually orthogonal there.*

(6) *If B is a half-rotation and A is not, the axis of B is orthogonal to the axis of A, or ends at the fixed point of A if A is parabolic, if and only if AB is a half-rotation. If this is the case, the axis of AB is also orthogonal to that of A, or ends at the fixed point of A if A is parabolic.*

(7) *The axes of A, B, C have a common perpendicular if and only if,*

$$\tau_{ABC} = \tau_{CBA}. \tag{7.7}$$

If A, B, C are all half-rotations, the condition for a common perpendicular becomes

$$\tau_{ABC} = 0. \tag{7.8}$$

(8) *If $aA + bB + cC = 0$ for nonzero scalars a, b, c while the matrices A, B, C represent nonparabolic elements, then the axes of the transformations corresponding to A, B, and C have a common perpendicular.*

In the special case that A and B preserve the upper half-plane UHP, they also preserve the vertical half-plane H in $\mathbb{H}^3$ based on $\mathbb{R}$. Their axes lie in H. If their axes intersect, then the axis of φ is orthogonal to H, and passes through the point of intersection; φ itself also preserves UHP and H. If instead the axes of A and B are disjoint, then the axis of φ lies in H as well and φ interchanges the upper and lower half-planes.

Example 7.2.2. Let $\mathcal{P}$ denote a regular hyperbolic octagon in $\mathbb{H}^2$ with vertex angles $\pi/4$. Going around $\partial\mathcal{P}$ in its positive direction, label its edges a, b, c, d, a^{-1}, b^{-1}, c^{-1}, d^{-1}. Let A denote the Möbius transformation that maps a to a^{-1}, sending $\mathcal{P}$ to the right side of a^{-1}, and similarly find transformations B, C, D. The group $G = \langle A, B, C, D \rangle$ is fuchsian and represents a genus 2 surface. Its generators satisfy the relation $ABCDA^{-1}B^{-1}C^{-1}D^{-1} = $ id, or $ABCD = DCBA$. Therefore ABC is conjugate to CBA and BCD to DCB. By property (7), the axes of A, B, C and of B, C, D have common perpendiculars in $\mathbb{H}^2$.

Example 7.2.3. Suppose that at the level of Möbius transformations, W is a word in the letters A, B with the property that negating the exponents of all the letters $(A \mapsto A^{-1}, B \mapsto B^{-1})$ changes W to W^{-1}. Then $\varphi W \varphi = W^{-1}$, in other words φ interchanges the fixed points of W, if W is loxodromic. The axis of W is orthogonal to the axis of φ. If A, B generate a fuchsian group, then the axes of A, B and W in $\mathbb{H}^2$ necessarily intersect at a fixed point of φ. Conclude with Jørgensen [1978] that on any hyperbolic Riemann surface, a point x which is at the intersection of two closed geodesics, or is at the intersection of a closed geodesic with itself, is at the intersection of infinitely many distinct closed geodesics.

Proof of Proposition 7.2.1. (1) This is verified by a direct matrix computation. One may assume that A is either a diagonal matrix (elliptic or loxodromic), or one with a zero in the lower left entry and trace two (parabolic). The second statement follows from (7.5). See also Lemma 1.5.2.

(2) It follows from (7.1) that

$$(AB - BA) - (BA^{-1} - A^{-1}B) = (A + A^{-1})B - B(A + A^{-1}) = 0.$$

Similarly,

$$(AB - BA) - (B^{-1}A - AB^{-1}) = A(B + B^{-1}) - (B + B^{-1})A = 0.$$

Consequently,

$$\varphi = AB - BA = BA^{-1} - A^{-1}B = B^{-1}A - AB^{-1}.$$

Since

$$(AB - BA)A^{-1} = A(BA^{-1} - A^{-1}B), \quad (AB - BA)B^{-1} = B(B^{-1}A - AB^{-1}),$$

Equations (7.6) follow.

In the remainder of the proofs, we have to be careful when switching between matrices and Möbius transformations.

(3) Since its trace is zero, φ is a half-rotation. The relations (7.6) show that φ interchanges the fixed points or fixes the fixed point of A (and B) according to whether there are two or one. In the former case, the rotation axis of φ is orthogonal to the axis of A. In the latter case, the fixed point is an endpoint of the axis of φ. In all cases, there is only one line in $\mathbb{H}^3$ with the properties of the axis of φ. The last statement follows from (7.3).

(4) We have $\det \psi = 4 - \tau_A^2 \neq 0$ since $A \neq \pm I$ is not parabolic. Now ψ has zero trace and $\{\psi, A\} = 0$.

(5) If A and B are themselves half-rotations and hence equal to their inverses, (7.6) implies that $\{\varphi, AB\} = 0$. Hence by (1), the fixed points of AB on $\partial\mathbb{H}^3$ are the same as those of φ. If the axes of A and B are known to intersect at a point $x \in \mathbb{H}^3$,

AB necessarily fixes x as well. Thus AB is elliptic with the same rotation axis as φ. If in addition AB is a half-rotation, then from (7.1), $AB = -BA$ so $\varphi = 2AB$. Correspondingly, $\{B, AB\} = 2A$ and $\{AB, A\} = 2B$. The three axes of A, B, AB intersect mutually orthogonally at x.

(6) If A is not parabolic, its matrix is conjugate to a diagonal matrix. The matrix for B has the form

$$\begin{pmatrix} a & b \\ c & -a \end{pmatrix}, \quad \text{with} - a^2 - bc = 1.$$

The axis of B is orthogonal to that of A if and only if $a = 0$. This is exactly the condition that $\tau_{AB} = 0$. In this case, the axis of AB is also orthogonal to that of A.

If A is parabolic, then $\tau_{AB} = 0$ if and only if the fixed point of A is an endpoint of the axis of B. In this case AB is a half-rotation fixing the fixed point of A.

(7) From (7.1),

$$\tau_{ABC} I - \tau_{BAC} I = (AB - BA)C + C^{-1}(B^{-1}A^{-1} - A^{-1}B^{-1}).$$

Moreover,

$$\tau_{AB} I = AB + B^{-1}A^{-1} = \tau_{BA} = BA + A^{-1}B^{-1},$$

and therefore,

$$B^{-1}A^{-1} - A^{-1}B^{-1} = BA - AB = -\varphi.$$

Consequently,

$$\varphi C - C^{-1}\varphi = (\tau_{ABC} - \tau_{CBA})I.$$

When $\tau_{ABC} = \tau_{CBA}$, the axis of φ, which is already known to be orthogonal to that of A and B, is also orthogonal to the axis of C, since $\varphi C \varphi^{-1} = C^{-1}$. Conversely, the only line orthogonal to the axis of both A and B is the axis of φ. If that is also orthogonal to the axis of C, then $\varphi C \varphi^{-1} = C^{-1}$ and $\tau_{ABC} = \tau_{CBA}$.

Equation (7.7) is satisfied if there is a linear relation between A, B, C.

Finally, if all three of A, B, C are half-rotations,

$$(ABC)^{-1} = C^{-1}B^{-1}A^{-1} = -CBA.$$

Therefore,

$$\tau_{ABC} - \tau_{CBA} = 2\tau_{ABC},$$

and $\tau_{ABC} = \tau_{CBA}$ if and only if $\tau_{ABC} = 0$, that is, if and only if ABC is a half-rotation. □

7.3 Square roots

The normalized matrix B is a *square root* of the normalized matrix A if $B^2 = A$. We will use the notation $B = \sqrt{A}$ or $B = A^{1/2}$ with the understanding that the roots are determined only up to the factor ± 1.

Lemma 7.3.1. *We have:*

$$\sqrt{A} = \pm\frac{A+I}{\sqrt{2+\tau_A}} \quad \text{if } \tau_A \neq -2,$$

$$\sqrt{-A} = \pm\frac{A-I}{\sqrt{2-\tau_A}} \quad \text{if } \tau_A \neq 2.$$

The square roots of $-I$ are the normalized matrices with zero trace.

At the level of Möbius transformations, if $A \neq$ id is not parabolic it has two square roots $A+I$ and $A-I$. If $\sqrt{A}$ denotes one of them, the other has the form $\boldsymbol{\alpha}\sqrt{A}$ where $\boldsymbol{\alpha}$ is the half-rotation about the axis of A.

If A is parabolic, at the level of Möbius transformations A has one root, namely either $A+I$ or $A-I$ depending on whether τ_A is $+2$ or -2. It is parabolic as well.

Proof. It follows from (7.1) that $(A\pm I)^2 = (\tau_A\pm 2)A$. It is also true that $\det(A\pm I) = 2 \pm \text{tr}_A$. In terms of normalized matrices,

$$\left(\frac{A+I}{\sqrt{2+\tau_A}}\right)^2 = A, \quad \left(\frac{A-I}{\sqrt{2-\tau_A}}\right)^2 = -A,$$

where one or the other formula holds if $\tau_A = \pm 2$. If $\tau = \tau_A \neq \pm 2$, again using (7.1),

$$\frac{A \pm I}{\sqrt{2 \pm \tau}} = \mp \frac{A - A^{-1}}{\sqrt{4 - \tau^2}} \frac{A \mp I}{\sqrt{2 \mp \tau}}.$$

Now interpret the matrices as the corresponding Möbius transformations. In view of the equation above and Proposition 7.2.1(4), application of the half-rotation ψ about the axis of A sends one root of A to the other.

Finally suppose $B^2 = -I$. Because $\text{tr}(B^2) = -2$ we may conjugate the matrix B so as to have the form $\begin{pmatrix} a & b \\ 0 & d \end{pmatrix}$, with $ad = 1$ and $a^2 + d^2 = -2$. These two equations

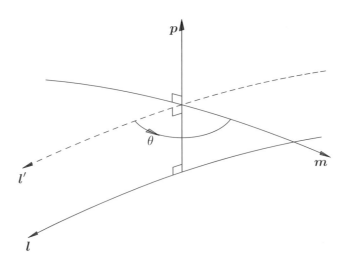

Fig. 7.1. Complex distance. The lines l and l' are coplanar. θ is the angle through which one must rotate l' counterclockwise, when looking along direction of the p.

imply that $\tau_B = 0$. We already know that any normalized matrix B of zero trace has the property that $B^2 = -I$. Conversely, any matrix of trace zero is a half-rotation. □

As an application we display the matrix formula

$$\{A, B\} = \sqrt{2 - \tau_K}\sqrt{-K}\, BA, \quad K = ABA^{-1}B^{-1}. \tag{7.9}$$

7.4 Complex distance

If $T : z \mapsto ke^{i\theta}z$, $k > 1$, the term *complex length* $\mathcal{L}$ of the resulting geodesic in $\mathbb{H}^3/\langle T \rangle$ refers to the number

$$\mathcal{L} = \log k + i\theta \quad \mathrm{mod}(2\pi i).$$

In invariant form the formula is

$$\tau_T = 2\cosh\frac{\mathcal{L}}{2} \quad \mathrm{mod}(\pm 1). \tag{7.10}$$

Let l, m be two *oriented* (hyperbolic) lines in $\mathbb{H}^3$ which do not have a common endpoint. The two lines have a unique common perpendicular line p. Orient p. The *complex distance* between l and m is the number

$$\chi(l, m) = d(l \cap p, m \cap p) + i\theta \quad (\mathrm{mod}\, 2\pi i).$$

Here d denotes the signed hyperbolic distance between $l \cap p$ and $m \cap p$; it is positive if the segment from $l \cap p$ to $m \cap p$ runs in the positive direction with respect to the positive direction along p. The distance is 0 if the lines intersect.

Thus in the upper half plane model $d(i, 2i) = \log 2 = -d(2i, i)$ if the vertical is oriented toward ∞. If the lines l and m are disjoint, the sign of $\mathrm{Re}\,\chi(l, m)$ is positive or negative depending on the orientation of the common perpendicular p. If the lines intersect, $\mathrm{Re}\,\chi(l, m) = 0$ and the orientation of p has no effect.

The angle θ is determined as follows. Assume first that the lines don't intersect. In the plane σ spanned by l and p, let $m' \subset \sigma$ denote the line through $m \cap p$, orthogonal to p, and oriented parallel to l. As sighted along the ray of p from $l \cap p$, let θ be the angle of *clockwise* rotation necessary to rotate the *positive* ray of m' onto the *positive ray* of m.

Changing the order (l, m) to (m, l) changes the angle from θ to $2\pi - \theta$. It also reverses the sign of the distance, the orientation of p being fixed. Therefore

$$\chi(m, l) = -\chi(l, m) \quad (\mathrm{mod}\, 2\pi i)$$

If l^-, m^- denote l, m with the opposite orientations,

$$\chi(l^-, m) = \chi(l, m^-) = \chi(l, m) + \pi i \quad (\mathrm{mod}\, 2\pi i).$$

The two lines l, m lie in a plane if and only if θ is 0 or π, mod 2π.

If the two lines intersect the angle determination is easier. Just look in the positive direction along p. and rotate m clockwise until its positive direction coincides with the positive direction of l.

Complex distance can also be expressed in terms of a cross ratio. Let l be oriented from endpoint r to endpoint s and m from u to v. Then

$$\cosh^2\left(\frac{\chi(l, m)}{2}\right) = (r, u, v, s).$$

It suffices to confirm this formula for two lines in the upper half-space model that are orthogonal to the vertical axis.

We see that $\exp \chi(l, m) = ke^{i\theta}$ is a continuous function of the triple of oriented lines (l, p, m), where l and m do not intersect and p is orthogonal to l and m.

7.5 Complex distance and line geometry

From Section 7.2 we know that the common perpendicular p to lines l, m is the axis of the half-rotation corresponding to the Lie product $\varphi = \{L, M\}$ of the half-rotations L, M about the lines. The line p is also the axis of ML. In fact, we see that the point $m \cap p$ is the midpoint of the segment of p between $l \cap p$ and $ML(l) \cap p$. Therefore

$$\chi(l, m) = \frac{1}{2}\big(d(l \cap p, ML(l \cap p)) + 2i\theta\big) = \frac{1}{2}\chi(l, ML(l)) \pmod{2\pi i},$$

where d is measured with respect to the positive direction along p.

Lemma 7.5.1. *The normalized matrices*

$$M_1 = \frac{ML + I}{\sqrt{\tau_{ML} + 2}}, \quad M_2 = \frac{ML - I}{\sqrt{\tau_{ML} - 2}}$$

correspond to the two transformations with axis p, related by a half-rotation, that send l onto m.

Proof. By Proposition 7.2.1(1), as Möbius transformations, M_1 and M_2 have the same pair of fixed points and therefore the same axis. Lemma 7.3.1 shows that $M_1^2 = M_2^2 = ML$. The three transformations ML, M_1, M_2 thus share the line p as axis. The orientation of p toward the attracting fixed point of ML agrees with its orientation from $l \cap p$ to $m \cap p$. Both M_1 and M_2 send l to m and $l \cap p$ to $m \cap p$, but they give m opposite orientations. □

We also record the following fact.

Lemma 7.5.2. *Given the axis p of a loxodromic transformation A and any line m orthogonal to p, there is a uniquely determined line l also orthogonal to p such that $A = ML$.*

Proof. Orient p toward the attracting fixed point of A and orient m arbitrarily. Set $y = m \cap p$ and find $x \in p$ such that the distance along p from x to y equals the distance from y to Ax. Find l orthogonal to p at x and orient it so that $\chi(l, m) = \frac{1}{2}\chi(l, Al)$.

Note that A automatically has the symmetries $A^{-1} = MAM$ and $A^{-1} = LAL$. □

Consider next the relation of the complex distance $\chi(l, m)$ to the trace of a loxodromic $A = ML$. Let λ, λ^{-1} denote the eigenvalues of A with $|\lambda| > |\lambda^{-1}|$ and $\operatorname{Re} \lambda \geq 0$. The Möbius transformation corresponding to A is conjugate to $z \mapsto \lambda^2 z$ and the complex distance along the axis of A, oriented towards the attracting fixed point, from a point x to MLx is $2 \log |\lambda| + 2\theta i$, where $\theta = \arg \lambda$, $-\pi/2 < \theta \leq \pi/2$. That is,

$$\theta = \arg \lambda, \quad -\pi/2 < \theta \leq \pi/2,$$
$$\chi(l, m) = \log |\lambda| + \theta i, \quad |\lambda| \geq 1, \tag{7.11}$$
$$\operatorname{tr}(A) = 2 \cosh \chi(l, m), \quad \operatorname{Re} \operatorname{tr}(A) \geq 0.$$

The transformation A as described by (7.11) is independent of the orientations of l and m.

Conversely, given lines l, m and the orthogonal p from l to m, define λ by $\log \lambda = \chi(l, m) \pmod{\pi i}$, which is independent of the orientations of l, m. Then

$$\tau_{ML} = (\lambda + \lambda^{-1}) = \pm 2 \cosh \chi(l, m). \tag{7.12}$$

As a consequence we obtain the two formulas which are independent of the order and orientations of l, m and the orientation of the axis of ML:

$$\tau_{ML}^2 = 4 \cosh^2 \chi(l, m), \tag{7.13}$$

$$\det(ML - LM) = -4 \sinh^2 \chi(l, m). \tag{7.14}$$

7.6 Exercises and explorations

7-1. When does $A^2 = 0$? Is $(AB)^{-1/2} = (B^{-1}A^{-1})^{1/2}$?

7-2. Suppose C and C' are orthogonal circles. Write down the equation of the half-rotation J that exchanges C and C' (for example you may choose the center of C to be 0 and the points of intersection to be $\pm ai$, $a > 0$). Suppose C^* is another circle orthogonal to C such that it and its interior is disjoint from C'. Prove that the radius of $J(C^*)$ is less than that of C^*.

7-3. *Ideal tetrahedra.* We will apply Equation (7.3) to the ideal tetrahedron with vertices at ∞ and 0, $P, Q \in \mathbb{C}$ (see Exercise (1-22)). Let $M_{0,\infty}, M_{PQ}$ denote half-rotations about the edges $[0, \infty], [P, Q]$, respectively. Show that

$$M_{0,\infty} M_{PQ} = \frac{1}{(P - Q)} \begin{pmatrix} -(P + Q) & 2PQ \\ 2 & -(P + Q) \end{pmatrix},$$

and denote its trace by τ. Writing $\tau = \lambda + \lambda^{-1}$, where λ denotes the larger eigenvalue, show that

$$\lambda = -\frac{\sqrt{P} \pm \sqrt{Q}}{\sqrt{P} \mp \sqrt{Q}}.$$

The axis of $M_{0,\infty} M_{PQ}$ is orthogonal to the two lines $[0, \infty]$, $[P, Q]$. Let λ denote the larger eigenvalue ($|\lambda| > 1$). Then

$$\log \lambda = \pm \log \frac{-\sqrt{P} - \sqrt{Q}}{\sqrt{P} - \sqrt{Q}} \pmod{\pi i}$$

is the complex distance between the lines $[0, \infty]$ and $[P, Q]$; compare 7.12. In terms of the cross ratio of the endpoints the distance is

$$\log(-\sqrt{P}, \sqrt{Q}, \sqrt{P}, \infty) \pmod{\pi i}.$$

Find the formulas for the distances between the other two pairs of opposite edges of the tetrahedron.

Check your formulas (or derive them in the first place using this case) by applying them to the case $P = 1$ and $Q = z^2$. Confirm that the eigenvalue $\lambda = \frac{z+1}{z-1}$; the complex distance between the lines is $\log \lambda$ (what is the ambiguity in these formulas?).

7-4. [Jørgensen 2000] For a loxodromic transformation A, (7.1) can be written $A + A^{-1} = (\lambda + \lambda^{-1})I$, where λ denotes the larger eigenvalue. Show that

$$A^k = -f_{k-1}(\lambda)I + f_k(\lambda)A,$$

where the coefficients are the polynomials

$$f_k(\lambda) = \frac{\lambda^k - \lambda^{-k}}{\lambda - \lambda^{-1}}.$$

Next consider the operator P defined as

$$PA := \lim_{k \to +\infty} \left(\frac{A}{\lambda}\right)^k = \frac{A - \lambda^{-1}I}{\lambda - \lambda^{-1}}.$$

The right side is a singular matrix. Also,

$$P(BAB^{-1}) = B(PA)B^{-1}, \qquad (PA)^2 = PN, \qquad PA + PA^{-1} = I,$$
$$(PA)(PA^{-1}) = 0, \qquad A = \lambda PA + \lambda^{-1}PA^{-1}.$$

In particular P has the properties of a projection.

At the level of Möbius transformations, if p_+, p_- denote the attracting and repelling fixed points of A, $\lim A^k(z)/\lambda^k = p_+$ for all $z \neq p_-$. In fact, by first confirming the formula when $p_+ = \infty$, $p_- = 0$, and then using conjugation, PA is the singular matrix

$$PA = \frac{1}{p_+ - p_-} \begin{pmatrix} p_+ & -p_+ p_- \\ 1 & -p_- \end{pmatrix}.$$

7-5. Prove that the transformation corresponding to the matrix B has the same axis as that corresponding to A if and only if $B = bI + aA$ for some scalars a, b. Conclude that all powers $B = A^k$ of A can be so represented.

Suppose the matrices A, B represent half-rotations. Then a matrix C also representing a half-rotation can be represented as $C = aA + bB$ for some scalars a, b if and only if the axis corresponding to C is perpendicular to the common perpendicular $\{A, B\}$ of the axes corresponding to A and B.

7-6. Suppose $G = \langle X, Y \rangle$ is a two-generator group. Suppose some word $W(X, Y)$ in the letters $X^{\pm 1}, Y^{\pm 1}$ satisfies $W(X, Y) = \mathrm{id}$. Then also $W(X^{-1}, Y^{-1}) = \mathrm{id}$. If X and Y are conjugate, that is if they have the same trace, then $W(Y, X) = \mathrm{id}$.

Hint: We have $W(\varphi X \varphi^{-1}, \varphi Y \varphi^{-1}) = \varphi W(X, Y) \varphi^{-1}$, for any Möbius transformation φ. Try $\varphi = \{X, Y\}$. Supose X and Y are conjugate and both are loxodromic or elliptic. Find the midpoint O of the segment of the axis of φ from the point that it crosses the axis of X to the point it crosses the axis of Y. Draw the line ℓ through O and orthogonal to the axis of φ so that the half-rotation φ_1 about ℓ interchanges the axes of X and Y and, if they are loxodromic, sends the attracting fixed point of X to that of Y. Then $\varphi_1 X \varphi_1^{-1} = Y$ and $\varphi_1 Y \varphi_1^{-1} = X$. If both X and Y are parabolic, the axis of φ runs between their fixed point. Find the point O on it such that the half-rotation φ_1 about a line ℓ through O and orthogonal to the axis of φ interchanges the fixed points of X and Y. (The point O satisfies $d(O, XO) = d(O, YO)$.)

Show that every nonelementary two-generator group $G = \langle A, B \rangle$ has an involution $A \mapsto A^{-1}$, $B \mapsto B^{-1}$. This is determined by the common orthogonal to their axes. If $\mathcal{M}(G)$ is a handlebody, the quotient under the involution is the complement of a 3-bridge knot, the singular set being the knot.

In fact, applying tameness and considering the compact core, show that for every (nonelementary) two-generator group G, either $\mathcal{M}(G)$ has finite volume, or G is a free group and $\mathrm{Int}(\mathcal{M}(G))$ is homeomorphic to the interior of a handlebody.

7-7. *A group G is determined by its traces.* Suppose A, B, C are loxodromic without fixed points in common. Show that the two-generator group $\langle A, B \rangle$ is determined up to conjugacy by the traces of $A, B, C = BA$. Then show with Wolpert [1981, Theorem 3.2], Culler and Shalen [1983, Prop. 1.5.3], that any finitely generated irreducible group (Exercise (2-1)) is determined up to conjugacy by the traces of a generating set, traces of pairs of distinct generators, and triples of distinct generators. The generators and all these products may be chosen to represent simple loops. *Hint:* Use the trace identities Lemma 1.5.6. Normalize A to have fixed points $0, \infty$ and B to have 1 as a fixed point. If the group is generated by $A, B, D, E, \ldots$ work in turn with A, D, DA, by temporarily conjugating so that D has fixed point 1, etc. Two parabolics A, B without a common fixed point can be conjugated so that A is the unit translation and B has fixed point at 0. Work out the precise requirements to carry out your argument. Discreteness is not needed for your proof.

When $S = \mathbb{H}^2/G$ is a n-punctured closed surface of genus with $n \geq 1$ and $g \geq 0$, it is shown in Seppälä and Sorvali [1988] both that S and $\mathfrak{T}(S)$ are determined by $6g + 2n - 6$ lengths of simple geodesics. These embed $\mathfrak{T}(G)$ into $\mathbb{R}^{6g+2n-6}$. For closed surfaces S, $\mathfrak{T}(G)$ cannot be embedded in $\mathbb{R}^{6g-6}$ by length functions [Hamenstädt 2003]. Instead, there is such an embedding into $\mathbb{R}^{6g-5}$, see Section 5.11.2 and Hamenstädt [2003] for this and prior work.

Knowing lengths of geodesics is essentially the same as knowing their traces by Equation (7.10).

Usually one works with the *trace field*, that is the squares of the traces. For a finite volume hyperbolic manifold, It is known that it is a finite extension of the rationals (cf. [Bowditch 2006, §5]).

7-8. *The length spectrum of a closed surface.* Let R be a closed surface of genus ≥ 2. The set of lengths of all distinct closed geodesics on R is called the *length spectrum.* Here multiplicity is counted; there may be many distinct geodesics with the same length. The length spectrum is independent of the orientation of R. Knowing all the lengths is equivalent to knowing the absolute values of the traces of all the closed geodesics. The length spectrum and corresponding trace values are functions on the extended moduli space $\mathfrak{Teich}(R)/\mathcal{MCG}^{\pm}(R)$.

It was a longstanding question: Does the length spectrum uniquely determine a point in $\mathfrak{Teich}(R)/\mathcal{MCG}^{\pm}(R)$? The answer is NO. However the point is uniquely determined if it lies in $(\mathfrak{Teich}(R)/\mathcal{MCG}^{\pm}) \setminus (V_g/\mathcal{MCG}^{\pm})$ where V_g is a proper real analytic subvariety of $\mathfrak{Teich}(R)$. This discovery was made by Scott Wolpert [1979].

7-9. *Quaternions again.* In addition to the identity matrix I, introduce the three normalized half-rotation matrices

$$E = \begin{pmatrix} i & 0 \\ 0 & -i \end{pmatrix}, \quad J = \begin{pmatrix} 0 & -i \\ -i & 0 \end{pmatrix}, \quad K = \begin{pmatrix} 0 & 1 \\ -1 & 0 \end{pmatrix},$$

as in Exercise (1-28). These matrices satisfy

$$EJ = K, \quad JK = E, \quad KE = J.$$

Furthermore their rotation axes are mutually orthogonal at their common point of intersection in $\mathbb{H}^3$.

The four matrices I, E, J, K form a basis of the four-dimensional complex vector space of complex 2×2 matrices. For any such matrix X write $X = x_1 I + x_2 E + x_3 J + x_4 K$ and correspondingly for matrices Y and Z. Here is a list of properties with respect to this linear structure:

(i) $\tau_X = 2x_1$ and $\det X = x_1^2 + x_2^2 + x_3^2 + x_4^2$.

(ii) $\varphi = \{X, Y\} = 2 \begin{vmatrix} I & J & K \\ x_2 & x_3 & x_4 \\ y_2 & y_3 & y_4 \end{vmatrix}$, $\operatorname{tr}(\varphi Z) = -4 \begin{vmatrix} z_2 & z_3 & z_4 \\ x_2 & x_3 & x_4 \\ y_2 & y_3 & y_4 \end{vmatrix}$, and

$\operatorname{tr}(XY) = 2(x_1 y_1 - x_2 y_2 - x_3 y_3 - x_4 y_4).$

Assume for the following that X, Y, Z are neither degenerate nor a multiple of I.

(iii) The axes of X and Y are the same if and only if X, Y, I are linearly dependent or, if $\tau_X = \tau_Y = 0$, if and only if X and Y are dependent. If X is parabolic then Y is too and with the same fixed point if and only if X, Y, I are dependent.

(iv) Z and the half-rotation $Z_0 = z_2 E + z_3 J + z_4 K$ have the same axis. Z is parabolic if and only if Z_0 is degenerate. If X and Y are half-rotations, then $\{X, Y\}$ and XY have the same axis.

(v) The axes of X, Y, Z have a common perpendicular if and only if X, Y, Z, I are linearly dependent, or, if $\tau_X = \tau_Y = \tau_Z = 0$, if and only if X, Y, Z are linearly dependent.

Also work out the formulas for XY and YX in terms of I, E, J, K. What is the condition for $XY = YX$?

7-10. Prove that lines $l, m \in \mathbb{H}^3$ lie in the same plane if and only if $\cosh \chi(l, m)$ is a real number.

7-11. Given an ideal tetrahedron (Exercise (1-22)) with edges $\alpha, \beta, \gamma, \delta$, construct the three common perpendiculars between opposite edges. A half-rotation about any one of them maps the tetrahedron onto itself. Consequently the three perpendiculars must meet at a point; the three half-rotations are the nonzero elements of the tetrahedral symmetry group $\mathbb{Z}_2 \otimes \mathbb{Z}_2$.

7-12. *Tubular neighborhood of a systole.* [Gabai et al. 2003] In a hyperbolic manifold suppose there is a shortest geodesic γ of length $l > 0$. Show that γ is necessarily a simple geodesic. Prove that there is an embedded tubular neighborhood (see Exercise (1-4)) about γ of radius at least $l/4$. *Hint:* Suppose the assertion is false. Then the tubular neighborhood of radius $l/4$ intersects itself at a point p. There are at least two perpendicular segments from p to γ. Construct a shorter curve by taking a segment of γ together with the two perpendiculars. The tubular neighborhood about γ is the projection of the tubular neighborhood about any lift $\ell \in \mathbb{H}^3$ of γ.

7-13. *A parametrization of two-generator groups.* Here is a way to parametrize two-generator groups $\langle A, B \rangle$ where A, B are loxodromic without a common fixed point. Let ℓ denote the axis of A. Assume first the two lines $\ell, B^{-1}(\ell)$ do not intersect. Find a line β, orthogonal to the common perpendicular of $\ell, B^{-1}(\ell)$, with the property that the half-rotation β exchanges ℓ and $B^{-1}(\ell)$ and sends the attracting fixed point $p \in \ell$ of A to $B^{-1}(p)$. The transformation $T = B\beta$ maps ℓ onto itself; ℓ is the axis of T and p is its attracting fixed point.

The aim is to parametrize groups $\langle A, B \rangle$ to be uniquely determined up to conjugation by the three complex distances with positive real parts: $\chi = \chi(B^{-1}(\ell), \ell)$, $\log \lambda_A$, and $\log \lambda_T$, where λ_A, λ_T are the larger eigenvalues of A, T, upon arranging matters so that $\operatorname{Re} \chi > 0$.

Show that the parametrization works even if the lines $\ell, B^{-1}(\ell)$ intersect in $\mathbb{H}^3$ and/or B is elliptic.

This parametrization was used in Gabai et al. [2003] to computationally explore hyperbolic manifolds with short geodesics. If $\langle A, B \rangle$ is discrete, and A, B arise from simple geodesics on the quotient, one can find elements $Y \in \langle A, B \rangle$ such that for the pair A, Y, $\operatorname{Re} \chi$ is as small as possible (but not 0). This means that in the quotient manifold, the geodesic γ resulting from projecting ℓ is contained in an embedded tubular neighborhood of radius Re χ and no larger radius. The authors explore the question: Is Re $\chi \geq (\ln 3)/2$ for all discrete groups $\langle A, B \rangle$? They reduce their problem to a study of short geodesics in discrete groups determined by parameters that lie in a certain box in six-dimensional euclidean space. This required a massive computation of about three CPU years.

7-14. Consider half-infinite polygonal arcs L in $\mathbb{H}^3$ made up of closed geodesic segments $\ell_0, \ell_1, \ldots$ with the property that each ℓ_i meets ℓ_{i+1} at 90°. Let $\boldsymbol{\alpha}_i$ denote the complex distance between ℓ_{i-1} and ℓ_{i+1}, for $i = 1, 2, \ldots$.

Given a sequence of numbers $\{\alpha_i\} \subset \mathbb{C}$ and a base point $O \in \mathbb{H}^3$, show that such a polygonal line L from O can be uniquely constructed. Investigate under what circumstances depending on L that the sequence $\{\alpha_i\}$ converges. When is the sequence dense in $\mathbb{H}^3$?

7-15. *The McShane identity and Mirzakhani's asymptotic formula for geodesic lengths.* Greg McShane [1998] made the following remarkable discovery. On any once-punctured torus $\mathbb{T}$,

$$\sum_{\sigma \in \mathcal{S}} \frac{1}{1 + e^{\ell(\sigma)}} = \frac{1}{2},$$

where $\mathcal{S}$ is the collection of simple closed geodesics on $\mathbb{T}$ and $\ell(\sigma)$ is the length of σ.

There is an interesting generalization in McShane [1998]; Bowditch [1996]. Take a quasifuchsian representation $\rho : \pi_1(\mathbb{T}) \to \Gamma$ onto a once-punctured torus group. For each $\sigma \in \mathcal{S}$, let $\ell(\rho(\sigma))$ now denote the complex length of $\rho(\sigma)$ modulo $2\pi i \mathbb{Z}$. Then

$$\sum_{\sigma \in \mathcal{S}} \frac{1}{1 + e^{\ell(\rho(\sigma))}} = \frac{1}{2},$$

where the sum converges absolutely.

Mirzakhani [2008] used McShane's formula to derive (in particular) the following amazing asymptotic formula for a hyperbolic surface X of genus g with n punctures. The mapping class group acts in particular on X. Given a simple closed geodesic $\gamma \subset X$, examine the infinite set of simple closed geodesics on X that are in the orbit $\mathcal{M}(\gamma)$ under the mapping class group $\mathcal{M}$. For given L, denote by $s_X(L; \gamma)$ the number of geodesics in this set with length $\leq L$. Then the following limit exists,

$$n_\gamma(X) := \lim_{L \to \infty} \frac{s_X(L; \gamma)}{L^{6g+2n-6}} \in \mathbb{R}_+.$$

Moreover $n_\gamma(X)$ is a continuous function of X in its moduli space $\mathfrak{M}(X)$.

7-16. *The* $\log(2k-1)$*-Theorem.* Similar appearing to McShane's, but in essence quite different, is a useful formula derived in Anderson et al. [1996]. There it is shown that for a torsion-free, freely generated kleinian group $G = \langle g_1, \ldots, g_k \rangle$, $k \geq 2$, and any point $z \in \mathbb{H}^3$,

$$\sum_{i=1}^{k} \frac{1}{1 + e^{d_i}} \le \frac{1}{2}, \quad d_i = d(z, g_i(z)).$$

In particular, for some index, $d_i \ge \log(2k - 1)$. This is known as the $\log(2k - 1)$-Theorem.

7-17. An orientation-reversing isometry J of $\mathbb{H}^3$ with $J^2 = \mathrm{id}$ uniquely determines a plane P_J such that J is the reflection in P_J, possibly followed by the half-rotation in a line in P_J. The latter case does not arise if it is known that J fixes three distinct points in the circle ∂P_J. In other words, J is conjugate on $\partial \mathbb{H}^3$ to $z \mapsto \bar{z}$ or to $z \mapsto -\bar{z}$; see Exercise (1-31).

As an application, suppose J is an orientation-reversing isometry of $\mathcal{M}(G)$ with $J^2 = \mathrm{id}$. Assume that $S \subset \mathcal{M}(G)$ is a properly embedded, compact, orientable, incompressible surface, possibly with boundary $\partial S \subset \partial \mathcal{M}(G)$, that is neither a disk nor a cylinder. Assume that there is an arc τ from $O \in S$ to $J(O)$ such that for all loops $\gamma \in \pi_1(S; O)$, $\tau^{-1}J(\gamma)\tau$ is homotopic to γ. Show that there is a totally geodesic surface $S^* \subset \mathcal{M}(G)$ that is pointwise fixed by J. Moreover S^* is homotopic to S.

7-18. *Extension of once-punctured torus groups* [Wada 2003]. Prove:

(i) Every quasifuchsian once-punctured torus group $G = \langle A, B \rangle$ with parabolic commutator $K^2 = [A, B]$ has an index two extension $G^* = \langle P, Q, R \rangle$ generated by three half-rotations P, Q, R satisfying $RQP = K$.

(ii) Every quasifuchsian twice-punctured torus group $G = \langle A, B, L, L' \rangle$ with L, L' both parabolic and $[A, B] = L'L$ has an index two extension $G^* = \langle P, Q, R, S \rangle$ generated by four half-rotations P, Q, R, S satisfying $SRQP = L$.

In the first case the quotient orbifold is $(2, 2, 2, \infty)$, while in the second it is $(2, 2, 2, 2, \infty)$. Here we are referring to punctured spheres with the indicated branching.

Hint: For the first case set $Q = AB - BA$ and define $R = AQ$, $P = BQ$. For the second case define Q, R, P by the same formulas, noting that $RQP = K$ where $K^2 = [A, B]$. Then apply Wada's lemma (Exercise (1-34)) to find a half-rotation S with $SLS = L'$.

7-19. *Conformal averaging on* $\mathbb{S}^1$ [Schwartz 2006]. Let $W^{(n)'} = \{w_1, w_2, \ldots, w_n\}$ be $n \ge 4$ distinct, cyclically arranged points on the unit circle $\partial \mathbb{H}^2$. The complementary intervals $\{(w_i, w_{i+1})\}$ will be subdivided as follows. Construct the common orthogonal ℓ to the two lines $[w_i, w_{i+1}]$, $[w_{i-1}, w_{i+2}]$. One of the endpoints w_i' of ℓ lies in the interval (w_i, w_{i+1}). When this is carried out for all intervals we end up with a new cyclically ordered set of distinct points $W^{(n)'} = \{w_1', w_2', \ldots, w_n'\}$ on the circle. Set up the interactive process $\{W^{(k+1)} = (W^{(k)})'\}$. Rich Schwartz proved that $\{W^{(2k)}\}$ converges exponentially fast to an ideal regular n-gon as $k \to \infty$. He interprets this

Fig. 7.2. The limit set of a twice-punctured torus quasifuchsian group computed using Wada's characterization. See Figure 6.2 for a differently constructed representation.

process as "conformal averaging". In the classical situation where w_i' is chosen as the *midpoint* of (w_i, w_{i+1}), the points $W^{(k)}$ become evenly spaced as $k \to \infty$.

7-20. [Brooks and Matelski 1981] Suppose T is a loxodromic with complex translation length δ; that is, if α denotes the axis of T oriented toward its attracting fixed point and ℓ is a line orthogonal to α then $\delta = \chi(\ell, T(\ell))$. Show that

$$\operatorname{tr}^2(T) = 4 \cosh^2 \frac{\chi(\ell, T(\ell))}{2},$$

and for any Möbius transformation S with β the axis of $S_1 = STS^{-1}$,

$$\text{tr}(STS^{-1}T^{-1}) - 2 = \left(1 - \cosh \chi(\ell, T(\ell))\right)\left(1 - \cosh \chi(\alpha, \beta)\right).$$

Their paper exploits that $\langle S, T \rangle$ is discrete only when $\{\cosh(\chi(\alpha, \beta_i))\}$ is a discrete set. Here β_i is the axis of the inductively defined $S_i = S_{i-1}TS_{i-1}^{-1}$. More extensive investigations generalizing Jørgensen's inequality are carried out along this line in [Gehring and Martin 1994].

7-21. Prove the Jacobi identity:

$$\{\{A, B\}, C\} + \{\{B, C\}, A\} + \{\{C, A\}, B\} = 0.$$

For applications to geometry see Ivanov [2011].

7-22. *Symmetry lines.* There are two notions of symmetry lines that have been important in studying the combinatorics of fundamental polyhedra: the first for studying cyclic groups [Jørgensen 1973], the second for studying once-punctured torus quasifuchsian groups [Jørgensen 2003].

Symmetry lines I. Fix a point $\mathcal{O} \in \mathbb{H}^3$ which we will call the basepoint. Given a Möbius transformation g which does not fix $\mathcal{O}$, let β denote the line through $\mathcal{O}$ which is orthogonal to the axis of g, if g not parabolic. If g is parabolic, let β be the line through $\mathcal{O}$ to its fixed point. Next construct the plane e_g which is the perpendicular bisector of the line segment $[\mathcal{O}, g^{-1}(\mathcal{O})]$. In particular, if $\mathcal{O}$ is the origin in the ball model or ∞ in the upper half-space model, $e_g = \mathcal{I}(g)$, the isometric plane.

If instead $\mathcal{O} \in \partial\mathbb{H}^3$, and g does not fix $\mathcal{O}$ construct e_g as follows. There is a unique horosphere $\mathcal{H}_\mathcal{O}$ at $\mathcal{O}$ such that the horosphere $g^{-1}\mathcal{H}_\mathcal{O}$ at $g^{-1}\mathcal{O}$ is externally tangent to $\mathcal{H}_\mathcal{O}$. The line between $\mathcal{O}$ and $g^{-1}\mathcal{O}$ necessarily passes through the point of tangency. (See Lemma 1.5.4.) Take e_g to be the plane through the point of tangency orthogonal to the line between $\mathcal{O}$ and $g^{-1}(\mathcal{O})$.

Recall from Lemma 7.3.1 that if $\text{tr}^2(g) \neq 4$, the square roots of $g^{\pm 1}$ are given as normalized matrices by

$$g^{1/2} = \frac{g \pm I}{\sqrt{2 \pm \tau_g}}, \qquad g^{-1/2} = \frac{g^{-1} \pm I}{\sqrt{2 \pm \tau_g}}.$$

If $\text{tr}^2(g) = 4$ ($g \neq \text{id}$), the roots of $g^{\pm 1}$ are given by the two expressions above that have nonvanishing denominators.

Lemma 7.6.1. *Set* $\alpha = g^{-1/2}(\beta)$. *Then* $g = \beta\alpha$ *and the line* $\alpha \subset e_g$. *Furthermore,* $e_\alpha = e_g$. *Let* β^* *denote the line through* $\mathcal{O}$ *orthogonal to* α. *The plane* e_g *can be alternately characterized as that plane orthogonal to* β^* *at its point of intersection with* α.

Proof. We begin by remarking that when g is not parabolic, the line α is independent of which square root of g is selected, for β is orthogonal to the axis of g.

Assume first that $\mathcal{O} \in \mathbb{H}^3$. In the ball model of $\mathbb{H}^3$, replace g by a conjugate so that $\mathcal{O}$ becomes the origin. Then e_g is the isometric plane $\{\vec{x} : |g'(\vec{x})| = 1\}$. Also $|\beta'(\vec{x})| = 1$ for all $\vec{x} \in \mathbb{H}^3$ since β is now a euclidean rotation about a diameter. It suffices to prove Lemma 7.6.1 under this normalization of $\mathcal{O}$.

Suppose first g is not parabolic. Recall from Lemma 7.5.2 and Lemma 7.5.1 that α can be alternately described as the line orthogonal to the axis of g such that g is the composition of the half-rotations $g = \beta\alpha$; the axis of g is the common perpendicular of α and β. We claim that the axis of α lies in e_g and is orthogonal to the segment $[\mathcal{O}, g^{-1}(\mathcal{O})]$ at its midpoint.

For any $\vec{x} \in \mathbb{H}^3$, $|g'(\vec{x})| = |\alpha'(\vec{x})|$. At any fixed point $\vec{x}$ of α, $|\alpha'(\vec{x})| = 1$. Thus the axis of α lies in the plane e_g. (Another argument is that α maps e_g onto itself, and its action in e_g is conjugate to the action $z \to \bar{z}$ in $\mathbb{C}$ so that $|\alpha'(\vec{x})| = 1$ for all $\vec{x} \in e_g$.)

Now $g^{-1}(\mathcal{O}) = \alpha\beta(\mathcal{O}) = \alpha(\mathcal{O})$. Since α maps the segment $[\mathcal{O}, g^{-1}(\mathcal{O})]$ onto itself switching the endpoints, it fixes the midpoint $\vec{p}$. Therefore $[\mathcal{O}, g^{-1}(\mathcal{O})]$ must be orthogonal to the axis of α at $\vec{p}$. Since $\mathcal{O}$ and $g^{-1/2}(\mathcal{O})$ have the same distance from the axis of g, and since $\alpha = g^{-1/2}(\beta)$, necessarily $\vec{p} = g^{-1/2}(\mathcal{O})$. Finally, e_α is the perpendicular bisector of the segment $[\mathcal{O}, \alpha(\mathcal{O}) = g^{-1}(\mathcal{O})]$ so that $e_\alpha = e_g$.

If instead g is parabolic, the line α also goes through the fixed point of g. It is also true that $g = \beta\alpha$. The rest of the argument is the same. Another way to confirm this is to assume g first is loxodromic but then allow it to converge to a parabolic and follow the geometry.

Upon reviewing the proof we can confirm that there is no essential difference if $\mathcal{O} \in \partial\mathbb{H}^3$. In this case however it is an endpoint of α that is $g^{-1/2}(\mathcal{O})$. The line α cuts the line $[\mathcal{O}, g^{-1}(\mathcal{O})]$ at its intersection with e_g. $\square$

The line $\alpha \subset e_g$ is called the *symmetry line* of the plane e_g.

Both the plane e_g and its symmetry line $\alpha = \alpha_g$ are uniquely determined by g once $\mathcal{O}$ is chosen (and is not a fixed point of g). In the ball model, if $\mathcal{O} = 0$, then e_g is the isometric plane. In the upper half-space model, when $\mathcal{O} = \infty$, e_g is the isometric plane. Therefore using the ball model, if $\mathcal{O} \in \mathbb{H}^3$ then g can be replaced by AgA^{-1} where $A\mathcal{O} = 0$ and Ae_g is the isometric plane for AgA^{-1}. Using the upper half-space model, if $\mathcal{O} \in \partial\mathbb{H}^3$, then g can be replaced by AgA^{-1} where $A\mathcal{O} = \infty$. These observations are important enough to record formally:

Lemma 7.6.2. *Let $\mathcal{O} \in \mathbb{H}^3 \cup \partial\mathbb{H}^3$ be a basepoint and g a Möbius transformation that does not fix $\mathcal{O}$. Let $\mathcal{O}_1 = A(\mathcal{O})$ be a new basepoint, also not fixed by g. Set $g_1 = AgA^{-1}$. Then the symmetry line α_1 and plane e_{g_1} for g_1 with respect to the basepoint $\mathcal{O}_1$ are related to those for g with respect to $\mathcal{O}$ as follows: $\alpha_1 = A(\alpha)$ and $e_{g_1} = A(e_g)$.*

Symmetry lines II. The basis for another notion of symmetry line is the following fact.

Lemma 7.6.3. *Suppose A and B are loxodromic while $K = AB^{-1}A^{-1}B$ is parabolic with fixed point $\mathcal{O} \in \partial\mathbb{H}^3$. At the level of Möbius transformations the following hold.*

- *The line $\alpha = \{AB, B^{-1}\} = K^{-1/2}A$ lies in the plane $e_A = e_\alpha$.*
- *The line $\beta = \{BA^{-1}B, B^{-1}A\} = K^{-1/2}B$ lies in the plane $e_B = e_\beta$.*
- *The line $\gamma = \{B^{-1}AB, B\} = K^{-1/2}AB$ lies in the plane $e_{AB} = e_\gamma$.*
- *The line $\gamma^\sharp = \{A, B^{-1}\} = K^{-1/2}AB^{-1} = K^{1/2}B^{-1}A$ lies in the plane $e_{AB^{-1}} = e_{B^{-1}A} = e_{\gamma^\sharp}$.*
- *The lines $\alpha^\sharp = \{AB^{-1}, A^{-1}BA^{-1}\} = K^{1/2}A^{-1}$, $\beta^\sharp = \{A, B^{-1}A^{-1}\} = K^{1/2}B^{-1}$ are the symmetry lines of $e_{A^{-1}}, e_{B^{-1}}$ respectively.*
- *$B = \alpha\gamma = \gamma^\sharp\alpha$, $K^{-1}A = \beta\gamma = \beta^\sharp\gamma^\sharp$, $\beta\gamma\alpha = \alpha\gamma^\sharp\beta^\sharp = K^{-1/2}$.*
- *The lines $\alpha, \beta, \gamma, \gamma^\sharp, \ldots$ are mutually disjoint in $\mathbb{H}^3 \cup \mathbb{S}^2$.*

Proof. Replace the transformations by conjugates if necessary so that the point $\mathcal{O}$ becomes ∞ in the upper half-space model. Necessarily $\tau_K = -2$, since $\{A, B^{-1}\}$ is nonsingular. At the level of transformations and referring to Lemma 7.3.1, $K^{1/2} = K - I$ and $K^{-1/2} = K^{-1} - I$; they are parabolic transformations fixing ∞. The formulas are verified by using $K - I = (AB^{-1} - B^{-1}A)A^{-1}B = AB^{-1}(A^{-1}B - BA^{-1})$ and correspondingly expressing $K^{-1} - I$. We also use the facts that the any half-rotation is identical to its inverse and that X, $K^{\pm 1/2}X$, and $K^{\pm 1}X$ all have the same isometric circle and plane.

Finally if α and γ, or any two distinct symmetry lines, intersected at $\vec{x} \in \mathbb{H}^3$, then B would fix x and could not be loxodromic. If instead the two half rotations α and γ had a common fixed point on S^2, then the composition $\alpha\gamma$ would have zero trace, which is impossible. $\qquad\square$

In the situation of Lemma 7.6.3, the lines α, β, γ will be called the *symmetry lines* for the planes $e_A, e_B, e_{AB}, \ldots$ respectively. Correspondingly, $\alpha, \beta^\sharp, \gamma^\sharp$ are the symmetry lines for $\mathcal{J}(A), \mathcal{J}(B^{-1}), \mathcal{J}(AB^{-1})$.

Lemma 7.6.2 is worth repeating to cover the present case.

Lemma 7.6.4. *Let $\mathcal{O} \in \mathbb{H}^3 \cup \partial\mathbb{H}^3$ be a basepoint not fixed by A, B and set $\mathcal{O}_1 = X(\mathcal{O})$ and $K_1 = XKX^{-1}$. Then Lemma 7.6.3 holds with respect to the basepoint $\mathcal{O}_1$ with A, B and the half-rotations replaced by their conjugates XAX^{-1}, XBX^{-1}.*

7-23. Find the analog of Lemma 7.6.3 in the case that A and B are loxodromic but their commutator is elliptic.

8

Right hexagons and hyperbolic trigonometry

In this chapter we will apply the line geometry developed in Chapter 7 to obtain many of the formulas of hyperbolic trigonometry. Good references are Beardon [1983] and Fenchel [1989].

Recall that $\cosh z = (e^z + e^{-z})/2$ and $\sinh z = (e^z - e^{-z})/2$.

8.1 Generic right hexagons

Let α, β, γ be lines in $\mathbb{H}^3$ no two of which have a common point or endpoint, and not all three are orthogonal to the same line. We will use the same notation to represent the half-rotations about these lines. When needed, we will use A, B, C to denote corresponding normalized matrices of zero trace.

The axis γ^* of the loxodromic transformation $C_0^* = BA$ is orthogonal to the lines α and β, the axis α^* of $A_0^* = CB$ is orthogonal to β and γ, and the axis β^* of $B_0^* = AC$ is orthogonal to γ and α. Note that

$$C_0^* B_0^* A_0^* = -I, \tag{8.1}$$

so that C_0^*, say, is automatically determined from A_0^* and B_0^*. The half-rotation matrices that correspond to $\alpha^*, \beta^*, \gamma^*$ are respectively (see Proposition 7.2.1(4)),

$$A^* = \frac{A_0^* - A_0^{*-1}}{\sqrt{4 - \tau^2(A_0^*)}}, \quad B^* = \frac{B_0^* - B_0^{*-1}}{\sqrt{4 - \tau^2(B_0^*)}}, \quad C^* = \frac{C_0^* - C_0^{*-1}}{\sqrt{4 - \tau^2(C_0^*)}}. \tag{8.2}$$

In terms of Möbius transformations we may write

$$A^* = \{C, B\}, \quad B^* = \{A, C\}, \quad C^* = \{B, A\}. \tag{8.3}$$

The rotation axes of A^*, B^*, C^* are the axes of the loxodromic CB, AC, BA respectively. The intermediate matrices A_0^*, B_0^*, C_0^* are quite useful, irrespective of the awkward notation.

Lemma 8.1.1. (i) *No two of the lines $\alpha^*, \beta^*, \gamma^*$ have a common endpoint.*
(ii) *The three lines $\alpha^*, \beta^*, \gamma^*$ do not have a common perpendicular.*

Proof. No two of the lines $\alpha^*, \beta^*, \gamma^*$ coincide because of our hypothesis that α, β, γ do not have a common perpendicular.

Consider for example α^* and β^*. Each one is orthogonal to γ. If α^* and β^* had a common endpoint then a right-angled hyperbolic triangle with two right angles would be formed. This is impossible since the angle sum must be less than π.

No line ℓ is orthogonal to all three $\alpha^*, \beta^*, \gamma^*$. For if ℓ were orthogonal to α^* and β^*, say, then they would have two common orthogonals, ℓ and γ. Because the common orthogonal is a unique line, $\ell = \gamma$. If $\ell = \gamma$ were also orthogonal to γ^*, then all of α, β, γ would be orthogonal to γ^*, a contradiction. $\qquad\square$

A *generic right hexagon* is one determined by a triple of lines α, β, γ such that

- no pair of lines have a common point or endpoint,
- the three lines do not have a common perpendicular, and
- no two of the dual lines $\alpha^*, \beta^*, \gamma^*$ intersect in $\mathbb{II}^3$.

For generic hexagons, the six segments cut off by pairwise intersections in the order

$$(\alpha, \gamma^*, \beta, \alpha^*, \gamma, \beta^*)$$

form a right-angled hexagon $\mathcal{H}ex(\alpha, \beta, \gamma)$, which is not planar unless α, β, γ all lie in a plane. Prescribing the cyclic sequence of intersections in the order indicated orients each of the six line segments of the hexagon. The orientation is consistent with the orientation of each line $\alpha^*, \beta^*, \gamma^*$ toward the attracting fixed point of each loxodromic transformation $\gamma\beta, \alpha\gamma, \beta\alpha$ respectively. Each side s is opposite its dual side s^*.

We stress that for generic hexagons, triples of lines (α, β, γ) and $(\alpha^*, \beta^*, \gamma^*)$ are interchangeable with each other: each triple is dual to the other.

Corollary 8.1.2 (Petersen–Morley Theorem). *The three altitudes of a generic right hexagon in $\mathbb{H}^3$ have a common perpendicular.*

Proof. In the notation we have been using, the altitudes are contained in the axes of the three half-rotations $h_\alpha = \{\alpha, \gamma\beta\}$, $h_\beta = \{\beta, \alpha\gamma\}$, and $h_\gamma = \{\gamma, \beta\alpha\}$. These are the common perpendiculars to the pairs of opposite lines (α, α^*), (β, β^*), (γ, γ^*), respectively. Now compute the three Lie products; their sum is zero. Apply Proposition 7.2.1(8). $\qquad\square$

Let $\mathcal{S}$ denote the space of ordered triples (α, β, γ) of unoriented lines in $\mathbb{H}^3$ which produce generic right hexagons.

Lemma 8.1.3. *The space $\mathcal{S}$ is connected. Once an initial choice of half-rotation matrices A, B, C corresponding to (α, β, γ) is made at one point of $\mathcal{S}$, then by continuity a choice is uniquely determined at all other points. In particular one generic right hexagon $\mathcal{H}ex(\alpha, \beta, \gamma)$ can be moved continuously through generic right hexagons to any other.*

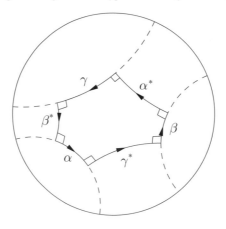

Fig. 8.1. Configuration of sides of the hexagon determined by α, β, γ.

Proof. A line $\ell \in \mathbb{H}^3$ can be moved into a small neighborhood of a point on $\partial \mathbb{H}^3$. Three lines can be moved close to any three distinct points on $\partial \mathbb{H}^3$ without intersecting each other. The three lines can be adjusted so that the hexagon they determine is generic. If in the course of the motion the three lines have a common perpendicular ℓ, then an arbitrarily small change in the position of any one of them (so long as it is not a rotation about ℓ) will destroy this property. Likewise pairwise intersections of the dual lines $\alpha^*, \beta^*, \gamma^*$ can be avoided. To be specific, if γ is taken as the vertical half-line rising in the upper half-space model from from $z = 0$, then the lines α^* with endpoints $u_1, u_2 \in \mathbb{C}$ and β^* with endpoints $v_1, v_2 \in \mathbb{C}$ have a common point of intersection with γ if and only if $u_2 = -u_1$, $v_2 = -v_1$ while $|u_1| = |v_1|$. To avoid the common intersection all that is needed is to move one endpoint slightly.

Thus the movement to the neighborhood of distinct points on $\partial \mathbb{H}^3$ can be made so that the hexagon remains generic at all intermediate points. $\qquad \square$

8.2 The sine and cosine laws

In Chapter 7 we introduced the notation of complex distance between *oriented* lines ℓ_1, ℓ_2 as

$$\chi(\ell_1, \ell_2) = \log \rho + i\theta \pmod{2\pi i}, \quad \rho > 0,$$

where $\log \rho$ is the distance from $\ell_1 \cap p$ to $\ell_2 \cap p$ along the oriented line p orthogonal to ℓ_1 and ℓ_2. We will continue to use the notation from Section 8.1 that α, β, γ are three lines which together with their dual lines $\alpha^*, \beta^*, \gamma^*$ form a generic right hexagon. The lines will be oriented as before by choosing the cyclic order of vertices. Since $\sinh(z + \pi i) = -\sinh z$ and $\cosh(z + \pi i) = -\cosh z$ the formulas below are independent of which orientation is chosen. The laws are:

Law of sines: $\qquad \dfrac{\sinh \chi(\beta, \gamma)}{\sinh \chi(\beta^*, \gamma^*)} = \dfrac{\sinh \chi(\gamma, \alpha)}{\sinh \chi(\gamma^*, \alpha^*)} = \dfrac{\sinh \chi(\alpha, \beta)}{\sinh \chi(\alpha^*, \beta^*)}.$ $\qquad$ (8.4)

Law of cosines: $\cosh \chi (\alpha, \beta) =$

$\cosh \chi (\beta, \gamma) \cosh \chi (\gamma, \alpha) + \sinh \chi (\beta, \gamma) \sinh \chi (\gamma, \alpha) \cosh \chi (\alpha^*, \beta^*).$ (8.5)

These formulas apply to any cyclic permutation of the six sides $(\alpha, \gamma^*, \beta, \alpha^*, \gamma, \beta^*)$.

Proof. For a start, we have to confirm the identity

$$4(\operatorname{tr}(XYX^{-1}Y^{-1}) - 2) = (2\operatorname{tr}(XY) - \operatorname{tr}(X)\operatorname{tr}(Y))^2 - (\operatorname{tr}^2(X) - 4)(\operatorname{tr}^2(Y) - 4). \quad (8.6)$$

This is done by applying Lemma 1.5.6(ii)-(i).

Let $\tilde{C}$ be a normalized matrix corresponding to the loxodromic transformation $\beta^*\alpha^*$. It has axis γ and complex translation length $2\chi(\alpha^*, \beta^*)$ (7.11). Using (8.3) and (7.14) we find that

$$\pm 4[\sinh \chi (\gamma, \alpha) \sinh \chi (\beta, \gamma)]\tilde{C} = (B_0^* - B_0^{*-1})(A_0^* - A_0^{*-1}). \quad (8.7)$$

For later application we will also record the result of using a different formula for the determinants of the right-hand side derived in the proof of Proposition 7.2.1(4), namely

$$\pm \left((\operatorname{tr}^2(B_0^*) - 4)(\operatorname{tr}^2(A_0^*) - 4)\right)^{1/2} \tilde{C} = (B_0^* - B_0^{*-1})(A_0^* - A_0^{*-1}). \quad (8.8)$$

Here A, B, C denote the normalized half-rotations corresponding to α, β, γ.

Expanding the right side of (8.7) and bringing in Lemma 1.5.6(i) we find that the trace of the right side is

$$2\operatorname{tr}(A_0^*B_0^*) - 2\operatorname{tr}(A_0^*B_0^{*-1}) = 4\operatorname{tr}(A_0^*B_0^*) - 2\operatorname{tr}(A_0^*)\operatorname{tr}(B_0^*).$$

Taking the trace of the left side as well,

$$\pm 8\big(\sinh \chi (\gamma, \alpha) \sinh \chi (\beta, \gamma)\big) \cosh \chi (\alpha^*, \beta^*) = 4\operatorname{tr}(A_0^*B_0^*) - 2\operatorname{tr}(A_0^*)\operatorname{tr}(B_0^*).$$

Squaring both sides gets rid of the $\pm$ ambiguity:

$$16\sinh^2 \chi (\alpha, \gamma) \sinh^2 \chi (\beta, \gamma) \cosh^2 \chi (\alpha^*, \beta^*) = \big(2\operatorname{tr}(A_0^*B_0^*) - \operatorname{tr}(A_0^*)\operatorname{tr}(B_0^*)\big)^2.$$

Now replace $\cosh^2$ by $1 + \sinh^2$. After doing so and after separating the terms on the left, replace $16\sinh^2 \chi (\alpha, \gamma) \sinh^2 \chi (\beta, \gamma)$ by the alternative expression for the determinants of the right-hand side of (8.7), as was used in (8.8). In doing so we get our chance to apply (8.6). Applying (8.1) in the process and representing the starred elements in terms of the unstarred ones, the result is:

$$\sinh^2 \chi (\alpha, \gamma) \sinh^2 \chi (\beta, \gamma) \sinh^2 \chi (\alpha^*, \beta^*) = \frac{\delta^2}{4},$$

$$\delta^2 = \operatorname{tr}(A_0^*B_0^*A_0^{*-1}B_0^{*-1}) - 2 = -\operatorname{tr}(CBA)^2 - 2 = -\operatorname{tr}^2(CBA).$$

Note that δ^2 is invariant under cyclic permutation of A, B, C.

Define δ by choosing the sign of $\pm i \operatorname{tr}(CBA)$ so that

$$\sinh \chi (\alpha, \gamma) \sinh \chi (\beta, \gamma) \sinh \chi (\alpha^*, \beta^*) = \frac{\delta}{2}. \quad (8.9)$$

We claim the definition of δ for one triple determines it by continuity for all triples of lines α, β, γ without a common point or endpoint. Start with an initial choice of α, β, γ and matrices A, B, C and let these range over the full space $\mathcal{S}$. The claim is valid simply because at no point can any of the terms in the left side of (8.9) vanish.

The law of sines follows from the identity

$$\frac{\sinh \chi(\alpha^*, \beta^*)}{\sinh \chi(\alpha, \beta)} = \frac{\delta/2}{\sinh \chi(\alpha, \beta) \sinh \chi(\beta, \gamma) \sinh \chi(\gamma, \alpha)},$$

since the right side is invariant under cyclic permutation.

We are also ready for the law of cosines. The starting point here is (8.8). Operating on the left side as before, and on the right side bringing in (7.13) we wind up with

$$\pm \sinh \chi(\beta, \gamma) \sinh \chi(\gamma, \alpha) \cosh \chi(\alpha^*, \beta^*)$$
$$= \cosh \chi(\alpha, \beta) \pm \cosh \chi(\beta, \gamma) \cosh \chi(\gamma, \alpha).$$

We have also used the fact that $B_0^* A_0^* = -AB$, $B_0^* = AC$ and $A_0^* = CB$.

It remains to settle the matter of signs. Once again we do this by continuity. We may assume that α, β, γ determine a planar, regular right hexagon. In this case all the complex distances $\chi(\cdot, \cdot)$ are equal to $d + \pi i$ (why?) where d is the common side length. We now have to look at the possibilities for the equation

$$\pm(1 - \cosh^2 d) = -1 \pm \cosh d.$$

The only situation for which there is a positive solution for $\cosh d$ occurs when the signs are in the order $(+, -)$. The only solution is $\cosh d = 2$ or $d = \log(2 + \sqrt{3})$. This choice gives the law as stated, and it remains true as stated for the whole space by continuity.

As we have seen, the triples of lines α, β, γ and $\alpha^*, \beta^*, \gamma^*$ are interchangeable. $\qquad \square$

The case of planar right hexagons is assigned as Exercise (8-6).

8.3 Degenerate right hexagons

In the notation of Section 8.1, it is possible that two of the lines $\alpha^*, \beta^*, \gamma^*$ intersect in $\mathbb{H}^3$. If for example β^* and γ^* intersect at $p \in \mathbb{H}^3$, they span a plane P. The line α is then orthogonal to P at p. This forces the side of the hexagon which lies on α to reduce to the single point p, which is a vertex. We will regard such a polygon as a "degenerate hexagon". The degeneration can be avoided by moving one or more of the lines α, β, γ slightly.

A line in $\mathbb{H}^3$ is determined by its endpoints on $\partial\mathbb{H}^3$. A single point $\zeta \in \partial\mathbb{H}^3$ can be regarded as the limit of a sequence of lines, it can be regarded as an *ideal line*. Upon adapting this point of view, it is natural by comparison with (7.3) to represent ζ by the projective equivalence class of the singular, zero-trace, nonzero matrix

$$\begin{pmatrix} \zeta & -\zeta^2 \\ 1 & -\zeta \end{pmatrix}, \quad \text{or} \quad \begin{pmatrix} 0 & -1 \\ 0 & 0 \end{pmatrix} \text{ when } \zeta = \infty.$$

The "common perpendicular" γ^* to two distinct lines α and β which have a common endpoint $\zeta \in \partial\mathbb{H}^3$ can be interpreted to be the *ideal line* ζ itself. This ideal line can also be interpreted to be "orthogonal" to the plane spanned by (α, β). This interpretation agrees with the construction of γ^* as a Lie product. Indeed, if Q_1, Q_2 denote the other endpoints of α, β respectively, we have from (7.3) again

$$AB - BA = -\frac{16(Q_1 - Q_2)^2}{(\zeta - Q_1)^2(\zeta - Q_2)^2} \begin{pmatrix} \zeta & -\zeta^2 \\ 1 & -\zeta \end{pmatrix},$$

where A, B denote half-rotation matrices for α, β.

This latter interpretation is the limiting case of two lines which intersect in $\mathbb{H}^3$. Indeed, suppose α and β intersect in $\vec{x} \in \mathbb{H}^3$. The common perpendicular γ^* is the line through $\vec{x}$ and perpendicular to the plane spanned by α and β. However unlike the case of the common perpendicular between disjoint lines, the ordering α, β no longer determines an orientation of γ^*. If we choose a sequence of disjoint lines α_n, β_n which converge to α, β, the common perpendicular γ_n^* of α_n and β_n converges to γ^*. There are two ways to orient γ_n^* depending on whether α_n is regarded as "over" or "under" β_n. The two choices induce opposite orientations on γ^*.

When α and β have a common endpoint the asymptotic distance between the lines is zero. The complex distance $\chi(\alpha, \beta)$ is either 0 or πi depending on the relative orientations of α and β.

When α and β intersect in $\mathbb{H}^3$, they span a plane. We have

$$\chi(\alpha, \beta) = \pm i\theta \text{ or } \pm i(\pi - \theta),$$

where θ is the acute angle formed by α and β with the sign in each term depending on how γ^* is oriented and the choice of term θ or $\pi - \theta$ depending on how α and β are oriented to each other. The bottom line is that, in all four cases,

$$\cosh^2 \chi(\alpha, \beta) = \cos^2 \theta, \quad \sinh^2 \chi(\alpha, \beta) = -\sin^2 \theta.$$

This also includes the cases of a common endpoint ($\theta = 0, \pi$).

In view of this discussion we will take the expression "common perpendicular" to include the case that one or both "lines" are ideal lines. The common perpendicular between two distinct ideal lines is interpreted as the ordinary line with those endpoints.

We now return to our construction of right hexagons. Suppose α, β, γ are distinct oriented lines in $\mathbb{H}^3$. We now allow that any two may have a common point or endpoint, but still exclude the possibility that all three have a common perpendicular line (that case is beyond the pale, as no closed figure results). In particular, not all three lines have a common point or endpoint.

Let γ^* denote the common perpendicular to α, β, α^* the common perpendicular to β, γ, and β^* the common perpendicular to γ, α. Any or all of these perpendiculars may be "ideal lines". At points of intersection in $\mathbb{H}^3$, orient the perpendiculars arbitrarily. The collection of the six oriented lines/ideal lines is called a generalized right hexagon. It is either a generic right hexagon or a it is a *degenerate right hexagon*. Degenerated hexagons have three, four, or five sides; the "degenerate sides" become vertices. If the degenerate side is on $\partial\mathbb{H}^3$, the vertex angle there is zero.

We will treat degenerate hexagons as limiting cases of generic right hexagons, and apply the laws as dictated by continuity.

For example, if (α, β, γ) form an ideal triangle, $(\alpha^*, \beta^*, \gamma^*)$ represent the ideal vertices. Conversely if (α, β, γ) represent three distinct points on $\partial\mathbb{H}^3$, $(\alpha^*, \beta^*.\gamma^*)$ are the edges of the associated ideal triangle.

8.4 Formulas for triangles, quadrilaterals, and pentagons

In this section we will present three typical examples of how the right hexagon formulas can be adapted to give formulas for degenerate hexagons, that is, polygons with less than six sides. The trick is to add "ideal" sides of zero length at some vertices so the polygon can then be interpreted as a degenerate case of a right hexagon. Other cases are presented in the exercises.

In each case the edges will be oriented so that the polygonal object lies to the left. In other words, the sides are labeled and oriented so that the vertices appear in the cyclic order $\alpha \cap \gamma^*, \gamma^* \cap \beta, \beta \cap \alpha^*, \alpha^* \cap \gamma, \gamma \cap \beta^*, \beta^* \cap \alpha$, with appropriate interpretation for degenerate sides.

Right triangles

In (say) the ball model of $\mathbb{H}^3$, move α, β, γ to lie in the equatorial plane forming there a right triangle with the hypotenuse contained in γ, orientation as shown in Figure 8.2. Then $\alpha^*, \beta^*, \gamma^*$ are orthogonal to the plane, Orient them so they are pointing toward the lower hemisphere. Denote the side lengths by a, b, c respectively where c is the hypotenuse. Let α, β denote the vertex angles opposite the side lengths a, b respectively.

Then $\chi(\alpha^*, \beta^*) = c$, $\chi(\beta^*, \gamma^*) = a$, and $\chi(\gamma^*, \alpha^*) = b$ while $\chi(\alpha, \beta) = \pi i/2$, $\chi(\beta, \gamma) = i(\pi - \alpha)$, and $\chi(\gamma, \beta) = i(\pi - \beta)$. To make this last computation,

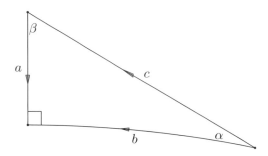

Fig. 8.2. A right (planar) triangle indicating the degenerate sides.

imagine say γ moved slightly out of the plane but parallel to its original position and orthogonal to β^*. Then looking along β^* in the direction of γ to α we see that a rotation of angle $\pi - \beta$ is required to rotate γ with its orientation onto α with its orientation. Putting these values in the laws gives

$$\cosh c = \cosh a \cosh b, \qquad \sinh c = \frac{\sinh b}{\sin \beta} = \frac{\sinh a}{\sin \alpha}. \qquad (8.10)$$

The length of two sides determines the length of the third and also determines the angles. we also find that

$$\tanh b = \sinh a \tan \beta, \qquad \operatorname{sech} c = \tan \alpha \tan \beta. \qquad (8.11)$$

Some more formulas for right triangles are given in Exercise (8-3). Formulas for the general triangle are presented in Exercise (8-2).

Planar pentagons with four right angles

Move α and β into a plane and orient γ^* by thinking of β slightly lower than α at their common vertex v so that γ^* is oriented so as to point from α to β. Then $\chi(\alpha, \beta)$ approaches θi. The resulting right hexagon has its side on γ^* degenerated to the vertex v. We obtain either a convex pentagon or a figure overlapping itself.

Here we will work out the formulas for the convex case; the case of self-intersection is in Exercise (8-5). Place the sides in order α, $[\gamma^*]$, β, α^*, γ, β^* where the side on γ^* reduces to the vertex v and the interior angle at v is θ. Denote the lengths of the sides contained in α, β, γ by a, b, c, and in α^*, β^* by a^*, b^*. Then $\chi(\alpha^*, \beta^*) = c + \pi i$, $\chi(\beta^*, \gamma^*) = a + \pi i/2$, and $\chi(\gamma^*, \alpha^*) = b + \pi i/2$ while $\chi(\alpha, \beta) = (\pi - \theta)i$. The law of cosines becomes

$$\cosh c = \sinh a \sinh b - \cosh a \cosh b \cos \theta,$$

or alternatively,

$$\cos \theta = \sinh a^* \sinh b^* \cosh c - \cosh a^* \cosh b^*, \qquad (8.12)$$

So c is determined by a, b^* and θ, and θ is determined by a^*, b^*, and c; two convex right pentagons whose corresponding side lengths are identical are isometric.

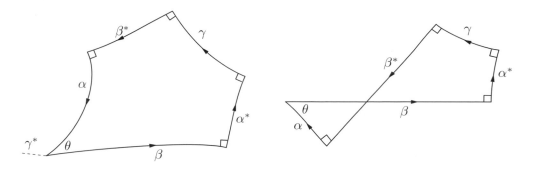

Fig. 8.3. Planar pentagons with four right angles.

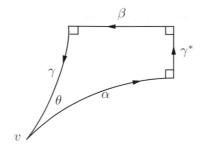

Fig. 8.4. A quadrilateral with three right angles.

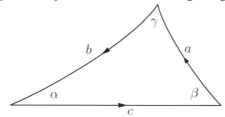

Fig. 8.5. A generic triangle.

The law of sines becomes

$$\frac{\cosh a}{\sinh a^*} = \frac{\cosh b}{\sinh b^*} = \frac{\sinh c}{\sin \theta}. \tag{8.13}$$

Specialize to the cases $\theta = 0$ (v is an ideal vertex) and $\theta = \pi i/2$.

Quadrilaterals with three right angles

Here we are not assuming that the quadrilateral is planar. Let v denote the vertex with angle θ. Label the sides in order as α, γ^*, β, $[\alpha^*]$, γ, $[\beta^*]$, where the brackets indicate the lines associated with degenerate sides—which correspond to vertices—and $[\beta^*]$ is associated with v.

The lines α^* and β^* are perpendicular to the planes determined by β, γ and γ, α respectively. Orient α^* and β^* to point into these planes, and interpret β to lie over γ on α^* and γ over α on β^*. Then $\chi(\beta, \gamma) = -\pi i/2$ and $\chi(\gamma, \alpha) = (\theta - \pi)i$. In addition, $\chi(\gamma^*, \alpha^*) = \chi(\gamma^*, \gamma) - \pi i/2$. Using this information the law of sines gives

$$\sinh \chi(\alpha, \beta) \cosh \chi(\gamma^*, \gamma) \;=\; \sin \theta \sinh \chi(\alpha^*, \beta^*).$$

The law of cosines tells us that

$$\cosh \chi(\alpha, \beta) \;=\; -\sin \theta \cosh \chi(\alpha^*, \beta^*).$$

From the two laws we conclude that

$$\tanh \chi(\alpha^*, \beta^*) = -\tanh \chi(\alpha, \beta) \cosh \chi(\gamma, \gamma^*), \tag{8.14}$$

and

$$\cos \theta \;=\; \sinh \chi(\alpha, \beta) \sinh \chi(\gamma^*, \gamma). \tag{8.15}$$

In particular, if the quadrilateral is planar (and necessarily convex) then

$$\cosh c^* = \sin\theta \cosh c, \qquad (8.16)$$

where c is the length of a side on γ and c^* the length of the opposite side on γ^*. Equation (8.15) becomes

$$\cos\theta = \sinh b \sinh c^*, \qquad (8.17)$$

where b is the length of the side on β. This latter equation is used to confirm that the sign is correctly chosen when taking square roots to obtain (8.15).

Two planar quadrilaterals with three right angles with corresponding side lengths identical are isometric. From (8.15) we deduce that $0 \leq \theta < \pi/2$; there are no hyperbolic rectangles, as we already know. On the other hand, regular quadrilaterals with 60° angles tessellate the plane.

8.5 Exercises and explorations

8-1. Show how to form a right hexagon from six edges of a hyperbolic cube.

8-2. *Law of sines and cosines for triangles.* Consider the general hyperbolic triangle with sides of length a, b, c and opposite angles labeled α, β, γ.

Derive the law of sines and cosines for the triangles, namely:

Law of cosines 8.5.1.

$$\cosh c = \cosh a \cosh b - \sinh a \sinh b \cos\gamma, \qquad (8.18)$$

$$\cosh c = \frac{\cos\alpha \cos\beta + \cos\gamma}{\sin\alpha \sin\beta}. \qquad (8.19)$$

Law of sines 8.5.2.

$$\frac{\sinh a}{\sin\alpha} = \frac{\sinh b}{\sin\beta} = \frac{\sinh c}{\sin\gamma}. \qquad (8.20)$$

Conclude that two triangles with the same angles are isometric.

Show that the quantity $\sinh a \sin\gamma$ is independent of the lengths b, c and the angles α, β that form the triangle. Here is a nice application of this fact:

Suppose $\sigma = \sigma(s) \subset \mathbb{H}^2$ is a geodesic parameterized by hyperbolic arc length and ζ is a point not on σ. At each point $\sigma(s)$ there is a geodesic segment from ζ to $\sigma(s)$; denote its length by $x(s)$. Let $\theta(s)$ be the angle of intersection at $\sigma(s)$ measured counterclockwise from σ. Deduce as in Epstein et al. [2004] the formula

$$\theta'(s) = -\coth x(s) \sin\theta(s).$$

Hint: You will need to know that $x'(s) = \cos\theta(s)$. This is seen using the fact that near $\sigma(s)$ the hyperbolic metric is almost euclidean, therefore $\Delta x \sim \Delta s \cos\theta$.

Finally consider the hyperbolic triangle with vertices x, y, z'. Let z be a point on the side (x, y). Derive the formula

$$\cosh d(z, z') \sinh d(x, y) = \cosh d(x, z') \sinh d(y, z) + \cosh d(y, z') \sinh d(x, z),$$
(8.21)

where as usual $d(\cdot, \cdot)$ is the hyperbolic distance.

Prove that if the shortest geodesic γ in a $\mathcal{M}(G)$ has length $\ell > 1.353$, then the tube of radius $\frac{1}{2} \log 3$ about it is embedded (Gabai–Meyerhoff–Thurston). *Hint:* Fix a lift γ^*. If γ_1^* is another of distance d away, then by the law of cosines applied to a right triangle with base along γ_1^* and sides of lengths, $d, \geq \ell, \leq \ell/2$ and $\cosh d \geq (\cosh \ell)/(\cosh \ell/2)$.

8-3. *Right-angled triangles.* Return to the case of a right-angled triangle labeled as in Section 8.4. Verify the additional formulas

$$\cosh c = \cot \alpha \cot \beta,$$

$$\cos \alpha = \tanh b \coth c, \qquad\qquad \cos \beta = \tanh a \coth c,$$

$$\sinh a = \cot \beta \tanh \beta, \qquad\qquad \sinh b = \cot \alpha \tanh a.$$

Specialize to the case that $b = c = \infty, \alpha = 0$.

8-4. *Convex planar quadrilateral with two adjacent right angles.* Let c denote the length of the base which has right angles at its endpoints. Let c' denote the length of the opposite side, and a, b the other two sides. Denote the vertex angle facing side b by β and that facing side a by α; necessarily $\alpha + \beta < \pi$.

Verify the formulas

$$\frac{\cosh a}{\sin \alpha} = \frac{\cosh b}{\sin \beta} = \frac{\sinh c'}{\sinh c},$$

$$\cosh c' = -\sinh a \sinh b + \cosh a \cosh b \cosh c$$

$$\cosh c = -\cos \alpha \cos \beta + \sin \alpha \sin \beta \cosh c'.$$

These give the exact formulas for the orthogonal projection in a plane of a line segment onto a geodesic (see Exercise (8-10) for a simpler version).

8-5. Complete the formulas for planar pentagons with four right angles by considering the self-intersecting case when the side along $\boldsymbol{\beta}^*$ crosses the side along b. The law

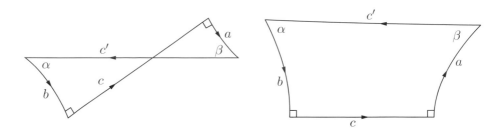

Fig. 8.6. Planar quadrilaterals with two right angles.

of sines remains the same except for some changes of sign. When applying the law of cosines, for instance $\chi(\boldsymbol{\beta}^*, \boldsymbol{\gamma}^*)$ changes to $a - \pi i/2$.

8-6. *Planar right hexagons.* Work out the formulas in the special case that the right hexagon lies in a plane. Confirm that in the convex case the two laws have the form

$$\frac{\sinh a}{\sinh a^*} = \frac{\sinh b}{\sinh b^*} = \frac{\sinh c}{\sinh c^*}, \tag{8.22}$$

$$\cosh c = -\cosh a \cosh b + \sinh a \sinh b \cosh c^*. \tag{8.23}$$

Derive the corresponding formulas in the self-intersecting case, forming two quadrilaterals each with three right angles.

How many side lengths uniquely determine the hexagon up to isometry, orientation preserving or reversing?

Show that given any $a, b, c > 0$ there exists a unique convex right hexagon with alternating sides of length a, b, c. Equivalently, there exist three mutually disjoint geodesics in $\mathbb{H}^2$ whose respective distances apart are exactly a, b, c.

Specialize to the cases that all sides have the same length, and then that all vertex angles are zero.

8-7. *Pants.* A pair of pants on a surface is a triply connected planar region bounded by three mutually disjoint closed geodesics. The three common perpendiculars in turn divide the pants into two right hexagons. Show that there is an orientation-reversing conformal map (a reflection) that pointwise fixes the common perpendiculars, and interchanges the two hexagons. That is, every pants is preserved by three orientation-reversing involutions each of which maps a boundary component onto itself.

Show that every closed Riemann surface of genus g can decomposed—and then built—by pants with geodesic boundaries. What about punctured surfaces, compact bordered surfaces?

In particular the area of a pants is 2π.

Show that the pants is uniquely determined by the lengths of the three boundary geodesics.

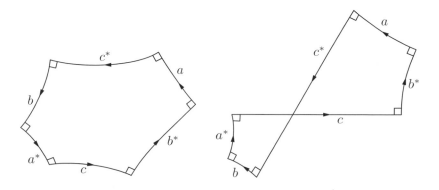

Fig. 8.7. Planar right hexagons.

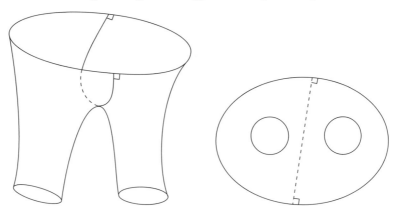

Fig. 8.8. A pair of pants with a seam.

In fact suppose the lengths of the three boundary curves c_1, c_2, c_3 of a pants are L_1, L_2, L_3. Denote the length of the seam that is the perpendicular between c_1 and c_2 by ℓ. Show that

$$\cosh \ell = \frac{\cosh L_3 + \cosh L_1 \cosh L_2}{\sinh L_1 \sinh L_2}.$$

See Fathi et al. [1979, §8] for a description of pants geometry.

Consider the limiting case that one boundary component shrinks to a single point (puncture) and the other two have lengths a, b. There is uniquely determined geodesic whose endpoints are at the puncture and which cuts the pants P into two annular regions P_1, P_2, where P_1 has the a-length side. Each P_i in turn can be cut into two quadrilaterals each with one ideal vertex and three right angles. The two quadrilaterals $Q_{1,i}, Q_{2,i}$ of each P_i are symmetric under reflection. One side of $Q_{1,i}$ has length $a/2, b/2$ respectively. Denote the lengths of the other two finite sides by a', b', respectively. Show that

$$\sinh(a/2) \sinh a' = 1, \quad \sinh(b/2) \sinh b' = 1.$$

Now assemble $2g - 2$ pants into closed surfaces $\{R\}$ of genus $g \geq 2$. Discuss how their boundary lengths yields a continuous map into $\mathfrak{T}(R)$. What additional information about the assembly do you need to do to get a homeomorphism?

8-8. *Collars.* It is a fact (a consequence of the Schwarz Lemma [Ahlfors 1978]) a nested pair of simply connected regions $\Omega_1 \subset \Omega_2$ with corresponding hyperbolic metrics $\rho_1(z)|dz|$ and $\rho_2(z)|dz|$ have the following property: $\rho_1(z) > \rho_2(z)$, $z \in \Omega_1$. From this, deduce the following sharp form of the collar lemma (see Buser [1992, Chapter 4]).

Collar Lemma 8.5.3. On a hyperbolic surface R suppose $\alpha, \beta \subset R$ are mutually disjoint simple geodesics of lengths a, b. Set

$$a' = \operatorname{arcsinh} \frac{1}{\sinh(a/2)}, \quad b' = \operatorname{arcsinh} \frac{1}{\sinh(b/2)}.$$

Then the distance-a' annular neighborhood of width $2a'$ about α is disjoint from the distance-b' annular neighborhood of width $2b'$ of β.

For more computational practice show that the length L_α of each boundary component of the collar of distance a' from the geodesic α of length a is

$$L_\alpha = a \cosh a' = \frac{a}{\tanh a/2}.$$

Note that $L_\alpha \to 2$ as $a \to 0$.

Prove [Beardon 1983, Theorem 8.3.1], which asserts that if X, Y generate a nonelementary fuchsian group without elliptic elements then

$$\sinh \frac{d(z, X(z))}{2} \sinh \frac{d(z, Y(z))}{2} \geq 1 \quad \text{for all } z \in \mathbb{H}^2.$$

This inequality is best possible.

8-9. *Polar, cylindrical, and horocyclic coordinates.* Show that for the hyperbolic metric ds in $\mathbb{H}^2$,

$$ds^2 = d\rho^2 + \sinh^2 \rho \, d\theta^2, \tag{8.24}$$
$$ds^2 = d\rho^2 + \cosh^2 \rho \, dt^2, \tag{8.25}$$
$$ds^2 = e^{-2\rho} dt^2 + d\rho^2. \tag{8.26}$$

Equation (8.24) is called the polar representation of the hyperbolic metric ds; ρ denotes hyperbolic distance from the origin and θ is the angle from the positive axis to the ray ρ. Equally θ can be interpreted as the angular measure on $\partial \mathbb{H}^2$.

Equation (8.25) is the metric representation in terms of geodesic coordinates; t is arclength along a geodesic α, for example the real diameter in the disk model, and ρ is the distance of a point from α, along an orthogonal line through t. The equation can be regarded as a two-dimensional form of cylindrical coordinates. It is often called the *Fermi coordinates*.

Equation (8.26) the metric representation in terms of *horocyclic coordinates*. Here t is arclength along a horocycle, and ρ is signed distance along a geodesic orthogonal at t. We choose the sign of ρ so that positive distance is toward the point on $\partial \mathbb{H}^2$ determined by the horocycle.

The first formula is derived from expressing the hyperbolic metric in (euclidean) polar coordinates,

$$ds^2 = \frac{4(dx^2 + dy^2)}{(1 - |z|^2)^2} = \frac{4(dr^2 + r^2 d\theta^2)}{(1 - r^2)^2},$$

together with the fact that

$$e^\rho = \frac{1 + r}{1 - r}.$$

Equation (8.26) is most easily derived in the upper half-plane model with the horocycle $\{y = a > 0\}$. The desired formula results from substituting $x = at$ and $y = ae^\rho$ into $ds^2 = (dx^2 + dy^2)/y^2$.

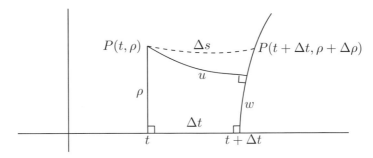

Fig. 8.9. Cylindrical coordinate approximation.

Equation (8.25) is more complicated to derive. Construct two quadrilaterals Q, Q^* as follows.

The base of Q is on a geodesic; the left endpoint of the base has coordinate t along the geodesic and the right endpoint is $t + \Delta t$, where Δt is a small deformation. The adjacent vertical sides are at right angles. The left side has length ρ and its top end is labeled $P(t, \rho)$; its coordinates are (t, ρ). The right side has length w. The top side of Q is orthogonal to the right side; denote its length by u. Q has three right angles; the nonright angle is subtended at the vertex $P(t, \rho)$. Now extend Q by taking a small deformation $\Delta\rho$ and extending the right side until it has length $\rho + \Delta\rho$; label the vertex $P(t + \Delta t, \rho + \Delta\rho)$. Let Δs denote the length of the segment from $P(t, \rho)$ to $P(t + \Delta t, \rho + \Delta\rho)$. Thus we now have a larger quadrilateral Q^* whose right side has length $w + v$ for some $v > 0$.

We first work with Q. Insert the diagonal from the upper left to the lower right and denote its length by d. It divides the right angle at the lower right vertex of Q into angles α, $\pi/2 - \alpha$, where we take α to be adjacent to the bottom.

The diagonal forms two right triangles. Therefore

$$\cosh d = \cosh \Delta t \cosh \rho = \cosh u \cosh w,$$

and also

$$\tanh \rho = \sinh \Delta t \tan \alpha; \quad \tanh u = \sinh w \tan(\pi/2 - \alpha).$$

Now $\tanh u \sim u$ and $\sinh \Delta t \sim \Delta t$, while $\cosh \Delta t \sim 1$ and $\cosh u \sim 1$. Hence $w \sim \rho$, and then

$$u \sim \cosh \rho \, \Delta t.$$

Now examine the right triangle $Q^* \setminus Q$. The side of length v satisfies $v = \rho + \Delta\rho - w \sim \Delta\rho$. For its hypotenuse, $\Delta s^2 \sim u^2 + v^2$, since the sides are very short. From these calculations Equation (8.25) follows.

The corresponding formulas for $\mathbb{H}^3$ are essentially the same:

$$ds^2 = d\rho^2 + \sinh^2 \rho \, d\theta^2, \tag{8.27}$$

$$ds^2 = d\rho^2 + d \sinh^2 \rho, \, d\theta^2 + \cosh^2 \rho \, dt^2 \geq d\rho^2 + \cosh^2 \rho \, dt^2 \tag{8.28}$$

$$ds^2 = d\rho^2 + e^{-2\rho} \, ds^2_{\mathcal{H}}. \tag{8.29}$$

Equation (8.29) is in terms of horocyclic coordinates about a horosphere $\mathcal{H}$, where $ds_{\mathcal{H}}$ denotes distance in $\mathcal{H}$ and ρ is distance along a line orthogonal to $\mathcal{H}$ with positive direction toward the associated point on $\partial \mathbb{H}^3$.

Equation (8.28) is in terms of cylindrical coordinates about a geodesic α; t denotes distances along α, $\rho > 0$ denotes distances from α so that (ρ, θ) are polar coordinates in the plane orthogonal to α at t. The volume form in these coordinates is

$$dV = \sinh \rho \cosh \rho \, dt \, d\rho \, d\theta. \tag{8.30}$$

Equation (8.28) invites the following interpretation. Denote the convex core of $\mathcal{M}(G)$ by $\mathcal{C}(G)$. If ρ represents the shortest distance of a point exterior to $\mathcal{C}(G)$ to $\partial \mathcal{C}(G)$, the hyperbolic metric restricted to the exterior of $\mathcal{C}(G)$ satisfies

$$ds^2 \geq \cosh^2 \rho \, ds^2_{\partial \mathcal{C}} + d\rho^2 > \frac{1}{4} e^{2\rho} ds^2_{\partial \mathcal{C}} + d\rho^2.$$

One says that the exterior of $\mathcal{C}(G)$ is *exponentially flaring*.

Boundary length estimates for triangles and cylinders. The hyperbolic area element in polar coordinates is

$$dA = \sinh \rho \, d\rho \, d\theta. \tag{8.31}$$

If Δ is a hyperbolic triangle, v is one of its vertices, and α is the opposite side, Area(Δ) can be expressed in polar coordinates about v as,

$$\text{Area}(\Delta) = \iint_{\Delta} \sinh \rho(\theta) \, d\rho \, d\theta = \int (\cosh \rho(\theta) - 1) \, d\theta.$$

The area of the triangular region in Δ *outside* the distance-r neighborhood of α is

$$\int (\cosh(\rho(\theta) - r) - 1) \, d\theta < e^{-r} \int \sinh \rho(\theta) \, d\theta < e^{-r} \, \text{Len}(\alpha),$$

since $r < \rho(\theta)$. In particular the area of a triangle is less than the length of its shortest side.

Following on in this vein, let $f : S^1 \times [0, 1] \to \mathcal{M}(G)$ be the embedding of a closed cylinder, that is a free homotopy between the simple loops $\gamma_1 = f(S^1 \times \{0\})$, $\gamma_2 = f(S^1 \times \{1\})$. Replace each arc $\{f(\theta) \times [0, 1])\}$ by a geodesic arc with the same endpoints, obtaining as a consequence a ruled cylinder C. The cylinder C can be approximated by a union of thin quadrilaterals. Each quadrilateral can be divided into two triangles. One of the triangles abuts γ_1, the other abuts γ_2. The total area of the latter triangles is less than the length of γ_2. Following Thurston [1979a, §9.3], prove from the estimate above that

$$\text{Area}(C \setminus N_r(\gamma_2)) \leq e^{-r} \, \text{Len}(\gamma_2) + \text{Len}(\gamma_1), \tag{8.32}$$

where $N_r(\gamma_2)$ is the distance-r annular neighborhood of γ_2. In particular, the area of C is less than the length of ∂C. This finds essential use in the theory of ending laminations for example in Lemma 5.7.2.

8-10. *Orthogonal projection strictly reduces distances.* Establish this often-used property: Let $\ell \in \mathbb{H}^3$ be a line and σ be a line segment of finite length which we may assume is disjoint from ℓ. Let $\sigma^* \in \ell$ denote the orthogonal projection of σ to ℓ. Show that the length of σ^* is strictly less than the length of σ (unless σ is itself a segment of ℓ).

That is, if x and y in $\mathbb{H}^3$ lie on the same side of ℓ and have distance $\geq r$ from ℓ, then the orthogonal projection of the two points x, y onto ℓ satisfies

$$d(\pi(x), \pi(y)) < d(x, y). \tag{8.33}$$

Hint: First verify that in a right triangle, the hypotenuse is strictly longer than either leg. Then verify that in a planar quadrilateral with three right angles, the length of a side with one end at the vertex with vertex angle $\theta \neq \pi/2$ is strictly greater than the length of the opposite side. Next show that the same property holds for a nonplanar quadrilateral with three right angles by taking its orthogonal projection to the plane formed by two orthogonal edges. Finally return to the original problem and drop a perpendicular from one end of σ to form a quadrilateral with three right angles and a right triangle.

Deduce from Equation (8.28) that it is also true that

$$d(\pi(x), \pi(y)) < 2e^{-r} d(x, y), \tag{8.34}$$

which is better than Equation (8.33) when $r > \log 2$. More precisely, if x, y are equal distance r from ℓ and the segment $[\pi(x), \pi(y)]$ has length L, then

$$2 \cosh d(x, y) = 1 + \cosh L + (\cosh L - 1) \cosh 2r.$$

8-11. *Riemann surfaces made out of pentagons.* Show that there exists a unique regular right-angled pentagon up to isometry; find a formula for the side length s.

Position it so that one side is the "bottom". We will refer to the sides directly on its right and its left as the "vertical" sides. Reflect across the bottom giving a 6-sided right polygon with its two vertical sides of length $2s$ and the others of length s. Then reflect this pair across say the right vertical side giving a right-angled 8-sided polygon P, the union of four pentagons, four of whose sides have length $2s$. From this you can construct a hyperbolic surface of genus two. From now on, P will be the fundamental unit.

Keep adjoining copies of P to the right side. Each time you attach a unit P, you will attach four pentagons. You will be subtracting an edge and adding five new edges. In short, you will be able to construct a closed surface of genus $g \geq 2$ out of $4(g - 1)$ pentagons comprising a $4g$-sided right polygon. We have proved:

Any closed surface of genus $g \geq 3$ is a covering surface of a closed surface of genus two.

8-12. *Riemann surfaces made out of equilateral triangles.* Consider a closed topological surface embedded in $\mathbb{R}^3$, say, of genus exceeding one. Triangulate it in any way. Then based on the combinatorics of your triangulation, build a homeomorphic surface R made out of euclidean equilateral triangles. It need not be embedded in $\mathbb{R}^3$. Still, R can be given a complex structure by flattening the vertex angles to 2π. What is remarkable about such a surface are its properties [Bowers and Stephenson 2004; Jones and Singerman 1996; Mulase and Penkava 1998; Schneps 1994; Stephenson 1999]:

The following properties are equivalent:

(i) R is composed of euclidean equilateral triangles.
(ii) There exists a meromorphic function $f : R \rightarrow \mathbb{S}^2$ such that its critical values, that is the image of its critical points C, lie in the set $\{0, 1, \infty\}$. (A critical point is a point ζ about which f is not a local homeomorphism; the corresponding critical value is $f(\zeta)$.)
(iii) $R \setminus C$ is a finite cover of the 3-punctured sphere; alternatively, $R \setminus C = \mathbb{H}^2/\Gamma$ where Γ has finite index in the modular group.
(iv) R can be represented by an algebraic curve whose coefficients lie in a finite extension to the field $\mathbb{Q}$ of rational numbers.

The first listed property can be characterized in terms of the existence of a special meromorphic quadratic differential $q \, dz^2$ on R whose critical trajectories divide R into equilateral triangles in the singular euclidean metric associated with $q \, dz^2$. In fact the f-preimage of the segments $[\infty, 0]$, $[0, 1]$, $[1, \infty]$ form a triangular graph on R. Such graphs were called by Grothendieck *dessins d'enfants*, although it seems unlikely a child would come up with one. It has the property that the vertices, the elements of C, can be labeled $+$ or $-$ so that the two endpoints of each edge have the opposite sign. On $\mathbb{S}^2$, the graph complement can be connected and simply connected, so it can just be a finite tree. For a wealth of information about dessins see Schneps [1994].

These Riemann surfaces, called *Belyĭ surfaces*, can be nicely uniformized by the circle packing technique.

Since there are infinitely many triangulations possible, it is natural to ask whether the Belyĭ surfaces are dense in the Teichmüller space. Compare with Exercise (4-17). After all, the coefficients of an algebraic curve can be approximated by algebraic numbers.

See also Exercise (2-26).

8-13. *Cutting a Riemann surface into euclidean polygons.* Let R be a closed (to simplify our discussion) hyperbolic surface of finite area and $\varphi(z) \, dz^2$ a quadratic differential (see Exercise (5-33)). Normalize φ to have unit area:

$$\int \int_R |\varphi z| \, dx dy = 1.$$

It determines a local euclidean structure on R by taking the preimage of its local mapping into $\mathbb{C}$,

$$w = \Phi(z) = \int^z \sqrt{\varphi} \, dz.$$

The local mapping is well defined away from zeros, up to an additive constant. The pull back of the euclidean metric from $\mathbb{C}$, gives a euclidean metric $ds = |dw|$ in R which is singular at the zeros of φ. The preimage of euclidean line segments is geodesic segments on R in the φ-metric.

Now the φ-metric $|dw|$ has a global extension to R, except at the $4g - 4$ zeros of φ. We can think of R as composed of a finite number of rectangles, for example, the preimage of rectangles in $\mathbb{C}$. If we choose a rectangle in $\mathbb{C}$ to have a vertex at the image of a zero, then its pullback to R will have the vertex angle $\frac{2\pi}{4(2k+1)}$, when the zero has order k.

Between any two points p, q of R there is a unique φ-geodesic γ. If γ does not go through any zeros then γ is the preimage of a straight line segment. If γ travels through m-zeros, it will be the union of $m + 1$ straight segments, meeting at angles determined by the orders. Similarly, in the free homotopy class of any simple closed geodesic on R there is a φ-geodesic which may well have a finite number of corners if it passes through zeros. A closed φ-geodesic may not be unique in its free homotopy class. Nonuniqueness will occur in the case in which there is a family of mutually disjoint, parallel φ-geodesics which fill an annulus on R.

A saddle connection is a φ-line between two of its zeros which contains no other zeros. By a process of induction, a finite set of saddle connections can be found which cut R into a finite number of polygons of unit total area: Φ maps each $P^* \subset R$ onto a euclidean polygon $P \subset \mathbb{C}$. The edges of the image polygons are arranged in parallel, equal length pairs (e, e'), each pair coming from the two sides of a saddle connection in R.

There is a euclidean translation $A : z \mapsto \pm z + a$ that maps e onto e'. However the preimage of the individual points $p \in e$ may or may not be the preimage of their images $A(p)$.

Special case: translation surfaces. This case arises in the case that $\varphi(z) \, dz^2 = (\omega(z) \, dz)^2$; here $\omega(z) \, dz$ is a holomorphic differential on R and is referred to as an *Abelian differential* after Abel. In this case all zeros of φ must have even orders and the number of zeros of ω is $2g - 2$. The simplest case is that of a torus with fundamental parallelogram in $\mathbb{C}$ and Abelian differential dz. In this case the translations referred to above are all of the form $A : z \mapsto +(z + a)$.

Currently, the study of dynamics on these surfaces is a very active area. For a survey, see Masur [2006]; the following geometric description is taken from this paper.

By using ω and cutting along saddle connections, R can be divided into a union of $n \geq 1$ euclidean polygons $\{\Delta_k\}$. Orient the boundary of each so that the polygon lies to its left. For each oriented edge of Δ_i, there is an oriented edge of some Δ_j such that the two edges are parallel of the same length. Glue these two edges together in the opposite orientation by a parallel translation $z \mapsto z + a$ so the glued edge becomes a common edge of the two polygons each with their original orientations.

In this manner, the polygons fit together to form a closed Riemann surface. Each vertex—zero of ω—becomes a cone point whose angle is an integral multiple of 2π.

On the other hand, if φ is not the square of an abelian differential, the linear mapping between at least some pairs of edges is given by $B(z) = -z + b$, a translation followed by a 180° degree rotation. In this case the preimages in R of points $p \in e$ and $B(p) \in e'$ are the same and R is not a translation surface.

Conversely, a Riemann surface R' and the corresponding differential can be constructed from a suitable assemblage of polygons in $\mathbb{C}$ coming from R with linear identifications between parallel edge pairs [Masur and Smillie 1993]. Such an assemblage, under the action of an affine map $A \in \mathrm{SL}(2, \mathbb{R})$, changes its shape. Using the corresponding boundary equivalences, the image can be reconstituted into a new Riemann surface R'. The map A determines a quasiconformal map $R \to R'$. In fact A determines a homeomorphism of Teichmüller space onto itself.

8-14. *Ptolemy relation for ideal quadrilaterals in* $\mathbb{H}^2$. The formula below was discovered by Robert Penner [Penner 1987, Prop. 2.6].

In the unit disk $\mathbb{D}$ mark four distinct points $(x_1, x_2, x_3.x_4)$ in counterclockwise order on $\partial\mathbb{D}$. Draw the geodesics (a, b, c, d) between each pair of successive points thereby forming an inscribed ideal quadrilateral. Now truncate the quadrilateral by horodisks at the vertices thereby forming a right-angled octahedron. Denote the truncated geodesics by (a_0, b_0, c_0, d_0). Denote their (hyperbolic) lengths by $(\ell(a_0), \ell(b_0), \ell(c_0), \ell(d_0))$.

Now take the diagonal geodesics u, v between the opposite vertices (x_1, x_3), (x_2, x_4), truncate by the given horodisks, and denote their truncated lengths by $\ell(u), \ell(v)$. Renormalize each of the six truncated lengths by taking exponentials:

$$\lambda^2(\cdot) = e^{\ell(\cdot)}.$$

Prove that

$$\lambda(u)\lambda(v) = \lambda(a_0)\lambda(c_0) + \lambda(b_0)\lambda(d_0).$$

This formula continues to hold if the horodisks about one or more successive pairs of vertices intersect. For each such case, the resulting segment lies in the intersection of the two adjacent horodisks. Its length is taken as the *negative* of the length of arc in the intersection.

Is there a corresponding theorem in $\mathbb{H}^3$?

8-15. *Hyperbolic Heron's formula.* The following formula for the area $|A|$ of a euclidean triangle is attributed to Heron of Alexandria, who lived almost 2000 years ago. Letting a, b, c denote the side lengths of a euclidean triangle and $s = (a + b + c)/2$ the half-perimeter,

$$|A|^2 = s(s - a)(s - b)(s - c).$$

An analogue of sorts for hyperbolic geometry is as follows [Fenchel 1989]. Let $A, B, C = -(BA)^{-1}$ be normalized matrices. Fix one of the eigenvalues of each of the matrices and denote the choices by $\lambda_A, \lambda_B, \lambda_C$. Set $\chi_A = \log \lambda_A$, $\chi_B = \log \lambda_B$, and $\chi_C = \log \lambda_C$. Finally set

$$s = \frac{\chi_A + \chi_B + \chi_C}{2}.$$

The formula states that

$$\text{tr}(ABA^{-1}B^{-1}) - 2 = 16 \cosh s \cosh(s - \chi_A) \cosh(s - \chi_B) \cosh(s - \chi_C). \quad (8.35)$$

Now prove this! (*Hint:* Start by expressing the left-hand side first in terms of the traces and then in terms of the eigenvalues.)

Confirm that the numerical value of the right side of (8.35) is invariant under action of the group G of Möbius transformations generated by $\langle A, B \rangle$. That is the generators A, B used in the formula can be replaced by generators $\varphi(A)\,\varphi(B)$ where φ is any automorphism of G. (*Hint:* The commutator $[A, B]$ is independent of φ.)

Assume that none of the eigenvalues is ± 1 (the traces are not ± 2). Show that there exist half-rotation matrices A^*, B^*, C^* with $A = C^*B^*$, $B = A^*C^*$, $C = B^*A^*$ such that the associated axes form a generalized right hexagon, after assigning orientations. Confirm that

$$-\frac{\text{tr}^2(C^*B^*A^*)}{16} = \cosh s \cosh(s - \chi_A) \cosh(s - \chi_B) \cosh(s - \chi_C).$$

In some of the following exercises the term "generator pair (A, B)" of a once-punctured torus group is used. This means that A, B are loxodromic with parabolic commutator $K = [A, B] = ABA^{-1}B^{-1}$ and generate a discrete group acting on $\mathbb{H}^3$. Thus if A, B are represented by matrices, $\text{tr}(K) = -2$. We can normalize so that $K = \left(\begin{smallmatrix} -1 & -2 \\ 0 & -1 \end{smallmatrix}\right)$.

8-16. Here is a construction, from Parker and Series [1995], associated with a special generator pair $\langle A, B \rangle$ that gives rise to a convex planar pentagon with four right angles and an ideal vertex.

Let α^* denote the axis of A oriented toward its attracting fixed point, and $\beta^* = B(\alpha^*)$ the axis of $BA^{-1}B^{-1}$ likewise oriented toward its attracting fixed point. We are going to assume that α^* and β^* lie in the same plane P and that P is preserved by A. Such a situation will arise in Exercise (8-21).

Denote the common perpendicular to α^* and β^* by γ. Find the line β, perpendicular to α^* such that $A^{-1}(\gamma) = \beta$. Find the line α perpendicular to β^* so that $BA^{-1}B^{-1}(\gamma) = \alpha$.

For the sixth line γ^* we would like to take the axis of $C = A \cdot BA^{-1}B^{-1} = K$. But this is parabolic. So we have instead an ideal line at the ideal vertex which is the fixed point of K.

Now for the side lengths, $\chi(\beta, \gamma) = a^* + \pi i$ and $\chi(\gamma, \alpha) = b^* + \pi i$, but $a^* = b^*$ (why?). Hence, from (8.12), we have

$$\cosh d = \frac{1 + \cosh^2(L/2)}{\sinh^2(L/2)},$$

where d is the distance between the axes of A and BAB^{-1} and L is the translation length of A: $L = 2 \log \lambda$ where λ is the larger eigenvalue of A. The transformation A can be assumed to have real eigenvalues since it preserves the plane P. This equation can be transformed by introducing the half-angle formula to become

$$\cosh \frac{d}{2} = \frac{1}{\tanh(L/2)}, \quad \tanh \frac{d}{2} = \frac{1}{\cosh(L/2)}. \tag{8.36}$$

8-17. [Minsky 1999] Suppose (A, B) is a generator pair of a once-punctured torus group. Let $\boldsymbol{\alpha}, \boldsymbol{\beta}, \boldsymbol{\gamma}$ denote the axes of $A, B, C = AB$ respectively. Prove for the right hexagon $\boldsymbol{\alpha}, \boldsymbol{\gamma}^*, \boldsymbol{\beta}, \boldsymbol{\alpha}^*, \boldsymbol{\gamma}, \boldsymbol{\beta}^*$ that

$$\sinh^2 \chi(\boldsymbol{\beta}^*, \boldsymbol{\gamma}^*) \sinh^2 \chi(\boldsymbol{\gamma}^*, \boldsymbol{\alpha}^*) \sinh^2 \chi(\boldsymbol{\alpha}, \boldsymbol{\beta}) = -1. \tag{8.37}$$

This equation does not depend on how the sides are oriented. *Hint:* To the law of cosines, introduce Lemma 1.5.6(iv) upon recalling Equation (7.13).

8-18. [Minsky 1999] Continuing with the situation of Exercise (8-17), let c^* denote the length of the side on $\boldsymbol{\gamma}^*$.

Consider the ray $\boldsymbol{\beta}_o^*$ of $\boldsymbol{\beta}^*$ that extends from the common endpoint of the sides on $\boldsymbol{\alpha}, \boldsymbol{\beta}^*$, runs along the side on $\boldsymbol{\beta}^*$, and ends at $\zeta_1 \in \partial \mathbb{H}^2$. Let ℓ_1 be the ray from ζ_1 that is orthogonal to $\boldsymbol{\gamma}^*$. Let d_1 be the length of segment of the side on $\boldsymbol{\gamma}^*$ cut off by ℓ_1, namely the length of the orthogonal projection of $\boldsymbol{\beta}_o^*$ to $\boldsymbol{\gamma}^*$. Correspondingly take the ray $\boldsymbol{\alpha}_o^*$ of $\boldsymbol{\alpha}^*$ and let d_2 denote the projection of $\boldsymbol{\alpha}_o^*$ onto $\boldsymbol{\gamma}^*$. Prove that

$$c^* \leq d_1 + d_2 + \log 3. \tag{8.38}$$

Hint: We have two quadrilaterals each with three right angles and one zero angle. From Equation (8.14)

$$\sinh^2 \chi(\boldsymbol{\alpha}, \ell_1) \sinh^2 \chi(\boldsymbol{\gamma}^*, \boldsymbol{\beta}^*) = 1,$$
$$\sinh^2 \chi(\boldsymbol{\beta}, \ell_2) \sinh^2 \chi(\boldsymbol{\gamma}^*, \boldsymbol{\alpha}^*) = 1,$$

and with the help of (8.37) confirm that

$$\sinh^2 \chi(\boldsymbol{\alpha}, \boldsymbol{\beta}) = - \sinh^2 \chi(\boldsymbol{\alpha}, \ell_1) \sinh^2 \chi(\boldsymbol{\beta}, \ell_2).$$

Now if $d = |\mathrm{Re}\, z|$ then

$$\frac{e^d - 1}{2} \leq \frac{e^d - e^{-d}}{2} < |\sinh z| < \frac{e^d + e^{-d}}{2} < e^d.$$

Applying this to the previous equation gives the desired result.

8-19. [Minsky 1999] Given a loxodromic element X define

$$Core_D(X) := \{\vec{x} \in \mathbb{H}^3 : d(\vec{x}, X\vec{x}) \leq L(X) + D\}.$$

Here $L(X)$ is the translation length $2 \log |\lambda|$ of X, where λ denotes the larger eigenvalue. Thus $Core_D(X)$ contains the tube of radius D about the axis of X, but it is larger, especially for small $L(X)$, as X becomes closer to parabolic. In the limit, the core becomes a horoball at the parabolic fixed point.

Prove that for

$$D = 4\operatorname{arcsinh} 1 + \log 3 = \log 3(1 + \sqrt{2})^4,$$

any generator pair (A, B) of any once-punctured torus group satisfies

$$Core_D(A) \cap Core_D(B) \neq \varnothing.$$

Hint: Apply Exercise (8-18) to the two quadrilaterals each with three right angles formed by the lines from endpoints of β^* and α^* orthogonal to γ^*. For $p \in \gamma^*$ on the projection of β^*, the distance of p in particular from the side on α does not exceed $2\operatorname{arcsinh} 1$ and therefore by triangle inequality $d(p, Ap) \leq 4\operatorname{arcsinh} 1 + L(A)$ (note $A^{-1} = \beta^*\gamma^*$). Likewise $d(p, Bp) \leq 4\operatorname{arcsinh} 1 + L(B)$. By Equation (8.38) the D-cores intersect on the side on γ^*.

Corollary 8.5.4. *Given a generator pair (A, B) there exists a point $p \in \mathbb{H}^3$ and $\rho < \infty$ such that both $d(p, Ap) < \rho$ and $d(p, Bp) < \rho$.*

The next four problems outline the development by John Parker and Caroline Series of explicit bending formulas for simple bending in once-punctured torus deformation spaces.

8-20. *The Parker–Series bending formula* [Parker and Series 1995]. Let's start with the following observation. Suppose P is a plane in $\mathbb{H}^3$ and ℓ is a given line in P. Suppose we are also given a Möbius transformation V such that the line $V^{-1}(\ell)$ also lies in P and does not have a common endpoint with ℓ. let $\ell^\perp$ be the line perpendicular to both ℓ and $V^{-1}(\ell)$. Set $\zeta = \ell^\perp \cap V^{-1}(\ell)$. The map V sends P onto a plane $V(P)$ which intersects P along ℓ. It sends ℓ to $V(\ell)$. The line $V(\ell^\perp)$ through $V(\zeta) \in \ell$ is the line perpendicular to ℓ and $V(\ell)$.

Find the midpoint $\zeta_m \in \ell$ between $\ell^\perp \cap \ell$ and $V(\zeta)$. Let ℓ_m be the line through ζ_m, perpendicular to ℓ, such that half-rotation ι_m about ℓ_m sends the line $\ell^\perp$ onto the line $V(\ell^\perp) \subset V(P)$. Necessarily ι_m exchanges the planes P and $V(P)$. Conclude that ι_m also sends $V^{-1}(\ell)$ onto $V(\ell)$.

Now construct the line ℓ_0 orthogonal to P and passing through the midpoint of the segment $[\ell^\perp \cap \ell, \ell^\perp \cap V^{-1}(\ell)]$ of $\ell^\perp$. Let ι_0 denote the half-rotation about ℓ_0. Then

$$\iota_m\iota_0(\ell) = \iota_m(V^{-1}(\ell)) = V(\ell),$$
$$\iota_m\iota_0(V^{-1}(\ell)) = \iota_m(\ell) = \ell.$$

That is, the transformation $X = \iota_m\iota_0$ sends the lines ℓ to $V(\ell)$ and $V^{-1}(\ell)$ to ℓ. Show that this implies that $X = V$. The axis of V must then be the common perpendicular to ℓ_m and ℓ_0.

Next consider $\ell_1 = V(\ell_0) = \iota_m(\ell_0)$ which is orthogonal to $V(\ell^\perp)$ midway between $V(\zeta)$ and $V(\ell)$. Then $\iota_1 = \iota_m\iota_0\iota_m$ is the corresponding half-rotation and we also have $\iota_1\iota_m = \iota_m\iota_0 = V$. Thus ℓ_1 is also orthogonal to the axis of V.

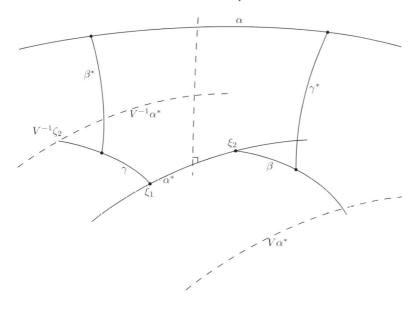

Fig. 8.10. A Parker–Series right hexagon expressing bending.

With this construction under our belt we can set up the following interesting right hexagon. Suppose U and V are loxodromic transformations whose axes have no common endpoint. Assume the axis α^* of U and $V^{-1}(\alpha^*)$ lie in a plane P. Show that the following six lines determine a right-angled hexagon:

$$\alpha^* = Axis(U),$$
$$\gamma = \text{common perpendicular to } \alpha^* \text{ and } V^{-1}(\alpha^*) = Axis(V^{-1}UV),$$
$$\beta^* = \text{line orthogonal to } P \text{ at midpoint of segment } [\gamma \cap \alpha^*, \gamma \cap V^{-1}(\alpha^*)],$$
$$\alpha = Axis(V),$$
$$\gamma^* = V(\beta^*),$$
$$\beta = V(\gamma).$$

To prepare for the law of cosines orient the sides of the hexagon in the usual way. Let $d > 0$ denote the hyperbolic distance between a^* and $V^{-1}(\alpha^*)$. Then

$$\chi(\alpha^*, \beta^*) = \chi(\beta^*\gamma^*) = \frac{d - i\pi}{2}.$$

Also $\chi(\beta^*, \gamma^*) = L + \pi i$ where $L = 2\log\lambda > 0$ and λ is the larger eigenvalue of V. Therefore

$$\cosh L = \sinh^2 \frac{d}{2} + \cosh^2 \frac{d}{2} \cosh \chi(\beta, \gamma).$$

Specialize to the case that (A, B) is a generator pair for a once-punctured torus group; this means that A, B are loxodromic generators with parabolic commutator $ABA^{-1}B^{-1}$. Set $U = A$ and $V = B$ so that in the formula above d is the distance between the axes of A and $B^{-1}AB$. Assume that A preserves the plane P so the eigenvalues of A are real; $\mathcal{L}_A$ will denote the larger eigenvalue. Let $\mathcal{L}_B$ denote the larger eigenvalue of B; $\mathcal{L}_B$ will not in general be real so the translation length of B is $|\mathcal{L}_B|$. However the real part $\mathcal{L}_B$ is positive. Now we can incorporate (8.36): apply the double angle formula to $\cosh \mathcal{L}_B$ and substitute for $\cosh \frac{d}{2}$ and $\sinh \frac{d}{2}$ to end up with the beautiful formula:

The Parker–Series Bending Formula 8.5.5.

$$\cosh^2 \frac{\chi(\boldsymbol{\beta}, \boldsymbol{\gamma})}{2} = \cosh^2 \frac{\mathcal{L}_A}{2} \tanh^2 \frac{\mathcal{L}_B}{2}. \tag{8.39}$$

We will digress in order to interpret $\chi(\boldsymbol{\beta}, \boldsymbol{\gamma})$. Consider the vertical half plane Q based on $\mathbb{R}$ in the upper half-space model of $\mathbb{H}^3$. Suppose a fuchsian once-punctured torus group is acting in Q. Let $\boldsymbol{\alpha}^*$ denote the positive vertical axis oriented toward ∞ and assume it is the axis a hyperbolic generator X. Now X corresponds to a simple loop α, not retractable to the puncture, on the quotient punctured torus and $\boldsymbol{\alpha}^*$ is a lift of α. Choose a simple loop σ which crosses α exactly once, and from left to right. Fix a lift σ^* which crosses the line $\boldsymbol{\alpha}^*$ from its left side to its right at a point ζ and is the axis of a Möbius transformation Y. We may arrange things so that (X, Y) is a generator pair.

We will deform the fuchsian group $\langle X, Y \rangle$ by a *quakebend*, also known as a *complex earthquake*, or *shearing and bending*. We start by giving a complex number of the form $\kappa = \log \rho + \phi i$, with $\rho > 0$ and $-\pi < \phi < \pi$.

Apply the transformation $S_\rho : z = x + iy \mapsto \rho z$ to the *right* half of P. In particular the point $\zeta \in \boldsymbol{\alpha}^*$ is moved up or down along $\boldsymbol{\alpha}^*$, depending on whether ρ is > 1 or < 1, signed hyperbolic distance $\log \rho$. Correspondingly the right half of the line σ^* is moved up or down and becomes a half line from $\rho\zeta$. The point $B(\zeta)$ on the right half of σ moves to $\rho Y(\zeta) \in P$. Actually S is the restriction of Möbius transformation of $\mathbb{H}^3$.

Next rotate the right half of Q about the vertical $\boldsymbol{\alpha}^*$ by angle ϕ, measured so that $\phi = 0$ corresponds to no bending. Denote this elliptic transformation by E_ϕ. The point $E_\phi Y(\zeta)$ rotates off P to a point $\hat{\zeta} \in \mathbb{H}^3$.

This process results in the *deformation* $Y \mapsto Y_\phi = E_\phi S_\rho Y$ and

$$G = \langle X, Y \rangle \mapsto G_\phi = \langle X, Y_\phi \rangle.$$

This will be a new punctured torus group provided κ is sufficiently small and that the deformed generators still satisfy the trace relation. Note that the quakebend depends on two real parameters; preservation of the commutator trace -2 gives rise to two real equations.

For the group G_ϕ, the lines $\boldsymbol{\alpha}^*$ and $Y_\phi(\boldsymbol{\alpha}^*)$ lie in the same plane Q_ϕ—the rotated right half of Q. The angle ϕ is the dihedral angle between $Y_\phi^{-1}(Q_\phi)$ and Q_ϕ.

Return to the situation preceding (8.39). Let X correspond to A and Y to B. In the Parker–Series Equation (8.39),

$$\chi(\boldsymbol{\beta}, \boldsymbol{\gamma}) = \kappa \log \rho + \phi i,$$

where $\log \rho > 0$ is the distance along α^* from $\alpha^* \cap \beta$ to $\alpha^* \cap \gamma$, and ϕ is the dihedral angle from P to $B(P)$.

Bend the right half-plane abutting α^* with respect to the left half-plane so that the two make the dihedral angle ϕ where $\phi = 0$ corresponds to no bending at all, $-\pi < \phi < \pi$. Let $x \in \mathbb{R}$ be a given number. The line σ is broken in two parts. The left part ends at $\zeta \in \alpha^*$. From ζ continue for x units along α^* (in the positive or negative direction depending on the sign of x) reaching a point $\zeta_1 \in \alpha^*$. Now continue in the right half plane from ζ_1, making the same angle with α^* as the original σ.

Corollary 8.5.6. *Suppose the sequence of generator pairs $\{(A_n, B_n)\}$ are the result of bending along the axis of A_n. Assume that $\lim A_n$ is a parabolic transformation. Then on a subsequence either $\lim |\mathcal{L}_{B_n}| = \infty$, or $\lim \rho_n = 1$ and $\lim \phi_n = \pi$.*

Can the second possibility occur?

8-21. *Real traces.* Suppose (A, B) is the generator pair of a fuchsian once-punctured torus group. Assume that the axis of B is orthogonal to the axis of A. Bend B along the axis of A getting a new element B_ϕ. Show that the trace of B_ϕ is real.

Hint: Assume A, B act in the vertical half plane along $\mathbb{R}$ in the upper half-space model and the axis of A is the vertical axis from the origin. Show that $B_\phi = E_\phi B$, where

$$E_\phi = \begin{pmatrix} e^{i\phi} & 0 \\ 0 & e^{-i\phi} \end{pmatrix}.$$

8-22. [Parker and Series 1995] Choose loxodromic Möbius transformations A and B so that the common perpendicular to their axes is the vertical half-line from $0 \in \mathbb{C}$ in the upper half-space model of $\mathbb{H}^3$. Show that their fixed points are necessarily symmetric with respect to 0. Normalize A to have fixed points $+1$. Write the fixed points of B as $\pm re^{i\theta}$. Show that if both A and B have real traces, which we may take to be ≥ 0, then

$$A = \begin{pmatrix} a & b \\ b & a \end{pmatrix}, \quad B = \begin{pmatrix} u & vre^{i\theta} \\ (v/r)e^{-i\theta} & u \end{pmatrix},$$
$$a^2 - b^2 = u^2 - v^2 = 1, \quad a, b, u, v > 0.$$

We will require as well that $\mathrm{tr}[A, B] = -2$. Using Proposition 7.2.1, show that necessarily $\theta \neq 0$ and the trace condition is satisfied if and only if either $r = 1$ or $\theta = \pi/2$. We will work only with the case that $\theta = \pi/2$ which holds if and only if

$$bv\left(r + \frac{1}{r}\right) = 2, \quad b, v > 0, \ r \geq 1. \tag{8.40}$$

Then the group $G = \langle A, B \rangle$ depends only on the two real numbers $b, v > 0$. When $r = 1$, that is when $bv \sin \theta = 1$, show that G is fuchsian, preserving the unit disk.

Show that each of $\mathfrak{I}(A)$ and $\mathfrak{I}(A^{-1})$ is tangent to $\mathfrak{I}(B)$ and to $\mathfrak{I}(B^{-1})$ at a point on the line segment between the centers. Therefore the common exterior of the four circles has two components, an inner one containing 0 and an outer one containing ∞.

Each circular polygon is bounded by four sides. Show that A and B pair the opposite sides of both components (see Section 1.6). Conclude that the points of tangency are parabolic fixed points of $[A, B]$ and its conjugates. Therefore the quotient consists of two once-punctured tori and (A, B) is a generator pair of a group in our space. In terms of complex probabilities introduced in Exercise (1-35), what are their coordinates? Then draw the fundamental region for some parameter values using Wada's computer program.

The common exterior of the planes determined by the four circles forms the Ford fundamental polyhedron, which therefore has only four faces.

8-23. *The single bend formula* [Parker and Series 1995]. Carry on with the two real parameter class of once-punctured torus groups $G = \langle A, B \rangle$ introduced in Exercise (8-22) above. These represent a two-dimensional slice through the four real dimensional deformation space $\mathfrak{T}$. Assume that $r > 1$ so that G is not fuchsian. We are going to explicitly describe how each of the two boundary components of the convex hull are bent. It will turn out that each component arises by bending along a single line and then its conjugates: the groups are obtained by pure bending along the axis of A, and automatically, along the axis of B. The two bending angles serve also as parameters for the class of groups G.

Start by establishing the symmetry with respect to the reflections J in the real axis and $J_\perp$ in the imaginary axis: $JAJ = A$, $J_\perp A J_\perp = A^{-1}$, and $JBJ = B^{-1}$, $J_\perp B J_\perp = B$. The symmetries preserve $\langle A, B \rangle$ as well. Therefore the limit set and the convex hull are also invariant under these symmetries.

Using symmetry show that $\Lambda(G)$ intersects the real and imaginary axes only in the four fixed points $\{\pm 1, \pm ri\}$. This sets the stage for showing the axes of A and B lie in the opposite boundary components of the convex hull.

Let $\mathcal{C}$ denote the convex hull of the limit set as in Section 3.10. First examine the vertical half plane P resting on $\mathbb{R}$. This plane contains the axis of A, and the axis α of A is automatically contained in $\mathcal{C}$. If α is not on the boundary $\partial\mathcal{C}$ then $P \cap \partial\mathcal{C}$ consists of two convex arcs σ_1, σ_2 which are separated in P by α, one on each boundary component Σ_1, Σ_2 of $\mathcal{C}$. By the paragraph above, the endpoints of these arcs must be the fixed points of A. If either is a geodesic it must agree with α and $\alpha \subset \partial\mathcal{C}$. At most one can be α. Each of σ_1, σ_2 which is not a geodesic has bends in it. For such to occur, there must be a (geodesic) line or lines l_1, l_2 in the same boundary component or components of $\mathcal{C}$ and which intersect P.

Now $J(l_1)$, say, must also be a line in Σ_1 yet it would cross l_1 unless $J(l_1) = l_1$. Hence l_1 is orthogonal to the plane P. Show that l_1 must in fact lie in the vertical plane $P_\perp$ resting on the imaginary axis. Once we know this, we can conclude that $l_1 = \beta$. Since this cannot also hold for l_2 it follows that $l_1 = \alpha \subset \Sigma_1$. Since the argument applies to *all* bending lines $l_1 \subset \Sigma_1$ that are transverse to P it shows that β is the only such line.

Repeating the argument deduce that $\beta \subset \Sigma_2$ and α is the only bending line that is transverse to $P_\perp$.

Summing up, the axis of A lies in Σ_1. All its conjugates in $G = \langle A, B \rangle$ do as well since Σ_1 is invariant under G. These are the totality of the bending lines on Σ_1, and all the bending angles are the same. Likewise the axis of B and its conjugates comprise the bending lines of Σ_2, all with the same bending angles. The conditions Equation (8.40) on the two real parameter groups $\langle A, B \rangle$ are *necessary and sufficient that the corresponding convex hull boundary components are bent along the axes of A and B*.

The endpoints of the bending lines on the two components separate each other on $\Lambda(G)$.

Apply to this case the Parker–Series bending formula. We first point out that the assumption made in Exercise (8-16) holds, namely that the axis α of A and the axis $B^{-1}(\alpha)$ of $B^{-1}AB$ lie in the same plane: For the bending lines in Σ_1 separate it into infinitely many components. Each component is contained in the hyperbolic plane determined by any two of its infinitely many boundary components.

Return to Equation (8.39). In the present case the translation lengths L_A, L_B have been taken to be both real and positive. The left side must be positive as well. Since $\chi(\alpha, \beta) = \log \rho + i\phi$, this is possible only if $\rho = 1$ so that the group G is obtained by pure bending. Therefore for Σ_1,

$$\cos \frac{\phi_A}{2} = \cosh \frac{L_A}{2} \tanh \frac{L_B}{2}.$$

Interchanging the roles of A and B for Σ_2,

$$\cos \frac{\phi_B}{2} = \cosh \frac{L_B}{2} \tanh \frac{L_A}{2}.$$

Here $\phi_A, \phi_B \in (0, \pi)$ are the bending angles at the axes of A and B. In terms of the matrix A, L_A has the expression

$$L_A = 2\log(a+b), \quad \cosh \frac{L_A}{2} = a, \quad \sinh \frac{L_A}{2} = b,$$

and the analogous equations in terms of u, v hold for L_B. Show that $\phi_A = 0$ if and only if $bv = 1$, that is, $r = 1$ and G is fuchsian. In this case also $\phi_B = 0$.

Solving for L_A, L_B when $r > 1$, we end up with

Theorem 8.5.7 (The Parker–Series Single Bend Formula).

$$b = \sinh \frac{L_A}{2} = \sin \frac{\phi_A}{2} \cot \frac{\phi_B}{2}, \quad v = \sinh \frac{L_B}{2} = \sin \frac{\phi_B}{2} \cot \frac{\phi_A}{2}. \qquad (8.41)$$

In particular, there is a homeomorphism between pairs of angles $\phi_A, \phi_B \in (0, \pi)$ and (nonfuchsian) quasifuchsian groups with convex hull boundaries bent along the axes of A, B at angles ϕ_A, ϕ_B respectively.

Bibliography

Abikoff, W. and Harvey, W., "Extremal Kleinian groups", *Proc. Amer. Math. Soc.* **140** (2012), 267–278.

Abikoff, W. and Maskit, B., "Geometric decompositions of Kleinian groups", *Amer. J. Math.* **99**:4 (1977), 687–697.

Abikoff, W., Earle, C. J. and Mitra, S., "Barycentric extensions of monotone maps of the circle", pp. 1–20 in *In the tradition of Ahlfors and Bers, III* Contemp. Math., 355, Amer. Math. Soc., Providence, RI, 2004.

Accola, R. D. M., "Invariant domains for Kleinian groups", *Amer. J. Math.* **88** (1966), 329–336.

Adams, C. C., "The noncompact hyperbolic 3-manifold of minimal volume", *Proc. Amer. Math. Soc.* **100**:4 (1987), 601–606.

Agard, S., "Distortion theorems for quasiconformal mappings", *Ann. Acad. Sci. Fennicae* Series A **423** (1968).

Agol, I., "Bounds on exceptional Dehn filling", *Geom. Topol.* **4** (2000), 431–449.

Agol, I., "Volume change under drilling", *Geom. Topol.* **6** (2002), 905–916.

Agol, I., "Tameness of hyperbolic 3-manfolds", preprint, 2004. Available at www.arXiv.org/abs/math.GT/0405568.

Agol, I., "Criteria for virtual fibering", *J. Topol.* **1** (2008), 269–284.

Agol, I., "The minimal volume orientable 2-cusped 3-manifolds", *Proc. A.M.S.* **138**(2010), 3723–3732.

Agol, I., "Bounds on exceptional Dehn filling II", *Geom. Topol.* **14** (2010), 1921–1940.

Agol, I., "The Virtual Haken Conjecture, with an appendix by Agol, Groves, and Manning", 2012, arXiv:1204.2810.

Agol, I. and Liu, Y., "Presentation length and Simon's conjecture", *J. Amer. Math. Soc.* **25** (2012), 151–187,

Agol, I., Storm, P. and Thurston, W. P., with an appendix by N. Dunfield, "Lower bounds on volumes of hyperbolic Haken 3-manifolds", *J. Amer. Math. Soc.* **20** (2007), 1053–1077.

Ahlfors, L. V., "Finitely generated Kleinian groups", *Amer. J. Math.* **86** (1964), 413–429.

Ahlfors, L. V., *Lectures on quasiconformal mappings*, Van Nostrand Mathematical Studies **10**, Van Nostrand, Toronto, 1966; Corrected 2nd edition with additional chapters by Earle and Kra, Shishikura, and Hubbard, A.M.S., 2006.

Ahlfors, L. V., *Conformal invariants: topics in geometric function theory*, McGraw-Hill, New York, 1973.

Ahlfors, L. V., *Complex analysis*, 3rd ed. McGraw-Hill, New York, 1978.

Ahlfors, L. V., *Möbius transformations in several dimensions*, Ordway Professorship Lectures in Mathematics, Univ. of Minnesota, Minneapolis, MN, 1981.

Ahlfors, L. V. and Sario, L., *Riemann surfaces*, Princeton Mathematical Series **26**, Princeton Univ. Press, Princeton, NJ, 1960.

Akiyoshi, H., Sakuma, M., Wada, M. and Yamashita, Y., "Ford domains of punctured torus groups and two-bridge knot groups", in *Knot Theory* (Proceedings of the workshop held in Toronto dedicated to 70th birthday of Prof. K. Murasugi), 1999.

Akiyoshi, H., Sakuma, M., Wada, M. and Yamashita, Y., "Jørgensen's picture of punctured torus groups and its refinement", pp. 247–273 in *Kleinian groups and hyperbolic 3-manifolds* (Warwick, 2001), edited by Y. Komori et al., London Math. Soc. Lecture Note Ser. **299**, Cambridge Univ. Press, Cambridge, 2003.

Akiyoshi, H., Sakuma, M., Wada, M. and Yamashita, Y., "Punctured torus groups and 2-bridge knot groups I", *Lecture Notes in Mathematics* **1909**, Springer, Berlin, 2007.

Anderson, J. W., "Intersections of analytically and geometrically finite subgroups of Kleinian groups", *Trans. Amer. Math. Soc.* **343**:1 (1994), 87–98.

Anderson, J. W., "The limit set intersection theorem for finitely generated Kleinian groups," *Math. Res. Lett.* **3**:5 (1996), 675–692.

Anderson, J. W. and Canary, R. D., "Algebraic limits of Kleinian groups which rearrange the pages of a book", *Invent. Math.* **126**:2 (1996a), 205–214.

Anderson, J. W., *Hyperbolic Geoometry, 2nd ed.* Springer-Verlag London, Ltd., London, 2005.

Anderson, J. W., Talk at Iberoamerican Congress, CUNY, May, 2014.

Anderson, J. W., and Canary, R. D., "Cores of hyperbolic 3-manifolds and limits of Kleinian groups", *Amer. J. Math.* **118**:4 (1996b), 745–779.

Anderson, J. W., and Maskit, B., "On the local connectivity of limit set of Kleinian groups", *Complex Variables Theory Appl.* **31** (1996), 177–183.

Anderson, J. W., Canary, R. D., Culler, M. and Shalen, P. B., "Free Kleinian groups and volumes of hyperbolic 3-manifolds", *J. Differential Geom.* **43**:4 (1996), 738–782.

Anderson, J. W., Canary, R. D. and McCullough, D., "The topology of deformation spaces of Kleinian groups", *Ann. of Math.* (2) **152**:3 (2000), 693–741.

Antonakoudis, S., "The bounded orbits conjecture for complex manifolds", Harvard preprint.

Arnoux, P. and Yoccoz, J.-C., "Construction de diff'eomorphismes pseudo-Anosov", *C. R. Acad. Sci. Paris* **292** (1981), 75–78.

Astala, K., "Area distortion of quasiconformal mappings", *Acta Math.* **173** (1994), 37–60.

Baba, S., "A Schottky decomposition theorem for complex projective structures", *Geometry and Topology* **14** (2010), 117–151.

Baba, S., "2π-Graftings and complex projective structures", *(Math GT) arXiv:1011.5051.*

Ballmann, W., Gromov, M. and Schroeder, V., "Manifolds of nonpositive curvature", *Progress in Mathematics* **61**, Birkhäuser, Boston, 1985.

Basmajian, A., "Universal length bounds for non-simple closed geodesics on hyperbolic surfaces", *J. Topol.* **6** (2013), 513–524.

Bass, H. "Groups of integral representation type", *Pacific J. Math.* **86**:1 (1980), 15–51.

Beardon, A. F. *The geometry of discrete groups*, Graduate Texts in Mathematics **91**, Springer, New York, 1983.

Beardon, A. F. and Jørgensen, T. "Fundamental domains for finitely generated Kleinian groups", *Math. Scand.* **36** (1975), 21–26.

Beardon, A. F. and Maskit, B. "Limit points of Kleinian groups and finite sided fundamental polyhedra", *Acta Math.* **132** (1974), 1–12.

Beardon, A. F. and Pommerenke, C. "The Poincaré metric of plane domains", *J. London Math. Soc.* (2) **18**:3 (1978), 475–483.

Beardon, A. F. and Stephenson, K. "The uniformization theorem for circle packings", *Indiana Univ. Math. J.* **39**:4 (1990), 1383–1425.

Benedetti, R. and Petronio, C. *Lectures on hyperbolic geometry*, Universitext, Springer, Berlin, 1992.

Bergeron, N., "La conjecture des sous-groups de surfaces", Séminar Bourbaki **no. 811** (2012), 1–28.

Bergeron, N. and Wise, D. T., "A boundary criterion for cubulation", *Amer. J. Math.* **134** no. 3 (2012), 843–859.

Bers, L., "Simultaneous uniformization", *Bulletin A.M.S.* **66** (1960), 94–97.

Bers, L., "On boundaries of Teichmüller spaces and on Kleinian groups, I", *Ann. of Math.* (2) **91** (1970a), 570–600.

Bers, L., "Spaces of Kleinian groups", pp. 9–34 in *Several Complex Variables* (College Park, MD, 1970), vol. I, edited by J. Horváth, Lecture Notes in Math. **155**, Springer, Berlin, 1970b.

Bers, L., "The action of the modular group on the complex boundary", pp. 33–52 in *Riemann Surfaces and Related Topics,* Ann. Math. Stud **97**, Princeton Univ. Press 1981.

Bers, L., "An inequality for Riemann surfaces", pp. 87–93 in *Differential geometry and complex analysis*, Springer, Berlin, 1985.

Bers, L., "On iterates of hyperbolic transformations of Teichmüller space", *Amer. J. Math.* **105** (1983), 1–11.

Bers, L., "An extremal problem for quasiconformal mappings and a theorem of Thurston", *Acta Math.* **141** (1978), 73–98.

Bers, L. and Greenberg, L., "Isomorphisms between Teichmüller spaces", pp. 53–79 in *Advances in the Theory of Riemann Surfaces* (Stony Brook, NY, 1969), edited by L. V. Ahlfors et al., Ann. of Math. Studies **66**, Princeton Univ. Press, Princeton, N.J., 1971.

Bessières, L., Besson, G., Boileau, M., Maillot, S. and Porti, J., *Geometrization of 3-Manifolds*, European Math. Soc., 2010.

Besson, G., Courtois, G. and Gallot, S., "Lemme de Schwarz réel et applications géométriques", *Acta Math.* **183**:2 (1999), 145–169.

Bestvina, M., "Degenerations of the hyperbolic space", *Duke Math. J.* **56**:1 (1988), 143–161.

Bestvina, M., "$\mathbb{R}$ trees in topology, geometry, and group theory", *arXiv/9712210* (2008).

Bestvina, M., "Geometric group theory and 3-manifolds hand in hand: the fulfillment of Thurston's vision", *Bull. Amer. Math. Soc.* **51** (2014), 53–70.

Bestvina, M. and Feighn, M., "A hyperbolic Out(Fn)-complex", *Groups Geom. Dyn.* **4** (2010), 31–58.

Biringer, I. and Souto, J., "Algebraic and geometric convergence of discrete representations into $\mathrm{PSL}_2(\mathbb{C})$", *Geom. & Top.* **14** (2010), 2431–2477.

Biringer, I. and Souto, J., "A finiteness theorem for hyperbolic 3-manifolds", *J. Lon. Math. Soc.* **84** (2011), 227–242.

Biringer, I. and Souto, J., "Rank of the fundamental group and topology of hyperbolic 3-manifolds", 2006 Souto preprint, to appear.

Birman, J. S., "Mapping class groups of surfaces: a survey", pp. 57–71 in *Discontinuous groups and Riemann surfaces* (College Park, MD, 1973), edited by L. Greenberg, Ann. of Math. Studies **79**, Princeton Univ. Press, Princeton, NJ, 1974.

Birman, J. S. and Series, C., "Geodesics with bounded intersection number on surfaces are sparsely distributed", *Topology* **24**:2 (1985), 217–225.

Bishop, C. J. and Jones, P. W., "Hausdorff dimension and Kleinian groups", *Acta Math.* **179**:1 (1997), 1–39.

Bleiler, S. A. and Hodgson, C. D., "Spherical space forms and Dehn filling", *Topology* **35**:3 (1996), 809–833.

Bobenko, A. I. and Springborn, B. A., "Variational principles for circle patterns and Koebe's theorem", *Trans. Amer. Math. Soc.* **356**:2 (2004), 659–689.

Boileau, M. and Porti, J., *Geometrization of 3-orbifolds of cyclic type*, Astérisque **272**, 2001.

Boileau, M., Maillot, S. and Porti, J., *Three-dimensional orbifolds and their geometric structures*, Panoramas et Synthèses **15**, Société Mathématique de France, Paris, 2003.

Boileau, M., Leeb, B. and Porti, J., "Geometrization of 3-dimensional orbifolds", *Ann. of Math.* (2) **162**:1 (2005), 195–290.

Bonahon, F., "Bouts des variétés hyperboliques de dimension 3", *Ann. of Math.* (2) **124**:1 (1986), 71–158.

Bonahon, F., "The geometry of Teichmüller space via geodesic currents", *Invent. Math.* **92**:1 (1988), 139–162.

Bonahon, F., "Shearing hyperbolic surfaces, bending pleated surfaces and Thurston's symplectic form", *Ann. Fac. Sci. Toulouse Math.* (6) **5**:2 (1996), 233–297.

Bonahon, F., "Geodesic laminations on surfaces", pp. 1–37 in *Laminations and foliations in dynamics, geometry and topology* (Stony Brook, NY, 1998), edited by M. Lyubich et al., *Contemp. Math.* **269**, Amer. Math. Soc., Providence, RI, 2001.

Bonahon, F., "Geometric structures on 3-manifolds", pp. 93–164 in *Handbook of geometric topology*, edited by R. J. Daverman and R. B. Sher, North-Holland, Amsterdam, 2002.

Bonahon, F., *Low-Dimensional Geometry,* A.M.S. Providence, 2009.

Bonahon, F. and Otal, J.-P., "Laminations mesurées de plissage des variétés hyperboliques de dimension 3", *Ann. Math.* **160** (2004), 1013–1055.

Bonk, M., "Singular surfaces and meromorphic functions", *Notices Amer. Math. Soc.* **49**:6 (2002), 647–657.

Bowditch, B. H. and Mess, G., "A 4-dimensional Kleinian group", *Trans. Amer. Math. Soc.* **344** (1994), 391–405.

Bowditch, B. H., "A proof of McShane's identity via Markoff triples", *Bull. London Math. Soc.* **28**:1 (1996), 73–78.

Bowditch, B. H., "Markoff triples and quasi-Fuchsian groups", *Proc. London Math. Soc.* (3) **77**:3 (1998), 697–736.

Bowditch, B. H., "A course on geometric group theory", *MSJ Memoirs* **16** Mathematical Society of Japan, Tokyo, 2006.

Bowditch, B. H., "Notes on tameness", *Enseign Math.* **56** (2010), 229-285.

Bowditch, B. H., "An upper bound for injectivity radii in convex cores", *Groups, Geometry, and Dynamics,* **7**, European Math. Soc, (2013), 109–126.

Bowditch, B. H., "The ending lamination theorem", Warwick preprint, preliminary draft (2011).

Bowditch, B. H. and Epstein, D. B. A., "Natural triangulations associated to a surface", *Topology*, **27**(1988), 91–117.

Bowers, P. L. and Stephenson, K., "The set of circle packing points in the Teichmüller space of a surface of finite conformal type is dense", *Math. Proc. Cambridge Philos. Soc.* **111**:3 (1992), 487–513.

Bowers, P. and Stephenson, K., " A branched Andreev-Thurston theom for circle packings of the sphere", *Proc. London Math. Soc.* **73** (1996), 185–215.

Bowers, P. L. and Stephenson, K., *Uniformizing dessins and Belyĭ maps via circle packing*, Mem. Amer. Math. Soc. **805**, Amer. Math. Soc., Providence, 2004.

Brendle, T. E. and Farb, B., "Every mapping class group is generated by 6 involutions", *J. Algebra* **278**:1 (2004), 187–198.

Bridgeman, M., "Average bending of convex pleated planes in hyperbolic three-space", *Invent. Math.* **132**:2 (1998), 381–391.

Bridgeman, M., "Bounds on the average bending of the convex hull boundary of a Kleinian group", *Michigan Math. J.* **51**:2 (2003), 363–378.

Bridgeman, M. and Canary, R., "The Thurston metric on hyperbolic domains and boundaries of convex hulls", *Geometric and Functional Analysis* **20** (2010), 1317–1353.

Bridgeman, M. and Kahn, J., "Hyperbolic volume of n-manifolds with geodesic boundary and orthospectra", *Geometric and Functional Analysis*, **20** (2010).

Bridson, M. R., and Haefliger, A., *Metric spaces of non-positive curvature,* Springer-Verlag, Berlin, 1999.

Brock, J. F., "Continuity of Thurston's length function", *Geom. Funct. Anal.* **10**:4 (2000), 741–797.

Brock, J. F., "Boundaries of Teichmüller spaces and end-invariants for hyperbolic 3-manifolds", *Duke Math. J.* **106**:3 (2001a), 527–552.

Brock, J. F., "Iteration of mapping classes and limits of hyperbolic 3-manifolds", *Invent. Math.* **143**:3 (2001b), 523–570.

Brock, J. F., "The Weil-Petersson metric and volumes of 3-dimensional hyperbolic convex cores", *J. Amer. Math. Soc.* **16** (2003), 495–535.

Brock, J. F. and Bromberg, K. W., "Cone-manifolds and the density conjecture", pp. 75–93 in *Kleinian groups and hyperbolic 3-manifolds*, London Math. Soc. Lecture Note Ser. **299**, Cambridge Univ. Press, Cambridge, 2003.

Brock, J. F. and Bromberg, K. W., "On the density of geometrically finite Kleinian groups", *Acta Math.* **192**:1 (2004), 33–93.

Brock, J. F. and Dumas, D., "Thurston's Jewel: A convex hull image", Online resource Available at http://dumas.io/jewel/.

Brock, J. F. and Dunfield, N., "Injectivity radii of hyperbolic integer homology 3-spheres", Available at www.arXiv.org/math.GT/1304.0391.

Brock, J. F., and Farb, B., "Curvature and rank of Teichmüller space", *Amer. J. Math.* **128** (2006), 1–22.

Brock, J. F. and Margalit, D., "Weil-Petersson isometries via the pants complex", *Proc. A.M.S.* **135** (2007), 795–803.

Brock, J. and Souto, J., "Algebraic limits of geometrically finite manifolds are tame", *Geom. Funct. Anal.* bf 16 (2006), 1–39.

Brock, J., Bromberg, K., Evans, R. and Souto, J., "Tameness on the boundary and Ahlfors' measure conjecture", *Publ. Math. Inst. Hautes Études Sci.* no. 98 (2003), 145–166.

Brock, J., Canary, R. D. and Minsky, Y., "Classification of Kleinian surface groups, II: The ending lamination conjecture", *Ann. of Math.* **176** (2012), 1–149.

Brock, J., Canary, R. D., Bromberg, K. and Minsky, Y., 'Local topology in deformation spaces of hyperbolic 3-manifolds", *Geom. & Topology* **15** (2011),1169–1224.

Bromberg, K., "Hyperbolic Dehn surgery on geometrically infinite 3-manifolds", preprint, 2000. Available at www.arXiv.org/math.GT/0009150.

Bromberg, K., "Hyperbolic cone-manifolds, short geodesics, and Schwarzian derivatives", *J. Amer. Math. Soc.* **17**:4 (2004), 783–826.

Bromberg, K., "Projective structures with degenerate holonomy and the Bers density conjecture", *Ann. of Math.* **166** (2007), 77–93.

Bromberg, K., "The space of Kleinian punctured torus groups is not locally connected", *Duke Math. J.*, **156** (2011), 349–385.

Bromberg, K. and Holt, J., "Self-bumping of deformation spaces of hyperbolic 3-manifolds", *J. Differential Geom.* **57**:1 (2001), 47–65.

Bromberg, K. and Souto, J., "The density conjecture, a prehistoric approach", in preparaton.

Brooks, R., "On the deformation theory of classical Schottky groups", *Duke Math. J.* **52**:4 (1985), 1009–1024.

Brooks, R., "Circle packings and co-compact extensions of Kleinian groups", *Invent. Math.* **86**:3 (1986), 461–469.

Brooks, R., "Twist surfaces", pp. 85–103 in *Random walks and discrete potential theory* (Cortona, 1997), edited by M. Picardello and W. Woess, Symposia Mathematica **39**, Cambridge Univ. Press, Cambridge, 1999.

Brooks, R. and Matelski, J. P., "The dynamics of 2-generator subgroups of PSL(2, $\mathbb{C}$)", pp. 65–71 in *Riemann surfaces and related topics* (Stony Brook, NY, 1978), edited by I. Kra and B. Maskit, Ann. of Math. Stud. **97**, Princeton Univ. Press, Princeton, NJ, 1981.

Brunner, A. M., Frame, M. L., Lee, Y. W. and Wielenberg, N. J., "Classifying torsion-free subgroups of the Picard group", *Trans. Amer. Math. Soc.* **282**:1 (1984), 205–235.

Burger, M. and Canary, R. D., "A lower bound on λ_0 for geometrically finite hyperbolic n-manifolds", *J. Reine Angew. Math.* **454** (1994), 37–57.

Burger, M., Gelander, T., Lubotzky, A. and Mozes, S., "Counting hyperbolic manifolds", *Geom. Funct. Anal.* **12**:6 (2002), 1161–1173.

Buser, P., "Geometry and spectra of compact Riemann surfaces", *Progress in Mathematics* **106**, Birkhäuser, Boston, 1992.

Button, J. O., "Fibred and virtually fibred hyperbolic 3-manifolds in the censuses", *Experiment. Math.* **14**:2 (2005), 231–255.

Calegari, D. and Gabai, D., "Shrinkwrapping and the taming of hyperbolic 3-manifolds", *J. Amer. Math. Soc.* **19** (2006), 385–446.

Callahan, P. J., Dean, J. C. and Weeks, J. R., "The simplest hyperbolic knots", *J. Knot Theory Ramifications* **8**:3 (1999), 279–297.

Canary, R. D., "The Poincaré metric and a conformal version of a theorem of Thurston", *Duke Math. J.* **64**:2 (1991), 349–359.

Canary, R. D., "On the Laplacian and the geometry of hyperbolic 3-manifolds", *J. Differential Geom.* **36**:2 (1992), 349–367.

Canary, R. D., "Ends of hyperbolic 3-manifolds", *J. Amer. Math. Soc.* **6**:1 (1993), 1–35.

Canary, R. D., "A covering theorem for hyperbolic 3-manifolds and its applications", *Topology* **35**:3 (1996), 751–778.

Canary, R. D. and Hersonsky, S., "Ubiquity of geometric finiteness in boundaries of deformation spaces of hyperbolic 3-manifolds", *Amer. J. Math.* **126**:6 (2004), 1193–1220.

Canary, R. D. and McCullough, D., *Homotopy equivalences of 3-manifolds and deformation theory of Kleinian groups*, Mem. Amer. Math. Soc. **812**, Amer. Math. Soc., Providence, 2004.

Canary, R. D. and Minsky, Y. N., "On limits of tame hyperbolic 3-manifolds", *J. Differential Geom.* **43**:1 (1996), 1–41.

Canary, R. D. and Storm, P., "Moduli spaces of hyperbolic 3-manifolds and dynamics of character varieties", *Comm. Math. Helv.* **88** (2013), 221–251.

Canary, R. D. and Storm, P., "The curious moduli space of unmarked Kleinian surface groups", *Amer. Jour. Math.* **134** (2012), 71–85.

Canary, R. D. and Taylor, E., "Kleinian groups with small limit sets", *Duke Math. J.* **73**:2 (1994), 371–381.

Canary, R. D., Epstein, D. B. A. and Green, P., "Notes on notes of Thurston", pp. 3–92 in *Analytical and geometric aspects of hyperbolic space* (Warwick and Durham, 1984), edited by D. B. A. Epstein, London Math. Soc. Lecture Note Ser. **111**, Cambridge Univ. Press, Cambridge, 1987. Reprinted in Canary et al. [2006].

Canary, R. D., Minsky, Y. N. and Taylor, E. C., "Spectral theory, Hausdorff dimension and the topology of hyperbolic 3-manifolds", *J. Geom. Anal.* **9**:1 (1999), 17–40.

Canary, R. D., Culler, M., Hersonsky, S. and Shalen, P. B., "Approximation by maximal cusps in boundaries of deformation spaces of Kleinian groups", *J. Differential Geom.* **64**:1 (2003), 57–109.

Canary, R. D., Marden, A. and Epstein, D. B. A., (editors), *Fundamentals of hyperbolic geometry: selected expositions*, edited by R. D. Canary et al., London Math. Soc. Lecture Note Ser. **328**, Cambridge Univ. Press, Cambridge, 2006.

Cannon, J. W., "The theory of negatively curved spaces and groups", pp. 315–369 in *Ergodic theory, symbolic dynamics, and hyperbolic spaces (Trieste, 1989)*, Oxford Sci. Publ., Oxford Univ. Press, New York, 1991.

Cannon, J. W., "Geometric group theory", pp. 261 305 in *Handbook of geometric topology*, North-Holland, Amsterdam, 2002.

Cannon, J. W., "Cannon's conjecture", in *McGraw-Hill 2011 yearbook of science and technology*, McGraw-Hill, New York (2011).

Cannon, J. W. and Cooper, D., "A characterization of cocompact hyperbolic and finite-volume hyperbolic groups in dimension three", *Trans. Amer. Math. Soc.* **330**:1 (1992), 419–431.

Cannon, J. W. and Thurston, W. P., "Group invariant Peano curves", *Geom. Topol.* **11** (2007), 1315–1355.

Cannon, J. W., Floyd, W. J., Kenyon, R. and Parry, W. R., "Hyperbolic geometry", pp. 59–115 in *Flavors of geometry*, Math. Sci. Res. Inst. Publ. **31**, Cambridge Univ. Press, Cambridge, 1997.

Cannon, J. W., Floyd, W. J. and Parry, W. R., "Twisted face pairing 3-manifolds", *Trans. A.M.S.* **354**, (2002), 2369–2397.

Cao, J. G., "The Bers-Nielsen kernels and souls of open surfaces with negative curvature", *Michigan Math. J.* **41**:1 (1994), 13–30.

Cao, C. and Meyerhoff, G. R., "The orientable cusped hyperbolic 3-manifolds of minimum volume", *Invent. Math.* **146**:3 (2001), 451–478.

Casson, A. J. and Bleiler, S. A., *Automorphisms of surfaces after Nielsen and Thurston*, London Mathematical Society Student Texts **9**, Cambridge Univ. Press, Cambridge, 1988.

Casson, A. and Jungreis, D., "Convergence groups and Seifert fibered 3-manifolds", *Invent. Math.* **118**:3 (1994), 441–456.

Charney, R., "An introduction to right-angled Artin groups", *Geom. Dedicata* **125** (2007), 141–158.

Chavel, I., *Riemannian geometry—a modern introduction*, Cambridge Tracts in Mathematics **108**, Cambridge Univ. Press, Cambridge, 1993.

Choi, Y.-E. and Series, C., "Lengths are coordinates for convex structures", *J. Differential Geom.* **73**:1 (2006), 75–117.

Chow, B. and Knopf, D., "The Ricci flow: an introduction", *Mathematical Surveys and Monographs* **110**, Amer. Math. Soc., Providence, RI, 2004.

Conder, M., Martin, G. and Torstensson, A., "Maximal symmetry groups of hyperbolic 3-manifolds", *New Zealand J. Math.* **35** (2006), 37–62.

Cooper, D., "The volume of a closed hyperbolic 3-manifold is bounded by π times the length of any presentation of its fundamental group", *Proc. Amer. Math. Soc.* **127**:3 (1999), 941–942.

Cooper, D. and Long, D. D. and Reid, A. W., "Essential closed surfaces in bounded 3-manifolds", *J. Amer. Math. Soc.* **10** (1997), no. 3, 553–563.

Cooper, D. and Lackenby, M., "Dehn surgery and negatively curved 3-manifolds", *J. Differential Geom.* **50**:3 (1998), 591–624.

Cooper, D., Hodgson, C. D. and Kerckhoff, S. P., *Three-dimensional orbifolds and cone-manifolds*, MSJ Memoirs **5**, Mathematical Society of Japan, Tokyo, 2000.

Coulson, D., Goodman, O. A., Hodgson, C. D. and Neumann, W. D., "Computing arithmetic invariants of 3-manifolds", *Experiment. Math.* **9**:1 (2000), 127–152.

Coxeter, H. S. M., *Introduction to geometry*, John Wiley & Sons, New York, 1961.

Culler, M., "Lifting representations to covering groups", *Adv. in Math.* **59**:1 (1986), 64–70.

Culler, M. and Dunfield, N., "Snappy", Available at http://www.uic.edu/t3m/SnapPy/doc/ (2012). Software to compute hyperbolic manifolds.

Culler, M. and Shalen, P. B., "Varieties of group representations and splittings of 3-manifolds", *Ann. of Math.* (2) **117**:1 (1983), 109–146.

Culler, M. and Shalen, P. B., "Margulis numbers for Haken manifolds", *math.GT/arXiv:1006. 3467* (2010).

Culler, M. and Vogtmann, K., "Moduli of graphs and automorphisms of free groups", *Invent. Math.* **84** (1986), 91–119.

Dahmani, F. Guirardel, V. Osin, D., "Hyperbolically embedded subgroups and rotating families in groups acting on hyperbolic spaces", *arXiv:1111.7048v3*.

De-Spiller, D. A., "Equimorphisms, and quasiconformal mappings of the absolute", *Dokl. Akad. Nauk SSSR* **194** (1970), 1006–1009.

Donaldson, S., *Riemann surfaces*, Oxford U. Press, New York, 2011.

Douady, A. and Earle, C. J., "Conformally natural extension of homeomorphisms of the circle", *Acta Math.* **157**:1-2 (1986), 23–48.

Doyle, P., "On the bass note of a Schottky group", *Acta Math* **160** (1988), 249–284.

Dumas, D., "The Schwarzian derivative and measured laminations on Riemann surfaces", *Duke Math. J.* **140** (2007), 203–243.

Dumas, D., "Complex projective structures", in *Handbook of Teichmüller space*, A. Papadoupoulos, ed., Eur. Math. Soc. 2009.

Dumas, D., "Skinning maps are finite-to-one", Math GT, arXiv:1203:0273.

Dumas, D. and Kent IV, R. P., "Slicing, skinning, and grafting", *Amer. J. Math.* **131** (2009), 1419–1429.

Dumas, D. and Wolf, M., "Projective structures, grafting and measured laminations", *Geometry and Topology* **12** (2008), 351–386.

Dunbar, W. D. and Meyerhoff, G. R., "Volumes of hyperbolic 3-orbifolds", *Indiana Univ. Math. J.* **43**:2 (1994), 611–637.

Dunfield, N. M., "Surfaces in finite covers of 3-manifolds: the Virtual Haken Conjecture", Video of lecture at GEAR retreat, Urbana, August 2012, www.math.uiuc.edu/~nmd.

Dunfield, N. M. and Thurston, W. P., "The virtual Haken conjecture: experiments and examples", *Geom. Topol.* **7** (2003), 399–441.

Dunfield, N. M. and Thurston, W. P., "Finite covers of random 3-manifolds", *Invent. Math.* **166** (2006), 457–521.

Duren, P. L., *Univalent functions*, Grundlehren der Mathematischen Wissenschaften **259**, Springer, New York, 1983.

Earle, C. J., "The infinite Nielsen kernels of some bordered Riemann surfaces", *Michigan Math. J.* **40**:3 (1993), 445–458.

Earle, C. J. and Kra, I., "On holomorphic mappings between Teichmüller spaces", pp. 107–124 in *Contributions to analysis (a collection of papers dedicated to Lipman Bers)*, Academic Press, New York, 1974.

Earle, C. J. and Marden, A., "Holomorphic plumbing coordinates", in *Quasiconformal mappings, Riemann surfaces, and Teichmueller spaces,* Y. Jiang, S. Mitra, eds., Contemporary Math. Series AMS, **573** (2012), 67–97.

Earle, C. J. and Marden, A., "Holomorphic plumbing coordinates on Teichmueller and compactified moduli space", to appear.

Earle, C. J. and Marden, A., "Disjointness of parabolic canonical regions", to appear.

Earle, C. J., Kra, I. and Krushkal, S. L., "Holomorphic motions and Teichmüller spaces", *Trans. Amer. Math. Soc.* **343**:2 (1994), 927–948.

Edmonds, A. L., "Deformation of maps to branched coverings in dimension two", *Annals of Math.* **110** (1979a), 113–125.

Edmonds, A. L., "Deformation of maps to branched coverings in dimension three", *Math. Ann.* **245** (1979b), 273–279.

Edmonds, A. L., Kulkarni, R. S., Stong, R. E., "Realizability of branched coverings of surfaces," *Trans. Amer. Math. Soc.* **282** (1984), 773–790.

Efremovič, V. A. and Tihomirova, E. S., "Equimorphisms of hyperbolic spaces", *Izv. Akad. Nauk SSSR Ser. Mat.* **28** (1964), 1139–1144. In Russian.

Epstein, C. L., "The hyperbolic Gauss map and quasiconformal reflections", *J. Reine Angew. Math.* **372** (1986), 96–135.

Epstein, D. B. A., "Curves on 2-manifolds and isotopies", *Acta Math.,* **115** (1966). 83–107.

Epstein, D. and Gunn, C., *Supplement to Not knot*, A K Peters, 1991.

Epstein, D. B. A. and Marden, A., "Convex hulls in hyperbolic space, a theorem of Sullivan, and measured pleated surfaces", pp. 113–253 in *Analytical and geometric aspects of hyperbolic space* (Warwick and Durham, 1984), edited by D. B. A. Epstein, London Math. Soc. Lecture Note Ser. **111**, Cambridge Univ. Press, Cambridge, 1987. Reprinted in Canary et al. [2006].

Epstein, D. B. A. and Markovic, V., "Extending homeomorphisms of the circle to quasiconformal homeomorphims of the disc", *Geom. Top.* **10** (2007), 517–595.

Epstein, D. B. A. and Penner, R. C., "Euclidean decompositions of noncompact hyperbolic manifolds", *J. Differential Geom.* **27** (1988), 67–80.

Epstein, D. B. A. and Petronio, C., "An exposition of Poincaré's polyhedron theorem", *Enseign. Math.* (2) **40**:1-2 (1994), 113–170.

Epstein, D. B. A., Cannon, J. W., Holt, D. F., Levy, S. V. F., Paterson, M. S. and Thurston, W. P., *Word processing in groups*, Jones and Bartlett Publishers, Boston, 1992.

Epstein, D. B. A., Marden, A. and Markovic, V., "Quasiconformal homeomorphisms and the convex hull boundary", *Ann. of Math.* (2) **159**:1 (2004), 305–336.

Epstein, D. B. A., Marden, A. and Markovic, V., "Complex earthquakes and deformations of the unit disk", *J. Differential Geom* **73** (2006), 119–166.

Eskin, A., Masur H. and Rafi, K., "Rigidity of Teichmueller space", *arXiv:1506.04774*.

Evans, R., "Weakly type-preserving sequences and strong convergence", *Geom. Dedicata* **108** (2004a), 71–92.

Evans, R. A., "Tameness persists in weakly type-preserving strong limits", *Amer. J. Math.* **126**:4 (2004b), 713–737.

Evans, R. A., "Uniformly bounded radii of balls in convex cores of hyperbolic 3-manifolds", *Auckland preprint* (2005).

Evans, R. A., "McMullen's conjecture for convex cores of hyperbolic 3-manifolds", *Auckland preprint* (2006).

Farb, B., "Relatively hyperbolic groups", *Geom. Funct. Anal.* **8**:5 (1998), 810–840.

Farb, B., "Some problems on mapping class groups and moduli space", *Problems on mapping class groups and related topics*, 11–55, in Proc. Sympos. Pure Math., A.M.S. **74**, 2006.

Farb, B. and Margalit, D., *A Primer on Mapping Class Groups*, Princeton University Press, 2012.

Farkas, H. M. and Kra, I., *Riemann surfaces, 2nd ed.* Graduate Texts in Mathematics **71**, Springer, New York, 1991.

Fathi, A., Laudenbach, F. and Poénaru, V., *Travaux de Thurston sur les surfaces*, Astérisque **66**, Société Mathématique de France, Paris, 1979. English translation, Princeton Univ. Press, 2012. Book review by D. Margalit, *Bull. A.M.S.* **51** (2013).

Fatou, P., "Fonctions automorphes", in *Théorie des fonctions algébriques et de leurs intégrales*, 2nd ed., vol. 2, Gauthier-Villars, Paris, 1930.

Feighn, P. and McCullough, D., "Finiteness conditions for 3-manifolds with boundary", *Amer. J. Math.* **109**:6 (1987), 1155–1169.

Feighn, M. and Mess, G., "Conjugacy classes of finite subgroups of Kleinian groups", *Amer. J. Math.* **113**:1 (1991), 179–188.

Fenchel, W., *Elementary geometry in hyperbolic space*, Studies in Mathematics **11**, de Gruyter, Berlin, 1989.

Fletcher, A. and Markovic, V., *Quasiconformal maps and Teichmüller theory*, Oxford U. Press, New York, 2007.

Floyd, W., Weber, B. and Weeks, J., "The Achilles' heel of O(3, 1)?", *Experiment. Math.* **11**:1 (2002), 91–97.

Ford, L. R., *Automorphic Functions*, McGraw-Hill, New York, 1929. Reprinted by Chelsea, New York, 1951.

Fox, R. H. and Artin, E., "Some wild cells and spheres in three-dimensional space", *Ann. of Math.* (2) **49** (1948), 979–990.

Freedman, M. H. and Gabai, D., "Covering a nontaming knot by the unlink", *Algebraic and Geometric Topology* **7** (2007), 1561–1578.

Frigerio, R., Martelli, B. and Petronio, C., "Small hyperbolic 3-manifolds with geodesic boundary", *Experiment. Math.* **13**:2 (2004), 171–184.

Fujii, M., "Hyperbolic 3-manifolds with totally geodesic boundary which are decomposed into hyperbolic truncated tetrahedra", *Tokyo J. Math.* **13**:2 (1990), 353–373.

Gabai, D., "On 3-Minifolds Finitely Covered by Surface Bundles", pp. 145–155 in *Low dimensional topology and Kleinian groups* (Warwick and Durham, 1984), edited by D. B. a. Epstein, London Math. Soc. Lecture Note Ser. **112**, Cambridge Univ. Press, Cambridge, 1986.

Gabai, D., "Convergence groups are Fuchsian groups", *Ann. of Math.* (2) **136**:3 (1992), 447–510.

Gabai, D., "On the geometric and topological rigidity of hyperbolic 3-manifolds", *J. Amer. Math. Soc.* **10** (1997), 37–74.

Gabai, D., "The Smale conjecture for hyperbolic 3-manifolds: Isom(M^3)≃Diff(M^3)", *J. Differential Geom.* **58** (2001a), 113–149.

Gabai, D., "Almost filling laminations and the connectivity of ending lamination space", *Geometry & Topology* **13** (2009), 1017–1041.

Gabai, D., "On the topology of ending lamination space", *Geom. Topol.* **18** (2014), 2683–2745.

Gabai, D., Meyerhoff, G. R. and Milley, P., "Volumes of tubes in hyperbolic 3-manifolds", *J. Differential Geom.* **57**:1 (2001b), 23–46.

Gabai, D., Meyerhoff, G. R. and Thurston, N., "Homotopy hyperbolic 3-manifolds are hyperbolic", *Ann. of Math.* (2) **157**:2 (2003), 335–431.

Gabai, D., Meyerhoff, G. R. and Milley, P., "Minimum volume cusped hyperbolic three-manifolds", *J. Amer. Math. Soc.* **22** (2009), 1157–1215.

Gabai, D., Meyerhoff, G. R. and Milley, P., "Mom technology and hyperbolic 3-manifold", pp. 84–107 in *In the tradition of Ahlfors-Bers*, Contemp. Math. **510**, Amer. Math. Soc., Providence, RI, 2010.

Gabai, D., Meyerhoff, G. R. and Milley, P., "Mom technology and volumes of hyperbolic 3-manifolds", *Comment. Math. Helv.* **86** (2011), 145–188.

Gallo, D., Kapovich, M. and Marden, A., "The monodromy groups of Schwarzian equations on closed Riemann surfaces", *Ann. of Math.* (2) **151**:2 (2000), 625–704.

Gardiner, F. R. and Lakic, N., "Comparing Poincaré densities", *Ann. of Math.* **154** (2001), 245–267.

Gaster, J., "A family of non-injective skinning maps with critical points", (2012) Available at www.arXiv.org/math.GT/1212.6210.

Gehring, F. W., "Rings and quasiconformal mappings in space", *Trans. Amer. Math. Soc.* **103** (1962), 353–393.

Gehring, F. W. and Martin, G. J., "Commutators, collars and the geometry of Möbius groups", *J. Anal. Math.* **63** (1994), 175–219.

Gilman, J., *Two-generator discrete subgroups of* PSL(2, **R**), Mem. Amer. Math. Soc. **561**, Amer. Math. Soc., Providence, 1995.

Gilman, J., "Algorithms, complexity and discreteness criteria in PSL(2, ℂ)", *J. Anal. Math.* **73** (1997), 91–114.

Gilman, J. and Waterman P., "Classical two-parabolic T-Schottky groups", *J. Anal. Math.* **98** (2006), 1–42.

Goldman, W. M., "Projective structures with Fuchsian holonomy", *J. Differential Geom.* **25**:3 (1987), 297–326.

Goldman, W. M., "Topological components of spaces of representations", *Invent. Math.* **93** (1988), 557–607.

Goldman, W. M., "The modular group action on real SL(2)-characters of a one-holed torus", *Geom. Topol.* **7** (2003), 443–486.

Goodman, O., "Snap", 2006 Available at http://sourceforge.net/projects/snap-pari. Software to computing arithmetic invariants of hyperbolic 3-manifolds.

Goodman, O., Heard, D. and Hodgson, C., "Commensurators of cusped hyperbolic manifolds", *Experiment Math.* **17** (2008), 283–306.

Gordon, C. M. and Luecke, J., "Knots are determined by their complements", *J. Amer. Math. Soc.* **2**:2 (1989), 371–415.

Gray, J., *Linear differential equations and group theory from Riemann to Poincaré*, Birkhäuser, Boston, 1986. 2nd edition, 2000.

Gray, J., 2002. Unpublished manuscript.

Gray, J., *Henri Poincaré: A Scientific Biography*, Princeton U. Press, 2013.

Greenberg, L., "Discrete groups of motions", *Canad. J. Math.* **12** (1960), 415–426.

Greenberg, L., "Discrete subgroups of the Lorentz group", *Math. Scand.* **10** (1962), 85–107.

Greenberg, L., "Fundamental polyhedra for kleinian groups", *Ann. of Math.* (2) **84** (1966), 433–441.

Greenberg, L., "Commensurable groups of Moebius transformations", pp. 227–237 in *Discontinuous groups and Riemann surfaces* (College Park, MD, 1973), edited by L. Greenberg, Ann. of Math. Studies **79**, Princeton Univ. Press, Princeton, NJ, 1974.

Greenberg, L., "Maximal groups and signatures", pp. 207–226 in *Discontinuous groups and Riemann surfaces* (College Park, MD, 1973), edited by L. Greenberg, Ann. of Math. Studies **79**, Princeton Univ. Press, Princeton, NJ, 1974.

Greenberg, L., "Finiteness theorems for Fuchsian and Kleinian groups", pp. 199–257 in *Discrete groups and automorphic functions* (Cambridge, 1975), edited by W. Harvey, Academic Press, London, 1977.

Greenberg, L., "Homomorphisms of triangle groups into PSL(2, **C**)", pp. 167–181 in *Riemann surfaces and related topics* (Stony Brook, NY, 1978), edited by I. Kra and B. Maskit, Ann. of Math. Stud. **97**, Princeton Univ. Press, Princeton, NJ, 1981.

Gromov, M., "Groups of polynomial growth and expanding maps", *Inst. Hautes Études Sci. Publ. Math.* no. 53 (1981a), 53–73.

Gromov, M., "Hyperbolic manifolds (according to Thurston and Jørgensen)", pp. 40–53 in *Bourbaki Seminar, 1979/80*, Lecture Notes in Math. **842**, Springer, Berlin, 1981b.

Gromov, M., "Hyperbolic groups", pp. 75–263 in *Essays in group theory*, Math. Sci. Res. Inst. Publ. **8**, Springer, New York, 1987.

Gromov, M. and Thurston, W., "Pinching constants for hyperbolic manifolds", *Invent. Math.* **89**:1 (1987), 1–12.

Gunn, C. and Maxwell, D., *Not Knot Video*, The Geometry Center, UMN; DVD distributed by A.K. Peters/Taylor and Francis, 1991.

Guo, R., "Characterizations of hyperbolic geometry among Hilbert geometries: A survey", in *Handbook of Hilbert Geometry*, A. Papadopoulos, ed., to appear.

Halpern, N., "A proof of the collar lemma", *Bull. London Math. Soc.* **13** (1981), 141–144.

Hamenstädt, U., "Length functions and parameterizations of Teichmüller space for surfaces with cusps", *Ann. Acad. Sci. Fenn.* **28** (2003), 75–88.

Hamenstädt, U., "Train tracks and the Gromov boundary of the complex of curves" pp. 187–207 in *Spaces of Kleinian groups*, London Math. Soc. Lecture Note Ser. **329**, Cambridge Univ. Press, Cambridge, 2006.

Hamenstädt, U., "Geometry of the mapping class group III: Quasiisometric rigidity", *arXiv: mathGT/0512429* (2007).

Hartshorn, K., "Heegaard splittings of Haken manifolds", *Pacific J. Math.* **204** (2002), 61–75.

Harvey, W. J., "Spaces of discrete groups", pp. 295–348 in *Discrete groups and Automorphic Functions* (Cambridge, 1975), edited by W. Harvey, Academic Press, London, 1977.

Harvey, W. J., "Boundary structure of the modular group", pp. 245–251 in *Riemann surfaces and related topics* (Stony Brook, NY, 1978), edited by I. Kra and B. Maskit, Ann. of Math. Stud. **97**, Princeton Univ. Press, Princeton, NJ, 1981.

Heard, D., "Orb," Available at http://www.ms.unimelb.edu.au/~snap/orb.html. software to compute hyperbolic structures on graph complements and orbifolds.

Hejhal, D. A., "Monodromy groups and linearly polymorphic functions", *Acta Math.* **135**:1 (1975), 1–55.

Hempel, J., *3-Manifolds*, Princeton Univ. Press, Princeton, NJ, 1976.

Hempel, J., "Residual finiteness for 3-manifolds", pp. 379–396 in *Combinatorial group theory and topology (Alta, Utah, 1984)*, Ann. of Math. Stud. **111**, Princeton Univ. Press, Princeton, NJ, 1987.

Hempel, J., "3-manifolds from the curve complex", *Topology* **40** (2001), 630–657.

Hidalgo, R. and Maskit, B., "On neoclassical Schottky groups", *Trans. Amer. Math. Soc.* **358** (2006), 4765–4792.

Hildebrand, M. and Weeks, J., "A computer generated census of cusped hyperbolic 3-manifolds", pp. 53–59 in *Computers and mathematics (Cambridge, MA, 1989)*, Springer, New York, 1989.

Hilden, H. M., Lozano, M. T. and Montesinos, J. M. "On knots that are universal", *Topology* **24**:4 (1985), 499–504.

Hilden, H. M., Lozano, M. T., Montesinos, J. M. and Whitten, W., "On universal groups and three-manifolds", *Invent. Math* **87** (1987), 441–456.

Hocking, J. G. and Young, G. S., *Topology*, Addison-Wesley, Reading (MA), 1961.

Hodgson, C. D. and Kerckhoff, S. P., "Rigidity of hyperbolic cone-manifolds and hyperbolic Dehn surgery", *J. Differential Geom.* **48**:1 (1998), 1–59.

Hodgson, C. D. and Weeks, J. R., "Symmetries, isometries and length spectra of closed hyperbolic three-manifolds", *Experiment. Math.* **3**:4 (1994), 261–274.

Hoffman, N., Ichihara, K., Kashiwagi, M., Masai, H., Oishi, S. and Takayasu, A., "Verified computations for hyperbolic 3-manifolds", *arXiv 1310.3410* (2013).

Holt, J., "Multiple bumping of components of deformation spaces of hyperbolic 3-manifolds", *Amer. J. Math.* **125**:4 (2003), 691–736.

Hou, Y., "Kleinian groups of small Hausdorff dimension are classical Schottky groups", *Geom. Topol.* **14** (2010), 473–519.

Hubbard, J. H., *Teichmüller Theory*, Matrix Editions, Ithaca, N.Y., 2006.

Imayoshi, Y. and Taniguchi, M., *An introduction to Teichmüller spaces*, Springer, Tokyo, 1992.

Ito, K., "Exotic projective structures and quasi-Fuchsian space", *Duke Math. J.* **105**:2 (2000), 185–209.

Ivanov, N. V., *Subgroups of Teichmüller modular groups*, Translations of Mathematical Monographs **115**, American Mathematical Society, Providence, RI, 1992.

Ivanov, N. V., "Automorphisms of complexes of curves and of Teichmüller spaces", in *Progress in Knot Theory and Related Topics*, Travaux en Cours **56**, Herman, Paris, 1997, also Internat. Math. Res. Notices 1997, no. 14, 651–666.

Ivanov, N. V., "Mapping class groups", in *Handbook of geometric topology*, North-Holland, Amsterdam, 2002.

Ivanov, N. V., "Arnol'd, the Jacobi identity, and orthocenters", *Amer. Math. Monthly* **118** (2011), 41–65.

Jaco, W., *Lectures on three-manifold topology*, CBMS Regional Conference Series in Mathematics **43**, Amer. Math. Soc., Providence, 1980.

Jaco, W. H. and Shalen, P. B., *Seifert fibered spaces in 3-manifolds*, Mem. A.M.S. **21** no. 220, Providence (1979).

Johannson, K., *Homotopy equivalences of 3-manifolds with boundaries*, Lecture Notes in Math. **761**, Springer, Berlin, 1979.

Jones, G. A., "Counting subgroups of non-Euclidean crystallographic groups", *Math. Scand.* **84 (1)** (1999), 23–39.

Jones, G. A. and Singerman, D., "Theory of maps on orientable surfaces", *Proc. London Math. Soc.* (3) **37**:2 (1978), 273–307.

Jones, G. and Singerman, D., "Belyĭ functions, hypermaps and Galois groups", *Bull. London Math. Soc.* **28**:6 (1996), 561–590.

Jørgensen, T., "On cyclic groups of Möbius transformations", *Math. Scand.* **33** (1973), 250–260.

Jørgensen, T., "On reopening of cusps", 1974a.

Jørgensen, T., "Some remarks on Kleinian groups", *Math. Scand.* **34** (1974b), 101–108.

Jørgensen, T., "On discrete groups of Möbius transformations", *Amer. J. Math.* **98**:3 (1976), 739–749.

Jørgensen, T., "Compact 3-manifolds of constant negative curvature fibering over the circle", *Ann. Math.* (2) **106**:1 (1977a), 61–72.

Jørgensen, T., "A note on subgroups of $SL(2, \mathbf{C})$", *Quart. J. Math. Oxford Ser.* (2) **28**:110 (1977b), 209–211.

Jørgensen, T., "Closed geodesics on Riemann surfaces", *Proc. Amer. Math. Soc.* **72**:1 (1978), 140–142.

Jørgensen, T., "Commutators in SL(2, $\mathbf{C}$)", pp. 301–303 in *Riemann surfaces and related topics* (Stony Brook, NY, 1978), edited by I. Kra and B. Maskit, Ann. of Math. Stud. **97**, Princeton Univ. Press, Princeton, NJ, 1981.

Jørgensen, T., "Composition and length of hyperbolic motions", pp. 211–220 in *In the tradition of Ahlfors and Bers* (Stony Brook, NY, 1998), Contemp. Math. **256**, Amer. Math. Soc., Providence, RI, 2000.

Jørgensen, T., "On pairs of once-punctured tori", pp. 183–207 in *Kleinian groups and hyperbolic 3-manifolds (Warwick, 2001)*, London Math. Soc. Lecture Note Ser. **299**, Cambridge Univ. Press, Cambridge, 2003.

Jørgensen, T. and Kiikka, M., "Some extreme discrete groups", *Ann. Acad. Sci. Fenn. Ser. A I Math.* **1**:2 (1975), 245–248.

Jørgensen, T. and Klein, P., "Algebraic convergence of finitely generated Kleinian groups", *Quart. J. Math. Oxford Ser.* (2) **33**:131 (1982), 325–332.

Jørgensen, T. and Marden, A., "Two doubly degenerate groups", *Quart. J. Math. Oxford Ser.* (2) **30**:118 (1979), 143–156.

Jørgensen, T. and Marden, A., "Algebraic and geometric convergence of Kleinian groups", *Math. Scand.* **66**:1 (1990), 47–72.

Jørgensen, T., Lascurain, A. and Pignataro, T., "Translation extensions of the classical modular group", *Complex Variables Theory Appl.* **19**:4 (1992), 205–209.

Kahn, J. and Markovic, V., "The good pants homology and a proof of the Ehrenpreis conjecture", *Ann. of Math.*, to appear.

Kahn, J. and Markovic, V., "Immersing almost geodesic surfaces in a closed hyperbolic 3-manifold", *Ann. of Math.* **175** (2012a), 1127–1190.

Kahn, J. and Markovic, V., "Counting essential surfaces in a closed hyperbolic 3-manifold", *Geometry & Topology* **16** (2012b). 601–624.

Kamishima, Y. and Tan, S. P., "Deformation spaces on geometric structures", pp. 263–299 in *Aspects of low-dimensional manifolds*, Adv. Stud. Pure Math. **20**, Kinokuniya, Tokyo, 1992.

Kapovich, M., "Hyperbolic manifolds and discrete groups", *Progress in Mathematics* **183**, Birkhäuser, Boston, 2001.

Kapovich, M., "Kleinian groups in higher dimensions", *Progress in Mathematics*, **265** (2007), 485–562.

Kapovich, M., "Dirichlet fundamental domains and topology of projective varieties", *Invent. Math.* **194** (2013), 631–672.

Kazhdan, D. A. and Margulis, G. A., "A proof of Selberg's conjecture", *Mat. Sb. (N.S.)* **75**:(117) (1968), 163–168. In Russian; translated in *Math. USSR Sbornik* **4** (1968), 147–152.

Keen, L., Rauch, H. E. and Vasquez, A. T., "Moduli of punctured tori and the accessory parameter of Lamé's equation", *Trans. A. M. S.* **255** (1979), 201–230.

Kellerhals, R., "Volumes of cusped hyperbolic manifolds", *Topology* **37**:4 (1998), 719–734.

Kellerhals, R. and Zehrt, T., "The Gauss-Bonnet formula for hyperbolic manifolds of finite volume", *Geom. Dedicata* **84**:1–3 (2001), 49–62.

Kent IV, R. P., "Skinning maps", *Duke Math. J.* **151** (2010), 279–336.

Kerckhoff, S. P., "The asymptotic geometry of Teichmüller space", *Topology* **19**:1 (1980), 23–41.

Kerckhoff, S. P., "The Nielsen realization problem", *Ann. of Math.* (2) **117**:2 (1983), 235–265.

Kerckhoff, S. P., "Earthquakes are analytic", *Comment. Math. Helv.* **60**:1 (1985), 17–30.

Kerckhoff, S. P., "The measure of the limit set of the handlebody group", *Topology* **29**:1 (1990), 27–40.

Kerckhoff, S. P., " Lines of minima in Teichmüller space", *Duke Math. J.* **65** (1992), 187–213.

Kerckhoff, S. P. and Thurston, W. P., "Noncontinuity of the action of the modular group at Bers' boundary of Teichmüller space", *Invent. Math.* **100**:1 (1990), 25–47.

Klarreich, E., "Semiconjugacies between Kleinian group actions on the Riemann sphere", *Amer. J. Math.* **121**:5 (1999a), 1031–1078.

Klarreich, E., "The boundary at infinity of the curve complex and the relative mapping class group," preprint (1999b). Available at http://www.nasw.org/users/klarreich/research.htm.

Klein, F., "Über die sogenannte nicht-euklidische Geometrie", *Math. Ann.* **4** (1871), 573–625. Page number cited in text is for the English translation, which appears in *Sources of hyperbolic geometry*, translated and edited by John Stillwell, Amer. Math. Soc., Providence, 1996.

Kleineidam, G. and Souto, J., "Algebraic convergence of function groups", *Comment. Math. Helv.* **77**:2 (2002), 244–269.

Kleineidam, G. and Souto, J., "Ending laminations in the Masur domain", pp. 105–129 in *Kleinian groups and hyperbolic 3-manifolds (Warwick, 2001)*, London Math. Soc. Lecture Note Ser. **299**, Cambridge Univ. Press, Cambridge, 2003.

Kleiner, B. and Lott, J., "Notes on Perelman's Papers", *Geometry and Topology* **12** (2008), 2587–2855.

Kojima, S., "Isometry transformations of hyperbolic 3-manifolds", *Topology Appl.* **29**:3 (1988), 297–307.

Kojima, S., "Nonsingular parts of hyperbolic 3-cone-manifolds", pp. 115–122 in *Topology and Teichmüller spaces* (Katinkulta, 1995), edited by S. Kojima et al., World Scientific, River Edge, NJ, 1996.

Kojima, S., "Deformations of hyperbolic 3-cone-manifolds", *J. Differential Geom.* **49**:3 (1998), 469–516.

Kojima, S., Mizushima, S. and Tan, S. P., "Circle packings on surfaces with projective structures: A survey", in *Spaces of kleinian groups*, edited by Y. Minsky et al., London Math. Soc. Lecture Notes **329**, Camb. Univ. Press, 2006.

Komori, Y. and Sugawa, T., "Bers embedding of the Teichmüller space of a once-punctured torus", *Conform. Geom. Dyn.* **8** (2004), 115–142.

Korkmaz, M., "Automorphisms of complexes of curves on punctured spheres and on punctured tori", *Topology Appl.* **95** (1999), 85–111.

Korkmaz, M., "Generating the surface mapping class group by two elements", *Trans. A.M.S.* **357** (2005), 3299–3310.

Koundouros, S., "Universal surgery bounds on hyperbolic 3-manifolds", *Topology* **43**:3 (2004), 497–512.

Kra, I., "On spaces of Kleinian groups", *Comment. Math. Helv.* **47** (1972), 53–69.

Kulkarni, R. S. and Shalen, P. B., "On Ahlfors' finiteness theorem", *Adv. Math.* **76**:2 (1989), 155–169.

Labourie, F., *Lectures on representations of surface groups*, Zurich Lectures in Advanced Mathematics, EMS, 2013.

Lackenby, M. and Meyerhoff, R., "The maximal number of exceptional Dehn surgeries", *Invent. Math* **191** (2013), 341–382.

Lamping, J., Rao, R. and Pirolli, P., "A focus + content technique based on hyperbolic geometry for viewing large hierarchies", pp. 401–408 in *Proceedings of the ACM SIGCHI Conference on Human Factors in Computing Systems*, ACM, New York, 1995.

Le, Thang, "Homology, torsion growth and Mahler measure", *Comment. Math. Helv.* **89** (2014), 719–757.

Le Calvez P., "A periodicity criterion and the section problem on the Mapping Class Group", *arXiv:1202.3106* (2012).

Lecuire, C., "Une caractérisation des laminations géodésiques mesurées de plissage des variétés hyperboliques et ses conséquences", pp. 103–115 in *Séminaire de Théorie Spectrale et Géométrie, vol. 21. Année 2002–2003*, Sémin. Théor. Spectr. Géom. **21**, Univ. Grenoble I, Saint, 2003.

Lecuire, C., "Bending map and strong convergence", Available at www.math.univ-toulouse.fr/~lecuire (2004).

Lecuire, C., "Plissage des vari étés hyperboliques de dimension 3", [Pleating of hyperbolic 3-manifolds] *Invent. Math.* **164** (2006), 85–141.

Lecuire, C., "Continuity of the bending map", *Ann. Fac. Sci. Toulouse Math. (6)***17** (2008), 93–119.

Lecuire, C., "An extension of Masur domain", *Spaces of Kleinian groups*, London Math. Soc. Lecture Note Ser., 329, Cambridge Univ. Press, Cambridge, 2006, 49–56.

Lehner, J., "Discontinuous groups and automorphic functions", *Mathematical Surveys* **8**, Amer. Math. Soc., Providence, 1964.

Lehto, O., *Univalent functions and Teichmüller spaces*, Graduate Texts in Mathematics **109**, Springer, New York, 1987.

Leininger, C. J. and McReynolds, D. B., "Seperable subgroups of mapping class groups", *Topology Appl.* **154** (2007), 1–10.

Leininger, C., Long, D. D. and Reid, A. W., "Commensurators of finitely generated nonfree Kleinian groups", *Algebr. Geom. Topol.* **11** (2011), 605–624.

Leung, N. C. and Wan, T. Y. H., "Harmonic maps and the topology of conformally compact Einstein manifolds", *Math. Res. Lett.* **8**:5-6 (2001), 801–812.

Li, T., "Heegard surfaces and measured laminations I: The Waldhausen conjecture", *Invent. Math.* **167** (2007), 135–177.

Li, T., "An algorithm to determine the Heegaard genus of a 3-manifold", *Geometry & Topology* **15** (2011), 1029–1106.

Li, T., "Rank and Genus of 3-manifolds", *J. Amer. Math. Soc.*, to appear.

Lickorish, W. B. R., "A representation of orientable combinatorial 3-manifolds", *Ann. of Math.* **76** (1962), 531–540.

Lickorish, W. B. R., *An introduction to knot theory*, Graduate Texts in Mathematics **175**, Springer, New York, 1997.

Liu, Y. and Markovic, V., "Homology of closed curves and surfaces in closed hyperbolic manifolds", *arXiv:mathGT/1309.7418v2* (2013).

Long, D. D., Reid, A. W. and Thistlehwaite, M., "Zariski dense surface subgroups in SL(3, $\mathbb{Z}$)", *Geometry and Topology* **15** (2011), 1–9.

Luo, F., "Automorphisms of the complex of curves I: Hyperbolicity", *Invent. Math.* **138** (1999), 103–149.

Lyndon R. C. and Schupp, P. E., *Combinatorial Group Theory* Ergebnisse der Mathematik und ihrer Grenzgebiete, **89** Springer-Verlag, Berlin, (1977).

Maclachlan, C. and Reid, A. W., *The arithmetic of hyperbolic 3-manifolds*, Graduate Texts in Mathematics **219**, Springer, New York, 2003.

Magid, A., "Deformation spaces of kleinian surface groups are not locally connected", *Geometry and Topology* **16** (2012), 1247–1320.

Magnus, W., "Residually finite groups", *Bull. A.M.S.* **75** (1969), 305–316.

Magnus, W., "Noneuclidean tesselations and their groups", *Pure and Applied Mathematics* **61**, Academic Press, New York, 1974.

Magnus, W., "Rings of Fricke characters and automorphism groups of free groups", *Math. Z.* **170**:1 (1980), 91–103.

Maher, J., "Random Heegaard splittings" *J. Topol.* **3** (2010), no. 4, 997–1025.

Maher, J., "Random walks on the mapping class group", *Duke Math. J.* **156** (2011), no. 3, 429–468.

Marden, A., "On finitely generated Fuchsian groups", *Comment. Math. Helv.* **42** (1967), 81–85.

Marden, A., "On homotopic mappings of Riemann surfaces", *Ann. of Math.* (2) **90** (1969), 1–8.

Marden, A., "An inequality for Kleinian groups", pp. 295–296 in *Advances in the Theory of Riemann Surfaces (Proc. Conf., Stony Brook, N.Y., 1969)*, Ann. of Math. Studies, No. 66, Princeton Univ. Press, Princeton, NJ, 1971.

Marden, A., "The geometry of finitely generated kleinian groups", *Ann. of Math.* (2) **99** (1974a), 383–462.

Marden, A., "Kleinian groups and 3-dimensional topology: a survey", pp. 108–121 in *A crash course on Kleinian groups* (San Francisco, CA, 1974), edited by L. Bers and I. Kra, Lecture Notes in Math. **400**, Springer, Berlin, 1974b.

Marden, A., "Schottky groups and circles", pp. 273–278 in *Contributions to analysis (a collection of papers dedicated to Lipman Bers)*, Academic Press, New York, 1974c.

Marden, A., "Universal properties of Fuchsian groups in the Poincaré metric", pp. 315–339 in *Discontinuous groups and Riemann surfaces* (College Park, MD, 1973), edited by L. Greenberg, Ann. of Math. Studies **79**, Princeton Univ. Press, Princeton, NJ, 1974d.

Marden, A., "Geometrically finite Kleinian groups and their deformation spaces", pp. 259–293 in *Discrete groups and automorphic functions* (Cambridge, 1975), edited by W. Harvey, Academic Press, London, 1977.

Marden, A., "Geometric relations between homeomorphic Riemann surfaces", *Bull. Amer. Math. Soc. (N.S.)* **3**:3 (1980), 1001–1017.

Marden, A., "A proof of the Ahlfors Finiteness Theorem", in *Spaces of kleinian groups*, edited by Y. Minsky et al., L.M.S. Lecture Notes **329**, Camb. Univ. Press, 2006.

Marden, A., "Deformations of Kleinian groups", pp. 411–446 in *Handbook of Teichmüller Theory I*, A. Papadopoulos, ed., European Math. Soc., Zürich, 2007.

Marden, A. and Markovic, V., "Characterization of plane regions that support quasiconformal mappings to their domes", *Bull. L.M.S.* **39** (2008), 962–972.

Marden, A. and Strebel, K., "The heights theorem for quadratic differentials on Riemann surfaces", *Acta Math.* **153**:3-4 (1984), 153–211.

Marden, A. and Strebel, K., "On the ends of trajectories", pp. 195–204 in *Differential geometry and complex analysis*, Springer, Berlin, 1985.

Marden, A. and Strebel, K., "Pseudo-Anosov Teichmueller mappings", *J. Analyse Math.* **46** (1986), 194–220.

Marden, A. and Strebel, K., "A characterization of Teichmüller differentials", *J. Differential Geom.* **37**:1 (1993), 1–29.

Margalit, D., "Automorphisms of the pants complex", *Duke Math. J.* **121** (2004), 457–479.

Margulis, G. A., *Discrete Subgroups of Semisimple Lie Groups*, Ergebnisse der Mathematik und ihrer Grenzgebiete, **17**, Berlin: Springer-Verlag, 1991.

Markovic, V., "Quasisymmetric groups", *J. Amer. Math. Soc.*, **19** (2006), 673–715.

Markovic, V., "Realization of the mapping class group by homeomorphisms", *Invent. Math.* **168** (2007), 523–566.

Markovic, V., "Criterion for Cannon's Conjecture", *Geom. Funct. Anal.* **23** (2013), 1035–1061.

Markovic, V., "Harmonic maps between 3-dimensional hyperbolic spaces", *Invent. Math.*, to appear.

Markovic, V., "Harmonic Maps and the Schoen Conjecture", CalTech preprint (2015).

Markovic, V. and Sarić, D., "Teichmüller mapping class group of the universal hyperbolic solenoid", *Trans. A.M.S.* **358** (2006), 2637–2650.

Marshall, T. H. and Martin, G. J., "Minimal co-volume hyperbolic lattices, II; Simple torsion in a Kleinian group", *Ann. of Math.*, **176** (2012), 1–41.

Maskit, B., "A characterization of Schottky groups", *J. Analyse Math.* **19** (1967), 227–230.

Maskit, B., "The conformal group of a plane domain", *Amer. J. Math.* **90** (1968), 718–722.

Maskit, B., "On boundaries of Teichmüller spaces and on Kleinian groups, II," *Ann. of Math.* (2) **91** (1970), 607–639.

Maskit, B., "Self-maps on Kleinian groups", *Amer. J. Math.* **93** (1971), 840–856.

Maskit, B., "Intersections of component subgroups of Kleinian groups", pp. 349–367 in *Discontinuous groups and Riemann surfaces* (College Park, MD, 1973), edited by L. Greenberg, Ann. of Math. Studies **79**, Princeton Univ. Press, Princeton, NJ, 1974.

Maskit, B., "Parabolic elements in Kleinian groups", *Ann. of Math.* (2) **117**:3 (1983), 659–668.

Maskit, B., *Kleinian groups*, Grundlehren der Mathematischen Wissenschaften **287**, Springer, Berlin, 1988.

Masur, H., "On a class of geodesics in Teichmüller space", *Ann. of Math.* (2) **102**:2 (1975), 205–221.

Masur, H., "Interval exchange transformations and measured foliations", *Ann. of Math.* (2) **115**:1 (1982), 169–200.

Masur, H., "Ergodic actions of the mapping class group", *Proc. A.M.S.* **94** (1985), 455–459.

Masur, H., "Measured foliations and handlebodies", *Ergodic Theory Dynam. Systems* **6**:1 (1986), 99–116.

Masur, H., "Hausdorff dimension of the set of nonergodic foliations of a quadratic differential", *Duke math. J.* **66**:3 (1992), 387–442.

Masur, H., "Ergodic theory of translation surfaces", pp. 527–547 in *Handbook of dynamical systems* **1B** Elsevier, Amsterdam, 2006.

Masur, H. A. and Minsky, Y. N., "Geometry of the complex of curves, I: Hyperbolicity", *Invent. Math.* **138**:1 (1999), 103–149.

Masur, H. A. and Minsky, Y. N., "Geometry of the complex of curves, II: Hierarchical structure", *Geom. Funct. Anal.* **10**:4 (2000), 902–974.

Masur, H. and Schleimer, S., "The pants complex has only one end", pp. 209–218 in *Spaces of Kleinian groups*, edited by Y. Minsky, M. Sakuma, C. Series, London Math. Soc. Lecture Notes **329**, Cambridge Univ. Press, Cambridge, 2006.

Masur, H. A. and Scheimer, S., "The geometry of the disk complex", *J. Amer. Math. Soc.* **26** (2013), 1–62.

Masur, H. A. and Smillie, J., "Quadratic differentials with prescribed singularities and pseudo-Anosov diffeomorphisms", *Comment. Math. Helvetici* **68** (1993), 289–307.

Masur, H. and Tabachnikov, S., "Rational billiards and flat structures", pp. 1015–1089 in *Handbook of dynamical systems*, vol. 1A, North-Holland, Amsterdam, 2002.

Masur, H. and Wolf, M., "Teichmüller space is not Gromov hyperbolic", *Ann. Acad. Sci. Fenn. Ser. A I Math.* **20** (1995), 259–267.

Masur, H. and Wolf, M., "The Weil-Petersson isometry group", *Geom. Dedicata* **93** (2002), 177–190.

Matsuzaki, K. and Taniguchi, M., *Hyperbolic manifolds and Kleinian groups*, Oxford Mathematical Monographs, The Clarendon Press Oxford Univ. Press, New York, 1998.

McCarthy, J. and Papadopoulos, A., "The mapping class group and a theorem of Masur-Wolf", *Topology Appl.* **96**(1996), 75–84.

McCarthy, J. and Papadopoulos, A., "The visual sphere of Teichmüller space and a theorem of Masur-Wolf", *Ann. Acad. Sci. Fenn. Math.* **24** (1999), 147–154.

McCullough, D., "Compact submanifolds of 3-manifolds with boundary", *Quart. J. Math. Oxford Ser.* (2) **37**:147 (1986), 299–307.

McCullough, D. and Miller, A., "Homeomorphisms of 3-manifolds with compressible boundary", *Mem. Amer. Math. Soc.* **344**, Amer. Math. Soc., Providence, 1986.

McCullough, D., Miller, A. and Swarup, G. A., "Uniqueness of cores of noncompact 3-manifolds", *J. London Math. Soc.* (2) **32**:3 (1985), 548–556.

McMullen, C., "Iteration on Teichmüller space", *Invent. Math.* **99**:2 (1990), 425–454.

McMullen, C., "Cusps are dense", *Ann. of Math.* (2) **133**:1 (1991), 217–247.

McMullen, C. T., "Renormalization and 3-manifolds which fiber over the circle", *Ann. of Math. Stud.* **142**, Princeton Univ. Press, Princeton, NJ, 1996.

McMullen, C. T., "Complex earthquakes and Teichmüller theory", *J. Amer. Math. Soc.* **11**:2 (1998), 283–320.

McMullen, C. T., "Hausdorff dimension and conformal dynamics, I: Strong convergence of Kleinian groups", *J. Differential Geom.* **51**:3 (1999), 471–515.

McMullen, C. T., "Local connectivity, Kleinian groups and geodesics on the blowup of the torus", *Invent. Math.* **146**:1 (2001a), 35–91.

McMullen, C. T., Program "lim" for computing limit sets of kleinian groups. Available at www.math.harvard.edu/~ctm/programs/ (2001b).

McMullen, C., "Rigidity of Teichmüller curves", *Math. Res. Lett.* 16 (2009), 647–649.

McMullen, C., "Riemann surfaces, dynamics, and geometry", *Harvard Course Notes, 2014.*

McShane, G., "Simple geodesics and a series constant over Teichmüller space", *Invent. Math.* **132**:3 (1998), 607–632.

Meeks, III, W. H. and Yau, S-T., "The equivariant Dehn's lemma and loop theorem", *Comment. Math. Helv.* **56**:2 (1981), 225–239.

Meyerhoff, R., "A lower bound for the volume of hyperbolic 3-manifolds", *Canad. J. Math.* **39**:5 (1987), 1038–1056.

Milnor, J., *Collected papers*, vol. 1, Publish or Perish, Houston, TX, 1994.

Milnor, J., "Towards the Poincaré conjecture and the classification of 3-manifolds", *Notices Amer. Math. Soc.* **50**:10 (2003), 1226–1233.

Minsky, Y. N., "Harmonic maps, length, and energy in Teichmüller space", *J. Differential Geom.* **35** (1992), 151–217.

Minsky, Y. N., "Teichmüller geodesics and ends of hyperbolic 3-manifolds", *Topology* **32** (1993), 625–647.

Minsky, Y. N., "On rigidity, limit sets, and end invariants of hyperbolic 3-manifolds", *J. Amer. Math. Soc.* **7**:3 (1994a), 539–588.

Minsky, Y. N., "On Thurston's ending lamination conjecture", in *Conf. Proc. Lecture Notes Geom. Topology* **3**, K. Johannson, ed., Internat. Press, Cambridge, MA, 1994b.

Minsky, Y. N., "The classification of punctured-torus groups", *Ann. of Math.* (2) **149**:2 (1999), 559–626.

Minsky, Y., "Kleinian groups and the complex of curves", *Geom. Top.* **4** (2000), 117–148.

Minsky, Y. N., "Bounded geometry for Kleinian groups", *Invent. Math.* **146**:1 (2001), 143–192.

Minsky, Y., "The Classification of Kleinian surface groups, I: Models and bounds," *Ann. of Math.*, **171** (2010), 1–107.

Minsky, Y. N., "Combinatorial and geometrical aspects of hyperbolic 3-manifolds", pp. 3–40 in *Kleinian groups and hyperbolic 3-manifolds (Warwick, 2001)*, London Math. Soc. Lecture Note Ser. **299**, Cambridge Univ. Press, Cambridge, 2003.

Mirzakhani, M., "Growth of the number of simple closed geodesics on hyperbolic surfaces", *Ann. Math.* **168** (2008), 97–125.

Miyachi, H., "Cusps in complex boundaries of one-dimensional Teichmüller space", *Conform. Geom. Dyn.* **7** (2003), 103–151.

Mj(Mitra), M., "Cannon-Thurston maps for hyperbolic group extensions", *Topology* **37**:3 (1998a), 527–538.

Mj(Mitra), M., "Cannon-Thurston maps for trees of hyperbolic metric spaces", *J. Differential Geom.* **48**:1 (1998b), 135–164.

Mj, M., "Ending laminations and Cannon-Thurston maps", *Geom. Funct. Anal.* **24** (2014), 297–321.

Mj, M., "Cannon-Thurston maps for kleinian groups", *arXiv:1002.0996v3[math.GT]* (2011), 22pp.

Mj, M., "The Cannon-Thurston maps for surface groups", *Ann. of Math.* **179** (2014), 1–80.

Montesinos, J. M., *Classical tessellations and three-manifolds*, Universitext, Springer, Berlin, 1987.

Moore, R. L., "Concerning upper semi-continuous collections of continuua", *Trans. A.M.S.* **27** (1927), 416–428.

Morgan, J. W., "On Thurston's uniformization theorem for three-dimensional manifolds", pp. 37–125 in *The Smith conjecture (New York, 1979)*, Pure Appl. Math. **112**, Academic Press, Orlando, FL, 1984.

Morgan, J. W., "Recent progress on the Poincaré conjecture and the classification of 3-manifolds", *Bull. Amer. Math. Soc. (N.S.)* **42**:1 (2005), 57–78.

Morgan, J. and Tian, G., *Ricci Flow and the Poincaré Conjecture*, **3V**, Clay Math. Monographs, A.M.S. (2007).

Morgan, J. and Tian, G., "Completion of the proof of the geometrization conjecture," arXiv:0809.4040.

Morgan, J. W. and Shalen, P. B., "Valuations, trees, and degenerations of hyperbolic structures, I," *Ann. of Math.* (2) **120**:3 (1984), 401–476.

Morgan, J. W. and Shalen, P. B., "Degenerations of hyperbolic structures, II: Measured laminations in 3-manifolds", *Ann. of Math.* (2) **127**:2 (1988), 403–456.

Morgan, J. W. and Shalen, P. B., "Degenerations of hyperbolic structures, III: Actions of 3-manifold groups on trees and Thurston's compactness theorem", *Ann. of Math.* (2) **127**.3 (1988), 457–519.

Mosher, L., "Geometric survey of subgroups of mapping class groups," pp. 387–410 in *Handbook of Teichmüller theory. Vol. I.* IRMA Lect. Math. Theor. Phys., **11**, Eur. Math. Soc., Zürich, 2007.

Mostow, G. D., *Strong rigidity of locally symmetric spaces*, Princeton Univ. Press, Princeton, NJ, 1973.

Mulase M. and Penkava, M., "Ribbon graphs, quadratic differentials on Riemann surfaces, and algebraic curves defined over $\overline{\mathbf{Q}}$", *Asian J. Math.* **2**:4 (1998), 875–919.

Mumford, D., "A remark on Mahler's compactness theorem", *Proc. Amer. Math. Soc.* **28** (1971), 289–294.

Mumford, D., Series, C. and Wright, D., *Indra's pearls*, Cambridge Univ. Press, New York, 2002.

Munkres, J., "Obstructions to the smoothing of piecewise-differentiable homeomorphisms", *Ann. of Math.* (2) **72** (1960), 521–554.

Munzner, T., "H3: Laying out large directed graphs in hyperbolic space", pp. 2–10 in *Proc. 1997 IEEE Symposium on Information Visualization*, 1997.

Myers, R., "Simple knots in compact, orientable 3-manifolds", *Trans. Amer. Math. Soc.* **273**:1 (1982), 75–91.

Myers, R., "End reductions, fundamental groups, and covering spaces of irreducible open 3-manifolds", *Geom. Topol.* **9** (2005), 971–990.

Namazi, H. and Souto, J., "Non-realizability and ending laminations; Proof of the Density Conjecture", *Acta Math.* **209** (2012), 323–395.

Nash, J., "The imbedding problem for Riemannian manifolds," *Ann. of Math.* **63** 1956, 20–63.

Neumann, W. D., "Notes on geometry and 3-manifolds", pp. 191–267 in *Low dimensional topology* (Eger, 1996, and Budapest, 1998), edited by J. Károly Böröczky et al., Bolyai Soc. Math. Stud. **8**, János Bolyai Math. Soc., Budapest, 1999.

Neumann, W. D. and Reid, A. W., "Arithmetic of hyperbolic manifolds", pp. 273–310 in *Topology '90* (Columbus, OH, 1990), edited by B. Apanasov et al., de Gruyter, Berlin, 1992.

Nicholls, P. J., *The ergodic theory of discrete groups*, London Mathematical Society Lecture Note Series **143**, Cambridge Univ. Press, Cambridge, 1989.

Ohshika, K., "Ending laminations and boundaries for deformation spaces of Kleinian groups", *J. London Math. Soc.* (2) **42**:1 (1990), 111–121.

Ohshika, K., "Geometrically finite Kleinian groups and parabolic elements", *Proc. Edinburgh Math. Soc.* (2) **41**:1 (1998a), 141–159.

Ohshika, K., "Rigidity and topological conjugates of topologically tame Kleinian groups", *Trans. Amer. Math. Soc.* **350**:10 (1998b), 3989–4022.

Ohshika, K., *Discrete groups*, Translations of Mathematical Monographs **207**, Amer. Math. Soc., Providence, RI, 2002.

Ohshika, K., *Kleinian groups which are limits of geometrically finite groups*, Mem. Amer. Math. Soc. **834**, Amer. Math. Soc., Providence, 2005.

Ohshika, K., "Constructing geometrically infinite groups on boundaries of deformation spaces," *J. Math. Soc. Japan* **61** (2009), 1261–1291.

Ohshika, K., "Realising end invariants by limits of minimally parabolic, geometrically finite groups," *Geometry & Topology* **15** (2011), 827–890.

Ohshika, K. and Soma, T., *Geometry and topology of geometric limits I* arXiv:1002.4266 (2010).

Ol'shanskiĭ, A. Y., "Almost every group is hyperbolic", *Internat. J. Algebra Comput.* **2**:1 (1992), 1–17.

Otal, J.-P., "Sur le nouage des géodésiques dans les variétés hyperboliques", *C. R. Acad. Sci. Paris Sér. I Math.* **320**:7 (1995), 847–852.

Otal, J.-P., *Le théorème d'hyperbolisation pour les variétés fibrées de dimension 3*, Astérisque **235**, 1996. English translation, *The hyperbolization theorem for fibered manifolds of dimension 3*, SMF/AMS Texts and Monographs **7**, Amer. Math. Soc., Providence, RI, 2001.

Otal, J.-P., "Thurston's hyperbolization of Haken manifolds", pp. 77–194 in *Surveys in differential geometry, vol. III* edited by C. C. Hsiung and S.-T. Yau, International Press, Boston, 1998.

Parker J. R. and Series, C., "Bending formulae for convex hull boundaries", *J. Anal. Math.* **67** (1995), 165–198.

Patterson, S. J., "Lectures on measures on limit sets of Kleinian groups", pp. 281–323 in *Analytical and geometric aspects of hyperbolic space* (Warwick and Durham, 1984), edited by D. B. A. Epstein, London Math. Soc. Lecture Note Ser. **111**, Cambridge Univ. Press, Cambridge, 1987. Reprinted in Canary et al. [2006]

Penner, R. C., "The decorated Teichmüller space of punctured surfaces", *Commun. Math. Phys.* **113** (1987), 299–339.

Penner, R. C., "Bounds on least dilatations", *Proc. A.M.S* **113** (1991), 443–450.

Penner, R. C., and Harer, J. L., *Combinatorics of train tracks*, Annals of Mathematics Studies **125**, Princeton Univ. Press, Princeton, NJ, 1992.

Perelman, G., "Finite extinction time for the solutions to the Ricci flow on certain three-manifolds", *arXiv:math.DG/0307245*, (2003a).

Perelman, G., "Ricci flow with surgery on three-manifolds", *arXiv:math.DG/0303109*, (2003b).

Petronio, C. and Porti, J., "Negatively oriented ideal triangulations and a proof of Thurston's hyperbolic Dehn filling theorem", *Expo. Math.* **18**:1 (2000), 1–35.

Petronio, C. and Weeks, J. R., "Partially flat ideal triangulations of cusped hyperbolic 3-manifolds," *Osaka J. Math.* **37**:2 (2000), 453–466.

Pollicott, M. and Sharp, R., "Length asymptotics in higher Teichmüller theory", *Proc. A.M.S.* **142** (2014), 101–112.

Pommerenke, C., "On uniformly perfect sets and Fuchsian groups", *Analysis* **4**:3-4 (1984), 299–321.

Pommerenke, C., *Boundary behaviour of conformal maps*, Grundlehren der Mathematischen Wissenschaften **299**, Springer, Berlin, 1992.

Prasad, G., "Strong rigidity of **Q**-rank 1 lattices", *Invent. Math.* **21** (1973), 255–286.

Przeworski, A., "Tubes in hyperbolic 3-manifolds", *Topology Appl.* **128**:2-3 (2003), 103–122.

Purcell, J. S. and Suoto, J., "Geometric limits of knot complements", *J. Topol* **3** (2010), 759–785.

Rafi, K. and Schleimer, S., "Curve complexes are rigid", *Duke Math. J.* **158** (2011), 225–246.

Ratcliffe, J. G., *Foundations of hyperbolic manifolds*, Graduate Texts in Mathematics **149**, Springer, New York, 1994.

Ratcliffe, J. G. and Tschantz, S. T., "The volume spectrum of hyperbolic 4-manifolds", *Experiment. Math.* **9**:1 (2000), 101–125.

Rees, M., "An alternative approach to the ergodic theory of measured foliations on surfaces", *Ergodic Theory Dynamical Systems* **1**:4 (1981), 461–488 (1982).

Reimann, H. M., "Invariant extension of quasiconformal deformations", *Ann. Acad. Sci. Fenn. Ser. A I Math.* **10** (1985), 477–492.

Riley, R., "A quadratic parabolic group", *Math. Proc. Cambridge Philos. Soc.* **77** (1975), 281–288.

Riley, R., "Applications of a computer implementation of Poincaré's theorem on fundamental polyhedra," *Math. Comp.* **40**:162 (1983), 607–632.

Rivin, I., "A characterization of ideal polyhedra in hyperbolic 3-space", *Ann. of Math.* (2) **143**:1 (1996), 51–70.

Rodin, B. and Sullivan, D., "The convergence of circle packings to the Riemann mapping", *J. Differential Geom.* **26**:2 (1987), 349–360.

Roeder, R. K. W., Hubbard, J. H. and Dunbar, D., "Andreev's theorem on hyperbolic polyhedra," *Ann. Inst. Fourier, Grenoble* **57** (2007), 825–882.

Rolfsen, D., *Knots and links*, Math. Lecture Series **7**, Publish and Perish, Berkeley, 1976. Second printing, 1990.

Royden, H. L., "Automorphisms and isometries of Teichmüller space", pp. 369–383 in *Advances in the theory of Riemann surfaces* (Stony Brook, NY, 1969), edited by L. Ahlfors et al., Ann. of Math. Studies **66**, Princeton Univ. Press, Princeton, NJ, 1971.

Rüedy, R. A., "Embeddings of open Riemann surfaces", *Comment. Math. Helv.* **46** (1971), 214–225.

Sageev, M., "Ends of group pairs and non-positively curved cube complexes", *Proc. London Math. Soc.* **71** (1995), 585–617.

Sageev, M., "CAT(0) cube complexes and groups," *IAS/Park City mathematics series*, A.M.S.

Scannell, K. P. and Wolf, M., "The grafting map of Teichmüller space", *J. Amer. Math. Soc.* **15**:4 (2002), 893–927.

Schafer, J. A., "Representing homology classes on surfaces", *Canad. Math. Bull.* **19**:3 (1976), 373–374.

Schleimer, S., lecture at a Columbia conference, Aug. 2013.

Schneps, L., (editor), *The Grothendieck theory of dessins d'enfants*, edited by L. Schneps, London Mathematical Society Lecture Note Series **200**, Cambridge Univ. Press, Cambridge, 1994.

Schwartz, R. E., "A conformal averaging process on the circle", *Geometriae Dedicata* **117** (2006), 19–46.

Scott, G. P., "Compact submanifolds of 3-manifolds", *J. London Math. Soc.* (2) **7** (1973a), 246–250.

Scott, G. P., "Finitely generated 3-manifold groups are finitely presented", *J. London Math. Soc.* (2) **6** (1973b), 437–440.

Scott, P., "Subgroups of surface groups are almost geometric", *J. London Math. Soc.* (2) **17**:3 (1978), 555–565.

Scott, P., "A new proof of the annulus and torus theorems", *Amer. J. Math.* **102**:2 (1980), 241–277.

Scott, P., "The geometries of 3-manifolds", *Bull. London Math. Soc.* **15**:5 (1983), 401–487.

Scott, P. and Tucker, T., "Some examples of exotic noncompact 3-manifolds", *Quart. J. Math. Oxford Ser.* (2) **40**:160 (1989), 481–499.

Selberg, A., "On discontinuous groups in higher-dimensional symmetric spaces", pp. 147–164 in *Contributions to function theory (internat. Colloq. Function Theory, Bombay, 1960)*, Tata Institute of Fundamental Research, Bombay, 1960.

Seppälä, M. and Sorvali, T., "Paramaterization of Teichmüller spaces by geodesic length functions," in *Holomorphic Functions and Moduli, Vol II*, Drasin, et al eds., MSRI publ., Springer-Verlag, 1988.

Sen, H. "Virtual domination of 3-manifolds," *http://msp.org/gt/2015/19-4/p10.xhtml.*

Series, C., "An extension of Wolpert's derivative formula", *Pacific J. Math.* **197** (2001), 223–239.

Series, C., "Thurston's bending measure conjecture for once punctured torus groups", pp. 75–89 in *Spaces of Kleinian groups*, L.M.S. Lecture Note **329**, Cambridge Univ. Press, Cambridge, 2006.

Shalen, P., "A generic Margulis number for hyperbolic 3-manifolds", in pp. 103–109 in *Topology and geometry in dimension three,* Contemp. Math. **560**, A.M.S., 2011.

Shiga, H. and Tanigawa, H., "Projective structures with discrete holonomy representations", *Trans. Amer. Math. Soc.* **351**:2 (1999), 813–823.

Shimizu, H., "On discontinuous groups operating on the product of the upper half planes", *Ann. of Math.* (2) **77** (1963), 33–71.

Slodkowski, Z., "Holomorphic motions and polynomial hulls", *Proc. Amer. Math. Soc.* **111** :2 (1991), 347–355.

Soma, T., "Geometric limits of quasi-fuchsian groups", math.GT/0702725.

Soma, T., "Existence of ruled wrappings in hyperbolic 3-manifolds", *Geom. Topol.* **10** (2006), 1173–1184.

Souto, J., "A note on the tameness of hyperbolic 3-manifolds", *Topology* **44**:2 (2005), 459–474.

Souto, J., "Short geodesics in hyperbolic compression bodies are not knotted," www.math.ubc.ca/~jsouto.

Springer, G., *Introduction to Riemann surfaces*, reprinted by AMS Chelsea Pub., 2002.

Stephenson, K., "The approximation of conformal structures via circle packing", pp. 551–582 in *Computational methods and function theory* (Nicosia, 1997), Ser. Approx. Decompos. **11**, World Scientific, River Edge, NJ, 1999.

Stephenson, K., "Circle packing: a mathematical tale", *Notices Amer. Math. Soc.* **50**:11 (2003), 1376–1388.

Stephenson, K., *Introduction to circle packing*, Cambridge Univ. Press, Cambridge, 2005.

Stillwell, J., *Sources of hyperbolic geometry*, History of Mathematics **10**, Amer. Math. Soc., Providence, RI, 1996.

Storm, P. A., "Minimal volume Alexandrov spaces", *J. Differential Geom.* **61**:2 (2002), 195–225.

Storm, P., "The barycenter method on singular spaces", *Comment. Math. Helv.* **82** (2007), 133–173.

Strebel, K., *Quadratic differentials*, vol. 5, Ergebnisse der Mathematik und ihrer Grenzgebiete (3) [Results in Mathematics and Related Areas (3)], Springer, Berlin, 1984.

Struik, D. J., *Lectures on classical differential geometry*, Addison-Wesley, Cambridge, MA, 1950.

Sturm, J. and Shinnar, M., "The maximal inscribed ball of a Fuchsian group", pp. 439–443 in *Discontinuous groups and Riemann surfaces* (College Park, MD, 1973), edited by L. Greenberg, Ann. of Math. Studies **79**, Princeton Univ. Press, Princeton, NJ, 1974.

Sullivan, D., "The density at infinity of a discrete group of hyperbolic motions", *Inst. Hautes Čtudes Sci. Publ. Math.* 50 (1979), 171–202.

Sullivan, D., "On the ergodic theory at infinity of an arbitrary discrete group of hyperbolic motions", pp. 465–496 in *Riemann surfaces and related topics* (Stony Brook, NY, 1978), edited by I. Kra and B. Maskit, Ann. of Math. Stud. **97**, Princeton Univ. Press, Princeton, NJ, 1981.

Sullivan, D., "Quasiconformal homeomorphisms and dynamics, II: Structural stability implies hyperbolicity for Kleinian groups", *Acta Math.* **155**:3-4 (1985), 243–260.

Sullivan, D., "Related aspects of positivity in Riemannian geometry", *J. Differential Geom.* **25**:3 (1987), 327–351.

Sun, H., "Virtual domination of 3-manifolds", *MathGT/arXiv:1401.7049* (2014).

Swarup, G. A., "Two finiteness properties in 3-manifolds", *Bull. London Math. Soc.* **12**:4 (1980), 296–302.

Tanigawa, H., "Grafting, harmonic maps and projective structures on surfaces", *J. Differential Geom.* **47**:3 (1997), 399–419.

Thurston, W. P., "The geometry and topology of three-manifolds", lecture notes, Princeton University, 1979a. The 2003 electronic edition is available at Available at http://msri.org/publications/books/gt3m.

Thurston, W., "Hyperbolic geometry and 3-manifolds", pp. 9–25 in *Low-dimensional topology (Bangor, 1979)*, London Math. Soc. Lecture Note Ser. **48**, Cambridge Univ. Press, Cambridge, 1979b.

Thurston, W. P., "Hyperbolic structures on 3-manifolds: overall logic", lecture notes, Bowdoin College, 1980.

Thurston, W. P., "Three-dimensional manifolds, Kleinian groups and hyperbolic geometry", *Bull. Amer. Math. Soc.* (*N.S.*) **6**:3 (1982), 357–381.

Thurston, W. P., "Earthquakes in two-dimensional hyperbolic geometry", pp. 91–112 in *Analytical and geometric aspects of hyperbolic space* (Warwick and Durham, 1984), edited by D. B. A. Epstein, London Math. Soc. Lecture Note Ser. **112**, Cambridge Univ. Press, Cambridge, 1986a. Reprinted in Canary et al. [2006].

Thurston, W. P., "Hyperbolic structures on 3-manifolds, I: Deformation of acylindrical manifolds", *Ann. of Math.* (2) **124**:2 (1986b), 203–246.

Thurston, W. P., "Hyperbolic structures on 3-manifolds, II: surface groups and 3-manifolds which fiber over the circle", preprint, 1986c. Available at www.arXiv.org/abs/math.GT/9801045.

Thurston, W. P., "Zippers and univalent functions", pp. 185–197 in *The Bieberbach conjecture (West Lafayette, Ind., 1985)*, Math. Surveys Monogr. **21**, Amer. Math. Soc., Providence, RI, 1986d.

Thurston, W. P., "On the geometry and dynamics of diffeomorphisms of surfaces", *Bull. Amer. Math. Soc.* (*N.S.*) **19**:2 (1988), 417–431.

Thurston, W. P., *Three-dimensional geometry and topology*, vol. 1, S. Levy, ed., Princeton Mathematical Series **35**, Princeton Univ. Press, Princeton, NJ, 1997.

Thurston, W. P. "Minimal stretch maps between hyperbolic surfaces", preprint, 1998. Available at www.arXiv.org/abs/math.GT/9801039.

Tucker, T. W., "Non-compact 3-manifolds and the missing-boundary problem", *Topology* **13** (1974), 267–273.

Tucker, T. W., "A correction to a paper of A. Marden (*Ann. of Math.* (2) **99** (1974), 383–462)", *Ann. of Math.* (2) **102**:3 (1975), 565–566.

Tukia, P., "On two-dimensional quasiconformal groups," *Ann. Acad. Sci. Fenn. Ser. A I Math.* **5** (1980), 73–78.

Tukia, P., "Differentiability and rigidity of Möbius groups", *Invent. Math.* **82**:3 (1985a), 557–578.

Tukia, P., "On isomorphisms of geometrically finite Möbius groups", *Inst. Hautes Études Sci. Publ. Math.* no. 61 (1985b), 171–214.

Tukia, P., "Quasiconformal extension of quasiisometric mappings compatible with a Möbius group", *Acta Math.* **154** (1985c), 153–193.

Tukia, P., 2005. private communication.

Van Vleck, E. B., "On the combination of non-loxodromic substitutions", *Trans. Amer. Math. Soc.* **20**:4 (1919), 299–312.

Vogtmann, K., "What is outer space?" *AMS Notices* **55** (7), Aug. 2008.

Vuorinen, M., *Conformal geometry and quasiregular mappings*, Lecture Notes in Math. **1319**, Springer, Berlin, 1988.

Wada, M., "Exploring the space of twice punctured torus groups", Aug. 3 2003. Lecture at the Cambridge workshop *Spaces of Kleinian Groups and Hyperbolic 3-Manifolds*.

Wada, M., "OPTi's algorithm for discreteness determination", *Experiment. Math.* **15**:1 (2006), 61–66.

Wada, M., "OPTi", A computer program to visualize quasiconformal deformations of once-punctured torus groups (2011). Available at http://www.math.sci.osaka-u.ac.jp/~wada.

Waldhausen, F., "On irreducible 3-manifolds which are sufficiently large", *Ann. of Math.* (2) **87** (1968), 56–88.

Wang, H. C., "Discrete nilpotent subgroups of Lie groups", *J. Differential Geometry* **3** (1969), 481–492.

Wang, H. C., "Topics on totally discontinuous groups", pp. 459–487 in *Symmetric spaces* (St. Louis, MO, 1969–1970), edited by W. M. Boothby and G. L. Weiss, Pure and Appl. Math. **8**, Dekker, New York, 1972.

Weeks, J. R., *The shape of space* **2nd ed.** Monographs and Textbooks in Pure and Applied Mathematics, **249**, Marcel Dekker, Inc., New York, 2002.

Weeks, J. R., "Convex hulls and isometries of cusped hyperbolic 3-manifolds", *Topology Appl.* **52**:2 (1993), 127–149.

Weeks, J., "The Poincaré dodecahedral space and the mystery of the missing fluctuations", *Notices Amer. Math. Soc.* **51**:6 (2004), 610–619.

Weeks, J., "SnapPea," Available at www.geometrygames.org/SnapPea/ (2012). Then click on "SnapPy", or go to www.math.uic.edu/t3m/SnapPy/doc/. Original software to compute hyperbolic manifolds.

Weeks, J., "Computation of hyperbolic structures in knot theory", in *Handbook of knot theory*, edited by W. W. Menasco and M. B. Thistlethwaite, Elsevier, Amsterdam, 2005.

Weiss, H., "Local rigidity of 3-dimensional cone-manifolds," *J. Differential Geom.* **71** (2004), 437–506.

Whitehead, G. W., *Elements of homotopy theory*, Graduate Texts in Mathematics **61**, Springer, New York, 1978.

Wielenberg, N. J., "Discrete Moebius groups: fundamental polyhedra and convergence", *Amer. J. Math.* **99**:4 (1977), 861–877.

Wielenberg, N., "The structure of certain subgroups of the Picard group", *Math. Proc. Cambridge Philos. Soc.* **84**:3 (1978), 427–436.

Wielenberg, N. J., "Hyperbolic 3-manifolds which share a fundamental polyhedron", pp. 505–513 in *Riemann surfaces and related topics* (Stony Brook, NY, 1978), edited by I. Kra and B. Maskit, Ann. of Math. Stud. **97**, Princeton Univ. Press, Princeton, NJ, 1981.

Wise, D. T., "The structure of groups with a quasiconvex hierarchy", www.math. mcgill.ca/wise/papers.html (2011) 187pp.

Wise, D. T., *From riches to raags: 3-manifolds, right angled Artin groups, and cubical geometry,* CBMS Lecture Notes, 2012.

Wolpert, S., "The length spectrum as moduli for compact Riemann surfaces", *Annals of Math.* **109** (1979), 323–351.

Wolpert, S., "An elementary formula for the Fenchel-Nielsen twist", *Comment. Math. Helvetici* **56** (1981), 132–135.

Wolpert, S. A., "The Weil-Petersson metric geometry", pp. 47–64 in *Handbook of Teichmüller theory, Vol. II*, IRMA Lect. Math. Theor. Phys. **13**, Eur. Math. Soc., Zürich, 2009.

Wright, D. J., "Searching for the cusp", in *Spaces of kleinian groups*, edited by Y. Minsky et al., London Math. Soc. Lecture Notes **329**, Camb. Univ. Press, 2006.

Yamada, A., "On Marden's universal constant of Fuchsian groups. II," *J. Analyse Math.* **41** (1982), 234–248.

Zhu, X. and Bonahon, F., "The metric space of geodesic laminations on a surface: I", *Topology and Geometry*, **8** (2004), 539–564.

Index

2π Lemma, 359

$\mathbb{H}^2$, $\mathbb{H}^3$; hyperbolic 2D, 3D space, 12

$\Lambda(G)$; limit set of group G, 62

$\mathcal{M}(G)$; hyperbolic manifold
$\qquad \mathbb{H}^3 \cup \Omega(G)/G$, 66

$\Omega(G)$; discontinuity set of group G, 64

δ-hyperbolic, δ-thinness, 109

λ-Lemma, 362

$\log(2k - 1)$ theorem on group actions, 272, 439

$\mathcal{MCG}(R)$; mapping class group of surface R, 91

$\mathfrak{R}(G)$; representation space of group G, 277

$\mathfrak{R}_{\mathrm{disc}}(G)$; discreteness locus $= \mathcal{AH}(G)$, 279

$\mathfrak{Teich}(R)$; Teichmüller space of surface R, 87

$\mathfrak{T}(R)$, $\mathfrak{T}(G)$; quasifuchsian space, surface R, group G, 280

Abikoff, William, 198

absolute measure of length, 7

accidental parabolic, 198, 239

Accola, Robert D. M., 65

acylindrical manifold, 198, 239, 382

Adams, Colin, 191, 252

Adams, Scot, xviii

Agard, Steve, 96

Agol, Ian, xvii, xviii, 84, 114, 248, 252, 293, 300, 358, 359, 363, 386, 389, 396, 405, 422

Ahlfors, Lars, xiii, 17, 25, 42, 61, 77, 91, 94, 96, 154, 186, 200, 204, 308, 332, 336, 368, 456
    Conjecture/Theorem, 184, 202, 295
    Finiteness Theorem, 64, 66, 115, 122, 192, 194, 234, 363

Akiyoshi, Hirotaka, 144

algebraic convergence, 219

algebraic surface, 77

Anderson, James, xvii, 81, 196, 209, 238, 284, 285, 329, 347, 349, 416, 439

Andreev-Thurston Theorem, 10

annulus, 25
    modulus of, 336

Anosov mappings of tori, 340

anti-Möbius transformation, 1, 44

Antonakoudis, Stergios, 381

Aougab, Terik, 103

area
    as a function of topology, 187
    of disk, ball, 16
    of tube boundary, 104

arithmetic kleinian group, 401

Arnoux, Pierre, 343

Artin, Emil, 295

Astala, Kari, 202

atoroidal manifold, 382
automatic group, 111
automorphism
  Dehn twist, 322
    iterated, 321
  discrete group on $\mathbb{S}^1$, 395
  extention from $\mathbb{S}^1$ to disk, 395
  inner and outer, 283, 286, 353
    of a 3-manifold, 370
  of fundamental group, 283
  pseudo-Anosov, 322
  reducible, 322
    pseudo-Anosov, 322
automorphism of a surface, 371, *see also*
    Dehn twist
  Anosov maps on tori, 102
  finite order, 92
    Nielsen Realization Problem, 92

B-groups, 309
Baba, Shinpei, 415
Baba, Shipei, xviii
ball
  circumference, 17
  volume, 17
ball, upper halfspace (UHS) models
  formulas for ball model, 30
Ballmann, Werner, 358
baseball, 11
Basmajian, Ara, xviii, 217
Bass, Hyman, 383
Beardon, Alan, 42, 133, 142, 145, 185,
    199, 267, 444, 457
Beltrami differential, 90, 288, 353, 368
Beltrami, Eugenio, xvi
  differential, 86
    for finitely generated kleinian
        group, 183
  equation, 85, 86
Belyĭ functions, 117, 461
bending
  angle, 168

lamination, 170, 178, 179, 296, 302,
    366
  lines, 172
  measure, 169
    existence theorem, 178
Benedetti, Riccardo, 250
Bergeron, Nicolas, 386, 389
Bers (analytic) boundary
  geometric limits, 321, 346
  limit of iteration, 322
  locally connected case, 311
Bers slice, 306
  Bers (analytic) boundary, 310, 353
  extended, 308, 413
    quasifuchsian locus, 413
  extened, 415
Bers, Lipman, 94, 145, 272, 281, 308,
    310, 336, 369
  conjecture, *see* Density Conjecture
Bessiéres, Laurent, 395
Besson, Gérard, 188, 395
Bestvina, Mladen, 258, 259, 261, 354,
    390
Betti number/rank of $H_1(M^3)$, 387
Bianchi groups, 400
bilipschitz map, 201
billiards, 116
Biringer, Ian, 239, 334
Birman, Joan, 160, 342
Bishop, Christopher, 201, 202, 334
Bleiler, Steven, 34, 80, 192, 358
Bobenko, Alexander, 10, 268
Boileau, Michel, 199, 395, 402, 403
Bólyai, János, xvi
Bonahon, Francis, xvii, 26, 158, 160,
    164, 165, 172, 178, 180, 195, 214,
    270, 282, 293, 295, 297, 298, 302,
    316, 385, 407
  Criteria A and B, 292, 294, 310, 313,
    357
Bonk, Mario, 77
Borromean rings, 398, 399, 401
  approximation of complement, 397

boundary
  groups, 309
  parallel embedded surface, 333
boundary component
  compressible (indecomposible), 355
  conformal, 73
  ideal, 76
  incompressible, 151, 382
bounded geometry, 303, 329
Bounded Image Theorem, 382
Bowditch
  manifold constant, 176
Bowditch, Brian, xviii, 24, 75, 110, 176,
    184, 215, 279, 298, 300, 438
Bowers, Philip, 10, 268
brain cortex, 268
branch cover, point, value, 68, 80
branch locus, 70
Brendle, Tara, 92
Bridgeman, Martin, 174, 178, 198, 378
Bridson, Martin R., 294
Brin, Matthew, 358
Brock, Jeffrey, xi, xviii, 238, 239, 293,
    299, 300, 302, 303, 305, 314, 316,
    321, 322, 352, 353, 373, 390, 406,
    416
Bromberg, Kenneth, xviii, 50, 238, 239,
    287, 289, 293, 305, 310, 314, 373,
    406, 410, 411, 416
Brooks, Robert, xi, 117, 264, 440
Brunner, Andrew, 401
bumping, 289, 416, 418, 419
  self-bumping, 289, 310
Burger, Marc, 253, 334
Buser, Peter, 133, 456
Button, Jack, 393

Calegari, Daniel, 293, 300
Callahan, David, 393
Canary, Richard, xviii, 11, 115, 158,
    160, 172, 173, 198, 201, 238, 279,
    282, 285, 286, 288, 292–295,

298–300, 303, 334, 337, 347, 349,
    354, 356–359, 363, 416, 439
Cannon Conjecture, 110
Cannon, James, 36, 38, 42, 110, 111, 199
Cannon-Thurston mappings, 327
  a sufficient condition, 370
  degenerate Schottky groups, 332
  of kleinian groups, 330
  singly or doubly degenerated
    quasifuchsian groups, 330
Cao, Chun, 251
Cao, Jian Guo, 212
Carathéodory convergence, 233
Casson, Andrew, 34, 80, 341, 386, 395
CAT(−1), 294
CAT(0), 393
Cauchy–Riemann equations, 85
Cayley graph, 108
  δ-thin, 109
  dual to polyhedra tessellation, 143
  geodesic, 108
Cayley–Hamilton identity, 425
census of manifolds, 393
cerebral cortex, 268
character variety, 362
Chavel, Isaac, 262
Cheeger constant, 335
Cheeger, Jeff, 34
Choi, Young-Eun, 179
Chow, Bennett, 395
circle packing, 10, 264, 413
  obtaining polyhedra, 268
circles
  euclidean and hyperbolic centers, 26
circumference of a disk, 16
closed manifold, 69
collapsing laminations, 329
collapsing map, 329, 339
Collar Lemma, 133, 456
combining groups, 152, 193
commensurable groups, 196
commensurator of a group, 196
commutator, 2, 19

subgroup, 67

compact core, 180, 197, 281, 284, 291, 359, 435
   relative compact core, 181, 281, 291
   ends of manifold, 291, 357

companion knot/link, 396, 397

complex length/distance, 431, 438, 441, 446
   between lines, 431

complex probabilities, 47

complex projective structure, 412
   grafting, 413
   monodromy (holonomy) group, 413

composition of Möbius t., 2

compressible/incompressible
   boundary component, 149
   surface, 151

compressible/incompressible surface, 151

compressing curve, 83

compressing disk, 151

compression body, 83, 194, 354
   embedding in $\mathbb{S}^3$, 397

computer software, *see also* Snap, SnapPi, OPTi
   for Bers slices, 415
   for cyclic loxodromic groups, 245

Conder, Marston, 252

cone
   angle/axis, 71, 254
   manifold, 255, 256, 406
   point, 69, 71, 76, 80

conformal
   averaging, 440
   boundary, 12, 73
   groups, 107
   map, 1, 77
   metric, 85
   model, 8

congruence subgroup, 96

conical limit point, 184, 198, 199

conjugate
   groups, 55

Möbius transformations, 2

convergence, *see also* algebraic, geometric, Hausdorff
   Carathéodory, 233
   Gromov-Hausdorff c. of metric spaces, 259
   of limit sets, 236
   of simple loops, 163
   type, 203

convex cocompact group, 177

convex core, 115, 175, 177, 256, 279
   bending measure, 178, 179
   boundary, 178, 303, 338, 363
   bounded embedded balls, 280
   bounded thickness, 176
   compact, 179
   in $\mathbb{H}^2$, 211
   maximal cusp, 319
   totally geodesic boundary, 377
   volume, 378

convex hull, 167
   bending measure, 169
   floor and dome, 169
   in $\mathbb{H}^2$, 419

Cooper, Daryl, xviii, 199, 262, 360, 402

corner in manifold boundary, 420

coset graph, 354

Coulson, David, 400

cover transformation, 68

covering surface
   branched, 68, 69
   normal, 68

Covering Theorem, 292

coverings of surfaces/3-manifolds
   normal coverings, 67
   regular coverings, 67
   Riemann surfaces, 67
   topological branched coverings, 393

Coxeter, H. M. S., 18

critical exponent, 202

cross ratio, 3, 5, 26, 35
   and distances, angles, 25
   convergence, 54

cube complex/hyperplanes, 393

Culler, Marc, 66, 246, 258, 277, 288,
    303, 354, 399, 439

curvature, *see also under* sectional
    Gaussian, 47
    of arcs, 45
    of circle, 46
    of equidistant arc, 46
    of equidistant surface, 47
    of horocycles, 45
    sectional, 19

curve complex, 350
    arc complex, 352
    disk complex, 352
    pants complex, 352

cusp, *see also under* maximal cusp
    cusp cylinders/cusp tori **Definitions**,
        125
    density, 288
    elimination, 359
    on deformation space boundary, 287,
        313, 338
    paired punctures, solid pairing tube
        Definition, 125
    rank of, 145
    solid cusp cylinders/solid cusp tori
        Definitions, 125

cyclic group, 56, 441

Cylinder Theorem, 150, 156, 194

cylindrical manifold, 198

Dahmani, Francois, 343

De-Spiller, D. A., 200

deck (=cover) transformation, 68

deformation, *see also under* Teichmüller
        space
    of kleinian groups, 281
    quasiconformal, 280, 287
    space, 276
        interior of closure, 363
    space boundary
        inclusiveness of groups, 305
        local connectivity, 416

degenerate group
    compression body, 304
    doubly, 304, 310, 373, 401
    partially, 313
    singly, 310, 313, 373

degenerate hexagon, 448

degree of map to closed manifold, 393

Dehn filling
    exceptional slopes, 249
    on link complements, 398

Dehn surgery, 248, 404

Dehn Surgery Theorem, 248, 404

Dehn twist, 91, 339, 343
    fixed point, 342
    iteration in $\mathfrak{T}(R)$, 344
    surface automorphism, 339
    variation of length, 120

Dehn's Lemma, xiii

Dehn's Lemma and Loop Theorem, 195
    applications, 151
    equivariant, 150

Dehn-Nielsen-Baer Theorem, 353

Delaunay triangulation, 206

Density Conjecture/Theorem, 290, 305,
        310

dessins d'enfants, 117, 461

developing map, 255, 413

dihedral group, 60, 119

dilatation, 85

Dirichlet fundamental polyhedron, 135
    generic:Jørgensen-Marden conjecture,
        208

discontinuity set $\Omega(G)$, 64

discrete group, 55
    with all real traces, 190

discreteness locus, 279
    in projective structure, 415

disk
    area, 17
    circumference, 17

diskbusting curves, 294, 357, 359, 405

diskbusting link, 357

divergence type, 203

dodecahedral group, 60, 119
dome over $\Omega \subset \mathbb{S}^2$, 168
  relation to geometry of Ω, 177
Donaldson, Simon, 77
Douady, Adrien, 154
double horocycle, 147
Double Limit Theorem, 304, 373
double of a surface, 81
doubling a manifold, 403
drilling out simple geodesics, 404
Drilling Theorem, 407
Dumas, David, xi, 382, 412, 414, 415, 420
Dunbar, William, 10, 252
Dunfield, Nathan, 386, 390, 393, 396, 399, 424
Duren, Peter, 233

Earle, Clifford, xvii, 94, 120, 141, 154, 212, 218, 274, 309, 325, 344, 362
Earle–Marden coordinates, 274, 275
earthquake, 210, 270
Earthquake Theorem, 211
edge cycle, 137
edge relation, 135, 137
Edmonds, Allan, 70
Edmonds, Allan L., 393
Efremovich, V., 199
Ehrenpreis Conjecture, 87
eigenvalues
  geometrically finite/infinite, 333
  of a 2×2 matrix, 4
  properties of $\lambda_1(\mathcal{M}(G))$, 335
  when $\lambda_0(\mathcal{M}(G)) = 0$, 334
electrification in geometric groups, 111
elementary group, 56, 61, 94
  all elements elliptic, 222
elementary representation, 277
elliptic transformation, 3
  axis, 13
  maximal order in closed surface, 190
end reduction, 358

end/relative end of a manifold, 290, 291, 294
  case of a surface, 76
  compressible/incompressible, 293
  geometrically (in)finite, 291
  indecomposable, 293
  tame end, 291
ending lamination, 297, 298
  Conjecture/Theorem, 286, 290, 296, 300, 304, 373, 407
  definition, 301
  existence, 298
endpoint of geodesic, 12
engulfing property, 358
Epstein, David B. A., xviii, 27, 111, 142, 158, 160, 168–174, 177, 197, 206, 211, 215, 217, 286, 398
equidistant curve/surface, 13, 46
ergodicity, 204
  and rigidity, 200
  unique, 166, 343
Eskin, Alex, 184
essential cylinder (annulus), 150, 198
  primitive, 349
essential disk, 149
ETH Zürich, xviii
Euclid, 6
Euler characteristic, 69, 187
Evans, Richard, xviii, 238, 239, 279, 293
excess angle, 51
exponential growth, 11
extended (quasi)fuchsian group, 194
extension $\partial \mathcal{M}(G) \to \mathcal{M}(G)$, 155, 240
extension to $\mathbb{S}^2$ from $\Omega(G)$, 212
extension to $\mathbb{H}^3$ of a univalent function, 50
extremal length, 365

Farb, Benson, 92, 93, 111, 190, 343, 353
Farey graph, 351
Farey sequence, 49, 103, 104
Farkas, Hershel, 77, 188

Fathi, Albert, 158, 163, 213, 316, 343, 372, 456
Fatou, Pierre, 133
Fay, John, 66
Feighn, Mark, 197, 354
Fenchel, Werner, 42, 44, 167, 208, 444
Fenchel-Nielsen coordinates of $\mathfrak{Teich}(R)$, 94
Ferguson, Helaman, 78, 379
Fermat curve, 77
Fermat's Last Theorem, 96
fibering over the circle, 374
figure-8 knot, 190, 401
filling/arational lamination, 166
    filling pair, 167, 373
finite group of Möbius transformations, 60
finitely generated kleinian groups, 197
finitely presented group, 74
finiteness theorem
    for cusps, 197
    for finite subgroups, 197
Fletcher, Alastair, 91
foliation, *see also* measured foliation
    (un)stable, 342
Ford region/polygon/polyhedron, 138, 140
Ford region/polyhedron, 245
    finite-sided, 145
    generalization, 217
Ford, Lester R., 21, 60, 61, 119
four-manifolds, 75
Fox, Ralph, 295
fractal, 64, 81, 114, 201
fractional linear transformations, 1
Frame, Michael, 401
free group, 74, 81, 82
    outer space, 354
    two generator, 335
free homotopy, 150
Freedman, Michael, 293
Fricke, Robert, 50
fuchsian centers, 414

fuchsian group, 50, 62, 80
    1st and 2nd kind, 81
    deformations, 88
    extended, 194
    finite index subgroups, 118
    finitely generated, 144
    geometric limits, 256
    least area, 189
    maximal, 189
    naming of, xvi
    Nielsen kernel, 211
    representation variety, 360
    triangle group, 61, 72, 98, 105, 189
    universal horodisks, 218
Fujii, Michihiko, 378
function group, 65, 194
fundamental group, 67
fundamental polyhedron, 105, 135, 441
    Dirichlet, 137
    Ford, 138
        generalized, 138
    not locally finite, 142

Gabai, David, 133, 252, 283, 293, 300, 341, 386, 395, 409, 454
    with Meyerhoff and N. Thurston, 130, 182, 437
Gallo, Daniel, 413, 415
Gardiner, Fred, 96
Gaster, Jonah, 376, 409
Gauss, Johann Friedrich, xv
Gauss, map, hyperbolic, 50
Gauss–Bonnet formula, 18, 179, 187, 188, 359
gaussian curvature, 17
gaussian integer, 105
Gehring, Fred, 153, 155, 186, 200, 441
Gelander, Tsachik, 253
geodesic, 159, 264, 428
    arclength, 32
    complex length, 431
    exiting sequence, 297
    length spectra of closed surfaces, 436

penetration of horodisk, 190
recurrent, 158
self-intersecting, 217
space of –s, 264
unknotted, 409
geodesic lamination, 158, 419
filling pair, 373
maximal, 167
measured, 161
projective, 162
total angle measure of transverse
arc, 214
uniquely ergodic, 167
minimal, 167
realizable, 173, 296
geometric convergence, 185, 225, 257
at Bers slice boundary, 344
at quasifuchsian boundary, 323
Benjamini-Schramm (BS)
convergence, 272, 390
by renormalization, 271
polyhedral, 226
geometric group theory, 108
geometric intersection number, 101, 162,
339
estimates, 213
two measured laminations, 165
geometric structures, 394
geometrically (in)finite end, 291
geometrically finite groups, 144, 149
definitions, 145
density on boundary, 305
essential compactness, 144
minimally parabolic, 284
Geometrization Conjecture/Theorem,
xiv, 394, 396
Geometry Center, 111, 379
Gilman, Jane, 56, 82, 115
Goldman, William, 24, 31, 411, 412, 420
Goodman, Oliver, 271, 400
Gordon, Cameron, 396
grafting, 410–412
2π-grafting, 411

Gray, Jeremy, xv, xvi
Green's formula, 19
Green's function, 204
Green, Paul, 158, 160, 172, 173
Greenberg, Leon, 41, 56, 65, 145, 147,
189, 196, 209, 235, 278, 369
Gromov, Mikhail, 34, 110, 249, 259, 358
–'s Theorem, 263
hyperbolicity, 109–111
a summary, 353
norm, 263
Grothendieck, Alexandre, 117, 461
group, *see* free, kleinian, quasifuchsian
etc.
δ-hyperbolic, 110
combination theory, 152, 193
complex conjugate, 103
containing only elliptics, 222
Gromov hyperbolic, 110
HNN extension, 293
hyperbolic, 110
indecomposable, 292
inverse limit, 113
Klein 4-group, 62
LERF, 114
marked, 277
normalizer, 68
presentation, 74
profinite completion, 113
relatively hyperbolic, 111
residually finite, 114
separable subgroup, 114
word hyperbolic, 110
group properties, summary, 114
Groves, Daniel, 386
Guirardel, Vincent, 343
Gunn, Charlie, 398, *see Not Knot*
Guo, Ren, xviii, 45

Haefliger, André, 294
Haglund, Frédéric, 386
Haken manifold, 282, 383
half-rotation, 425, 434, 444

Halpern, Naomi, 141, 218
Hamenstädt, Ursula, 316, 351, 436
Hamilton, Richard, 395
handle, 76
handlebody, 83, 340, 341
harmonic (hyperbolic) maps, 154, 332
Hartshorn, Kevin, 423
Harvey, William, 65, 256, 350
Hausdorff
  convergence
    definition, 232
    of limit sets, 238, 239, 254
  dimension, 201
    of limit sets, 333
    union of simple geodesics, 160
  measure, 201
Heard, Damian, 400
Heegaard splitting, 84
  Heegaard genus of $\mathcal{M}(G)$, 84
  splitting distance, 423
Hejhal, Dennis, 415, 418
Hempel, John, 84, 112, 149, 152, 153,
    195, 282, 382, 384, 394, 403
Hersonsky, Sa'ar, 288, 303
hexagonal, *see also* right hexagon
  packing, 191, 268
  punctured torus, 103, 207
  torus, 190, 308, 309
Hidalgo, Rubén, 409
hierarchy, 383
Hilbert, David, 18
  metric, 44
Hildebrand, Martin, 393
Hilden, H M, 399
HNN-extension, 383
Hocking, John, 65
Hodgson, Craig, 41, 192, 248, 255, 358,
    393, 400, 402, 405, 406
holomorphic motion, 362
holonomy group, 413
holonomy map, 255
Holt, John, 289, 310

homeomorphisms between manifolds,
  282
homology
  $\mathbb{S}^3$, 390, 424
  basis, 101, 116
  group, 67
  Torelli group, 423
homotopy, 183, 205
homotopy equivalence, 187, 281, 284
  between manifolds, 282
  between surfaces, 282
  homeomorphisms, 282
  primitive shuffle, 285
  shuffle of rolodex, book pages, 285
horizon, 32
horocycle, 14
  double, 147
  foliation by –s, 214
horodisk, 14
  general form in $\Omega(G)$, 140
  in a torsion free fuchsian group, 218
  in simply connected region, 218
  penetration by geodesics, 190
horosphere and horoball, 14, 21, 123,
  127
  maximal, 192
  penetration by planes, 190
Hubbard, John H., 10, 77, 91, 94
hyperbolic
  cone manifold, 255, 256, 406
    based on unknotted geodesic, 411
    deformations, 407
  cube, 453
  Gauss map, 50
  geometry, 6
  group, 111
  harmonic maps, 154, 332
  law of (co)sines
    for hexagons, 446
    for pentagons, 451
    for quadrilaterals, 452
    for triangles, 453
  manifold, 66, *see also under* manifold

boundary area, 350
Bowditch constant, 176
covering, 70
cubulation, 393
diameter bound, 262
manifold double, 378
Minsky Model Theorem, 299
noncuspidal part, 294
random choices, 423
totally geodesic, 377
volume bound, 262
with corners, 420
zero first homology, 397
metric
simply connected domain, 42
orbifold, 71, 256
existence theorem, 402
structure of singular set, 72
quadrilaterals, 454
Ptolemy relation, 463
right hexagon, 446, 453, 455
degenerate, 448, 450
generic, 444
right triangle, 451
space, 8
transformation, 4
trigonometry, 446
hyperbolic group, 109, 110
hyperbolic knots, 396
hyperbolic metric, 30
annulus, 95
cylindrical (Fermi) coordinates, 457
horocyclic coordinates, 457
polar coordinates, 457
punctured disk, 95
solid angle, 13
hyperbolic plane, 8
disk, upper half-plane (UHP) models, 8
hyperbolic space, 8
ball, upper halfspace (UHS) models, 8
polyhedra, 10

Hyperbolization Theorem, 110, 284, 338, 348, 360, 371, 378, 382, 385
for surfaces, 79
hyperboloid model, 37
imaginary length, 37
light cone, 37
timelike, lightlike, spacelike, 37
hyperelliptic involution, 100, 106

I-bundle, 81
icosahedral group, 60, 119
ideal
bigon, 162
boundary component, 76
line, 448, 449
point, 14
tetrahedra, 34, 191, 433, 437
triangle, 7, 14
triangulation, 269
vertex, 7
incomplete hyperbolic metric, 254
incompressible surface, 151, 382
doubly incompressible, 173
indecomposable group, 151
injectivity radius, 126, 397
positive lower bound, 303
interval exchange transformations, 214
invariant spiral, 4
involution
conjugation by, 47
hyperelliptic, 100, 106
irreducible manifold, 382
irreducible representations, 278
isometric circle, 20, 22
excess, 51
isometric plane, 20
isoparametric inequality, 111, 262
isothermal coordinates, 85
isotopy, 183, 205
mapping class group, 91
of metrics, 314
Ito, Kentaro, 418
Ivanov, Nikolai, 93, 343, 351

Jørgensen, Troels, xiv, xvii, 33, 47, 49, 51, 56, 105, 134, 143–145, 219, 225, 231, 234, 235, 237, 245, 284, 308, 310, 319, 425, 434, 441
  complex probabilities, 308
  inequality, 56, 57, 105, 127, 143, 220, 223, 225, 226, 229, 441
    cases of equality, 105
    new parabolics, 237
Jaco, William, 84, 149, 152, 363, 382, 384, 394
Jacobi identity, 441
jet, 50
Johannson, Klaus, 240, 281, 394
Jones, Gareth, 117, 188
Jones, Peter, 201, 202, 334
Jungreis, Douglas, 341, 386, 395

Kahn, Jeremy, xvii, 87, 378, 389
Kamishima, Yoshinobu, 419, 420
Kapovich, Michael, 123, 184, 208, 248, 253, 259, 277, 333, 363, 385, 402, 413, 415, 420
Kapovich, Misha, xviii
Keen, Linda, 103, 133
Kellerhals, Ruth, 252
Kent IV, Richard P., 382, 408
Kerckhoff, Steven, xviii, 92, 164, 165, 211, 214, 248, 255, 314, 325, 341, 344, 346, 356, 365, 402, 404–406
Kiikka, Maire, 105
Klarreich, Erica, 329, 351
Klein, Felix, 7
  bottle, 403
  model, 37
  surface, 78
Klein, Peter, 219
Klein–Maskit combination theory, 152, 193
Kleineidam, Gero, 304, 354–356
Kleiner, Bruce, 395
kleinian group, 62
  arithmetic, 401

convergence, 231
determined by its traces, 435
doubly degenerate, 374
finitely generated, 197
higher dimensional, 253
infinitely generated, 66, 326, 421
naming of, xvi
presentation, 74
quaternion representation, 42
two-generator, 49, 363, 435
Knopf, Dan, 395
knot
  complement, 105
    hyperbolic structure, 396
  figure-8, 190, 375
  longitude and meridian, 397
  Seifert surface, 397
Knotted Wye, 379
knotted/unknotted geodesic, 409
Kojima, Sadayoshi, xviii, 255, 404, 413
Komori, Yohei, 308, 415
Korkmaz, Mustafa, 351
Koundouros, Stelios, 397
Kra, Irwin, 94, 188, 281, 369
Kulkarni, Ravi, 70, 197

Labourie, Francois, 68
Lackenby, Marc, 249, 360
Lakes of Wada, 65
Lakic, Nikola, 96
lamination, *see also under* geodesic lamination
  (un)stable, 342, 372
  ending, 301
  minimal, 161
Lamping, John, 11
laplacian (hyperbolic), 332, 333
Lascurain, Antonio, 105
lattice, 99
Laudenbach, Francois, 158, 163, 213, 316, 343, 372
Le Calvez, Patrice, 92
Le, Thang, 390

Lecuire, Cyril, 178, 354
Lee, Youn, 401
Leeb, Bernhard, 402, 403
Lehner, Joseph, 57, 96
Lehto, Olli, 91, 107
Leininger, Chris, 114
LERF, (all f.g. subgroups separable),
 112, 114
Leung, Naichung, 182
level k congruence subgroup, 96
Levy, Silvio, xi, xviii
Li, Peter, 155
Li, Tao, 84
Lickorish, Raymond, 397
Lie product, 426
Lieninger, Chris, 196
lifting to a matrix group, 66
limit set, 63, *see also under* conical limit
 set
 convergence, 236
 Hausdorff dimension, 201
 Hausdorff distance, 238
 locally connected, 331
 tangents, 107
line geometry, 425
link
 complement, 105
 Dehn surgery along, 249
 diskbusting, 358, 359
 indecomposable, 397
linked/unlinked geodesics, 409
Liouville measure of geodesics, 26
Liu, Yi, 262, 389
Lobachevsky, function, 35
Lobachevsky, Nikolai Ivanovich, xvi
local connectivity, 287
 quasifuchsian discreteness locus, 287
 of limit sets, 327
Long, Darren D., 196
longitude and meridian, 397
Loop Theorem, *see* Dehn's Lemma
Lott, John, 395
loxodromic

curve, 4
transformation, 3
 axis, 13
Lozano, M T, 399
Lubotzky, Alexander, 253
Luecke, John, 396
Luo, Feng, 351
Lyndon, Roger C., 354

Möbius strip, 264
Möbius transformation, 1
 axis, 13, 427, 434
 composition, 2
 composition of reflections, 27
 convergence, 53
 eigenvalues and eigenvectors, 4
 extension to 3-space, 6
 half-rotation, 425
 images of horizontal lines, 121
 in ≥ 3 dimensions (*Liouville's
  theorem*), 28
 normalized, 2
 normalized matrix representation, 2
 square roots, 429
 standard forms, 4
Maclachlan, Colin, 383, 401
Magid, Aaron, 287
magnetic resonance imaging, 268
Magnus, Wilhelm, 50, 98
Maher, Joseph, 111, 423
Maillot, Sylvain, 395
Maloni, Sara, xviii
Mané, Ricardo, 362
Mandelbrot, Benoît, 81, 201
manifold, *see also under* hyperbolic
  manifold
 aspherical, 386
 boundary topology, 152
 containing knots/unknots, 409
 from face pairing of polyhedra, 424
 geometrically atoroidal, 383
 graph of manifolds, 354
 Haken, 383

incompressible, 383
  boundary incompressible, 383
irreducible, 383
pared, 383
random choice, 424
toroidal=homotopically t., 383
manifold vs orbifold, 67
manifolds
higher dimensional, 253
Manning, Jason, 386
mapping class group, 111, 372
  5-punctured sphere, 343
  action on Thurston boundary, 316
  classification of elements, 341
  closed hyperbolic 3-manifold, 183
  definition, 91
  exceptional cases, 92
  extended, 92
  extension to $\mathcal{M}(G)$, 364
  finite index subgroup, 93
  finite subgroups, 341
  not realizable by homeos, 92
  puncture fixing subgroup, 93
  random walk, 424
  rigidity, 187
  torsion free, finite index normal
      subgroup, 341
mapping torus, 374
Marden, Albert, 27, 80, 82, 123, 141,
    152, 168–171, 174, 177, 195, 197,
    211, 218, 234, 235, 237, 274, 280,
    281, 286, 313, 325, 343, 344, 364,
    365, 369, 377, 413, 415
  Conjecture, *see* Tameness Conjecture
  Isomorphism (or Rigidity) Theorem,
      182, 183, 346
Margalit, Dan, 93, 190, 353
Margulis, Grigori, 200, 400
  constant, 134
Markov identity and conjecture, 23, 24
Markovic, Vlad, 155

Markovic, Vladimir, xvii, xviii, 87, 91,
    92, 110, 118, 154, 168–170, 174,
    177, 208, 212, 389
Marshall, Timothy, 252
Martin, Gaven, 252, 441
Maskit, Bernard, xiv, 65, 82, 107, 152,
    157, 185, 193, 195, 198, 199, 281,
    290, 309, 310, 313, 329, 338, 369,
    409
  Planarity Theorem, 82, 115, 153
Masur, Howard, xvii, xviii, 111, 116,
    184, 214, 300, 316, 350, 353, 356,
    364, 463
  domain, 354–356
Matelski, Peter, 440
Mathematical Sciences Research
    Institute, 78
matrix group from kleinian group, 66
Matsuzaki, Katsuhiko, 73, 202, 204, 240
maximal cusp, 287, 289, 313, 338
  density of –s, 314
maximal lamination, 166
Maxwell, Delle, *see Not Knot*
McCullough, Darryl, 180, 195, 197, 282,
    285, 286, 355
McMullen, Curtis, xi, 96, 154, 215, 254,
    279, 288, 303, 308, 314, 329, 336,
    338, 363, 373, 377, 381, 400, 416
McReynolds, David, 114
McShane, Gregory
  Mcshane identity, 438
measured
  foliation, 165
    from interval exchange, 214
    horocyclic, 214
    quadratic differentials, 366
  lamination, 161
    arational/filling, 166
    by sequence of lengths, 164
    finite, 163
    length, 164
    quadratic differentials, 364
    sequence convergence, 163

uniquely ergodic, 166, 167
Meeks III, William H., 150
meromorphic
  function, 117, 411, 419
    locally injective, 412
  lamination, 419
  quadratic differential, 461
Mess, Geoffrey, 75, 197
Meyerhoff, Robert, 129, 130, 133, 134,
    182, 249, 251, 252, 283, 409, 454
Miller, Andrew, 180, 195, 355
Milley, Peter, 133, 252
Milnor, John, xvi, 35, 395
minimal lamination, 161, 166
minimally parabolic, 284
Minkowski space, 38
  light cone, 40
  timelike, lightlike, spacelike, 40
Minsky Model Theorem, 299
Minsky, Yair, xi, xviii, 111, 173, 293,
    298–300, 303, 308, 328, 334, 339,
    350, 353, 365, 375, 410, 416
Mirzakhani, Maryam, 160
  geodesic length formula, 438
Mitra, Sudab, 154
Miyachi, Hideki, 308
Mj, Mahan, xvii, 330, 375
Möbius transformation
  orientation reversing, *see* anti-Möbius
modular group, 96
  =mapping class group, 92
  extended, 92, 94
  Farey sequence, 104
modular transformation, 100
moduli space, 94
  compactifications, 94
  definition, 93
  manifold cover, 94
  of a 3-manifold, 370
  triangulation and compactification,
    215
modulus
  of annulus, 336

of circular quadrilateral, 265
monodromy group, 413
Montel's Theorem, 53
Montesinos, José Maria, 394, 399
Moore, R.L., 328, 332
Mordell Conjecture, 77
Morgan, John, 258, 385, 395
Mosher, Lee, 343
Mostow, George, xiii, 187, 200, 377
  Rigidity Theorem, 182, 186, 283, 378
  history, 200
Mozes, Shasar, 253
MSRI, 78
multicurve, 162, 412
  intersection numbers, 162
Mumford, David, xvii, 49, 81, 104, 115,
    142, 253, 257, 287
Munkres, James, 153
Munzner, Tamara, 11, 66, 96
Myers, Robert, 358, 405, 408

Namazi, Hossein, 305
Nash, John, 76
navigation, 4
nearest point retraction, 168, 169, 175,
    419
negative curvature, *see also* pinching
    and hyperbolic
  characterizations, 17
  discrete, 17
  of groups, 109, 111
nerve, 265
Neumann, Walter, 191, 271, 400
new parabolics, 237
Nicholls, Peter, 202, 204, 205
Nielsen Realization Problem, 92, 211,
    341
Nielsen, Jakob, 167, 208
  kernel, 211
  transformation, 335
non-Euclidean geometry, xv
normal subgroup, 68
normalizer of a subgroup, 68

Not Knot, 66, 96, 398
number theory, 96

octagon, hyperbolic, 427
octahedral group, 60, 119
Ohshika, Ken'ichi, xvii, xviii, 110, 238, 258, 293, 295, 305, 323, 337, 422
Ol'shanskiĭ, A. Yu., 111
OPTi, 308
orbifold, 67, 70, 402
    cover, 73
    euclidean, 120
    minimum volume, 252
    spherical, 119
    theorem, 402
Orbifold Theorem, 403
orbital counting function, 203
ordinary set $\Omega(G)$, 64
oriented lines, 431, 446
orthogonal projection, 11, 460
Osin, Denis, 343
Otal, Jean-Pierre, 164, 178, 258, 259, 355, 356, 374, 385, 407, 409, 420
outer circles, 20, 21
Outer space, 354

page shuffling, *see* rolodex
paired punctures, 147
    joining, 274
    opening up, 284
pants, 455
    all medium sized, 272
    cuff lengths, 456
    decomposition, 165, 269, 288, 304, 421
pants complex, 352
Papakyriakopoulos, Christos, xiii
parabolic group, 98, 190
    discrete extension, 102
    horosphere and horoball, 14
    intrinsic horosphere euclidian metric, 124
    least (translation) length, 126

parabolic transformation, 3
    accidental, 198
    associated geometric structures, 145
    formulas for, 427
    new parabolics, 237
parallel loops, 337
pared manifold, 382
Parker, John, xi, 180, 468, 471
Parker–Series bending formulas, 468, 471
Patterson, Samuel J., 202, 205
Patterson-Sullivan measure on limit sets, 204
Peano curve
    equivariant construction, 327, 332
Penner, Howard, xi
Penner, Robert C., 217, 343, 463
pentagon, 451, 454, 460
Perelman, Grigori, xiv
    Geometerization Theorem, 385
Petersen–Morley Theorem, 445
Petronio, Carlo, 142, 207, 248, 250, 270
Picard group, 105
Pignataro, Thea, 105
pinching
    curvature bounds, 358
    estimate, 288, 336
    limiting process, 253, 287, 289, 304, 309, 312, 319, 337
    loops, 313, 323
    Theorem (Ohshika), 337
Pirelli, Peter, 11
planar
    covering surface, 81
    pentagon, 451, 455
    quadrilateral, 454
    Riemann surface, 82
    right hexagon, 455
pleated surfaces, 171, 173
    uniform injectivity, 173
pleating locus, 172
plumbing coordinates, 274
PNC manifolds, 358, 360

Poénaru, Valentin, 158, 163, 213, 316, 343, 372
Poincaré, Henri, xiii, xvi, 12, 38, 135
  Conjecture, 12, 84, 382
  dodecahedral space, 12
  Polyhedron Theorem, 142
  series, 205
point of approximation, *see also under* conical limit set
Poisson integral formula, 332
Pollicott, Mark, 369
polygon, hyperbolic, 7
polyhedral
  convergence, 226
  deformation, 36
  group, 60, 116, 118
  surface, 17, 76
polyhedron
  hyperbolic, 10
  volume, 36
Pommerenke, Christian, 201, 212, 233, 329, 331
Porti, Joan, 248, 250, 395, 402, 403
Prasad, Gopal, 182, 187
presentation, 74
  length, 262
primitive curve, 271
primitive group element, 59
profinite completion, 118
projective model, 37
projective structure
  discreteness locus, 415, 419
  extended Bers slice, 414
  quasifuchsian locus, 414
  Thurston coordinates, 420
properly discontinuous, 56, 64, 78
Przeworski, Andrew, 133, 252
pseudo-Anosov mappings, 342, 372
pseudosphere, 18
puncture, 76, 78
punctured torus, 103, 335, 439, 441
  group, 103, 144, 313, 319, 439, 441, 465

hexagonal, 103, 207
Purcell, Jessica S., 397

quadratic differential, 90, 205, 364
  Abelian differential, 463
  bundle over $\mathfrak{T}(R)$, 415, 419
  singular euclidean metric, 463
quadratic differentials, 90
quadratic forms, 96
quadrilateral
  circular, 264
  marked, 265
  planar with two right angles, 454
  with three right angles, 452
quasiconformal
  definition, 85
  deformation, 86, 463
  deformation space, 280
  extension $\Omega(G)$ to $\mathbb{S}^2$, 213
  metric definition, 85, 200
quasifuchsian, *see also* degenerate group
  deformation space, 305, 373
  group, 155, 194, 305
    illustration, 311, 312
quasifuchsian space, 308
  Bers slices, 308
  Earle slice, 309
  Maskit slice, 308
  nonlocal connectivity of closure, 416
quasiisometry, 155, 199, 200
quasisymmetric boundary mapping, 91
quaternions, 42, 436

RAAG: right-angled Artin group, 392
Rafi, Kasra, 184, 351
ramification, *see* branch
rank of cusp, 145
rank two subgroups, 272
Rao, Ramana, 11
Ratcliffe, John, 35, 73, 191, 252
real $\mathbb{R}$-trees, 258, 259, 362
real projective structure, 411
realizable lamination, 173, 296

recurrent geodesic, 158
recurrent ray, 185, 199
reducible, *see also* irreducible
reducible automorphism, 341
reducible group, 95
    representation, 277
Rees, Mary, 165
reflection
    in a point, 44
    in a sphere or plane, 1, 25, 29, 44
regular exhaustion, 290
regular set $\Omega(G)$, 64
Reid, Alan, 196, 271, 383, 401
Reimann, Hans Martin, 154
relatively hyperbolic group, 111
    with respect to subgroup, 111
relator, 74
representation variety, 276
    character variety, 362
    discreteness locus, 279, 289, 416, 418
    fuchsian groups, 360
    local coordinates, 179
    quasiconformal deformation space,
        280
residually finite, 112
residually finite group, 112, 114
retraction
    in hyperbolically convex set, 197
    nearest point, 168, 169, 175, 419
    of $\mathbb{H}^3$ to line, 11
rhumb line, 4
Ricci flow, 395
Riemann Mapping Theorem, 77
Riemann surface, 75
    (integral) grafting, 412
    Belyĭ functions, 117, 461
    built from pants, 455
    canonical hyperbolic triangulations,
        215
    compact bordered, 81
    conformal embedding in $\mathbb{R}^3$, 76
    cut into polygons, 463
    Ehrenpreis Conjecture, 87

    from equilateral triangles, 461
    from ideal triangles, 269
    from interval exchange, 214
    from pentagons, 460
    genus 2, 107
    isometric embeddings in $\mathbb{R}^{17}$, $\mathbb{R}^{51}$, 76
    length of closed geodesics, 133
    marked, 88
    polygon decompositions, 217
    projective structure, 415
    spine, 216
    translation surface, 463
Riemann–Hurwitz formula, 69, 118
riemannian 3-manifolds
    pinched negative sectional curvature,
        386
right triangle, 450, 454
rigidity
    3-manifolds with boundary, 183
    mapping class group, 187
    of homotopies, 182
    of homotopy equivalences, 282
    quasiconformal, 281
    topological, 303
Riley, Robert, xiv, 143, 375, 400, 401
Rips, Eliahu, 109
Rivin, Igor, 10, 268
Rodin, Burt, 266
Roeder, Roland K.W., 10, 268
Rogness, Jonathan, xi
Rolfsen, Dale, 399
rolodex, 285, 347, 349
Royden, Halsey, 93
Rüedy, Reto, 76

Sad, Ricardo, 362
Sageev, Michah, 386, 390
Sakai, Tsuyoshi, 404
Sakuma, Makoto, 144
Sarić, Dragomir, 118
Sario, Leo, 77, 82
satellite knot/link, 396, 397
Scannell, Kevin P., 412

Schafer, James, 117
Schläfli formula, 36
Schleimer, Saul, xviii, 351, 352
Schneps, Leila, 461
Schoen Conjecture, 155
Schottky group, 81, 87, 114, 115, 144,
        152, 195, 257
    boundary cusp, 289
    classical, 82
    degenerated, 331
    dimension of limit set, 201
    illustration, 307
    simply degenerate, 332
Schroeder, Viktor, 358
Schupp, Paul E., 354
Schwartz, Rick, 439
Schwarz Lemma, 456
schwarzian derivative, 50, 413
schwarzian differential equations, 413
Scott, Peter, xiv, xviii, 74, 75, 112, 180,
        194, 295, 385, 394, 405
Scott–Shalen theorem, 276
sectional curvature, 19, 182
    pinched, 358
Seifert Conjecture, 386, 395
Seifert fiber space, 394, 395
Seifert–Weber dodecahedral space, 139
Selberg's Lemma, 57, 73, 148, 184, 197,
        359
separable subgroup, 112, 114
Seppälä, Mika, 436
Series, Caroline, xi, 49, 81, 104, 115,
        120, 142, 160, 179, 180, 257, 287,
        468, 471
Shalen, Peter, 74, 133, 180, 197, 246,
        258, 277, 288, 303, 363, 394, 439
Sharp, Richard, 369
Shiga, Hiroshige, 414
Shimizu, Hideo, 134
Shinnar, Meir, 134
short geodesics, 127, 438
    drilling out, 407
shrinkwrapping, 294

shuffle, 285, 347, 349
Siegel, Carl Ludwig, 57, 96
Sierpiński gasket (carpet), 378, 379
simplicial volume, 264
simultaneous uniformization, 157, 305,
        306
Singerman, David, 117
singular fiber, 394
singular set of an orbifold, 71, 402
Skinning Lemma, 323, 376, 408
skinning map, 376
Skora, Richard, 259
Slodkowski, Zbigniew, 362
Smillie, John, 463
SnapPea/SnapPy, 270, 399
solenoid, 118
solid torus, 83
    longitude and meridian, 397
Soma, Teruhiko, xvii, 294, 323, 408, 422
Sorvali, Tuomas, 436
soul, 212
Souto, Juan, xviii, 238, 239, 280, 293,
        294, 305, 334, 354–356, 397, 405,
        409
sphere at infinity, 12
sphere theorem, 382
spherical manifold, 12
spine, 135
spinning, 269
Springborn, Boris, 10
Springer, George, 77
square root of Möbius t., 429
stabilizer, 55
standard form of Möbius t., 4
Stephenson, Kenneth, xi, 10, 267, 268
stereographic projection, 1, 28, 36
    $\mathbb{B}^3 \to$ UHS., 29
Stillwell, John, xvi
Stong, Robert, 70
Storm, Peter, 188, 378, 396, 403
Strebel, Kurt, 89, 343, 364, 365
strong convergence, 239
strong stability, 280

Strong Torus Theorem, 385

structural stability of groups, 363

Struik, Dirk, 17, 19

Sturm, Jacob, 134

subgroup separable, 114

Sugawa, Toshiyuki, 308, 415

Sullivan, Dennis, 110, 170, 197, 204,
    205, 266, 334, 362, 363
  group stability, 363
  K-theorem, floor to dome, 170
  Rigidity Theorem, 86, 183, 303, 308,
    310

Sun, Hongbin, 390

surface area and volume, 262

surface automorphims
  finite order, 341

surface automorphisms
  Dehn twists
    iterates, 316, 325
  pseudo-Anosov, 327, 341, 372
    axis, 342
    fixed points, 372
    iterates, 322, 325
    rank and abelian group, 372
  reducible, 322, 341

Surface Subgroup Conjecture/Theorem,
  387
  counting immersed surfaces, 388

Swarup, Ananda, 180

symmetry lines, 442

systole, 103, 437

Tam, Luem-Fai, 155

tame
  end, 291, 386
  manifold, 291, 292, 422, *see also*
    Ahlfors' Conjecture, Bonahon's
    Criteria, untameness

Tameness Conjecture (Theorem), 360

Tameness Conjecture/Theorem, 238,
  239, 289, 293, 294, 299, 334, 363

Tan, Ser Peow, 413, 419, 420

tangent bundle of a hyperbolic surface,
  33

tangents to limit sets, 107

Tanigawa, Harumi, 414

Taylor, Edward, 201, 334

Teichmüller lemma, 87

Teichmüller mapping, 89
  extremal, 90

Teichmüller modular group
  = mapping class group, 92

Teichmüller space
  Bers (analytic) boundary, 310
  Bers slice, 307
  biholomorphic automorphisms, 94
  bounded orbits, 381
  comparison Bers and Thurston
    boundaries, 316
  complex structure, 94
  definitions, 88
  dimension, 89
  geodesic rigidity, 215
  geodesics, 90
  global complex analytic coordinates,
    94
  global real analytic coordinates, 94
  higher Teichmüller space, 369
  isometries, 94
  metric, 89
  natural tessellation, 215
  pseudo-Anosov action, 373
  quasi-isometric rigidity, 184
  ray, 364
  relative hyperbolicity, 111
  surface with cone points, 369
  Thurston (geometric) boundary, 316
    convergent sequences, 315
    pseudo-Anosov fixed points, 373

Teichmüller, Oswald, 87

tetrahedral group, 60, 119

tetrahedron, flattened, 270

tetrahedron, ideal, 433, 437
  thinness, 36
  volume, 36

Theorema Egregium, 18
thick/thin decomposition, 134, 173, 192
Thickstun, Thomas, 358
thin part, *see* thick/thin decomposition
Thurston, Nathaniel, 130, 182, 283, 409, 454
Thurston, William, xiv, xvii, 50, 111, 112, 154, 163, 167, 169, 187, 191, 199, 215, 246, 248, 249, 253, 268, 290, 292, 293, 295, 297, 298, 314, 325, 334, 339, 341, 343, 344, 346, 358, 372, 373, 375, 377, 379, 382, 386, 393–396, 398, 401–403, 419, 424
   orbifold/reflection trick, 420
   Compactness Theorem, 240, 261, 279, 280, 338
   coordinates, 420
   earthquakes, 210, 211
   geometric finiteness, 149, 192
   Gluing Theorem, 381
   Hyperbolization Theorem, 385
   pleated surfaces, 173
   thick/thin, 173
Tihomirova, E., 199
topological rigidity, 182, 282
Torelli group
   homology spheres, 424
torsion-free, 55, 66
Torstensen, Anna, 252
torus, 99, 190, *see also* punctured torus
   hexagonal, 190, 308, 309
   knot, 396
   marked, 100
   slope of simple loop, 100
   square, 103
Torus Theorem, 405
totally geodesic boundary, 179, 377
trace
   –s determine group, 435
   and Dehn twist, 343
   calculations for cyclic groups, 32
   definition, 2

identities, 23, 32, 47, 50
   signed, 24
train tracks, 366
   switch condition, 367
   weighted, 367
Tranah, David, xviii
triangle
   area, 7, 19, 459
   area and side length, 27
   group, 61, 98
   uniform thinness, 15
tripod, 258
Tschantz, Steven, 252
tubular neighborhood
   of geodesic, 13, 26, 123, 130, 133
      volume/area, 104
   of systole, 437
   universal, 134
Tucker, Thomas, 295
Tukia, Pekka, 154, 183, 184, 186, 209
twisted I-bundle over Klein bottle, 403
type preserving, 88, 277

UHP, 28, *see* upper half-plane
uniform injectivity, 269
Uniformization Theorem, 77
uniformization, simultaneous, 157, 305
uniformly perfect set, 43, 272
uniquely ergodic, 166, 343
universal
   ball, 127
   ball $\subset \mathcal{M}(G)$, 127
   constants, 127
   cover, 67
      PSL$(2, \mathbb{R})$, 68
   elementary neighborhood, 127
   horoball/horosphere, 127
   horodisk, 190
      extended form in $\Omega(G)$, 141
   hyperbolic solenoid, 118
   isolation of cone axes, 127
   tubular neighborhoods, 127, 134
universe, curvature of, 12

University of Minnesota, xviii, 379
University of Warwick, xviii
Unknottedness Theorem, 409
untameness, 296

van Kampen's Theorem, 151, 383
Van Vleck, Edward, 27
vector space of 2×2 matrices, 436
Virtual Domination
    Conjecture/Theorem, 390
Virtual Fibering Conjecture/Theorem,
    387
Virtual Haken Conjecture/Theorem, 386
visual angle, 32
visual sphere, 12
Vogtmann, Karen, 354
volume
    3-manifold minimums, 251
    3-orbifold minimums, 251
    estimated by thick part, 104
    geometrically finite, 149
    higher dimensions, 252
    manifold bound, 262
    of ball, 16
    of convex core, 378
    of hyperbolic manifolds, 245, 396
    of maximal horoball, 192
    of polyhedra, 36
    of tetrahedra, 36
    of tubes, 104
    simplicial (of a manifold), 264
    well ordering, 251
Voronoi diagram, 206
Vuorinen, Matti, 28

Wada, Masaaki, 47, 144, 308, 415, 439
Waldhausen, Friedhelm, xiv, 84, 152,
    186, 282, 383, 386
Wan, Tom, 182
Wang, Hsien-chung, 134, 253
Waterman, Peter, 82, 115
Weeks manifold, 401
Weeks, Jeffrey, xvii, 12, 41, 207, 270,
    393, 399
Weil-Petersson metric, 352, 353
Weiss, Hartmut, 256
Whitehead link, 252, 401
Whitehead, George, 282
Whitten, W C, 399
Wielenberg, Norbert, 105, 143, 149, 400,
    401
wild embedding, 295
Wiles, Andrew, 96
Wise, Daniel T., xvii, 389
Wolf, Michael, 353, 412
Wolpert, Scott, 120, 353, 436
word-hyperbolic, 109
wormhole, 379
wrapping around a loop, 417
Wright, David, xi, xvii, 49, 81, 104, 115,
    142, 257, 287, 378

Yamada, Akira, 217
Yamashita, Yasushi, 144, 308
Yau, Shing-Tung, 150
Yoccoz, Jean-Christophe, 343
Young, Gail S., 65

Zhu, Xiaodong, 302